CANCER BIOLOGY

FOURTH EDITION

Raymond W. Ruddon, M.D., Ph.D.

University of Michigan Medical School
Ann Arbor, Michigan

OXFORD

UNIVERSITY PRESS

2007

OXFORD

UNIVERSITY PRESS

Oxford University Press, Inc., publishes works that further
Oxford University's objective of excellence
in research, scholarship, and education.

Oxford   New York
Auckland   Cape Town   Dar es Salaam   Hong Kong   Karachi
Kuala Lumpur   Madrid   Melbourne   Mexico City   Nairobi
New Delhi   Shanghai   Taipei   Toronto

With offices in
Argentina   Austria   Brazil   Chile   Czech Republic   France   Greece
Guatemala   Hungary   Italy   Japan Poland   Portugal   Singapore
South Korea   Switzerland   Thailand   Turkey   Ukraine   Vietnam

Published by Oxford University Press, Inc.
198 Madison Avenue, New York, New York 10016
www.oup.com

Oxford is a registered trademark of Oxford University Press

Library of Congress Cataloging-in-Publication Data
Ruddon, Raymond W., 1936–
Cancer biology / Raymond W. Ruddon.— 4th ed.
    p. ; cm.
Includes bibliographical references and index.
ISBN-13: 978-0-19-517543-1 (cloth)
ISBN-13: 978-0-19-517544-8 (pbk.)
1. Cancer.   2. Carcinogenesis.   3. Molecular biology.   I. Title.—
[DNLM: 1. Cell Transformation, Neoplastic.   2. Neoplasms
Etiology.   QZ202R914C 2007]
RC261.R85 2007
616.99'4071—dc22   2006010326

1005223384

9 8 7 6 5 4 3 2 1

Printed in the United States of America
on acid-free paper

I dedicate this book to my spouse, Lynne Ruddon,
who has been my best friend and the love of my life for over 45 years.
Her continual and unflagging patience and support
have made possible whatever success I have experienced in my professional career.

Preface

There have been a significant number of advances in the field of cancer research since the first edition of *Cancer Biology*, which was published in 1981. These include advances in defining the genetic and phenotypic changes in cancer cells, the genetic susceptibility to cancer, molecular imaging to detect smaller and smaller tumors, the regulation of gene expression, and the "-omics" techniques of genomics, proteomics, and metabolomics, among others. Yet, the goals of the fourth edition of *Cancer Biology* remain the same as those of the earlier editions, namely to provide a historical perspective on key developments in cancer research as well as the key advances of scientific knowledge that will lead to a greatly increased ability to prevent, diagnose, and treat cancer. Unfortunately, many aspects of the exciting breakthroughs in our knowledge of basic cancer biology have yet to be translated into standard care for patients. This will require an expanded ability of basic scientists and clinical researchers to learn to speak each other's language and to collaborate on bringing basic research findings to the bedside. A goal for this book, which may seem overly ambitious if not a bit pompous, is to provide part of the lingua franca for these groups of experimentalists to better communicate. Now more than ever it has become clear that to achieve real breakthroughs in improving much needed diagnosis and treatment of cancer and other multifaceted chronic diseases, an interaction is required among researchers in many fields, including molecular biologists, chemists, computational scientists, biomedical engineers, epidemiologists, and health services researchers, as well as dedicated physicians, nurses, and other health care professionals.

I would like to thank the many investigators who have allowed me to use data from their own research to illustrate key points in the text. I would also like to thank the numerous colleagues who have read the earlier editions and used them in their teaching. Their comments have been helpful in revising the text. I am especially gratified by the feedback from some individuals who have said that *Cancer Biology* was their first exposure to the field of cancer research and that reading it inspired them to seek a career in the field.

I want to thank Denise Gonzalez for preparation of some of the early chapters of the book. I am greatly indebted to Paulette Thomas for her diligent and patient work on the preparation of the illustrations and on other technical components of the book. I am especially indebted to Kathy Christopher for her careful preparation and preliminary editing of the text. Without her, the book could not have been completed. I also want to thank the editors and production staff at Oxford University Press who made the book happen.

Contents

CANCER BIOLOGY

1

Characteristics of Human Cancer

WHAT EVERYONE WANTS TO KNOW ABOUT CANCER

Patients

During my career as a cancer scientist, I have frequently received calls from individuals who recently heard a physician tell them the ominous words "You have cancer," or from people who have heard that statement about a family member or close friend. The first question usually is "What can you tell me about this kind of cancer?" They may have already visited several Internet sites and have some information, not always accurate or scientifically based. If the patient is a child and the inquiry comes from parents, they frequently have a great feeling of guilt and want to know what they did wrong, or they may lash out at some perceived environmental agent that they think is the cause, such as water pollutants or electromagnetic fields from high-power lines in their neighborhood. Individuals or their family members then want to know what caused the cancer, what the meaning of the test results is, what the treatment options are, and, if the tumor has spread, if there are any preventive measures that can be taken to stop further spread of the cancer. If cancer is in the family, they may ask what their chances are of getting cancer. These are questions that are always difficult to answer. One of the goals of this book is to try to provide the scientific basis for approaching these questions.

Physicians and Health Care Professionals

The members of the health care team who take care of cancer patients have a different set of questions. These may include the following: What are the most appropriate diagnostic tests with low false negatives and false positives? What are the differential diagnoses that need to be ruled out? And once the diagnosis is made, what is the stage and histological grade? Is the disease local, regional, or metastatic? What is the likely prognosis and the best therapeutic approach? How often is follow-up of the patient required and for how long? If the disease progresses, how may the treatment approaches change? Some of the data that relate to answering these questions will also be discussed in the book.

Cancer Researchers

Basic scientists and clinicians working in the field of cancer research, by contrast, have yet another set of fundamental questions: What are the basic mechanisms of malignant transformation of cells? What causes of cancer can be identified? Knowing that, what preventive measures can be taken? Are there genetic profiles, hereditary or induced by spontaneous mutations, that correlate with susceptibility or progression of cancer? Can the gene expression patterns of cancer cells be used to identify targets for cancer diagnosis or therapy? What proof-of-principle studies are needed

3

to verify these targets? What type of clinical trials is needed to determine the toxicity and efficacy of a new therapeutic modality? These questions will also be addressed.

WHAT IS CANCER?

A few years ago I was at a small meeting with a group of distinguished cancer biologists and clinicians. It was an interesting meeting because there were also distinguished scientists from other fields. The idea of the meeting was to stimulate cross-fertilization of ideas from different scientific disciplines, with the hope that new paradigms for approaching the causes of cancer and its course would be conceived.

One of the first questions that one of the non-cancer researchers asked was, what is the definition of cancer? It was somewhat startling to hear the vigorous discussion and even squabbling among the distinguished cancer scientists in their attempt to define cancer. Although most could agree on a few key characteristics, everyone had their own caveats or additional variations to add. So, like all good academic groups, they appointed a committee to come up with a consensus definition. As the most gullible person there, I agreed to chair the committee. After many phone calls and E-mails going back and forth, we came up with the definition and more detailed description below. I should note that the definition is the sort of thing that would appear in a dictionary and the description contains some of the points and caveats thought crucial for taking into account the characteristics of this multifaceted disease.

Definition of Cancer

Cancer is an abnormal growth of cells caused by multiple changes in gene expression leading to dysregulated balance of cell proliferation and cell death and ultimately evolving into a population of cells that can invade tissues and metastasize to distant sites, causing significant morbidity and, if untreated, death of the host.

Description of Cancer

Cancer is a group of diseases of higher multicellular organisms. It is characterized by alterations in the expression of multiple genes, leading to dysregulation of the normal cellular program for cell division and cell differentiation. This results in an imbalance of cell replication and cell death that favors growth of a tumor cell population. The characteristics that delineate a malignant cancer from a benign tumor are the abilities to invade locally, to spread to regional lymph nodes, and to metastasize to distant organs in the body. Clinically, cancer appears to be many different diseases with different phenotypic characteristics. As a cancerous growth progresses, genetic drift in the cell population produces cell heterogeneity in such characteristics as cell antigenicity, invasiveness, metastatic potential, rate of cell proliferation, differentiation state, and response to chemotherapeutic agents. At the molecular level, all cancers have several things in common, which suggests that the ultimate biochemical lesions leading to malignant transformation and progression can be produced by a common but not identical pattern of alterations of gene readout. In general, malignant cancers cause significant morbidity and will be lethal to the host if not treated. Exceptions to this appear to be latent, indolent cancers that may remain clinically undetectable (or in situ), allowing the host to have a standard life expectancy.

Some points in the description may not seem intuitively obvious. For example, cancer doesn't just occur in humans, or just mammals for that matter. Cancer (or at least tumorous growths—these may or may not have been observed to metastasize) has been observed in phyla as old as Cnidaria, which appeared almost 600 million years before the present, and in other ancient phyla such as Echinodermata (> 500 million years old), Cephalopoda (500 million years old), Amphibia (300 million years old), and Aves (150 million years old). Curiously, cancer has never been seen (or at least reported) in a number of phyla such as Nematoda, Tradigrada, and Rotifera. It is intriguing to consider that these organisms may have some protective mechanisms that prevent them from getting tumors. If so, it would be important to find out what these mechanisms are.

One thing is clear, though, which is that cancer is a disease of multicellular organisms. This trait implies that there is something inherent in the ability of cells to proliferate in clumps or to differentiate into different cell types and move around in the body to sites of organogenesis that is key to the process of tumorigenesis. Problems occur when these processes become dysregulated.

One might also argue that evolution itself has played some tricks on us because some of the properties selected for may themselves be processes that cancer cells use to become invasive and metastatic. Or to phrase it differently: Is cancer an inevitable result of a complex evolutionary process that has advantages and disadvantages? Some of these processes might be the following:

1. The mechanism of cell invasiveness that allows the implantation of the early embryo into the uterine wall and the development of a placenta.
2. Cell motility that allows neural cells, for example, to migrate from the original neural crest to form the nervous system.
3. The development of a large, complex genome of up to 40,000 genes that must be replicated perfectly every time a cell divides.
4. The large number of cells in a human or higher mammal that must replicate and differentiate nearly perfectly every time (some can be destroyed if they become abnormal).
5. The long life span of humans and higher mammals, increasing the chance for a genetic "hit" to occur and lead a cell down a malignant path.

As we shall see in later chapters of this book, cancer cells take advantage of a number of these events and processes.

Other questions that arose at the gathering above from scientists not in the field of cancer were the following:

1. Is there a single trait or traits that all cancer cells have?
2. How many genetic "hits" does it take to make a cancer cell?
3. What kinds of genes are involved in these hits?

These questions are all dealt with in later chapters. Suffice it to say here that for a cell to become cancerous or at least take the first steps to becoming cancerous, at least two genetic hits are required. One may be inherited and another accrued after birth or both may be accrued after birth (so-called somatic, or spontaneous, hits). The kinds of genes involved are oncogenes, which when activated lead to dysregulated cell proliferation, and tumor suppressor genes, which become inactivated or deleted, producing a loss of the cell's checks and balances controlling cell proliferation and differentiation.

The single most common, if not universal, trait that occurs in all cancers is genetic drift. or the ability of cells to lose the stringent requirement for precise DNA replication and to acquire the ability to undergo sequential progressive changes in their genome, through mutations, gene rearrangement, or gene deletion. This has sometimes been called the acquisition of a "mutator phenotype."

WHAT SIGNIFICANT EVENTS HAVE HAPPENED IN CANCER RESEARCH IN THE LAST 25 YEARS?

As I was beginning to gather my thoughts for the fourth edition of *Cancer Biology*, one of my colleagues mentioned that he thought it would be of interest to describe the significant things that have happened in cancer biology in the 25 years since the first edition was published (1981). Many things have happened since then, of course, and everyone has their favorite list. But looking back at the table of contents for the first edition and at the outline for this edition, several things struck me, as listed below.

1. Cancer susceptibility genes. In 1981 we knew that familial clustering of some cancers occurred, for example, with colon cancer, but the genes involved in this hadn't been determined. The APC, BRCA-1, BRCA-2, and p53 inherited mutations, for example, were not known at that time. Research in this area has identified a number of genes involved in cancer susceptibility, and with modern cloning techniques, more are identified every few months.
2. The techniques of modern molecular biology were in their infancy at that time. Polymerase chain reaction (PCR), DNA

microarrays, protein chips, and bioinformatics were not terms in anybody's dictionary.

3. Genes involved in cancer initiation and promotion were very poorly defined. Although we knew that chemicals and irradiation could damage DNA and initiate cancer in animals and humans, the specific genes altered were almost completely unknown. We now know a lot about the genes involved at various stages of a number of cancers. For example, the work of Bert Vogelstein and colleagues has defined a pathway sometimes called the "Vogelgram" for the progression of colon cancer (see Chapter 5). We knew that DNA repair was important and that heritable conditions of defective DNA repair (e.g., xeroderma pigmentosum) could lead to cancer, but the ideas about the mechanisms of DNA repair were primitive.

4. The identification of oncogenes didn't really start until the early 1980s. The *src* gene was identified in 1976 by Stehelin et al., and *erb*, *myc*, and *myb* oncogenes were identified in the late 1970s, but this was about the limit of our knowledge (see Chapter 5).

5. The term *tumor suppressor gene* wasn't even coined until the early 1980s, although their existence had been implied from the cell fusion experiments of Henry Harris, (Chapter 5) who showed that if a normal cell was fused with a malignant cell, the phenotype was usually nonmalignant. The RB gene was the first one cloned, in 1983 by Cavenee et al. (Chapter 5) p53 was originally thought of as an oncogene. It wasn't realized until 1989 that wild-type p53 could actually suppress malignant transformation. A number of tumor suppressor genes have, of course, been identified since then.

6. Starting in the 1970s, cell cycle checkpoints were identified in yeast by Lee Hartwell and colleagues, but the identification of human homologs of these genes didn't occur until the late 1980s (see Chapter 4).

7. Tumor immunology was still poorly understood in 1981—both the mechanism of the immune response and the ability to manipulate it with cytokines, activated dendritic cells, and vaccines. Such manipulation was not in the treatment armamentarium.

8. The first treatment of a patient with gene therapy occurred in 1990. Several gene therapy clinical trials for cancer are under way and some gene therapy modalities will likely be approved in the next few years.

9. The viral etiology of cancer was still being widely debated in 1981. The involvement of Epstein-Barr virus in Burkitt's lymphoma and of hepatitis B virus in liver cancer was becoming accepted, but the role of viruses in these diseases and in cervical cancer, Kaposis' sarcoma, and in certain T-cell lymphomas became clearer much later.

10. Although some growth factors that affect cancer cell replication, such as IGF-1 and IGF-2, FGF, NGF, PDGF, and EGF, were known in 1981, knowledge about their receptors and signal transduction mechanisms was primitive indeed. Tumor growth factor α was known as sarcoma growth factor (SGF), and the existence of its partner, TGF-β, was only implied from what was thought to be a contaminating HPLC peak from the purification procedure. The explosion of knowledge about signal transduction mechanisms and how these pathways interact has been a tremendous boon to our understanding of how cells respond to signals in their environment and communicate with each other.

11. Knowledge about the regulation of gene expression has greatly increased in the past 25 years, on the basis of our current information on the packaging of chromatin, transcription factors, coinducers and corepressors, and inhibitory RNA (siRNA).

12. While not topics discussed in detail in the earlier editions of *Cancer Biology*, advances in diagnostic imaging such as magnetic resonance imaging (MRI), computed tomography (CT), and positron emission tomography (PET) have significantly im-

proved cancer diagnosis. Improved radiation therapy, combined modality therapy, bone marrow transplant, and supportive care have also improved significantly.

BASIC FACTS ABOUT CANCER

Cancer is a complex family of diseases, and carcinogenesis, the events that turn a normal cell in the body into a cancer cell, is a complex multistep process. From a clinical point of view, cancer is a large group of diseases, perhaps up to a hundred or more, that vary in their age of onset, rate of growth, state of cellular differentiation, diagnostic detectability, invasiveness, metastatic potential, response to treatment, and prognosis. From a molecular and cell biological point of view, however, cancer may be a relatively small number of diseases caused by similar molecular defects in cell function resulting from common types of alterations to a cell's genes. Ultimately, cancer is a disease of abnormal gene expression. There are a number of mechanisms by which this altered gene expression occurs. These mechanisms may occur via a direct insult to DNA, such as a gene mutation, translocation, amplification, deletion, loss of heterozygosity, or via a mechanism resulting from abnormal gene transcription or translation. The overall result is an imbalance of cell replication and cell death in a tumor cell population that leads to an expansion of tumor tissue. In normal tissues, cell proliferation and cell loss are in a state of equilibrium.

Cancer is a leading cause of death in the Western world. In the United States and a number of European countries, cancer is the second-leading killer after cardiovascular disease, although in the United States since 1999 cancer has surpassed heart disease as the number one cause of death in people younger than 85.[1] Over 1.3 million new cases of cancer occur in the United States each year, not including basal cell and squamous cell skin cancers, which add another 1 million cases annually. These skin cancers are seldom fatal, do not usually metastasize, and are curable with appropriate treatment, so they are usually considered separately. Melanoma, by contrast, is a type of skin cancer that is more dangerous and can be fatal, so it is considered

with the others. The highest mortality rates are seen with lung, colorectal, breast, and prostate cancers (Fig. 1–1). Over 570,000 people die each year in the United States from these and other cancers. More people die of cancer in 1 year in the United States than the number of people killed in all the wars in which the United States was involved in the twentieth century (Fig. 1–2).

In many cases the causes of cancer aren't clearly defined, but both external (e.g., environmental chemicals and radiation) and internal (e.g., immune system defects, genetic predisposition) factors play a role (see Chapter 2). Clearly, cigarette smoking is a major causative factor. These causal factors may act together to initiate (the initial genetic insult) and promote (stimulation of growth of initiated cells) carcinogenesis. Often 10 to 20 years may pass before an initiated neoplastic cell grows into a clinically detectable tumor.

Although cancer can occur at any age, it is usually considered a disease of aging. The average age at the time of diagnosis for cancer of all sites is 67 years, and about 76% of all cancers are diagnosed at age 55 or older. Although cancer is relatively rare in children, it is the second-leading cause of death in children ages 1–14. In this age group leukemia is the most common cause of death, but other cancers such as osteosarcoma, neuroblastoma, Wilms' tumor (a kidney cancer), and lymphoma also occur.

Over eight million Americans alive today have had some type of cancer. Of these, about half are considered cured. It is estimated that about one in three people now living will develop some type of cancer.

There has been a steady rise in cancer death rates in the United States during the past 75 years. However, the major reason why cancer accounts for a higher proportion of deaths now than it did in the past is that today more people live long enough to get cancer, whereas earlier in the twentieth century more people died of infectious disease and other causes. For example, in 1900 life expectancy was 46 years for men and 48 years for women. By 2000, the expectancy had risen to age 74 for men and age 80 for women. Thus, even though the overall death rates due to cancer have almost tripled since 1930 for men and gone up over 50% for women,

Estimated New Cases*

Males   Females

Prostate	232,090	33%		Breast	211,240	32%
Lung and Bronchus	93,010	13%		Lung and Bronchus	79,560	12%
Colon and Rectum	71,820	10%		Colon and Rectum	73,470	11%
Urinary Bladder	47,010	7%		Uterine Corpus	40,880	6%
Melanoma of the Skin	33,580	5%		Non-Hodgkin Lymphoma	27,320	4%
Non-Hodgkin Lymphoma	29,070	4%		Melanoma of the Skin	26,000	4%
Kidney and Renal Pelvis	22,490	3%		Ovary	22,220	3%
Leukemia	19,640	3%		Thyroid	19,190	3%
Oral Cavity and Pharynx	19,100	3%		Urinary Bladder	16,200	2%
Pancreas	16,100	2%		Pancreas	16,080	2%
All Sites	**710,040**	**100%**		**All Sites**	**662,870**	**100%**

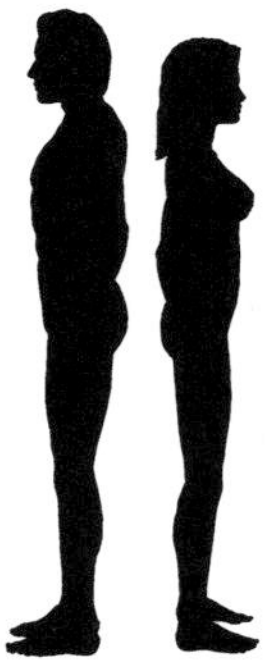

Estimated Deaths

Males   Females

Lung and Bronchus	90,490	31%		Lung and Bronchus	73,020	27%
Prostate	30,350	10%		Breast	40,410	15%
Colon and Rectum	28,540	10%		Colon and Rectum	25,750	10%
Pancreas	15,820	5%		Ovary	16,210	6%
Leukemia	12,540	4%		Pancreas	15,980	6%
Esophagus	10,530	4%		Leukemia	10,030	4%
Liver and Intrahepatic Bile Duct	10,330	3%		Non-Hodgkin Lymphoma	9050	3%
Non-Hodgkin Lymphoma	10,150	3%		Uterine Corpus	7310	3%
Urinary Bladder	8970	3%		Multiple Myeloma	5640	2%
Kidney and Renal Pelvis	8020	3%		Brain and Other Nervous System	5480	2%
All Sites	**295,280**	**100%**		**All Sites**	**275,000**	**100%**

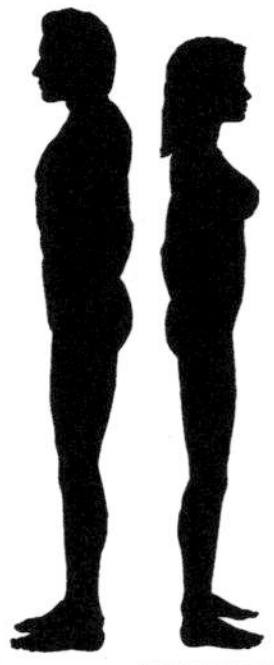

Figure 1–1. Ten leading cancer types for estimated new cancer cases and deaths, by sex, United States, 2005. *Excludes basal and squamous cell skin cancers and in situ carcinoma except urinary bladder. Estimates are rounded to the nearest 10. Percentage may not total 100% due to rounding. (From American Cancer Society, Surveillance Research, 2005. *CA Cancer J Clin* 2005; 55:10–30, with permission.)

the age-adjusted cancer death rates in men have only increased 54% in men and not at all for women.[2]

The major increase has been in deaths due to lung cancer. Thus, cigarette smoking is a highly suspect culprit in the observed increases. In addition, pollution, diet, and other lifestyle changes may have contributed to this increase in cancer mortality rates (Chapter 3). The mortality rates for some cancers has decreased in the past 50 years (e.g., stomach, uterine cervix); however, the mortality rates have been essentially flat for many of the major cancers such as breast, colon, and prostate, although 5-year survival rates have improved for these cancers (see Chapter 3).

It is instructive to examine the trends in cancer mortality over time to get some clues about the causes of cancer. For males, lung cancer remains the number one cancer killer (Fig. 1–3). With a lag of about 20 years, its rise in mortality parallels the increase in cigarette smoking among men, which has an almost identical curve starting in the early 1900s. Lung cancer mortality rates for men have decreased somewhat since 1990, and death rates for colorectal cancer have dropped slightly in recent years, whereas prostate cancer mortality has increased somewhat. Stomach cancer mortality has dropped significantly since the early 1900s, presumably because of better methods of food preservation

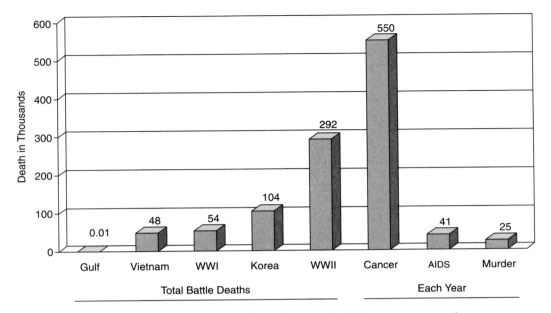

Figure 1–2. Total battle deaths from all wars with U.S. involvement in the twentieth century, compared to number of deaths each year from cancer, AIDS, and murder in the United States. (Personal communication from Don Coffey, Johns Hopkins University, with permission.)

(e.g., better refrigeration, less addition of nitrate and nitrate preservatives). Cancer of the gastroesophageal junction, however, has risen significantly in recent years, perhaps due to obesity and increased incidence of gastric reflux into the esophagus in the U.S. population.

Somewhat surprising, perhaps, is the fact that lung cancer has overtaken breast cancer as the number one cancer killer in women (Fig. 1–4). This increase occurred in the late 1980s and, as was the case for males, parallels the rise in the percentage of women who smoke. Smoking started to increase dramatically during World War II. Rosie the Rivetter picked up some bad male habits along with increased access to traditionally male jobs.

Breast cancer mortality rates have remained stubbornly stable, although a small decrease (5%) has occurred since 1990. Uterine cancer death rates have been going down, primarily through earlier detection and treatment of cervical cancer. Female colon cancer mortality has been decreasing, but the reasons for this aren't clear. As in males, stomach cancer mortality in women has been going down for many years.

The good news is that more and more people are being cured of their cancers today. In the 1940s, for example, only one in four persons diagnosed with cancer lived at least 5 years after treatment; in the 1990s that figure rose to 40%. When normal life expectancy is factored into this calculation, the relative 5-year survival rate is about 64% for all cancers taken together.[1] Thus, the gain from 1 in 3 to 4 in 10 survivors means that almost 100,000 people are alive now who would have died from their disease in less than 5 years if they had been living in the 1940s. This progress is due to better diagnostic and treatment techniques, many of which have come about from our increasing knowledge of the biology of the cancer cell.

HALLMARKS OF MALIGNANT DISEASES

Malignant neoplasms or cancers have several distinguishing features that enable the pathologist or experimental cancer biologist to characterize them as abnormal. The most common

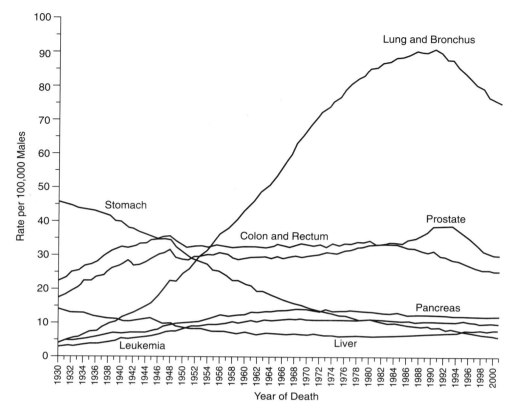

Figure 1–3. Annual age-adjusted cancer death rates° among males for selected cancer types, United States, 1930 to 2001. °Rates are age adjusted to the 2000 U.S. standard population. Because of changes in ICD coding, numerator information has changed over time rates for cancers of the lung and bronchus, colon and rectum, and liver are affected by these changes. (From U.S. Mortality Public Use Data Tapes, 1960 to 2001, *U.S. Mortality Volumes*, 1930 to 1959, National Center for Health Statistics, Centers for Disease Control and Prevention, with permission.)

types of human neoplasms derive from *epithelium*, that is, the cells covering internal or external surfaces of the body. These cells have a supportive stroma of blood vessels and connective tissue. Malignant neoplasms may resemble normal tissues, at least in the early phases of their growth and development. Neoplastic cells can develop in any tissue of the body that contains cells capable of cell division. Though they may grow fast or slowly, their growth rate frequently exceeds that of the surrounding normal tissue. This is not an invariant property, however, because the rate of cell renewal in a number of normal tissues (e.g., gastrointestinal tract epithelium, bone marrow, and hair follicles) is as rapid as that of a rapidly growing tumor.

The term *neoplasm*, meaning new growth, is often used interchangeably with the term *tumor* to signify a cancerous growth. It is important to keep in mind, however, that tumors are of two basic types: benign and malignant. The ability to distinguish between benign and malignant tumors is crucial in determining the appropriate treatment and prognosis of a patient who has a tumor. The following are features that differentiate a malignant tumor from a benign tumor:

1. Malignant tumors invade and destroy adjacent normal tissue; benign tumors grow by expansion, are usually encapsulated, and do not invade surrounding tissue. Benign tumors may, however, push aside

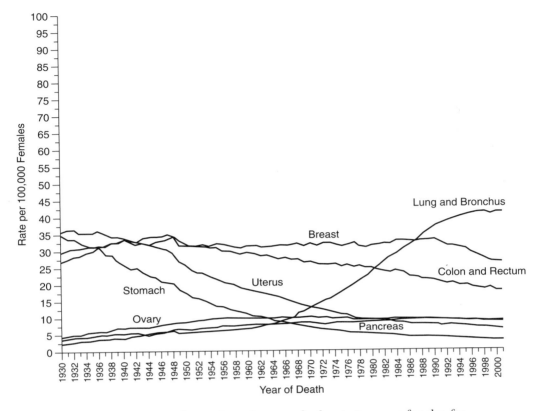

Figure 1–4. Annual age-adjusted cancer death rates° among females for selected cancer types, United States, 1930 to 2001. °Rates are age adjusted to the 2000 U.S. standard population. Because of ICD coding, numerator information has changed over time, rates for cancers of the uterus, ovary, lung and bronchus, and colon and rectum are affected by these changes. Uterus cancers are for uterine cervix and uterine corpus combined. (From U.S. Mortality Public Use Data Tapes, 1960 to 2001, *U.S. Mortality Volumes,* 1930 to 1959, National Center for Health Statistics, Centers for Disease Control and Prevention, with permission.)

normal tissue and may become life threatening if they press on nerves or blood vessels or if they secrete biologically active substances, such as hormones, that alter normal homeostatic mechanisms.

2. Malignant tumors metastasize through lymphatic channels or blood vessels to lymph nodes and other tissues in the body. Benign tumors remain localized and do not metastasize.

3. Malignant tumor cells tend to be "anaplastic," or less well differentiated than normal cells of the tissue in which they arise. Benign tumors usually resemble normal tissue more closely than malignant tumors do.

Some malignant neoplastic cells at first structurally and functionally resemble the normal tissue in which they arise. Later, as the malignant neoplasm progresses, invades surrounding tissues, and metastasizes, the malignant cells may bear less resemblance to the normal cell of origin. The development of a less well-differentiated malignant cell in a population of differentiated normal cells is sometimes called *dedifferentiation.* This term is probably a misnomer for the process, because it implies that a differentiated cell goes backwards in its developmental process after carcinogenic insult. It is more likely that the anaplastic malignant cell type arises from the progeny of a tissue "stem cell" (one that still has a capacity for renewal and is not yet fully differentiated), which has been blocked or diverted in its pathway to form a fully differentiated cell.

Examples of neoplasms that maintain a modicum of differentiation include islet cell tumors of the pancreas that still make insulin, colonic adenocarcinoma cells that form glandlike epithelial structures and secrete mucin, and breast carcinomas that make abortive attempts to form structures resembling mammary gland ducts. Hormone-producing tumors, however, do not respond to feedback controls regulating normal tissue growth or to negative physiologic feedback regulating hormonal secretion. For example, an islet cell tumor may continue to secrete insulin in the face of extreme hypoglycemia, and an ectopic adrenocorticotropic hormone (ACTH)-producing lung carcinoma may continue to produce ACTH even though circulating levels of adrenocortical steroids are sufficient to cause Cushing's syndrome (see Chapter 6). Many malignant neoplasms, particularly the more rapidly growing and invasive ones, only vaguely resemble their normal counterpart tissue structurally and functionally. They are thus said to be "undifferentiated" or "poorly differentiated."

4. Malignant tumors usually, though not invariably, grow more rapidly than benign tumors. Once they reach a clinically detectable stage, malignant tumors generally show evidence of significant growth, with involvement of surrounding tissue, over weeks or months, whereas benign tumors often grow slowly over several years.

Malignant neoplasms continue to grow even in the face of starvation of the host. They press on and invade surrounding tissues, often interrupting vital functions; they metastasize to vital organs, for example, brain, spine, and bone marrow, compromising their functions; and they invade blood vessels, causing bleeding. The most common effects on the patient are cachexia (extreme body wasting), hemorrhage, and infection. About 50% of terminal patients die from infection (see Chapter 8).

Differential diagnosis of cancer from a benign tumor or a nonneoplastic disease usually involves obtaining a tissue specimen by biopsy, surgical excision, or exfoliative cytology. The latter is an examination of cells obtained from swabbings, washings, or secretions of a tissue suspected to harbor cancer: the "Pap test" involves such an examination.

CLASSIFICATION OF HUMAN CANCERS

Although the terminology applied to neoplasms can be confusing for a number of reasons, certain generalizations can be made. The suffix *oma*, applied by itself to a tissue type, usually indicates a benign tumor. Some malignant neoplasms, however, may be designated by the *oma* suffix alone; these include lymphoma, melanoma, and thymoma. Rarely, the *oma* suffix is used to describe a nonneoplastic condition such as granuloma, which is often not a true tumor, but a mass of granulation tissue resulting from chronic inflammation or abscess. Malignant tumors are indicted by the terms *carcinoma* (epithelial in origin) or *sarcoma* (mesenchymal in origin) preceded by the histologic type and followed by the tissue of origin. Examples of these include adenocarcinoma of the breast, squamous cell carcinoma of the lung, basal cell carcinoma of skin, and leiomyosarcoma of the uterus. Most human malignancies arise from epithelial tissue. Those arising from stratified squamous epithelium are designated *squamous cell carcinomas*, whereas those emanating from glandular epithelium are termed *adenocarcinomas*. When a malignant tumor no longer resembles the tissue of origin, it may be called *anaplastic* or *undifferentiated*. If a tumor is metastatic from another tissue, it is designated, for example, an adenocarcinoma of the colon metastatic to liver. Some tumors arise from pluripotential primitive cell types and may contain several tissue elements. These include mixed mesenchymal tumors of the uterus, which contain carcinomatous and sarcomatous elements, and teratocarcinomas of the ovary, which may contain bone, cartilage, muscle, and glandular epithelium.

Neoplasms of the hematopoietic system usually have no benign counterparts. Hence the terms *leukemia* and *lymphoma* always refer to a malignant disease and have cell-type designations such as acute or chronic myelogenous leukemia, Hodgkin's or non-Hodgkin's lymphoma, and so on. Similarly, the term *melanoma* always refers to a malignant neoplasm derived from melanocytes.

MACROSCOPIC AND MICROSCOPIC FEATURES OF NEOPLASMS

The pathologist can gain valuable insights about the nature of a neoplasm by careful examination of the overall appearance of a surgical specimen. Often, by integrating the clinical findings with macroscopic characteristics of a tumor, a tentative differential diagnosis can be reached. Also, notation of whether the tumor is encapsulated, has extended through tissue borders, or reached to the margins of the excision provides important diagnostic information.

The location of the anatomic site of the neoplasm is important for several reasons. The site of the tumor dictates several things about the clinical course of the tumor, including (1) the likelihood and route of metastatic spread, (2) the effects of the tumor on body functions, and (3) the type of treatment that can be employed. It is also important to determine whether the observed tumor mass is the primary site (i.e., tissue of origin) of the tumor or a metastasis. A primary epidermoid carcinoma of the lung, for example, would be treated differently and have a different prognosis than an embryonal carcinoma of the testis metastatic to the lung. It is not always easy to determine the primary site of a neoplasm, particularly if the tumor cells are undifferentiated. The first signs of a metastatic tumor may be a mass in the lung noted on CT scan or a spontaneous fracture of a vertebra that had been invaded by cancer cells. Because the lungs and bones are frequent sites of metastases for a variety of tumors, the origin of the primary tumor may not be readily evident. This is a very difficult clinical situation, because to cure the patient or to produce long-term remission, the oncologist must be able to find and remove or destroy the primary tumor to prevent its continued growth and metastasis. If histologic examination does not reveal the source of the

primary tumor, or if other diagnostic techniques fail to reveal other tumor masses, the clinician has to treat blindly, and thus might not choose the best mode of therapy.

Another consideration is the accessibility of a tumor. If a tumor is surgically inaccessible or too close to vital organs to allow complete resection, surgical removal is impossible. For example, a cancer of the common bile duct or head of the pancreas is often inoperable by the time it is diagnosed because these tumors invade and attach themselves to vital structures early, thus preventing curative resection. Similarly, if administered anticancer drugs cannot easily reach the tumor site, as is the case with tumors growing in the pleural cavity or in the brain, these agents might not be able to penetrate in sufficient quantities to kill the tumor cells.

The site of the primary tumor also frequently determines the mode of, and target organs for, metastatic spread. In addition to local spread, cancers metastasize via lymphatic channels or blood vessels. For example, carcinomas of the lung most frequently metastasize to regional lymph nodes, pleura, diaphragm, liver, bone, kidneys, adrenals, brain, thyroid, and spleen. Carcinomas of the colon metastasize to regional lymph nodes, and by local extension, they ulcerate and obstruct the gastrointestinal tract. The most common site of distant metastasis of colon carcinomas is the liver, via the portal vein, which receives much of the venous return from the colon and flows to the liver. Breast carcinomas most frequently spread to axillary lymph nodes, the opposite breast through lymphatic channels, lungs, pleura, liver, bone, adrenals, brain, and spleen.

Some tissues are more common sites of metastasis than others. Because of their abundant blood and lymphatic supply, as well as their function as "filters" in the circulatory system, the lungs and the liver are the most common sites of metastasis from tumors occurring in visceral organs. Metastasis is usually the single most important criterion determining the patient's prognosis. In breast carcinoma, for example, the 5-year survival rate for patients with localized disease and no evidence of axillary lymph node involvement is about 85%; but when more than four axillary nodes are involved, the 5-year survival is about 30%, on average.[3]

The anatomic site of a tumor will also determine its effect on vital functions. A lymphoma growing in the mediastinum may press on major blood vessels to produce the superior vena caval syndrome, manifested by edema of the neck and face, distention of veins of the neck, chest, and upper extremities, headache, dizziness, and fainting spells. Even a small tumor growing in the brain can produce such dramatic central nervous system effects as localized weakness, sensory loss, aphasia, or epileptic-like seizures. A lung tumor growing close to a major bronchus will produce airway obstruction earlier than one growing in the periphery of the lung. A colon carcinoma may invade surrounding muscle layers of the colon and constrict the lumen, causing intestinal obstruction. One of the frequent symptoms of prostatic cancer is inability to urinate normally.

The cytologic criteria that enable the pathologist to confirm the diagnosis, or at least to suspect that cancer is present (thus indicating the need for further diagnostic tests), are as follows:

1. The morphology of cancer cells is usually different from and more variable than that of their counterpart normal cells from the same tissue. Cancer cells are more variable in size and shape.
2. The nucleus of cancer cells is often larger and the chromatin more apparent ("hyperchromatic") than the nucleus in normal cells; the nuclear-to-cytoplasmic ratio is often higher; and the cancer cell nuclei contain prominent, large nucleoli.
3. The number of cells undergoing mitosis is usually greater in a population of cancer cells than in a normal tissue population. Twenty or more mitotic figures per 1000 cells would not be an uncommon finding in cancerous tissue, whereas less than 1 per 1000 is usual for benign tumors or normal tissue.[4] This number, of course, would be higher in normal tissues that have a high growth rate, such as bone marrow and crypt cells of the gastrointestinal mucosa.
4. Abnormal mitosis and "giant cells," with large, pleomorphic (variable size and shape) or multiple nuclei, are much more common in malignant tissue than in normal tissue.
5. Obvious evidence of invasion of normal tissue by a neoplasm may be seen, indicating that the tumor has already become invasive and may have metastasized.

GRADE AND STAGE OF NEOPLASMS

Histologic Grade of Malignancy

The histologic grading of malignancy is based on the degree of differentiation of a cancer and on an estimate of the growth rate as indicated by the mitotic index. It was generally believed that less differentiated tumors were more aggressive and more metastatic than more differentiated tumors. It is now appreciated that this is an oversimplification and, in fact, not a very accurate way to assess the degree of malignancy for certain kinds of tumors. However, for certain epithelial tumors, such as carcinomas of the cervix, uterine endometrium, colon, and thyroid, histologic grading is a fairly accurate index of malignancy and prognosis. In the case of epidermoid carcinomas, for example, in which keratinization occurs, keratin production provides a relatively facile way to determine the degree of differentiation. On the basis of this criterion, and others like it, tumors have been classified as grade I (75% to 100% differentiation), grade II (50% to 75%), grade III (25% to 50%), and grade IV (0% to 25%).[4] More recent methods of malignancy grading also take into consideration mitotic activity, amount of infiltration into surrounding tissue, and amount of stromal tissue in or around the tumor. The chief value of grading is that it provides, for certain cancers, a general guide to prognosis and an indicator of the effectiveness of various therapeutic approaches.

Tumor Staging

Although the classification of tumors based on the preceding descriptive criteria helps the oncologist determine the malignant potential of a tumor, judge its probable course, and determine the patient's prognosis, a method of discovering the extent of disease on a clinical basis and a universal language to provide standardized criteria among physicians are needed. Attempts to develop an international language for describing the extent of disease have been carried out by two major agencies—the *Union Internationale Contre le Cancer* (UICC) and the *American Joint Committee for Cancer Staging and End Results Reporting* (AJCCS). Some of the objectives of the classification system developed by these groups are (1) to aid oncologists in planning treatment; (2) to provide categories for estimating prognosis and evaluating results of treatment; and (3) to facilitate exchange of information.[5] Both the UICC and AJCCS schemes use the T, N, M classification system, in which T categories define the primary tumor; N, the involvement of regional lymph nodes; and M, the presence or absence of metastases. The definition of extent of malignant disease by these categories is termed *staging*. Staging defines the extent of tumor growth and progression at one point in time; four different methods are involved:

1. Clinical staging: estimation of disease progression based on physical examination, clinical laboratory tests, X-ray films, and endoscopic examination.
2. Tumor imaging: evaluation of progression based on sophisticated radiography—for example, CT scans, arteriography, lymphangiography, and radioisotope scanning; MRI; and PET.
3. Surgical staging: direct exploration of the extent of the disease by surgical procedure.
4. Pathologic staging: use of biopsy procedures to determine the degree of spread, depth of invasion, and involvement of lymph nodes.

These methods of staging are not used interchangeably, and their use depends on agreed-upon procedures for each type of cancer. For example, operative findings are used to stage certain types of cancer (e.g., ovarian carcinomas) and lymphangiography is required to stage Hodgkin's disease. Although this means that different staging methods are used to stage different tumors, each method is generally agreed on by oncologists, thus allowing a comparison of data from different clinical centers. Once a tumor is clinically staged, it is not usually changed for that patient; however, as more information becomes available following a more extensive workup, such as a biopsy or surgical exploration, this information is, of course, taken into consideration in determining treatment and estimating prognosis. Staging provides a useful way to estimate at the outset what a patient's clinical course and initial treatment should be. The actual course of the disease indicates its true extent. As more is learned about the natural history of cancers, and as more sophisticated diagnostic techniques become available, the criteria for staging will likely change and staging should become more accurate (see Chapter 7).

It is important to remember that staging does not mean that any given cancer has a predictable, ineluctable progression. Although some tumors may progress in a stepwise fashion from a small primary tumor to a larger primary tumor, and then spread to regional nodes and distant sites (i.e., progressing from stage I to stage IV), others may spread to regional nodes or have distant metastases while the primary tumor is microscopic and clinically undetectable. Thus, staging is somewhat arbitrary, and its effectiveness is really based on whether it can be used as a standard to select treatment and to predict the course of disease.

Although the exact criteria used vary with each organ site, the staging categories listed below represent a useful generalization.[6]

Stage I ($T_1 N_0 M_0$): Primary tumor is limited to the organ of origin. There is no evidence of nodal or vascular spread. The tumor can usually be removed by surgical resection. Long-term survival is from 70% to 90%.

Stage II ($T_2 N_1 M_0$): Primary tumor has spread into surrounding tissue and lymph nodes immediately draining the area of the tumor ("first-station" lymph nodes). The tumor is operable, but because of local spread, it may not be completely resectable. Survival is 45% to 55%.

Stage III ($T_3 N_2 M_0$): Primary tumor is large, with fixation to deeper structures. First-station lymph nodes are involved; they may be more than 3 cm in diameter and fixed to underlying tissues. The tumor is not usually resectable, and part of the tumor mass is left behind. Survival is 15% to 25%.

Stage IV (T_4 N_3 M_+): Extensive primary tumor (may be more than 10 cm in diameter) is present. It has invaded underlying or surrounding tissues. Extensive lymph node involvement has occurred, and there is evidence of distant metastases beyond the tissue of origin of the primary tumor. Survival is under 5%.

The criteria for establishing lymph node involvement (N categories) are based on size, firmness, amount of invasion, mobility, number of nodes involved, and distribution of nodes involved (i.e., ipsilateral, contralateral, distant involvement): N_0 indicates that there is no evidence of lymph node involvement; N_1 indicates that there are palpable lymph nodes with tumor involvement, but they are usually small (2 to 3 cm in diameter) and mobile; N_2 indicates that there are firm, hard, partially movable nodes (3 to 5 cm in diameter), partially invasive, and they may feel as if they were matted together; N_3 indicates that there are large lymph nodes (over 5 cm in diameter) with complete fixation and invasion into adjacent tissues; N_4 indicates extensive nodal involvement of contralateral and distant nodes.

The criteria applied to metastases (M categories) are as follows: M_0, no evidence of metastasis; M_1, isolated metastasis in one other organ; M_2, multiple metastases confined to one organ, with minimal functional impairment; M_3, multiple organs involved with no to moderate functional impairment; M4, multiple organ involvement with moderate to severe functional impairment. Occasionally a subscript is used to indicate the site of metastasis, such as M_p, M_h, M_o for pulmonary, hepatic, and osseous metastases, respectively.

Diagnostic procedures are getting more sophisticated all the time. Improved CT, MRI, and PET scanners, as well as ultrasound techniques, are being developed to better localize tumors and determine their metabolic rate. One can visualize the day when "noninvasive biopsies," based on the ability to carry out molecular and cellular imaging by means of external detection of internal signals, may at least partially replace the need for biopsy or surgical specimens to get diagnostic information (see Chapter 7). There will always be the need, however, for clinical pathologists to examine tissue specimens to confirm noninvasive procedures, at least for the foreseeable future. The ultimate diagnosis, prognosis, and selection of a treatment course will depend on this.

Although the TNM system is useful for staging malignant tumors, it is primarily based on a temporal model that assumes a delineated progression over time from a small solitary lesion to one that is locally invasive, then involves lymph nodes, and finally spreads through the body. While this is true for some cancers, the linearity of this progression model is an oversimplification. For example, some patients have aggressive tumors almost from the outset and may die before lymph node involvement becomes evident, whereas others may have indolent tumors that grow slowly and remain localized for a long time, even though they may become large.

In addition, the TNM staging system does not take into account the molecular markers that we now know can more clearly define the status of a cancer, e.g., its gene array and proteomic profiles (see Chapter 7). Nor does the TNM system, as a prognostic indicator, take into account the varied responsiveness of tumors to various therapeutic modalities. Thus, treatment choices and prognostic estimates should be based more on the molecular biology of the tumor than the tumor's size, location, or nodal status at the time of diagnosis.[7]

References

1. A. Jemal, T. Murray, E. Ward, A. Samuels, R. C. Tivari, A. Ghafoor, E. J. Feuer, and M. J. Thun: Cancer statistics, 2005. *CA Cancer J Clin* 55:10, 2005.
2. P. A. Wingo, C. J. Cardinez, S. H. Landis, R. T. Greenlee, A. G. Ries, R. N. Anderson, and M. J. Thun: Long-term trends in cancer mortality in the United States, 1930–1998. *Cancer* 97:3133, 2003.
3. I. C. Henderson and G. P. Canellos: Cancer of the breast—The past decade. *N Engl J Med* 302:17, 1980.
4. S. Warren: Neoplasms. In W. A. D. Anderson, ed.: *Pathology*. St. Louis: C. V. Mosby, 1961, pp. 441–480.
5. P. Rubin: A unified classification of cancers: An oncotaxonomy with symbols. *Cancer* 31:963, 1973.
6. P. Rubin: Statement of the clinical oncologic problem. In P. Rubin, ed.: *Clinical Oncology*. Rochester: American Cancer Society, 1974, pp. 1–25.
7. H. B. Burke: Outcome prediction and the future of the TNM staging system. *J Natl Cancer Inst* 96:1408, 2004.

2

Causes of Cancer

Perhaps the most important question in cancer biology is what *causes* the cellular alterations that produce a cancer. The answer to this question has been elusive. If the actual cause of these alterations were known, the elimination of factors that produce cancer and the development of better treatment modalities would likely follow. Cancer prevention might become a reality.

A cancerous growth has a number of predictable properties. The incidence rates of various cancers are strongly related to environmental factors and lifestyle, and cancers have certain growth characteristics, among which are the abilities to grow in an uncontrolled manner, invade surrounding tissues, and metastasize. Also, when viewed microscopically, cancer cells appear to be less well differentiated than their normal counterparts and to have certain distinguishing features, such as large nuclei and nucleoli. Most cancers arise from a single clone of cells, whose precursor may have been altered by insult with a carcinogen. In most cases cancer is a disease of aging. The average age at diagnosis is over 65 and malignant cancers arise from a lifetime accumulation of "hits" on a person's DNA. These hits may result from genetic susceptibility to environmental agents such as chemicals; radiation; or viral, bacterial, or parasitic infections; or from endogenously generated agents such as oxygen radicals. It is often said that we would all get cancer if we lived long enough.

There is frequently a long latent period, in some cases 20 years or more, between the initiating insult and the appearance of a clinically detectable tumor. During this time, cellular proliferation must occur, but it may originally be limited by host defenses or lack of access to the host's blood supply. During the process of tumor progression, however, escape from the host's defense mechanisms and vascularization of the growing tumor ultimately occur.

The genetic instability of cancer cells leads to the emergence of a more aggressively growing tumor frequently characterized by the appearance of poorly differentiated cells with certain properties of a more embryonic phenotype. During tumor progression, considerable biochemical heterogeneity becomes manifest in the growing tumor and its metastases, even though all the neoplastic cells may have arisen originally from a single deranged cell. Any theory that seeks to explain the initiation of cancer and its progression must take these observations into consideration.

In this chapter, we will examine what is known about various chemical, physical, and viral carcinogenic agents and discuss the putative mechanisms by which they cause cancer.

THE THEORY OF "HITS"

As noted above, with the exception of childhood malignancies such as leukemias and sarcomas that occur in children, cancer incidence increases with age. Most of the common adult solid tumors begin to increase after age 45 and go up logarithmically with age after that, as shown for colorectal cancer (Fig. 2–1).[1] This has led to the idea that it takes multiple cellular hits to explain the age-related incidence of malignancy. Most

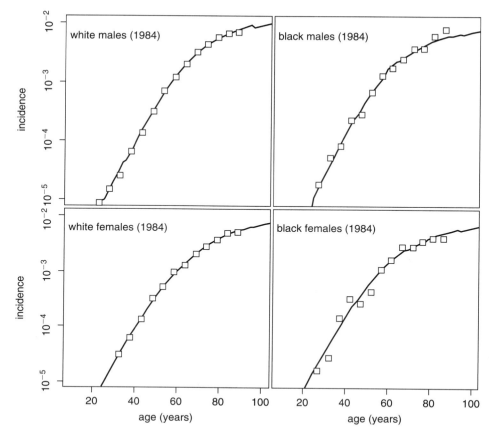

Figure 2–1. Observed (squares) and predicted (lines) incidence of colorectal cancer by race and gender in the Surveillance, Epidemiology, and End Results (SEER) registry (1984). (From Luebeck and Moolgavkar,[1] with permission.)

of these hits are thought to be mutational in origin and to result from chromosomal damage or base changes in DNA. The number of hits needed to produce the initiation of a malignant event may vary from one to six or more. However, progression to a full-blown invasive metastatic cancer almost always requires multiple hits. A few examples will make the point.

In chronic myleogenous leukemia (CML), there is an inciting chromosomal translocation that involves a piece of chromosome 22 being lost. This was first observed by Nowell and Hungerford,[2] who named this small chromosome the Philadelphia chromosome. It was later shown by Rowley[3] that this was a reciprocal translocation between chromosomes 9 and 22 (Fig. 2–2), which produces a chimeric protein called Bcr/Abl that is a constitutively active tyro-

sine kinase promoting cell proliferation (see Chapter 4). Thus, CML appears to be triggered by this one-hit event and is probably the reason why the drug Gleevec, which targets this kinase, is effective as a single agent in CML.

A second example is retinoblastoma. There are two forms of this disease, hereditary and spontaneous. Both forms appear to require two initiating genetic events, leading Knudson, who studied this disease in detail, to postulate the two-hit hypothesis.[4] In the hereditary form, one genetic mutation is inherited at birth and a second one occurs later (Fig. 2–3). This must be the case, since every cell in the eye contains the hereditary mutation, but only three to four tumors on average develop in a retinal cell population of several million cells in affected individuals.

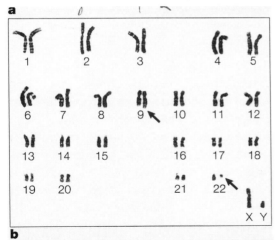

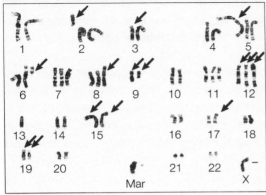

Figure 2–2. A comparison of karyotypes. *a.* Chronic myelogenous leukemia, showing the typical 9;22 translocation and an otherwise normal karyotype. *b.* Non–small cell carcinoma of the lung, showing abnormalities of both number and structure. The arrows indicate aberrant chromosomes. (From Knudson,[4] reprinted by permission from Macmillan Publishers Ltd.)

Most adult solid cancers (e.g., colon, lung, breast, prostate) likely require several hits to achieve a full malignant state. The best example of this is colon cancer, for which at least five hits appear to be required to produce an invasive carcinoma (Fig. 2–4). Because of genetic instability, a characteristic of most solid cancers, many more genetic alterations are frequently seen in later stages of cancer progression.[5] This has been ascribed to a "mutator phenotype" observed in many cancers.[6] In contrast to single genetic defect cancers such as CML, the prospect of finding effective single therapeutic agents is unlikely for most solid tumors. Most likely, multiple aberrant cell signaling pathways will need to be inhibited for effective chemotherapeutic regimens to be achieved. However, if there are identifiable time intervals between the multiple hits that lead to cancer, perhaps detectable by early screening for surrogate markers of progression, there may be a window of opportunity for preventive agents (see Chapter 9).

CHEMICAL CARCINOGENESIS

Historical Perspectives

Carcinogenic chemicals and irradiation (ionizing and ultraviolet) are known to affect DNA and to be mutagenic under certain conditions. Thus, one of the long-standing theories of carcinogenesis is that cancer is caused by a genetic mutation; however, it is now known that epigenetic mechanisms are also involved.

Evidence that chemicals can induce cancer in humans has been accumulating since the sixteenth century (reviewed in Reference 7). In 1567, Paracelsus described a "wasting disease of miners" and proposed that exposure to something in the mined ores caused the condition. A similar condition was described in 1926 in Saxony and was later identified as the "lung cancer of the Schneeberg mines." It was realized much later that the cause of this was probably exposure to radon. Nevertheless, Paracelsus could probably be called "the father of occupational carcinogenesis." It is Bernadini Ramazzini, however, who published a systematic account of work-related diseases in 1700, who is more logically considered the founder of occupational medicine.[7]

Later in the eighteenth century, the first direct observation associating chemicals was made by John Hill, who in 1761 noted that nasal cancer occurred in people who used snuff excessively. In 1775, Percival Pott reported a high incidence of scrotal skin cancer among men who had spent their childhood as chimney sweeps. One hundred years later, von Volkman, in Germany, and Bell, in Scotland, observed skin cancer in workers whose skin was in continuous contact with tar and paraffin oils, which we now know contain polycyclic aromatic hydrocarbons. In 1895, Rehn

Hereditary

Nonhereditary

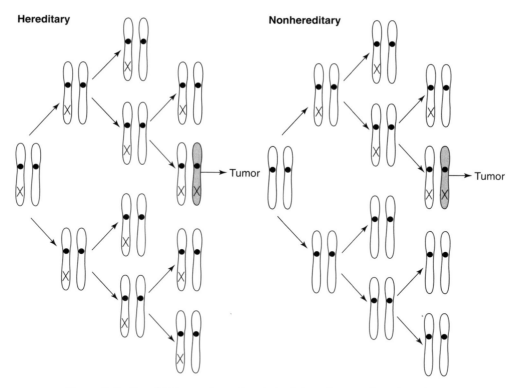

Figure 2–3. Two-hit tumor formation in both hereditary and nonhereditary retinoblastoma. A "one-hit" clone is a precursor to the tumor in nonhereditary retinoblastomas, whereas all retinoblasts (indeed, all cells) are one-hit clones in hereditary retinoblastoma. (From Knudson,[4] reprinted by permission from Macmillan Publishers Ltd.)

reported the development of urinary bladder cancer in aniline dye workers in Germany. Similar observations were later made in a number of countries and established a relationship between heavy exposure to 2-naphthylamine, benzidine, or 4-aminobiphenyl and bladder cancer. Thus, the first observations of chemically induced cancer were made in humans. These observations led to attempts to induce cancer in animals with chemicals. One of the first successful attempts was made in 1915, when Yamagiwa and Ichikawa induced skin carcinomas by the repeated application of coal tar to the ears of rabbits. This and similar observations by other investigators led to a search for the active carcinogen in coal tar and to the conclusion that the carcinogenic agents in tars are the polycyclic aromatic hydrocarbons. Direct evidence for that came in the 1930s from the work of Kennaway and Heiger, who demonstrated that synthetic 1,2,5,6-dibenzanthracene

is a carcinogen, and from the identification of the carcinogen 3,4-benzpyrene in coal tar by Cook, Hewitt, and Hieger. Induction of tumors by other chemical and hormonal carcinogens was described in the 1930s, including the induction of liver tumors in rats and mice with 2', 3-dimethyl-4-aminoazobenzene by Yoshida, of urinary bladder cancer in dogs with 2-naphthylamine by Hueper, Wiley, and Wolfe, and of mammary cancer in male mice with estrone by Lacassagne. The list of known carcinogenic chemicals expanded in the 1940s with the discovery of the carcinogenicity of 2-acetylaminofluorene, halogenated hydrocarbons, urethane, beryllium salts, and certain anticancer alkylating agents. Since the 1940s, various nitrosamines, intercalating agents, nickel and chromium compounds, asbestos, vinyl chloride, diethylstilbestrol, and certain naturally occurring substances, such as aflatoxins, have been added to the list of

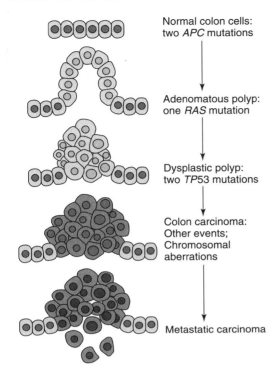

Normal colon cells: two *APC* mutations

Adenomatous polyp: one *RAS* mutation

Dysplastic polyp: two *TP53* mutations

Colon carcinoma: Other events; Chromosomal aberrations

Metastatic carcinoma

Figure 2–4. A possible five-hit scenario for colorectal cancer, showing the mutational events that correlate with each step in the adenoma–carcinoma sequence. (From Knudson,[4] reprinted by permission from Macmillan Publishers Ltd.)

known carcinogens. A list of some known human carcinogens is found in Table 2–1, and the structures of some known carcinogens are shown in Figure 2–5.

Metabolic Activation of Chemical Carcinogens

As studies on the reactions of carcinogens with cellular macromolecules progressed, it became apparent that most of these interactions resulted from covalent bond formation between an electrophilic form of the carcinogen and the nucleophilic sites in proteins (e.g., sulfur, oxygen, and nitrogen atoms in cysteine, tyrosine, and histidine, respectively) and nucleic acids (e.g., purine or pyrimidine ring nitrogens and oxygens). Frequently, the parent compound itself did not interact in vitro with macromolecules until it had been incubated with liver homogenates or liver microsomal fractions. These studies led to the realization that metabolic activation of certain carcinogenic agents is necessary to produce the "ultimate carcinogen" that actually reacts with crucial molecules in target cells. With the exception of the very chemically reactive alkylating agents, which are activated in aqueous solution at physiologic pH (e.g., *N*-methyl-*N*-nitrosourea), and the agents that intercalate into the DNA double helix by forming tight noncovalent bonds (e.g., daunorubicin), most of the known chemical carcinogens undergo some metabolic conversions that appear to be required for their carcinogenic action. Some examples of these metabolic conversions are given next.

Donors of Simple Alkyl Groups

Included in this group are the dialkylnitrosamines, dialkylhydrazines, aryldialkyltriasenes, alkylnitrosamides, and alkylnitrosimides. The alkylnitrosamides and alkylnitrosimides do not require enzymatic activation because they can react directly with water or cellular nucleophilic groups. The alkylnitrosamines, alkylhydrazines, and alkyltriazenes, however, undergo an enzyme-mediated activation step to form the reactive electrophile (Fig. 2–6). These agents are metabolically dealkylated by the mixed-function oxidase system in the microsomal fraction (endoplasmic reticulum) of cells, primarily liver cells. The monoalkyl derivatives then undergo a nonenzymatic, spontaneous conversion to monoalkyldiazonium ions that donate an alkyl to cellular nucleophilic groups in DNA, RNA, and protein.[8]

Cytochrome P-450–Mediated Activation

A number of carcinogenic chemicals are chemically inert nucleophilic agents until they are converted to active nucleophiles by the cytochrome p-450–dependent mixed function oxidases, or CYPs So far, 57 genes encoding these enzymes have been identified in the human genome. The CYPs most involved in carcinogen activation are CYP1A1, 1A2, 1B1, 2A6, and 3A4. A wide variety of chemical carcinogens such as aromatic and heterocyclic amines, aminoazo dyes, polycyclic aromatic hydrocarbons, *N*-nitrosamines, and halogenated olefins are activated by one or more of

Table 2–1. Selected Human Chemical Carcinogens

Compounds	Main Sources and Uses	Affected Organs and Cancer Type
AMINOAZO DYES		
o-Aminoazotoluene	Pigments, coloring oils; immunosuppressant	Liver, lung, bladder
N,N-dimethyl-4-aminoazobenzene	Color polishes, waxes (no longer in use)	Lung, liver
ANTICANCER DRUGS		
Melphalan	Chemotherapy	Leukemia
Thiotepa	Chemotherapy (no longer in use)	Leukemia
AROMATIC AMINES AND AMIDES		
2-Naphthylamine	Dyes; antioxidant (no longer in use)	Bladder
4-Aminobiphenyl	Dyes; antioxidant (no longer in use), research tool	Bladder
2-Acetylaminofluorene	Model compound; tested as a pesticide	Liver, bladder
AROMATIC HYDROCARBONS		
Benzo[a]pyrene	Coal tar, roofing, cigarette smoke	Skin, lung, stomach
2,3,7,8-Tetrachlorodibenzo-p-dioxin	No commercial use; tested as a pesticide	Lung, lymphoma, liver
Polychlorinated biphenyls	Flame retardants, hydraulic fluids	Liver, skin
METALS (AND COMPOUNDS)		
Arsenic	Natural ores, alloys; pharmaceutical agent	Skin, lung, liver
Cadmium	Natural ores; pigments, batteries, ceramics	Lung, prostate, kidney
Nickel	Natural ores; alloys, electrodes, catalysts	Lung, nasal cavity
NATURAL CARCINOGENS		
Aflatoxin B$_1$	A mycotoxin (found in contaminated food)	Liver
Asbestos (fibrous silicates)	Thermal insulation, gaskets (declining usage)	Lung, mesothelioma
N-NITROSO COMPOUNDS		
N-Nitrosodimethylamine	Polymers, batteries, nematocide (no longer in use)	Liver, lung, kidney
4-(Methylnitrosamino)-1-(3-pyridyl)-1-butanone	Cigarette smoke; research tool	Lung, liver
OLEFINS		
Ethylene oxide	Glycol and polyester production; sterilization	Leukemia, lymphoma
Vinyl chloride (VC)	Plastics (PVC), copolymers	Liver (angiosarcoma)
Trichloroethylene	Degreasing operations, adhesives, lubricants	Liver, kidney
PARAFINS AND ETHERS		
1,2-Dichloroethane	VC production, solvent, degreaser (no longer in use)	Liver, lung, breast
Bis(chloromethyl)ether	Technical applications (rarely used)	Lung
Mustard gas (sulphur mustard)	Chemical warfare in World War I; research	Lung
Nitrogen mustard	Limited application as antineoplastic agent	Lung, skin, lymphoma

these CYPs (Fig. 2–7). Some of these compounds are further activated by subsequent steps; for example, 2-acetylaminofluorene (AAF) is further modified by a sulfotransferase to form the ultimate DNA-binding moiety.

Somewhat surprisingly, glutathione-S-transferase (GST), which had been thought to be involved only in detoxifying carcinogens, has been shown to activate some industrial chemicals,[7] so GST appears to have a dual role, depending on the chemical.

2-Acetylaminofluorene

The metabolic interconversions of this compound were studied in detail by the Millers and colleagues.[9,10] In 1960, it was shown that AAF is converted to a more potent carcinogen, N-hydroxy-AAF, after the parent compound was fed to rats. Although both AAF and N-hydroxyl-AAF are carcinogenic in vivo, neither compound reacted in vitro with nucleic acids or proteins, suggesting that the ultimate carcinogen was another, as-yet

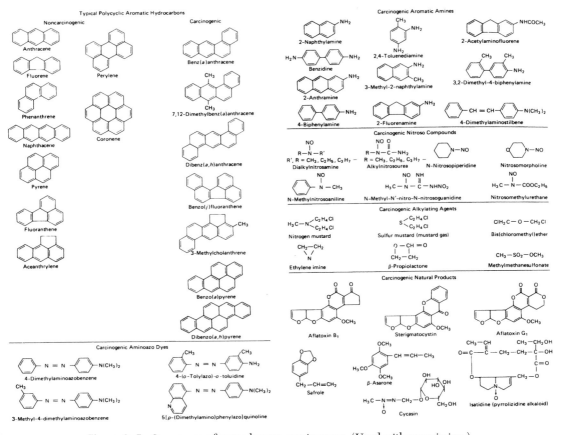

Figure 2–5. Structures of some known carcinogens. (Used with permission.)

unidentified metabolite. Subsequent studies showed that N-hydroxy-AAF is converted in rat liver to a sulfate, N-sulfonoxy-AAF, by means of a cytosol sulfotransferase activity (Fig. 2–7). This compound reacts with nucleic acids and proteins and appears to be the ultimate carcinogen in vivo. It is also highly mutagenic, as determined by assays of DNA-transforming activity (see below).

Other enzymatic conversions of AAF occur in rat liver, for example, N-hydroxy-AAF is converted to N-acetoxy-AAF, N-acetoxy-2-aminofluorene and the O-glucuronide (conjugate with glucoronic acid). These enzymatic reactions may also be involved in the conversion of AAF to carcinogenic metabolites, especially in nonhepatic tissues, which often have low sulfotransferase activity for N-hydroxy-AAF. The acetyltransferase-mediated activity converts N-hydroxy-AAF to N-acetoxy-2-aminofluorene, which is also a strong electrophile

and may be the ultimate carcinogen in nonhepatic tissues.

Other Aromatic Amines

Electrophilic forms of the aromatic amines result from their metabolic activation, and the positively charged nitrenium ion formed from naphthylamine and aminobiphenyl compounds has been implicated as the ultimate urinary bladder carcinogen in dogs and humans. Hydroxylamine derivatives of these compounds are formed in the liver and then converted to a glucuronide. The glucuronide conjugate is excreted in the urine, where the acid pH can convert it back to hydroxylamine and subsequently to a protonated hydroxylamine, which rearranges to form a nitrenium ion by a loss of water. The electrophilic nitrenium ion can then react with nucleophilic targets in the urinary bladder epithelium.

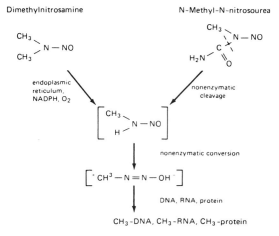

Figure 2–6. The enzymatic and nonenzymatic activations of dimethylnitrosamine and *N*-methyl-*N*-nitroso reactive nucleophiles. (Used with permission.)

Polycyclic Aromatic Hydrocarbons

In 1950, Boyland[11] suggested that the carcinogenicity of polycyclic aromatic hydrocarbons (PAH) was mediated through metabolically formed epoxides. It was originally thought that the key epoxide formation involved the K region of the hydrocarbon ring structure.[12] However, subsequent studies demonstrated that K-region epoxides had little carcinogenecity in vivo. An extensive amount of work has gone on since the 1950s to characterize the metabolism and carcinogenic potential of the PAH (reviewed in Reference 7). It is now generally accepted that conversion of PAH to dihydrodiol epoxides is a crucial pathway in the formation of the ultimate carcinogen. For instance, studies from a number of laboratories have indicated that 7β,8α-dihydroxy-9α, 10αepoxy-7,8,9,10-tetrahydrobenzo(a)pyrene is an ultimate mutagenic and carcinogenic metabolite of benzo(a)pyrene. The evidence is strong that the analogous metabolites of other PAHs that are similar to benzo(a)pyrene are also the ultimate carcinogens of these compounds.

It should be noted that although a number of interactions of chemical carcinogens with DNA and other cellular macromolecules have been observed, there is still no formal proof that the major reaction products detected in cells exposed to these agents are the ones actually involved in carcinogenesis. It could be that some minor or as-yet undetected reaction is the cru-

cial one. Moreover, because carcinogenesis is a multistage process involving initiation, a lag time, promotion, and tumor progression, multiple actions of a carcinogen—or alternatively, the actions of multiple carcinogens—appear to be necessary to produce a clinically detectable malignant neoplasm. An important point to note, however, is that although the PAH diol-epoxides vary considerably in their biological reactivity, the level of mutations in cells is quantitatively related to the level of diol-epoxide-DNA adducts, and the carcinogenicity of different PAHs correlates with the DNA adducts in lung tissue.[7]

Of particular interest is the association of CYP1A1 levels with cigarette smoking. CYP1A1 is inducible in various extrahepatic tissues by the PAH contained in cigarette smoke. This has led a number of investigators to examine the relationship between inducibility of CYP1A1 and susceptibility to lung cancer. Early studies of Kellerman et al.[13] showed a correlation between aryl hydrocarbon hydroxylase induction in peripheral blood lymphocytes and the incidence of lung cancer. More recent studies have shown the formation of DNA-benzo(a)pyrene adducts in pulmonary tissue from cigarette smokers and a higher level of CYP1A1 expression in lung tissue from cigarette smokers than in nonsmokers (89% vs. 0% respectively).[14] In addition, CYP1A1 was elevated in about half the lung cancers from smokers, compared to only 25% of lung cancers from nonsmokers. A genetic polymorphism of CYP1A1 combined with a genetic deficiency in GST (which detoxifies the electrophilic metabolites of PAH) is associated with an increased risk of cigarette smoking–induced lung cancer.[15]

Cigarette smoke contains other toxic chemicals in addition to PAH. One of the most deadly is a carcinogenic nitrosamine, nitrosamine 4-(methylnitrosamino)-1-(3-pyridyl)-1-butanone (NNK), which is metabolized in several steps by a cytochrome P-450, cyclooxygenase, or lipoxygenase to produce metabolites that bind DNA.[16] One of the DNA adducts formed is an O^6-methylguanine that causes a GC-AT transitional base mispairing that has been associated with an activating point mutation in the K-*ras* oncogene. This mutation has been observed in NNK-induced pulmonary adenocarcinomas in mice and rats.[17]

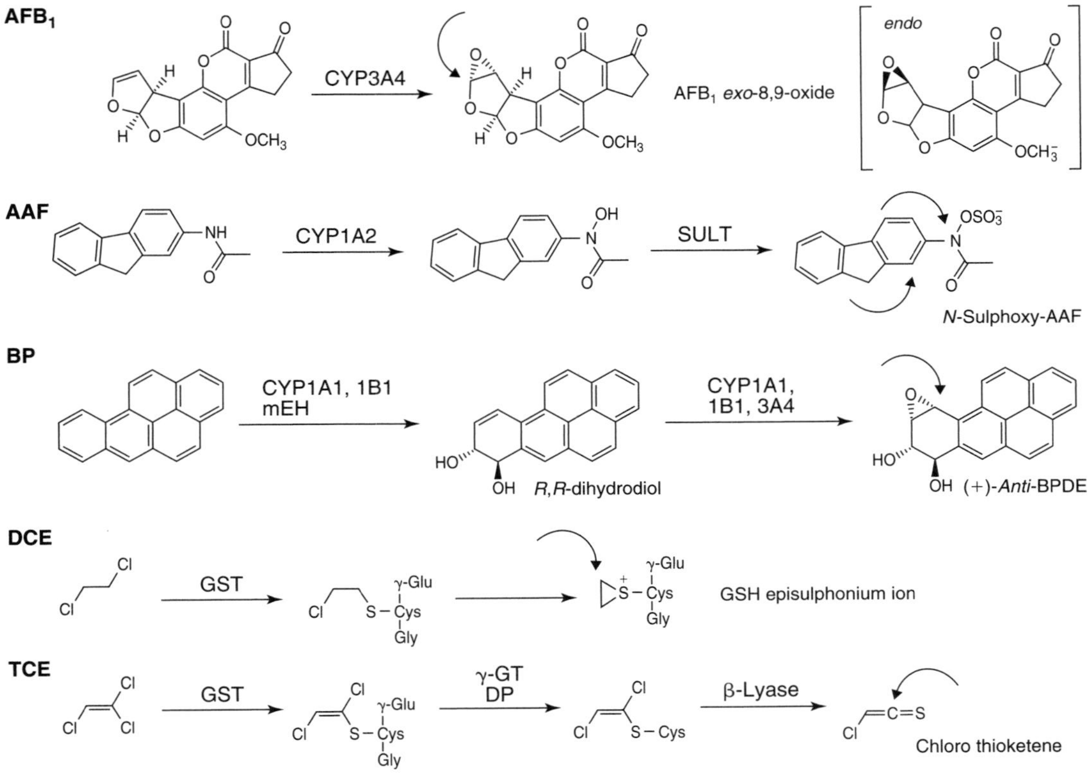

Figure 2–7. Enzymatic conversion of some selected human carcinogens toward their ultimate DNA-reactive metobolites. Activation of aflatoxin B[1] (AFB[1]), 2-acetylaminofluorene (AAF), and benzo[a]pyrene (BP) requires the activity of cytochrome P450–dependent monooxygenases (CYPs). CYP3A4 activates AFB, at its 8,9-bond, resulting in the AFB, exo-8,9-oxide. The *endo*-diasteromer is not formed by CYP3A4, but might be formed in small amounts by CYP1A2. AAF is converted by CYP1A2 into N-hydroxy-AAF, which subsequently might undergo sulphotransferase (SULT)-catalysed esterification into the ultimate genotoxic form, the N-sulphoxy-AAF, BP is initially converted mainly by CYP1A1 or CYP1B1 into the 7,8-epoxide. This epoxide is a substrate of microsomal epoxide hydrolase (mEH), which produces the 7,8-dihydrodiol. Both reactions together stereoselectively form the R,R-dihydrodiol. Further epoxidation at the vicinal double bond catalyzed by CYP1A1, CYP1B1, and CYP3A4 generates the ultimate genotoxic diol-epoxide of BP (BPDE). Of the four possible resulting diastereomers, the (+)-anti-BPDE is formed at the highest levels. 1,2-Dichloroethane (DCE) is activated by glutathione-S-transferases (GSTs) into glutathione (GSH) half-mustard and GSH episulphonium electrophiles, which can bind directly to DNA. GST-catalyzed conjugation of trichloroethylene (TCE) produces GSH adducts. Cleavage of the terminal amino acids by γ-glutamyltransferase (γ-GT) and cysteinylglycine dipeptidase (DP) activity give rise to cysteine (Cys) adducts that can be converted into genotoxic thioketenes by the kidney-specific cysteine conjugate β-lyase. The red arrows point to the position of the nucleophile (DNA, protein, GSH) attack. GSH conjugates of AFB, oxide, or PAH diol-epoxides are detoxification products. (From Luch,[7] reprinted by permission from Macmillan Publishers Ltd.)

Another important enzyme in the carcinogen activation pathway is the microsomal epoxide hydrolase that catalyzes the stereoselective hydration of alkene and arene oxides to *trans*-dihydrodiols. This enzyme is inducible by various xenobiotics, including some chemical carcinogens and phenobarbital, and its level is increased in hepatic nodules and hepatomas induced by chemical carcinogens.[18]

DNA Adduct Formation

Since most chemical carcinogens react with DNA and are mutagenic, interactions with DNA have been viewed as the most important reactions of these agents with cellular macromolecules. Reaction of chemical carcinogens with DNA is the simplest mechanism that explains the induction of a heritable change in a cell leading to malignant transformation; thus many investigators view this as the most plausible mechanism for initiation of carcinogenesis. Representative agents from virtually all classes of chemical carcinogens have been shown to affect DNA in some way, and a number of distinct biochemical-reaction products have been identified after treatment of cells in vivo or in culture with carcinogenic agents.

The principal reaction products of the nitrosamines and similar alkylating agents with DNA are N-7 and O^6 guanine derivatives. However, the extent of O^6 alkylation of DNA guanine residues correlates better with mutagenic and carcinogenic activity than the quantitatively greater N-7 alkylation of guanine residues (see below). Reactions also occur with other DNA bases, and these may be important in subsequent mutagenic or carcinogenic events. Aflatoxin forms adducts of guanine at the N-7 position after metabolic activation. The principal reaction product of AAF with cellular DNA is the C-8 position of guanine, just as it is for RNA. Other carcinogenic aromatic amines, such as N-methyl-4-aminoazobenzene, also produce C-8 substituted guanine residues as their major nucleic acid reaction product (adduct). Polycyclic aromatic hydrocarbons, after activation, also react with DNA and RNA, forming adducts involving the 2-amino group of guanine, but other reaction products derived from guanine, adenine, and cytosine have been observed as well.[7]

The potential biological consequences of DNA base–adduct formation by chemical carcinogens are several. In some cases, it may stabilize an intercalation reaction in which the flat planar rings of a polycyclic hydrocarbon are inserted between the stacked bases of double-helical DNA and distort the helix, leading to a frame-shift mutation during DNA replication past the point of the intercalation.[19] Alkylated bases in DNA can mispair with the wrong base during DNA replication—for example, O^6 methylguanine pairs with thymine instead of cytosine during DNA replication, leading to a base transition (i.e., GC→AT) type of mutation during the next round of DNA replication.[20] Many of the base adducts formed by carcinogens involve modifications of N-3 or N-7 positions on purines that induce an instability in the glucosidic bond between the purine base and deoxyribose, resulting in loss of the base and creation of an apurinic site in DNA.[21] This "open" apurinic site can then be filled by any base, but most commonly by adenine, during subsequent DNA replication. This substitution can result in a base transition (purine–pyrimidine base change, but in the same orientation, e.g., GC→AT) or a base transversion (inverted purine–pyrimidine orientation, e.g., GC→TA). Finally, interaction with some carcinogens has been shown to favor a conformational transition of DNA from its usual double-helical B form to a Z-DNA form. This could alter the transcribability of certain genes, since B→Z conformational transitions are thought to be involved in regulating chromatin structure.

Another interesting point is that interaction of chemical carcinogens with DNA or chromatin does not appear to be a random process. For example, when the ultimate carcinogen of benzo[a]pyrene, that is, its diol epoxide metabolite, is reacted with cloned chicken β-globin DNA, it preferentially binds in a 300–base pair sequence immediately 5′ to the RNA cap site.[22] Since this region is thought to contain sequences involved in regulating gene transcription, its alteration by a chemical carcinogen could change the function of genes downstream from the regulatory sequences. Moreover, treatment of the large polythene chromosomes of *Chironomus* with the ultimate carcinogen benzo(a)pyrene diol epoxide in vitro or administration of the parent unmetabolized compound in vivo to *Chironomus* larvae demonstrates that the carcinogen binds preferentially to areas most active in gene transcription.[23] DNA in transcribing regions associated with the nuclear matrix also appears to be a preferential target for carcinogen binding. Taken together, these data indicate that the specificity of carcinogen binding is determined to some extent by the base sequence of DNA, its location within the nucleus (e.g., association with nuclear matrix), and the structure of chromatin, with active, "open" sites being favored.

Interaction of Chemical Carcinogens with Oncogenes and Tumor Suppressor Genes

Cellular oncogenes and tumor suppressor genes are two of the critical DNA targets for chemical carcinogens, leading to activation of oncogenes and the inactivation of suppressor genes. This will be discussed further in Chapter 5, but a few examples will be given here.

Carcinogens can activate cellular oncogenes (proto-oncogenes) by a variety of mechanisms including base substitution (point) mutations, chromosomal translocations, and gene amplification. One fairly common example is the activation of *ras* proto-oncogenes by chemical and physical carcinogens in both cultured mammalian cells and animal models (reviewed in Ref. 24). H-*ras* and K-*ras* proto-oncogene mutations, for example, have been observed in rodent models of skin, liver, lung, and mammary carcinogenesis. The observed mutations in the tumors correlate with expected base adducts formed by the carcinogen: G→A base transitions with alkylating agents (e.g., NMU and MNNG), G→T transversions for benzo(a)pyrene, A→T transversions for 7,12 dimethylbenzanthracene, G→T transversions and G→A transitions for aflatoxin B1.[24] These mutations appear to reflect similar base substitution mutations in human tumors.

The best documented example of a tumor suppressor gene being inactivated during carcinogenesis is the *p53* gene. Mutations of the *p53* gene have been observed in animal tumors and in a wide variety of human cancers. Most of the mutations are point mutations involving "hot spots" in exons 5 through 8. Interestingly, these are the most highly conserved domains of these exons. In human colon tumors, the majority of the mutations are G→A transitions (just as for *ras*); however, other types of base alterations of *p53* are seen in other human cancers.

Perhaps the most interesting observation is the finding of a high incidence of *p53* point mutations in hepatocellular carcinomas in patients from parts of China and southern Africa where exposure to aflatoxin B1 is endemic.[25,26] Most of these are at a single site, the third base of codon 249, and are G→T transversions. Moreover, hepatocellular carcinomas from low-aflatoxin exposure areas appear to only rarely have this mutation.

Carcinogen-Induced Epigenetic Changes

Even though the application of Ockham's (or Occam's) razor to the effects of chemical carcinogens leads to the concept that the genotoxic results of carcinogen-DNA binding are the simplest, most straightforward explanation for their carcinogenicity, a number of important epigenetic effects are also observed. For example, changes in gene expression patterns caused by carcinogen-induced epigenetic alterations such as changes in DNA methylation or histone acetylation have been observed after exposure of cells to carcinogens. This pattern has been observed, for example, during cells' exposure to the carcinogenic metals nickel, cadmium, or arsenic.[7] The carcinogenic effects of nickel have been linked to DNA hypermethylation and histone deacetylation, both of which can alter chromatin structure and cause epigenetic silencing of tumor suppressor genes (see Chapter 5).

Tumor Initiation, Promotion, and Progression

The idea that development of cancer is a multistage process arose from early studies of virus-induced tumors and from the discovery of the cocarcinogenic effects of croton oil. Rous and colleagues found that certain virus-induced skin papillomas in rabbits regressed after a period of time and that papillomas could be made to reappear if the skin was stressed by punching holes in it or by applying such irritant substances as turpentine or chloroform. These findings led Rous and his associates to conclude that tumor cells could exist in a latent or dormant state and that the tumor induction process and subsequent growth of the tumor involved different mechanisms, which they called "initiation and promotion."[27] The term *cocarcinogen* was coined by Shear, who discovered that a basic fraction of creosote oil enhanced the production of mouse skin tumors by benzo(a)pyrene.[28] In 1941, Berenblum[29] reported that among mice receiving a single skin painting of a carcinogen, such as methylcholanthrene, only a small number of animals developed papillomas, but if the same area of skin was later painted repeatedly with croton oil, which by itself is not carcinogenic, almost all the animals developed skin carcinomas. Taken together, the data of these investigators suggested a multistage mechanism for carcinogenesis.

Studies of the events involved in the initiation and promotion phases of carcinogenesis were greatly aided by the identification of agents that have primarily an initiating activity, such as urethane or a low dose of a "complete" carcinogen (see below), and by the purification of the components of croton oil that have only a promoting activity. Diesters of the diterpene alcohol phorbol were isolated from croton oil and found to be the tumor-promoting substances.[30,31] Of these, 12-*O*-tetradecanoylphorbol-13-acetate (TPA) is the most potent promoter.[32]

A scheme used to study the initiation–promotion phases of mouse skin carcinogenesis is depicted in Figure 2–8. Typically, tumor initiation is brought about by the single application of an initiator, such as urethane, or a subcarcinogenic dose of an agent with both initiating and promoting activity, such as the polycyclic hydrocarbon benzo(a)pyrene; promotion is carried out by repeated application of a phorbol ester, such as TPA (e.g., three times a week).[31,33] Benign papillomas begin to appear at 12 to 20 weeks and by about 1 year, 40% to 60% of the animals develop squamous cell carcinomas. If the promoting agent is given alone, or before the initiating agent, usually no malignant tumors occur.

The progression stage of carcinogenesis is an extension of the tumor promotion stage and results from it in the sense that the cell proliferation caused by promoting agents allows the cellular damage inflicted by initiation to be propagated, and the initiated cells are clonally expanded. This propagation of damaged cells in which genetic alterations have been produced leads to the production of more genetic alterations. This genetic instability is the hallmark of the progression phase of carcinogenesis and leads to the chromosomal translocations and aneuploidy that are frequently seen in cancer cells.[34] Such alterations in the genome of the neoplastic cell during the progression phase lead to the increased growth rate, invasiveness, and metastatic capability of advanced neoplasms. Some of the gene expression alterations that occur during tumor initiation and promotion are shown in Figure 2–9 (see color insert).

Evidence for multistage induction of malignant tumors has also been observed for mammary gland, thyroid, lung, and urinary bladder and in cell culture systems (reviewed in Reference 9), thus it seems to be a general phenomenon. This experimental evidence is consistent with the observed clinical history of tumor development in humans after exposure to known carcinogens—that is, initial exposure to a known chemical or physical carcinogen, a long lag period during which exposure to promoting agents probably occurs, and finally the appearance of a malignant tumor.

Several characteristics of tumor initiation, promotion, and progression provide some insight into the mechanisms involved in these processes. Initiation can occur after a single, brief exposure to a potent initiating agent. The actual initiation events leading to transformation into a dormant tumor cell appear to occur within one mitotic cycle, or about 1 day for the mouse skin system.[32] Furthermore, initiation appears to be irreversible; the promoting agent can be given for up to a year later and a high percentage of tumors will still be obtained. Thus, the initiation phase only requires a small amount of time, it is irreversible, and it must be heritable because the initiated cell conveys the malignant alteration to its daughter cells. All these properties are consistent with the idea that the initiation event involves a genetic mutation, although other "epigenetic" explanations are possible (see above). The promotion phase, by contrast, is a slow, gradual process and requires a more prolonged exposure to the promoting agent. Promotion

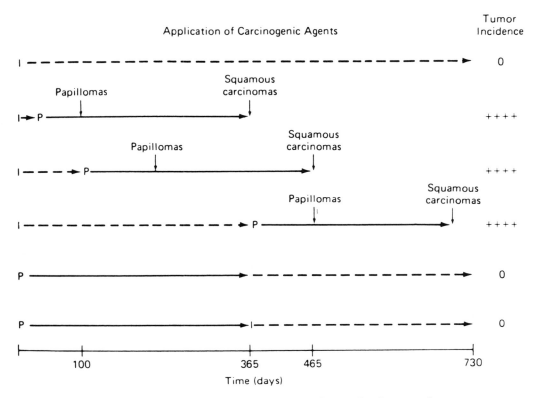

Figure 2–8. Scheme of initiation–promotion phases of induction of carcinogenesis in mouse skin. Initiation is caused by the single application of a subcarcinogenic dose of an agent such as 7,12-dimethylbenz[a]anthracene, benzo[a]pyrene, or urethane. Promotion is carried out by repeated application (e.g., k three times a week) of an agent such as the phorbol ester TPA. Papillomas develop within 12 to 20 weeks, squamous carcinomas in about 1 year. Solid lines indicate continual application of agent; dotted lines indicate the duration of time without exposure to agents. Note that promoter may be added up to 1 year after a single application of the initiating agent and tumors still occur. I, initiator; P, promoter. (Used with permission.)

occupies the greater part of the latent period of carcinogenesis, is at least partially reversible, and can be arrested by certain anticarcinogenic agents (see Chapter 9). Tumor promotion is a cell proliferation phase that propagates the initiated damage and leads to the emergence of an altered clone of cells. Most promoting agents are mitogens for the tissue in which promotion occurs. Tumor progression requires continued clonal proliferation of altered cells, during which a loss of growth control and an escape from host defense mechanisms become predominant phenotypic traits. This process allows growth to progress to a clinically detectable tumor.

The later events in the tumor progression phase are also thought to be irreversible because of the pronounced changes in the genome that have occurred leading into this phase. Agents that are "pure" progression-causing agents are hard to identify, but the free radical–generating agent benzoylperoxide appears to be a progression-inducing agent during experimental epidermal carcinogenesis.[35]

It should be noted that some potent carcinogens are "complete carcinogens" in that at certain doses they can by themselves induce a cancer. Such agents include polycyclic aromatic hydrocarbons, nitrosamines, certain aromatic amines, and aflatoxin B1. When these agents are

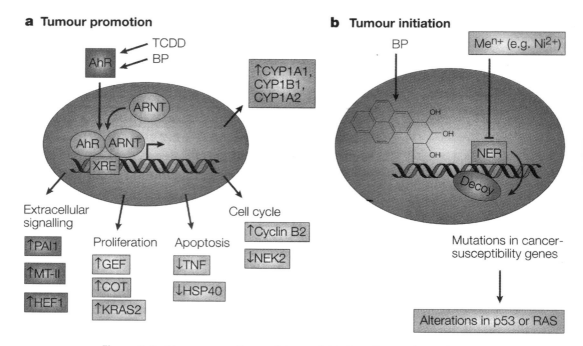

Figure 2–9. Tumor promotion and tumor initiation. Genotoxic carcinogens can induce damage in tumor suppressors or oncogenes in different ways, all of which contribute to the transformation of normal cells into tumor cells—this is known as the *tumor initiation* stage in carcinogenesis. Some chemical carcinogens are also capable of promoting the outgrowth of those transformed cell clones and of contributing to the generation of visible tumor cell masses—this is known as the *tumor promotion* stage in carcinogenesis. *a.* Chemical compounds such as 2,3,7,8-tetrachlorodibenzo-*p*-dioxin (TCDD) or benzo[a]pyrene (BP) result in tumor promotion through arylhydrocarbon receptor (AhR)–mediated signal transduction. Binding of TCDD or BP to AhR leads to activation and translocation of the complex into the nucleus. After heterodimerization with the AhR nuclear translocator (ARNT), the complex binds to xenobiotic-responsive elements (XREs) and induces the expression of a variety of different genes involved in carcinogen metabolism, including CYP forms 1A1, 1B1, and 1A2. It also changes the expression pattern of several factors involved in cellular growth and differentiation, such as plasminogen-activator inhibitor type 1 (PAI1), metallothionein II (MT-II), human enhancer of filamentation 1 (HEF1), guanine nucleotide exchange factor (GEF), COT, and KRAS. Pro-apoptosis factors such as tumor necrosis factor (TNF; superfamilies 3, 6, 8, 9, 10) and heat-shock protein 40 (HSP40) are down-regulated, and cell cycle genes can either be up-regulated (such as cyclin B2) or down-regulated (such as NEK2). The factors shown here are only a few examples of the great number of factors that have been shown by gene expression analysis to be altered following carcinogen exposure. COT and NEK2 are serine/threonine kinases.[70,71] *b.* Tumor initiation occurs through DNA adduct–derived mutations in cancer susceptibility genes. DNA binding by genotoxic carcinogens such as activated BP leads to the induction of base pair or frameshift mutations in cancer susceptibility genes such as *TP53* or *RAS*. The mutagenic potency of such polycyclic aromatic hydrocarbon diol-epoxide-DNA adducts can be increased because of inhibition of nucleotide excision repair (NER) by metal ions (Me$^+$; for example, Ni^{2+}), or as a result of NER factor immobilization at repair-resistant DNA-adduct sites, also known as decoy adducts. (From Luch,[7] reprinted by permission from Macmillan Publishers Ltd.)

given in sufficient dose to animals during cancer-causing protocols, they can cause DNA damage and produce tissue necrosis, which is itself enough to stimulate several rounds of cell proliferation in response to the tissue damage. In this situation, the promotion–progression phases are often collapsed in time, resulting in the production of aneuploid malignant cells.[34]

Mechanisms of Tumor Initiation

Initiation of malignant transformation of normal cells by a carcinogenic agent involves a permanent, heritable change in the gene expression of the transformed cell. This could come about by either direct genotoxic or mutational events, in which a carcinogenic agent reacts directly with DNA, or by indirect or "epigenetic" events that modulate gene expression without directly reacting with the base sequence of DNA. Most investigators favor the mutational theory of carcinogenesis—that is, that the initiating events involve a direct action on the genome.

The mutational theory depends on three kinds of evidence:

1. Agents that damage DNA are frequently carcinogenic. As discussed previously, chemical carcinogens are usually activated to form electrophilic agents that form specific reaction products with DNA. The extent of formation of some of these reaction products, for example, alkyl-O^6-guanine, has been shown to correlate with mutagenicity and carcinogenicity of certain chemical agents. Ultraviolet and ionizing radiation also interact with DNA at doses that are carcinogenic.

2. Most carcinogenic agents are mutagens. A number of in vitro test systems using mutational events in microorganisms have been developed to rapidly screen the mutagenic potential of various chemical agents. One of the best known of these, the Ames test, is based on certain characteristics of specially developed strains of the bacterium *Salmonella typhimurium*. The tester strain, a mutant line that requires exogenous histidine for its growth (hisauxotroph), has a poor excision repair mechanism and an increased permeability to exogenously added

chemicals. Using this system, together with a liver microsomal fraction that has the capacity to activate most chemical carcinogens metabolically, Ames and colleagues have shown that about 90% of all carcinogens tested are also mutagenic.[36] Moreover, few noncarcinogens show significant mutagenicity in this test system. Malignant transformation can be induced in a variety of cultured mammalian cells by agents that are mutagenic for the same cells. For example, carcinogenic polycyclic hydrocarbons cause mutations, as measured by induction of resistance to 8-azaguanine, ouabain, or elevated temperature, in Chinese hamster V79 cells if the cells are cocultured with lethally irradiated rodent cells that can metabolize the hydrocarbons to their electrophilic, active metabolite.[37,38] In these studies, mutagenicity was obtained with the carcinogenic hydrocarbons 7,12-dimethylbenz(a)anthracene, benzo(a)pyrene, and 3-methylcholanthrene. There was no mutagenicity with a noncarcinogenic hydrocarbon, and the degree of mutagenicity was related to the degree of carcinogenicity of the chemicals in vivo.

3. Incidence of cancer in patients with DNA-repair deficiencies is increased. In individuals with certain recessively inherited disorders, the prevalence of cancer is significantly higher than in the general population.[39] The connecting link between these disorders is the inability to repair certain kinds of physical or chemical damage to DNA. The high incidence of cancer in these diseases constitutes the best available evidence for a casual relationship between mutagenicity and carcinogenicity in humans.

One example of xeroderma pigmentosum (XP) is characterized by extreme sensitivity of the skin to sunlight and is the most widely studied of the repair-deficient human diseases. Virtually 100% of affected individuals will eventually develop some form of skin cancer. In addition, heterozygotes who carry the XP gene but do not have the disease appear to have a higher incidence of nonmelanoma skin cancer.[40] All individuals with XP are defective in repair of ultraviolet damage to DNA, and most of them have a defect in the excision repair pathway (see below). The repair defect ranges from 50% to 90% repair efficiency in cells from different patients, and there is good correlation between the severity of the molecular defect and the extent of the disease. The defect in most patients appears to be at the nicking or incision step of excision repair, although patients in one complementation group have normal excision repair and are defective in postreplication repair.[39] The XP cells are also less efficient at repairing chemically induced damage to their DNA.

Other examples of enhanced susceptibility to cancer in individuals with DNA repair deficiencies are ataxia telangiectasia (AT), Fanconi's anemia (FA), and Bloom's syndrome (BS). Cell lines derived from AT patients are defective in repair replication following exposure to irradiation or carcinogenic chemicals.[40] Patients with AT are more prone to develop leukemia and some other cancers. FA cells have a defect in the repair of cross-linked bases in DNA; they also appear to have a slight deficiency in the repair of γ-ray- or ultraviolet-induced damage, and patients with FA are at increased risk to develop cancer.[39] Lymphoblastoid cell lines from patients with a type of BS characterized by a high rate of sister chromatid exchange (SCE) are more highly tumorigenic after brief exposure in vitro to 4-nitroquinoline-N-oxide and N-methyl-N'-nitro-N-nitrosoguanidine (MNNG) in the nude mouse assay than are lymphoblastoid cells from normal individuals.[41]

Although the mutational theory is the simplest explanation of a heritable change in a cell that could produce a cancer-initiating event, not all initiating agents are mutagenic in the test systems used, and some do not react directly with DNA. For example, malignant tumors can be produced by agents that do not interact with DNA or by certain cells placed in abnormal tissue locations. Prolonged administration of the estrogenic compound diethylstilbestrol (DES)

has been shown to produce renal adenocarcinomas in male hamsters. The estrogen-dependent induction of these tumors is inhibited by the simultaneous administration of testosterone or progesterone.[42] Malignant neoplasms arise when rat ovarian tissue is transplanted into normal rat spleen, presumably because of the hormonal imbalance thus induced.[43] Tumors can also be induced in rats by the insertion of plastic or metal films, depending only on the physical state (i.e., solid versus porous or fibrous form) of the inserted material. In any of these cases, it is unlikely that mutational events resulting from interaction with nuclear DNA could have produced the tumors, although it has been shown that an oxidative metabolite of DES can induce SCE in cultured hamster embryo cells.[44] SCE was also observed in human hepatoma cells that metabolize DES but not in cell lines that do not.[45] Presumably, this phenomenon occurs as the result of the formation of phenoxy radical intermediates from DES metabolites by a peroxidase-mediated reaction and the DNA strand breakage that is produced by these oxygen radicals.

In cases in which direct interaction with DNA does not seem to occur, it is likely that regulation of DNA *expression* is changed by epigenetic alterations that result in malignant transformation stemming from chromatin conformational changes. These changes can be induced by DNA methylation or histone modifications as noted above or by other mechanisms, such as (1) interaction of carcinogenic agents with membrane proteins that regulate cell surface receptors for growth factors or that control feedback regulation of cell proliferation in the cell's microenvironment; (2) reaction with RNA molecules involved in translation of proteins; (3) binding to regulatory proteins that control gene transcription; or (4) interaction with proteins involved in the cell's mitotic apparatus. These changes could be "heritable," at least for several cell generations, because each daughter cell receives a certain complement of the parent cell's RNA and protein. In any case, the ultimate result is a cell that is genetically unstable, since tumor progression leads to the appearance of cells that are genetically different from cells of the tumor's tissue of origin.

Endogenous Carcinogenesis

An important question that arises is, what is the source of mutations in the human genome that leads to cancer? One might argue that the answer is obvious. We live in a sea of carcinogens: PAHs from automobile exhaust, industrial pollution, pesticide residues in foods, chlorinated organic compounds in drinking water, etc. Furthermore, epidemiologists argue that almost 30% of human cancers are related to cigarette smoking. Yet, a significant amount of cancers occur in people with no clear evidence of exposure to clearly defined carcinogens. For a number of cancers of the pancreas, ovary, kidney, and breast, for example, there are in most cases no clear geographic or genetic risk factors (although heritable genetic changes may account for 5% to 10% of some cancers such as breast cancer). Thus, if cancer is initiated through a mutation or a series of mutations, how might these arise?

One possibility is that "spontaneous" mutations arising from an inherent error rate in the fidelity of DNA replication and/or repair could give rise to mutations, some of which by chance could be in key genes involved in regulation of cell proliferation and differentiation. The spontaneous or background mutation rate in human somatic and germline cells has been estimated to be 1.2 to 1.4×10^{-10} mutations per base pair per cell division[6] (i.e., one mistake in 10 billion base pairs per each cell division). Since there are about 10^{14} cells in the adult human (with a genome of 3×10^9 base pairs) and they undergo an estimated 10^{16} cell division cycles in a normal life span, about 3.6×10^{15} single base (point) mutations could arise in a lifetime.[6] If a single mutation could produce a cancer, this would lead to 3.6×10^{15} spontaneously arising cancer cells in a lifetime, a highly unlikely proposition. Several considerations moderate this wildly excessive number. For example, most base changes are repaired; not all base changes are in coding regions and some are silent (not producing an altered protein); not all mutations produce a cancer cell (they may not occur in key oncogenes or tumor suppressor genes); and more than one mutation is necessary to produce a cancer cell (in one study, 25% to 50% of human colon cancers, for example, contained nine or more mutations).[46]

If at least two mutations are required (e.g., as the Knudson model predicts), the required mutation rate would be the square of 1.4×10^{-10}. In this case, the expected number of spontaneously arising cancer cells in an individual's lifetime would be about 300,000. Even this number seems high. And even if it were true, most of these cells would die or be eliminated by immune mechanisms. If more than two mutations are required to produce a cancer, then the number becomes much smaller, e.g., 5.5 cancers per 100,000 individuals for three mutations. Thus, if this latter assumption is correct, i.e., that three or more mutations are required to produce a cancer, then the spontaneous mutation rate could not by itself explain the number of cancers in the human population. Moreover, the type and distribution of spontaneous mutations differ from that of chemically induced mutations in cells and from that of a number of mutations found in human cancers.[47] Thus, although spontaneous mutations may contribute to the causation of human cancer, they are unlikely by themselves to cause the initiation and progression events that lead to most invasive, metastatic neoplasms.

Several potential mechanisms exist for spontaneous mutations in human cells. These include depurination, deamination, damage to DNA by oxygen radicals, and errors in DNA replication. Depurination is the most common potentially mutagenic event, occurring at a rate of about 10,000 depurination events per cell per day.[48] This results from breakage of the N-glycosidic bond connecting a purine base to the deoxyribose-phosphate backbone of DNA and creates a gap in the base sequence. When DNA polymerase encounters such a gap during DNA replication, it may insert the wrong base, usually an adenine, in place of the missing base. Obviously, this couldn't happen very often or the mutation rate in the human genome would be higher than it is.

Deamination of cytidine to uridine occurs at 1/500 the rate of depurination, or about 20 events per cell per day. Since uridine base pairs with adenine during DNA replication, this could lead to a G→A transition. Also, deamination of methylcytosine can occur producing thymidine, which if not repaired, could produce a G→A transition.

The rate of damage to DNA produced by oxygen radicals, which are continuously generated in cells by normal metabolism, isn't clear, but it could be as high or higher than depurination. Measurements in human urine of 8-hydroxydeoxyguanosine and thymine glycol, which are oxidative breakdown products of DNA, suggest that 10,000 oxygen radical–induced alterations in DNA could occur per cell per day.[49] However, cells have stringent mechanisms to protect themselves against free radicals generated by cellular metabolism. These include superoxide dismutase (SOD), catalase, and glutathione generating systems. Oxygen radical damage, if unrepaired, can produce single-strand and double-strand breaks in the DNA backbone.

Errors in DNA replication or repair could result from any of the mechanisms described above. In addition, errors introduced by the DNA replicating machinery itself can occur. DNA polymerases, though usually incredibly accurate, can make some mistakes. Using a ϕ bacteriophage DNA replication system, Kunkel and Loeb[50,51] have determined the error rate of mammalian DNA polymerases. DNA pol α, the major DNA replicating polymerase in eukaryotic cells, has an error rate of 1/30,000 to 1/200,000 bases, depending on the method of purification; pol β, a major repair enzyme has a 1/5000 error rate; and pol δ, a polymerase with proof reading ability, has an error rate of 1/500,000. The most frequent error for all polymerases are single base substitutions and "minus one" base frameshifts. These error rates, determined in vitro, probably overestimate the error rate in vivo, however, because purified enzymes were used, and repair enzymes that are part of the intracellular DNA replication complex may have been lost. Nevertheless, errors in DNA replication are another potential source for spontaneous mutations leading to cancer.

The facts that the actual cancer rate in the population can't be readily explained by the background spontaneous mutation rate and that genetic instability increases with tumor progression have led to the hypothesis that malignant cells have a way to increase their error rate and that they gain some selective advantage from this. The term *mutator phenotype* has been used to describe this phenomenon,[6] and refers to the ability of tumor cells to direct their own mutation rate or, more precisely, to allow a rapid accumulation of errors that favor their survival. One way this could occur is through alterations in DNA polymerases involved in DNA replication and repair, although evidence for such altered enzymes is not conclusive.

One thing that is clear, however, is that cancer is, in general, a disease of aging. As noted above, the average age at time of diagnosis of a malignant tumor is about 65. Moreover, the incidence of a number of adult solid cancers increases with the fourth to sixth power of age.[52] This finding, plus the fact that aneuploidy and other genetic alterations increase during tumor progression, which may occur over many years (up to 15 or 20 years for some cancers), supports the notion that it is the accumulation of genetic errors over time that is most dangerous for the human genome.

Mechanisms of Tumor Promotion and Progression

Tumor-initiating agents most likely act by interacting with DNA to induce mutations, gene rearrangements, or gene amplification events that produce a genotypically altered cell. What happens next is that the initiated cells undergo a clonal expansion under the influence of promoting agents that act as mitogens for the transformed cell type. As will be discussed later, these promoting actions appear to be mediated by cell membrane events, although a direct action of promoters on DNA has also been proposed. It is important to note that multiple clones of cells are likely to be initiated by a DNA-damaging agent in vivo and that, through a rare second event, one or a small number of these clones progresses to malignant cancer.

It may be useful to think of the *promotion phase* as the stage of cell proliferation and clonal expansion induced by mitogenic stimuli and of the *progression phase* as the gradual evolution of genotypically and phenotypically altered cells that occurs due to genetic instability of the progressing cells. This process leads to the development of cell heterogeneity within a tumor, an idea first described by Foulds[53] and later expanded by Nowell.[54] During the progression phase, which can take many years in humans, individual tumors develop heterogeneity with respect to their invasive and metastatic characteristics, antigentic specificity, state of cellular differentiation, and responsiveness to hormones, drugs, and immune-modulating agents. Presumably, some powerful selection process goes on to favor the growth of one progressing cell type over another. This preferential selection may be due to a certain cell type developing a growth advantage in the host's tissues over its peers, as proposed by Nowell, or to the host's immunologic defense system being able to recognize and destroy some cell types better than others, thus providing the selection pressure for expansion of one clone over another, or to a combination of these factors. Experimental evidence supports such a selection of tumor cells growing in vivo. For example, Trainer and Wheelock[55] have shown that during the growth of L5178Y lymphoma cells in mice, a continual selection of cells with a decreasing ability to be killed by cytolytic T lymphocytes (CTL) "armed" against the tumor occurs, until an "emergent phenotype" appears that is highly resistant to the CTL cells.

CENTRAL DOGMA OF TUMOR PROGRESSION

The standard concept of how cancer starts is that malignant tumors arise from a single cell transformed by a chemical carcinogen, oncogenic virus, radiation damage, endogenous genetic damage caused by oxidative insult to DNA, or any of a host of other potential ways (e.g., chronic infections with a bacteria such as *H. pylori* or with a parasite such as schistosomiasis, or hormonal imbalance). Once the initiated cell starts to undergo clonal expansion, it undergoes multiple genetic changes, due to genetic instability, leading to an invasive metastatic cancer. This progression is thought to occur sequentially, as exemplified by the work of Vogelstein and colleagues on colon cancer.[46] The idea here is that colon cancer goes through a series of "evolutionary" changes from hyperplasia, to early-stage adenoma, to late-stage adenoma, to carcinoma, and finally to metastatic cancer.

There is, however, another point of view proposed by Weinberg and colleagues.[56,57] This hypothesis, for which there are supportive clinical data (see below), states that the genes involved in driving invasiveness and metastasis may be expressed early in the progression pathway and

actually be the same genes involved in a selective growth advantage for these cells. These cells may be lurking even in early-stage cancers. That is, some cancers are predestined almost from the beginning to evolve into invasive, metastatic tumors and some are not. This possibility has huge implications for cancer screening, diagnosis, and choice of therapy. Numerous women receive a diagnosis of ductal carcinoma in situ of the breast based on mammography screening, and many men receive a diagnosis of prostate cancer based on a prostate-specific antigen (PSA) test and subsequent biopsy. And yet many of these patients have indolent tumors that would not affect their overall life expectancy, and they still often undergo significant surgical and drug treatments. The problem is that we are only beginning to be able to tell (e.g., by gene expression arrays) which of these so-called early-stage cancers will be lethal and which ones won't.

Another point of the Weinberg theory is that the genetic alterations that occur during tumor progression do not necessarily occur in a given sequence and are probably different for different cancers.[56] One might even suggest that they may be different in different patients who have the same histological tumor type. Ultimately, however, these genetic and phenotypic changes lead to a similar loss of cell proliferation control and expression of a panoply of genes (maybe not the identical ones) that make some tumors invasive and metastatic.

There are clinical data supporting some of these concepts. In a study by van de Vijver et al.,[58] it was determined that the gene expression profile of breast cancers was a much better predictor of disease outcome in patients with breast cancer than standard clinical and histopathological staging. Indeed, they could restratify patients listed as low risk or high risk by clinical staging into a more accurate prognostic outcome category (based on actual metastasis-free survival) through gene expression arrays. In addition, Al-Hajj et al.[59] were able to identify and isolate the more tumorigenic cells from a heterogeneous population of breast tumors in eight of nine patients. These more aggressive cell types were identified by their cell surface markers and by repeated passage in nude mice. Each time the more aggressive cells were injected into nude mice they produced tumors, whereas the marker-negative cells did not grow. These data suggest that the aggressive tumorigenic cells can be prospectively identified in initial tumor biopsies containing mixed populations of cells and can be used to discriminate patients with potentially more aggressive tumors.

MECHANISMS OF TUMOR-PROMOTING AGENTS

The terms *tumor promotion*, *tumor progression*, and *multistage carcinogenesis* are overlapping and somewhat redundant. Some people use these terms interchangeably and some use them to define discreet steps in the carcinogenesis process. Mechanistically, tumor promotion and progression are a continuum, even though they appear to be "multistage." *Promotion* involves a clonal expansion (proliferative phase), and *progression* usually refers to the genetic alteration phase. But as was noted above, the genes involved in these steps are overlapping or similar. Nevertheless, studies of chemical carcinogenesis models have been used to define and discriminate initiation events and promotion or progression events, and these studies have been useful in determining the genetic and biochemical steps involved in these steps, as well as providing targets for drug therapy and chemoprevention.

The isolation and characterization of tumor-promoting agents have provided the tools to study the mechanisms of tumor promotion in vitro and in vivo. The reader is reminded that these agents are primarily defined by their ability to promote skin carcinogenesis in the mouse skin-painting assay, and the mechanisms by which they do this may or may not be relevant to the mechanism of tumor promotion and progression during carcinogenesis in other organs in experimental animals or in humans. Nevertheless, the study of these compounds has been extremely useful in determining the biochemical actions of tumor promoters. Of the promoting agents examined, the phorbol esters have been the most widely studied. Still, one must ask: what the "phorbol esters" are in human carcinogenesis. Most likely they are factors to which we are continually exposed through our diet, cigarette smoke, and other kinds of environmental agents. This answer leads to a second question: Do all these agents act through the same receptor or, if not, through the same biochemical steps? The answer is not known, but the list of potential promoters in the human environment is so large that it seems unlikely that they would all act by means of the same proximal ("receptor") mechanism. More likely, they act through different steps in a cascade leading to the same end point—namely, clonal expansion of initiated cells and progressive selection of genetically variant populations of tumor cells.

Tumor-promoting phorbol esters produce a wide variety of biochemical changes in cells. A number of these changes may be related to the ability of these agents to promote the growth of initiated tumor cells in vivo. Many of the cellular changes induced by phorbol esters are reminiscent of characteristics of the transformed phenotype (see Chapter 4). The effects of phorbol esters on cultured cells include (1) induction of ornithine decarboxylase, 5′-nucleotidase, ATPase, and plasminogen activator activities; (2) stimulation of sugar transport, DNA synthesis, and cell proliferation; and (3) alteration of cell morphology with a loss of cell surface fibronectin and the appearance of a diffuse pattern of actin-containing cytoskeletal elements (reviewed in Reference 60). In addition, phorbol esters stimulate anchorage-independent growth of adenovirus-transformed cells[61] and inhibit the terminal differentiation of chicken myoblasts[62] and chondroblasts,[63] murine lipocytes,[64] erythroleukemia cells,[65] and neuroblastoma cells.[66] Tumor-promoting phorbol esters also transform mouse embryo fibroblasts treated with ultraviolet light[67] and enhance the transformation of human lymphocytes by Epstein-Barr virus.[68] These cell culture effects are exerted by low concentrations (nonmolar range) of phorbol esters, and there is generally a correlation between the potencies of phorbol esters for the cell culture effects and their potencies as promoters in mouse skin carcinogenesis. Phorbol esters share a number of biological properties with epidermal growth factor (EGF) and may act by mechanisms similar to EGF.

An interesting observation suggests that TPA can induce neoplastic transformation of fibroblasts from humans genetically predisposed to cancer.[69] In these experiments, fibroblasts derived from individuals with familial adenomatosis of the colon and rectum were treated with TPA in culture and then injected into athymic mice. Cultures treated with TPA produced tumors in the mice, whereas untreated cultures did not. These results indicate that the fibroblasts from adenomatosis patients exist in an "initiated" state due to the dominant mutation that produces the disease, and that this dominantly inherited trait can be induced to undergo malignant progression by treatment with promoting agents alone. This observation supports the idea that initiation of cancer is a mutagenic event and has profound implications for human cancer. For example, if the promoting agents present in our environment could be identified and exposure to them eliminated or significantly diminished, could human cancer be prevented? This approach could conceivably be more effective than eliminating exposure to initiating agents, since exposure to them need be only very short and is irreversible. Completely preventing exposure to initiating agents over a lifetime is not practical; however, if the promotion phase takes 15 to 20 years, expanding it to 30 to 40 years would mean that most individuals could have a life expectancy approaching normal before they developed a fatal cancer.

Chronic application of TPA to mouse skin indicates that a number of target cells may be involved in the tumor promotion phase of skin carcinogenesis.[70] When TPA is applied twice weekly to the skin of mice for several weeks, cell damage, edema, and acute inflammation occur in both the epidermis and dermis during the first week. By 3 weeks, epidermal hyperplasia occurs, accompanied by chronic inflammation in the dermis and hyperplasia of the hair follicles. These features remain until TPA treatment ceases. Although many of these changes regress within 2 weeks after TPA application ceases, an increased number of hair follicles, capillary vessels, and mast cells in the dermis and an increased dermal thickness remain. This finding indicates that the target cells for TPA are not confined to the epidermis and suggests that chronic tissue irritation, as evidenced by the hyperplastic and inflammatory responses, plays a role in tumor promotion. Although the mechanism of this damage is not totally clear, it has been demonstrated that tumor-promoting phorbol esters produce a number of changes in cell membranes and can indirectly damage chromatin by generating oxygen radicals.

TPA and other tumor-promoting phorbol esters stimulate Na^+ outside/H^+ inside exchange across cell membranes, leading to a transient intracellular alkalinization that precedes mitogenic events and appears to be coupled to at least some of the phorbol ester–induced changes in gene expression in target cells.[71–73] TPA-induced Na^+/H^+ exchange also produces cell swelling,[74] and this could lead to further changes in cell functions and integrity. Other membrane effects of tumor-promoting phorbol esters are inhibition of gap-junctional intercellular communication;[75] phosphorylation of cell surface receptors for EGF, insulin-like growth factor 1 (IGF-1), insulin, and transferrin, leading to decreased ligand–receptor binding and increased receptor internalization;[76] reorganization of actin and vinculin elements in the cytoskeleton;[77] and structural rearrangement of the nuclear matrix–intermediate filament scaffold.[78]

Tumor-promoting phorbol esters also appear to be able to alter cellular gene expression by indirectly altering DNA structure and chromosomal proteins by generating oxygen radicals (reviewed in Reference 79). TPA has been shown to induce chromosomal alterations in a variety of human cell types, and these effects are inhibited by the addition of antioxidants. TPA also stimulates poly-ADP-ribosylation of chromosomal proteins in human monocytes, an effect frequently stimulated by DNA strand breakage and one that could modify gene expression. Further evidence for the role of oxygen radicals in tumor promotion comes from the observations that O_2^-, H_2O_2, and certain organic hydroperoxides promote carcinogenesis in chemical- or irradiation-initiated cells. In contrast, antioxidants such as butylated hydroxytoluene and butylated hydroxyanisole inhibit transformation of initiated, TPA-treated mouse cells. Some of the changes in gene expression induced by TPA may be due to oxygen-radical generation, since induction of ornithine decarboxylase by TPA is blocked by the antioxidant enzymes catalase and superoxide dismutase. These DNA-damaging effects of tumor promoters would be expected to induce chromosomal breaks and gene rearrangements. It has been demonstrated that during a single-step selection assay for methotrexate (MTX) resistance in cultured mouse fibroblasts, TPA causes a 100-fold increase in the incidence of MTX-resistant colonies, an effect shown to be due to MTX-gene amplification in these cells.[80] Thus, tumor promoters could also alter gene expression by favoring gene rearrangements and gene amplification events in initiated cells that already have damaged DNA and a propensity for genetic instability.

The DNA-damaging effects of tumor-promoting agents seem to be incompatible with the view that the tumor-promotion phase of carcinogenesis is at least partially reversible. However, most cells have mechanisms to protect themselves against the generation of oxygen radicals, and the ability of agents like TPA to produce oxygen radical–mediated damage in normal cells may be relatively low. This effect would be expected to be increased in cells whose DNA was already damaged or in cells whose oxygen radical–scavenging mechanisms are compromised. Moreover, the effects of oxygen radical–induced damage may be cumulative over time, thus explaining the long duration of the tumor promotion and progression phase. It could also explain, at least in part, why cancer is a disease of aging, since aged individuals would have accumulated many more "hits" on their genetic material over time, and there is some evidence that the ability to scavenge free radicals decreases in senescent cells. It is interesting that cells from patients with hereditary diseases, such as ataxia telangiectasia, Fanconi's anemia, and Bloom's syndrome, which are all characterized by increased cancer incidence, are hypersensitive to damage by agents that induce oxygen-radical formation. For example, increased oxygen tension causes an excessive amount of chromosomal aberrations in cells from patients with Fanconi's anemia.[79] Moreover, the serum of patients with ataxia telangiectasia or Bloom's syndrome contains DNA-breaking ("clastogenic") factors; this effect, which can be observed when this serum is added to cultures of normal human cells, is inhibited by addition of superoxide dismutase to the cultures.

In addition to stimulating cell proliferation and altered gene expression, phorbol ester tumor promoters induce the secretion of plasminogen activator and type IV collagenase by human fibroblasts.[81] Because proteases and collagenases released by tumor cells would foster degradation of the growth-limiting basal lamina, release of such enzymes may be another way in which tumor promoters foster tumor expansion and ultimately invasion into underlying tissues. TPA also induces angiogenesis, at least in vitro, as evidenced by its ability to cause cultured endothelial cells to infiltrate into an underlying collagen matrix and form an extensive network of capillary-like structures.[82]

Many of the effects of tumor-promoting phorbol esters are thought to be due to their ability to activate a calcium-dependent protein kinase known as protein kinase C (PKC). The mechanism for TPA activation of PKC has been worked out (reviewed in References 83 and 84). Interaction of TPA with its receptor kinase favors binding of the inactive cytosolic form to the cell membrane, where it is activated. TPA acts as diacylglycerol (DAG) does, and can substitute for it by increasing the affinity of PKC for Ca^{2+} and phosphatidylserine, thereby fostering the translocation of PKC from cytosol to plasma membrane and causing its activation. Part of TPA chemically resembles DAG. TPA thus acts synergistically with Ca^{2+}-mobilizing agents, such as those that activate the inositol phospholipid turnover cascade (see Chapter 4). Unlike DAG, TPA and similar phorbol esters have a long half-life in cellular membranes, which may explain how they can provide a prolonged signal for cell proliferation, unregulated by the normal feedback mechanism provided by turnover of DAG and the subsequent inactivation of PKC.

Experimental Models for the Study of Carcinogenesis

A number of models for the study of carcinogenesis have been developed over the years. Historically, two of the most useful ones have been the initiation-promotion model of mouse skin carcinogenesis (the "skin-painting" model) and the induction of liver cancers in rats.

The classic model of carcinogenesis is the single application of an initiating agent such as a polycyclic aromatic hydrocarbon followed by the continuous application of a promoting agent like TPA to the backs of shaved mice. Much of what we know about tumor initiation, promotion, and progression has come from this model system.

Initiation and promotion during mouse skin carcinogenesis produce multiple benign squamous papillomas. A few squamous cell carcinomas eventually arise from the papillomas over many months. However, malignant conversion can be speeded up by exposure of papilloma-bearing mice to mutagens, which activates oncogenes such as H-*ras* and causes loss of tumor suppressor genes such as *p53*, as noted above.

The mouse skin carcinogenesis model is also a useful one in which to study the role of diet and chemopreventive agents in carcinogenesis (see also Chapter 9). For example, calorie-restricted diets have been shown to reduce the number and size of papillomas during and following promotion with TPA in DMBA-initiated SENCAR mice.[85] Furthermore, the latency period for occurrence of carcinomas was increased and the total number of carcinomas was decreased. Application of apigenin, a plant alkaloid,[86] retinoic acid,[87] and prostratin, a nonpromoting phorbol ester[88] have been shown to inhibit the promotion phase (appearance of papillomas) of mouse skin carcinogenesis.

Multistage carcinogenesis has also been observed for liver tissue. For example, Peraino et al.[89] observed that a 3-week exposure of rats to AAF in the diet produced only a small number of hepatomas after several months, but if the animals were subsequently treated with phenobarbital for several months after carcinogen feeding was discontinued, a high incidence of hepatomas was noted. Similar results have been obtained by Kitagawa et al.,[90] who fed rats a nonhepatocarcinogenic dose of 2-methyl-N,N-dimethyl-4-aminoazobenzene for 2 to 6 weeks, and then a dietary administration of phenobarbital for 70 weeks. By 72 weeks, many large hepatocellular carcinomas had developed in the phenobarbital-treated animals, whereas only a few small tumor nodules were observed in the rats not given phenobarbital. Thus, the action of phenobarbital appears to be analogous to that of TPA in the mouse skin system—that is, it "fixes" the damage to cells induced by an initiating agent and causes a clone of cells arising from a damaged cell to proliferate. However, whereas TPA stimulates DNA synthesis and hyperplasia in skin, phenobarbital produces only a transient and relatively small increase in DNA synthesis in liver. Perhaps that is all that is needed to fix

the carcinogenic damage and to allow for the initial proliferation of a damaged clone of cells. Once the damaged clone is present, it could undergo alteration due to its genetic instability and gradually progress to a detectable malignant tumor. This idea is supported by the experiments of Pitot et al.,[91] who treated rats with a single dose of diethylnitrosamine by intubation 24 hours after partial hepatectomy (partial removal of the liver), which stimulates DNA synthesis and cell proliferation in the remaining tissue. If the animals were then treated, starting 8 weeks later, with phenobarbital in the diet for 6 months, many small, phenotypically heterogeneous foci characterized by glucose-6-phosphatase–deficient areas, ATPase-deficient areas, and γ-glutamyltranspeptidase-containing areas developed in the liver. Many of these animals also had hepatomas, for which the enzyme-altered foci appear to represent the early stage of neoplastic development. Thus in this case, phenobarbital appears to have stimulated the replication of dormant initiated cells, which, in the absence of the promoter, would not have proliferated. If each enzyme-altered focus observed in these experiments were a clone derived from a single cell, about 10^4 to 10^5 cells in the liver were "initiated" by diethylnitrosamine, and a very small number of these subsequently underwent clonal proliferation during phenobarbital feeding.[91] Thus the conversion of these abnormal foci, or early nodules, as they have been called, to a malignant neoplasm is a rare event.

Newer models of carcinogenicity have involved the use of knock-out or knock-in rodent models, in which various oncogenes, tumor-suppressor genes, or susceptibility genes have been engineered into or out of rodent embryos (usually mice). This process has enabled the definition of some of the genes that are key to various steps in the tumor-initiation promotion and progression steps. These tumor models are now being superceded by conditional genetic knock-out models in mice that allow for the controlled expression of oncogenes or tumor suppressor genes in a way that more closely mimics "spontaneously" arising human cancers (Table 2–2).

Conditional gene expression in the mouse has been achieved by mutations induced by FLP/FRT or Cre/lox P site-specific recombination

Table 2–2. Conditional and Inducible Mouse Tumor Models

Tumor Type	Conditional or Inducible Gene*
TUMOR MODELS THAT USE CRE/lOXP OR FLP/FRT RECOMBINATION SYSTEMS	
Colorectal adenomas	Apc^{loxP}
Mammary adenocarcinomas	$Brca1^{loxP} + Trp53^{+/-}$
Mammary adenocarcinomas	$Brca2^{loxP} + Trp53^{loxP}$
Mammary adenocarcinomas	$Brca2^{loxP}$
Schwannomas	$Nf2^{loxP}$
Lung adenocarcinomas	$Stop^{loxP} + Kras^{G12D}$
Pituitary tumours	Rb^{loxP} or Rb^{FRT}
Medulloblastomas	$Rb^{loxP} + Trp53^{loxP}$
Liver haemangiomas	Vhl^{loxP}
TUMOR MODELS THAT USE SPONTANEOUS RECOMBINATION	
Lung tumors, thymic lymphomas, skin papillomas	$Kras^{G12D}$
TUMOR MODELS THAT USE RETROVIRAL GENE DELIVERY	
Gliomas	$ErbB2 + Cdk4$ or $ErbB2 + Cdkn2a^{-/-}$
Gliomas	$Pdfg$ or $Pdfg + Cdkn2a^{-/-}$
Gliomas	PyV-mT
Glioblastomas	$Kras^{G12D} + Akt$
Ovarian carcinomas	Combinations of $Kras^{G12D}$, Myc, and Akt in wildtype or $Trp53^{-/-}$ background
TUMOR MODELS THAT USE REGULATABLE ONCOGENES	
B-cell leukemia	Bcr-Abl1
Skin tumors	ErbB2
Lung hyperplasias	$Fgf7$
Papillomatosis	Myc
T-cell lymphomas, AML	Myc
T-cell lymphomas	Myc
Melanomas	$Hras^{V12G} + Cdkn2a$
Lung adenocarcinomas	$Kras^{G12D} + Cdkn2a^{-/-}$ or $Kras^{G12D} + Trp53^{-/-}$
Salivary gland hyperplasia	SV40 Tag

*Combinations with conventional tumor suppressor gene knockouts (for example, $Cdkn2a^{-/-}$ or $Trp53^{-/-}$) are included. *Apc*, adenomatous polyposis coli; *Bcr-Abl1*, breakpoint cluster region Abelson 1; *Brca1*, breast cancer gene 1; *Brca2*, breast cancer gene 2; *Cdk4*, cyclin-dependent kinase 4; *Cdkn2a*, cyclin-dependent kinase inhibitor 2a (which encodes the Ink4a and Arf tumor suppressors); *ErbB2*, avian erythroblastic leukemia viral oncogene homologue 2 (which encodes an epidermal growth factor receptor homologue); *Fgf7*, fibroblast growth factor 7; *Hras*, Harvey rat sarcoma viral oncogene homologue; *Kras*, Kirsten rat sarcoma viral oncogene homologue; *Myc*, avian myelocytomatosis viral oncogene homologue; *Nf2*, neurofibromatosis type 2; *Pdgf*, platelet-derived growth factor; *Pyv-mT*, polyomavirus middle T antigen; *Rb*, retinoblastoma; SV40 TAg, simian virus 40 large T antigen; *Trp53*, transformation-related protein 53; *Vhl*, Von Hippel-Landau. (From Jonkers and Berns,[92] reprinted by permission from Macmillan Publishers Ltd.)

systems, regulatable oncogene expression, and retroviral gene transfer in transgenic mice that express an avian retroviral receptor (reviewed in Reference 92). These models provide for the induction of somatic mutations in tissue-specific and time-sequenced way. These models more closely mimic human cancer development, which involves the activation of oncogenes and inactivation of tumor suppressor genes over time. They enable investigators to determine the contribution that individual mutations make to

the various stages of tumor development. These models also provide a way to validate various targets for anticancer drug development.

Validity of Tests for Carcinogenicity

There is quite a bit of debate among scientists and regulatory agencies about how to assess the carcinogenic hazards of chemicals, both manmade and natural, in our environment. Much of this debate has spilled over into the media, generating a sort of "carcinogen-of-the-month club" and much confusion among the public. Indeed, as one observer put it, "Cancer news is a health hazard."[93]

For many years, the prevailing view among cancer epidemiologists has been that 60% to 90% of human cancers are attributable to environmental and lifestyle factors, including cigarette smoking, diet, ultraviolet irradiation, sexual practices, parasitic and viral infections, industrial pollution, and, more recently, pesticides.[94,95]

The implication of this attribution is that most cancers are preventable. Thus the prevailing view, adopted by federal regulatory agencies, is that, as much as possible, all carcinogens should be eliminated from the environment. One outcome of this view is the famous (or infamous, depending on your point of view) Delaney clause to the Food and Drug Act.[96] The Delaney clause interdicts the use of any food additive in processed foods (interestingly, it pays no attention to pesticide residue on non-processed foods) that is found to be carcinogenic at any dose in one or more animal species. Methods to detect carcinogens have since become a major issue for the food and pharmaceutical industries. The question, then, is how can carcinogens be identified before they are found retrospectively to cause human cancer?

The classic approach has been long-term studies in rodents, exposing them to a "maximum tolerated dose" (MTD) for the life span of the animal. An MTD is defined as the highest dose that can be given without causing severe weight loss or other signs of life-threatening toxicity. This kind of testing is very expensive (the estimated range is up to $1 million to $2 million per compound) and time consuming (3 to 4 years). In addition, the doses used in these tests are usually orders of magnitude higher than those that most humans would ever be exposed to. Moreover, these tests fail to take into account the differences in pharmacokinetics, drug metabolism, and excretion mechanisms between mice or rats and humans. This situation has led to some fascinating snafus. Saccharin is a good example. Sodium saccharin, the artificial sweetener, was shown to increase the incidence of bladder cancer in rats when administered beginning at conception or at birth and continuing through an animal's lifetime. Hence, it was banned as a food additive.

The problem is that (1) humans would have to consume 25 kilograms of sodium saccharin a day to achieve the cancer-causing dose in male rats, and (2) humans don't have the same excretion patterns as those of male rats.[97] It was later found that male rats are considerably more susceptible to the saccharin-caused bladder cancer than female rats, mice, hamsters, monkeys, and, according to epidemiological studies, humans. The main difference among these species is that male rats excrete large amounts of protein in the urine, unlike the other species, and silicate microcrystals form in the male rats' urinary bladder as a result. This leads to a chronic irritation of the uroepithelium, increased cell proliferation, and a hyperplasia eventually leading to carcinoma.[97] These effects are not observed in humans. The realization of this species and dose-reality difference eventually resulted in the re-release of saccharin as a sweetener, albeit with warnings being required on the package.

The example of saccharin and certain other chemical carcinogens raises several important questions about carcinogenicity testing. What is an appropriate dose and time frame to use in animal carcinogenicity assays? What effects might lower dose but lifetime exposure to a chemical have on humans? Is there a "threshold" dose for a given agent, below which human exposure is safe? Are there cheaper, more accurate, faster ways to find out if a chemical is carcinogenic? How does one evaluate the relative risk of exposure to natural chemicals in food and water compared to man-made chemicals? Is there an acceptable risk–benefit ratio for economically important chemicals? Can society afford the cost of removing every trace of a potentially carcinogenic substance from the environment? Are there dietary or other factors that can be used to supplement diets to prevent or delay cancer even if one is exposed to known or unknown carcinogens? Can highly susceptible individuals be identified so that they can be advised to avoid certain employment or activities? Can exposure levels be determined that could lead to removal of substances from an environment and/or careful follow-up of individuals so exposed?

These are all difficult questions to answer, and there is considerable debate about the answers, but some approaches have been taken to answer some of them. The Ames test for muta-genesis in bacteria *Salmonella* is one of the widely used short-term tests that have been developed. It is fast, inexpensive, and reasonably accurate. However, the Ames test doesn't detect certain kinds of chemical carcinogens, e.g., those that don't bind to DNA such as chloro-form and dibromochloromethane, and it over-predicts carcinogenicity for others. In one sur-vey of 224 chemicals, the Ames test had a sensitivity of only 54% (percentage of true car-cinogens identified) and a specificity of 70% (true negatives, i.e., chemicals correctly identi-fied as noncarcinogens).[98]

Other short-term tests include induction of resistance to antimetabolites in cultured cell lines, and chromosomal breakage in exposed cultured cells. Using the data from short-term tests, particularly the Ames test, in combination with determination of a "chemically alerting" structure (i.e., a chemical containing structural elements known to be carcinogenic, such as electrophile-producing side chains), Ashby and Tennant[99] showed a high correlation between those compounds that were structurally alerting and those that were mutagenic. When taken to-gether, a reasonable correlation with carcino-genicity was predicted, based on carcinogenicity in animal tests.

One of the problems with the use of the MTD approach to carcinogenicity testing in animals is that in addition to being a dose often an order of magnitude more than a dose to which humans are likely to be exposed, the MTD is a dose that is often sufficient to kill cells and induce a prolif-erative, tissue-repair response in target organs. This increased cell proliferation puts cells at risk for propagating an unrepaired lesion produced by the spontaneous background mutation rate (see above) or by exposure to an environmental agent. This has led Cohen and Ellwein[97] to propose a biological model for carcinogenesis based on whether a chemical is genotoxic or non-genotoxic. The key feature of this model is that an agent can increase the incidence of cancer by either damaging a cell's DNA (genotoxic) or stimulating cell proliferation (non-genotoxic), increasing the likelihood for a spontaneous genetic error to occur during DNA replication. They provide as evidence for this a number

of examples of genotoxic chemicals, e.g., 2-acetylaminofluorene (2-AAF), that cause differ-ent effects in different tissues depending on the proliferative response they produce in a given tissue. The authors also cite non-genotoxic chem-icals that can act at high enough doses to induce cell damage and a mitogenic response that in-creases the genetic error rate. In the example of 2-AAF, even though the dose–response effect may depend on the chemical's effect on cell prolifer-ation in a given tissue, a threshold for the carci-nogenic effect is unlikely. In the case of non-genotoxic chemicals, a threshold effect is likely.

Weinstein and others[100,101] have argued that there is definitive evidence that spontaneous mutation or endogenous DNA damage is car-cinogenic and that there is no consistent corre-lation between the inherent growth fraction (percent proliferating cells) in a tissue and cancer incidence in that tissue. It seems, how-ever, that cell proliferation has to occur for a cancer to develop and that the self-renewal stem cells of a tissue are the most likely targets for carcinogens. Unfortunately, there is no well-defined way to simulate whole-organism carci-nogenicity testing other than in whole organ-isms, where the parameters of absorption, tissue distribution, metabolism (activation and deacti-vation), pharmacokinetics, and excretion can be evaluated. Furthermore, it can be argued that a full dose-range, including doses that stimulate cell proliferation, of potential carcinogens should continue to be tested in animals.[101] The diffi-culty comes in translating these data to realistic human exposure levels. Thus, computers will still not be able to replace good judgment and common sense.

The debate about linear versus nonlinear models for estimating the human carcinogeni-city of chemicals continues.[102] The linear, no-threshold model (Fig. 2–10) assumes that all carcinogens act similarly and that there is no threshold for their carcinogenic action. The logical conclusion from this assumption is that some risk of carcinogenicity exists for any dose of a carcinogen. This notion defies what is known from all other toxicological and phar-macological events, i.e., that there is a dose below which no toxicity or drug response occurs (Fig. 2–11). It also implies that cancer is a one-hit

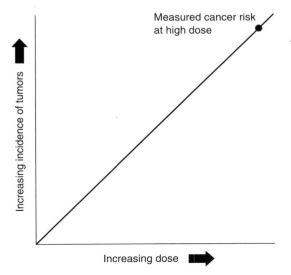

Figure 2–10. Linear curve. Curves with this appearance are not usually found experimentally in dose–response assays, and the idea that a dose–response curve could take such a form is now considered obsolete. (From *America's War on "Carcinogens": Reassessing the Use of Animal Tests to Predict Human Cancer Risk*, p. 52, with permission.)

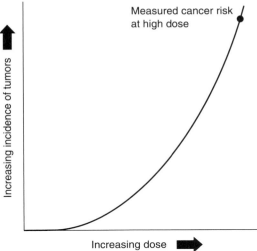

Figure 2–11. Nonlinear threshold. (From *America's War on "Carcinogens": Reassessing the Use of Animal Tests to Predict Human Cancer Risk*, p. 53, with permission.)

event. The no-threshold risk calculation is derived from the extrapolation of high–dose exposure situations. For example, the incidence of lung cancer in deep mineworkers exposed to radon is plotted versus radon exposure linearly back through 0/0 on the plot (Fig. 2–10). In other words, some infinitismal exposure to radon will still cause some cancers. From this sort of extrapolation, it is concluded that no level of exposure to radon is safe. This model is still the one most commonly used for cancer risk assessment by government agencies and its use has led to the propagation of a number of "cancer myths" (see Chapter 3).

Ames and colleagues[103,104] have argued that natural chemicals are as likely as synthetic chemicals to be carcinogenic in various tests and that on the basis of actual levels of human exposure (other than industrial workers or farmers, for example, who might have high exposure rates to industrial chemicals or pesticides), natural chemicals are at least as dangerous as synthetic ones. In high-dose tests, 30% to 50% of both natural and synthetic chemicals are carcinogens,

mutagens, teratogens, and clastogens (DNA-damaging agents).[104] Thus the authors concluded that natural and synthetic chemicals are equally likely to be positive in animal cancer tests and that at the low doses of most human exposures, "the comparative hazards of synthetic pesticide residues are insignificant."[103]

IRRADIATION CARCINOGENESIS

A number of the points made about chemical carcinogenesis are also true for radiation-induced carcinogenesis. Both X-rays and ultraviolet (UV) radiation, for example, produce damage to DNA. As with chemical carcinogens, this damage induces DNA repair processes, some of which are error prone and may lead to mutations. The development of malignant transformation in cultured cells after irradiation requires cell proliferation to "fix" the initial damage into a heritable change and then to allow clonal proliferation and expression of the typical transformed phenotype.[105] Fixation appears to be complete after the first postirradiation mitotic cycle. In the case of mouse C3H/10 T 1/2 cells, expression of radiation-induced transformation requires an additional 12 rounds of cell division. Thus, as in the case of chemical carcinogenesis, a promotion phase is required for full expression of the initiated malignant alteration. Moreover, when low doses of chemical carcinogens and X-rays are used together, these two types of agents act synergistically to produce malignant transformation.[105]

When cells are exposed to UV light in the 240 to 300 nm range, the bases acquire excited energy states, producing photochemical reactions between DNA bases (reviewed in Reference 106). The principal products in DNA at biologically relevant doses of UV light are cyclobutane dimers formed between two adjacent pyrimidine bases in the DNA chain. Both thymine–thymine and thymine–cytosine dimers are formed. That formation of these dimers is linked to mutagenic events (see below).

Heavy exposure to sunlight induces similar changes in human skin, and the degree of exposure to sunlight is closely related to the incidence of skin cancer (see Chapter 3). Whether continuing exposure to UV rays in sunlight is the promoting agent in skin cancer or additional promoting events are required is not clear, but it seems that UV irradiation is a complete carcinogen, just as some chemicals are—that is, it has both initiating and promoting activities. Patients who cannot efficiently repair UV-induced damage, such as those with xeroderma pigmentosum, have a much higher risk of developing malignant skin tumors.

Ionizing Radiation

The history of radiation carcinogenesis goes back a long way (reviewed in Reference 107). The harmful effects of X-rays were observed soon after their discovery in 1895 by W. K. Röntgen. The first observed effects were acute, such as reddening and blistering of the skin within hours or days after exposure. By 1902, it became apparent that cancer was one of the possible delayed effects of X-ray exposure. These cancers, which included leukemia, skin cancers, lymphomas, and brain tumors, were usually seen in radiologists only after long-term exposure before adequate safety measures were adopted, thus it was thought that there was a safe threshold for radiation exposure. The hypothesis that small doses of radiation might also

cause cancer was not adopted until the 1950s, when data from atomic bomb survivors in Japan and certain groups of patients treated with X-rays for noncancerous conditions, such as enlarged thyroids, were analyzed. These and other data led to the concept that the incidence of radiation-induced cancers might increase as a linear, nonthreshold function of dose. Thus the debate about whether there is a safe threshold pertains to radiation carcinogenesis, just as it does to chemical carcinogenesis.

In radiation carcinogenesis, the damage to DNA, and hence its mutagenic and carcinogenic effect, is due to the generation of free radicals as the radiation passes through tissues. The amount of radical formation and ensuing DNA damage depend on the energy of the radiation. In general, X-rays and gamma rays have a low rate of linear energy transfer, generate ions sparsely along their tracks, and penetrate deeply into tissue. This profile contrasts with that of charged particles, such as protons and α particles, which have a high linear energy transfer, generate many more radical ions locally, and have low penetration through tissues. The damage to DNA can include single- and double-strand breaks, point mutations due to misrepair deletions, and chromosomal translocations.[107–109] The molecular genetic events that follow radiation damage to cells include (1) induction of early-response genes such as c-jun and Egr-1; (2) induction of later-response genes such as tumor necrosis factor-α (TNF-α), fibroblast growth factor (FGF), and platelet-derived growth factor-α (PDGF-α); (3) activation of interleukin-1 (IL-1) PKC[110]; and (4) activation of oncogenes such as c-myc and K-ras.[111] Induction of these genes may be involved in the cellular responses to irradiation and in the longer-range effects that lead to carcinogenesis. At any rate, the production of clinically detectable cancers in humans after known exposures generally occurs after long latent periods. Estimates of these latent periods are 7 to 10 years for leukemia, 10–15 years for bone, 27 years for brain, 20 years for thyroid, 22 years for breast, 25 years for lung, 26 years for intestinal, and 24 years for skin cancers.

A more recent example of nuclear fallout leading to environmental exposure to radiation is the Chernobyl accident, which happened on

April 26, 1986. A steam explosion blew the lid off the reactor. The graphite core caught fire and over 10^{19} becquerels (Bqs) of radioisotopes were released, producing a fallout that covered much of Belarus, Northern Ukraine, and part of the Russian Federation. Estimates are that 10–20 million people were exposed to significant fallout. There were some deaths due to acute radiation sickness from high levels of exposure. However, the long-term effects are still being recorded. So far, the reliable reports of increases in cancer incidences are mostly limited to thyroid cancer.[112] This finding is in contrast to cancer incidence among atomic bomb survivors in Japan, some of whom developed cancers of various types, including cancers of the thyroid, breast, lung, stomach, esophagus, bladder, leukemia, and lymphoma (although the incidence of cancers in Japanese atomic bomb survivors was less than would have been predicted by radiation exposure). The reason for this discrepancy is most likely that those exposed to the Chernobyl fallout received primarily dosage from β-emitters, mostly isotopes of iodine, which concentrates in the thyroid. Atomic bomb survivors, by contrast, received whole-body irradiation from neutrons and gamma rays.

Another interesting point about the Chernobyl survivors is that the type of thyroid cancer they developed, mostly among those under 2 years of age if they were exposed, were 98% papillary, many with an unusual morphology, whereas in non-exposed populations, only 67% of childhood thyroid cancers are papillary.[112]

Expression of two families of oncogenes, the c-*ret* and *ras* families, has been shown to be involved in papillary thyroid cancers. The oncogene c-*ret* is a receptor tyrosine kinase activated by gene rearrangement, and two of these, *ret-ptc* 1 and *ret-ptc* 3, are activated in papillary carcinomas. Since c-*ret* is activated by rearrangement, the high proportion of double-strand DNA breaks seen in radiation-induced papillary carcinomas of the thyroid may explain its activation.

Since the thyroid is not the only tissue that concentrates iodine, malignancies of other tissues that also concentrate iodine, such as the breast, salivary gland, and stomach, may appear in higher incidence as time goes on. Moreover, other isotopes including cesium were present in the fallout, and inhabitants of parts of the Ukraine and Belarus are still exposed to low levels of radioactive cesium. The long-term effects, if any, of such exposure is not yet clear.

Ultraviolet Radiation

Ultraviolet radiation–induced lesions, generated by UV-B (280–320 nm wavelength) or UV-A (320–400 nm wavelength), result from DNA damage, which is converted to mutations during cellular repair processes. UB-B and UV-A generate different types of DNA damage and DNA repair mechanisms (reviewed in Reference 113). Irradiation with UV-B produces cyclobutane pyrimidine dimers that are repaired by nucleotide excision repair. If left unrepaired, C→T and CC→TT base transitions occur. UV-A-induced DNA damage produces mostly oxidative lesions via photosensitization mechanisms and is repaired by base excision repair. UV-B and UV-A also produce different effects on the immune system and elicit different transcriptional and inflammatory responses. While the specific mechanisms by which UV radiation induces basal cell or squamous cell carcinomas or melanoma are not clear, a number of signal transduction pathways are affected that can either lead to apoptosis or to increased cell proliferation (Fig. 2–12). UV irradiation activates receptor tyrosine kinases and other cell surface receptors. It also enhances phosphorylation by ligand-independent mechanisms via inhibition of protein tyrosine phosphatase activity. Ligand-dependent cell surface receptor activation can also occur by activation of autocrine or paracrine release of growth factors from keratinocytes, melanocytes, or neighboring fibroblasts. It is clear, however, that better animal models are needed to clearly define the mechanisms by which UV light causes human cancer.

OXYGEN FREE RADICALS, AGING, AND CANCER

The diseases of aging include cardiovascular disease, decline in function of the immune system, brain dysfunction, and cancer. People living in the United States who are 65 or older have

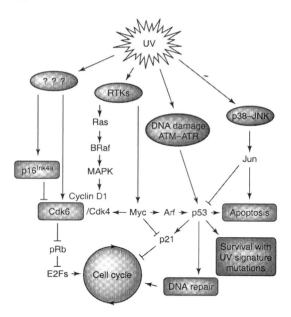

Figure 2–12. Immediate and long-term effects of ultraviolet (UV) radiation on skin cells and their genomes are complex. Immediate cellular responses to UV can occur through stress pathways (p38, mitogen-activated protein kinase [MAPK] and Jun N-terminal kinase [JNK]), cell surface receptors (receptors tyrosine kinase [RTKs]) and direct DNA damage. The response to DNA damage is largely mediated by p53, which can arrest growth and facilitate DNA repair or, if the damage is too extensive, induce apoptosis. Mutation-bearing melanocytes that escape these fates and survive are the seeds of potential future melanomas. Proteins of the Rb pathway, such as p16[Ink4a] and cyclin-dependent kinase 6 (Cdk6), are important biological targets of UV. However, the genetic changes observed (loss of *p16[Ink4a]* and amplification of *Cdk6* [green]) are not characteristic of UV-induced mutagenesis and, hence, these consequences are probably indirect, with stochastic mutations selected for in cells that go on to become melanomas. Arrows do not necessarily represent direct interactions. ATM, ataxia telengiectasia mutated; ATR, ataxia telengiectasia and Rad3 related. (From Merlino and Noonan,[113] with permission.)

10 times the risk of those under age 65 for developing cancer.

Part of the increase in cancer incidence with aging could be due to an accumulation of damage to DNA over a lifetime of exposure to carcinogenic substances. Another, perhaps more likely, possibility is that cellular damage produced by endogenous oxidants accumulates over time and the body's ability to repair this damage decreases with age. There is a fair amount of circumstantial evidence to support this latter hypothesis (reviewed in References 114–116). Oxidative damage to DNA, proteins, lipids, and other macromolecules accumulates with age. Oxidation products formed during normal metabolic processes in cells include superoxide (O_2^-), hydrogen peroxide (H_2O_2), and hydroxyl radical (OH). These are also produced in cells by radiation, and they are capable of damaging DNA and producing mutagenesis. Of these, the hydroxyl radical appears to be the primary DNA-damaging species, but it has a short half-life and high reactivity, so it must be generated in close proximity to DNA.[117] This may occur in the cell nucleus by an interaction of H_2O_2 with chromatin-bound metals such as Fe^{+2} by the following reaction:

$$Fe^{+2} + H_2O_2 \rightarrow Fe^{+3} + .OH + OH.^-$$

Singlet oxygen, which is produced by lipid peroxidation or by the respiratory bursts from neutrophils, is also mutagenic and has a much longer half-life than the hydroxyl radical. Lipid peroxidation can also give rise to mutagenic products such as lipid epoxides, hydroperoxides, alkanyl and peroxyl radicals, and α, β unsaturated aldehydes.[114]

Cells have multiple mechanisms to protect themselves from oxidative damage, including superoxide dismutase, catalase, glutathione peroxidase, and glutathione-S-transferases. In addition, DNA damaged by oxidants is subject to repair (see below), oxidized proteins are degraded by proteases, and lipid peroxides are destroyed by glutathione peroxidase. Nevertheless, some oxidative damage and misrepair may persist, and the ability to carry out these repair mechanisms decreases with aging. It is estimated that the human genome suffers about 10,000 "oxidative hits" to DNA per cell per day.[114] Mutations accumulate with age in the rat so that an "old" rat (2 years old) has twice as many DNA lesions per cell as a young rat. Furthermore, the frequency of somatic mutations found in human lymphocytes is about ninefold higher in the aged than in neonates.[118] How much of this mutation frequency is due to oxidative damage of DNA isn't clear, but a number of altered bases have been observed in cells undergoing oxidative stress. These include hydroxymethyl uracil, thymine glycol, 8-hydroxyguanine,

8-hydroxyadenine, and formamido derivatives of altered purines.[119] Some of these products appear in the urine and may be an index of oxidative damage. They are also produced by exposure of DNA to ionizing radiation and oxygen-radical generators. 8-Hydroxyguanine appears to be the most frequently altered base to result from oxidative damage to DNA, and if this base is left unrepaired in DNA it produces G→T transversion.[119]

Endogenous oxidants can also damage proteins and lipids. Oxygen free-radicals catalyze the oxidative modification of proteins leading to an age-related increase in carbonyl content of cellular proteins.[115] For example, there is a significant increase in carbonylated proteins in human erythrocytes from older individuals, and the carbonyl content of proteins in cultured human skin fibroblasts increases exponentially with the age of the fibroblast donor.[115] In addition, the protein carbonyl content of fibroblasts from individuals with premature aging (progeria and Werner's syndrome) is higher than that in age-matched controls. There is also an age-related decrease in neutral alkaline protease activity that degrades oxidized proteins. The end result is an increased retention of damaged proteins with aging. The degree to which oxidative protein damage contributes to diseases of aging and cancer isn't clear, but treatment of gerbils with the radical trapping agent tert-butyl-α-phenylnitrone inhibits age-related increases in oxidized protein in the brain and blocks age-related memory loss (as measured by a radial-arm maze test).[120] Oxygen-radical damage to lipids leads to fluorescent lipid oxidation products that appear to result from cross-links between proteins and lipid peroxidation products, and these also increase with age.[114]

Caloric or protein restriction in the diet slows oxidative damage to proteins and DNA and decreases the rate of formation of neoplasms in rodents. Similar results are seen by dietary supplementation with antioxidants such as tocopherol (vitamin E), ascorbate (vitamin C), and carotenoids such as β-carotene, leading to the hypothesis, supported by epidemiological data in humans, that dietary intake of such substances could decrease the incidence of human cancer (see Chapter 9). It should also be noted that one of the major sources of exogenous oxidant exposure is the oxides of nitrogen found in cigarette smoke.[114]

GENETIC SUSCEPTIBILITY AND CANCER

As was noted above, there are a number of inherited cancer susceptibility gene mutations, such as xeroderma pigmentosum, Fanconi's anemia, and ataxia telangiectasia. These types of inherited defects that lead to cancer are generally caused by a deficiency in DNA repair pathways. Almost certainly we have only scratched the surface of inherited cancer susceptibility genes that make an individual more prone to developing cancer. Other susceptibility genes may include alterations in the metabolic enzymes that metabolize drugs and environmental toxins, polymorphisms in genes that regulate utilization of certain essential nutrients such as folic acid, or inherited mutations in tumor suppressor genes.

The completion of the Human Genome Project allows a systematic approach to discovering the genetic alterations that make individuals prone to developing various diseases. The Environmental Genome Project is producing a catalogue of variation in genes involved in catabolizing toxins, nutrient metabolism, and DNA repair.[121] These data, which will be largely generated by detection of single nucleotide polymorphisms (SNPs), will enable toxicologists and cancer biologists to predict individual susceptibility to diseases triggered or promoted by environmental pollutants, diet, and other lifestyle factors. Some examples of this SNP analysis approach are the increased susceptibility of individuals with altered folate metabolism genes to develop leukemia after benzene exposure and the ethnic variation in the BRCA1 gene SNPs that affect susceptibility to breast cancer.

MULTIPLE MUTATIONS IN CANCER

In most cases, it takes years for a full-blown invasive, metastatic cancer to develop from a small clone of initiated cells. This process might take 20 years or more, during which time an initiated clone of cells undergoes clonal expansion via multiple cell doublings. As these clones expand, various cells in the population accumulate multiple genetic alterations, some of which facilitate dysregulated cell proliferation and some of which lead to cell death. These genetic alterations can include point mutations, chromosomal translocations, gene deletions, gene amplifications, loss of genetic heterozygosity (LOH), and loss of genetic imprinting (LOI). These will be discussed in detail in Chapter 5. This accumulation of genetic defects that occurs during clonal expansion of transformed cells is due to "genetic instability." The cause of this genetic instability is not clearly understood, but it includes defects in cell replication checkpoint controls and decreased ability to repair DNA damage.

There is evidence for the accumulation of thousands of mutations in cancer cells derived from human tumors. For example, examination of the colon tumor–derived DNA from patients with hereditary non-polyposis colon cancer (HNPCC) reveals that as many as 100,000 repetitive DNA sequences are altered from the mismatch DNA repair defects that these patients' cells harbor (reviewed in Reference 122). Mismatch repair defects have also been noted in "sporadic" (not known to be hereditary) cancers.

As noted earlier, one hypothesis explaining the genetic instability of transformed cells is the mutator phenotype hypothesis, championed by Loeb and colleagues.[122] This hypothesis states that an "initial mutator [gene] mutation generates further mutations including mutations in additional genetic stability genes, resulting in a cascade of mutations throughout the genome." The molecular defect that could provide this phenotype could be a mutation in DNA polymerases that leads to error-prone DNA replication. The mutator phenotype would have to be generated early in tumorigenesis for this hypothesis to be valid. There are a number of arguments against this idea, such as observations that there is not necessarily an increased mutation rate in cancer cells over that of normal cells[123] and that a similar "evolution" of genetically altered cancer cells could arise by clonal selection followed by clonal expansion of cells with a genetic alteration that provides a proliferative advantage.[124]

DNA REPAIR MECHANISMS

Not all interactions of chemicals and irradiation with DNA produce mutations. In fact, all cells have efficient repair mechanisms that repair such lesions. DNA repair mechanisms include sets of enzymes that survey DNA for specific kinds of damage, remove the altered portion of DNA, and then restore the correct nucleotide sequence. The important role of DNA repair in human cancer has been established by the finding that a number of inherited defects in DNA repair systems predispose individuals to getting cancer. These diseases include xeroderma pigmentosum, ataxia telangiectasia, Fanconi's anemia, Bloom's syndrome, Cokayne's syndrome, and hereditary retinoblastoma.[125]

There are several types of DNA repair systems, a number of which have been preserved from bacteria to humans. These include[125–127] (1) abnormal precursor degradation, e.g., the hydrolysis of the oxidized nucleotide triphosphate 8-hydroxy-dGTP to its nucleotide 8-OH-dGMP, preventing incorporation into DNA; (2) a visible light-activated photoreactivation repair mechanism for removal of UV-induced cyclobutane pyrimidine dimmers; (3) strand break repair via an action of DNA ligase, exonuclease, and polymerase activities; (4) base excision repair that recognizes simple base alterations such as cytosine deamination to uracil and requires the action of (a) a purine or pyrimidine glycosylase that breaks the deoxyribose-base bond, (b) an endonuclease to cleave at the abasic site, (c) a phosphodiesterase to clip away the "naked" abasic site, (d) DNA polymerase, and (e) DNA ligase to refill and reclose the site; (5) nucleotide excision repair that recognizes bulky DNA base adducts, pyrimidine dimers, and base crosslinks and requires the concerted action of enzymes and recognition factors (see below); and (6) 0^6-alkyguanine-DNA alkyltransferase that recognizes and removes small alkyl adducts from DNA. In mammalian cells, key repair mechanisms are base excision repair, nucleotide excision repair, transcription-coupled repair, homologous recombination and end joining, and mismatch repair.[128]

Excision repair is the most general DNA repair mechanism in higher organisms. Base excision repair removes damage such as deaminated bases, oxidized or ring-opened bases generated by hydroxyl or superoxide radicals, and abnormally methylated bases such as 3-methyladenine.[126] Nucleotide excision repair requires sequential steps of (1) preincision recognition of damage; (2) incision of the damaged DNA strand at or near the damaged site; (3) excision of the damaged site and local removal of nucleotides in both directions from the defect in the affected DNA strand; (4) repair replication to replace the excised region, using the undamaged strand as a template; and (5) ligation to join the repaired sequence of nucleotides at its 3′ end to the contiguous DNA strand.[125]

DNA repair is usually very accurate, but if repair cannot occur prior to or during DNA replication it may be error prone. This error-prone, post-replication repair seems to be brought into play by certain types of agents or when a cell is overwhelmed by damage that it cannot handle by excision repair before the cell enters S phase during the next round of cell division. In this case, the new DNA is synthesized on templates that still contain damaged bases, leading to mispairing or recombinational events that transfer damaged bases to daughter strands. For example, in mammalian cells, 5% to 30% of UV-induced thymidine dimers are transferred from parental to daughter strands during postreplication repair.[129]

Nucleotide excision repair (NER) of DNA in eukaryotic cells requires several gene products. Some of these gene products appear to be identical or highly homologous in yeast, rodents, and humans.[130,131] A number of defects in the NER system have been found by studying mutations in cells from patients with xeroderma pigmentosum, in whom at least nine different kinds of mutations (i.e., nine different complementation groups) have been found.[125] Some of these XP genes have been cloned and found to be highly homologous to yeast *RAD* genes that are required for excision repair in *Saccharomyces cerevisiae*.[130–133] Some of the cloned human genes also correct repair defects in mutant rodent cells and are called *excision repair cross-complementing* (ERCC) genes.

The *RAD2* gene in *S. cerevisiae* shares remarkable sequence homology to the XP-G gene of the G complementation group of xeroderma, which is the same as the ERCC 5 gene for correcting repair defects in humans. The RAD2 gene product has been shown to have the ability to act as a single-stranded DNA endonuclease, directly implicating the *RAD2* gene and its XPG human homologue as an important component in the incision of a damaged DNA strand during excision repair.[131] Two other human DNA repair defect diseases, Cockayne's syndrome (CS) and PIBIDS, which is a photosensitive form of the brittle-hair disease trichothiodystrophy (TTD), also have genetic defects that may correlate to genes in yeast. It is peculiar, however, that patients with PIBIDS have all of the symptoms of CS as well as some of those of TTD, and patients with CS and PIBIDS do not appear to have a higher than expected incidence of cancer. This is one of the curious examples of a remarkable clinical heterogeneity among patients with the same apparent genetic defects, that is, the same mutation in different individuals giving rise to different clinical syndromes. Such heterogeneity has been observed in cystic fibrosis as well. In the case of NER defects, mutations in one gene may give rise to symptoms of XP, combined XP and CS, or PIBIDS.[130] These kinds of results indicate that identifying mutations is only a first step in understanding the mechanism of the disease process. A mutant gene may act differently in one cell type than in another cell type, depending on its interaction with cell-specific transcription factors or *cis*-regulatory elements, interaction of its gene product with other gene products expressed at different levels in different cell types, or post-transcriptional and post-translational mechanisms in different cell types that regulate the expression and/or function of the protein coded for by the mutant gene in question. One candidate for this latter point is the way protein folding is regulated. Proteins must fold into particular conformations to have biological activity, and some mutations observed in human genetic diseases appear to affect protein folding and intracellular translocations processes. Such defects are seen in certain forms of cystic fibrosis, α_1-anti-trypsin deficiency, and certain glycogen storage diseases.

It has become increasingly important to understand the mechanisms of DNA repair and the defects associated with repair processes, because gene mutations identified in association with hereditary forms of various cancers have been found to be mutations in DNA repair enzymes (Table 2–3). This intriguing discovery highlights the importance of DNA repair processes in maintaining the integrity of the human genome and in protecting the organism from genetic alterations that can lead to cancer. It also highlights the importance of studying basic biochemical and genetic functions in lower phylogenetic systems, because often that is where the important functions of genes are first discovered. The discovery of the human hereditary non-polyposis colon cancer gene is a case in point. Without the knowledge of how DNA mismatch repair genes work in yeast and bacteria and a somewhat serendipitous finding based on a gene bank search, identification of the function of the colon cancer genes as mutated DNA repair genes may have taken many months or years longer.[134]

Two other points about DNA repair systems should be made here. One is that the DNA damage recognition and repair complex contains proteins that function in concert with transcription factors (e.g., the repair protein complex XPBC/ERCC3 interacts with the basal transcription factor TFIIH) and RNA polymerase II.[130] For example, 25 or more proteins participate in the machinery involved in nucleotide excision repair. These are assembled in step-wise fashion at the site of the lesion and then disassembled after the repair event is completed.[128] A second point is that actively transcribed genes are repaired more rapidly than inactive genes. For example, repair of lesions in the active dihydrofolate reductase (DHFR) gene that are induced by UV damage or alkylating agents is done much more rapidly and completely than overall genome repair in the same cells.[125] Moreover, actively transcribed proto-oncogenes or tumor suppressor genes, although perhaps more subject to damage by carcinogens, are also avidly repaired on the transcribing strand. Thus mutations may accumulate on the non-transcribed strand. Mutations in proto-oncogenes that are not actively transcribed in a given cell type may not become evident until such time as they become expressed. This delay might explain, in part, the long latency time of certain forms of human cancer.

The last repair mechanism mentioned above, namely O^6-alkylguanine-DNA alkyltransferase (AGT), is important for the repair of alkyl

Table 2–3. Human Syndromes with Defective Genome Maintenance

Syndrome	Affected Maintenance Mechanism	Main Type of Genome Instability	Major Cancer Predisposition
Xeroderma pigmentosum	NER (±TCR)	Point mutations	UV-induced skin cancer
Cockayne syndrome	TCR	Point mutations	None°
Trichothiodystrophy	NER, TCR	Point mutations	None°
Ataxia telangiectasia (AT)	DSB response and repair	Chromosome aberrations	Lymphomas
AT-like disorder	DSB response and repair	Chromosome aberrations	Lymphomas
Nijmegen breakage syndrome	DSB response and repair	Chromosome aberrations	Lymphomas
BRCA 1 and BRCA 2	HR	Chromosome aberrations	Breast (ovarian) cancer
Werner syndrome	HR? TLS?	Chromosome aberrations	Various cancers
Bloom syndrome	HR?	Chromosome aberrations (SCE↑)	Leukemia, lymphoma, others
Rothmund-Thomson syndrome	HR?	Chromosome aberrations	Osteosarcoma
Ligase IV deficiency†	EJ	Recombination fidelity	Leukemia(?)
HNPCC	MMR	Point mutations	Colorectal cancer
Xeroderma pigmentosum variant	TLS‡	Point mutations	UV-induced skin cancer

°Defect in transcription-coupled repair triggers apoptosis, which may protect against UV-induced cancer.

†One patient with leukemia and radiosensitivity was described with active-site mutation in ligase IV.

‡Specific defect in relatively error-free bypass replication of UV-induced cyclobutane pyrimidine dimmers.

Abbreviations: BER, base-excision repair; DSB, double-strand break; EJ, end joining; HNPCC, hereditary non-polyposis colorectal cancer; HR, homologous recombination; MMR, mismatch repair; NER, nucleotide-excision repair; SCE, sister-chromatid exchange; TCR, transcription-coupled repair; TLS, translesion synthesis; UV, ultraviolet. (From Heijmakers[128])

(Reprinted by permission from Macmillan Publishers Ltd. Nature 411:366–374)

adducts in chemical carcinogen–damaged DNA.[127,135] AGT is both an alkyltransferase and an alkyl acceptor protein that transfers, for example, a methyl group from O^6-methylguanine to an internal cysteine, forming S-methylcysteine in the protein and regenerating unalkylated G in the DNA strand. This transferase can repair O^6-methyl G and O^4-methyl T and is inhibited by O^6-benzylguanine.[136] This inhibitory effect has been employed to show the importance of O^6-alkylations in carcinogen-induced DNA damage and mutations and to augment the cytoxicity of alkylating anticancer drugs such as bischloroethylnitrosourea (BCNU). AGT is most active in the repair of smaller alkyl adducts; for example, O^6-methyl G is repaired about three times faster than O^6-ethyl G, and it is likely that even larger adducts are repaired primarily by excision repair.[127] The importance of AGT in preventing chemical carcinogenesis in vivo has been shown by the prevention of N-methyl-N-nitrosourea-induced thymic lymphomas in transgenic mice bearing the human AGT gene.[135]

VIRAL CARCINOGENESIS

Historical Perspectives

It has long been suspected that various forms of cancer, particularly certain lymphomas and leukemias, are caused or at least "co-caused" by transmissible viruses. This theory has had its ups and downs during the first half of this century, and it was not generally accepted until the 1950s that viruses can cause malignant tumors in animals. The known carcinogenic effects of certain chemicals, irradiation, chronic irritation, and hormones did not fit with the idea of an infectious origin of cancer. In early experiments, the basic assay to determine whether cancer could be induced by a transmissible agent involved transmititng malignant disease by inoculation of filtered extracts prepared from diseased tissues. If the disease occurred in animals inoculated with such filtrates, it was assumed to be caused by a virus. In 1908, Ellermann and Bang[137] transmitted chicken leukemia by cell-free, filtered extracts and thus were among the first to demonstrate the viral etiology of this disease. In 1911, Rous[138] induced sarcomas in chickens by filtrates obtained by passing tumor extracts through filters that were impermeable to cells and bacteria. These findings remained dormant for two decades until Shope showed, in 1933, that the common cutaneous papillomas of wild rabbits in Kansas and Iowa were caused by a filterable agent.[139] It was later found that when these tumors were transplanted subcutaneously they became invasive squamous cell carcinomas.[140] In 1934, Lucké observed that kidney carcinomas commonly found in frogs in New England lakes could be transmitted by lyophilized cell-free extracts.[141] Two years later, Bittner demonstrated the transmission of mouse mammary carcinoma through the milk of mothers to offspring.[142] This was the first documented example of transmission of a tumor-inducing virus from one generation to another.

Drawing on the experiments of Bittner, Gross postulated that mouse leukemia was also caused by a virus and that occurrence of the disease in successive generations of mice was due to transmission of virus from parents to offspring.[143] The proof of this hypothesis eluded Gross for a number of years until he was prompted, by evidence based on transmission of Coxsackie viruses to newborn mice, to attempt inoculation of mice less than 48 hours old. Using this approach, he successfully transmitted mouse leukemia by injecting filtered extracts prepared from organs of inbred AK or C58 mice, which have a high incidence of "spontaneous" leukemia, or from embryos of these mice, into newborn C3H mice, which have a very low incidence of leukemia. These experiments demonstrated for the first time that mouse leukemia is caused by a virus and that the virus is transmitted in its latent form through embryos. This led to the isolation of a mouse leukemia virus.[144] The isolated virus was also found to induce leukemias and lymphomas in inbred strains of mice. Electron-microscopic studies[145] showed that the mouse leukemia virus is spheroid, has a diameter of about 100 nm, and contains a dense, centrally located "nucleus" separated from the external envelope by a clear circular zone. The Gross mouse leukemia virus was classified as a *type C* virus, a term now used to describe a wide variety of RNA-containing oncogenic viruses of similar morphology.

The RNA oncoviruses have been classified by morphological criteria. Intracytoplasmic type A particles were initially observed in early embryos of mice and in certain murine tumors. These A particles are noninfectious, bud into intracellular membranes rather than through the plasma membrane, and thus stay within the cell. They have an active reverse transcriptase and exist as a proviral form in chromosomal DNA. Type B viruses have spikes on their outer envelope, bud from cells, and have been identified primarily in murine species, mouse mammary tumor virus (MMTV) being an example. Type C viruses have been found widely distributed among birds and mammals, can induce leukemias, sarcomas, and other tumors in various species, and have certain gene sequences that are homologous to "transforming" sequences isolated from various human tumors (see below). Another subgroup, type D RNA oncoviruses, has been isolated from primate species but their oncogenic potential is not well established. The subtypes of RNA tumor viruses, known as *Retroviridiae*, share a genetically related genome containing a *gag-pol-env* gene sequence coding for virus internal structural proteins, the special type of RNA-directed DNA polymerase called *reverse transcriptase* and *viral envelope proteins*, respectively. Thus, they most likely share a common evolutionary heritage.[146] However, distinct subclasses of retrovirus evolution, based on *pol* gene sequence homologies, have been found; one major pathway gives rise to mammalian type C viruses and a second to A, B, D, and avian type C oncoviruses.[146] A more recent addition to the retrovirus classification is the human T-cell leukemia virus (HTLV), isolated from patients with certain forms of adult T-cell leukemias (discussed later). The *pol* gene of HTLV appears to have evolved from a progenitor common to the types A, B, D, and avian C oncoviruses rather than from the mammalian C type.[146] If true, this would be unusual because most mammalian type C viruses share antigenic determinants among several *gag, pol,* and *env* gene products, suggesting a common progenitor for this subclass of retroviruses.

Unlike most infectious viruses, oncoviruses can be transmitted through the germline of animal species, and thus these viral genes can be passed from one generation to the next, often in a silent form. The widespread distribution of gene sequences homologous to those of oncoviruses throughout the animal kingdom suggests that these sequences are evolutionarily very old. One of the most important observations of Bittner and Gross was that an oncogenic virus could be transmitted vertically from parent to offspring. In the case of mouse leukemia, it became apparent that a mouse born to AK or C58 parents receives at birth the genetic information for malignant disease. These findings suggest that later in the life of a carrier animal, the expression of virally coded genes, perhaps triggered by exposure to chemicals, irradiation, hormone imbalance, or chronic irritation, becomes activated, causing leukemia. In the case of the mouse (and perhaps of humans), the activation of latent oncogenic viruses may not occur during the life span of the carrier animal, and the animal may remain disease-free, even though it carries and transmits the viral genome to its progeny. The way in which this type of vertical transmission could occur is explained by later findings that the genome of oncogenic viruses is integrated into the host cells' genome prior to cellular transformation (see Chapter 5).

Another oncogenic virus was discovered in the tissue extracts of leukemic AK mice after it was noted that when newborn C3H mice were injected with such extracts, some of the C3H mice developed parotid gland tumors rather than leukemia.[143] Some of these mice also developed cutaneous sarcomas, mammary carcinomas, and other malignancies. Eddy, Stewart, and colleagues found that virus produced by cultured cells after infection with extracts from leukemic AK mice organs caused parotid gland tumors and a variety of other neoplasms in mice, hamsters, and rats. They suggested that this multiple-tumorigenic agent be called *polyoma virus.*[147]

Since the early pioneering work in this field, a number of other oncogenic viruses have been identified and characterized. These include (1) feline leukemia virus, shown by inoculation of cell-free extracts from leukemic cats into newborn kittens; (2) SV40 virus, shown to be latent and harmless in the rhesus monkey but to induce leukemias and sarcomas after inoculation into newborn hamsters; (3) adenoviruses, which cause the common cold in humans and induce sarcomas in newborn hamsters and rats; and (4) such herpesviruses as *Herpes saimiri*, which is indigenous in the New World squirrel monkey and may induce lymphosarcomas and leukemias when inoculated into certain species of monkeys. Table 2–4 lists some of the different types of RNA and DNA oncogenic viruses.

Role of Viruses in the Causation of Human Cancer

To prove a causal relationship between a putative cancer-causing virus and human cancer is not a simple task. Such proof relies on evidence that is to a fair extent circumstantial. This evidence includes (1) epidemiological data showing a correlation between living in an area of endemic viral infection and a type of cancer; (2) serological evidence of antibody titers to viral antigens in patients with a given cancer type; (3) evidence for insertion of viral DNA into a cancer-bearing host's cell genome; (4) evidence for a consistent chromosomal translocation, particularly those involving an oncogene, in virally infected patients; (5) data showing that viral infection of cells in culture or transfection of viral genes into cells causes cell transformation and the ability of such cells to produce tumors in nude mice; and (6) development of cancers of the suspected target organ in transgenic mice produced by embryonic gene transfer of viral genes.

On the basis of this sort of evidence, some human cancers are considered to be caused by viral infection either directly or indirectly. By "directly," I mean that the viral gene(s) can themselves cause cells to become malignant (sometimes also requiring the loss of a tumor suppressor gene). By "indirectly," I mean that viral infection may simply cause the progression of malignant cell growth by producing an immunodeficiency state (e.g., the occurrence of non-Hodgkin's lymphoma in HIV-infected patients) or by stimulating the proliferation of already transformed cells. Sometimes viral infection acts in concert with other infectious agents or chemical carcinogens. Such is the case for malarial infection of Epstein-Barr virus (EBV)–infected patients and for aflatoxin exposure of individuals bearing the hepatitis B viral genome in their liver cells (see below). The types of

Table 2–4. Examples of Oncogenic Viruses

Virus	Species of Isolation
I. Oncogenic RNA viruses	
A. Acute-acting type°	
Rous sarcoma	Chickens
Fujinami sarcoma	Chickens
Retculoendotheliosis	Chickens and turkeys
Avian erythroblastosis	Chickens
Avian myeloblastosis	Chickens
Avian myelocytomatosis	Chickens
Moloney sarcoma	Mice
Abelson leukemia	Mice
FBJ osteosarcoma	Mice
Harvey/Kirsten sarcoma[†]	Rat
Rat sarcoma	Rat
Feline sarcoma	Cat
Woolly monkey sarcoma	Woolly monkey
B. Chronic type[‡]	
Avian leucosis	Chickens
Mouse leukemia°°	Mice
Feline leukemia	Cat
Bovine leukemia	Cow
Gibbon ape leukemia[††]	Gibbon ape
Mouse mammary tumor	Mice
Human T lymphotropic viruses	Human
II. Oncogenic DNA viruses	
A. Papovaviruses	Rabbit, man, dog, cow, and others
Papilloma	Mouse
Polyoma	Monkey
SV40	Human
JC	Human
BK	
B. Adenoviruses	Human,[‡‡] monkey, birds, cow
C. Herpes viruses	
Epstein-Barr	Human
Lucke carcinoma	Frog
Marek's disease	Chicken

°These viruses are acute, transform cells in vitro, have rapid disease induction in vivo, and carry "transforming *onc* gene" related to cell gene. Most are replication defective but can be isolated free of helper virus.

[†]These hybrid viruses, created experimentally, contain mouse helper virus and rat "src" sequences.

[‡]These viruses are chronic, have no transformation in vitro, long latency period in vivo, and no evidence of transforming gene. These all appear to be horizontally transmitted; in some cases, related sequences are found in cell DNA.

°°The Friend leukemia virus complex contains a defective genome, codes for a small envelope glycoprotein not incorporated into virions, does not transform cells in vitro, and perhaps should be placed in a separate category.

[††]Viruses show a distant relationship to mouse DNA, but not that of primates; this indicates "ancient" horizontal transmission.

[‡‡]There are 31 members of the human adenovirus group and at least 12 induce tumors in newborn animals and/or transform cells in vitro.

human cancer thought to be caused by viral infection and the strength of epidemiological associations are shown in Table 2–5.[148]

Table 2–5. Human Cancer Viruses and Associated Cancers, Strength of Association, and Necessary Preconditions

Virus	Cancer	Strength of Epidemiologic Association	Required Precondition
HBV	Hepatocellular carcinoma	Strong	None
HTLV-I	T-cell lymphoma	Strong	None
EBV	Burkitt's lymphoma	Strong	Chronic malaria
EBV	High-grade lymphoma	Strong	HIV
HPV	Cervical cancer	Consistent	?None
EBV	Nasopharyngial carcinoma	Inconsistent	—
HSV-2	Cervical cancer	Inconsistent	—

EBV, Epstein-Barr virus; HBV, hepatitis B virus; HSV-2, herpes simplex virus 2; HTLV-1, human T-lymphotropic virus 1. (From Henderson[148])

Association of Epstein-Barr Virus and Human Cancers

Epstein-Barr virus has been linked to four different types of human cancer: Burkitt's lymphoma (BL), nasopharyngeal carcinoma (NPC), B-cell lymphomas in immunosuppressed individuals such as HIV-infected patients, and some cases of Hodgkin's lymphoma.[149] The evidence is strongest for an association with BL and NPC.

Infection with EBV does not by itself cause cancer. On average, across the world, about 90% of the population may be infected by the time they reach adulthood. In some endemic areas, the incidence rate approaches 100%. In developing countries, EBV infection often occurs in young childhood. In more affluent societies, EBV infection tends to occur as the "kissing age" of adolescence or young adulthood is reached, and manifests itself as infectious mononucleosis. In developing countries, particularly in equatorial Africa, concomitant or subsequent infection with the malarial parasite induces B-cell proliferation and an immunodeficiency state that leads to malignant transformation and progression. There is a consistent chromosomal translocation involving immunoglobulin genes, usually on chromosome 14, and sequences within or adjacent to the c-*myc* gene locus on chromosome 8 (see Chapters 5 and 7).

The role of EBV in NPC is less well characterized, but the evidence for an association includes high serum antibody titers against EBV antigens and the presence of EBV DNA in NPC cells. Similar evidence suggests an association between EBV infection and induction of some B-cell lymphomas and some Hodgkin's disease cases in immunosuppressed individuals, al-

though the exact role of EBV remains to be elucidated.[149]

Hepatitis Virus and Hepatocellular Carcinoma

Epidemiological evidence strongly points to a link between chronic hepatitis B virus (HBV) infection and hepatocellular carcinoma (HCC). In areas where HBV infection is endemic, such as Taiwan, Senegal, South Africa, Hong Kong, China, and the Philippines, the incidence of HCC is much higher than in countries where HBV infection is less common (reviewed in Reference 149). Hepatocellular carcinoma usually appears after decades of chronic liver involvement due to HBV-induced liver cell damage and regeneration. HBV DNA can be found in the majority of liver cancers from patients in high-risk areas and a specific piece of HBV DNA called *HBx*, which encodes a transcription factor, is found in HCC cells and can induce liver tumors in transgenic mice.[150] Some chromosomal modifications are also observed in HCC. These include alterations of the short arm of chromosome 11, deletions in the long arm of chromosome 13, and point mutations of the *p53* gene on chromosome 17. These latter mutations are particularly interesting because such mutations are seen in areas where there is concomitant exposure to aflatoxins in foods, suggesting a joint role of HBV infection and chemical carcinogens causing HCC.

There is also some evidence that chronic infection with hepatitis C virus, which may cause chronic liver injury and regeneration, may be a causative agent for HCC and be a factor particularly in HBV-negative cases. Currently there are a number of countries in which widespread vaccination against HBV is occurring, and time will tell whether such vaccination lowers the incidence of HCC in these areas.

Papillomaviruses and Cervical Cancer

A large class of papillomaviruses that are pathogenic for humans (HPV) have been identified. More than 60 genotypic subtypes have been isolated.[148] Only two HPV subtypes have been closely associated with cervical cancer, HPV 16 and HPV 18. The evidence for this association is the following (reviewed in Reference 149): (1) viral DNA is found in about 90% of cervical cancers; (2) in most cases a specific piece of the viral DNA is integrated into the host's genome; (3) the vast majority of all HPV-positive cervical cancers contain cells that express two specific gene transcripts, E6 and E7; (4) E6 and E7 genes of high-risk HPV (e.g., HPV 16 and 18), but not of low-risk HPV subtypes, immortalize human cells in culture; (5) E6- and E7-expressing cells frequently undergo a progression to aneuploidy and gene amplification in culture; (6) the E6 and E7 oncoproteins bind to and inactivate or degrade the p53 and RB tumor suppressor gene proteins, respectively; and (7) uterine cervical dysplasia can be induced in mice by inoculation of a recombinant retrovirus bearing the E6 and E7 genes of HPV 16 into the vagina of mice.[151]

HTLV-I and Adult T-Cell Leukemia

The retrovirus HTLV-I was first identified in interleukin-2 (IL-2)-stimulated T lymphocytes from two patients, one with a T-cell lymphoma and the other with T-cell leukemia.[152,153] This was the first demonstration that a retrovirus could cause malignancy in humans, although it had long been suspected that retroviruses might do so, based on numerous examples in animals.

Since the original isolation of a human T-cell leukemia and lymphoma virus, several similar isolates have been made from patients with T-cell neoplasms in different parts of the world, including Japan, Africa, the Caribbean basin, England, and the Netherlands.[154] The isolates

from patients with adult T-cell leukemia-lymphoma (ATLL), named *HTLV-I*, were found to have several characteristics in common. In 1982, a new subgroup of HTLV, called *HTLV-II*, was identified.[155] The DNA sequence of the two subgroups is clearly different, but there is significant homology. Moreover, the two types of virus genomes encode a very similar p24 core protein and share a common mechanism of gene activation, indicating that they are members of the same family. HTLV-I and -II are retroviruses with reverse transcriptase and an RNA genome of about 9 kilobases. DNA sequences homologous to HTLV DNA are not found in the genome of normal human cells; thus, they are exogenous, not endogenous, genomic sequences, in contrast to the c-*onc* genes that are homologous to v-*onc* genes of retroviruses. HTLVs do not appear to carry their own *onc* gene. In this respect they appear to be similar to the chronic-acting retroviruses such as avian leukosis virus, and they have an LTR-*gag-pol-env*-LTR gene arrangement typical of other retroviruses. HTLV-I and -II can transform normal human T cells in culture and the transformed cells contain at least one proviral DNA copy, transcribe viral DNA, and make low levels of viral proteins.[155]

In addition to the genes common to all retroviruses, the HTLV-I genome contains a 1.6 kilobase sequence at the 3′ terminal region that encodes at least two *trans*-acting regulator proteins: a 40,000 Dalton protein product of a gene called *tax* and a 27,000 Dalton protein product of a gene called *rex*. Transgenic mice bearing the *tax* gene develop multiple mesenchymal tumors at about 3 months of age.[156]

A high percentage of patients with ATLL and certain T-cell lymphomas have antigens to HTLV-I proteins in their serum, and their tumor cells contain one or more copies of the HTLV genome.[154,157,158] But patients with childhood cancers, non–T-cell leukemias and lymphomas, myeloid leukemias, Hodgkin's disease, and solid tumors do not have evidence of HTLV antibodies in their serum, and healthy individuals in nonendemic areas are also antibody negative.[157] However, almost 50% of relatives of ATLL patients and about 12% of healthy blood donors in endemic areas have been reported to be antibody positive. These data indicate the T-cell specificity of the neoplastic transformation process induced by HTLV and suggest horizontal spread of the virus among people. Major clusters of HTLV-related T-cell cancers are found in areas of high endemic infection with HTLV, such as southwestern Japan, in the Caribbean basin, and in certain areas of South America and Africa.[154] As is the case with Burkitt's lymphoma, however, only certain people infected with the virus get ATLL. Thus, other predisposing factors must exist.

Horizontal transmission among individuals appears to require prolonged and intimate contact with an HTLV-positive person. Cell-to-cell transmission can also be demonstrated in cell culture systems when HTLV-producing cells are cocultured with normal T lymphocytes.[154] The HTLV-infected, transformed cells often produce infectious HTLV, but virus-nonproducing transformed cells, containing the integrated viral genome as a provirus, are also observed.

Only about 1 out of 25 to 30 infected individuals will eventually develop ATLL, and HTLV-I DNA is consistently demonstrated in the ATL cells from these patients. There appears to be a latency period between primary infection and leukemia development of several decades.[149]

From these findings, the following question obviously arises: Where did these viruses come from and are they a recently evolved class of viruses? Although, there is no definitive evidence for HTLV infections in humans before the late 1970s, HTLV viruses have probably been around for a long time, perhaps hundreds or thousands of years. Most likely, there was an animal vector originally, and recently the virus may have undergone some evolutionary change that made it more infectious or more cytopathic for humans. Analysis of serum samples from subhuman primates, for example, shows that several Old World monkey species, including Japanese and Chinese macaques, African green monkeys, and baboons, are seropositive for HTLV.[159] In Japan, where HTLV is endemic in humans in certain areas, the distribution of virus in primates is much more widespread, suggesting independent entry of HTLV-like virus into the primate and human populations and arguing against current transmission between primates and humans. Because of the widespread infection of African Old World primates with HTLV-like viruses, it has been proposed that the origin of HTLV was in Africa and that spread to other countries may have occurred by means of explorers who introduced infected primates or had contact with infected primates in Africa.[159]

In conclusion, it is fair to say that infections with oncogenic viruses are clearly associated with certain kinds of human cancer. However, even in those cases in which viral infection appears to be a predisposing factor, viral infection itself is insufficient to cause cancer. In all cases, there are other contributing factors, which include cell type–specific mitogenic stimulation, suppression of the immune response, and, possibly, genetic factors. It is also clear, though, that a combination of infection with certain oncogenic viruses, chronic mitogenic stimulation of the virus-infected cells, and a concomitant immune deficiency state have a high propensity to induce the cancerous process in human beings.

It should be noted that the human immunodeficiency virus (HIV) that causes AIDS is also a slow infectious retrovirus virus (lentivirus) that is a T-cell lymphotropic virus with some genomic similarities to HTLV. HIV infects CD4+ cells and causes disease by its immunosuppressive effects. Patients with AIDS are at high risk to develop Kaposi's sarcoma and non-Hodgkin's lymphoma. These cancers most likely arise because of the immunosuppressive effects of the AIDS virus rather than the direct transforming activity of the virus, although some direct cell-transforming effects have been observed.

References

1. E. G. Luebeck and S. H. Moolgavkar: Multistage carcinogenesis and the incidence of colorectal cancer. *Proc Natl Acad Sci USA* 99:15095, 2002.
2. P. C. Nowell and D. A. Hungerford: A minute chromosome in human chronic granulocytic leukemia. *Science* 132:1497, 1960.
3. J. D. Rowley: A new consistent chromosomal abnormality in chronic myelogenous leukemia identified by quinacrine fluorescence and Giemsa staining. *Nature* 243:290, 1973.
4. A. G. Knudson: Two genetic hits (more or less) to cancer. *Nat Rev Cancer* 1:157, 2001.
5. C. Lengaur, K. W. Kinzler, and B. Vogelstein: Genetic instabilities in human cancers. *Nature* 396:643, 1998.

6. L. A. Loeb: Mutator phenotype may be required for multistage carcinogenesis. *Cancer Res* 51: 3075, 1991.

7. A. Luch: Nature and nuture—Lessons from chemical carcinogenesis. *Nat Rev Cancer* 5: 113, 2005.

8. P. D. Lawley: Carcinogenesis by alkylating agents. In C. E. Searle. ed.: *Chemical Carcinogens, Monograph* No. 173. Washington, DC: American Chemical Society, 1976, pp. 83–244.

9. E. C. Miller: Some current perspectives on chemical carcinogenesis in humans and experimental animals: Presidential address. *Cancer Res* 38:1479, 1978.

10. E. C. Miller and J. A. Miller: Searches for ultimate chemical carcinogens and their reactions with cellular macromolecules. *Cancer* 47:2327, 1981.

11. E. Boyland: The biological significance of metabolism of polycyclic compounds. *Biochem Soc Symp* 5:40, 1950.

12. A. Pullman and B. Pullman: Electronic structure and carcinogenic activity of aromatic molecules. *Adv Cancer Res* 3:117, 1955.

13. G. Kellerman, C. R. Shaw, and M. Luyten-Kellerman: Aryl hydrocarbon hydroxylase inducibility and bronchogenic carcinoma. *N Engl J Med* 289:934, 1973.

14. T. L. McLemore, S. Adelberg, M. C. Liu, N. A. McMahan, S. J. Yu, W. C. Hubbard, et al.: Expression of CYP1A1 gene in patients with lung cancer: Expression for cigarette smoke–induced gene expression in normal lung tissue and for altered gene regulation in primary pulmonary carcinomas. *J Natl Cancer Inst* 82:1333, 1990.

15. K. Nakachi, K. Imai, S.-I. Hayashi, and K. Kawajiri: Polymorphisms of the CYP1A1 and glutathione S-transferase genes associated with susceptibility to lung cancer in relation to cigarette dose in a Japanese population. *Cancer Res* 53:2994, 1993.

16. H. M. Schuller: Mechanisms of smoking-related lung and pancreatic adenocarcinoma development. *Nat Rev Cancer* 2:455, 2002.

17. S. A. Belinsky, T. R. Devereux, and M. W. Anderson: Role of DNA methylation in the activation of proto-oncogenes and the induction of pulmonary neoplasia by nitrosamines. *Mutat Res* 233:105, 1990.

18. V. D.-H. Ding, R. Cameron, and C. B. Pickett: Regulation of microsomal, xenobiotic epoxide hydrolase messenger RNA in persistent hepatocyte nodules and hepatomas induced by chemical carcinogens. *Cancer Res* 50:256, 1990.

19. M. E. Hogan, N. Dattagupta, and J. P. Whitlock, Jr.: Carcinogen-induced alteration of DNA structure. *J Biol Chem* 256:4504, 1981.

20. J. S. Eadie, M. Conrad, D. Toorchen, and M. D. Topal: Mechanism of mutagenesis by O^6-methylguanine. *Nature* 308:201, 1984.

21. L. A. Loeb: Apurinic sites as mutagenic intermediates. *Cell* 40:483, 1985.

22. T. C. Boles and M. E. Hogan: Site-specific carcinogen binding to DNA. *Proc Natl Acad Sci USA* 81:5623, 1984.

23. P. D. Kurth and M. Bustin: Site-specific carcinogen binding to DNA in polytene chromosomes. *Proc Natl Acad Sci USA* 82:7076, 1985.

24. C. C. Harris: Chemical and physical carcinogenesis: Advances and perspectives for the 1990s. *Cancer Res* 51:5023, 1991.

25. I. C. Hsu, R. A. Metcalf, T. Sun, J. Welsh, N. J. Wang, and C. C. Harris: p53 gene mutational hotspot in human hepatocellular carcinomas from Qidong, China. *Nature* 350:427, 1991.

26. B. Bressac, M. Kew, J. Wands, and M. Ozturk: Selective G to T mutations of p53 gene in hepatocellular carcinoma from southern Africa. *Nature* 350:429, 1991.

27. W. F. Friedewald and P. Rous: The initiating and promoting elements in tumor production: An analysis of the effects of tar, benzpyrene, and methylcholanthrene on rabbit skin. *J Exp Med* 80:101, 1944.

28. R. D. Sall and M. J. Shear: Studies in carcinogenesis. XII. Effect of the basic fraction of creosote oil on the production of tumors in mice by chemical carcinogens. *J Natl Cancer Inst* 1:45, 1940.

29. I. Berenblum: The mechanism of carcinogenesis: A study of the significance of carcinogenic action and related phenomena. *Cancer Res* 1:807, 1941.

30. E. Hecker: Isolation and characterization of the cocarcinogenic principles from croton oil. *Methods Cancer Res* 6:439, 1971.

31. B. L. Van Duuren: Tumor-promoting agents in two-stage carcinogenesis. *Prog Exp Tumor Res* 11:31, 1969.

32. I. Berenblum: Sequential aspects of chemical carcinogenesis: Skin. In F. F. Becker, ed.: *Cancer: A Comprehensive Treatise*. New York: Plenum Press, 1975, pp. 323–344.

33. R. K. Boutwell: Some biological aspects of skin carcinogenesis. *Prog Exp Tumor Res* 4:207, 1964.

34. H. c. Pitot and Y. P. Dragan: Facts and theories concerning the mechanisms of carcinogenesis. *FASEB J* 5:2280, 1991.

35. J. F. O'Connell, A. J. P. Klein-Szanto, D. M. DiGiovanni, J. W. Fries, and T. J. Slaga: Enhanced malignant progression of mouse skin tumors by the free-radical generator benzoyl peroxide. *Cancer Res* 46:2863, 1986.

36. J. McCann, E. Choi, E. Yamasaki, and B. N. Ames: Detection of carcinogens as mutagens in the *Salmonella*/microsome test: Assay of 300 chemicals. *Proc Natl Acad Sci USA* 72:5135, 1975.

37. E. Huberman and L. Sachs: Cell-mediated mutagenesis of mammalian cells with chemical carcinogens. *Int J Cancer* 13:326, 1974.

38. E. Huberman and L. Sachs: Mutability of different genetic loci in mammalian cells by metabolically activated carcinogenic polycyclic hydrocarbons. *Proc Natl Acad Sci USA* 73:188, 1976.

39. R. B. Setlow: Repair deficient human disorders and cancer. *Nature* 271:713, 1978.

40. M. Swift and C. Chase: Cancer in families in xeroderma pigmentosum. *J Natl Cancer Inst* 62:1415, 1979.

41. Y. Shiraishi, T. H. Yosida, and A. A. Sandberg: Malignant transformation of Bloom syndrome B-lymphoblastoid cell lines by carcinogens. *Proc Natl Acad Sci USA* 82:5102, 1985.

42. H. Kirkman: Estrogen-induced tumors of the kidney in the Syrian hamster. *National Cancer Institute Monograph* No. 1, 1959, p. 137.

43. M. S. Biskind and G. S. Biskind: Development of tumors in the rat ovary after transplantation into the spleen. *Proc Soc Exp Biol Med* 55:176, 1944.

44. T. Tsutsui, G. H. Degan, D. Schiffman, A. Wong, H. Maizumi, J. A. McLachlan, and J. C. Barrett: Dependence on exogenous metabolic activation for induction of unscheduled DNA synthesis in Syrian hamster embryo cells by diethylstilbestrol and related compounds. *Cancer Res* 44:184, 1984.

45. S. K. Buenaventura, D. Jacobson-Kram, K. L. Dearfield, and J. R. Williams: Induction of sister chromatid exchange by diethylstilbestrol in metabolically competent hepatoma cells but not in b/fibroblasts. *Cancer Res* 44:3851, 1984.

46. E. R. Fearon and B. Vogelstein: A genetic model for colorectal tumorigenesis. *Cell* 61:759, 1990.

47. B. S. Strauss: The origin of point mutations in human tumor cells. *Cancer Res* 52:249, 1992.

48. L. A. Loeb: Endogenous carcinogenesis: Molecular oncology into the twenty-first century—Presidential address. *Cancer Res* 49:5489, 1989.

49. C. Richter, J-W. Park, and B. N. Ames: Normal oxidative damage to mitochondrial and nuclear DNA is extensive. *Proc Natl Acad Sci USA* 85:6465, 1988.

50. T. A. Kunkel and L. A. Loeb: Fidelity of mammalian DNA polymerase. *Science* 213:765, 1981.

51. T. A. Kunkel: The mutational specificity of DNA polymerase-β during in vitro DNA synthesis: production of frameshift, base substitution and deletion mutations. *J Biol Chem* 260: 5787, 1985.

52. B. N. Ames, R. L. Saul, E. Schwiers, R. Adelman, and R. Cathcart: Oxidative DNA damage as related to cancer and aging: The assay of thymine glycol, thymidine glycol, and hydroxymethyluracil in human and rat urine. In R. S. Sohal, L. S. Birnbaum, and R. G. Cutler, eds.: *Molecular Biology of Aging: Gene Stability and Gene Expression*. New York: Raven Press, 1985, pp. 137.

53. L. Foulds: The experimental study of tumor progression: A review. *Cancer Res* 14:327, 1954.

54. P. C. Nowell: The clonal evolution of tumor cell populations. *Science* 194:23, 1976.

55. D. L. Trainer and E. F. Wheelock: Characterization of L5178 Y cell phenotypes isolated during progression of the tumor-dormant state in DBA/2 mice. *Cancer Res* 44:2897, 1984.

56. D. Hanahan and R. A. Weinberg: The hallmarks of cancer. *Cell* 100:57, 2000.

57. R. Bernards and R. A. Weinberg: A progression puzzle. *Nature* 418:823, 2002.

58. M. J. van de Vijver, Y. D. He, L. J. van't Veer, H. Dai, et al.: A gene-expression signature as a predictor of survival in breast cancer. *N Engl J Med* 347:1999, 2002.

59. M. Al-Hajj, M. S. Wicha, A. Benito-Hernandez, S. J. Morrison, and M. F. Clarke: Prospective identification of tumorigenic breast cancer cells. *Proc Natl Acad Sci USA* 100:3983, 2003.

60. D. B. Rifkin, R. M. Crowe, and R. Pollack: Tumor promoters induce changes in the chick embryo fibroblast cytoskeleton. *Cell* 18:361, 1979.

61. P. B. Fisher, J. H. Bozzone, and I. B. Weinstein: Tumor promoters and epidermal growth factor stimulate anchorage-independent growth of adenovirus-transformed rat embryo cell. *Cell* 18:695, 1979.

62. R. Cohen, M. Pacifici, N. Rubernstein, J. Biehl, and H. Holtzer: Effect of tumor promoter on myogenesis. *Nature* 226:538, 1977.

63. M. Lowe, M. Pacifici, and H. Holtzer: Effects of phorbol-12-myristate-13-acetate on the phenotypic program of cultured chrondroblasts and fibroblasts. *Cancer Res* 38:2350, 1978.

64. L. Diamond, T. O'Brien, and A. Rovera: Inhibition of adipose conversion of 3Y3 fibroblasts by tumor promoters. *Nature* 269:247, 1977.

65. E. Fibach, R. Gambari, P. A. Shaw, G. Maniatis, R. C. Reuben, S. Sassa, R. A. Rifkind, and P. A. Marks: Tumor promoter–mediated inhibition of cell differentiation: Suppression of the expression of erythroid functions in murine erythroleukemia cells. *Proc Natl Acad Sci USA* 76: 1906, 1979.

66. D. Ishii, E. Fibach, H. Yamasaki, and I. B. Weinstein: Tumor promoters inhibit morphological differentiation in cultured mouse neuroblastoma cells. *Science* 200:556, 1978.

67. S. Mondal and C. Heidelberger: Transformation of C3H/10T 1/2 CL8 mouse embryo fibroblasts by ultraviolet irradiation and a phorbol ester. *Nature* 260:710, 1976.

68. N. Yamamoto and H. zur Hausen: Tumor promoter TPA enhances transformation of human leukocytes by Epstein-Barr virus. *Nature* 280:244, 1979.

69. L. Kopelovich, N. E. Bias, and L. Helson: Tumor promoter alone induces neoplastic transformation of fibroblasts from human

genetically predisposed to cancer. *Nature* 282: 619, 1979.

70. C. M. Aldaz, C. J. Conti, I. B. Gimenez, T. J. Slaga, and A. J. P. Klein-Szanto: Cutaneous changes during prolonged application of 12-O-tetradecanoylphorbol-13-acetate on mouse skin and residual effects after cessation of treatment. *Cancer Res* 45:2753, 1985.

71. P. M. Rosoff, L. F. Stein, and L. C. Cantley: Phorbol esters induce differentiation in a pre-B-lymphocyte cell line by enhancing Na^+/H^+ exchange. *J Biol Chem* 259:7056, 1984.

72. J. M. Besterman, W. S. May Jr., H. LeVine III, E. J. Cragoe Jr., and P. Cuatrecasas: Amiloride inhibits phorbol ester–stimulated Na^+/H^+ exchange and protein kinase C: An amiloride analog selectively inhibits Na^+/H^+ exchange. *J Biol Chem* 260:1155, 1985.

73. I. Sussman, R. Prettyman, and T. G. O'Brien: Phorbol esters and gene expression: The role of rapid changes in K^+ transport in the induction of ornithine decarboxylase by 12-O-tetradecanoyl-13-acetate in BALB/c 3T3 cells and a mutant cell line defective in $Na^+K^+Cl^-$cotransport. *J Cell Biol* 101:2316, 1985.

74. S. Grinstein, S. Cohen, J. D. Goetz, A. Rothstein, and E. W. Gelfand: Characterization of the activation of Na^+/H^+ exchange in lymphocytes by phorbol esters: Change in cytoplasmic pH dependence of the antiport. *Proc Natl Acad Sci USA* 82:1429, 1985.

75. C.-C. Chang, J. E. Trosko, H.-J. Kung, D. Bombick, and F. Matsumura: Potential role of the src gene product in inhibition of gap-junctional communication in NIH/3T3 cells. *Proc Natl Acad Sci USA* 82:5360, 1985.

76. W. S. May, N. Sahyoun, S. Jacobs, M. Wolf, and P. Cuatrecasas: Mechanism of phorbol diester–induced regulation of surface transferring receptor involves the action of activated protein kinase C and an intact cytoskeleton. *J Biol Chem* 260:9419, 1985.

77. M. Schliwa, T. Nakamura, K. R. Porter, and U. Euteneuer: A tumor promoter induces rapid and coordinated reorganization of actin and vinculin in cultured cells. *J Cell Biol* 99:1045, 1984.

78. E. G. Fey and S. Penman: Tumor promoters induce a specific morphological signature in the nuclear matrix–intermediate filament scaffold of Madin-Darby canine kidney (MDCK) cell colonies. *Proc Natl Acad Sci USA* 81:4409, 1984.

79. P. A. Cerutti: Pro-oxidant states and tumor promotion. *Science* 227:375, 1985.

80. J. Barsoum and A. Varshavsky: Mitogenic hormones and tumor promoters greatly increase the incidence of colony-forming cells bearing amplified dihydrofolate reductase genes. *Proc Natl Acad Sci USA* 80:5330, 1983.

81. T. Solo, T. Turpeenniemi-Hujanen, and K. Tryggvason: Tumor-promoting phorbol esters and cell proliferation stimulate secretion of basement membrane (type IV) collagen-degrading metalloproteinase by human fibroblasts. *J Biol Chem* 260:8526, 1985.

82. R. Montesano and L. Orci: Tumor-promoting phorbol esters induce angiogenesis in vitro. *Cell* 42:469, 1985.

83. Y. Nishizuka: The role of protein kinase C in cell surface signal transduction and tumor production. *Nature* 308:693, 1984.

84. J. E. Niedel, L. J. Kuhn, and G. R. Vandenbark: Phorbol diester receptor copurifies with protein kinase C. *Proc Natl Acad Sci USA* 80:36, 1983.

85. D. F. Birt, J. C. Pelling, L. T. White, K. Dimitroff, and T. Barnett: Influence of diet and calorie restriction on the initiation and promotion of skin carcinogenesis in the SENCAR mouse model. *Cancer Res* 51:1851, 1991.

86. H. Wei, L. Tye, E. Bresnick, and D. F. Birt: Inhibitory effect of apigenin, a plant flavonoid, on epidermal ornithine decarboxylase and skin tumor promotion in mice. *Cancer Res* 50:499, 1990.

87. A. Leder, A. Kuo, R. D. Cardiff, E. Sinn, and P. Leder: v-Ha-ras transgene abrogates the initiation step in mouse skin tumorigenesis: Effects of phorbol esters and retinoic acid. *Proc Natl Acad Sci USA* 87:9178, 1990.

88. Z. Szallasi and P. M. Blumberg: Prostratin, a nonpromoting phorbol ester, inhibits induction by phorbol 12-myristate 13-acetate of ornithine decarboxylase, edema, and hyperplasia in CD-1 mouse skin. *Cancer Res* 51:5355, 1991.

89. C. Peraino, R. J. M. Fry, E. Staffeldt, and W. E. Kisielski: Effects of varying the exposure to phenobarbital on its enhancement of 2-acetylaminofluorene-induced hepatic tumorigenesis in the rat. *Cancer Res* 33:2701, 1973.

90. T. Kitagawa, H. C. Pitot, E. C. Miller, and J. A. Miller: Promotion by dietary phenobarbital of hepatocarcinogenesis by 2-methyl-N, N-dimethyl-4-aminoazobenzene in the rat. *Cancer Res* 39:112, 1979.

91. H. C. Pitot, L. Barsness, T. Goldsworthy, and T. Kitagawa: Biochemical characterization of stages of hepatocarcinogenesis after a single dose of diethylnitrosamine. *Nature* 271:456 1978.

92. J. Jonkers and A. Berns: Conditional mouse models of sporadic cancer. *Nat Rev Cancer* 2:251, 2002.

93. S. R. Lichter: Why cancer news is a health hazard. *Wall Street Journal* November 12, 1993.

94. J. Higginson: Changing concepts in cancer prevention: Limitations and implications for future research in environmental carcinogenesis. *Cancer Res* 48:1381, 1988.

95. F. Perera and P. Boffetta: Perspectives on comparing risks of environmental carcinogens. *J Natl Cancer Inst* 80:1282, 1988.

96. K. Müster, E. M. Whelan, G. L. Ross, and A. N. Stimola, eds.: Animal cancer testing in laws and regulations. In *America's War on*

"Carcinogens". New York: American Council on Science and Health, 2005, pp. 107–115.

97. S. M. Cohen and L. B. Ellwein: Genetic errors, cell proliferation, and carcinogenesis. *Cancer Res* 51:6493, 1991.

98. A. Hay: Testing times for the tests. *Nature* 350: 555, 1991.

99. J. Ashby and R. W. Tennant: Definitive relationships among chemical structure, carcinogenicity and mutagenicity for 301 chemicals tested by the U.S. NTP. *Mutat Res* 257:229, 1991.

100. I. B. Weinstein: Mitogenesis is only one factor in carcinogenesis. *Science* 251:387, 1991.

101. R. L. Melnick: Does chemically induced hepatocyte proliferation predict liver carcinogenesis? *FASEB J* 6:2698, 1992.

102. K. Meister, E. M. Whelan, G. L. Ross, and A. N. Stimola, eds.: Limitations and results of animal carcinogen testing. In *America's War on "Carcinogens".* New York:: American Council on Science and Health, 2005, pp. 45–57.

103. B. N. Ames, M. Profet, and L. S. Gold: Dietary pesticides (99.99% all natural) *Proc Natl Acad Sci USA* 87:7777, 1990.

104. B. N. Ames, M. Profet, and L. S. Gold: Nature's chemicals and synthetic chemicals: Comparative toxicology (carcinogens/mutagens/teratofens/dioxin). *Proc Natl Acad Sci USA* 87:7782, 1990.

105. J. B. Little: Radiation carcinogenesis in vitro: Implications for mechanisms. In H. H. Hiatt, J. D. Watson, and J. A. Winsten, eds.: *Origins of Human Cancer.* Cold Spring Harbor, NY: Cold Spring Harbor Laboratory, 1977, pp. 923–939.

106. W. A. Haseltine: Ultraviolet light repair and mutagenesis revisited. *Cell* 33:13, 1983.

107. A. C. Upton: The biological effects of low-level ionizing radiation. *Sci Am* 246:41, 1982.

108. A. J. Grosovsky, J. G. De Boer, P. J. De Jong, E. A. Drobetsky, and B. W. Glickman: Base substitutions, frameshifts, and small deletions constitute ionizing radiation-induced point mutations in mammalian cells. *Proc Natl Acad Sci USA* 85:185, 1988.

109. L. H. Lutze, R. A. Winegar, R. Jostes, F. T. Cross, and J. E. Cleaver: Radon-induced deletions in human cells: Role of nonhomologous strand rejoining. *Cancer Res* 52:5126, 1992.

110. R. R. Weichselbaum, D. E. Hallahan, Z. V. Sukhatme, A. Dritschilo, M. L. Sherman, et al.: Review: Biological consequences of gene regulation after ionizing radiation exposure. *J Natl Cancer Inst* 83:480, 1991.

111. S. J. Garte, F. J. Burns, T. Ashkenazi-Kimmel, M. Felber, and M. J. Sawey: Amplification of the c-*myc* oncogene during progression of radiation-induced rat skin tumors. *Cancer Res* 50:3073, 1990.

112. D. Williams: Cancer after nuclear fallout: Lessons from the Chernobyl accident. *Nat Rev Cancer* 2:543, 2002.

113. G. Merlino and F. P. Noonan: Modeling gene–environment interactions in malignant melanoma. *Trends Mol Med* 9:102, 2003.

114. B. N. Ames, M. K. Shigenaga, and T. M. Hagen: Oxidants, antioxidants, and the degenerative diseases of aging. *Proc Natl Acad Sci USA* 90:7915, 1993.

115. E. R. Stadtman: Protein oxidation and aging. *Science* 257:1220, 1992.

116. L. D. Youngman, J.-Y. K. Park, and B. N. Ames: Protein oxidation associated with aging is reduced by dietary restriction of protein or calories. *Proc Natl Acad Sci USA* 89:9112, 1992.

117. H. Joenje: Genetic toxicology of oxygen. *Mutat Res* 219:193, 1989.

118. S. A. Grist, M. McCarron, A. Kutlaca, D. R. Turner, and A. A. Morley. In vivo human somatic mutation: Frequency and spectrum with age. *Mutat Res* 266:189, 1992.

119. T. M. Reid and L. A. Loeb: Mutagenic specificity of oxygen radicals produced by human leukemia cells. *Cancer Res* 52:1082, 1992.

120. J. M. Carney, P. E. Starke-Reed, C. N. Oliver, R. W. Lundum, M. S. Cheng, et al.: Reversal of age-related increase in brain protein oxidation, decrease in enzyme activity, and loss in temporal and spatial memory by chronic administration of the spin-trapping compound N-tertbutyl-alpha-phenylnitrone. *Proc Natl Acad Sci USA* 88:3633, 1991.

121. J. Kaiser: Tying genetics to the risk of environmental diseases. *Science* 300:563, 2003.

122. L. A. Loeb, K. R. Loeb, and J. P. Anderson: Multiple mutations and cancer. *Proc Natl Acad Sci USA* 100:776, 2003.

123. I. Tomlinson and W. Bodmer: Selection, the mutation rate and cancer: Ensuring that the tail does not wag the dog. *Nat Med* 5:11, 1999.

124. P. C. Nowell: The clonal evolution of tumor cell populations. *Science* 194:23, 1976.

125. V. A. Bohr, D. H. Phillips, and P. C. Hanawalt:: Heterogeneous DNA damage and repair in the mammalian genome. *Cancer Res* 47:6426, 1987.

126. T. Lindahl: Instability and decay of the primary structure of DNA. *Nature* 362:709, 1993.

127. A. E. Pegg: Mammalian O^6-alkylguanine-DNA alkyltransferase: Regulation and importance in response to alkylating carcinogenic and therapeutic agents. *Cancer Res* 50:6119, 1990.

128. J. H. J. Hoeijmakers: Genome maintenance mechanisms for preventing cancer. *Nature* 411: 366, 2001.

129. R. B. Setlow, F. E. Ahmed, and E. Grist: Xeroderma pigmentosum: Damage to DNA is involved in carcinogenesis. In H. H. Hiatt, J. D. Watson, and J. A. Winsten, eds.: *Origins of Human Cancer.* Cold Spring Harbor, NY: Cold Spring Harbor Laboratory, pp. 889–202, 1977.

130. D. Bootsma and J. H. J. Hoeijmakers: Engagement with transcription. *Nature* 363:114, 1993.

131. Y. Habraken, P. Sung, L. Prakash, and S. Prakash: Yeast excision repair gene RAD2 encodes a single-stranded DNA endonuclease. *Nature* 366:365, 1993.

132. D. Scherly, T. Nouspikel, J. Corlet, C. Ucla, A. Bairoch, et al.: Complementation of the DNA repair defect in xeroderma pigmentosum group G cells by a human cDNA related to yeast RAD2. *Nature* 363:182, 1993.

133. E. C. Friedberg: Xeroderma pigmentosum, Cockayne's syndrome, helicases, and DNA repair: What's the relationship? *Cell* 71:887, 1992.

134. J. E. Cleaver: It was a very good year for DNA repair. *Cell* 76:1, 1994.

135. L. L. Dumenco, E. Allay, K. Norton, and S. L. Gerson: The prevention of thymic lymphomas in transgenic mice by human O^6-alkylguanine-DNA alkyltransferase. *Science* 259:219, 1993.

136. S. M. O'Toole, A. E. Pegg, and J. A. Swenberg: Repair of O^6-methylguanine and O^4-methylthymidine in F344 rat liver following treatment with 1,2-dimethylhydrazine and O^6-benzylguanine. *Cancer Res* 53:3895, 1993.

137. V. Ellermann and O. Bang: Experimentelle Leukämie bei Hühnern. *Zentralbl f Bacteriol* 46:595, 1908.

138. P. Rous: A sarcoma of the fowl transmissible by an agent separable from the tumor cells. *J Exp Med* 13:397, 1911.

139. R. E. Shope: Infectious papillomatosis of rabbits. *J Exp Med* 58:607, 1933.

140. P. Rous and J. W. Beard: The progression to carcinoma of virus-induced rabbit papillomas (Shope). *J Exp Med* 62:523, 1935.

141. B. A. Lucké: A neoplastic disease of the kidney of the frog, *Rana pipiens. Am J Cancer* 20:352, 1934.

142. J. J. Bittner: Some possible effects of nursing on the mammary gland tumor incidence in mice. *Science* 84:162, 1936.

143. L. Gross: Viral etiology of cancer and leukemia: A look into the past, present, and future—G. H. A. Clowes memorial lecture. *Cancer Res* 38:485, 1978.

144. L. Gross: Development and serial cell-free passage of a highly potent strain of mouse leukemia virus. *Proc Soc Exp Biol Med* 94:767, 1957.

145. L. Dmochowski: Viruses and tumors in the light of electron microscope studies: A review. *Cancer Res* 20:997, 1960.

146. I.-M. Chiu, R. Callahan, S. R. Tronick, J. Schlom, and S. A. Aaronson: Major pol gene progenitors in the evolution of oncoviruses. *Science* 223:364, 1984.

147. B. E. Eddy, S. E. Stewart, and W. Berkeley: Cytopathogenicity in tissue cultures by a tumor virus from mice. *Proc Soc Exp Biol Med* 98:848, 1958.

148. B. E. Henderson: Establishment of an association between a virus and a human cancer. *J Natl Cancer Inst* 81:320, 1989.

149. H. zur Hausen: Viruses in human cancers. *Science* 254:1167, 1991.

150. C.-M. Kim, K. Koike, I. Saito, T. Miyamura, and G. Jay: HBx gene of hepatitis B virus indices liver cancer in transgenic mice. *Nature* 351:317, 1991.

151. T. Sasagawa, M. Inoue, H. Inoue, M. Yutsudo, O. Tanizawa, et al.: Induction of uterine cervical neoplasias in mice by human papillomavirus type 16 E6/E7 genes. *Cancer Res* 52:4420, 1992.

152. B. J. Poiesz, F. W. Ruscetti, A. F. Gazdar, P. A. Bunn, J. D. Minna, et al.: Detection and isolation of type-C retrovirus particles from fresh and cultured lymphocytes of a patient with cutaneous T-cell lymphoma. *Proc Natl Acad Sci USA* 77:7415, 1980.

153. B. J. Boiesz, F. W. Ruscetti, M. S. Reitz, V. S. Kalyanaraman, and R. C. Gallo: Isolation of a new type-C retrovirus (HTLV) in primary uncultured cells of a patient with Sézary T-cell leukemia. *Nature* 295:268, 1981.

154. R. C. Gallo and F. Wong-Staal: Current thoughts on the viral etiology of certain human cancers. *Cancer Res* 44:2743, 1984.

155. V. S. Kalyanaraman, M. G. Sarngadharan, M. Robert-Guroff, D. Blayney, D. Golde, et al.: A new subtype of human T-cell leukemia virus (HTLC-II) associated with a T-cell variant of hairy cell leukemia. *Science* 215:571, 1982.

156. M. Nerenberg, S. H. Hinrichs, R. K. Reynolds, G. Khoury, and G. Jay: The *tat* gene of human T-lymphotropic virus type I induces mesenchymal tumors in transgenic mice. *Science* 237:1324, 1987.

157. R. C. Gallo, V. S. Kalyanaraman, M. G. Sarngadharan, A. Sliski, E. C. Vonderheid, et al.: Association of the human type C. retrovirus with a subset of adult T-cell cancers. *Cancer Res* 43:3892, 1983.

158. F. Wong-Staal, B. Hahn, V. Manzari, S. Colombini, G. Franchini, et al.: A survey of human leukemias for sequences of a human retrovirus. *Nature* 302:626, 1983.

159. H.-G. Guo, F. Wong-Staal, and R. C. Gallo: Novel viral sequences related to human T-cell leukemia virus in T-cells of a seropositive baboon. *Science* 223:1195, 1984.

3

The Epidemiology
of Human Cancer

TRENDS IN CANCER INCIDENCE AND MORTALITY

U.S. Data

Long-range trends in the incidence of various cancers in different populations provide clues to the causes of cancer. Because of the long latency period between the first exposure to carcinogenic agents and the appearance of clinically detectable cancer, which may be up to 20 or 30 years, current trends probably reflect carcinogen exposure that began decades earlier. Genetic predisposition plays a role here in that individuals carrying genetic susceptibility genes may develop cancer more rapidly or at an earlier age. Another major factor that affects the overall incidence of cancer is the change in the average age of the population. The average age at the time of diagnosis (averaged for all tumor sites) is 67,[1] and as a higher proportion of the population reaches age 60 and above the incidence of cancer will go up as a result of this factor alone. Moreover, with the long-term downward trends in other causes of death, primarily infectious and cardiovascular diseases, more people live to an age when the risk of developing cancer becomes high. It is projected, for example, that about one in four males and one in five females born in 1985 in the United States will eventually die of cancer.[2] This is up from about 18% for males and 16% for females born in 1975. If current trends continue, about one in three Americans now living will develop some form of cancer.[3]

The U.S. age-adjusted cancer death rates for selected cancer types from 1930 through 2001, the latest year for which complete data are available, were discussed in Chapter 1. A number of points stand out from these data; for example, the alarming increased mortality rate for lung cancer in both males and females. Even though this steep rise of mortality rate for females was slower to occur than in males, the death rate due to lung cancer surpassed that of breast cancer in the late 1980s. The mortality rate for lung cancer in males has declined to some extent since the 1990s, probably due to a decreased rate of smoking in young males that started much earlier. Unfortunately, the incidence of smoking among young adults appears to be on the rise again since the year 2000.

Of over 1.3 million new cases of cancer each year in the United States, about 570,000 patients die every year. The overall mortality rates since the 1970s are disconcerting (Fig. 3–1); they are basically flat. There is a small downward trend in mortality rates for men since the mid- to late 1990s, most likely because of better treatment, although diagnosticians will argue that this is due to earlier diagnosis, primarily for prostate cancer. While this conjecture may be partly correct, the overall survival data do not suggest the concept that prostate-specific antigen (PSA) levels have made a large difference in long-term overall survival. This is a controversial area and will be discussed further in Chapter 7. The data for prostate cancer indicate a spike in cancer incidence for men from 1990 to 1995, due to the introduction of large-scale PSA testing (Fig. 3–2).

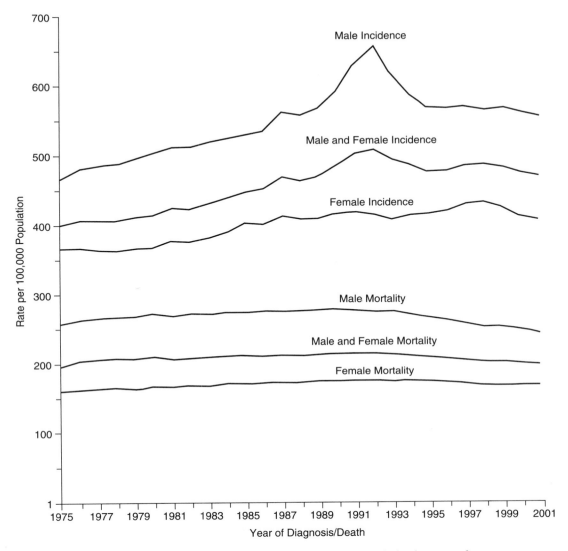

Figure 3–1. Annual age-adjusted cancer incidence and death rates° for all sites, by sex, United States, 1975 to 2001. °Rates are age adjusted to the 2000 U.S. standard position. (*Source*: Incidence data from Surveillance, Epidemiology, and End Results (SEER) Program, nine oldest registries, 1975 to 2001, Division of Cancer Control and Population Sciences, National Cancer Institute, 2004. Mortality data from U.S. Mortality Public Use Data Tapes, 1960 to 2001, National Center for Health Statistics, Centers for Disease Control and Prevention, 2004. From Jemal et al.,[3] with permission.)

If a significant number of these men were going to die of prostate cancer, the mortality rate should have fallen more dramatically than it has in recent years. That is not to say that some men's lives have not been saved by early diagnosis of prostate cancer, but on a population basis, it is by no means clear that the current PSA test has been a huge success.

The overall number of yearly cancer cases by state in the United States is shown in Figure 3–3.[4] These figures are not normalized by population, so they reflect the absolute numbers only. It should be noted that basal cell and squamous cell carcinoma of the skin are excluded from these data and those discussed above, because these cancers are almost always curable, even

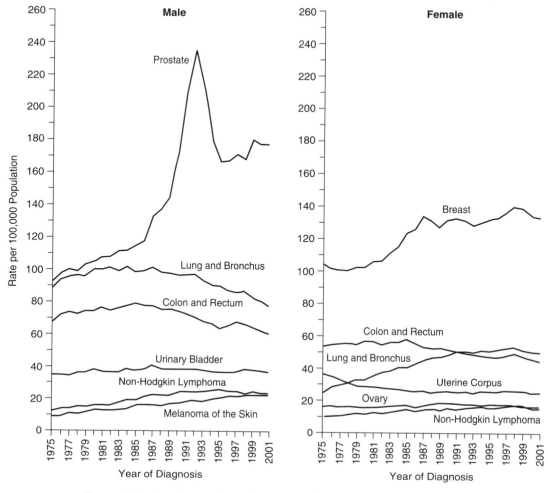

Figure 3–2. Annual age-adjusted cancer incidence rates° among males and females for selected cancer types, United States, 1975 to 2001. °Rates are age adjusted to the 2000 U.S. standard population. (*Source*: Surveillance, Epidemiology, and End Results (SEER) Program, nine oldest registries, 1975 to 2001, Division of Cancer Control and Population Sciences, National Cancer Institute, 2004. From Jemal et al.,[3] with permission.)

though the number of cases is enormous. The incidence rate for these nonmelanoma skin cancers is over one million new diagnoses per year.[3]

Cancer Is a Global Problem

Cancer is clearly a worldwide problem. The incidence and mortality rates for various cancers are similar, though not identical, among developed countries. In the developing world, as countries become more westernized and their populations achieve longer life expectancy, cancer rates are increasing. Although there are differ-

ences among developing and developed countries in the incidence rates of certain cancers, lung cancer is the most common cancer among men in both regions of the world and breast cancer is the most common cancer in women (Fig. 3–4).[5]

There are, however, regional differences in the distribution of various cancers in different regions of the world that reflect differing etiologic factors. For example, infectious etiology plays a greater role in certain parts of the world, e.g., the role of schistosomiasis infections in causing bladder cancer in parts of Africa and that of hepatitis B infections in liver cancer in China

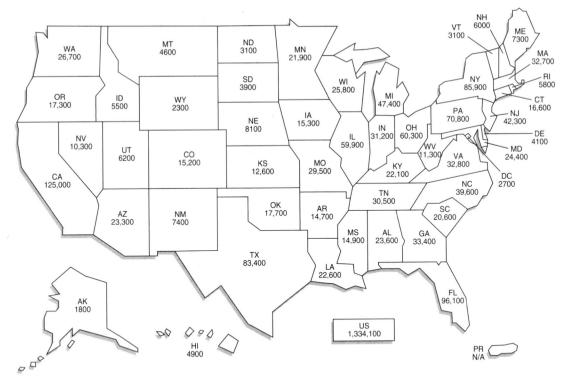

Figure 3–3. Estimated number of new cancer cases in the United States, by state, for 2003. Rates are age adjusted to the 2000 U.S. standard population. Excluded are basal and squamous cell skin cancers and in situ carcinomas except urinary bladder. (From *Cancer Facts and Figures—2003*,[4] with permission from American Cancer Society, Inc., 2006.)

and other parts of the Southeast Asia. Cancers of the stomach and uterine cervix are higher in parts of the developing world, whereas colorectal and prostate cancers are higher in the developed world.[5] The global death rate due to cancer is estimated to be more than 6 million people annually, with about 10 million new diagnoses every year.[5] Over 22 million individuals have been diagnosed with cancer worldwide. Thus, there is a global and growing cancer problem.

DATA FOR SOME PREVALENT HUMAN CANCERS

Lung Cancer

Lung cancer is the most common cancer worldwide and the leading type of cancer mortality in men. The incidence of lung cancer closely tracks the incidence of cigarette smoking, with a lag of about 20 years, in both men and women (Fig. 3–5).[6a] Since the mid-1900s in men and 1950s for women, the rate of lung cancer has risen dramatically. The highest rates include the United States, United Kingdom, Japan, and Australia and the lowest rates are in Africa and Southern Asia.[6] Female incidence rates are highest in the United States, Canada, Denmark, and the United Kingdom but lower in countries such as France, Japan, and Spain, where the prevalence of smoking among women has been low until recently.

The etiology of lung cancer is predominantly related to cigarette smoking. The data for this are overwhelming.[6] Other proposed causes include environmental pollution, occupational exposure (e.g., asbestos, coal mining), passive smoke inhalation, and radon exposure. These are all minor players and, in fact, the role of passive smoke and radon in the home as causes are insignificant, in

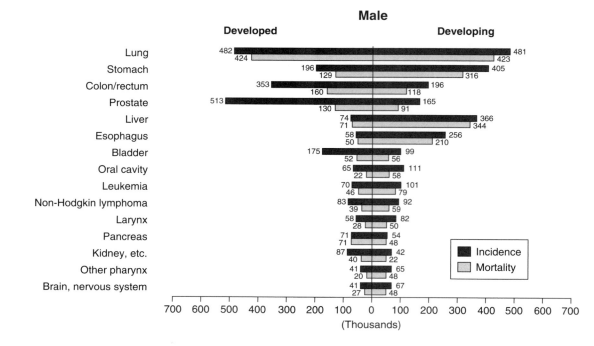

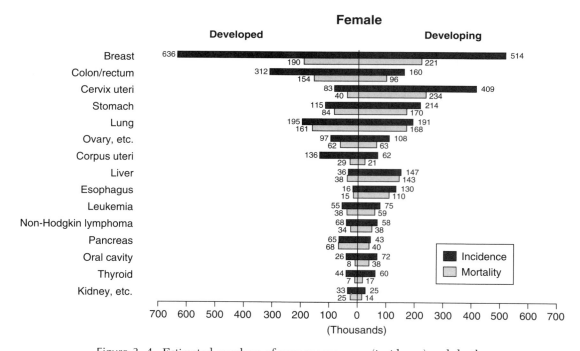

Figure 3–4. Estimated numbers of new cancer cases (incidence) and deaths (mortality) in 2002. Data shown in thousands for developing and developed countries by cancer site and sex. (From Parkin et al.,[5] with permission.)

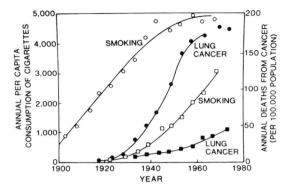

Figure 3–5. Trends in smoking prevalence and lung cancer in British males and females. The data for this chart are from England and Wales. In men, smoking (open circles) began to increase at the beginning of the twentieth century, but the corresponding trend in deaths from lung cancer (filled circles) did not begin until after 1920. In women, smoking (open squares) began later, and the increase in lung cancer deaths in women (filled squares) has appeared only recently. (From Loeb et al.,[6a] with permission.)

spite of environmentalist hype (see Cancer Myths, below).

Lung cancer is still classified by histological cell type into squamous cell carcinoma, adenocarcinoma, and large- and small-cell carcinomas. The first three are often lumped together as non–small cell lung carcinomas. Squamous cell carcinoma is strongly associated with smoking and is the most prevalent type. Adenacarcinomas are less strongly associated with smoking. These tumors are gland forming and mucous producing. Small cell lung carcinomas are usually aggressive and invasive and are often metastic at the time of diagnosis. They are in general more sensitive to chemotherapy than other lung cancers, but usually relapse and are fatal. Small cell lung carcinomas (SCLC) also tend to produce a variety of hormones and are associated with paraneoplastic syndromes (see Chapter 8).

There are a number of genetic alterations observed in progressive lung cancers (Table 3–1). Mutations in p53 are common, and begin to be observed in early tumors, particularly squamous cell carcinomas. K-ras activation occurs more commonly in adenocarcinomas than the other cell types. Loss of heterozygosity at chromosomal locus 3p and of fragile histidine triad (FHIT) occur early in neoplastic transformation and are associated with smoking. Silencing of the cyclin-dependent kinase inhibitor by gene methylation or deletion has been observed in 30% or more of adenocarcinomas and squamous cell carcinomas. The implications of these genetic changes are described in Chapter 5. Other genes have been associated with lung carcinogenesis, including erb b1 and b2, the myc family of genes, c-myc, N-myc, and L-myc. Gene expression profiling has shown that a variety of genes are up- or down-regulated in non–small cell lung cancers (see Chapter 7). Overall 5-year survival is poor for all types of lung cancer. It is only 15%, but it is 49% if diagnosed when the disease is still localized (about 15% of cases).

Breast Cancer

Over 210,000 new cases of invasive breast cancer and over 40,000 deaths due to breast cancer

Table 3–1. Genetic Alterations in Lung Tumors

Gene	Locus	Alteration	Frequency (% of tumors)		
			Small Cell Carcinoma	Adenocarcinoma	Squamous Cell Carcinoma
p53	17p13	Deletion, mutation (G:C > T:A), (overexpression)	70–90	30	50
KRAS	12p21	Mutation (GGT > TGT)	<1	15–60	8–9
CDKN2A/p16[INK4]	9p21	Deletion, mutation, hypermethylation	<1	27–59	33–40
LOH 3p	3p	Deletion (loss of heterozygosity)	100	50–85	
FHIT	3p14.2	Delection (loss of heterozygosity), transcriptional dysregulation	76	40–76	

From World Cancer Report,[6] with permission.

occurred in 2003 in the United States, which is a decrease of about 2,000 annual deaths from the peak year of 1995.[3,3a] Breast cancer is also the most common malignancy of women worldwide, with more than one million new cases occurring annually. Although breast cancer is more common in developed Western societies, with an incidence rate greater than 80 per 100,000 population per year, incidence rates are increasing in the developing world. From 1975 to 1990, the largest increases (1%–5%) were seen in Asia, Africa, and parts of Europe that previously had low incidence rates.[7]

Risk factors include duration of exposure to female hormones (early menarche and late menopause); reproductive factors (nulliparity, late age of first pregnancy); dietary and low physical activity factors (obesity; high-fat diet); ionizing radiation during breast development; chronic use of hormone replacement therapy; and genetic inheritance (family history) of breast cancer such as brca1, brca2, or p53 germline mutations.

Although high dietary fat intake has been associated with an increased risk of breast cancer in animal studies, international population comparisons, and some case–control studies, this association has not been found in some prospective studies. Case–control studies have supported a positive risk correlation with high intake of saturated fat, but a pooled analysis of prospective studies found only a weakly positive association (reviewed in Reference 8).

There have been some issues with the way these studies were carried out. For example, earlier prospective studies on the association of fat intake and breast cancer included relatively few premenopausal women who later developed breast cancer. Since one would expect that high fat intake during premenopausal years may be the time at risk for later tumorigenesis, this is an important group to study. Cho et al.[8] studied the dietary fat intake and breast cancer risk among 90,655 women of ages 26–46 years enrolled in the Nurses Health study in 1991. Fat intake was assessed with a food-frequency questionnaire at baseline in 1991 and again in 1995. Breast cancers were confirmed by review of pathology reports. The conclusion from this study was that intake of animal fat, mostly from red meat and high-fat dairy products, during premenopausal

years is associated with an increased risk of breast cancer.

In contrast to the somewhat conflicting results on the role of dietary fat, there is a general consensus that obesity is a significant risk factor for breast cancer. For example, a study of the relationships between body mass index (BMI) with serum estrogen levels and breast cancer incidence found that there was an increased risk of breast cancer with increasing BMI among postmenopausal women.[9] This result was largely associated with increased bioavailable serum concentration of estrogen in the women with high BMIs. The mechanism of this is postulated to result from elevated production of estrogen by aromatase in adipose tissue and a decrease in the serum concentration of sex hormone–binding globulin.[10]

Early detection of breast cancer is key to survival rates. The 5-year survival rate for localized breast cancer is >95% but drops to 78% for regional spread, and 23% for metastic disease.[4] Screening mammography, though subject to debate regarding the women's age to start this and its role in overall survival, is still the most widely used screening tool. Improved methods such as MRI may enhance the accuracy of diagnosis.

Another issue related to mammography screening is the high incidence of ductal carcinomas in situ (DCIS) that is detected, over 30% in some screening centers. The catch is that some of these lesions will progress to invasive disease and some will not,[11] and currently there is no good way to tell which ones will progress and which will not. Gene expression microarrays may make this determination possible in the future.[12]

Germline mutations, including brca1 and brca2, account for only 15%–20% of familial-related breast cancers and only 5% of all breast cancers.[13] These genetic susceptibility genes have variable penetrance in various individuals, which most likely reflects the expression of hormone metabolizing genes,[14] DNA repair genes, immune surveillance, H-ras, and androgen receptor genes.[13] The lifetime risk of developing breast cancer for women with the brca1 mutation varies from 36% to 80% depending on the population studied.[13] The risk increases wih age: by age 80, the relative risk is 80% for individuals with a brca1 mutation. Thus, finding the additional genes that modulate breast cancer susceptibility is a key area for research.

Some of the key genes involved in the pathogenesis and progression of breast cancer have been identified. These include loss of heterozygosity on chromosomal loci 13q, 9p, and 16q that involve *rb* (the retinoblastoma gene), CDKN2 (encoding the p16 protein), and CDH1 (encoding the E-cadherin protein).[15] Other common types of genetic alteration are the amplification of *erb* B2, *c-myc*, cyclin D1, and insulin-like growth factor genes.[16]

Colorectal Cancer

In the United States, 105,000 cases of colon cancer and 40,000 cases of rectal cancer were expected in 2005, and an estimated 56,290 deaths for both combined.[3] This number accounts for about 10% of all cancer deaths in the United States. Over 940,000 cases of colorectal occur annually worldwide.[17] Although cancers of the colon and rectum are relatively rare in developing countries, they are the second-most frequent malignancies in the developed world. This discrepancy appears to be largely due to

the conditions of an affluent lifestyle, because the major risk factors are a diet abundant in fat, refined carbohydrates, and animal protein and low in fiber, combined with physical inactivity.[18] However, the primary risk factor is age, with over 90% of cases diagnosed in people over age 50. This profile may be due to a lifetime of biological and chemical insults resulting from the above associated risk factors.

There is a clear overlay of genetic predisposition for colorectal cancer. Two of the genetically inherited syndromes are familial adenomatous polyposis (FAP) and hereditary non-polyposis colon cancer (HNPCC). The genes involved in these syndromes have been identified (see below).

The molecular genetic changes that occur during malignant transformation and tumor progression have been well studied by Vogelstein and colleagues.[19] Although the genetic alterations that occur during colon cancer progression suggest that the changes occur sequentially, this is probably not the case. It is more likely that it is the accumulation of the changes that occur in the *APC*, K-*ras*, *DCC*, *p53* genes, and other genes that results in invasive colon carcinoma (Fig. 3–6). Moreover, Smith et al.[20] have reported

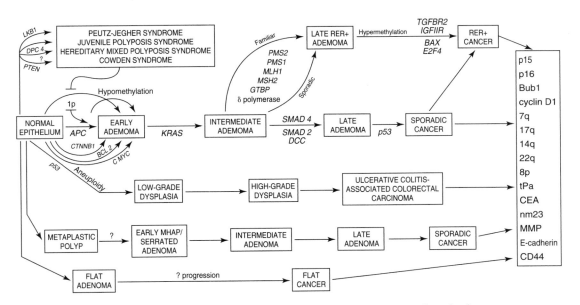

Figure 3–6. Putative genetic pathways in colorectal cancer. It is thought that most tumors develop according to the original Vogelstein model (bold arrows). MHAP, mixed hyperplastic adenomatous polyps. (From *World Cancer Report*,[17] with permission.)

that in contrast to the predictions of the sequential model of mutation accumulation, only 6.6% of colon cancers from a cohort of 106 colorectal cancer patients had mutations in all three genes: *APC*, K-*ras*, and *p53*, whereas 38.7% of tumors had mutations in only one of these genes. The most common combination of mutations was *APC* plus *p53* (27.1%). Mutations in both K-*ras* and *p53* were rare. These data suggest that colorectal cancers are a heterogeneous family of tumors that can arise from alternative pathways.

Treatment is primarily surgical and is often curative if disease is localized. Chemotherapy or chemotherapy plus irradiation is used for deeply invasive or metastatic disease. Usually metastasis occurs first in the liver, for which local hepatic radiation plus hepatic artery infusion improves survival.[21] However, 5-year survival for metastatic colon cancer is dismal, 9%, in contrast to 90% 5-year survival for patients with localized disease (Dukes A). Unfortunately, only about 37% of patients are diagnosed at that stage.[4] Recently, clinical studies have shown chemopreventive effects with aspirin and nonsteroidal anti-inflammatory drugs (NSAIDS), particularly COX-2 inhibitors.[22]

Liver Cancer

Although not a major cause of cancer in Western societies, primary hepatocellular carcinoma is a huge problem in a number of areas of the world, especially Southeast Asia and Africa. Worldwide, about 560,000 new cases occur annually.[23] Liver cancer is extremely difficult to treat and overall, 5-year survival (all stages) is only about 7% in the United States and even lower in developing countries. The primary risk factor in parts of the world where liver cancer is prevalent is hepatitis B infection. Other risk factors include infection with hepatitis C virus, chronic liver cirrhosis, alcohol abuse, aflatoxin exposure, and parasitic infections.

The molecular genetic changes in liver cancers are diverse and probably reflect the various causative agents. For example, *p53* mutations are an early event in high-incidence areas (e.g., China) but a late progression-related event in Western countries.[23] K-*ras* mutations and cyclin D1 amplification are only observed in a small subset of patients. Other observed genetic alterations include CDKN2A, M6P/IGF2R, SMAD gene family members, and cyclin D and A genes.

Pancreatic Cancer

Even though pancreatic cancer is the tenth leading cancer by incidence in men and ninth in women, it is the fourth leading cause of cancer deaths in the United States. Overall survival is poor. For all stages combined, 1-year survival is 21% and 5-year survival is 4%.[4] It is a "silent" disease symptomatically and difficult to diagnose early, but even for those patients diagnosed with so-called early disease, 5-year survival is only 17%.

Pancreatic cancer is primarily a disease of developed countries. About 216,000 cases occur annually worldwide. Risk factors include smoking, obesity, physical inactivity, chronic pancreatitis, diabetes, liver cirrhosis, and high-fat diet.[4] Familial genetic risk appears to account for about 10% of cases. K-*ras* and *p53* gene alterations are the most common ones observed. As can be surmised from the poor survival rates, no effective treatment currently exists.

Cancers of the Female Reproductive Tract

The most common cancers of the female reproductive tract are cervical, ovarian, and endometrial arcinomas Annually worldwide, there are 470,000 new cases of cervical cancer (with 230,000 deaths; 80% in developing countries), 190,000 cases of ovarian cancer (mostly in postmenopausal women in developed countries), and 188,000 new cases of endometrial cancer (also mainly in postmenopausal women in developed countries).[24]

Cervical Cancer

This is almost two different diseases: one in the developing world and a different one in the developed world, although the etiology is similar. For example, in the United States and other developed nations, routine PAP smears and gynecologic examinations detect the majority of cases early and the cure rate is close to 100% for patients with pre-invasive lesions (cervical intraepithelial neoplasia, CIN). The cure rate is 92% for localized carcinomas that are minimally invasive.[4] However, it is a different story in developing parts of the world, where 80% of the 470,000 annual worldwide cases occur.

The primary risk factor is sexually transmitted infection with certain species of human papilloma virus and is linked to having early sex and multiple sex partners. Additional risk factors are HIV infection and a weak immune system.

Ovarian Cancer

About 190,000 new cases and 114,000 deaths from ovarian cancer occur annually in the world.[24] The highest case loads are in Scandinavia, Eastern Europe, the United States, and Canada. Low incidence rates occur in Africa and Asia. As with most cancers, the risk increases with age. Risk factors include obesity, a history of pelvic inflammatory disease, polycystic ovary syndrome, and endometriosis. Hormone replacement therapy increases risk, whereas oral contraceptives and tubal ligation decrease risk. A family history of breast or ovarian cancer also increases risk and involves mutations in brca1 and brca2 genes. Hereditary non-polyposis colon cancer has also been associated with ovarian cancer.

Symptoms of ovarian cancer are vague and may include abdominal bloating, vague digestive disturbances, and, rarely, abnormal vaginal bleeding. For this reason, early diagnosis is very difficult and only about 25% of cases are detected at localized stage.[4] Overall 5-year survival for all stages is 53%, and in the minority of cases with localized disease, 5-year survival is over 90%. The tumor marker CA-125 has had some usefulness in diagnosis, but it is not sufficiently sensitive or specific to be used for routine screening.[25] Treatment involves surgery, radiation, and chemotherapy.

Genetic alterations observed during disease progression are overexpression of Her2/neu and c-myc. K-ras activation is often observed and p53 mutations occur in 50% of cases.[24] Ovarian cancer appears to develop through multiple chromosomal abnormalities.[26] Ovarian carcinomas develop through three or more phases of karyotypic derangements. In the first phase, karyotypic abnormalities proceed in a step-wise fashion involving either hyperdiploid imbalances (with additions of genetic material at chromosomes +7/+8/+12 or deletions at 6q−/1q−. A second phase involves increased chromosomal instability

and phase three is characterized by triploid formation linked to the 6q−/1q− pathway.

Endrometrial Cancer

Cancers of the uterine corpus are the seventh-most common cancer of women worldwide, with 189,000 new cases and 45,000 deaths annually worldwide. Sixty percent of these cases are in developed countries, with the United States and Canada having the greatest prevalence.[24] The etiology is linked to reproductive history: highest in nulliparous women and those with late menopause. The cause is linked to duration and type of estrogen exposure. Estrogen replacement therapy, tamoxifen treatment, early menarche, and late menopause are examples of this. Addition of progesterone to estrogen in hormone replacement regimens lowers the risk of ovarian cancer, but not of breast cancer, in which both estrogen and progesterone are cell proliferation promoting.

Endometrial cancer is often diagnosed early because abnormal uterine bleeding is a common symptom. Treatment is by hysterectomy, followed by radiation and/or chemotherapy if the disease has become invasive or metastatic. Five-year survival for local, regional, and distant stage disease are 96%, 64%, and 26%, respectively. Common genetic alterations are seen in K-ras, PTEN, p53, Her 2/neu, cell cycle regulatory genes (e.g. cyclin D1, p16/INK4a), and E-cadherin.[24]

Prostate Cancer

An estimated 232,000 new cases of prostate cancer and 30,000 deaths are expected in the United States in 2005.[3] About 200,000 deaths occur worldwide annually, mostly affecting men over age 65 in developed countries.[27] African-American men have the highest prostate cancer incidence rates in the world, for reasons that aren't clear but may relate to genetic polymorphisms in the synthesis and metabolism of androgenic hormones.[5]

The only well-established risk factors are age, race, and heredity. Environmental factors are involved, as indicated by incidence rates that change as populations migrate. For example, first-generation Japanese American men (Nisei) born in the United States experience an increase in prostate cancer that more closely resembles the incidence rate of the Caucausian U.S. male population (this increase for risk is also true for breast and colorectal cancers). Dietary risk factors are a diet rich in red meat and high fat. Dietary antioxidants such as selenium and lycopene lower risk.[28]

Family history of prostate cancer is one of the most consistent risk factors: up to 40% of prostate cancer patients hava a familial association.[28] However, although some candidate genes have been identified, no smoking gun similar to *brca*1 and *brca*2 for breast cancer has been found.

Several molecular genetic alterations associated with prostate carcinogenesis and progression have been identified.[28] These involve chromosomes 8, 10, 13, 16 and 17. Loss of heterozygosity on chromosome 8p appears to be an early event. Interestingly, hypermethylation of the promoter region of the glutathione S-transferase P1 (GST-P1) gene has been observed in about 90% of prostate carcinomas and 70% of high-grade prostatic intraepithelial neoplasias (PIN). Since GST-P1 is involved in detoxification of carcinogens and promoter methylation has been implicated in gene silencing (see Chapter 5), the hypothesis is that silencing of this gene reduces a protective effect in the prostate gland. Methylated DNA sequences can be detected in the urine and this may provide a tumor marker that is more sensitive and specific than PSA (see Chapter 5). Measurement of age-related, oxygen radical–induced DNA damage, as measured by 8-hydroxyl adenine and guanine levels in normal vs. tumor prostatic tissue,[29] suggests that age-accumulated damage to prostatic tissue is important for the neoplastic transformation in the gland. This may also explain why antioxidants such as lycopene and selenium have a protective effect.

The origin of cancer in human tissues is thought to be clonal, in that it arises from a single clone of transformed cells that undergo expansion, and through cellular evolution accompanied by genetic instability, these cells eventually become a full-blown malignant neoplasm. Prostate cancer may be an exception to that. Through microdissection of tumors in different locations from the same patients and detemination of four DNA microsatellite polymorphic markers, it was concluded that different tumor foci in a given patient have an independent origin.[30] Similarly, Macintosh et al.,[31] employing precise microdissection of different tumor foci in individual patients,

found that phenotypically similar tumor foci had different genotypes, providing additional evidence for the multifocality of tumor development in the prostate. This finding has important clinical implications, because these different tumor foci may have different degress of invasiveness, androgen-dependence, and drug responsiveness.

Screening for prostate cancer by the PSA test has led to a large increase in the detection of early prostate lesions. The number of these that would actually develop into invasive cancer that would kill men is subject to intense debate. The huge increase in prostate cancer incidence seen in the United States between 1990 and 1995, as detected by the PSA test, has not translated into a great change in overall mortality (Fig. 3–2). Moreover, in the United Kingdom, the rise in incidence of prostate cancer, though shower and more delayed, also has not led to very great changes in overall prostate cancer mortality.[32] PSA, originally thought to be a specific prostate epithelial marker, has been observed in other tissues, including liver, colon, lung, kidney, breast, ovarian, and parotid tumors as well as in breast milk and amniotic fluid.[32] Transiently increased serum PSA has also been observed in patients infected with hepatitis A virus.[33] Thus, more sensitive and specific diagnostic tests for prostate cancer are needed. Also, methods to determine which prostate lesions are likely to progress to invasive, metastatic disease and which are likely to remain indolent for a man's life span are badly needed.

Treatment for prostate cancer includes surgery or radiation therapy for localized lesions, followed by androgen deprivation therapy, and finally by chemotherapy if the disease progresses. The high incidence of impotence and urinary incontinence of postsurgical resection is still a problem, though now much less so with nerve-sparing surgical approaches.[34] Five-year survival for all stages is close to 97%. However, 10-year and 15-year survival rates are 79% and 57%, respectively.[4]

Urinary Bladder Cancer

Bladder cancer is the ninth-most common malignancy in the world.[35] There are 330,000 new cases and 130,000 deaths each year. In the United States alone there are over 63,000 new cases annually and 13,000 deaths.[3]

Smoking is the greatest risk factor and is estimated to be a causative factor in 65% of males and 30% females in some developed countries. Historically, some types of bladder cancer were associated with abuse of analgesic combinations containing phenacetin and occupational exposure in the aniline dye industry (e.g., exposure to 2-naphthylamine). In Egypt and some other African nations, chronic bladder infections with *Shistosoma haematodium* are a risk factor.

Bladder cancer is treated by surgical removal if disease is invasive. Superficial, localized cancers can be treated by local instillation of immuno-modulatory agents (e.g., bacilli Calmette-Guerin [BCG]) or chemotherapy. Chemotherapy and/or radiation therapy have been used as an adjuvant or neo-adjuvant (before surgery) to cystectomy.

Five-year survival in the United States and other developed countries is over 90% for localized disease, 48% for regional disease, and 6% for metastatic disease.[4] In developing countries, overall 5-year survival is only 30%–50%[35] because detection and treatment are usually more delayed.

Lymphoma

Lymphomas are generally classified as Hodgkin's and non-Hodgkin's (NHL) lymphomas, although there are a number of subtypes of NHL that differ in their cellular morphology, response to chemotherapy, and prognosis. The new cases annually in the United States are in the range of 7300 for Hodgkin's disease and 56,000 for non-Hodgkin's.[3] Deaths due to NHL are about 20,000 annually, whereas only about 1400 deaths occur from Hodgkin's disease in the United States. Globally, about 62,000 cases of Hodgkin's disease occur annually, but over 280,000 cases of NHL occur annually, predominantly in more developed countries.[36]

The strongest known risk factors for NHL are chromosomal translocations, the inciting cause for which isn't usually clear. Viral infection, e.g., with Epstein-Barr virus, human herpes virus 8, or human T-lymphotropic virus-1 (HTLV-1), and acquired immunodeficiencies due to AIDS or immunosuppressive drugs, for example, have been suggested as causative.

The molecular pathogenesis of lymphomas has been well studied.[37] Because tissue is readily available and distinct genetic events such as chromosomal translocations are clearly related to disease progression, it has been easier to study such events in lymphomas and leukemia than in solid tumors such as lung, breast, and colon. Chromosomal translocations are often the inciting events in lymphomas, in contrast to solid tumors, where gene deletions are more common. The translocation events often involve the immunoglobin (Ig) loci and a proliferative or anti-apoptotic gene such as BCL-2. About one-sixth of all NHLs have translocations of the BCL-6 gene that encodes a transcriptional repressor of normal B-lymphocyte differentiation.[37] This favors cell proliferation and decreased cell senescence. Some NHLs have translocations that lead to over-expression of c-*myc* and D-type cyclins, which favor cell proliferation. A type of NHL called *mantle cell lymphoma* (MCL) exhibits a genomic deletion of the cell cycle checkpoint gene *p16* (INK4A) or a genomic amplification of the gene *bmi-1* that codes for a repressor of the p16 (INK4A) locus. Both of these alterations lead to loss of cell cycle checkpoint control. Still other lymphomas lose cell genome integrity by deletion or mutation of the *ATM* or *p53* genes.

These genetic alterations are summarized in Figure 3–7. A number of these genetic lesions involve pathways that will be seen again in other cancers.

Gene expression microarrays are now being employed to molecularly categorize a number of human cancers. One of the first practical demonstrations of this was for NHL. Alizadeh et al.[38] showed that diffuse large B-cell lymphomas (DLBCL) can be categorized by prognosis using gene arrays. Although clinical parameters can also predict survival, gene expression arrays are independent and perhaps more reliable predictors of prognosis. Furthermore, gene expression profiles of subgroups of DLBCL demonstrate that they are pathogenetically distinct diseases.

Lymphomas are as a class generally responsive to chemotherapy. The advent of the Mustargen (nitrogen mustard), Oncovin (vincristine), prednisone, procarbazine (MOPP) regimen by De Vita and colleagues[39] and subsequent variations on this theme have led to a high cure rate for Hodgkin's lymphoma. In general, NHLs are also responsive to combination chemotherapy, although somewhat less so than Hodgkin's

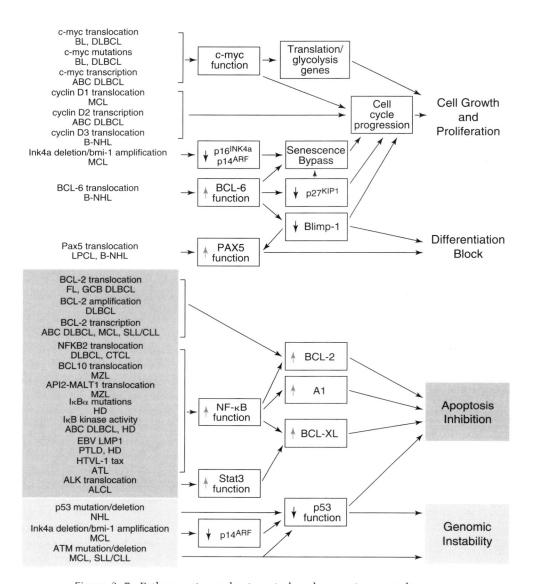

Figure 3–7. Pathogenetic mechanisms in lymphomas. Arrows indicate presumptive target genes and pathways affected by oncogenic events in various lymphoma types. An enhanced version of this figure with references is available as Supplemental Figure S1 at http://www.cancercell.org/cgi/content/full/2/5/363/DC1. Abbreviations: ALCL, anaplastic large-cell lymphomia; ATL, adult T-cell lymphoma; BL, Burkitt's lymphoma; B-NHL, B-cell non-Hodgkin's lymphoma; CLL, chronic lymphocytic leukemia; CTCL, cutaneous T-cell lymphoma; DLBCL, diffuse large B-cell lymphoma; EBV, Epstein-Barr virus; FL, follicular lymphoma; GCB, germinal-center B cell; HD, Hodgkin's disease; LPCL, lymphoplasmacytoid lymphoma; MZL, marginal zone lymphoma; PTLD, post-transplant lymphoproliferative disorder; SLL, small lymphocytic lymphoma. (From Staudt and Wilson,[37] reprinted by permission from Elsevier.)

lymphoma. Development of monoclonal anti-bodies such as rituximab target lymphoma cell surface receptors and are showing responses in chemoresistant NHLs.

Leukemia

Leukemias are classified on the basis of their cell type and chronicity of the disease. Thus, there are acute and chronic myeloid leukemias (AML and CML), and acute and chronic lymphocytic leukemias (ALL and CLL). Lymphocytic leukemias can be either of the B-cell or T-cell type. Rarer forms include monocytic, basophilic, eosinophilic, and erythroid leukemias. Multiple myeloma is an immunoglobulin-producing subtype of B-cell leukemia. Survival rates vary greatly depending on the cell type and chronicity. ALL is largely a disease of young childhood and thanks to effective combination chemotherapy has a 70% or higher cure rate. AML, however, is substantially less curable and 5-year survival is only 20%–30%.

Worldwide, about 250,000 new cases and 195,000 deaths occur each year.[40] The causes of leukemia are largely unknown, although some cases occur as secondary to earlier chemotherapy or exposure to radiation (e.g., atomic bomb survivors) or chemicals (e.g., benzene). Infection with the virus HTLV-1 is associated with adult T-cell leukemia in tropical countries and Japan.

Peak incidence of ALL is in the first 4 years of life, and while leukemia is often thought of as a childhood disease, it is diagnosed 10 times more often in adults. After infancy, there is a decline in leukemia incidence, but from ages 25 to 85 there is a steep increase in incidence (Fig. 3–8). ALL accounts for 25% of all childhood malignancies. AML is the most common leukemia in young adults up to about age 45, whereas CLL is the most common form in older adults.

With a better understanding of the molecular basis of leukemias[41] has come the development of so-called targeted chemotherapy, which is based on the determination of altered signal transduction pathways required and sufficient to cause (and maintain) the malignant phenotype. STI-571 (Gleevec) is the first example of such successful therapy (see below under Chronic Myelogenous Leukemia).

Leukemias result from clonal expansion of immature hematopoietic cells that are blocked in their differentiation to mature, functional blood cells. These cells accumulate in the bone marrow and spill over into the peripheral blood, crowding out their functional counterparts and leading to symptoms such as fatigue, weight loss, repeated infections, and excessive bruising and nosebleeds that bring patients to the physician. Through examination of the peripheral blood leukocytes and bone marrow a cell type of origin of the disease can be detected. Although histological examination is often sufficient to classify the leukemia type, gene expression microarrays are now able to stratify patients on the basis of molecular genetic profile, drug sensitivity, and prognosis.[42]

A variety of chromosomal abnormalities occur in various leukemias. The classic example is the Philadelphia chromosome in CML, which was originally detected by Nowell and Hungerford and later found by Rowley to be a translocation involving a piece of chromosome 22 translocated to chromosome 9 (see Chapter 2). This t(9;22) translocation creates a fusion gene *bcr/abl* that codes for a proliferation-promoting tyrosine kinase. It is this activated kinase that is the target of Gleevec, which induces complete remissions and increased survival in CML patients.[43] One reason for this targeted therapy working so well in CML is that the *bcr/abl* translocation is a key, if not *the* key, event initiating and maintaining the neoplastic character of CML cells. There are very few such examples in cancer biology, and this may be the reason it is so difficult to achieve therapeutic success with targeted therapy as a monotherapy for other cancers. A number of other gene-activating translocations occur in other leukemias, and these may also provide important targets for drug development (Table 3–2).

As for lymphomas, the leukemias as a class are generally responsive to chemotherapeutic agents, but overall survival rates differ markedly. For example, the 5-year survival rate for ALL is about 70%, whereas it is only 20%–30% for AML and CML. Patients with CLL tend to have 5-year survival rates in the 70% range. Five-year survival for multiple myeloma is 30%.[40]

Skin Cancer

Over one million cases of nonmelanoma skin cancers occur in the United States annually. These are mostly basal cell or squamous cell cancers and

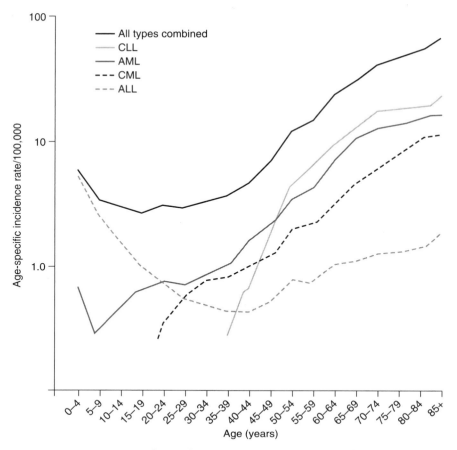

Figure 3–8. Age-specific incidence rates in the United States of leukemia overall and of different subtypes. ALL, acute lymphoblastic leukemia; AML, acute myeloid leukemia; CLL, chronic lymphocytic leukemia; CML, chronic myelogenous leukemia. Note the high incidence of ALL among children. (From *World Cancer Report*,[40] with permission.)

are highly curable by local excision (the Mohs procedure). Melanoma is a different story. Over 59,000 new cases of melanoma and over 7500 deaths occur every year in the United States, primarily in fair-skinned people.[3] Chronic accumulation of ultraviolet (UV)-induced skin damage is thought to be the primary cause of basal and squamous cell carcinomas, whereas multiple, intermittent severe burns, particularly during the young years of life, are thought to be the carcinogenic event for melanoma. There is also a strong hereditary link for melanoma, in addition to skin pigmentation.[44]

Melanoma is mostly a disease of people with light skin pigmentation. It is 10 times higher among Caucasians than African Americans, and it varies among Caucasian populations in relation

to proximity to the equator. Of the 133,000 new cases annually in the world, 80% occur in North America, Europe, and Australia.[45]

A large proportion of persons diagnosed with melanoma are cured by surgical excision, because the lesions are usually diagnosed early while the cancer is still localized. However, once melanoma has metastasized, it is extremely difficult to treat and almost always fatal. Melanoma is one of the most immunogenic of human cancers and immunomodulatory approaches have shown responses.

Alterations of chromosomes 1, 6, 7, and 10 are prevalent in melanoma but seem to be late in tumor progression.[44] Deletion of the tumor suppressor gene *p16 INK4A* are frequently observed. Epigenetic inactivation of the *p16*

Table 3–2. Common Recurrent Chromosomal Abnormalities in Acute Leukemia

Type and Subtype of Leukemia		Abnormality	Genes Involved
ACUTE MYELOID LEUKEMIA			
M0	Acute myeloblastic leukemia with minimal differentiation	inv(3q26) t(3;3)	EVI1
M2	Acute myeloblastic leukemia with maturation	t(8;21) t(6;9)	AML1-ETO° DEK-CAN
M3	Acute promyelocytic leukemia	t(15;17) t(11;17)	PML-RARα° PLZF-RARα° NPM-RARα°
M4	Acute myelomonocytic leukemia	11q23 t(3;3) inv(3q26) t(6;9)	MLL[†] EVI1 DEK-CAN
M4Eo	Acute myelomonocytic leukemia with abnormal eosinophils	inv(16) t(16;16)	CFBβ-MYH11°
M5	Acute monocytic leukemia	11q23 t(8;16)	MLL[†] MOZ-CBP
Secondary		5q° 7q° 11q23	Unknown[†] Unknown[†] MLL[†]
ACUTE LYMPHOBLASTIC LEUKEMIA			
		t(12;21)	TEL-AML1°
		t(1;19)	E2A-PBX1°
		t(4;11) t(11;19) t(1;11)	MLL[†]
		t(9;22)	BCR-ABL[†]
		t(8;14) t(2;8) t(8;22)	MYC[†]

°Confers good prognosis.

[†]Confers poor prognosis.

(From Stewart and Schuh,[41] reprinted by permission from Elsevier.)

INK4A gene by promoter methylation has also been observed. Cyclin D1 amplification has been detected, but curiously, this has only been observed in melanomas in relatively sun-protected areas of the body, suggesting a different mechanism of melanoma initiation in these areas compared to sun-exposed areas of the body. Interestingly, about 5% if melanoma cases clinically present as metastatic lesions with no known primary site, thus the immune system may have dealt with the primary skin location but had no effect on the metastases.[44]

Cancers of the Central Nervous System

"Brain tumors" are in reality a very diffuse collection of tumors of the central nervous system (CNS) (Table 3–3). Each of these tumors has a distinct biology, prognosis, and therapeutic approach. In the United States, there are about 17,000 new intracranial tumors diagnosed each year and about 13,000 deaths. Brain metastasis from other primary sites is much more common, with about 100,000 patients dying from this complication per year in the United States.[46]

Tumors of the CNS account for less than 2% of all global malignancies (about 175,000 cases), and the incidence does not vary widely among regions of the world.[47] The cause of CNS tumors is largely unknown. Certain hereditary syndromes produce a susceptibility for CNS malignancies. These include neurofibromatosis, von Hippel-Lindau disease, tuberous sclerosis, and Li Fraumeni syndrome. The location, age of clinical manifestation, 5-year survival, and associated genetic alterations are shown in Table 3–4. Glioblastomas are the

Table 3–3. Histologic Classification of Tumors of the Central Nervous System°

TUMORS OF NEUROEPITHELIAL TISSUE	
Astrocytic tumors	Pineal parenchymal tumors
Astrocytoma	Pineocytoma
Anaplastic astrocytoma	Pineoblastoma
Glioblastoma multiforme	
Pilocytic astrocytoma	Embryonal tumors
Pleomorphic xanthoastrocytoma	Medulloblastoma
Subependymal giant-cell astrocytoma	Primitive neuroectodermal tumor
Oligodendroglial tumors	**MENINGEAL TUMORS**
Oligodendroglioma	Meningioma
Anaplastic oligodendroglioma	Hemangiopericytoma
Mixed gliomas	Melanocytic tumor
Oligoastrocytoma	Hemangioblastoma
Anaplastic oligoastrocytoma	**PRIMARY CENTRAL NERVOUS SYSTEM**
Ependymal tumors	**LYMPHOMAS**
Ependymoma	Germ Cell Tumors
Anaplastic ependymoma	
Myxopapillary ependymoma	Germinoma
Subependymoma	Embryonal carcinoma
	Yolk-sac tumor (endodermal-sinus tumor)
Choroid-plexus tumors	Choriocarcinoma
Choroid-plexus papilloma	Teratoma
Choroid-plexus carcinoma	Mixed germ-cell tumors
Neuronal and mixed neuronal–glial tumors	**TUMORS OF THE SELLAR REGION**
Gangliocytoma	Pituitary adenoma
Dysembryoplastic neuroepithelial tumor	Pituitary carcinoma
Ganglioglioma	Craniopharyngioma
Anaplastic ganglioglioma	
Central neurocytoma	**METASTATIC TUMORS**

°This table has been abridged and modified from the World Health Organization classification.

(From De Angelis,[46] reprinted with permission from the Massachusetts Medical Society.)

most common brain tumors in adults, and they are really bad actors, because they are essentially incurable by surgery and are highly resistant to radiation therapy and chemotherapy. Neuroblastomas are the most common neural-related tumors in children, and the outlook for this disease is much better. A number of these tumors regress with age and the cure rate is over 90% for infants with the disease. A number of genetic pathways have been implicated in the progression of CNS tumors,[48] holding out the hope that better therapies will be developed in the future.

ROLE OF VARIOUS FACTORS IN THE DEVELOPMENT OF CANCERS

Most cancer types vary in incidence and mortality among different populations in different parts of the world. When populations move from one country to another, the rates for many cancers tend toward that of the local population rather than that of their country of origin. A classic example is the incidence rates among Japanese individuals living in Osaka, Japan, in contrast to those who have moved to Hawaii (Fig. 3–9).[49] Within a generation, the incidence for prostate, colon, and breast cancer begin to approach those of the United States population, whereas the incidence of stomach cancer, more prevalent in Japan, decreases. Another interesting point from the data in Figure 3–9 is that from 1970–71 to 1988–92, some of the "more Western cancers" became more prevalent in Japan, presumably because of a more Westernized diet and lifestyle.

Some specific causes of cancer are known, the most prominent of which is cigarette smoking (Table 3–5). However, the causes for the large

Table 3–4. Summary of Epidemiological Data on Intracranial Tumors

Tumor (WHO Grade)	Typical Location	Age at Clinical Manifestation (% of Cases)			5-Year Survival Rate (% of Patients)	Genetic Alterations
		0–20 years	20–45 years	>45 years		
Pilocytic astrocytoma (grade I)	Cerebellum, optic nerve	74	20	6	>85	*NF1* (neurofibromatosis cases)
Low-grade diffuse astrocytoma (grade II)	Cerebral hemispheres	10	61	29	>50	*p53* mutation
Glioblastoma (grade IV)	Cerebral hemispheres	3	25	72	>3	*EGFR* amplification, *PTEN* mutation, p16 deletion, LOH chromosome 10
Oliogodendroglioma (grade II/III)	Cerebral hemispheres	8	46	46	>50	LOH 1p, 19q
Ependymoma (grade IV)	Ventricles, spinal cord	37	38	25	<30	*NF1* (spinal tumors)
Medulloblastoma (grade IV)	Cerebellum	74	23	3	>50	Isochromosome 17, mutations of *p53*, *PTCH*, β-catenin
Neuroblastoma (grade IV)	Abdomen	>95			>90 (<1 year old) 20–50 (>1 year)	LOH 1p, 11q, *MYCN* amplification, trisomy 17q

LOH, loss of heterozygosity.

(From World Cancer Report,[47] with permission.)

global variation in the most common cancers such as breast, prostate, and colon-rectum remain unclear. In any case, the inescapable conclusion from these data is that environmental and lifestyle factors play the predominant role in cancer causation. That is not to say that genetic susceptibility factors, many of which remain to be elucidated, are not important (see below). Yet the implication here is that a high percentage of cancers are preventable, or at least "delayable." If it is true as some say, that if we lived long enough we'd all get cancer, then *delayable* may be a better term than *preventable*. Since there are three billion nucleoside bases in every cell's DNA and there are about 10^{14} cells in the human body, the chances of one or a few of these bases being misreplicated during cell division is enormous. Even if DNA editing and repair mechanisms were to take care of most of these replication errors, the chances that over a lifetime at least some of these would lead to malignant transformation in one or a few cells seems very likely indeed.

Table 3–6 illustrates the agents or circumstances for which there is good evidence of carcinogenicity in humans. These agents are generally divided into three categories: occupational, medical, and social. The data are obtained from the International Agency for Research on Cancer (IARC), based on studies of a working group that periodically accumulates and evaluates the evidence for human carcinogenesis. Although the information presented in Table 3–6 suggests that most human cancer is caused by occupational or medical exposures, this is not the case. Rather, the cause of the bulk of human cancers is unknown, and only after certain discrete exposures are well documented can cancer causation be laid at the feet of a particular agent. Sometimes rare cancers such as the hepatic angiosarcomas associated with occupational exposure to vinyl chloride call attention to certain agents. Most epidemiologists ascribe only about 2% to 5% of human cancers to occupational exposures. The best estimates of the proportion of cancer deaths due to various factors are shown in Table 3–5. Surprisingly, some authors ascribe only about 1% to 5% of cancer deaths to pollution.[51] There is considerable debate about the role of environmental exposure to air and water pollutants in cancer causation; this is discussed in more detail later. Obviously, the ranges cited for each of the potential causes are very broad in some cases,

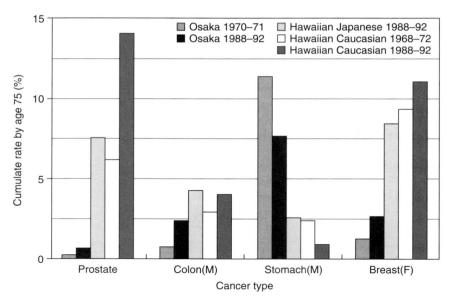

Figure 3–9. Cancer rates in migrants become similar to those in the local population. Cancer rates from 1988 to 1992 among Japanese migrants to Hawaii, and around 1970 and 1990 in Japan (Osaka) and in Hawaiian Caucasians. Local rates for prostate, colon, and breast cancer increased over time (partly because of increased completeness of diagnosis and registration, particularly for prostate cancer in Hawaiian Causcasians) and stomach cancer decreased; but the effects of migration were larger. (From Peto,[49] reprinted by permission from Macmillan Publishers Ltd.)

reflecting the degree of uncertainty about the whole business of ascribing definitive causes at the current state of our knowledge.

Cigarette Smoking

As noted above, epidemiologists have attributed as many as 30% of all cancer deaths to tobacco use, primarily cigarette smoking. In 2005, for example, approximately 163,000 deaths due to lung cancer were expected to occur in the United States (90,000 men and 73,000 women), making up about 31% of all deaths from cancer in men and 27% in women.[3] Although the most direct correlation is between cigarette smoking and lung cancer, tobacco use has also been implicated in cancers of the mouth, pharynx, larynx, esophagus, urinary bladder, pancreas, kidney, and, more recently, stomach and liver[49] and perhaps colorectal cancer.[52] Smoking of pipes or cigars has been implicated in the occurrence of cancers of the mouth, pharynx, larynx, and esophagus, but

this form of tobacco use is generally considered much less dangerous because the smoke is usually not inhaled. A number of studies have also suggested a correlation between "passive

Table 3–5. Estimated Percentage of Total Cancer Deaths Attributable to Established Causes of Cancer

Risk Factor	Percentage
Tobacco	30
Adult diet and obesity	30
Sedentary lifestyle	5
Occupational factors	5
Family history of cancer	5
Viruses and other biologic agents	5
Perinatal factors and growth	5
Reproductive factors	3
Alcohol	3
Socioeconomic status	3
Environmental pollution	2
Ionizing and ultraviolet radiation	2
Prescription drugs and medical procedures	1
Salt, other food additives, and contaminants	1

*Estimates according to Harvard Center for Cancer Prevention.
(From America's War on Carcinogens[50])

Table 3–6. Established Human Carcinogenic Agents and Circumstances

Agent or Circumstance	Type of Exposure°			Site of Cancer
	Occupational	Medical	Social	
Aflatoxin			+	Liver
Alcoholic drinks			+	Mouth, pharynx, larynx, esophagus, liver
Alkylating agents				
Cyclophosphamide		+		Bladder
Melphalan		+		Marrow
Aromatic amines				
4-Aminodiphenyl	+			Bladder
Benzidine	+			Bladder
2-Naphthylamine	+			Bladder
Arsenic†	+	+		Skin, lung
Asbestos	+			Lung, pleura, peritoneum
Benzene	+			Marrow
Bis(chloromethyl) ether	+			Lung
Busulfan		+		Marrow
Cadmium†	+			Prostate
Chewing (betel, tobacco, lime)			+	Mouth
Chromium†	+			Lung
Chlornaphazine		+		Bladder
Furniture manufacturer (hardwood)	+			Nasal sinuses
Immunosuppressive drugs		+		Recticuloendothelial system
Ionizing radiations‡	+	+		Marrow and probably all other sites
Isopropyl alcohol manufacture	+			Nasal sinuses
Leather goods manufacture	+			Nasal sinuses
Mustard gas	+			Larynx lung
Nickel†	+			Nasal sinuses, lung
Estrogens				
Unopposed		+		Endometrium
Transplacental (DES)		+		Vagina
Overnutrition (causing obesity)			+	Endometrium, gallbladder
Phenacetin		+		Kidney (pelvis)
Polycyclic hydrocarbons	+	+		Skin, scrotum, lung
Reproductive history:				
Late age at first pregnancy			+	Breast
Zero or low parity			+	Ovary
Parasites				
Schistosoma haematobium			+	Bladder
Chlonorchis sinensis			+	Liver (cholangioma)
Sexual promiscuity			+	Cervix uteri
Steroids				
Anabolic (oxymetholone)		+		Liver
Contraceptives		+		Liver (hamartoma)
Tobacco smoking			+	Mouth, pharynx, larynx, lung, esophagus, bladder
Ultraviolet light	+		+	Skin, lip
Vinyl chloride	+			Liver (angiosarcoma)
Virus (hepatits B)			+	Liver (hepatoma)

°A plus sign indicates that evidence of carcinogenicity was obtained.

†Certain compounds or oxidation states only.

‡For example, from X-rays, thorium, Thorotrast, some underground mining, or other occupations.

(From Doll and Peto[51])

smoking"—that is, exposure to smoke and others' cigarettes in the home or workplace—and lung cancer, but the data for this are questionable (see The Great Cancer Myths, below). An enormous amount of research on the relationship between tobacco smoking and cancer has been carried out over a number of years. The vast majority of these studies indicate cigarette smoking as a major cause of lung cancer. Epidemiologic studies, autopsy reports, and experimental animal data reviewed in the original U.S. Surgeon General's Report of 1964 and in subsequent U.S. Department of

Health, Education, and Welfare reports strongly support a casual relationship. The data can be summarized as follows.[53]

1. A strong relationship between cigarette smoking and lung cancer mortality in men has been demonstrated in numerous prospective and retrospective studies, with risks for all smokers as a group ranging from 11 to 22 times those of nonsmokers.[54]

2. A dose–response relationship between cigarette consumption and risk of development of lung cancer for both men and women has been demonstrated in numerous studies, with risk being much higher for men and women who are heavy smokers than for nonsmokers. Light smokers have an intermediate risk.[54,55]

3. Mortality from lung cancer directly attributable to cigarette smoking is increased in the presence of urbanization and such occupational hazards as uranium mining and exposure to asbestos.

4. Cessation of smoking results in lowered risk or mortality from lung cancer in comparison with continuation of smoking.[54,55]

5. Results from autopsy studies show that changes in the bronchial mucosa that are thought to precede development of bronchogenic carcinoma are more common in smokers than in nonsmokers, and there is a dose–response relationship for these changes.

6. Chronic inhalation of cigarette smoke or the intratracheal instillation of various fractions of tobacco smoke produce lung cancer in such experimental animals as dogs and hamsters.

7. Cell culture studies show that various constituents found in tobacco and cigarette smoke condensate produce malignant transformation of cells.

8. Numerous complete carcinogens and co-carcinogens (tumor promoters) have been isolated from cigarette smoke condensation.

Numerous mutagens and carcinogens have been identified in the particulate or vapor phases of tobacco smoke; these include benzo(a)-pyrene, dibenza(a)anthracene, nickel, cadmium, radioactive polonium[210]Po, hydrazine, urethan, formaldehyde, nitrogen oxides, and nitrosodiethylamine.[56] Moreover, mutagenic activity is 5- to 10-fold higher in the urine of smokers than that of nonsmokers.[57]

An increased incidence of chromosomal abnormalities has been observed in smokers' peripheral blood lymphocytes compared to lymphocytes of nonsmokers.[58,59] A meta-analysis of DNA adducts present in peripheral white blood cells or tissue of smokers who have cancer compared to smokers that don't showed that current smokers with high levels of adducts have an increased risk of lung and bladder cancers.[60] However, only a fraction of smokers develop lung cancer, thus suggesting individual variability in susceptibility. A study by Wei et al.[61] found that a low DNA repair capacity correlated with increased risk of lung cancer in a population of smokers.

There is a worldwide epidemic in lung cancer that reflects the increasing amount of tobacco use in the world. Over 900,000 new cases of lung cancer are reported yearly by the World Health Organization (WHO).[6] From 1960 to the 1980s, mortality rates from lung cancer in women increased more than 100% in Japan, Norway, Poland, Sweden, and the United Kingdom; more than 200% in Australia, Denmark, and New Zealand; and over 300% in Canada and the United States.[62] A WHO survey in the late 1980s also showed that one in three males about age 15 in developed countries smoke cigarettes and one-half of such individuals smoke in developing countries. The rates for females are lower but rising rapidly. It is estimated that the incidence of lung cancer in developing countries will be the same as in developed nations in 40 years.[62] And this is just for lung cancer. When one considers the other tobacco-related cancers, the number of preventable new cancer cases could be more than double that estimated for lung cancer alone. Clearly, unless smoking habits and other lifestyle and pollution factors change, the future looks bleak for attempts to prevent cancer.

An encouraging trend, however, has been noted in the United States, United Kingdom, Australia, and a few other countries, which may encourage other nations to reduce tobacco consumption. Lung cancer rates in men under age

54 started to decline in the 1980s in the United States, and there has been a leveling off or decline of mortality rates reflecting that. In the oldest age groups, however, a steady increase in lung cancer mortality rates continues, because of a cohort of individuals who started smoking in high numbers at an early age. The numbers for women are less optimistic since they started smoking later in the twentieth century as a group than men, and the 20-year lag in lung cancer is still catching up with them. Thus, lung cancer mortality rates in women are expected to climb for a number of years to come in much of the world. Although the decline in lung cancer incidence and mortality rates in some developed countries is encouraging, there is a disturbing upward trend in smoking rates among adolescents that doesn't bode well for the future.[63]

Alcohol

Alcohol is thought to interact with smoking in the causation of certain cancers, particularly oral and esophageal cancers. Alcohol appears to be synergistic with tobacco in causing cancers of the mouth, pharynx, larynx, and esophagus, but not that of the lung.[64,65] In liver cancer, there is good evidence that alcohol consumption sufficient to cause cirrhosis of the liver increases the incidence of liver cancers, perhaps secondary to the chronic damage to the liver caused by alcohol abuse. Pure alcohol is not by itself carcinogenic in animals and may exert its carcinogenic effect secondarily to tissue damage, as in the case of hepatic cirrhosis, or by facilitating uptake of carcinogens by exposed tissues, as may be the case for oral and esophageal cancer.[51] Other potential mechanisms include (1) a carcinogenic effect of other chemicals such as N-nitrosamines in alcoholic beverages; (2) a solvent action that facilitates absorption of carcinogens found in tobacco smoke; and (3) a carcinogenic effect due to acetaldehyde, a major metabolite of ethanol. Support for the latter concept includes data from Japanese populations in which a genetic polymorphism that results in high circulating acetaldehyde levels is associated with increased cancer risk.[66]

A small positive association between alcohol and breast cancer risk has been seen in some but not all epidemiological studies.[67] In some studies, even moderate consumption (less than three drinks per week) was associated with increased risk. However, considering all the data, a clear relationship between alcohol consumption and breast cancer at such low doses is not clear. An increased risk of breast cancer has been reported in women who consume one or more drinks per day, but this increased risk is alleviated by high folate intake.

Diet

Dietary factors are postulated to account for about 30% of cancers in developed countries and about 20% in developing countries.[68] This would make diet the second leading risk factor after tobacco. However, the exact dietary components that either increase or decrease cancer risk are not clear, despite numerous case–control and prospective studies on dietary factors. The best-known link between diet and cancer is the correlation of obesity with the increased incidence of various cancers such as esophagus, colorectum, breast, endometrium, and kidney.[68] Nevertheless, the large differences in cancer rates among various countries and regions of the world suggest that diet and lifestyle are key to explaining these differences. Some dietary risk factors and dietary protective factors are listed in Table 3–7. A number of probabilities and possibilities are listed in the table because the data are for the most part speculative or inconclusive. Some of the data for various dietary components are discussed below.

Dietary fat intake has long been thought to be a factor in the causation of various cancers including breast, prostate, and colon cancer. This link has been supported by comparison of average fat intake among high-incidence and low-incidence countries of the world, particularly for breast cancer. In a large Nurses' Health Study that examined the relationship between dietary fat and breast cancer risk over a 14-year period in 3000 women, there appeared to be a weak overall increased risk due to total dietary fat.[69] However, the results of six other large prospective studies that examined the relationship between fat intake and breast cancer all showed a weak or no association. A pooled analysis of all these studies indicated no significant

Table 3–7. Dietary Risk Factors, Dietary Protective Factors, and Other Major Risk Factors for Common Cancers

Cancer	Dietary and Diet-Related Risk Factors	Dietary Protective Factors	Other Major Risk Factors
Oral cavity, pharynx and esophagus	Alcohol Very hot drinks Obesity (adenocarcinoma of the oesophagus) Chinese-style salted fish (nasopharyngeal cancer)	Probably fruit and vegetables	Smoking
Stomach	Probably high intake of salt-preserved foods and salt	Probably fruit and vegetables	Infection by *Helicobacter pylori*
Colorectum	Obesity Possible red and processed meat	Probably fruit and vegetables and other plant foods rich in fiber	Sedentary lifestyle
Liver	High alcohol intake Foods contaminated with aflatoxins	None established	Hepatitis viruses
Pancreas	None established	None established	Smoking
Larynx	Alcohol	None established	Smoking
Lung	None established	Possibly fruit & vegetables	Smoking
Breast	Obesity after menopause Alcohol	None established	Reproductive and hormonal factors
Endometrium	Obesity	None established	Low parity
Cervix	None established	None established	Human papillomavirus
Prostate	None established	None established	None established
Kidney	Obesity	None established	None established

From Key et al.,[68] with permission.

relationship between fat intake and breast cancer. Similar conclusions have been reached for colon cancer risk.[70]

Another widely held belief is that high dietary fiber, particularly from grains, reduces colon cancer risk. Again, a number of studies do not support this hypothesis.[69] Furthermore, no evidence has been found that dietary fiber is related to the occurrence of colon adenomas in either men or women. An inverse relationship, however, was found between high dietary fiber and the risk of diverticular disease of the colon as well as coronary heart disease.

High consumption of fruits and vegetables has also been thought to have beneficial effects in reducing cancer risk, according to case–control studies. In recent cohort studies the association was much weaker. In a Nurses' Health Study and Health Professionals follow-up study, no overall association between dietary intake of fruits and vegetables and colon cancer was observed.[69] Some studies, however, have found that individuals who consume very low amounts of fruits

and vegetables have the greatest risk of colorectal cancer[71] and that high consumption of cruciferous vegetables is associated with reduced risk of prostate cancer.[72] In addition, serum levels of selenium have been shown to correlate with decreased risk of esophageal and gastric cancers.[73] Selenium[74] and vitamin E[75] have been reported to decrease risk for prostate cancer.

One relationship of cancer to diet that is clearer than a lot of other data is the role of folic acid. A number of studies have indicated that the intake of folic acid reduces cancer risk (reviewed in Reference 69). For example, a 15-year follow-up of the Nurses' Health Study showed a reduction in colon cancer risk with folate supplementation. The mechanism for this effect isn't clear, but two theories are (1) there is a folate-mediated production of methyl donors (via methionine) for DNA methylation, which regulates gene expression (see Chapter 5); and (2) there is a facilitated conversion of uracil to thymine for DNA synthesis and repair. High folate intake was found to protect against some of the carcinogenic effects of alcohol consumption; protection was seen for colon cancer in men and breast cancer in women.[69]

Because there are no clear dietary factors that explain the global diversity of cancer incidence and mortality, one might then ask what lifestyle factors are involved. Environmental pollution and occupational hazards aside (because they only contribute a relatively small percentage to overall risk), there are some correlations that may explain some of this diversity. These are obesity and physical activity, or to put it another way, energy intake versus output. Studies in animals suggest that energy (caloric) restriction has a powerful influence on tumor formation. For example, a 30% restriction in caloric intake can reduce mammary tumors in rats by 80% (reviewed in Reference 69). Although an exact measurement of energy balance in humans is hard to come by, there are significant data correlating obesity and lack of physical activity as cofactors in colon cancer in men and women and in breast cancer.[69]

Sexual Development, Reproductive Patterns, and Sexual Behavior

The duration of hormonal exposure appears to play a role in the susceptibility to breast cancer in women. The carcinogenic effects of hormones were first demonstrated in animals. In 1932, Lacassagne reported the induction of mammary carcinomas in mice injected repeatedly with an ovarian extract containing estrogen. Later, he also showed that the synthetic estrogen diethylstilbestrol produced mammary tumors in susceptible strains of mice.[76] Furthermore, ovariectomized mice and rats have a decreased frequency of breast cancer, whereas rodents subjected to increased levels of estrogen, progesterone, and prolactin have an increased frequency of breast cancer, although timing of exposure to individual hormones appears to be crucial.[77,78] Similarly in humans, a role of hormones in the development of breast cancer has been deduced from the known risk factors associated with the disease. These factors include early age of menarche, delayed age of first pregnancy, and delayed menopause, suggesting longer duration of exposure to hormonal stimulation as an etiologic

agent in breast cancer. Studies in blood levels and urinary excretion patterns of hormones in patients with breast cancer or women at risk to develop breast cancer have yielded conflicting results. However, in a study of women whose mothers had bilateral breast cancer and who thus had a high familial risk of also developing the disease, it was concluded that this increased risk was associated with elevated plasma levels of prolactin, progesterone, and estrogen.[79]

In men, late descent of the testes is associated with an increased susceptibility to testicular cancer, for reasons that are not clear, although this may reflect some faulty differentiation response in the testicular tissue. This cancer usually occurs in younger men (average age at diagnosis is 32 years).

Reproductive patterns are related to cancers of the uterine endometrium, ovary, and breast, all three of which are less common in women who have had children, particularly if early in their reproductive lives, than in women who have not had children.

Cancer of the uterine cervix is associated with early and frequent sexual contact with a variety of partners. There is strong evidence for an association of a transmissible virus, human papilloma virus, with cervical cancer (see Chapter 2).

Industrial Chemicals and Occupational Cancers

The chemicals and industrial processes that have a known or suspected etiologic role in the development of cancer are listed in Table 3–6. As noted above, about 2%–5% of all cancer deaths are attributed to occupational hazards. Of those agents listed as carcinogenic for humans, a number were identified because of their close association between an abnormal clustering of certain cancers and exposure to an industrial chemical or process. For example, epidemiologic studies of workers occupationally exposed to industrial levels of 4-aminobiphenyl have a higher incidence of bladder cancer.[80] Occupational exposure to asbestos fibers results in a higher incidence of lung cancer, mesotheliomas, gastrointestinal tract cancers, and laryngeal cancers.[81] As mentioned earlier, cigarette smoking and occupational exposure to asbestos act synergistically to increase

the incidence of lung cancer. Several epidemiologic studies have shown increased frequency
of leukemia in workers exposed to benzene.[81–83]
Two studies of workers exposed to bis(chloromethyl) ether have indicated an increased risk of
lung cancer, primarily small-cell carcinoma.[81]
There is also an increased risk of lung cancer
among workers in chromium industries. Occupational exposure to 2-naphthylamine has long
been known to be associated with urinary bladder
cancer.[81] Since the time of Percival Pott and his
study of chimney sweeps (see Chapter 2), coal
soot has been known as a cause of skin cancer.
Since that time, occupational exposure to soot,
coal tar, pitch, coal fumes, and some crude shale
and cutting oils has been shown to be associated
with cancers of the skin, lung, bladder, and gastrointestinal tract. The carcinogenicity of these
latter agents is probably related to their content of
polycyclic aromatic hydrocarbons (PAH).

The highest levels of human exposure to PAH
occur in industrial processes involving the use of
coal tar and pitch and in the production of coke
from coke ovens.[84] Epidemiological data indicate that ambient coal tar and pitch in iron and
steel foundries contains carcinogenic substances
and may lead to an increased incidence of lung
cancer, particularly in smokers who work in
such environments. A study by van Schooten
et al.[84] showed that coke-oven workers were
exposed to substantial concentrations of PAH in
the air, including benzo[a]pyrene and pyrene.
Forty-seven percent had detectable levels of
PAH-DNA adducts in their white blood cells,
compared to 30% of control subjects who worked
in another part of the plant. In both groups,
smokers had significantly higher levels of PAH
adducts than nonsmokers. Since the carcinogenic mechanisms of PAH involves metabolic
activation and alteration of DNA function (see
Chapter 2), these data suggest that exposure
to various carcinogens and/or susceptibility to
their DNA-damaging effects could be monitored by measuring DNA-adduct formation in
peripheral white blood cells or perhaps in urine,
if sensitive and specific enough assays could be
developed. It should be noted that such measurements would be subject to individual variations relating to variation in daily exposure levels, genetic differences in metabolic activation

of PAH and DNA repair mechanisms, smoking
habits, amount of air pollution with PAH in the
place of residence, and amount of PAH in the
drinking water and diet.

Herbicides

Herbicides are a heterogeneous class of chemicals widely used in agriculture, forestry, and
gardening to kill undesirable weeds and foliage.
Although agricultural workers and workers in the
plants that manufacture them are exposed to
the highest concentrations, the entire population is probably exposed to some level of herbicide contamination, albeit a low level. This
could come about from residual contamination
of foodstuffs, runoff into ground water used for
drinking supplies, or airborne contamination in
areas of heavy spraying.

Herbicides that have been used commercially
include the phenoxy compounds 2,4-dichloro-
phenoxyacetic acid (2,4-D) and 2,4,5-trichloro-
phenoxyacetic acid (2,4,5-T; also known as Agent
Orange), triazines, amides, benzoics, carbamates,
trifluralin, and uracils. 2,4,5-T, and to some extent 2,4-D, preparations were contaminated
with dioxins and furans, particularly prior to
1975 when government manufacturing restric-
tions limited the amount of allowable contamination. One contaminant, 2,3,7,8-tetrachlorodi-
benzo-p-dioxin (TCDD), is a potent mutagen
and a powerful carcinogen in animal studies.
Prior to 1975, TCDD concentrations of 1 part
per million (ppm) were observed in commercial
phenoxy herbicide preparations, whereas current levels are below 0.1 ppm.[85]

A study of phenoxy herbicide applicators has
shown detectable blood levels of TCDD.[85] Because TCDD is stored in fat tissue, its half-life
in the body may be as long as 7 years. Calendar
period of exposure and intensity of use of 2,4,5-
T were determinants of serum levels of TCDD.
TCDD serum levels were also associated with
intensity of exposure to 2,4-D, but this was
confined to individuals exposed before 1975.
Based on the assumed half-life of 7 years, some
workers would have had serum TCDD levels of
up to 329 parts per trillion (ppt); a maximum
of 26 ppt has been reported in the general population. In a Vietnam veteran heavily exposed to

Agent Orange, serum levels of 1530 ppt were estimated to have been reached during the time of his peak exposure. Adipose tissue concentrations of 540 ppt have been shown to cause thyroid cancers in animals.[85]

There is considerable controversy over the long-range health effects of exposure to 2,4,5-T, but it is clear that some individuals heavily exposed to 2,4,5-T prior to 1975 would most likely have achieved body concentrations of TCDD that have been shown to be carcinogenic in some animals.[85] As yet, however, there is not conclusive evidence for TCDD carcinogenicity in humans.[86] The manufacture of 2,4,5-T has been discontinued in most Western countries and its use banned in the United States since 1983.

The association of herbicide use with cancer in humans has been reviewed.[87] Review of studies from several countries and states in the United States shows significant evidence supporting a relationship between non-Hodgkin's lymphoma in farmers and exposure to phenoxy herbicides. Several studies have also reported a relationship to increased risk of soft tissue sarcomas with exposure to phenoxy herbicides. Although also implicated in some studies to be related to increased risk of cancers of the colon, lung, nasal passages, prostate, and ovary as well as leukemia and multiple myeloma, there have been too few definitive studies to demonstrate an exposure–risk relationship. An increased risk of non-Hodgkin's lymphoma has been associated with chronic exposure to a number of pesticides including 2,4-D, mecoprop, dicamba, and malathione.[88]

Another dilemma is that experimental animal studies don't convincingly demonstrate the carcinogenicity of 2,4-D and 2,4,5-T.[89] Thus it is difficult to develop a consistent public policy on the use of such substances, particularly when most regulatory decisions are made on carcinogenicity testing in animals. Nevertheless, it seems only prudent to carefully monitor exposure to herbicides and other occupationally related chemicals and to promote minimal exposure safety practices, particularly among farmers and other workers who may experience high exposure. In the United States, the National Cancer Institute, in collaboration with the Environmental Protection Agency, is undertaking a long-term cohort study of pesticide-exposed farmers.

Air and Water Pollutants

Air, water, and soil pollution is estimated to account for only 1%–4% of all cancers. A small percentage of lung cancer (less than 5%) may be due to chronic inhalation of outdoor air pollutants such as industrial or engine exhaust chemicals. Indoor air pollutants such as secondhand smoke and radon are thought to be contributors, but this risk is most likely exaggerated (see below). In China and some other Asian countries, chronic inhalation of cooking oil smoke may be a causative agent of lung cancer.[90] The contamination of the atmosphere by chlorofluorocarbons (whose production is now banned in developed countries) in refrigerant and propellants has been implicated in destruction of the ozone layer and a resultant increase in skin cancer due to a lower filtering of UV irradiation from the sun. Occupational exposure to inhaled asbestos, such as occurred in Liberty Ship building in World War II, has been clearly linked to mesothelioma.

Regarding water pollution, high exposure to arsenic in drinking water in certain countries (e.g., Bangladesh) and areas of the United States (Alaska) and South America (Argentina, Chile) appears to be related to an increased risk of bladder and skin cancers.[90] A number of other groups of water pollutants have been investigated as possible sources of cancer risk, but the data are not conclusive, even though a popular myth is that the contents of our drinking water are causing cancer.

Evidence that potential carcinogens in the air or water might cause cancer is based on several assumptions as well as on epidemiologic data. One of the assumptions is that there is a linear, nonthreshold dose–response relationship between the given dose of carcinogen and the number of cases of cancer. This assumption is based primarily on dose–response studies in experimental animals. Such a dose–response relationship carries the implication that there is no such thing as a safe level of exposure to a carcinogen (discussed in Chapter 2). Taking the nonthreshold approach to evaluation of exposure to environmental agents is, of course, the most conservative policy; it tends to predict the largest response (i.e., the largest number of cancers) for any given level of low-dose exposure. Since the possible consequences of exposure to carcinogens in the general environment are so enormous, a number of investigators think that it is appropriate to use this approach. Although this approach seems reasonable to environmentalists, currently only limited evidence supports it. Evidence from air pollution studies, for example, indicates that estimates of cancer risk by extrapolation of dose–response relationships may be an oversimplification of the problem. Large metropolitan areas have a substantially higher level of atmospheric carcinogens, such as benzo[a]pyrene, resulting from combustion of fossil fuels, than rural areas, yet some studies[91] show that nonsmokers in urban areas do not have a significantly higher risk of lung cancer than that of rural nonsmokers. However, urban *smokers* do have a significantly higher incidence of lung cancer than comparably heavy smokers in rural areas. These observations and others, such as the potentiation of lung cancer in uranium miners[92] and asbestos workers[93,94] who smoke, support the idea that a *combination* of urban air pollution and smoking is the most carcinogenic.

Numerous potential carcinogens have been found in air and water, particularly in areas near or downstream from large industrial complexes. For example, nitrosamines, a class of chemicals that are among the most potent carcinogens known from experimental animal studies, are present in the environment, albeit usually at very low concentrations.

In addition to industrial sources, domestic sewage treatment plant effluents may contain carcinogenic substances that may find their way into drinking water supplies. More than 50 chlorinated hydrocarbons have been identified in domestic sewage effluents.[95] This same study estimated that over 1000 tons of chlorinated organic compounds are discharged by sewage treatment plants into the nation's waterways annually. Chlorinated hydrocarbons result from the chlorination of water heavily polluted with organic chemicals.[96] Some of these chlorinated compounds are known to be carcinogenic in animals.

Although discharges from industrial and municipal waste treatment plants may be continuous sources of pollution, spills resulting from industrial operations, transportation accidents, or dumping of chemical wastes on or near bodies of water can contribute significant levels of hazardous substances to public water supplies.

The Environmental Protection Agency and other groups have undertaken studies of several large metropolitan areas to evaluate the level of contamination of public drinking water supplies and to assess the carcinogenic risk associated with this contamination. In a survey of 80 cities, a number of potentially dangerous trihalomethanes, including chloroform, bromodichloromethane, dibromochloromethane, and bromoform, were detected.[97] Chloroform, a known carcinogen in animals, was found in the drinking water of 80 cities. Carbon tetrachloride, also a known carcinogen, was found in the drinking water of 10 cities. In one survey[98], 325 organic chemicals were identified in the drinking water of various cities. Only about 10% of these have been adequately tested for carcinogenicity. Among the known or suspected carcinogens identified in drinking water are benzene, bis(chloromethyl) ether, carbon tetrachloride, chloroform, dieldrin, polychlorinated biphenyls, 1,1,2-trichloroethylene, and vinyl chloride. Thus, it is evident that the general public is exposed to a wide variety of environmental chemical carcinogens. Since there is a 20- to 30-year latent period between exposure to certain carcinogenic agents and the development of clinically detectable cancer, it will probably take several decades to fully evaluate the impact of our contaminated environment. It should be pointed out, however, that the expected correlations between exposure to a given carcinogen in the drinking water and the type of cancer expected to result from such exposure have not been established. For example, even though chloroform, a hepatocarcinogen, is the predominant organic contaminant in the drinking water of certain communities in Louisiana that take their water from the Mississippi, there is no increased mortality from hepatic cancers in those communities.[99]

Nevertheless, a considerable debate over the role of environmental pollutants in human cancer continues. On the basis of studies of cancer incidence in various regions in Africa, Higginson and Oettlé[100] provided some definitive data on the impact of environmental factors in the causation of human cancer. The work of Higginson and colleagues has generally been credited with establishing the fact that about two-thirds of all human cancers have an environmental cause and thus, theoretically at least, are preventable. This has led many people to believe that the environmental agents responsible for cancer are chemicals that we inhale or ingest. However, as Higginson himself has reiterated,[101,102] what he meant by "the environment" is the total milieu in which people live, including cultural habits, diet, exposure to various infectious agents, average age of menarche, number of children a woman bears, age of menopause—in short, the cultural as well as the chemical environment.

Although we have seen that clear correlations between excess occupational exposure to carcinogenic chemicals and some cancers can be made, the contribution of these occupationally related cancers to the total incidence of cancer in industrialized nations is small. Furthermore, although urban–rural differences in cancer incidence have been reported in several countries, these differences tend to disappear when homogeneous populations with a similar lifestyle, for example, the Mormons, are studied.[103] And, although in England and Wales, certain occupations have been associated with a different risk of cancer from that of the general population, nearly 90% of such variation is eliminated if individual groups of similar social class and habits are compared.[104] Other inconsistencies also occur. For example, bladder cancer is linked to certain chemical and allied industries in United States, but no clear industrial association has been found in Japan. Prostate cancer is higher in blacks than in whites living in the same counties in the United States, and both black and white males in the United States have a higher incidence of prostate cancer than men in the industrialized United Kingdom and Japan. However, the differences in incidence between the United States and the United Kingdom may be mainly due to PSA screening because PSA screening is not widely used in the United Kingdom. Interestingly, the mortality rates are similar between the two countries.[32]

Thus, the overall distribution patterns of cancer observed in North America and Western Europe, with high frequencies of lung, colon, breast, and uterine cancer, suggest some common factors in the environment of these regions in comparison with regions in Africa in which there is a much lower incidence of these malignant diseases. At present, however, it is unjustified to link these differences in incidence directly to recent food additives or chemical pollutants. Lifestyle differences appear to play a large role in the causation of these and other cancers.[101] For example, the varying incidence of cancer of the breast, ovary, and uterus can be related at least partly to differences in average age at onset of menarche, sexual behavior, and reproductive patterns among different population groups. Taken together, all the data accumulated to date suggest that cancer distribution patterns represent a variety of differences in lifestyle, with exposure to chemical pollutants in the ambient environment of industrialized societies contributing to some but an as-yet unclear percent of the total number of cancer deaths.

Radiation

Ultraviolet

It has been known for a long time that exposure to UV or ionizing irradiation can cause cancer in humans. The association between skin cancer and exposure to sunlight was observed more than 100 years ago, and in 1907 William Dubreuilh, a French dermatologist, reported epidemiologic evidence implicating sunlight as a cause of skin cancer, supporting the earlier observation.[105] In 1928, George Findlay, a British pathologist, experimentally verified this by inducing skin cancer in mice exposed to ultraviolet radiation.[106]

Ultraviolet radiation is a low-energy emission and does not penetrate deeply. Hence the skin absorbs most of the radiation and is the primary carcinogenic target. Because nonmelanomatous skin cancer is the most easily detectable and curable human cancer, the fact that it is also the most clearly identifiable is often overlooked. The fear of skin cancer, however, is apparently not sufficient to prevent people from overexposing themselves to this carcinogenic agent.

The evidence for the association of skin cancer and UV radiation is compelling and can be summarized as follows:[107]

1. Skin cancer occurs primarily on exposed areas, that is, the head, neck, arms, hands, and legs (in women).

2. Skin cancer is relatively rare in dark-skinned races in whom skin pigment filters out UV radiation, whereas it is common in fair-skinned people.
3. The incidence of skin cancer and the amount of exposure to sunlight are related.
4. Skin cancer frequency and the intensity of solar radiation are related. Going toward the equator, the prevalence of skin cancer in Caucasians increases in proportion to the intensity of UV radiation.
5. Skin cancer can be induced in laboratory animals by repeated exposure to UV radiation.
6. The inability to repair DNA damaged by UV radiation is associated with skin cancer. Thus, individuals with xeroderma pigmentosum, an inherited disease with a DNA-repair defect, almost always develop skin cancer.

Both malignant melanoma and nonmelanoma skin cancers are associated with exposure to UV radiation, although the dose–response curve is less steep for melanoma.[107] The most common types of skin cancer are basal cell carcinoma, which may be locally invasive but is almost never metastatic; squamous cell carcinoma, which is more aggressive than basal cell carcinoma, invades locally, and may rarely metastasize; and melanoma, which is less common than the other forms of skin cancer but is often highly malignant and rapidly metastatic, with an average 5-year survival rate of over 90% if detected and treated early but only 14% if metastatic.

Ionizing Radiation

The carcinogenic effects of ionizing radiation were discovered from studies of pioneer radiation workers who were occupationally exposed, individuals who were exposed to diagnostic or therapeutic radiation, and atomic bomb survivors. Malignant epitheliomas of the skin were observed in the earliest experimenters with X-rays and radium within a few years after their discovery in 1895 and 1898, respectively. By 1914, a total of 104 case reports of radiation cancers had been noted and analyzed.[108] In 1944, the role of ionizing radiation in the in-creased incidence of leukemia among radiologists was recognized.[109]

In more recent times, follow-up studies of atomic bomb survivors and people exposed to the aftereffects of the Chernobyl disaster have shown that exposure to radioactive fallout is linked to an increased susceptibility for thyroid cancer, breast cancer, and leukemia. Both the age at exposure and the dose of radiation received are key factors in determining risk of cancer.

The types of neoplasms produced in individuals exposed to ionizing radiation depend on a number of factors, including dose of radiation, age at time of exposure, and sex of the individual. Within 25 to 30 years after whole-body or trunk irradiation, there is an increased incidence of leukemia and cancers of the breast, thyroid, lung, stomach, salivary glands, other gastrointestinal organs, and lymphoid tissues. Other malignant neoplasms have been observed in tissues that were locally exposed to high doses of radiation. A number of unfortunate but striking examples are available. During the 1920s, in a factory in Orange, New Jersey, watch dials were painted with radium and mesothorium to make them luminescent. To get a fine tip on the brushes used to paint the dials, the workers wetted the brush tips on their tongues, leaving a deposit of radioactive material on the tongue. Approximately 800 young women were exposed to the radioactive materials in this manner. Radium is a radioisotope that becomes deposited in bone, and several years later, a high incidence of osteogenic sarcoma became evident in these workers.[110] Another radioisotope that becomes deposited in internal organs is thorium, which was used in the preparation of a radiocontrast solution called Thorotrast, once used for diagnostic purposes. The overall incidence of malignant diseases in patients who received this material has been found to be twice the expected incidence, with liver tumors and leukemias being about sixfold higher than expected.[111] Another example of local irradiation producing cancer a number of years later is the observation of thyroid cancers in individuals who had been irradiated over the neck during childhood for either a so-called enlarged thymus gland or hypertrophied tonsils and adenoids. An 83-fold increased risk for thyroid cancer has been noted in these cases.[112]

The period between irradiation and the appearance of cancer depends to some extent on the age at irradiation.[113] Juvenile tumors and leukemias associated with prenatal irradiation become evident in the first 2 to 3 years after birth, with a peak incidence at 5 years of age. The latency period following postnatal irradiation, however, is 5 to 10 years for leukemia and more than 20 years for most solid tumors. The increased incidence of leukemia and solid tumors also appears to be higher after prenatal than after postnatal irradiation. In general, the data suggest that the relative risk for cancers other than leukemia decreases with increasing age at the time of irradiation.

In the case of breast cancer, for example, women who received a radiation dose before age 40 have increased risk of developing breast cancer.[114] After age 40, radiation has a small effect on breast cancer risk. Women who were below age 10 at the time of exposure have an increased risk that does not become apparent until these women reach the age at which breast cancer usually occurs.[114] This increased risk persists for at least 35 years and may remain throughout life. Current evidence indicates that a very small number, probably less than 1%, of breast cancer cases result from diagnostic radiography.[114]

Radon

Radon is a radioactive gas that is ubiquitous in earth's atmosphere. It is formed from the radioactive decay of radium-236. Radium is found in substantial but varying amounts in soil and rocks and ends up in some building materials. Various parts of the country have varying amounts, as do certain localities within a small geographic area. There is extensive epidemiologic evidence that exposure to high levels of radon produces bronchogenic carcinoma (reviewed in Reference 115), most of which comes from studies of workers involved in deep mining of uranium and other ores. Because this epidemiologic evidence is quite compelling and because radon is so widespread, the potential that large numbers of people might develop lung cancer from such exposure has produced a radon scare in the United States, not unlike the asbestos scare.

An increased incidence of lung cancer in deep-well miners was observed in uranium and other ore miners in eastern Germany and western Czechoslovakia over 60 years ago. Exposure to radon among these miners was very high, approaching 3000 picocuries (pCi) per liter of air. In the early 1950s, an increase in lung cancer incidence was noted among uranium miners in Colorado. Later, an increased rate of lung cancer was also noted among miners working in iron, zinc, tin, and fluorspar mines. In these mines, radon levels were also high. Although these miners were also exposed to other potentially carcinogenic dusts, the common feature was exposure to radon. The excess number of lung cancer deaths in these miners (compared to nonminers) ranges from 0.3% to 13% and varies depending on the ambient air concentration.[115] This risk goes up in more than an additive manner for individuals who are also smokers.

Monitoring of homes began in a rather haphazard fashion in the 1980s. Nevertheless, some regions with high indoor levels were found, including the Reading Prong geological region extending from Pennsylvania to New York. Estimates of risk for lung cancer from radon exposure in residences is based on extrapolations from miner risk data. These estimates may or may not be realistic, as there is some evidence both ways. In a case–control study of 400 women with lung cancer, performed by the Department of Health in New Jersey, an increased risk was found for exposure levels of 2 pCi/L of air, but the results were not statistically significant.[116] A study done in China, in which median household radon levels ranged from 2.3 to 4 pCi/L of room air, no positive associations between radon levels and lung cancer was found,[117] a finding suggesting that projections of lung cancer risk from surveys of miners exposed to high-radon levels are overestimates.

Estimates of increased risk due to radon residential exposure very widely. For an average lifetime exposure to 1 pCi/L, estimates vary from 5000 to 20,000 excess lung cancer deaths per year in the United States.[115] These estimates uphold the conservative tradition of radiation protection. To put this in some perspective, a lifetime exposure to 4 pCi/L is estimated to cause a 1% increase in lung cancer, whereas the risk of smoking cigarettes increases the risk of lung cancer at least 10-fold over that of nonsmokers. Most homes in the United States have indoor radon levels less than 2 pCi/L and these are levels usually found in basements.[118]

Drugs

Drugs that have been associated with human cancers are listed in Table 3–8. Some of these have been used as therapeutic or diagnostic agents in medical practice. Among these, the anticancer drugs in particular have been implicated. A number of these (e.g., cyclophosphamide, melphalan, and busulfan) are alkylating agents, known to interact with DNA in a manner similar to that of known chemical mutagens and carcinogens (see Chapter 2).

Second cancers arising later in life from the effects of treatment of childhood cancers are a particular concern. It is estimated that from 3% to 12% of children treated for cancer will develop a new cancer within 20 years from the time of first diagnosis.[120] This is a 10-fold higher risk than that for age-matched controls. Exposure to therapeutic radiation and anticancer drugs such as alkylating agents nitrogen mustard, cyclophosphamide, procarbazine, and nitrosoureas are known risk factors for second cancers. About 25% of those patients who develop second cancers are known to have some genetic susceptibility such as Li-Fraumeni syndrome, retinoblastoma, neurofibromatosis, or a sibling with cancer[120] (see below). The most common malignant familial condition predisposing to second neoplasm is retinoblastoma, in which osteosarcomas and soft tissue sarcoma are most common.

The risk of second cancers in children surviving ALL, however, is relatively low. In a study of 9720 children treated for ALL from 1972 to 1988, 43 second cancers were seen.[121] Second cancers may occur within 5 years from first diagnosis. These are usually acute myelogenous leukemia (AML), CML, or non-Hodgkin's lymphoma. Secondary solid tumors, the most common of which are brain tumors, may be seen 5 to 15 years later. Brain tumors are most often seen

Table 3–8. Medicinal Drugs Classified as Carcinogenic to Humans

Drug or Drug Combination IARC Group I	Cancer Site and Cancer Type
Analgesic mixtures containing phenacetin	Kidney, bladder
Azathioprine	Lymphoma, skin, liver and bile ducts, soft connective tissue
N,N-bis(2-chloroethyl)-2-naphthylamine (Chlornaphazine)	Bladder
1,4-Butanediol dimethane-sulfonate (Myleran; Busulfan)	Leukemia
Chlorambucil	Leukemia
1-(2-Chloroethyl)-3-(4methyl-cyclohexyl)-1-nitrosourea (Methyl-CCNU)	Leukemia
Ciclosporin	Lymphoma, Kaposi's sarcoma
Cyclophosphamide	Leukemia, bladder
Diethylstilbestrol	Cervix, vagina
Etoposide in combination with cisplatin and bleomycin	Leukemia
Fowler's solution (inorganic arsenic)	Skin
Melphalan	Leukemia
8-Methoxypsoralen (Methoxsalen) plus ultraviolet radiation	Skin
MOPP and other combined (anticancer) chemotherapy including alkylating agents	Leukemia
Estrogen therapy, postmenopausal	Breast, uterus
Estrogen, non-steroidal	Cervix, vagina
Estrogens, steroidal	Uterus, breast
Oral contraceptives, combined°	Liver
Oral contraceptives, sequential	Uterus
Tamoxifen†	Uterus
Thiotepa	Leukemia
Treosulfan	Leukemia

°There is also conclusive evidence that these agents have a protective effect against cancers of the ovary and endometrium.

†There is conclusive evidence that tamoxifen has a protective effect against second breast tumors in patients with breast cancer.

(From World Cancer Report,[119] with permission.)

in patients who received cranial irradiation. Most other secondary solid cancers in this patient population also appear to result from irradiation. Of those not treated with irradiation, most will have received alkylating agents and a number appear to have some genetic predisposition. A recent study has shown a 1.4-fold increased risk of breast cancer following radiotherapy and alkylating agent chemotherapy in young women (age 30 years or younger).[122] Somewhat surprisingly, patients treated with alkylating agents alone had a lower risk of breast cancer compared to case–controls, presumably because of ovarian damage that inhibits estrogen production.

A higher incidence of lymphomas has been seen in patients who have received organ transplants for which they were treated with immunosuppressive drugs, some of which are also used in cancer chemotherapy. The most widely used immunosuppressive drug, cyclosporine, which has revolutionized the organ transplant field, has also been noted to cause lymphoproliferative disease, including lymphomas. This was particularly a problem in the earlier clinical trials when higher doses were being employed in combination with high doses of corticosteroids and antithymocyte globulin.[123] More recently, the incidence of lymphoproliferative disease has been reduced to about 1% in transplant patients treated with lower doses of cyclosporine.

Other drugs are also suspected of causing cancer in humans. For example, 10 cases of liver cell tumors have been reported in patients with blood disorders treated for long periods with the androgenic steroid oxymetholone.[124] Several studies indicate that chronic abuse of analgesics containing phenacetin leads to papillary necrosis of the kidney. It has been suggested that this is related to the subsequent development of transitional cell carcinoma of the renal pelvis in a number of these cases.[124]

Hormones

As noted earlier, the risk factors associated with breast cancer include age at menarche, age at the time of the first full-term pregnancy, and age at menopause. These factors suggest a role for estrogens and progesterone in breast cancer. Production of these hormones increases near menarche and starts to decrease in the perimenopausal period. Prolactin levels have been reported to decrease in women after full-term pregnancy[125] and this may provide some protective effect. Some studies indicate that the rate of cell proliferation is greater in nulliparous women than in parous women (reviewed in Reference 125), and this may reflect the lower hormonal levels in the latter group. Moreover, mitotic activity of breast epithelium varies during the menstrual cycle and peaks during the luteal phase, suggesting that progesterone also has a role in regulating the mitotic rate in breast tissue. Presumably, these hormones could act as promoters for cells initiated by some carcinogens, and the amount of duration of exposure to these hormones could then increase the risk in a woman who has a propensity to develop breast cancer. That other risk factors are also involved is evident from comparative data from United States and Japanese women. In these two groups, data on the first birth, nulliparity, and age at menopause show that the lower rate in Japanese women is not accounted for by these factors.[125] The remaining difference may be related to dietary fat and total body weight, both of which are, on the average, higher in United States women. The breast cancer risk associated with body weight is thought to operate through the increased levels of conversion of adrenal androgens to estrogen and lower levels of sex-hormone binding globulin in obese women.[126]

Estrogens have been used extensively in the treatment of postmenopausal symptoms and for the prevention of osteoporosis. There is a clear association between use of "unopposed" estrogen therapy (i.e., without progestins) and increased risk of endometrial cancer. Combination of lower doses of estrogen and a progestin protects against the estrogen-alone effect on the endometrium. The problem is that estrogen–progestin combined hormone replacement therapy (CHRT) enhances breast cancer risk.[127,128] In a case–control study of over 2500 postmenopausal women, CHRT was associated with a 10% higher risk of breast cancer for each 5 years of use.[127]

This risk association has been confirmed by "a Million Woman" study in the United Kingdom that reported a 1.74-fold increase in breast cancer in current HRT users who had taken hormones for 1–4 years and a 2-fold increase risk in 5- to 9-year current users.[129] Curiously, no increased risk was observed in women who had used HRT in the past, even if use was for greater than 10 years. Overall, current users were at a 1.66-fold increased risk of developing breast cancer and 1.22-fold increased risk of dying from it. To put this in some context one should consider the fact that the risk of developing breast cancer for a 50-year-old woman who has not had breast cancer is 11% in her lifetime.[130] Thus, the increased risk for a post-menopausal woman taking CHRT is 10% of 11%, or 1.1%, for each additional 5 years of her life if she continues to take CHRT. If she lives to be 80, the increased risk is 6.6% (age 50 to 80 = six 5-year periods; $6 \times 1.1\% = 6.6\%$).

Many epidemiologic studies have shown that oral contraceptives do not significantly affect the risk of breast cancer. However, a small increase has been reported in some studies in certain groups of women. These include women who have used contraceptives for several years before age 25 and/or before the first full-term pregnancy; women who continue to use oral contraceptives at age 45 and older; women with a history of benign breast disorders; multiparous, premeno-pausal women with early menarche; and women with a family history of breast cancer (reviewed in Reference 114).

Several studies have reported that use of diethylstilbestrol (DES) during pregnancy is associated with an overall risk of about 1.5-fold for developing breast cancer.[114] The well-documented appearance of vaginal adenocarci-nomas in women whose mothers had been treated with DES in early pregnancy with the intent of preventing abortion is another example of hormonally induced neoplasm.[131]

Infection

Cancer is not an infectious disease in the usual sense of the term. Doctors, nurses, and spouses who come into close contact with cancer patients do not have a higher risk of developing cancer than the rest of the population.[51] However, there is now known to be a clear association between infection with certain types of viruses and neoplastic disease. Infection with certain viruses probably acts in concert with other carcinogenic agents or processes. This is discussed in detail in Chapter 2. Suffice it to say

here that an association between infection with Epstein-Barr virus and Burkitt's lymphoma, hepatitis B and C viruses and liver cancer, human T-cell lymphotropic virus (HTLV) and leukemia, and human papilloma virus and cervical cancer are examples of this linkage. A number of neoplasms have also been associated with HIV infection in patients with AIDS. These include Kaposi's sarcoma and non-Hodgkin's lymphoma primarily, but central nervous system tumors and Hodgkin's disease are also seen in patients with AIDS.[132]

Infection with certain parasites also seems to be able to initiate a cascade of events culminating in malignant neoplastic disease in certain populations. The high incidence of bladder cancer in patients whose urinary bladders are infected with the shistosome parasite indigenous to Egypt and other parts of Africa as well as the occurrence of a type of liver cancer (cholangio-sarcoma) in patients with clonorchiasis, a parasitic infection of the liver, common in parts of China, are examples of this association. Infection with the bacterium *Helicobacter pylori* (*H. pylori*) is associated with gastric cancer, but other agents in the diet are likely to be cofactors.

AGING AND CANCER

Cancer is a disease of aging. The average age at diagnosis is 67 and the median age of patients with cancer in the United States is 70 years.[133] The incidence of cancer rises exponentially with age from ages 40 to 80 (Fig. 3–10). This age-related increase in cancer probably relates to the combined effects of accumulated genetic alterations (mutations, translocations, etc.), increased epigenetic gene silencing, telomere dysfunction, and altered tissue stroma as tissues age. There is evidence for each of these factors playing a role (reviewed in Reference 134).

Increased somatic mutations have been observed in aged cells and tissues from humans and mice. This process most likely occurs as a result of accumulated DNA damage due to exogenous and endogenous agents such as oxygen free radical–forming agents (see Chapter 2). Such mutations can also result from error-prone repair during DNA replication. Whether decreasing DNA repair capacity with aging is the

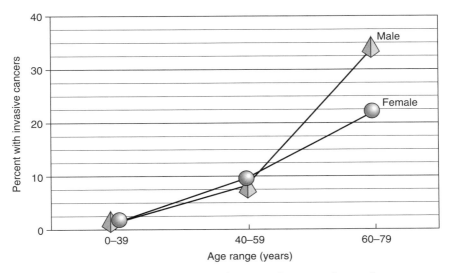

Figure 3–10. Cancer incidence as a function of age. Incidence of invasive cancer plotted against age ranges reveals exponential increase from age 40 to 80 years.[1] Note that beyond age 80, incidence of cancers plateaus. (From De Pinho,[134] reprinted by permission from Macmillan Publishers Ltd.)

culprit is not clear. However, in cells from aged mice and humans, an increased level of chromosomal abnormalities has been observed. In addition, an age-related decline in repair of UV-induced DNA damage has been found in cultured primary skin fibroblasts and lymphoblastoid cell lines, when comparing normal donors up to 10 years of age with normal donors in their 80s or 90s (reviewed in Reference 134). There is also an age-associated decrease in cellular levels of proteins involved in DNA repair, such as ERCC3 for excision repair, replication protein A, and p53. However, other investigators have not observed significant differences in DNA repair machinery of human keratinocytes in response to UV damage, so age-dependent changes in DNA repair capability is apparently not a sine qua non of the aging process. Germline defects in mismatch repair genes such as MSH6, by contrast, are associated with late-onset colon cancer (see Genetic Factors, below). Thus a variety of mechanisms may be involved in the deterioration of genome maintenance in the elderly. Epigenetic mechanisms may also be involved in the age-related increase in cancer incidence. These mechanisms involve DNA methylation and histone deacetylation reactions that regulate chromatin structure and gene

transcription (see Chapter 5). DNA methylation and histone deacetylation are involved in silencing genes, and if tumor suppressor genes are targets for these reactions one can predict the consequences. There is evidence for an age-related progressive increase in CpG island DNA methylation, a finding leading to the concept that this methylation, which often involves promoter sequences in tumor suppressor genes, is responsible for the age-related alterations leading to cancer.

Telomere dysfunction has also been implicated in the increased incidence of cancer among the elderly. Telomeres are shorter in some human cancer cells than in normal cells in the tissue of origin, which suggests that telomere shortening occurs during part of the carcinogenic process.[134] Telomerase, the enzyme that maintains telomere length (see Chapter 5), is then often reactivated. This reactivation is part of what provides the "immortality" of cancer cells. With the accumulation of somatic mutations over time, loss of cell cycle checkpoint controls, and activated telomerase, transformed cells are well on their way to progression to a dysregulated, genetically unstable population of cells.

It is now well known that the mesenchymal stroma, on which epithelial cells grow, divide, and differentiate, plays a key role in maintaining a normal tissue (see Chapter 4). It is also clear that the "malignant stroma" has a lot to do with fostering the carcinogenic process. For example, only prostate cancer–associated fibroblasts were able to sustain the growth of malignant prostatic cells in vivo, whereas fibroblasts from nonmalignant prostate tissue were not.[135] Some of the changes in cancer stroma are similar to some age-related changes in senescent dermal fibroblasts—i.e., their increased production of cytokines, proteases, and other extracellular matrix–degrading enzymes.[136]

An interesting model for the aging process is the roundworm *C. elegans*. This organism lives for only a few weeks, thus changes in gene expression and protein levels are telescoped over a short time, allowing for correlation with the aging process. It has been found that the insulin and insulin-like growth factor (IGF-1) pathways regulate the aging process by up-regulating longevity-favoring genes, such as antioxidant enzymes, stress-response proteins involved in protein folding, and antimicrobial genes, and down-regulating specific life-shortening genes in *C. elegans*.[137] These same pathways appear to affect life span in fruit flies and mice as well, and probably in humans.

DNA microarray technology has been used to measure mRNA levels in dividing fibroblasts isolated from young, middle-aged, and elderly normal humans and patients with progeria, an inherited genetic disorder characterized by premature aging.[136] Genes whose lower expression correlated with aging include genes involved in cell cycle checkpoints, chromosomal segregation, chromatin structure, and proteasome function. Ly et al.[136] propose "that the underlying mechanism of the aging process involves increasing errors in the mitotic machinery of dividing cells . . . and that this dysfunction leads to chromosomal pathologies that result in misregulation of genes involved in the aging process." Of all the factors that have been associated with increasing longevity in animals, caloric restriction is the best documented. One reason why caloric restriction appears to be a key to longer survival is that it enhances gene expression linked to suppression of DNA damage caused by mitotic recombination.[138]

Damage due to oxidation appears to underlie much of the age-associated effects involved in an increased risk of cancer. This phenomenon may be due to formation of reactive oxygen species that cause DNA strand breaks or base changes in DNA and also to protein oxidation. The oxygen free-radical hypothesis of aging postulates that the progressive decline in functional activity and chromosomal integrity with age results from the accumulation of oxidative damage by reactive oxygen species produced by cells' normal metabolic activity. There is support for this hypothesis in that steady-state levels of carbonylated proteins, produced by protein oxidation, increase with age.[139] These oxidized proteins are then targeted for intracellular proteolysis. It is also possible that the reverse is true, that transcriptional (e.g., frameshift or base change mutations) or translational errors produce misfolded proteins that are then oxidized, and it is the oxidation reaction that targets proteins for degradation. Indeed, the age-related cellular lesion in the latter case would be the accumulation of misfolded proteins, which then become oxidized.[139] Thus, it may be aged cells' inability to fold proteins correctly, due to transcriptional or translational errors (as a result of chromosomal changes over time), rather than an age-related decrease in oxidative defense systems that is the central defect. In any case, antioxidants have been reported to decrease the aging process and extend life span in some organisms. For example, exposure of *C. elegans* to small molecule synthetic superoxide dismutase and catalase mimetics increased life span by 44%.[140] Superoxide dismutase and catalase are cellular enzymes that protect many cell types, including mammalian cells, from oxidative damage by neutralizing reactive oxygen species, and the ability to pharmacologically enhance these activities would seem to be a good target for anti-aging drugs.

GENETIC FACTORS IN CANCER

A number of inherited traits are related to causation of cancer. A few cancers have a definite inheritance, whereas others may arise in individuals with a genetic defect that makes them more susceptible to potentially carcinogenic agents. There are really two different aspects to the genetics and cancer issue. First, the initiation and promotion–progression events that occur in the body over time are due to changes in the structure and function of the genome in adult cells (or in the case of pediatric cancers, children's cells). These are called *somatic mutations* and usually involve activation of oncogenes and inactivation of tumor suppressor genes. These changes accumulate over time and may be progression related, as described in the "Vogelgram" for colorectal cancer, discussed in Chapter 5. The second type of genetic basis for cancer is inherited defects. These are called *germline mutations* and they increase cancer suspectibility, usually by some interaction with the environment.

Inherited Cancers

A list of inherited cancer syndromes is shown in Table 3–9. These neoplasms represent a small fraction, perhaps 1% to 2%, of total cancers.[142] A high percentage of certain tumors, however, are genetically determined. For example, dominant genetic inheritance accounts for about 40% of retinoblastomas and 20% to 40% of Wilms' tumors (embryonal renal tumors) and neuroblastomas. Familial multiple polyposis of the large bowel is another example of a disease that is transmitted as a Mendelian-dominant trait, with about an 80% penetrance rate. Cancer of the large bowel will eventually occur in nearly 100% of untreated patients with familial multiple polyposis. There is also a predisposition to develop a variety of other neoplasms, particularly subcutaneous tumors and osteomas, in these latter patients.

The probability that an individual carrying the retinoblastoma gene will develop a tumor is about 95% and an average of three to four tumors occur in such a gene carrier. A child born without the gene has only 1 chance in 30,000 of developing retinoblastoma. This amounts to an about 100,000-fold increased risk in the gene carrier group (assuming an average of three tumors per gene carrier with an incidence of 95 per 100, compared to 1 tumor per 30,000 in the general population). Although the presence of the retinoblastoma gene virtually ensures that the carrier will develop such a tumor, oncogenesis at the cellular level must be a rare event because only three to four tumors, on average, develop in a retinal cell population of several million (this assumes that the cancer arises from the progeny of one or a small number of precursor cells, which appears to be the case for most types of cancer; see Chapter 2). Thus, the genetically dominant inherited mutation is not in itself sufficient to ensure that a retinal cell bearing the gene will become a cancer cell.

Approximately 50 forms of hereditary cancers have been reported, some of which are listed in Table 3–9. For example, for years breast cancer has been considered to have a familial association. Similar associations have been noted for ovarian cancer. Genetic studies have also associated occurrence of breast cancer with a variety of other tumors in the same families,[143] including associations between breast cancer and ovarian cancer, gastrointestinal tract cancer, soft tissue sarcoma, brain tumors, or leukemia. In a study of 12 pedigrees that had a clustering of breast and ovarian cancer among female relatives, the data suggested that a genetic factor was transmitted from affected mothers to half of their daughters and, in some families, father-to-daughter transmission appeared to occur. These observations suggest the possibility of X-chromosome linkage in the transmission of breast and ovarian cancer in these families. In a number of these family-associated cancers, specific chromosomal abnormalities or genetic mutations have been observed. These will be discussed in detail in Chapter 5. The following are few examples.

Table 3–10 lists some of the high-risk cancer susceptibility genes, their chromosomal location, and the associated cancers. Some of these genes are involved in genome integrity (*brca1* and *brca2*), some are cell cycle regulator genes (*p16* and *CDK4*), and some are DNA mismatch repair genes (e.g., *hMLH1*, *hMSH2*, and *hMSH6*). These genetic mutations usually demonstrate incomplete pentrance. For example, a woman carrying a mutated *brca1* gene has about a 70%-80% lifetime risk for developing breast cancer. It is clear that other cofactors are involved in this risk. The so-called high-risk susceptibility genes may also be involved in sporadic cancers for which no

Table 3–9. Inherited Cancer Syndromes Caused by a Single Genetic Defect°

Syndrome	Gene	Location	Cancer Site and Cancer Type
Familial retinoblastoma	*RB1*	13q14	Retinoblastoma, osteosarcoma
Multiple endocrine neoplasia II	*RET*	10q11	Medullary thyroid carcinoma, pheochromocytoma
Multiple endocrine neoplasia I	*MEN1*	11q13	Adrenal, pancreatic islet cells
Neurofibromatosis type I	*NF1*	17q11	Neurofibromas, optic gliomas, pheochromocytoma
Neurofibromatosis type II	*NF2*	22q2	Bilateral acoustic neuromas, meningiomas, cerebral astrocytomas
Bloom syndrome	*BLM*	15q26	Leukemia, lymphoma
Familial adenomatous polyposis	*APC*	5q21	Colorectal, thyroid
Von Hippel-Lindau	*VHL*	3p25	Renal cell carcinoma, pheochromocytoma
Familial Wilm's tumor	*WT1*	11q	Wilms tumor (kidney)
Xeroderma pigmentosum	*XP(A–D)*	9q,3p,19q,15p	Basal cell carcinoma, squamous cell carcinoma, melanoma (skin)
Fanconi anemia	*FAC*	16q, 9q, 3p	Acute leukemia
Li-Fraumeni syndrome	*p53*	17p13	Breast and andrenocortical carcinomas, bone and soft-tissue sarcomas, brain tumors, leukemia
Cowden syndrome	*PTEN*	10q22	Breast, thyroid
Gorlin syndrome	*PTCH*	9q31	Basal cell carcinoma
X-linked proliferative disorder	*XLP*	Xq25	Lymphoma
Peutz-Jeghers syndrome	*LKB1*	19p	Breast, colon
Ataxi telangiectasia	*ATM*	11q22	Leukemia, lymphoma

°The lifetime risk of cancer is high. There are usually recognizable phenotypic features that make the syndromes easy to identify clinically.
(From World Cancer Report,[141] with permission)

clear gene association has been found. The inherited cancer *syndrome* genes usually have a very high degree of penetrance and are relatively rare (e.g., the incidence of *p53* gene mutations involved in the Li-Fraumeni syndrome is 1 in 10,000 individuals).[141] The inherited *susceptibility* gene mutations are more common and are seen in common types of cancer and, as noted, in sporadic cancers. However, the distinction between the rarer inherited cancer syndromes and those mutations found in the more common cancers is somewhat arbitrary. For example, *rb1*, *apc*, *p53*, and *PTEN* mutations are involved in both inherited and sporadic cancers.

Table 3–10. High-risk Susceptibility Genes and Their Chromosomal Location°

Gene	Location	Associated Tumors
BRCA1	17q	Breast, ovary, colon, prostate
BRCA2	13q	Breast, ovary, pancreas, prostate
p16 INK4A	9p	Melanoma, pancreas
CDK4	6q	Melanoma, other tumors (rarely)
hMLH1	3p	Colorectal, endometrial, ovarian cancer
hMSH2	2p	Colorectal, endometrial, ovarian cancer
hMSH6	2p	Colorectal, endometrial, ovarian cancer
PMS1	2q	Colorectal cancer, other tumors (rarely)
PMS2	7p	Colorectal cancer, other tumors (rarely)
HPC2	17p	Prostate (rarely)

°Inherited mutations in these genes are associated with some common cancers.
(From World Cancer Report,[141] with permission)

Gene–Environment Interactions

Another way in which inherited susceptibility to cancer may be expressed is the way in which an individual can handle carcinogenic insults from the environment. For example, some individuals have a reduced capacity to metabolize carcinogens such as arylamines because of a slow acetylator phenotype, related to polymorphisms in the N-acetyltransferase-2 gene. Others may have a decreased ability to detoxify a number of carcinogenic agents due to polymorphisms in the glutathione S-transferase gene *GSTM*-1 or cytochrome P-450 genes *CYP2A6* or *CYP2D6*. Genes that regulate metabolism of drugs and other xenobiotics are often discussed under the heading of *pharmacogenetics*. A number of genetic polymorphisms that are related to human pharmacogenetic disorders are listed in Table 3–11. Evidence from a study of monozyotic

Table 3–11. Classification of Some Human Pharmacogenetic Disorders

LESS ENZYME OR DEFECTIVE PROTEIN

Succinylcholine apnea
Acetylation polymorphism
  Isoniazid-induced neurotoxicity
  Drug-induced lupus erythematosus
  Phenytoin-isoniazid interaction
  Isoniazid-induced hepatitis
  Arylamine-induced bladder cancer
Increased susceptibility to drug-induced hemolysis
  Glucose-6-phosphate dehydrogenase deficiency
  Other defects in glutathione formation or use
  Hemoglobinopathies
Hereditary methemoglobinemia
Hypoxanthine-guanine phosphoribosymtransferase-
  (HPRT)-deficiency
P450 mono-oxygenase polymorphisms
  Debrisoquine 4-hydroxylase deficiency
  Vitamin D-dependent rickets type I
  C21-Hydroxylase polymorphism
Enzymes of methyl conjugation
Hyperbilirubinemia
  Crigler-Naijar syndrome type II
  Gilbert's disease
Fish-odor syndrome

INCREASED RESISTANCE TO DRUGS

Inability to taste phenyithiourea
Coumarin resistance
Possibility of (or proven) defective receptor
  Steroid hormone resistance
  Cystic fibrosis

Trisomy 21
Dysautonomia
Leprechaunism
Defective absorption
  Juvenile pernicious anemia
  Folate absorption-conversion
Increased metabolism
  Succinylcholine resistance
  Atypical liver alcohol dehydrogenase
  Atypical aldehyde dehydrogenase

CHANGE IN DRUG RESPONSE DUE
TO ENZYME INDUCTION

The porphyrias
The *Ah* locus

ABNORMAL DRUG DISTRIBUTION

Thyroxine (hyperthyroidism or hypothyroidism)
Iron (hemochromatosis)
Copper (Wilson's disease)

DISORDERS OF UNKNOWN ETIOLOGY

Corticosteroid-induced glaucoma
Malignant hyperthermia associated with general
  anesthesia
Halothane-induced hepatitis
Chloramphenicol-induced aplastic anemia
Phenytoin-induced gingival hyperplasia
Thromboembolic complications caused by anovulatory
  agents

From Nebert and Weber[144]

(MZ) and dizygotec (DZ) twins has shown that susceptibility to carcinogens and mutagens is highly heritalde. Mutagen sensitivity was measured by exposing peripheral blood lymphocytes to mutagens in vitro and the correlation coefficients were all significantly higher in MZ than in DZ twins.[144a]

AVOIDABILITY OF CANCER

If, as a number of cancer epidemiologists contend, lifestyle accounts for about 80% of all malignant cancers, then presumably the same proportion of cancers should be avoidable. To be more specific, about 30% of all cancers are thought to be related to smoking, 3% to alcohol consumption, 30% to diet, 7% to sexual and reproductive patterns, and another 5% to occupational hazards and industrial products.[50,51] Moreover, about 1% are estimated to be related to drugs and medical procedures (primarily X-rays)

and 3% to geophysical factors (mostly exposure to sunlight). Thus, about 84% of all cancers should be avoidable, if these estimates are correct. The "ideal" man, then, should not smoke or drink; should eat a diet low in fat, rich in fiber and yellow vegetables; should protect himself from hazardous chemicals in the workplace and home; should minimize intake of drugs and avoid unneeded X-rays; and should protect himself from sunlight. The "ideal" woman (from the point of view of avoiding cancer) should do all this and, in addition, have at least one child early in her reproductive life and avoid multiple sex partners. Assuming that we cannot totally avoid the pollution in our environment, which epidemiologists tell us accounts for no more than 2% of cancers, exposure to certain infectious agents (5% of cancers), and certain other unknown factors including genetic determinants (about 4%), we should be able to decrease our cancer mortality rate by about 84% by simple, direct actions as individuals. This conclusion is almost certainly overly sanguine but appears to be worth the experiment.

Evidence to support the idea that most cancers are avoidable comes from different sorts of data: (1) differences in the incidence of cancer in various areas of the world (Table 3–12); (2) differences in the incidence rates for various types of cancer between residents of a country and those of the same ethnic group who have emigrated to another country (Fig. 3–9); (3) variations over time in the incidence of cancers within a given society or community; and (4) the identification of specific causes of cancer and preventive measures resulting from this (e.g., aniline dyes in bladder cancer, vinyl chloride in hepatic angiosarcomas, asbestos in mesotheliomas, etc.).

Risk Assessment

As was discussed more fully in Chapter 2, most chemical carcinogens are also mutagens. Because mutagenicity is much easier to measure experimentally than carcinogenicity, many of the tests used to assess the carcinogenic risk potential of substances in our environment are based on mutagenicity assays. Exposure of human beings

to mutagens occurs from chemicals in our diet, water, and air, from products that we use as cosmetics and drugs, and from cigarette smoking. As noted before, several mutagenic substances have been identified in cigarette smoke. Mutagens are also found among the natural products contained in foods such as products elaborated by molds (e.g., aflatoxin) or by edible plants that synthesize a variety of toxins, presumably to ward off insects, as well as among synthetic chemicals such as pesticides, industrial pollutants, and weed killers.[146] In short, we live in a sea of mutagens and carcinogens. Identification of potentially mutagenic substances in our environment is a major public health and political issue. Of the myriad of potential mutagens and carcinogens in our diet, only a few have been studied in detail. In addition, more than 65,000 synthetic chemicals are produced in the United States, and about 1000 new chemicals are introduced each year.[147] Only a small number of these were examined for mutagenic and carcinogenic potential before being marketed. Obviously, this is a major epidemiologic problem and one that has a major economic impact on private industry as well as on the consumer and taxpayer. What is needed are accurate, rapid,

Table 3–12. Worldwide Variation in Incidence of Common Cancers, with Range of Variation Expressed for Ages 35–64

Type of Cancer	High-Incidence Area	Low-Incidence Area	Range of Variation
Skin	Australia (Queensland)	India (Bombay)	>200
Buccal cavity	India	Denmark	>25
Nasopharynx	Singapore°	England	40
Bronchus	England	Nigeria	35
Esophagus	Iran	Nigeria	300
Stomach	Japan	Uganda	25
Liver	Mozambique	Norway	70
Colon	U.S.[†] (Connecticut)	Nigeria	10
Rectum	Denmark	Nigeria	20
Pancreas	New Zealand[‡]	Uganda	5
Breast	U.S.[†] (Connecticut)	Uganda	5
Uterine cervix	Colombia	Israel	15
Uterine corpus	U.S.[†] (Connecticut)	Japan	10
Ovary	Denmark	Japan	8
Bladder	U.S.[†] (Connecticut)	Japan	4
Prostate	U.S.°°	Japan	30
Penis	Uganda	Israel	300

°Chinese.

[†]The U.S. data are taken from the Connecticut Tumor Registry because it is the oldest continued cancer registry based on a defined population in this country.

[‡]Maori.

°°African Americans.

(From Doll[145])

and economically feasible tests to predict the mutagenic and carcinogenic potential of the numerous chemicals in our environment. In practice, however, it has not been possible to develop a "perfect" short-term test. A number of false positives and false negatives result from using these tests.

More than 100 short-term tests for mutagenicity and carcinogenicity have been developed. Some of the most widely used are bacterial mutagenesis (the Ames test), mutagenesis in cell culture systems, direct measurement of damage to DNA or chromosomes in exposed cells, and malignant transformation of cell cultures.[148] One of the most popular of the short-term tests is the Ames test, developed by Bruce Ames and colleagues.[147] The basis of this assay is the ability of a chemical agent to induce a genetic reversion of a series of *Salmonella typhimurium* tester strains, which contain either a base substitution or a frameshift mutation, from histidine requiring (his$^-$) to histidine nonrequiring (his$^+$). These strains have been specially developed for this assay by selecting clones that have a decreased cell surface barrier to uptake of chemicals and a decreased excision repair system. Other advantages of this system are the small genome of the bacteria (4×10^6 base pairs), the large number of cells that can be exposed per culture dish (about 10^9), and the positive selection of the mutated organisms (i.e., only the mutated organisms will grow under the test conditions). This system has great sensitivity: only about 1 in 1000 to 1 in 10,000 of the mutated bacteria need to be detected to give a positive test, and nanogram amounts of a potent mutagen can be detected as a positive. Both base substitution and frameshift mutagens can be detected, and, using the appropriate tester strains, the type of mutagen can be deduced because frameshift mutagens usually revert only frameshift mutations of the tester strains and not base substitution mutations, and vice versa. Because many mutagens must be metabolized to be active, a liver homogenate fraction containing microsomes is usually added to the incubation to provide the drug metabolizing enzymes.

The potential of various chemical agents to mutagenize mammalian cells has also been used as a short-term test. Frequently, mutation at the hypoxanthine-guanine phosphoribosyltransfer-ase (HGPRT) locus is used as a marker; the end point of the assay is loss of sensitivity to purine antimetabolites that must be activated by HGPRT to be effective, thus leading to the selection of HGPRT$^-$ clones. Cultured fibroblast cell lines such as Chinese hamster V79 or ovary (CHO) cells are frequently used in this way.

Agents that damage DNA can often be detected by examining an index of genotoxicity, such as unscheduled DNA synthesis, sister chromatid exchange, or chromosome breakage in cultured cells exposed to the agents in question.

Carcinogenic potential has also been estimated by the ability of chemicals to "transform" smooth, well-organized monolayers of normal diploid fibroblasts into cells that grow piled up on one another (transformed foci) or into a cell type that can grow suspended in soft agar (normal fibroblasts do not usually grow on soft agar). Sometimes the putative malignant cells are then injected into immunosuppressed or immunodeficient ("nude") mice to further demonstrate that they are malignant. All of these estimates of carcinogenic potential are fraught with danger in that a significant number of false negatives or false positives can occur.

No single short-term test is foolproof; however, if definitive evidence of genotoxicity has been obtained in more than one test, a chemical is highly suspect. An agent found to be mutagenic, DNA damaging, and a chromosome breaker is almost certain to also be carcinogenic.[148] Final proof of mutagenicity and carcinogenicity involves the chronic exposure of whole animals to the test chemical. Although the short-term in vitro tests have several advantages, a number of important components, such as absorption, pharmacokinetics, tissue distribution, metabolism, age or sex effects, and species specificity, cannot be duplicated in vitro. Tests in whole animals take a long time and, unfortunately, are very expensive.

One key question remains: how does one estimate the danger of low-dose exposures? More importantly, how does one estimate the risk of low-dose exposure over a lifetime? These are extremely difficult questions to answer, but in practical terms, as long as an individual's DNA-repair enzymes are working (see Chapter 2), there probably is some low level of exposure below which DNA lesions can be removed efficiently without permanent damage. The low level of mutation in human genes seems to argue in support of this conclusion.[149] If one could, in fact, measure the amount of DNA-adduct formation (i.e., the amount of DNA bases bound to carcinogen) after exposure to various doses of carcinogen, one could probably get a much better estimate of the risks involved in exposure to various amounts of carcinogenic agents.[150] A shift in the dose–response curve with low doses of carcinogen could occur for several reasons, all of which make linear extrapolations of dose–response data from animal studies tenuous. Some of these reasons relate to differences in metabolism, distribution, and overall pharmacokinetics among species.

THE GREAT CANCER MYTHS

Cancer is a dreadful disease. In most polls, it is the most feared disease of all. Coupled with this are the almost daily media reports of another carcinogen or cancer risk being found in our environment that produce a setting for the sometimes hysterical fear that cancer lurks around every corner. Epidemiological pronouncements that one out of eight women will die of breast cancer or one of every four men will get prostate cancer, while perhaps having some statistical validity if everyone would reach age 80 and die of nothing else, belies the real risk of getting and dying of cancer. A study published by Woloshin et al.[151] puts this rate in a more rational context.

These authors have developed charts for men and women that show the chance of dying from various causes based on age and smoking history (Figs. 3–11 and 3–12). Instead of giving risks in terms of population percentages, these data show risk in terms of individual risks. For example, their data indicate that a 60-year-old woman, even one who smokes, has a 4.5% chance of dying of a heart attack in the next decade, a 6.5% chance of dying of lung cancer, and a 0.7% chance of dying of breast cancer. Or to look at it another way, for every 1000 60-year-old women who are smokers, 45 will die of heart attacks, 65 of lung cancer, and 7 of breast cancer in the next 10 years. For 60-year-old women who have

never smoked, 14 of 1000 will die of heart disease, 5 of lung cancer, and 7 of breast cancer by the time they reach 70 years of age. For 60-year-old men who are smokers, 84 of 1000 will die of heart disease and 98 of lung cancer, but only 4 of 1000 will die of prostate cancer.

A few years ago there was debate about whether the incidence of childhood cancers is going up, with the Environmental Protection Agency and the National Cancer Institute taking opposite points of view.[152] Data obtained between 1975 and 1995 showed a slight increase in cancers of children, which appeared to be due to an increase in brain cancer. However, the rates of leukemia and lymphoma, which together account for about 35% of all childhood cancers, did not change. Since there are only about 1800 new cases of brain cancer in the United States per year, a small number of patients being diagnosed in any given year could skew the numbers. This is not to belittle the devastating effects of childhood cancer, but it must be kept in mind that cancer in children is a rare disease, about one-third of which is due to leukemia, and for which the overall 5-year survival rate for all childhood cancers combined is 70% to 94%.

While there is a tendency to blame environmental causes for cancer in children, this probably plays a small role. Hereditary gene mutations probably play a larger role. Most experts agree that a mother's smoking during pregnancy, electromagnetic fields from power lines, or other environmental toxicants play little role.[152]

There are always debates about what is or is not a human carcinogen. Many of them have been identified by occupation, a rare medical exposure, atomic bomb fallout, or viral or other infections. Determination of whether a chemical is a human carcinogen by high-dose exposure in rodents, frequently at doses that no human being would ever be exposed to, is notoriously inaccurate and has led to many false-positive claims. Another point of view is expressed by Bruce Ames, who has said that he is a "contrarian in the hysteria over tiny traces of chemicals that may or may not cause cancer. If you have thousands of hypothetical risks that you are supposed to pay attention to, that completely drives out the major risks you should be aware of,"[153] which I would add include

Risk chart for women who have never smoked*

Find the line with your age. The numbers next to your age tell how many of 1000 women will die in the next 10 years from each cause of death.

Age, y	Vascular disease		Cancer type					Infection			Accidents	Any cause
	Heart attack	Stroke	Lung	Breast	Colon	Ovarian	Cervical	Pneumonia	Influenza	AIDS		
20											2	4
25					Fewer than 1 death					1	1	5
30				1						2	2	7
35				1						2	2	11
40	1	1		2	1					2	2	17
45	2	1	1	4	1	1	1			1	2	26
50	4	2	2	5	2	1	1	1		1	2	42
55	7	2	3	6	3	2	1	1		1	2	66
60	14	4	5	7	4	3	1	2			2	105
65	30	10	7	9	6	3	1	4			3	158
70	52	19	10	10	8	4	1	7			5	247
75	90	36	11	11	11	4	1	15			7	381
80	153	62	11	12	14	4	1	30	1		11	581
85	221	86	8	12	15	3	1	48	3		14	771
90	272	96	5	10	14	2	1	64	5		15	879

Figure 3–11. Risk chart for women who have never smoked. The chart indicates the number of women per 1000 who will die from various diseases and for any reason during the next 10 years, beginning at the indicated age. (*The numbers of each row do not add up to the chance of dying from any reason because there are many other causes of death in addition to the ones listed here.) Shaded area indicates age group and disease combinations with fewer than 1 death per 1000. (From Woloshin et al.,[151] with permission.)

cigarette smoking, obesity, lack of antioxidants in the diet, sun overexposure, and inadequate access to health care in large parts of the United States and the world.

It has often been said that we live in a sea of carcinogens. Indeed, every time you're stuck in traffic or behind an exhaust-belching truck, you are inhaling a lung-full of potential carcinogens. If you live in a city and and drink chlorinated water, you are exposing yourself to a host of potential carcinogens. Most of the modern conveniences that we take for granted contain carcinogenic substances. The chair you sit in probably has polyurethane, another carcinogen, in the cushions. Another point is that modern technology such as high-sensitivity mass spectrometers can detect parts per billion of chemical substances. Thus, one must ask, is the detection of any level of a carcinogen dangerous? Perhaps it is worth keeping in mind an old adage in pharmacology: "a tiny amount of something doesn't necessarily cause anything and enough of something can cause anything."

Described below are some commonly held myths about agents that cause cancer.

Passive Smoking

Inhalation of smoke in an enclosed space, especially for prolonged periods of time, is not healthy. It may trigger an asthmatic attack in a child, for example. There are studies that show if nonsmokers are in a 10×10 square foot room with smokers, within a short time of exposure, effects on the cardiovascular system, e.g., heart rate and blood pressure, can be observed. This is not surprising. Nicotine, after all, is a drug that can cause cardiovascular and central nervous system effects. The data for passive smoking causing lung cancer, however, are skimpy at

Risk chart for men who have never smoked*

Find the line with your age. The numbers next to your age tell how many of 1000 men will die in the next 10 years from each cause of death.

Age, y	Vascular disease		Cancer type			Infection			Accidents	Any cause
	Heart attack	Stroke	Lung	Colon	Prostate	Pneumonia	Influenza	AIDS		
20									5	10
25					Fewer than 1 death			1	5	11
30	1							2	4	13
35	2							2	5	18
40	4	1		1				2	5	27
45	7	1	1	1		1		1	5	39
50	12	2	2	2	1	1		1	4	62
55	20	4	3	4	2	2		1	4	93
60	32	6	5	6	4	3			5	146
65	61	12	7	9	8	6			6	221
70	93	21	12	11	14	11			7	336
75	142	36	15	14	23	22	1		11	494
80	196	53	13	16	32	38	1		15	652
85	241	67	10	16	37	56	1		19	770
90	223	73	6	14	36	74	1		21	828

Figure 3–12. Risk chart for men who never have smoked. The chart indicates the number of men per 1000 who will die from various diseases and for any reason during the next 10 years, beginning at the indicated age. (°The numbers in each row do not add up to the chance of dying from any reason because there are many other causes in death in addition to the ones listed here.) Shaded area indicates age group and disease combinations with fewer than 1 death per 1000. (From Woloshin et al.,[151] with permission.)

best. In a multicenter case–control study of exposure to environmental tobacco smoke (ETS) and lung cancer in Europe, no association between childhood exposure to ETS and lung cancer risk was found.[154] There was weak evidence of a dose–response relationship between risk of lung cancer in spouses and workplace ETS, but no detectable risk after cessation of exposure. Similarly, no association between exposure to ETS and female breast cancer mortality was found in two large cohort studies, one involving over 146,000 women[155] and one involving over 116,000 women.[156]

One of the most definitive studies involved 118,094 adults in California enrolled in late 1959 and followed until 1998.[157] Of these, 35,561 were never smokers who had a spouse in the study with known smoking habits. No significant associations were found for current or former exposure to environmental tobacco smoke and

coronary heart disease or lung cancer, even before or after taking into consideration seven potential confounding factors and before or after excluding participants with pre-existing disease. This was true for follow-up periods 1960–65, 1966–72, 1973–85, and 1973–98. The authors concluded that "the association between exposure to environmental tobacco smoke and coronary heart disease and lung cancer may be considerably weaker than generally believed."

Radon in the Home

As noted above in the sections on the role of various factors in cancer development, radon is a well-established occupational carcinogen. There are significant data indicating an increased risk of lung cancer in deep-well miners, particularly among miners who smoke, and there is a dose–response relationship to this risk. Since radon in the soil can seep into homes in areas where there is a high natural soil context of radon gas, there has been concern that this exposure presents a potential risk of lung cancer. These risk assessments used the exposure dose–response relationship from studies of uranium miners and miners of other ores and extended this relationship by a linear nonthreshold model down to zero exposure.[158] Based on assumptions from this model and extrapolations from occupational data, it has been stated that radon is the second leading cause of lung cancer after cigarette smoking,[159] although these authors admit that "the effect of smoking on lung cancer risk appears to be an order of magnitude greater than the effect of radon." The difference between the linear nonthreshold extrapolation model and the threshold model for cancer risk estimation was discussed in Chapter 2.

A significant amount of data does not support this claim of the association of radon exposure in the home and lung cancer. A large case–control study of 1055 case subjects and 1544 controls in Finland did not indicate an increased risk of lung cancer from indoor radon exposure.[160] A meta-analysis of lung cancer risk from residential radon in eight epidemiological studies reported that in four of the eight studies the data for an association were positive or weakly positive and the remaining four showed no increased risk.[161] Other major studies also showed results that were equivocal.[161] In a small case–control study (138 cases, 291 controls) in an Italian alpine valley with high radon levels, an increased risk was observed, but the association was confined to male smokers.[162] A similar study done in Sweden indicated some risk due to radon exposure, but again it was higher among smokers.[163] A recent study done in the United Kingdom indicated that radon is not a risk factor for childhood cancers.[164]

Cell Phones

The use of cell phones has increased rapidly in the past few years. They are found in most parts of the world, even in remote areas of developing countries. Cell phones emit radiofrequency (RF) signals in a range between 800 and 2000 MHz, which puts it in the microwave range of the electromagnetic spectrum. RF radiation at sufficiently high levels can produce heat by inducing small electric currents. A typical cell phone operates with a power output that could only cause, at a maximum, a rise $0.1°$ C.[165] This amount could not be expected to have any significant biological effect. In addition, RF does not possess sufficient energy to remove electrons from atoms or molecules and thus does not produce ionizing radiation,[165] which is the kind that could damage DNA.

Because of the weak thermal and ionizing potential of RF from cell phones, it seems highly unlikely that this RF would cause cancer. A smattering of reports, however, from rodent studies have suggested some associated risk between RF in the potential range of a cell phone's MHz and tumors, a range that could damage DNA (reviewed in Reference 165). A number of human epidemiological and occupational exposure studies do not support any association of cell phone risk and cancer (reviewed in Reference 165). A study of 250,000 cell phone users in the United States did not show any increased cancer risk, and a case–control study from Sweden indicated no increase in brain tumors. In a study of 195,775 workers engaged in manufacturing and testing of cell phones, no association between RF exposure and brain, other nervous system cancers, or leukemia was found. A nationwide cohort study involving 420,000 cell phone users in Denmark found no association between cell phone use and tumors of the brain or salivary glands, leukemia, or other cancers.

Electromagnetic Fields

There have been some studies suggesting a link between magnetic fields generated by electrical power lines and childhood leukemia. For example, an excess incidence of leukemia in Swedish children was associated with the estimated electric current flow, based on historical records of local power companies;[166] however, that risk of childhood leukemia did not correlate with residential measurements of magnetic field made shortly after the time of diagnosis (reviewed in Reference 167). In fact, there have been a number of shortcomings in earlier epidemiological studies on this topic, such as discrepancies between results based on proxy measurements and those based on direct magnetic field measurements, the absence of supportive laboratory data, and lack of a substantive biological explanation for causing cancer.[167]

A very careful study done by National Cancer Institute investigators and their collaborators directly measured magnetic fields in cases' and controls' bedrooms, three or four other rooms, and the front doors of their houses.[167] In addition, they measured magnetic fields in homes where case subjects' and controls' families lived during their mothers' pregnancies. The results of this study showed that the risk of acute lymphoblastic leukemia (the most common malignancy of childhood) was not increased among children who lived in homes with the highest exposure to magnetic fields, and there was also no significant associated risk with magnetic-field levels of the homes where the mothers resided when pregnant.

Alcohol

Chronic alcohol abuse is associated with increased risk of certain cancers such as liver and oral pharyngeal cancers, in which there is usually some associated tissue toxicity as a prodromal factor. There have been a number of studies of the potential risk of alcohol consumption and breast cancer. One such study suggested that a woman who had as few as three alcoholic drinks a week had an increased risk of breast cancer. This is one of those conclusions that doesn't pass the "common sense" test, which should be used for all reports, particularly those in the media, about what causes cancer or what cures it.

In a meta-analysis of 28 case–control and 10 cohort studies, comparing drinkers and non-drinkers, the risk of breast cancer increased 24% with consumption of two drinks per day.[168] Another study, pooling data from six prospective studies, reported a 9% increase in breast cancer incidence with each 10 grams of alcohol consumed per day.[169] These data were mostly obtained from postmenopausal women. In another case–control study, recent alcohol consumption of 13 grams per day (about equivalent to 3 drinks of 100 proof whiskey) was associated with a 21% increased risk of breast cancer, but in the age group less than 30 years of age there was no increased risk (reviewed in Reference 170). In

another cohort study,[171] women who consumed 15 or more (!) grams of ethanol per day had a 26% increased risk. To put this in context, if the risk of a woman over age 50 in contracting breast cancer is 11%,[130] an increased risk of 25% is $0.11 \times 0.25 = 2.7\%$, or 2 to 3 of every 100 women. If the risk is 9%,[169] then the increased risk is 1%, or 1 of 100 women. Given the notorious underreporting of personal alcohol use, the odds are that at least in some of these studies the amount of alcohol actually consumed was underestimated.

A hypothesis for an association of alcohol use and breast cancer is that alcohol increases circulating levels of estrogen. Here again the data are conflicting. In one study of premenopausal women, who consumed 30 grams per day of ethanol, increased blood levels of estrogen were reported, whereas in another study, alcohol intake was not associated with plasma estrogen levels, but was associated with increased levels of androstenedione (reviewed in Reference 170).

Organochlorine Compounds, Polycyclic Aromatic Hydrocarbons, and Breast Cancer

Environmental exposure to organochlorine compounds such as polychlorinated biphenyls (PCBs), 2,2'-bis (p-chlorophenyl)-1, 1, 1-trichloroethane (DDT) and its metabolite DDE, and organochloro pesticides has been suggested as a risk factor for breast cancer. The basis for this claim is that some of these are carcinogenic in animals, have estrogenic activity, and are inducers of cytochrome P-450 enzymes that metabolize drugs, hormones, and various xenobiotics. Some epidemiological studies have suggested an association between this class of compounds and breast cancer risk, but these studies have been contradictory and inconclusive (reviewed in Reference 172).

Because of a purported clustering of breast cancer on Long Island, New York, and a fair amount of political pressure, a $30 million Long Island breast cancer study project was launched to examine the relationship between exposure to environmental agents and breast cancer incidence. The study was carried out under the auspices of the National Cancer Institute and the National Institute of Environmental Health

Sciences. (In fact, the incidence of breast cancer on Long Island is not significantly different from, and in many instances less than, other locales in New York State).[173] The published results of this study indicated that there was no association between increased rates of breast cancer and exposure to PCBs, DDT, or pesticides (reviewed in Reference 174). Studies done in other parts of the United States and of the world also show no significant correlation between serum or plasma levels of PCBs or other organochlorine compounds and breast cancer.[172,175–177] The only caveat to this is a report from a small cohort study that women living within one mile of hazardous waste sites containing organochlorines had a higher incidence of breast cancer (reviewed in reference 177a). However, when the data were pooled in a combined analysis, there was no association between breast cancer risk and blood levels of PCBs or DDE.[177a]

An argument has been made that since PCBs and DDTs have been banned since the 1970s, these may not have been the correct chemicals to look at or exposure of female babies in utero may be the key factor here. While there may be some truth to these assumptions since organochlorines do persist in the environment and can remain in the body for more than a decade. However, a case–control study based on cohorts of women who donated blood in 1974, 1989, or both and who were matched on age, race, menopausal status, and month and year of blood donation showed that even after 20 years of follow-up after exposure to relatively high concentrations of DDE or PCBs there was no association with an increased risk of breast cancer.[178]

One might argue that a better way to assess risk is to look at damage to the target in the body to which environmental agents might bind. This was done in a study that looked at polycyclic aromatic hydrocarbon–DNA adduct levels in blood mononuclear cells of women who live on Long Island. Samples from 576 breast cancer cases and 427 age-matched controls were assayed for PAH–DNA adducts by ELISA.[179] The levels of PAH–DNA adducts were slightly, though not significantly, higher among cases than controls. Also, there was no consistent association between adduct levels and passive cigarette smoke exposure or consumption of grilled or smoked foods. The authors also concluded that there

was not a dose-dependent relationship between exposure and adduct formation, suggesting that there is a threshold effect. This latter point is interesting because there are known polymorphisms in the enzymes that activate and detoxify PAHs, thus a pharmacogenetic analysis could reveal who may be at higher risk.

Finally, it is worth noting that a careful analysis of all risk factors for breast cancer must be done before one can conclude that a "cluster" of breast cancer cases is related to some local environmental factor. For example, a study done in the San Francisco Bay area, involving both Caucasian and African-American women, found that the elevated breast cancer incidence in the Bay area could be completely accounted for by regional differences in known risk factors, e.g., parity, age at first pregnancy, months of breast feeding, and ages at menarche and menopause.[180]

Antiperspirants

Recently, a rumor that underarm antiperspirants or deodorants caused breast cancer was widely circulated on the Internet and picked up in the media. This is another example of widely disseminated urban myth. In a case control study of 813 women with breast cancer diagnosed between 1992 and 1995, compared with 793 women without breast cancer, there was no link between breast cancer and regular use of antiperspirants or deodorants, even when applied after underarm shaving.[181] Both the American Cancer Society and the National Cancer Institute posted notices on their Web sites to assure the public that there is no scientific basis for this rumor.

Water Chlorination

Some of the compounds used to disinfect water in urban drinking water systems are carcinogenic at high doses in rodents. A study by Komulainen et al.[182] reported that administration of 3-chloro-4-(dichloromethyl)-5-hydroxy-2 (5H)-furanone (MX) to rats in their drinking water produced thyroid tumors and bile-duct neoplasms (cholangiomas). Based on linear extrapolation of the data on dose-exposure for induction of cholangiomas, the upper-bound cancer risk per unit dose for lifetime exposure to MX was estimated to be 100% per milligram MX per kilogram body weight per day. The cancer-causing dose estimates for humans are based on assumptions that rats and humans absorb, metabolize, and excrete MX in the same manner, that rats and human tissues have the same carcinogenic target or response, and that the extrapolation model for estimating low-exposure dose-response is an accurate reflection of human risk. None of these assumptions is likely to be true. Furthermore, MX levels in United States and Finland water supplies have been reported to range from 3 to 67 parts per trillion (ppt). Based on the highest estimate of 67 ppt, daily human exposures would be several orders of magnitude lower than those employed by Komulainen et al.[183] Using the extrapolation model, this could potentially produce two cancers per one million people.[183] Similar conclusions can be reached for other water disinfectants such as trihalomethanes. Thus, the overall cancer risk due to disinfectants in drinking water is very small and the risk–benefit ratio is very large. Disastrous consequences could result if water chlorination were stopped. For example, a cholera epidemic involving 300,000 people occurred in Peru as a consequence of inadequate disinfection of drinking water supplies.[183]

Abortion or Miscarriage and Breast Cancer

Some reports have suggested that incomplete pregnancies, terminated either by induced abortion or miscarriage, increases the risk of breast cancer (reviewed in Reference 184). A number of other studies have not shown an increased risk of breast cancer in women who have undergone induced abortions.[184] A well-controlled study of the effects of induced abortion and miscarriage on breast cancer incidence, involving age-, parity-, and race-matched cases and controls, showed that neither induced abortion nor miscarriage increased breast cancer risk.[184] This claim appears to be more of an issue of politics and religious beliefs than science.

Asbestos

It has been known for several decades, since the follow-up of Liberty Ship builders in World War II, that occupational exposure to asbestos at high levels can cause lung cancer and mesothelioma of the pleura and peritoneum. Mesotheliomas in asbestos workers can be induced by exposure to asbestos alone, whereas lung cancers are more likely to be caused by exposure to asbestos in smokers. However, since mesotheliomas are quite rare, lung cancer cases are much more common among asbestos workers.

Since the 1980s, regulations on occupational exposure have greatly reduced exposure to asbestos. Based on the known carcinogenic effect of occupational exposure to asbestos and its wide use as fire retardant material in homes and public buildings, a ground swell of public and political pressure has been mounted to remove every scrap of asbestos from schools, homes, and public buildings, creating a whole new industry. The validity of such overwrought concerns has been questioned for several reasons.[185] First of all, estimates of risk were based on extrapolations from occupational exposure to environmental exposure that are 1/100,000 of those to which asbestos workers were exposed in the past. Second, dose–response estimates used to extrapolate the data varied by a factor of 1000 among various studies. Third, estimates of risk of asbestos-induced cancer have not been validated in nonoccupational exposed populations. Finally, in a study of nonoccupationally exposed women in two chrysotile-asbestos mining regions, no increased risk of lung cancer was observed.[185] The authors concluded that the Environmental Protection Agency's extrapolation model overestimates the risk of asbestos-induced lung cancer by at least a factor of 10-fold.

Saccharin

As discussed in Chapter 2, there have been a number of reports that sodium saccharin, either in the diet or by direct installation, causes urinary bladder cancer in rats (reviewed in Reference 186). On the basis of these observations, the United States Food and Drug Administration proposed to ban saccharin from human use. This proposal was overturned by a moratorium passed by the United States Congress, but saccharin use was banned in Canada. It turns out that the bladder carcinogen effect is seen in certain species of male rats (but not in mice, hamsters, or guinea pigs) in which calcium phosphate–sodium saccharin–containing precipitates form in the urine after high-dose administration of the artificial sweetener (higher than humans would likely ever ingest). Dietary administration of high doses of other sodium salts such as sodium ascorbate, sodium citrate, or even sodium chloride caused the same effect. Even more conclusive is a study in monkeys using doses of sodium saccharin 5 to 10 times the allowable daily intake for humans that showed no carcinogenic effect on the primate urinary bladder.[186]

Acrylamide in Foods

A report from a Swedish group in April 2002 announced that there were high levels of acrylamide in certain cooked or fried foods such as French fries and potato chips. Since acrylamide is listed by the World Health Organization (WHO) as a probable human carcinogen, this raised a state of alarm. A WHO official stated: "Given that we know acrylamides are cancer-causing in animals and probably in humans, it is intolerable that they are in foods at the levels found, and we have to find a remedy."[187] A U.S. Food and Drug Administration (FDA) representative concurred, stating that "it is clear that acrylamide is a problem. It doesn't need to be in food." Some food safety advocacy groups started planning lawsuits against fast-food companies.

There is no consensus on the risk posed by this "new threat" to human health. Indeed, skeptics point out that acrylamide is also found at some level in breads, meat, and vegetables that have been eaten by people for thousands of years. Even given the huge increase in fast-food consumption in the last two to three decades, there have not been significant increases in cancers that could be associated with acrylamide.[187] In addition, there are no data proving that acrylamide is a human carcinogen. The amount given to rats to produce tumors would require that a person consume 35,000 potato chips or 182 pounds of French fries per day.

Further studies are under way to determine how acrylamide is generated in foods during cooking. One hypothesis is that in foods heated above 120° C, acrylamide can be generated by a reaction between amino acids (such as asparagine, which is present in potatoes and cereals) and reducing sugars. However, the amount of acrylamide produced can vary depending on the food, temperature and duration of heating, water, and starch content. Of course, as Bruce Ames points out, if we eliminated all the foods that contain potential human carcinogens (including peanut butter and a number of fruits and vegetables), there wouldn't be much left to eat. This was probably summed up best by Walter Willett, Professor of Epidemiology and Nutrition at the Harvard School of Public Health, who is quoted as saying, "We don't even know if it [acrylamide] is a carcinogen in humans based on the amounts we eat. Out of the 100 things you should worry about, I'd put this at 200."[187]

Alar

Alar was the trade name for a formulation containing daminoxide, a hormone-like substance that can slow the growth of certain varieties of apples that tend to rot as soon as they ripen. Alar was approved by the FDA in 1968 and was used by apple growers to improve the efficiency of the harvest. Its safety testing included a 2-year animal carcinogenicity study in rats that were negative (reviewed in Reference 188). During the 1970s, however, Alar and its byproduct showed increases in certain tumors in mice fed enormous doses of either substance (larger than the maximum tolerated dose [MTD]). Additional studies based on feeding mice doses exceeding the MTD showed similar results. This caused a media uproar. In one well-publicized instance, a famous actress interviewed on TV stated that Alar was going to give children cancer. A 60 Minutes TV segment in 1989 described Alar as "the most potent cancer-causing agent in the food supply."

In fact, there has never been any evidence that Alar causes childhood or any other type of cancer.[188] Nevertheless, sales of apples decreased, parents poured apple juice down the drain, and some schools removed all apple products from their lunch menus. Even though the FDA, Environmental Protection Agency, and U.S. Department of Agricultur tried to assure the public, the Alar scare continued, and the manufacturer halted sale of all Alar products for use on food crops. Since 1989, a number of agencies, including a British government advisory group, a United Nations panel, and the American Medical Association, have reviewed the scientific data and concluded that Alar did not pose a risk to human health.[188]

SV40 Virus in Early Polio Vaccines

Polio vaccines have been blamed for everything from initiating the AIDS epidemic to being a Western plot to subvert the developing world. Poliovirus vaccines that were used during the late 1950s and early 1960s were contaminated with simian virus 40 (SV40), a monkey virus that came from the monkey cells in which early batches of the vaccine were grown. A survey done in 1961 indicated that about 90% of U.S. citizens younger than 20 years of age (those born between 1941 and 1961) had received at least one immunization with poliovirus vaccine that may have contained SV40 virus.[189] It is difficult to determine the average exposure because the titers of live SV40 in different vaccine lots varied from undetectable to high.

Most epidemiological studies of populations who were immunized with polio vaccine potentially containing live SV40 during early childhood, which is presumed to be the highest at-risk age group for a lifetime risk of developing an exogenous agent-induced cancer, have failed to show any association of polio vaccination with increased risk of cancer more than 30 years following vaccination (reviewed in Reference 189). A population-based study in Denmark, comparing cancer incidence data from 1943 to 1997 with exposure to SV40-contaminated poliovirus vaccine, found that there was no associated increased incidence of mesothelioma, ependymoma, choroid plexus tumor, non-Hodgkin's lymphoma, or leukemia.[190] However, SV40 DNA sequences have been reported to be present in certain human cancers such as childhood brain tumors, osteosarcomas, and pleural mesotheliomas.

Pleural mesotheliomas are the cancers most often reported to contain SV40 DNA. Yet age-specific trends in the U.S. pleural mesothelioma incidence rates are not consistent with an effect caused by exposure to SV40-contaminated poliovirus vaccine.[189] To be totally comfortable about this issue, given the reports of SV40 DNA in human tumors, monitoring of people who received potentially SV40-contaminated poliovirus vaccinations should probably continue.

References

1. B. F. Hankey, L. A. Gloeckler-Ries, B. A. Miller, and C. L. Kosary: Overview. In *Cancer Statistics, 1973–1989*, NIH, Publ. No. 92–2789, I.1–17, 1992.
2. H. S. Seidman, M. H. Mushinski, S. K. Gelb, and E. Silverberg: Probabilities of eventually developing or dying of cancer—United States, 1985. *CA Cancer J Clin* 35:36, 1985.
3. A. Jemal, T. Murray, E. Ward, A. Samuels, R. C. Tiwari, A. Ghafoor, E. J. Feuer, and M. T. Thun: Cancer statistics—2005. *CA Cancer J Clin* 55:10, 2005.
3a. A. Jemal, R. Siegel, E. Ward, T. Murray, J. Xu, et al. Cancer statistics-2006. *CA-A Cancer Journal for Clinicians* 56:106, 2006.
4. *Cancer Facts and Figures—2003*. Atlanta, GA: American Cancer Society, 2003.
5. D. M. Parkin, F. Bray, J. Ferlay, and P. Pisani: Global cancer statistics, 2002. *CA Cancer J Clin* 55:74, 2005.
6. Lung cancer. In B. W. Stewart and P. Kleihues, eds.: *World Cancer Report*, World Health Organization. Lyon: IARC Press, 2003, pp. 182–187.
6a. L. A. Loeb, V. L. Ernster, K. E. Warner, J. Abbotts, and J. Laszlo: Smoking and lung cancer: An overview. *Cancer Res* 44:5940, 1984.
7. Breast cancer. In B. W. Stewart and P. Kleihues, eds.: *World Cancer Report*, World Health Organization. Lyon: IARC Press, 2003, pp. 188–193.
8. E. Cho, D. Spiegelman, D. J. Hunter, W. Y. Chen, M. J. Stampfer, G. A. Colditz, and W. C. Willett: Premenopausal fat intake and risk of breast cancer. *J Natl Cancer Inst* 95:1079, 2003.
9. Endogenous hormones and breast cancer collaborative group: Body mass index, serum sex hormones, and breast cancer risk in postmenopausal women. *J Natl Cancer Inst* 95:1218, 2003.
10. H. L. Judd, I. M. Shamonki, A. M. Frumar, and L. D. Lagasse: Origin of serum estradiol in postmenopausal women. *Obstet Gynecol* 59:680, 1982.
11. V. L. Ernster, R. Ballard- Barbask, W. E. Barlow, Y. Zheng, D. L. Weaver, et al.: Detection of ductal carcinoma in situ in women undergoing screening mammography. *J Natl Cancer Inst* 94:1546, 2002.
12. M. J. van de Vijver, Y. D. He, L. J. van't Veer, H. Dar, et al.: A gene-expression signature as a predictor of survival in breast cancer. *N Engl J Med* 347:1999, 2002.

13. K. N. Nathanson, R. Wooster, and B. L. Weber: Breast cancer genetics: What we know and what we need. *Nat Med* 7:552, 2001.

14. H. S. Feigelson, R. McKean-Cowdin, G. A. Coetzer, D. O. Stram, L. K. Kolonel, and B. E. Henderson: Building a multigenic model of breast cancer susceptibility: CYP17 and HSD17B1 are two important candidates. *Cancer Res* 61:785, 2001.

15. J. Baselga and L. Norton: Focus on breast cancer. *Cancer Cell* 1:319, 2002.

16. R. B. Dickson and M. E. Lippman. Cancer of the breast: Molecular biology of breast cancer. In V. T. J. DeVita, S. Heklman, and S. A. Rosenberg, eds.: *Principles and Practice of Oncology*. Philadelphia: Lippincott Williams and Wilkins, 2001, pp. 1633–1645.

17. Colon cancer. In B. W. Stewart and P. Kleihues, eds.: *World Cancer Report, World Health Organization*. Lyon: IARC Press, 2003, pp. 198–202.

18. J. D. Potter: Colorectal cancer: Molecules and populations. *J Natl Cancer Inst* 91:916, 1999.

19. E. R. Fearon and B. A. Vogelstein: A genetic model for colorectal tumorigenesis. *Cell* 61:759, 1990.

20. G. Smith, F. A. Carey, J. Beattie, M. J. V. Wilkie, T. J. Lightfoot, et al.: Mutations in APC, Kirsten-ras, and p53-alternative genetic pathways to colorectal cancer. *Proc Natl Acad Sci USA* 99:9433–9438, 2002.

21. N. Kemeny, Y. Huang, A. M. Cohen, W. Shi, J. A. Conti, et al.: Hepatic arterial infusion of chemotherapy after resection of hepatic metastases from colorectal cancer. *N Engl J Med*, 341:2039, 1999.

22. S. D. Markowitz, D. M. Dawson, J. Willis, and J. K. V. Willson: Focus on colon cancer. *Cancer Cell* 1:233, 2002.

23. Liver cancer. In B. W. Stewart and P. Kleihues, eds.: *World Cancer Report, World Health Organization*. Lyon: IARC Press, 2003, pp. 203–206.

24. Cancers of the female reproductive tract. In B. W. Stewart and P. Kleihues, eds.: *World Cancer Report, World Health Organization*. Lyon: IARC Press, 2003, pp. 215–222.

25. C. Crump, M. W. McIntosh, N. Urban, G. Anderson, and B. Y. Karlan: Ovarian cancer tumor marker behavior in asymptomatic healthy women: Implications for screening. *Cancer Epidemiol Biomarkers Prev* 9:1107, 2000.

26. M. Höglund, D. Gisselsson, G. B. Hansen, T. Säll, and F. Mitelman: Ovarian carcinoma develops through multiple modes of chromosomal evolution. *Cancer Res* 63:3378, 2003.

27. Cancers of the male reproductive tract. In B. W. Stewart and P. Kleihues, eds.: *World Cancer Report, World Health Organization*. Lyon: IARC Press, 2003, pp. 208–214.

28. W. Isaacs, A. De Marzo, and W. G. Nelson: Focus on prostate cancer. *Cancer Cell* 2:113, 2002.

29. D. C. Malins, P. M. Johnson, T. M. Wheeler, E. A. Barker, N. L. Polissar, and M. A. Vinson: Age-related radical-induced DNA damage is linked to prostate cancer. *Cancer Res* 61:6025, 2001.

30. L. Cheng, S-Y. Song, T. G. Pretlow, F. W. Abdul-Karin, H-J. Kung, et al.: Evidence of independent origin of multiple tumors from patients with prostate cancer. *J Natl Cancer Inst* 90:233, 1998.

31. C. A. Macintosh, M. Stower, N. Reid, and N. J. Maitland: Precise microdissection of human prostate cancers reveals genotypic heterogeneity. *Cancer Res* 58:23, 1998.

32. A. Shibata and A. S. Whittemore: Correspondence RE: Prostate cancer incidence and mortality in the United States and the United Kingdom. *J Natl Cancer Inst* 93:1109, 2001.

33. X. Bosch and O. Bernadich: Increased serum prostate-specific antigen in a man and a woman with hepatitis A. *N Engl J Med* 337:1849, 1997.

34. J. L. Stanford, Z. Feng, A. S. Hamilton, F. D. Gilliland, R. A. Stephenson, et al.: Urinary and sexual function after radical prostatectomy for clinically localized prostate cancer—The prostate cancer outcomes study. *JAMA* 283:354, 2000.

35. Bladder cancer. In B. W. Stewart and P. Kleihues, eds.: *World Cancer Report, World Health Organization*. Lyon: IARC Press, 2003, pp. 228–231.

36. Lymphoma. In B. W. Stewart and P. Kleihues, eds.: *World Cancer Report, World Health Organization*. Lyon: IARC Press, 2003, pp. 237–241.

37. L. M. Staudt and W. H. Wilson: Focus on lymphomas. *Cancer Cell* 2:363, 2002.

38. A. A. Alizadeh, M. B. Eisen, R. W. Davis, C. Ma, I. S. Lossos, et al.: Distinct types of diffuse large B-cell lymphoma identified by gene expression profiling. *Nature* 403:503, 2000.

39. V. T. De Vita, Jr., A. A. Serpick, and P. P. Carbone: Combination chemotherapy in treatment of advanced Hodgkin's disease. *Ann Intern Med* 73:881, 1970.

40. Leukemia. In B. W. Stewart and P. Kleihues, eds.: *World Cancer Report, World Health Organization*. Lyon: IARC Press, 2003, pp. 242–247.

41. A. K. Stewart and A. C. Schuh: White cells 2: Impact of understanding the molecular basis of haematological malignant disorders on clinical practice. *Lancet* 355:1447, 2000.

42. T. R. Golub, D. K. Slonim, P. Tamazo, et al.: Molecular classification of cancer: Class discovery and class prediction by gene expression monitoring. *Science* 286:531, 1999.

43. B. J. Drucker, S. Tamura, E. Breckdunger, et al.: Effects of a selective inhibitor of the *Abl* tyrosine kinase on the growth of *Bcr-Abl* positive cells. *Nat Med* 2:561, 1996.

44. A. N. Houghton and D. Polsky: Focus on melanoma. *Cancer Cell* 2:275, 2002.

45. Melanoma.In B. W. Stewart and P. Kleihues, eds.: *World Cancer Report, World Health Organization*. Lyon: IARC Press, 2003, pp. 253–256.

46. L. M. De Angelis: Brain tumors. *N Engl J Med* 344:114, 2001.

47. Tumours of the nervous system. In B. W. Stewart and P. Kleihues, eds.: *World Cancer Report, World Health Organization*. Lyon: IARC Press, 2003, pp. 265–269.

48. A. Behin, K. Hoang-Xuan, A. F. Carpentier, and J-Y. Delattre: Primary brain tumours in adults. *Lancet* 361:323, 2003.

49. J. Peto: Cancer epidemiology in the last century and the next decade. *Nature* 411:390, 2001.

50. The known causes of cancer. In K. Meister, E. M. Whelan, G. L. Ross, and A. N. Stimola, eds.: *America's War on "Carcinogens"*. New York: American Council on Science and Health, 2005, pp. 59–73.

51. R. Doll and R. Peto: The causes of cancer. *J Natl Cancer Res* 66:1191, 1981.

52. J. Bradbury: Add colorectal cancer to list of smoking-associated cancers, say experts. *Lancet* 356:2072, 2000.

53. *The Health Consequences of Smoking*, HEW Publ. No. (CDC) 74–8704. Washington, DC: U.S. Department of Health, Education, and Welfare, 1974, pp. 35–37.

54. D. R. Shopland, H. J. Eyre, and T. F. Pechacek: Smoking-attributable cancer mortality in 1991: Is lung cancer now the leading cause of death among smokers in the United States? *J Natl Cancer Inst* 83:1142, 1991.

55. L. Garfinkel and E. Silverberg: Lung cancer and smoking trends in the United States over the past 25 years. *CA Cancer J Clin* 41:137, 1991.

56. United States Public Health Service: *The Health Consequences of Smoking—Cancer: A Report of the Surgeon General*. Rockville, MD: U.S. Department of Health and Human Services, Office on Smoking and Health, 1982.

57. R. M. Putzrath, D. Langley, and E. Eisenstadt: Analysis of mutagenic activity in cigarette smokers' urine by high performance liquid chemotherapy. *Mutat Res* 85:97, 1981.

58. G. Obe, H. J. Voght, S. Madle, A. Fahning, and W. D. Heller: Double-blind study on the effect of cigarette smoking on the chromosomes of human peripheral blood lymphocytes in vivo. *Mutat Res* 92:309, 1982.

59. Vijayalaxmi and H. J. Evans: In vivo and in vitro effects of cigarette smoke on chromosomal damage and sister-chromatid exchange in human peripheral blood lymphocytes. *Mutat Res* 92: 321, 1982.

60. F. Veglia, G. Matullo, and P. Vineis: Bulky DNA adducts and risk of cancer: A meta-analysis. *Cancer Epidemiol Biomarkers Prev* 12:157, 2003.

61. Q. Wei, L. Cheng, C. I. Amos, L. E. Wang, Z. Guo, et al.: Repair of tobacco carcinogen–induced DNA adducts and lung cancer risk: a molecular epidemiologic study. *J Natl Cancer Inst* 92:1764, 2000.

62. J. Stjernswärd: Battle against tobacco. *J Natl Cancer Inst* 81:1524, 1989.

63. P. A. Wingo, L. A. G. Ries, G. A. Giovino, D. S. Miller, H. M. Rosenberg, et al.: Annual report to the nation on the status of cancer, 1973–1996, with a special section on lung cancer and tobacco smoking. *J Natl Cancer Inst* 91:675, 1999.

64. A. J. Tuyns, G. Pequinot, and O. M. Jensen: Les cancers de l'oesophage en Ille-et-Villaine en function des niveaux de consummation d'alcool et de tabac. Des risques qui se multiplient. *Bull Cancer (Paris)* 64:45, 1977.

65. B. Herity, M. Moriarity, G. J. Bourke, and L. Daly: A case–control study of head and neck cancer in the Republic of Ireland. *Br J Cancer* 43:177, 1981.

66. K. Matsuo, N. Hamajima, M. Shinoda, S. Hatooka, M. Inoue, et al.: Gene–environment interaction between an aldehyde dehydrogenase-2 (ALDH-2) polymorphism and alcohol consumption for the risk of esophageal cancer. *Carcinogenesis* 22:913, 2001.

67. J. L. Kelsey and M. D. Gammon: The epidemiology of breast cancer. *CA Cancer J Clin* 41:146, 1991.

68. T. J. Key, N. E. Allen, E. A. Spencer, and R. C. Travis: The effect of diet on risk of cancer. *Lancet* 360:861, 2002.

69. W. C. Willett: Diet and cancer: One view at the start of the millennium. *Cancer Epidemiol Biomarkers Prev* 10:3, 2001.

70. E. Giovannucci and W. C. Willett: Dietary factors and risk of colon cancer. *Ann Med* 26: 443, 1994.

71. P. Terry, E. Giovannucci, K. B. Michels, L. Bergkvist, H. Hansen, et al.: Fruit, vegetables, dietary fiber, and risk of colorectal cancer. *J Natl Cancer Inst* 93:525, 2001.

72. J. H. Cohen, A. R. Kristal, and J. L. Stanford: Fruit and vegetable intakes and prostate cancer risks. *J Natl Cancer Inst* 92:61, 2000.

73. S. D. Mark, Y-L. Qiao, S. M. Dawsey, Y-P. Wu, H. Katki, et al.: Prospective study of serum selenium levels and incident esophageal and gastric cancers. *J Natl Cancer Inst* 92:1753, 2000.

74. L. C. Clark, B. Dalkin, A. Krongrad, et al.: Decreased incidence of prostate cancer with selenium supplementation: Results of a double-

blind cancer prevention trial. *Br J Urol* 81:730, 1998.

75. O. P. Heinonen, D. Albanes, J. Virtamo, et al.: Prostate cancer and supplementation with α-tocopherol and β-carotene: Incidence and mortality in a controlled trial. *J Natl Cancer Inst* 90:440, 1998.

76. A. Lacassagne: Apparition d'adénocarcinomes mammaires chez des souris mâles, traiteés par une substance oestrogéne synthétique. *Compt Rend Soc Biol* 129:641, 1938.

77. J. E. Bruni and D. G. Monetmurro: Effect of pregnancy, lactation, and pituitary isografts on the genesis of spontaneous mammary gland tumors in the mouse. *Cancer Res* 31:1903, 1971.

78. C. J. Bradley, G. S. Kledzik, and J. Meites: Prolactin and estrogen dependency of rat mammary cancers at early and late stages of development. *Cancer Res* 36:319, 1976.

79. B. E. Henderson, M. C. Pike, V. R. Gerkins, and J. T. Casagrande: The hormonal basis of breast cancer: Elevated plasma levels of estrogen, prolactin, and progesterone. In H. H. Hiatt, J. D. Watson, and J. A. Winsten, eds.: *Origins of Human Cancer*. Cold Spring Harbor, NY: Cold Spring Harbor Laboratory, 1977, pp. 77–86.

80. M. R. Melamed: Diagnostic cytology of urinary tract carcinoma. *Cancer* 8:287, 1972.

81. International Agency for Research on Cancer, Report of an IARC Working Group: An evaluation of chemicals and industrial processes associated with cancer in humans based on human and animal data: IARC Monographs Volumes 1 to 20. *Cancer Res* 40:1, 1980.

82. M. Aksoy, E. Erdem, and G. Dincol: Leukemia in shoe-workers exposed chronically to benzene. *Blood* 44:837, 1974.

83. P. F. Infante, J. K. Wagoner, R. A. Rinsky, and R. J. Young: Leukaemia in benzene workers. *Lancet* 2:76, 1977.

84. F. J. van Schooten, F. E. van Leeuwen, M. J. X. Hillebrand, M. E. deRijke, A. A. M. Hart, et al.: Determination of benzo[a]pyrene diol epoxide-DNA adducts in white blood cell DNA from coke-oven workers: the impact of smoking. *J Natl Cancer Inst* 82:927, 1990.

85. E. S. Johnson, W. Parsons, C. R. Weinberg, D. L. Shore, J. Mathews, et al.: Current serum levels of 2,3,7,8-tetrachlorodibenzo-p-dioxin in phenoxy acid herbicide applicators and characterization of historical levels. *J Natl Cancer Inst* 84:1648, 1992.

86. E. S. Johnson: Human exposure to 2,3,7,8-TCDD and risk of cancer. *Crit Rev Toxicol* 21:451, 1992.

87. H. I. Morrison, K. Wilkins, R. Semenciw, Y. Mao, and D. Wigle: Herbicides and cancer. *J Natl Cancer Inst* 84:1866, 1992.

88. H. H. McDuffie, P. Pahwa, J. R. McLaughlin, J. J. Spinelli, S. Fincham, et al.: Non-Hodgkin's lymphoma and specific pesticide exposures in men: Cross-Canada study of pesticides and health. *Cancer Epidemiol Biomarkers Prev* 10: 11155, 2001.

89. A. Blair: Herbicides and non-Hodgkin's lymphoma: New evidence from a study of Saskatchewan farmers. *J Natl Cancer Inst* 82:544, 1990.

90. Environmental pollution. In B. W. Stewart and P. Kleihues, eds.: *World Cancer Report, World Health Organization*. Lyon: IARC Press, 2003, pp. 39–42.

91. W. Haenszel, D. B. Loveland, and M. G. Sirken: Lung-cancer mortality as related to residence and smoking histories. I. White males. *J Natl Cancer Inst* 28:947, 1962.

92. V. E. Archer, J. D. Gillam, and J. K. Wagoner: Respiratory disease mortality among uranium workers. *Ann N Y Acad Sci* 271:280, 1976.

93. W. J. Nicolson: Asbestos—The TLV approach. *Ann N Y Acad Sci* 271:152, 1976.

94. E. C. Hammond, I. J. Selikoff, and H. Seidman: Asbestos exposure, cigarette smoking, and death rates. *Ann N Y Acad Sci* 330:473, 1979.

95. R. L. Jolley: Chlorination effects on organic constituents in effluents from domestic sanitary sewage treatment plants. Publication No. 565. Oak Ridge, TN: Environmental Science Division, Oak Ridge National Laboratory, 1973.

96. J. J. Rook: Formation of haloforms during chlorination of natural waters. *J Soc Water Treat Exam* 23:234, 1974.

97. U.S. Environmental Protection Agency (EPA): Preliminary assessment of suspected carcinogens in drinking water. Report to Congress. Washington, D.C.: EPA, 1975.

98. G. A. Junk and S. E. Stanley: Organics in drinking water. Part I: Listing of identified chemicals. Springfield, VA: National Technical Information Services, 1975.

99. T. A. DeRouen and J. E. Diem: Relationships between cancer mortality in Louisiana drinking water source and other possible causative agents. In H. H. Hiatt, J. D. Watson, and J. A. Winsten, eds.: *Origins of Human Cancer*. Cold Spring Harbor, NY: Cold Spring Harbor Laboratory, 1977, pp. 331–345.

100. J. Higginson and A. G. Oettlé: Cancer incidence in the Bantu and "Cape Colored" races of South Africa: Report of a cancer survey in the Transvaal (1953–55). *J Natl Cancer Inst* 24:589, 1960.

101. J. Higginson and C. S. Muir: Environmental carcinogenesis: Misconceptions and limitations to cancer control. *J Natl Cancer Inst* 63:1291, 1979.

102. J. Higginson: Perspectives and future developments in research on environmental carcinogenesis. In A. C. Griffin and C. R. Shaw, eds.: *Carcinogens: Identification and Mechanism of Action*. New York: Raven Press, 1979, pp. 187–208.

103. J. L. Lyon, M. R. Klauber, J. W. Gardner, M. P. H. Melville, and C. R. Smart: Cancer incidence in Mormans and non-Mormans in Utah, 1966–1970. *N Engl J Med* 294:129, 1976.

104. A. J. Fox and A. M. Adelstein: Occupational mortality: Work or way of life? *J Epidemol Community Health* 32:73, 1978.

105. W. Dubreuilh: Epithéliomatose d'origins solaire. *Ann Dermatol Syphiliq* 8:837, 1907.

106. G. M. Findlay: Ultra-violet light and skin cancer. *Lancet* 2:1070, 1928.

107. E. L. Scott and M. L. Straf: Ultraviolet radiation as a cause of cancer. In H. H. Haitt, J. D. Watson, and J. A. Winsten, eds.: *Origins of Human Cancer*. Cold Spring Harbor, NY: Cold Spring Harbor Laboratory, 1977, pp. 529–546.

108. S. Feygin: *Du Cancer Radiologique*. Paris: Rousset, 1915.

109. H. C. March: Leukemia in radiologists. *Radiology* 43:275, 1944.

110. H. S. Martland and R. E. Humphries: Osteogenic sarcoma in dial painters using luminous paint. *Arch Pathol* 7:406, 1929.

111. J. S. Horta, L. C. Da Motta, and M. H. Tavares: Thorium dioxide effects in man. Epidemiological, clinical, and pathological studies. *Environ Res* 8:131, 1974.

112. L. H. Hempelmann, W. J. Hall, M. Phillips, R. A. Cooper, and W. R. Ames: Neoplasms in persons treated with X-rays in infancy: Fourth survey in 20 years. *J Natl Cancer Inst* 55:519, 1975.

113. A. C. Upton: Radiation effects. In H. H. Haitt, J. D. Watson, and J. A. Winsten, eds.: *Origins of Human Cancer*. Cold Spring Harbor, NY: Cold Spring Harbor Laboratory, 1977, pp. 477–500.

114. J. L. Kelsey and M. D. Gammon: The epidemiology of breast cancer. *CA Cancer J Clin* 41:146, 1991.

115. N. H. Harley and J. H. Harley: Potential lung-cancer risk from indoor radon exposure. *CA Cancer J Clin* 40:265, 1990.

116. J. Schoenberg and J. Klotz: A case–control study of radon and lung cancer among New Jersey women. New Jersey State Department of Health Technical Report, Phase I. Trenton, NJ: New Jersey State Department of Health, 1989.

117. W. J. Blot, Z.-Y. Xu, J. D. Boice, Jr., D.-Z. Zhao, B. J. Stone, et al.: Indoor radon and lung cancer in China. *J Natl Cancer Inst* 12:1025, 1990.

118. A. V. Nero, M. B. Schwehr, W. W. Nazaroff, and K. L. Revzan: Distribution of airborne radon-222 concentrations in United States homes. *Science* 234:992, 1986.

119. Medicinal drugs. In B. W. Stewart and P. Kleihues, eds.: *World Cancer Report*, World Health Organization. Lyon: IARC Press, 2003, pp. 48–50.

120. C. A. DeLaat and B. C. Lampkin: Long-term survivors of childhood cancer: Evaluation and identification of sequelae of treatment. *CA Cancer J Clin* 42:263, 1992.

121. J. P. Neglia, A. T. Meadows, L. L. Robison, T. H. Kim, W. A. Newton, F. et al.: Second neoplasms after acute lymphoblastic leukemia in childhood. *N Engl J Med* 325:1330, 1991.

122. L. B. Travis, D. A. Hill, G. M. Dores, M. Gospodarowicz, F. E. van Leeuwen, et al.: Breast cancer following radiotherapy and chemotherapy among young women with Hodgkin disease. *JAMA* 290:465, 2003.

123. P. E. Oyer, E. B. Stinson, S. W. Jamieson, S. A. Hunt, M. Billingham, et al.: Cyclosporin-A in cardiac allografting: A preliminary experience. *Transplant Proc* 15:1247, 1983.

124. International Agency for Research on Cancer, Report on an IARC Working Group: An evaluation of chemicals and industrial processes associated with cancer in humans based on human and animal data: IARC Monographs, Vols. 1–20. *Cancer Res* 40:1, 1980.

125. M. C. Pike, M. D. Krailo, B. E. Henderson, J. T. Casagrande, and D. G. Hoel: "Hormonal" risk factors, "breast tissue age," and the age-incidence of breast cancer. *Nature* 303:767, 1983.

126. P. C. MacDonald, C. D. Edman, D. L. Hemsell, J. C. Porter, and P. K. Siiteri: Effect of obesity on conversion of plasma androstenedione to estrone in postmenopausal women with and without endometrial cancer. *Am J Obstet Gynecol* 130:448, 1978.

127. R. K. Ross, A. Paganini-Hill, P. C. Wan, and M. C. Pike: Effect of hormone replacement therapy on breast cancer risk: Estrogen versus estrogen plus progestin. *J Natl Cancer Inst* 92:328, 2000.

128. C. Schairer, J. Lubin, R. Troisi, S. Sturgeon, L. Brinton, et al.: Menopausal estrogen and estrogen–progestin replacement therapy and breast cancer risk. *JAMA* 283:485, 2000.

129. Million Women Study Collaborators: Breast cancer and hormone-replacement therapy in the Million Women Study. *Lancet* 362:419, 2003.

130. K. Armstrong, A. Eisen, and B. Weber: Primary care—Assessing the risk of breast cancer. *N Engl J Med* 342:564, 2000.

131. A. L. Herbst, P. Cole, T. Colton, R. E. Scully, and S. J. Robboy: Age-incidence and risk of DES-related clear cell adenocarcinoma of the vagina and cervix. *Am J Obstet Gynecol* 128:43, 1976.

132. B. Safai, B. Diaz, and J. Schwartz: Malignant neoplasms associated with human immunodeficiency virus infection. *CA Cancer J Clin* 42:74, 1992.

133. W. B. Ershler and D. L. Longo: Aging and cancer: Issues of basic and clinical science. *J Natl Cancer Inst* 89:1489, 1997.

134. R. A. DePinho: The age of cancer. *Nature* 408:248, 2000.

135. A. F. Olumi, G. D. Grossfeld, S. W. Hayward, P. R. Carroll, T. D. Tlsty, and G. R. Cunha: Carcinoma-associated fibroblasts direct tumor

progression of initiated human prostatic epithelium. *Cancer Res* 59:5002, 1999.

136. D. H. Ly, D. J. Lockhart, R. A. Lerner, and P. G. Schultz: Mitotic misregulation and human aging. *Science* 287:2486, 2000.

137. C. T. Murphy, S. A. McCarroll, C. I. Bargmann, A. Fraser, R. S. Kamath, et al.: Genes that act downstream of DAF-16 to influence the lifespan of *Caenorhabditis elegans*. *Nature* 424: 277, 2003.

138. J. Campisi: Aging, chromatin, and food restriction—Connecting the dots. *Science* 289: 2062, 2000.

139. S. Dukan, A. Farewell, M. Ballesteros, F. Taddei, M. Radman, et al.: Protein oxidation in response to increased transcriptional or translational errors. *Proc Natl Acad Sci USA* 97:5746, 2000.

140. S. Melov, J. Ravenscroft, S. Malik, M. S. Gill, D. W. Walker, et al.: Extension of life-span with superoxide dismutase/catalase mimetics. *Science* 289:1567, 2000.

141. Genetic susceptibility. In B. W. Stewart and P. Kleihues, eds.: *World Cancer Report*, *World Health Organization*. Lyon: IARC Press, 2003, pp. 71–75.

142. A. G. Knudson, Jr.: Genetic predisposition to cancer. In H. H. Hiatt, J. D. Watson, and J. A. Winstein, eds.: *Origins of Human Cancer*. Cold Spring Harbor, NY: Cold Spring Harbor Laboratory, 1977, pp. 45–52.

143. H. T. Lynch, R. E. Harris, H. A. Giurgis, K. Maloney, L. L. Carmody, et al.: Familial association of breast/ovarian carcinoma. *Cancer* 41:1543, 1978.

144. D. W. Nebert and W. W. Weber: Chapter 7. In W. B. Pratt and P. Taylor, eds.: *Pharmacogenetics in Principles of Drug Action: The Basis of Pharmacology*. New York: Churchill Livingstone, 1990, pp. 469–531.

144a. X. Wu, M. R. Spitz, C. I. Amos, J. Lin, L. Shao, et al.: Mutagen sensitivity has high heritability: Evidence from a twin study. *Cancer Res* 66: 5993, 2006.

145. R. Doll: Introduction, In H. H. Hiatt, J. D. Watson, and J. A. Winstein, eds.: *Origins of Human Cancer*. Cold Spring Harbor, NY: Cold Spring Harbor Laboratory, 1977, pp. 1–12.

146. B. N. Ames: Identifying environmental chemicals causing mutations and cancer. *Science* 204:587, 1979.

147. B. N. Ames, W. E. Durston, E. Yamasaki, and F. D. Lee: Carcinogens are mutagens: A simple test system combining liver homogenates for activation and bacteria for detection. *Proc Natl Acad Sci USA* 70:2281, 1973.

148. J. H. Weisburger and G. M. Williams: Carcinogen testing: Current problems and new approaches. *Science* 214:401, 1981.

149. R. D. Kuick, J. V. Neel, J. R. Strahler, E. H. Y. Chu, R. Bargal, et al.: Similarity of spontaneous germinal and in vitro somatic cell mutation rates in humans: Implications for carcinogenesis and for the role of exogenous factors in "spontaneous" germinal mutagenesis. *Proc Natl Acad Sci USA* 89:7036, 1992.

150. D. G. Hoel, N. L. Kaplan, and M. W. Anderson: Implication of nonlinear kinetics on risk estimation in carcinogenesis. *Science* 219:1032, 1983.

151. S. Woloshin, L. M. Schwartz, and H. G. Welch: Risk charts: Putting cancer in context. *J Natl Cancer Inst* 94:799, 2002.

152. J. Kaiser: No meeting of minds on childhood cancer. *Science* 286:1832, 1999.

153. R. Twombly: Federal carcinogen report debuts new list of nominees. *J Natl Cancer Inst* 93: 1372, 2001.

154. P. Boffetta, A. Agudo, W. Ahrens, E. Benhamou, S. Benhamou, et al.: Multicenter case–control study of exposure to environmental tobacco smoke and lung cancer in Europe. *J Natl Cancer Inst* 90:1440, 1998.

155. D. Wartenberg, E. E. Calle, M. J. Thun, C. W. Heath, Jr., C. Lally, et al.: Passive smoking exposure and female breast cancer mortality. *J Natl Cancer Inst* 92:1666, 2000.

156. P. Reynolds, S. Hurley, D. E. Goldberg, H. Anton-Culver, L. Bernstein, et al.: Active smoking, household passive smoking, and breast cancer: Evidence from the California Teachers Study. *J Natl Cancer Inst* 96:29, 2004.

157. J. E. Entrom and G. C. Kabat: Environmental tobacco smoke and tobacco related mortality in a prospective study of Californians, 1960–98. *BMJ* 326:1–10, 2003.

158. J. M. Samet: Indoor radon exposure and lung cancer: Risky or not?—All over again. *J Natl Cancer Inst* 89:4, 1997.

159. H. Frumkin and J. M. Samet: Radon. *CA Cancer J Clin* 51:337, 2001.

160. A. Auvinen, I. Mäekeläinen, M. Hakama, O. Castrén, E. Pukkala, et al.: Indoor radon exposure and risk of lung cancer: A nested case–control study in Finland. *J Natl Cancer Inst* 88:966, 1996.

161. J. H. Lubin and J. D. Boice, Jr.: Lung cancer risk from residential radon: Meta-analysis of eight epidemiologic studies. *J Natl Cancer Inst* 89:49, 1997.

162. F. Pisa, F. Barbone, A. Betta, et al.: Residential radon and risk of lung cancer in an Italian alpine area. *Arch Environ Health* 56:208, 2001.

163. F. Lagarde, G. Axelsson, L. Damber, et al.: Residential radon and lung cancer among never smokers in Sweden. *Epidemiology* 12:396, 2001.

164. J. S. Neuberger and T. F. Gesell: Childhood cancers, radon, and γ radiation. *Lancet* 360:1437, 2002.

165. C. Johansen, J. D. Boice, Jr., J. K. McLaughlin, and J. H. Olsen: Cellular telephones and cancer—A nationwide cohort study in Denmark. *J Natl Cancer Inst* 93:203, 2001.

166. M. Feychting and A. Ahlbom: Magnetic fields and cancer in children residing near Swedish

high-voltage power lines. *Am J Epidemiol* 138:
467, 1993.

167. M. S. Linet, E. E. Hatch, R. A. Kleinerman, L. L.
Robison, W. T. Kaune, et al.: Residential exposure to magnetic fields and acute lymphoblastic
leukemia in children. *N Engl J Med* 337:1, 1997.

168. M. Longnecker: Alcoholic beverage consumption in relation to risk of breast cancer: Meta-
analysis and review. *Cancer Causes Control*
5:73, 1994.

169. S. A. Smith-Warner, D. Spiegelman, S. S. Yaun,
P. A. van den Brandt, A. R. Folsom, et al.:
Alcohol and breast cancer in women: A pooled
analysis of cohort studies. *JAMA* 279:535, 1998.

170. C. Byrne, P. M. Webb, T. W. Jacobs, G. Peiro,
S. J. Schnitt, et al.: Alcohol consumption and
incidence of benign breast disease. *Cancer Epidemiol Biomarkers Prev* 11:1369, 2002.

171. H. S. Feigelson, C. R. Jonas, A. S. Robertson,
M. L. McCullough, M. J. Thun, et al.: Alcohol,
folate, methionine and risk of incident breast
cancer in the American Cancer Society Cancer
Prevention Study II Nutrition Cohort. *Cancer
Epidemiol Biomarkers Prev* 12:161, 2003.

172. T. Zheng, T. R. Holford, S. T. Mayne, J. Tessari,
B. Ward, et al.: Risk of female breast cancer
associated with serum polychlorinated biphenyls
and 1,1-Dichloro-2,2'-bis(*p*-chlorophenyl)ethyl-
ene. *Cancer Epidemiol Biomarkers Prev* 9:167,
2000.

173. Breast cancer in New York State by zip code,
1993–1997. New York State Department of
Health, 2000. Access at http://www.health.state
.ny.us.

174. R. Trombley: Long Island study finds no link
between pollutants and breast cancer. *J Natl
Cancer Inst* 94:1348, 2002.

175. A. Demers, P. Ayotte, J. Brisson, S. Dodin, J.
Robert, et al.: Risk and aggressiveness of breast
cancer in relation to plasma organochlorine
concentrations. *Cancer Epidemiol Biomarkers
Prev* 9:161, 2000.

176. F. Laden, G. Collman, K. Iwamoto, A. J. Alberg,
G. S. Berkowitz, et al.: 1,1-Dichloro-2,2-bis
(*p*-chlorophenyl)ethylene and polychlorinated
biphenyls and breast cancer: Combined analysis
of five U.S. studies. *J Natl Cancer Inst* 93:768,
2001.

177. E. M. Ward, P. Schulte, B. Grajewski, A. Ander-
sen, D. Patterson, Jr., et al.: Serum organochlorine levels and breast cancer: A nested case–
control study of Norwegian women. *Cancer
Epidemiol Biomarkers Prev* 9:1357, 2000.

177a. D. M. Winn: The Long Island Breast Cancer
Study Project. *Nature Rev Cancer* 5:986,2005.

178. K. J. Helzlsouer, A. J. Alberg, H-Y. Huang,
S. C. Hoffman, P. T. Strickland, et al.: Serum concentrations of organochlorine compounds and the subsequent development of
breast cancer. *Cancer Epidemiol Biomarkers
Prev* 8:525, 1999.

179. M. D. Gammon, R. M. Santella, A. I. Neugut,
S. M. Eng, S. L. Teitelbaum, et al.: Environmental toxins and breast cancer on Long Island.
I. Polycyclic aromatic hydrocarbon DNA adducts. *Cancer Epidemiol Biomarkers Prevention* 11:677, 2002.

180. A. S. Robbins, S. Brescianini, and J. L. Kelsey:
Regional differences in known risk factors and
the higher incidence of breast cancer in San
Francisco. *J Natl Cancer Inst* 89:960, 1997.

181. D. K. Mirick, S. Davis, and D. B. Thomas:
Antiperspirant use and the risk of breast cancer.
J Natl Cancer Inst 94:1578, 2002.

182. H. Komulainen, V. M. Kosma, S. L. Vaitti-
nen, T. Vartianen, E. Kaliste-Korhonen, et al.:
Carcinogenicity of the drinking water mutagen
3-chloro-4-(dichloromethyl)–5-hydroxy-2(5H)-
furanone in the rat. *J Natl Cancer Inst* 89:848,
1997.

183. R. L. Melnick, G. A. Boorman, and V. Dellarco:
Water chlorination, 3-Chloro-4-(dichloromethyl)-
5-hydroxy-2(5H)-furanone (MX), and potential
cancer risk. *J Natl Cancer Inst* 89:832, 1997.

184. M. Mahue-Giangreco, G. Ursin, J. Sullivan-
Halley, and L. Bernstein: Induced abortion, miscarriage, and breast cancer risk of young women.
Cancer Epidemiol Biomarkers Prev 12:209,
2003.

185. M. Camus, J. Siemiatycki, and B. Meek: Nonoccupational exposure to chrysotile asbestos and
the risk of lung cancer. *N Engl J Med* 338:1565,
1998.

186. S. Takayama, S. M. Sieber, R. H. Adamson, U. P.
Thorgeirsson, D. W. Dalgard, et al.: Long-term
feeding of sodium saccharin to nonhuman
primates: Implications for urinary tract cancer.
J Natl Cancer Inst 90:19, 1998.

187. M. Mitka: Fear of frying: Is acrylamide in foods
a cancer risk? *JAMA* 288:2105, 2002.

188. Health scares based on animal carcinogen
testing. In K. Meister, E. M. Whelan, G. L. Ross,
and A. N. Stimola, eds.: *America's War on
Cancer*. New York: American Council on Science and Health, 2005, pp. 117–137.

189. H. D. Strickler, J. J. Goedert, S. S. Devesa,
J. Lahey, J. F. Fraumeni, et al.: Trends in U.S.
pleural mesothelioma incidence rates following simian virus 40 contamination of early
poliovirus vaccines. *J Natl Cancer Inst* 95:38,
2003.

190. E. A. Engels, H. A. Katki, N. M. Nielsen, J. F.
Winther, H. Hjalgrim, et al.: Cancer incidence
in Denmark following exposure to poliovirus
vaccine contaminated with simian virus 40. *J
Natl Cancer Inst* 95:532, 2003.

4

The Biochemistry and Cell
Biology of Cancer

HISTORICAL PERSPECTIVES

The development of knowledge about the biochemistry and cell biology of cancer comes from a number of disciplines. Some of this knowledge has come from research initiated a century or more ago. There has been a flow of information about genetics into a knowledge base about cancer, starting with Gregor Mendel and the discovery of the principle of inherited traits and leading through Theodor Boveri's work on the chromosomal mode of heredity and chromosomal damage in malignant cells[1] to Avery's discovery of DNA as the hereditary principle,[2] Watson and Crick's determination of the structure of DNA,[3] the human genome project, DNA microarrays, and proteomics. Not only has this information provided a clearer picture of the carcinogenic process, it has also provided better diagnostic approaches and new therapeutic targets for anticancer therapies.

Once cell culture techniques were developed[4] it became possible to test which genes are involved in malignant transformation and progression. This field of research led to the discovery of oncogenes[5] and tumor suppressor genes.[6] Hereditary studies led to the two-hit theory[7] and the concept of the hereditary nature of some cancers.[8] Chromosomal staining techniques enabled Nowell and Hungerford[9] and Rowley[10] to identify chromosomal translocation as a tumor initiating event.

Studies in yeast produced the concept of cell cycle checkpoints,[11] and investigations with C. elegans found genes involved in apoptosis.[12] The cell cycle began to be studied in great detail in lower organisms, and organisms such as clams, yeast, and fruit flies have contributed greatly to our understanding of the cell cycle events.[13]

The findings that simple molecules like cyclic AMP could direct a whole panoply of cellular functions[14] led to the discovery of signal transduction pathways, which are now becoming favored molecular targets for anticancer drug discovery.

Much of what we originally knew about the biochemical differences between normal and malignant cells, however, was discovered in their patterns of enzymatic activity. In the 1920s, Warburg studied glycolysis in a wide variety of human and animal tumors and found that there was a general trend toward an increased rate of glycolysis in tumor cells.[15] He noted that when normal tissue slices were incubated in a nutrient medium containing glucose, but without oxygen, there was a high rate of lactic acid production (anaerobic glycolysis); however, if they were incubated with oxygen, lactic acid production virtually stopped. The rate of lactic acid production was higher in tumor tissue slices in the absence of oxygen than in normal tissues, and the presence of oxygen slowed, but did not eliminate, lactic acid formation in the tumor slices. Warburg concluded that cancer cells have an irreversible injury to their respiratory mechanism, which increases the rate of lactic acid production even in the presence of oxygen (aerobic glycolysis). He regarded the persistence of this type of

117

glycolysis as the crucial biochemical lesion in neoplastic transformation. This old idea still has some credence in that there are hypoxic areas in the core of tumors, where anaerobic metabolism predominates. This has clinical implications because hypoxic cells do not respond as well to certain anticancer drugs or radiation therapy. The ability of lactate and pyruvate, end points of glycolysis, to enhance tumor progression appears to be mediated by the activation of hypoxia inducible factor-1 (HIF-1).[16] In addition to increased activity of enzymes of the glycolytic pathway, such as hexokinase, phosphofructokinase, and pyruvate kinase in cancer cells, hypoxia is also a common feature of many human solid cancers. These effects have been linked to tumor progression, metastasis, and multidrug resistance.[17] Interestingly, oncogenes such as *ras*, *src*, and *myc* enhance aerobic glycolysis by increasing the expression of glucose transporters and glycolytic enzymes (reviewed in Reference 16).

Cancer cells react to hypoxic conditions by upregulating expression of HIF-1, which is a transcription factor that in turn up-regulates expression of genes involved in glycolysis, glucose transport (GLUT-1), angiogenesis (VEGF), cell survival, and erythropoiesis. HIF-1 expression has been observed in cancers of the brain, breast, colon, lung, ovary, and prostate and their metastases but not in the corresponding normal tissues. Its expression in tumors correlates with poor prognosis.

Interest in tumor metabolism has been stimulated once again by modern techniques such as position emission tomography (PET), sensitive mass spectrometry (MS), and high-resolution nuclear magnetic resonance spectroscopy (NMR). PET uses fluorine-18 labeled fluorodeoxyglucose (FdG) to detect tissue regions of high glucose uptake, which is indicative of up-regulated glycolysis and increased metabolic rate. FdG PET imaging has shown that most primary and metastatic human cancers have increased glucose uptake.[17] This finding is indicative of a "glycolytic switch" in cancer cells and may be a precursor of tumor angiogenesis and metastasis.[17]

NMR and MS can now be used to measure mestatic profiles of cancer cells and the metabolic phenotype of tissues and organs. This so-called science of "metabolomics" can provide metabolic biomarkers of tumors such as pro-duction of the end products of glycolysis, lipid levels indicative of cell membrane turnover, and alterations in amino acids and nucleotide levels.[18]

Since mitochondria contain the enzymatic cascades for oxidative metabolism, it has been suggested that damage to mitochondria may be involved in the disruptions of oxidative metabolism seen in malignant tumors. Mutations of mitochondrial DNA (mtDNA) has been observed in a variety of human cancers, including bladder, head and neck, lung,[18] and ovarian[19] cancers. Interestingly, in the bladder cancers, the mutation hot spots were primarily in a nicotinamide adenine dinucleotide dehydrogenase subunit, a key component of the electron transfer machinery. This suggests a mechanism for the alterations in oxidative metabolism seen in malignant cells. Because mitochondrial DNA is exposed to high levels of reactive oxygen species generated during oxidative phosphorylation, it is not surprising that mtDNA is highly susceptible to mutational events. The mutational rate of mtDNA has been estimated to be 10 times higher than that of nuclear DNA.[19] Mitochondria also play a key role in apoptosis (see section on apoptosis below), and alterations in those mitochondria-mediated events are seen in cancer cells.

In the early 1950s, Greenstein formulated the "convergence hypothesis" of cancer, which states that the enzymatic activity of malignant neoplasms tends to converge to a common pattern.[20] Although he recognized some exceptions to this rule, he considered the generalization, based mostly on repeatedly transplanted tumor models, to be valid. It is now more fully appreciated that even though cancer cells do have some commonly increased metabolic pathways, such as those involved in nucleic acid synthesis, there is tremendous biochemical heterogeneity among malignant neoplasms, and that there are many fairly well-differentiated cancers that do not have the common enzymatic alterations he suggested. Thus, cancers do not have a universally uniform malignant phenotype as exemplified by their enzyme patterns.

On the basis of work of about 60 years ago, which evolved from studies on the production of hepatic cancer by feeding aminoazo dyes, the Millers advanced the "deletion hypothesis" of cancer.[21] This hypothesis was based on the observation that a carcinogenic aminoazo dye covalently bound liver proteins in animals undergoing carcinogenesis, whereas little or no dye binding occurred with the protein of tumors induced by the dye. They suggested that carcinogenesis resulted from "a permanent alteration or loss of protein essential for the control of growth."

About 10 years later, Potter suggested that the proteins lost during carcinogenesis may be involved in feedback control of enzyme systems required for cell division,[22] and he proposed the "feedback deletion hypothesis."[23] In this hypothesis, Potter postulated that "repressors" crucial to the regulation of genes involved in cell proliferation are lost or inactivated by the action of oncogenic agents on the cell, either by interacting with DNA to block repressor gene transcription or by reacting directly with repressor proteins and inactivating them. This prediction anticipated the discovery of tumor suppressor proteins, such as p53 and RB, by about 25 years.

Biochemical studies of cancer were also aided by the so-called minimal-deviation hepatomas developed by Morris and colleagues.[24] These tumors were originally induced in rats by feeding them the carcinogens fluorenylphthalamic acid, fluorenylacetamide compounds, or trimethylaniline. These hepatocellular carcinomas are transplantable in an inbred host strain of rats and have a variety of growth rates and degrees of differentiation. They range from slowly growing, well-differentiated, karyotypically normal cells to rapidly growing, poorly differentiated, polyploid cells. All these tumors are malignant and eventually kill the host. The term "minimal deviation" was coined by Potter[23] to convey the idea that some of these neoplasms differ only slightly from normal hepatic parenchymal cells. The hypothesis was that if the biochemical lesions present in the most minimally deviated neoplasm could be identified, the crucial changes defining the malignant phenotype could be determined. As Weinhouse[25] indicated, studies of these tumors greatly advanced our knowledge of the biochemical characteristics of the malignant phenotype, and they have ruled out many secondary or nonspecific changes that relate more to tissue growth rate than to malignancy.

The extensive biochemical analyses of the Morris minimal-deviation hepatomas led Weber to formulate the "molecular correlation concept" of cancer, which states that "the biochemical strategy of the genome in neoplasia could be identified by elucidation of the pattern of gene expression as revealed in the activity, concentration, and isozyme aspects of key enzymes and their linking with neoplastic transformation and progression."[26] Weber proposed three general types of biochemical alterations associated with malignancy: (1) transformation-linked alterations that correlate with the events of malignant transformation and that are probably altered in the same direction in all malignant cells; (2) progression-linked alterations that correlate with tumor growth rate, invasiveness, and metastatic potential; and (3) coincidental alterations that are secondary events and do not correlate strictly with transformation or progression. Weber maintained that key enzymes, that is enzymes involved in the regulation of rate and direction of flux of competing synthetic and catabolic pathways, would be the enzymes most likely to be altered in the malignant process. In contrast, "non-key" enzymes, that is, enzymes that are not rate limiting and do not regulate reversible equilibrium reactions, would be of lesser importance. As one would expect, a number of enzyme activities that Weber and others have found to be altered in malignant cells are those involved in nucleic acid synthesis and catabolism. In general, the key enzymes in the de novo and salvage pathways of purine and pyrimidine biosynthesis are increased and the opposing catabolic enzymes are decreased during malignant transformation and tumor progression. Weber noted that the degree of neoplasia was related to the concentrations of certain regulators of key metabolic pathways. The question of why anaplastic, rapidly growing tumors tend to be biochemically alike, whereas more well-differentiated tumors display a vast array of phenotypic characteristics, was approached by Knox.[27] He thought that the vast bulk of biochemical components in tumor tissues are "normal," in the sense that they are produced by certain specialized adult normal cells or by normal cells at some stage of their differentiation. In cancer cells, it is the combination and proportions of these normal components that are abnormal. The biochemical diversity of cancer cells, then, would depend on the cell of origin of the neoplasm and its degree of neoplasticity.[22] All too frequently, even now, in the histopathologic or biochemical characterization of cancer, a biochemical component that is present or absent or increased or decreased is not considered in relation to the particular cell of origin of a tumor, its differentiation state, or its degree of neoplasticity.

Taken together, the data on enzyme patterns of cancer cells indicate that undifferentiated, highly malignant cells tend to resemble one another and fetal tissues more than their adult normal counterpart cells, whereas well-differentiated tumors tend to resemble their cell of origin more than other tumors. Of course, between these two extremes several levels of neoplastic gradation occur, leading to the vast biochemical heterogeneity of tumors. This heterogeneity also exists for tumors of the same tissue type arising in different patients or even in the same patient at different stages of the disease.

The fact that more undifferentiated tumors tend to converge to a more fetal-like state is evidenced by a frequently observed production of oncodevelopmental gene products. A number of cancer cell characteristics, such as invasiveness and "metastasis," are also seen in embryonic tissue. For example, the developing trophoblast invades the uterine wall during the implantation step of embryonic development. During organogenesis, embryonic cells dissociate themselves from the surrounding cells and migrate to new locations, a process not unlike metastasis.

GROWTH CHARACTERISTICS
OF MALIGNANT CELLS

As will be discussed more in Chapter 5, most cancers (other than those for which there is a dominantly inherited cancer susceptibility gene) are an acquired molecular genetic disease in which a single (or a few) clone(s) of cells accumulate cellular genetic changes that progress to the full-blown cancer phenotype. Cancer can be characterized as a disease of genetic instability, altered cellular behavior, and altered cell–extracellular matrix interactions.[28] These alterations lead to dysregulated cell proliferation, and ultimately to invasion and metastasis. There are interactions between the genes involved in these steps. For example, the genes associated with loss of control of cell proliferation may also

be involved in genetic instability (rapidly proliferating cells have less time to repair DNA damage) and tumor vascularization that leads to dysregulated proliferation of cells, which in turn eats up more oxygen, creates hypoxia, and turns on HIF-1 and additional angiogenesis. Similarly, genes involved in tumor cell invasion may also be involved in loss of growth control (invasive cells have acquired the skills to survive in "hostile" new environments) and evasion of apoptosis (less cell death even in the face of a normal rate of cell proliferation produces more cells). As will be described in more detail below, the molecular genetic alterations of cancer cells lead to cells that can generate their own growth-promoting signals, are less sensitive to cell cycle checkpoint controls, evade apoptosis, and thus have almost limitless replication potential. The signal transduction mechanisms involved in this replication potential will also be discussed in this chapter. As will become clearer, these signaling pathways are interlinked. As was not initially realized, cancer cells have multiple proliferative pathways and can bypass an interdiction of one or more of these. This redundancy makes design of effective signal transduction-targeted chemotherapeutic drugs that target a single pathway very difficult indeed.

Cancer cells can also subvert the environment in which they proliferate. Alterations in both cell–cell and cell–extracellular matrix interactions also occur, leading to creation of a cancer-facilitating environment. For example, a common alteration in epithelial carcinomas is alteration of E-cadherin expression. E-cadherin is a cell–cell adhesion molecule found on all epithelial cells.

Cancer cells exhibit remarkable plasticity. Malignant cells have the ability to mimic some of the characteristics of other cell types as they progress and became less well differentiated. For example, cancer cells may assume some of the structure and function of vascular cells.[28] As cancer cells metastasize, they may eventually take on a new phenotype such that the tissues of origin may become unclear—so-called cancers of unknown primary site.

Phenotypic Alterations in Cancer Cells

Treatment of animals or cells in culture with carcinogenic agents is a means of studying discrete biochemical events that lead to malignant transformation. Studies of cell transformation in vitro, however, have many pitfalls. These "tissue culture artifacts" include overgrowth of cells not characteristic of the original population of cultured cells (e.g., overgrowth of fibroblasts in cultures that were originally primarily epithelial cells), selection for a small population of variant cells with continued passage in vitro, or appearance of cells with an abnormal chromosomal number or structure (karyotype). Such changes in the characteristics of cultured cell populations can lead to "spontaneous" transformation that mimics some of the changes seen in populations of cultured cells treated with oncogenic agents. Thus, it is often difficult to sort out the critical malignant events from the noncritical ones. Although closer to the carcinogenic process in humans, malignant transformation induced in vivo by treatment of susceptible experimental animals with carcinogenic chemicals or oncogenic viruses or by irradiation is even more difficult because it is hard to discriminate toxic from malignant events and to determine what role a myriad of factors, such as the nutritional state of the animal, hormone levels, or endogenous infections with microorganisms or parasites, might have on the in vivo carcinogenic events. Moreover, tissues in vivo are a mixture of cell types, and it is difficult to determine in which cells the critical transformation events are occurring and what role the microenvironment of the tissue plays. Thus, most studies designed to identify discrete biochemical events occurring in cells during malignant transformation have been done with cultured cells, since clones of relatively homogeneous cell populations can be studied and the cellular environment defined and manipulated. The ultimate criterion that establishes whether cells have been transformed, however, is their ability to form a tumor in an appropriate host animal. The generation of immortalized "normal" cell lines of a given differentiated phenotype from human embryonic stem cells has enhanced the ability to study cells of a normal genotype from a single source.[29] Such cell lines may also be generated by transfection of the telomerase gene into cells to maintain chromosomal length.

Over the past 60 years, much scientific effort has gone into research aimed at identifying the phenotypic characteristics of in vitro–transformed cells that correlate with the growth of a cancer in vivo. This research has tremendously increased our knowledge of the biochemistry of cancer cells. However, many of the biochemical characteristics initially thought to be closely associated with the malignant phenotype of cells in culture have subsequently been found to be dissociable from the ability of those cells to produce tumors in animals. Furthermore, individual cells of malignant tumors growing in animals or in humans exhibit marked biochemical heterogeneity, as reflected in their cell surface composition, enzyme levels, immunogenicity, response to anticancer drugs, and so on. This has made it extremely difficult to identify the essential changes that produce the malignant phenotype. Hahn et al.[30] have shown that ectopic expression of the human telomerase catalytic subunit (hTERT) in combination with the oncogenes h-ras and SV40 virus large-T antigen can induce tumorigenic conversion in normal human epithelial and fibroblast cells, suggesting that disruption of the intracellular pathways regulated by these gene products is sufficient to produce a malignant cell.

Since hyperproliferative conditions in patients, such as inflammatory bowel disease or psoriasis, may mimic some of the characteristics of malignant cells, it is important to use a number of criteria that define the malignant state. The evidence that these phenotypic properties found in transformed cells in culture are related to malignant neoplasia in vivo is discussed below.

Immortality of Transformed Cells in Culture

Most normal diploid mammalian cells have a limited life expectancy in culture. For example, normal human fibroblast lines may live for 50 to 60 population doublings (the "Hayflick index"), but then viability begins to decrease rapidly unless they transform spontaneously or are transformed by oncogenic agents. However, malignant cells, once they become established in culture, will generally live for an indefinite number of population doublings, provided the right nutrients and growth factors are present. It is not clear what limits the life expectancy of normal diploid cells in culture, but it may be related to the continual shortening of chromosomal telomeres each time cells divide. Transformed cells are known to have elevated levels of telomerase that maintain telomere length. Transformed cells that become established in culture also frequently undergo karyotypic changes, usually marked by an increase in chromosomes (polyploidy), with continual passage. This suggests that cells with increased amounts of certain growth-promoting genes are generated and/or selected during continual passage in culture. The more undifferentiated cells from cancers of animals or patients also often have an atypical karyology, thus the same selection process may be going on in vivo with progression over time of malignancy from a lower to a higher grade.

Decreased Requirement for Growth Factors

Other properties that distinguish transformed cells from their nontransformed counterparts are decreased density-dependent inhibition of proliferation[31] and the requirement for growth factors for replication in culture. Cells transformed by oncogenic viruses have lower serum growth requirements than do normal cells.[32] For example, 3T3 fibroblasts transformed by SV40, polyoma, murine sarcoma virus, or Rous sarcoma virus are all able to grow in a culture medium that lacks certain serum growth factors, whereas uninfected cells are not (reviewed in Reference 33).

Cancer cells may also produce their own growth factors that may be secreted and activate proliferation in neighboring cells (paracrine effect) or, if the same malignant cell type has both the receptor for a growth factor and the means to produce the factor, self-stimulation of cell proliferation (autocrine effect) may occur. One example of such an autocrine loop is the production of tumor necrosis factor-alpha (TNF-α) and its receptor TNFR1 by diffuse large cell lymphoma.[34] Co-expression of TNFα and its receptor are negative prognostic indicators of survival, suggesting that autocrine loops can be powerful stimuli for tumor aggressiveness and thus potentially important diagnostic and therapeutic targets.

Loss of Anchorage Dependence

Most freshly isolated normal animal cells and cells from cultures of normal diploid cells do not grow well when they are suspended in fluid or a semisolid agar gel. If these cells make contact with a suitable surface, however, they attach, spread, and proliferate. This type of growth is called *anchorage-dependent growth*. Many cell lines derived from tumors and cells transformed by oncogenic agents are able to proliferate in suspension cultures or in a semisolid medium (methylcellulose or agarose) without attachment to a surface. This is called *anchorage-independent growth*. This property of transformed cells has been used to develop clones of malignant cells.[35] This technique has been widely used to compare the growth properties of normal and malignant cells. Another advantage that has been derived from the ability of malignant cells to grow in soft agar (agarose) is the ability to grow cancer cells derived from human tumors to test their sensitivity to chemotherapeutic agents and to screen for potential new anticancer drugs.[36]

Loss of Cell Cycle Control and Resistance to Apoptosis

This topic will be discussed in more detail below. Suffice it to say here that normal cells respond to a variety of suboptimal growth conditions by entering a quiescent phase in the cell division cycle, the G_0 state. There appears to be a decision point in the G_1 phase of the cell cycle, at which time the cell must make a commitment to continue into the S phase, the DNA synthesis step, or to stop in G_1 and wait until conditions are more optimal for cell replication to occur. If this waiting period is prolonged, the cells are said to be in a G_0 phase. Once cells make a commitment to divide, they must continue through S, G_2, and M to return to G_1. If the cells are blocked in S, G_2, or M for any length of time, they die. The events that regulate the cell cycle, called *cell cycle checkpoints*, are defined in more detail below. This loss of cell cycle check point control by cancer cells may contribute to their increased susceptibility to anticancer drugs. Normal cells have mechanisms to protect themselves from exposure to growth-limiting conditions or toxic agents by calling on these check point control mechanisms. Cancer cells, by contrast, can continue through these checkpoints into cell cycle phases that make them more susceptible to the cytotoxic effects of drugs or irradiation. For example, if normal cells accrue DNA damage due to ultraviolet (UV) or X-irradiation, they arrest in G_1 so that the damaged DNA can be repaired prior to DNA replication. Another check point in the G_2 phase allows repair of chromosome breaks before chromosomes are segregated at mitosis. Cancer cells, which exhibit poor or absent check point controls, proceed to replicate the damaged DNA, thus accounting for persisting and accumulating mutations.

Changes in Cell Membrane Structure and Function

The cell surface membrane (plasma membrane) plays an important role in the "social" behavior of cells, that is, communication with other cells, cell movement and migration, adherence to other cells or structures, access to nutrients in the microenvironment, and recognition by the body's immune system. Alterations of the plasma membrane in malignant cells may be inferred from a variety of properties that characterize their growth and behavior, for example, the loss of density-dependent inhibition of growth, decreased adhesiveness, loss of anchorage dependence, and invasiveness through normal tissue barriers. In addition, a number of changes in the biochemical characteristics of malignant cells' surfaces have been observed. These include appearance of new surface antigens, proteoglycans, glycolipids, and mucins, and altered cell–cell and cell–extracellular matrix communication.

Alterations in Cell Surface Glycolipids, Glycoproteins, Proteoglycans, and Mucins

Aberrant glycosylation was first suggested as the basis for the tumor-associated changes in glycolipids by the finding of a remarkable accumulation of fucose-containing glycolipids found in human adenoarcinomas, some of which were identified as lactofucopentaose-III-ceramide, lactofucopentaose-II-ceramide (Lewis A blood group glycolipid), and lactodifucohexose and lactodifucooctose ceramide (Lewis B glycolipid).[37] These identifications were confirmed once monoclonal antibodies (mAbs) were used to identify antigens definitively. A number of mAbs with preferential reactivity for tumor cells over nor-

mal cells have been shown to react with Lewis blood group antigens, such as Le^x, Le^a, Le^b, or their analogues.[37]

The biochemical characterization or the aberrant glycosylation of glycoproteins was also demonstrated in earlier studies. The presence of high-molecular-weight glycopeptides with altered glycosylation patterns was detected on transformed cells in early studies before they were clearly chemically identified.[38,39] Later, the chemical basis for some of the changes in tumor cell glycoproteins was attributed to the fact that the N-linked oligosaccharides of tumor cells contain more multiantennary structures than the oligosaccharides derived from normal cells.[40]

Tumor-associated carbohydrate antigens can be classified into three groups[37]: (1) epitopes expressed on both glycolipids and glycoproteins, (2) epitopes expressed only on glycolipids, and (3) epitopes expressed only on glycoproteins. To the first group belongs the lacto-series structure that is found in the most common human cancers, such as lung, breast, colorectal, liver, and pancreatic cancers. The common backbone structure for these epitopes is $Gal\beta1 \rightarrow 3GlcNAc\beta1 \rightarrow 3Gal$ (type 1 blood group) or $Gal\beta1 \rightarrow 4 GlcNac \beta1 \rightarrow 3 Gal$ (type 2 blood group). The second group of epitopes, expressed exclusively on glycolipids, is mostly on the ganglio- or globo-series structures. This series of epitopes is expressed abundantly only on certain types of human cancers, such as melanoma, neuroblastoma, small cell lung carcinoma, and Burkitt's lymphoma. The third group of epitopes, seen only on glycoproteins, consists of the multiantennary branches of N-linked carbohydrates and the alterations of O-linked carbohydrate chains seen in some mucins.

Tumor-associated carbohydrate antigens can also be classified by the cell types expressing them, as those (1) expressed on only certain types of normal cells (often only in certain developmental stages) and greatly accumulated in tumor cells; (2) expressed only on tumor cells, for example, altered blood group antigens or mucins; and (3) expressed commonly on normal cells but present in much higher concentrations on tumor cells, for example, the GM ganglioside in melanoma and Le^x in gastrointestinal cancer.[37]

A variety of chemical changes that can explain the altered glycosylation patterns in tumor cells have been identified. These result from three kinds of altered processes: (1) incomplete synthesis and/or processing of normally existing carbohydrate chains and accumulation of the resulting precursor form; (2) "neosynthesis" resulting from activation of glycosyltransferases that are absent or have low activity in normal cells; and (3) organizational rearrangement of tumor cell membrane glycolipids.[41]

Moreover, the glycosyl epitopes found in glycolipids and glycoproteins make up microdomains that are involved in cell adhesion and signal transduction events. They function as a "glycosynapse" (analogous to the "immunological synapse") in mediating these events.[42] The cell motility, altered adhesive properties, and invasiveness observed in cancer cells are regulated by these glycosynapse complexes.[42]

Interest in the carbohydrate components of cell surface glycolipids, glycoproteins, and proteoglycans has been heightened by the fact that many of the monoclonal antibodies developed to tumor cell–associated antigens recognize these carbohydrate moieties or peptide epitopes exposed by altered glycosylation. Moreover, many of these have turned out to be blood group–specific antigens or modifications of blood group–specific antigens, some of which are antigens seen at certain stages of embryonic development and thus fit the definition of oncodevelopmental antigens. Thus, the field of chemical glycobiology is making significant contributions to our understanding of the cell surface biochemistry of normal and malignant cells. The aberrantly expressed glycans on tumor cells regulate a number of aspects of tumor progression, including cell proliferation, tumor invasion, angiogenesis, and metastasis.[43] Some of these altered glycans are detectable in the bloodstream and can be used as biomarkers of tumor burden and response to therapy.[43] In addition, there are some glycan-targeted anticancer therapeutics being tested in clinical trials. There is some evidence that tumor glycosphingolipids shed into the bloodstream may impair host immunity to some tumors.[43]

Role of Glycosyl Transferases and Oligosaccharide Processing Enzymes

The substitution of additional carbohydrate moieties on blood group–related structures is not the only aberrant modification of glycoproteins or glycolipids observed in cancer cells. Increased branching of asparagine-linked oligosaccharides and incomplete processing of these oligosaccharides have also been noted in certain cell-surface as well as secretory glycoproteins.[44,45] The increased activity of specific N-acetylglucosaminyl transferases in tumor cells appears to be responsible for the appearance of tri- and tetra-antennary structures, whereas the analogous glycoprotein in normal cells is often a biantennary structure. Unusually high expression of N-acetylglucosaminyltransferase-IVa has been observed in human choriocarcinoma cell lines and may be the enzymatic basis for the formation of abnormal biantennary sugar chains on human chorionic gonadotropin (hCG) produced by these cells.[46] Similarly, the extra fucosylations that appear on membrane glycoproteins and glycolipids have been associated with the induction of an unusual α-fucosyltransferase in chemical carcinogen–induced precancerous rat liver and in the resulting hepatomas.[47] These investigations strongly suggest that the regulation of glycosyltransferase genes is important in malignant transformation. Other changes in glycosyl transferase activities include a decrease in β1, 3-galactosyl transferase β3 Gal-T5 in human adenocarcinomas compared to normal colon.[48]

All these data strongly support the idea that glycosylation patterns change during transformation of normal cells into malignant ones. Because cell–cell interactions, adhesion to extracellular matrices, regulation of cell proliferation, and recognition by the host's immune system are all profoundly affected by the composition of the cell surface, the entire social behavior of a cell could be altered by such changes.

Additional evidence for the importance of glycosylation patterns of cell surface glycoproteins and glycolipids in the malignant phenotype comes from the use of glycosylation inhibitors and oligosaccharide-processing inhibitors. For example, tunicamycin, an inhibitor of addition of N-linked glycans to nascent polypeptide chains, castanospermine, an inhibitor of glucosidase, and KI-8110, an inhibitor of sialyltransferase activity, all reduce the number of lung metastases in murine experimental tumor models.[49-51] In addition, swainsonine, an inhibitor of mannosidase II, was shown to reduce the rate of growth of human melanoma xenografts in athymic nude mice,[52] and castanospermine was observed to inhibit the growth of *v-fms* oncogene-transformed rat cells in vivo.[53] These results support the hypothesis that the synthesis of highly branched complex-type oligosaccharides are associated with the malignant phenotype and may provide tumor cells with a growth advantage.

Mucins

Mucins are a type of highly glycosylated glycoproteins that a variety of secretory epithelial cells produce. They are 50% to 80% carbohydrate by weight and function to lubricate and protect ductal epithelial cells. They contain O-linked glycans (serine- and threonine-linked) of various lengths and structures, depending on the tissue type in which they are produced. They are made in a wide variety of tissues, including the gastrointestinal tract, lung, breast, pancreas, and ovary. Tumors arising in these organs may have an altered glycosylation pattern that distinguishes them from the normal mucins and renders them immunogenic.

Total expression of the mucins is increased in many cancers and up-regulated in some normal tissues under different physiologic states (e.g., lactating mammary gland).[54] Increased expression of the mucin 1 gene (*muc1*) has been observed in most adenocarcinomas of the breast, lung, stomach, pancreas, prostate, and ovary. Although *muc1*-encoded mucin has been the most extensively studied, cancer-related alterations in other mucins have been observed. Moreover, it appears that some cells, both normal and cancer, can express more than one mucin. Focal aberrant expression of *muc2* and *muc3* has been frequently observed in a variety of adenocarcinomas.[55] However, in general, mucin genes appear to be independently regulated and their expression is organ and cell type specific.[55]

There is evidence for host immune recognition of the breast cancer mucin, in that cytotoxic T lymphocytes isolated from breast cancer patients recognize a mucin epitope expressed on the breast cancer cells.[56] The immune-recognized epitope involves the core protein that appears to be selectively exposed on breast, ovarian, and other carcinomas. It has also been demonstrated that patients can produce antibodies to cancer mucins,[54] and this is the basis for the proposal that glycopeptides, because of the aberrantly processed mucins of cancer cells, may have some utility as tumor vaccines. Clinical trials of mucin-derived vaccines have been initiated.[57,58] Some mucin antigens are shed from tumor cells and can be detected in the sera of patients with pancreatic, ovarian, breast, and colon cancers. These include CA19-9, CA125, CA15-3, SPan-1, and DuPan-2, which are currently being used as tumor markers.[58]

The membrane-associated mucins of tumor cells have multiple roles in cancer biology.[58] They interact with and modify the microenvironment in which tumors grow. They provide an intermediate signaling pathway by exposing, through conformational changes, active growth factor domains and recruitment of signal-transducing molecules such as epidermal growth factor. Muc1 (and Muc4) play a role in regulation of cancer cell proliferation and differentiation. There is some evidence that the overexpression of Muc1 on tumor cells decreases cell–cell and cell–substratum interactions and that may promote tumor invasion and metastasis. Finally, there is evidence that tumor-derived mucins can modulate the immune response and may be immunosuppressive for T-lymphocyte responses.

Proteoglycans

The *proteoglycans* are high-molecular-weight glycoproteins that have a protein core to which are covalently attached large numbers of side chains of sulfated glycosaminoglycans as well as N-linked and/or O-linked oligosaccharides. They are categorized on the basis of their glycosaminoglycans into several types, including heparan sulfate, chondroitin sulfate, dermatan sulfate, and keratan sulfate.[59] The glycosaminoglycans have different repetitive disaccharide units bound to the core protein through a common glycosaminoglycan linkage region: GlcNAcβ1→3Galβ1→3Galβ1→4Xylβ1-O-Ser. The structure of the sulfated glycopeptides from the carbohydrate-protein linkage region of some of the proteoglycans has been determined.[60]

Proteoglycans interact via their multiple binding domains with many other structural macromolecules, giving them the capacity "to function as a multipurpose 'glue' in cellular interactions."[61] They bind together extracellular matrix (ECM) components, such as hyaluronic acid, collagen, laminin, and fibronectin; mediate binding of cells to the ECM; act as a reservoir for growth factors; and "present" growth factors to growth factor receptors on cells. The proteoglycans also act as cell adhesion factors by promoting organization of actin filaments in the cell's cytoskeleton. Proteoglycans have been shown to undergo both quantitative and qualitative changes during malignant transformation, and alterations have been reported in breast, colon, and liver carcinomas, in glioma cells, and in transformed murine mammary cells and 3T3 fibroblasts.

Two putative tumor suppressor genes are glycosyl transferases required for the biosynthesis of the proteoglycan heparan sulfate.[62] Mutations of these genes, called *ext1* and *ext2*, have been associated with the development of skeletal dysplasias, and these findings suggest that alterations in the synthesis of heparan sulfate precursor polysaccharide are involved in dysregulation of heparan sulfate production and function in tumor formation.

Modification of Extracellular Matrix Components

The ECM plays a key role in regulating cellular proliferation and differentiation. In the case of tumors, it is now clear that development of a blood supply and interaction with the mesenchymal stroma on which tumor cells grow are involved in their growth, invasive properties, and metastatic potential. This supporting stromal structure is continuously remodeled by the interaction between the growing tumor and host mesenchymal cells and vasculature. About 80% of the cells within a tumor are stromal cells, including fibroblasts, non-tumor epithelial cells, mast cells, and macrophages.

The ECM components include collagen, proteoglycans, and glycoproteins, such as fibronectin, laminin, and entactin. The ECM forms the milieu in which tumor cells proliferate and provides a partial barrier to their growth. Basement membranes are a specialized type of ECM. These membranes serve as a support structure for cells, act as a "sieving" mechanism for transport of nutrients, cellular metabolic products, and migratory cells (e.g., lymphocytes), and play a regulatory role in cell proliferation and differentiation.[63] Basement membranes also prevent the free passage of cells across them, but there are mechanisms that permit the passage of inflammatory cells. It is also clear that basement membranes act as regulators of cell attachment, through cellular receptors called integrins (see below). There is also "cross talk" between epithelial cells and their ECM to create a microenvironment for accurate signal transduction for growth factors and other regulatory molecules. It has been shown, for example, that exogenous reconstituted basement membranes stimulate specific differentiation of a variety of cell types, including mammary cells, hepatocytes, endothelial cells, lung alveolar cells, uterine epithelial cells, Sertoli cells, and Schwann cells.[64]

The basement membrane barrier can be breeched by tumor cells that release a variety of proteases, glycosidases, and collagenases that have the ability to degrade various components of the matrix and thus allow tumor cells to invade through tissue barriers and blood vessel and lymph channel walls. In addition, malignant cells themselves have receptors for and/or can produce certain components of the matrix; this capability enables them to bind to the vascular endothelium and may be involved in their ability to metastasize. Tumor cells may also release polypeptide factors that can modulate the type of proteoglycans produced by host mesenchymal cells. The tumor stromal cells, in turn, can release factors that favor tumor cell proliferation and invasiveness. For example, activated fibroblasts in the tumor stroma release a number of growth factors that stimulate cell proliferation, inhibit apoptosis, and alter cell differentiation and that up-regulate proteases involved in degrading the ECM (reviewed in References 65 and 66). These factors include hepatocyte growth factor (HGF), insulin-like growth factors (IGF)-1 and -2, EGF, TGF-α, TGF-β, interleukin-6, fibroblast growth factors (FGF)-2 and -10, and matrix metalloproteases-1 and -7.[65] These multiple effects of the tumor stroma on cancer growth and progression provide a number of potential targets for anticancer therapy.[67]

Cell-Extracellular Matrix and Cell-Cell Adhesion

Cells in tissues are attached to one another and to the ECM. Disruption of these adhesion events leads to increased cell motility and potential invasiveness of cells through the ECM. In addition, most cell types require attachment to the ECM for normal growth, differentiation, and function. This attachment is responsible for what is termed *anchorage dependence*. Normal cells that are detached from their binding to the ECM undergo apoptosis, whereas tumor cells that are less dependent on this attachment are free to proliferate, wander, and invade tissues.

Cell adhesion to the ECM is mediated by cell surface receptors called *integrins*. Integrins are a family of proteins consisting of $\alpha\beta$ heterodimers that are integral membrane proteins with a specific arginine, glycine, aspartic acid (RGD) amino acid sequence involved in binding to the ECM.[68] Integrins also link the external ECM cytoskeleton to the intracellular actin cytoskeleton, and via this connection a linkage to control of gene expression in the cell nucleus is established. In this way, cell-ECM interactions can control gene read-out involved in cell differentiation and function. Cell–ECM interactions occur via focal adhesions that consist of clusters of ECM-bound integrins, and these in turn connect to actin fibrils and the signal transduction machinery inside the cell. These signaling pathways include the focal adhesion kinase (FAK) pathway that participates in the control of anchorage dependence, and growth factor signaling pathways, such as the *ras-raf*-mitogen-activated kinase, protein kinase C, and phosphatidylinositol 3-kinase pathways.[69] Thus, integrins cooperate with growth factors to enhance mitogenic signaling. Alterations in integrin receptor expression have been observed in chemically transformed human cells and in human colon and breast cancer tissue.[70]

Cell-cell interactions are also important for the normal regulation of cell proliferation and differentiation. These interactions are mediated by a family of molecules called *cell adhesion molecules* (CAMs), which act as both receptors (on one cell) and ligands (for another cell). The expression of CAMs is programmed during development to provide positional and migratory information for cells. A large family of CAMs has been identified. One group of these, called *cadherins*, comprise a superfamily of Ca^{2+}-dependent transmembrane glycoproteins that play an essential role in the initiation and stabilization of cell-cell contacts. Regulation of cadherin-mediated cell–cell adhesion is important in embryonic development and maintenance of normal tissue differentiation.[71,72]

The extracellular domain of various cadherins is responsible for cell–cell homotypic binding (a given cadherin domain for a given cell type), and the conserved cytoplasmic domains interact with cytoplasmic proteins called *catenins*. Each cadherin molecule can bind to either β-catenin or γ-catenin, which in turn bind α-catenin. α-Catenin links the cadherin complex to the actin cytoskeleton. Cell lines that lack α-catenin lose normal cell-cell adhesiveness, and tumor cells with mutated or down-regulated α-catenin have increased invasiveness.[73]

E-cadherin is the predominant type of cadherin expressed in epithelial tissue. Alterations of E-cadherin expression and function have been observed in human cancers.[74] In addition, down-regulation of E-cadherin correlates with increased invasiveness, metastasis, and poor prognosis in cancer patients. Suppression of this invasive phenotype can be achieved by transfection of E-cadherin cDNA into carcinoma cells, and contrarily, invasiveness of E-cadherin gene-transfected cells can be restored by exposure of the cells to E-cadherin antibodies or an E-cadherin antisense RNA.[74] Germline mutations of the E-cadherin gene (*cdh1*) have been found in New Zealand Maori families with a dominantly inherited susceptibility to gastric cancer.[75]

The cell surface receptor for E-cadherin is β-catenin. Early mutations in the human colon cancer progression pathway affect the cellular distribution of β-catenin. In patients with colon cancer, the normal colonic epithelial cells adjacent to neoplastic lesions had mostly cell surface membrane expression of β-catenin, whereas cytoplasmic expression of β-catenin was observed in aberrant crypt foci.[76] Nuclear expression was observed in more advanced dysplasias and increased as adenomas progressed to carcinomas. These latter changes are also observed in less well-differentiated areas of tumors and are accompanied by loss of E-cadherin expression at the invasive front of breast carcinomas, possibly due to hypermethylation of the E-cadherin promoter.[77]

CELL PROLIFERATION VERSUS DIFFERENTIATION

A cancer develops from cells that are capable of dividing. All tissues in the body contain some cells that can divide and renew themselves. A subset of the cell population in any tissue can differentiate into the functional cells of that tissue. The normal process of cellular differentiation ultimately leads to an adult, fully differentiated, "dead-end" cell that cannot, under ordinary circumstances, divide again. These fully differentiated cells are the workhorse cells in most tissues of the body. They are the neurons in the brain controlling ideation and behavior; the liver cells that manufacture enzymes to metabolize substrates needed for growth, produce plasma proteins, and clear the blood of potentially toxic substances; the pancreatic cells that manufacture insulin and the enzymes necessary for digestion; the kidney cells that filter, secrete, or reabsorb substances and fluid in the formation of urine; the polymorphonuclear white blood cells that phagocytize and destroy bacteria; and so on. Under circumstances that are not clearly understood, cells that have the potential to divide can be changed by interaction with carcinogenic agents into a cell type that is capable of continued proliferation and thereby is prevented from achieving the normal state of complete differentiation. The carcinogen-altered cell is said to have undergone malignant transformation. Somehow the genes controlling cell proliferation are locked in the "on" position when they should be in the "off" position, and the genes controlling differentiation are either not expressed or are expressed only imperfectly. What we need to know to understand carcinogenesis and to develop ways of preventing or curing cancer, then, is contained in the mechanisms of normal cellular differentiation. Only by understanding these mechanisms can the manner in which cells are altered during malignant transformation be ascertained.

Differentiation is the sum of all the processes by which cells in a developing organism achieve their specific set of structural and functional characteristics. By the acquisition of these special traits, progeny cells are distinguishable from their parent cells and from each other. Somatic cells that share a set or a subset of structural and functional characteristics become organized into tissues in higher organisms. Tissues are arranged as organs, and organs make up the organism. Indeed, cellular differentiation is the sine qua non of multicellular life.

Differentiation requires a progressive restriction of genomic expression in the pathway from the totipotential fertilized ovum to the unipotent cells of specialized organs (Fig. 4–1).[78] The totipotentiality of cells starts to change very early after fertilization as the developing embryo

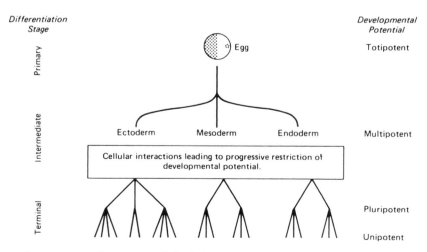

Figure 4–1. Schematic model of the various stages of differentiation in multicellular organisms. (Adapted from Rutter et al.,[78] with permission.)

proceeds through the blastula stage, where it is already evident that certain cells are pre-destined to form certain tissues. This process appears to occur by means of a combination of internal reprogramming (which perhaps occurs as early as the first cell division) and external stimuli by neighboring cells, proximity to the maternal circulation, and gradients of growth factors, oxygen, nutrients, ions, and so on that result from the position in which a cell finds itself as the early embryonic cells continue to divide. The whole microenvironment of the cell determines its developmental destiny. The process of differentiation appears to be fairly permanent in that, as tisues develop, some cells retain the capacity to divide, whereas others divide and then differentiate into cells with a more restricted phenotype. These latter cells are then said to be *pluripotent* rather than totipotent, that is, they are now committed to develop into one of the cell types peculiar to their tissue of origin. For example, a pluripotent stem cell of the bone marrow may differenti-ate into an erythrocyte, a polymorphonuclear leukocyte, or a megakaryocyte. It used to be thought that only a few proliferative tissues in the body contained stem cells that had unlim-ited proliferative capacity, e.g., the bone marrow hematopoietic stem cells, basal cells of the skin and crypt cells of the gastrointestinal tract. It is now known that, in fact, all tissues in the body, including the central nervous system, have stem cells that can re-enter the cell proliferation cycle if a tissue is damaged. These tissue stem cells are probably the ones most susceptible to carcino-genic stimuli.

An organism could not develop, of course, without vigorous cell replication. Nor could it survive without continued cell division. How-ever, there is normally a well-controlled balance between cell division, cell differentiation, and cell death. It is this delicate balance that is disrupted in cancer tissue. Cell differentiation usually pro-duces, ultimately, a cell that no longer has the capacity to divide, but many cells in the process of differentiating continue to divide. Hence, the two processes are not mutually exclusive.

The period following fertilization up to late blastula is a period of intense cell division with very little cell growth between mitoses. In am-phibians, for example, the cell cycle prior to the blastula stage is abbreviated; there is no inter-mitotic G_1 phase and G_2 is short. During blas-tulation, the G_2 period is prolonged, and a short G_1 appears. Thereafter, the G_1 period between S (DNA replication) phases lengthens until the tail-budding stage is reached, at which time the cell cycle approaches that of adult proliferating cells. The cell cytoplasm appears to exert control over the timing of nuclear DNA synthesis since nuclei from adult tissues in which DNA syn-thesis is rare can be induced to synthesize DNA by injection into unfertilized ova.

In mammalian cells, also, growth arrest is coordinated with expression of the differenti-ated phenotype—for example, in hematopoi-etic cells and in epithelial cells of the skin and gastrointestinal tract as well as in such cell cul-ture systems as the preadipocyte mouse 3T3 lines that can be induced to differentiate into fat cells. Cells transformed by carcinogenic agents or oncogenic viruses lose this ability to become growth arrested and to become terminally dif-ferentiated.

Mechanisms of Cellular Differentiation

Much of what we have learned about the cel-lular mechanism of differentiation has come from studies of lower organisms, including yeast, slime molds, round worms, sea urchins, fruitflies, zebrafish, and chickens. As more has become known about the genomes of these or-ganisms and of humans, we have also learned that there are orthologous genes and proteins that have similar sequence and function and that can be traced back in evolution. A number of these orthologs are involved in cell differentia-tion processes in all species. As an example, 50% of the genes of fruitflies have human equivalents and almost every human gene has a counterpart in the mouse.

The nematode (round worm) *Caenorhabditis elegans* (*C. elegans*) was the second eukaryotic organism that had its genome completely se-quenced; the first was the yeast *Saccharomyces cerevisiae* (*S. cerevisiae*). This sequencing al-lowed for the first time a direct comparison of orthologous genes from widely divergent organ-isms and provided background information on

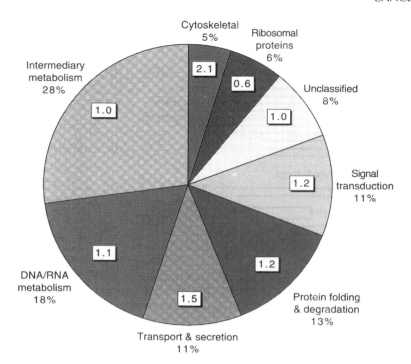

Figure 4–2. Distribution of core biological functions conserved in both yeast and worm. Yeast and worm protein sequences were clustered into closely related groups (BLASTP $P < 1 \times 10^{-50}$, with the >80% aligned length constraint). Each sequence group (including groups with two or more sequences) was assigned into a single functional category, relying primarily on the functional annotations for the yeast genes in Saccharomyces Genome Database (SGD) when available. The unclassified category contains groups of sequences without annotation. The boxed number within each category reflects the ratio of worm to yeast proteins for that category. (From Chervitz et al.,[79] reprinted with permission from the American Association for the Advancement of Science.)

how similar genes in humans may function.[79] These shared genes carry out fundamental biological processes such as intermediary metabolism; DNA and RNA synthesis and processing; and protein folding, trafficking, and degradation. The percent conservation of these functions between yeast and nematode is shown in Figure 4–2 (see color insert). There has, of course, been some sequence and functional divergence during evolution, yet a surprising number of similar gene sequences related to similar functions have been passed down to humans. A few examples will demonstrate this.

Functional conservation of proteins from widely divergent species was first shown experimentally by demonstrating that the mammalian RAS protein could substitute for yeast RAS in a RAS-deficient yeast strain (reviewed in Reference 79). The C. elegans ras gene is homologous to two yeast ras genes (ras1 and ras2). The cell cycle checkpoint genes cdc 28 in yeast and ncc-1 in C. elegans are orthologous pairs of the cyclin-dependent kinase family that are functionally interchangeable. Mammalian cells have similar cell cycle checkpoint control proteins.

A Hedgehog-related protein in yeast (HINT) has binding domains similar to nematode Hedgehog, although the yeast gene has a different function. Hedgehog is a key regulator of positional orientation in insect and vertebrate development (see below). The DNA binding domains of these yeast and worm proteins are specifically related to the helix-turn-helix domains of the transposases of animal and bacterial transposons, which provides an idea of how evolutionarily old these protein domains are.

Thus, the comparison of the complete protein sets of organisms like C. elegans and S. cerevisiae, based on correlation of sequence and function data, provides a powerful tool for understanding the functional role of their orthologous genes in human cells.

If we start with the lowest level and work our way up to mammals and humans, we can see what this comparitive study can teach us about how differentiation and development can go awry along the way. In addition, some interesting targets for drug discovery may come to light. Even though the genome sequences of humans and several lower organisms have established a fairly detailed list of genes required for key functions, an important next step is to determine the functions of the large pool of genes for which no function is currently known and to learn about how genes are networked to carry out specific biological processes. Gene expression data from DNA microarrays gives clues about this, but a more powerful tool is to determine how genes are co-regulated and whether this co-regulation has been conserved over evolution.

The availability of genomic sequence and DNA microarray data for humans and organisms such as yeast, C. elegans, and Drosophila allows determination of evolutionarily conserved expression over a large phylogenetic scale. Stuart et al.[80] have identified pairs of genes coexpressed over 3182 DNA microarrays from humans, Drosophila, C. elegans, and S. cerevisiae. They found 22,163 co-expression relationships, each of which have been conserved across these species. This finding implies that there is a selective advantage for these gene sets and that these genes are functionally related. One interesting gene set is comprised of five metagenes involved in cell proliferation and cell cycle regulation that were not previously known to be so involved.[80] A metagene is defined as a set of genes across multiple organisms whose protein sequences reflect the best reciprocal connection, based on a function. Expression levels of these genes were measured in human pancreatic cancer cells and normal pancreas and in loss-of-function mutants of C. elegans. All five gene sets were overexpressed in pancreatic cancers compared to normal tissue. In C. elegans loss-of-function mutants, germline cell proliferation was also suppressed, providing evidence that these genes are functionally linked in two widely divergent species. In addition, this network of genes contained many cell proliferation and cell cycle genes across multiple species. These data not only help to define a previously unknown function of genes but also help to define what other genes are linked to a functional network. This type of study will be a great tool to help define upstream and downstream targets for therapeutic intervention against malignant disease.

Slime Molds

In 1942, Ernest Runyon discovered that if he put samples of the slime mold *Dictyostelium discoidium* on both sides of a piece of porous cellophane, the individual cells began to move "tail to head" in streams toward a central point (reviewed in Reference 81). Remarkably, the cells on the other side of the cellophane lined up and moved in the same way, forming mirror-image streams. Runyon hypothesized that a diffusible molecule penetrated the cellophane and attracted the cells on the other side. This was the first example of a "chemoattractant" that could regulate cell movement and migration. This chemoattractant turned out to be cyclic AMP.[81] The tail-to-head orientation is achieved by the distribution of adenylyl cyclase that produces cAMP at the rear of the migrating cells, thus explaining how the cells line up—i.e., cAMP is produced at one end of the migrating cells, is secreted into the environment, and binds to cAMP receptors in the trailing cells. This process in turn leads to the accumulation of factors involved in cellular movement in the membrane at the front edge of the trailing cells. The induced ability "to march in close-order files" also facilitates the laying down of an extracellular matrix over which the cells can move smoothly. Cellular movement also requires an intact internal cellular skeleton of actin fibers. Mutations in adenylyl cyclase prevent this smooth concatenation of events. Thus, even this primitive organism processes properties similar to certain human developmental and functional properties—cell movement, response to chemoattractants, deposition and modulation of an extracellular matrix, and signaling through an internal cytoskeleton.

Because cAMP appears to play a key role in a large number of cellular processes (Table 4–1)

Table 4–1. Effects of cAMP on Various Cellular Processes

Enzyme or Process Affected	Tissues or Organism	Change in Activity or Rate*
Protein kinase[†]	Several	+
Phosphorylase	Several	+
Glycogen synthetase	Several	−
Phosphofructokinase	Liver fluke	+
Lipolysis	Adipose	+
Clearing factor lipase	Adipose	−
Amino acid uptake	Adipose	−
Amino acid uptake	Liver and uterus	+
Synthesis of several enzymes	Liver	+
Net protein synthesis	Liver	−
Gluconeogenesis	Liver	+
Ketogenesis	Liver	+
Steroidogenesis	Several	+
Water permeability	Epithelial	+
Ion permeability	Epithelial	+
Calcium resorption	Bone	+
Renin production	Kidney	+
Discharge frequency	Cerebellar Purkinje	−
Membrane potential	Smooth muscle	+
Tension	Smooth muscle	−
Contractility	Cardiac muscle	+
HCI secretion	Gastric mucosa	+
Fluid secretion	Insect salivary glands	+
Amylase release	Parotid glad	+
Insulin release	Pancreas	+
Thyroid hormone release	Thyroid	+
Calcitonin release	Thyroid	+
Research of other hormones	Anterior pituitary	+
Histamine release	Mast cells	−
Melanin granule dispersion	Melanocytes	+
Aggregation	Platelets	−
Aggregation	Cellular slime molds	+
Messenger RNA synthesis	Bacteria	+
Synthesis of several enzymes	Bacteria	+
Proliferation	Thymocytes	+
Cell growth	Tumor cells	−

*+, increase; −, decrease.

[†]Stimulation of protein kinase is known to mediate the effects of cAMP on several systems, such as the glycogen synthetase and phosphorylase systems, and may be involved in many or even most of the other effects of cAMP. (From Sutherland[82])

including cell proliferation and differentiation, a number of investigators have speculated that alterations in the cyclic nucleotide–generating or response systems may be altered during malignant transformation. A considerable amount of work has been done to show this, and there does appear to be some correlation, but the exact role of cyclic nucleotides in this process has not yet been defined. A number of studies have shown that transformed fibroblasts regain a number of the characteristics of untransformed cells after treatment with cAMP analogues. These characteristics include a more flattened morphology, an increased adhesion to the substratum, a decreased agglutinability by lectins,

and a decreased rate of cell proliferation. However, not all transformed cells respond to cAMP treatment in this way. The response is determined by the cell of origin of the transformed cell line. Fibroblastic cells, in general, tend to respond to the cAMP in the previously described manner, whereas epithelial cells often do not. Experiments with various clones of rat kidney cells, for example, showed that in a fibroblastic clone, intracellular levels of cAMP rose as the cells reached confluency, but this was not the case in an epithelial clone.[83] Murine sarcoma virus–transformed fibroblastic clones of rat kidney cells did not have elevated intracellular cAMP at confluency, and they responded to treatment with exogenous cAMP analogues by exhibiting a slower growth rate and a flattened cell morphology. Neither the growth nor the morphology of the epithelial clone was affected by exogenous cAMP.

It appears that certain parts of the response system for cAMP differ in certain kinds of cells. In support of this idea, experiments with the S49 lymphoma cell line showed that the proliferation of these cells was inhibited by cAMP but was not inhibited in a mutant S49 cell line defective in cAMP-dependent protein kinase.[84] Similar results have been obtained for mutant Chinese hamster ovary cells with a variant cAMP-dependent protein kinase. There may also be alterations in the cAMP-binding regulatory subunit of cAMP-dependent protein kinase or in the translocation step involved in the nuclear uptake of protein kinase, a step that appears to be required for response to cAMP in some cells. Decreased binding of cAMP and an altered cAMP-binding protein have been demonstrated in a cAMP-unresponsive line of Walker 256 carcinosarcoma cells compared with the responsive parent line.[85] The nuclear translocation of cAMP-binding proteins and protein kinase was also markedly diminished in the unresponsive tumor cells after treatment with dibutyryl camp.[86]

It is clear that cAMP affects the proliferation rate of some normal and transformed cultured cells and that cAMP levels are lower in some transformed cell lines. It is not clear if changes in cAMP lebels or the cAMP-response system are *responsible* for the appearance of the transformed phenotype and, more importantly, for the loss of normal growth control. In some cells, alterations in intracellular cAMP appear to be more closely related to the morphologic characteristics of the transformed phenotype than to growth control; in fact, the two events are clearly dissociable in certain cell types. Nevertheless, it is clear that induction of cAMP in several types of cultured neoplastic cells induces a more differentiated, less transformed phenotype.

Several lines of evidence implicate G protein–coupled receptors in malignant transformation.[87] Overexpression of acetylcholine or serotonin receptors in NIH 3T3 cells causes ligand-dependent transformation. Bombesin-like peptides are secreted by some small-cell lung carcinoma cells and stimulate their growth, and antibodies to bombesin inhibit tumor cell proliferation. Some pituitary, adrenal cortical, and ovarian tumors have point mutations in G proteins coupled to adenylyl cyclase that could lead to constitutive overproduction of cAMP.

The $\alpha_{-1\beta}$-adrenergic receptor is a member of the G protein–coupled receptor superfamily and activates PI hydrolysis, a signaling pathway that is activated by a number of growth factors and that plays a crucial role in mitogenesis. Mutation of three amino acids residues in the third intracellular loop (see Fig. 4-3)[88] increases the binding affinity of norepinephrine and its ability to stimulate PI hydrolysis by two to three orders of magnitude.[89] Moreover, this activating mutation renders the receptor constitutively active, stimulating PI turnover even in the absence of ligand. When the wild-type gene for the $\alpha_{1\beta}$ receptor is transfected into rat or NIH 3T3 fibroblasts, the cells express high levels of this receptor, become transformed in response to norepinephrine, and form tumors when injected into nude mice. When the mutated gene is transfected into fibroblasts, the cells spontaneously form transformed foci in the absence of ligand and have an enhanced ability to form tumors in nude mice. Thus, the $\alpha_{1\beta}$ adrenergic

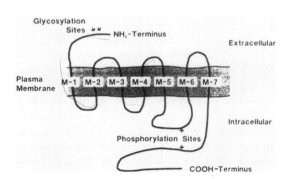

Figure 4–3. Schematic representation of the membrane organization of plasma membrane receptors (such as adrenergic receptors, substance K receptors, or opsins) that are linked to G proteins. An extracellular amino-terminal region with sites of glycosylation on asparagines residues is followed by seven membrane-spanning domains (M1 to M7) interspersed with three intercellular and three extracellular loops and then an intracellular carboxy terminus. The consensus sequences expected at sites for phosphorylation are found in the third intracellular loop and carboxyl-terminal regions. (From Taylor and Insel,[88] with permission.)

receptor gene acts like a proto-oncogene and when activated or overexpressed is a transforming oncogene. These data suggest that other G protein–coupled receptors of this type can act as oncogenes in certain cell types. This further suggests a host of strategies for chemotherapeutic interdiction of this system, for example, the design of specific antagonists of the G protein–coupled receptors that may be activated or overexpressed in tumor cells.

There is also evidence that alteration of G protein subunits themselves can cause alterations in fibroblast growth characteristics. For example, transfection and overexpression of a mutated G protein $\alpha_{1\beta}$ subunit gene, a gene shown to be involved in proliferation of fibroblasts and differentiation of myeloid cells, in fibroblasts produces increased cell proliferation and anchorage-independent growth, indicating a role for this G-protein subunit in regulation of fibroblast cell proliferation and in transformation events.[90]

Yeast

Since the yeast genome has been sequenced, a number of functional and evolutionary correlations have become more evident. RNA interference (RNAi) has been used to block gene expression, thus allowing the "knock out" of specific sequences and the determination of what the key functional defect is. Silencing of genes to detect function has also been used in yeast, *C. elegans*, and mammalian cells (see Chapter 5).

Recent advances in detecting protein-protein interacting networks, in addition to complete genome sequencing, provide another tool to look at evolutionary changes over time and speciation. Fraser et al.[91] have used this concept to catalog 3541 interactions between 2445 yeast proteins to estimate evolutionary rates for these protein sets. Comparison of orthologous sequences between *S. cerevisiae* and *C. elegans* provided a method to compare differences and hence delineate conservation of protein interacting networks during evolution. Although protein-protein interacting networks evolve more slowly than individual proteins, such interacting proteins appear to co-evolve, in that substitutions in one protein results in selective pressure for reciprocal changes in interacting

partners.[91] Moreover, the proteins in interacting sets presumably act in the same functional pathways in different organisms, and although not clearly demonstrated across widely divergent species, these sets most likely have similar functional duties in different organisms.

Some of the striking correlates between yeast and higher organisms involve the cell cycle checkpoint pathways, which were originally demonstrated in yeast and now have been shown to be analogous in mammalian, including human, cells (see Cell Cycle Regulation, below).

Sea Urchin

Development of a "body plan" for multicellular organisms requires a precise interaction of multiple factors and is controlled by what Eric Davidson and colleagues call a "gene regulatory network" (GRN).[92] Expression of genes in this complex network is in turn replicated by "waves" of transcription factors and repression that turn genes on and off in an exquisitely orchestrated pattern. This complex array of events is not easily studied in complex animal systems such as mammals, but in simple organisms such as sea urchins it can be analyzed in reasonable detail.

The "heart" of developmentally regulated gene networks is the genes encoding transcription factors and *cis*-regulatory elements that control the expression of these genes. Each of the *cis*-regulatory elements gets multiple inputs from transcription factors (TFs) encoded by other genes in the network. These TFs in turn recognize specific gene sequences in the *cis*-regulatory elements. These *cis*-regulatory systems at the center of a developmental GRN receive carefully timed input information based on the rise and fall of the TFs to which they respond. Another key element is the signal transduction machinery that produces the signals for TF expression. This machinery is turned on by chemostimulants in the local environment, which are also present in kinetic waves of production and release from cells in the local cellular milieu. What turns on the expression of these chemoregulants at the appropriate time still remains to be completely understood. The chemoregulants that specify developmentally regulated events must of necessity operate along gradients in the cellular environment of an embryo. Otherwise, all cells would be exposed to the same signals and no cell type specific differentiation could occur, given that all the cells in a given organism carry the same genome.

This differentiation is achieved by the release of cellular differtiating signals, called *morphogens*, by certain "pacemaker" cells in the local embryonic environment, similar to the way that cyclic AMP regulates cell orientation, movement, and differentiation in slime molds as described above. *Morphogens* are defined as "signaling molecules that are produced in a restricted region of a [developing] tissue and move away from their source to form a long-range concentration gradient."[93] Cells then differentiate in response to local morphogen concentration depending on how far away they are from the morphogen source. Examples of morphogens are Wingless, Hedgehog, and Dpp, which form gradients and elicit differentiation responses across distances during wing and leg development in *Drosophila*. Multiple target genes are induced depending on the local concentration of these ligands and binding to their responsive (receptor-bearing) cells.

The reason that sea urchin development is an ideal system in which to study these programs is because the most closely examined examples of a *cis*-regulatory information-processing system is the developmental expression of the *endo 16* gene of the sea urchin.[94] *Endo 16* encodes a polyfunctional protein that is secreted into the lumen of the embryonic and larval midgut. It is expressed first in progenitor cells of the endomesoderm, then throughout the gut, and finally only in the midgut. Early- and late-expression phases are controlled by two different regulatory gene sequences, each several hundred base-pairs long.

Control of gene expression regulated by these sequences is subject to input from nine DNA-sequence specific TFs as time-varying regulatory signals. The end result of all this is the development and differentiation of the primitive gut in the sea urchin. The steps in this process are instructive about what goes on in the development of higher organisms and is worth cataloging here (Table 4–2).

Another key point about developmentally regulated GRNs is that they are subject to both positive and negative regulation. The developmental process is moved forward by intergenic feedback loops, both positive and negative, that lock in a given developmental state. These seldom go backwards once stabilized (with the possible

Table 4–2. Phenomenological Aspects of Endomesoderm Specification in Sea Urchin Embryos: Developmental Process.

1. *Autonomouse cues of maternal origin*
   Nuclearization of β-catenin in micromeres (by fourth cleavage) and veg₂ cells (from sixth cleavage on)
   Exclusion of ectodermal transcription factors from vegetal-most cell nuclei
   Nuclearization of Otx factor in micromeres at fourth cleavage

2. *Early micromere signal*
   Micromere signal to veg₂ (fourth through sixth cleavage) required for normal endomesodermal specification

3. *Wnt8/Tcf loop*
   Wnt8 ligand expressed throughout endomesodermal domain maintains and strengthens β-catenin/Tcf input in these nuclei
   B-catenin/Tcf input required for endomesoderm specification

4. *Late micromere signal*
   Expression of Delta ligand in micromeres
   Activation of Notch signal transduction in veg₂ descendants adjacent to micromeres that receive Delta signal

5. *Skeletogenesis*
   Skeletogenic functions expressed after ingression of skeletogenic cells in late blastula

6. *Specification of veg₂ mesoderm and endoderm*
   Segregation of cell type precursors within vegetal plate complete by late blastula
   Mesoderm cells turn off endoderm genes, leaving endoderm genes expressed in peripheral veg₂ cells

7. *Specification of veg₁ endoderm*
   Wnt8 signal from veg₂ to veg₁ and activation of β-catenin nuclearization in abutting veg₁ cells

8. *Invagination of archenteron*
   veg₂ mesoderm carried inward at tip of archenteron on gastrulation
   Followed by roll-in of veg₁ endoderm, contributing mainly hindgut

From Davidson et al.,[94] reprinted with permission from the America Association for the Advancement of Science exception of stem cell transdifferentiation or carcinogenic insult (see Stem Cells, below). This feedback network for intergenic loops used to stabilize a developmental phenotype is not peculiar to sea urchins. It is also seen, for example, in the *hox* gene network that controls rhombomere development in the mouse hindbrain (reviewed in Reference 92). Thus, mammals use similar developmentally regulated systems.

Once the *cis*-regulatory network is used to achieve a partially complete differentiated state, later epigenetic processes such as genetic imprinting, DNA methylation, and regulation of chromatin packaging contribute to further differentiation and stabilization of a differentiated tissue phenotype (see Chapter 5).

Drosophila Melanogaster

Regulation of cellular polarity and "positionality" is key to determining the cell motility, positioning, and orientation of body axes (e.g., anterior-posterior [A-P], head-to-tail). Genes involved in this key developmental process are evolutionarily conserved to a large extent. Recently, it was found that homologs of genes that control polarity of hairs on the epidermal cells of *Drosophila* (fruit fly) wings also control polarized cell motility that determines cellular movements underlying body shape orientation in vertebrates.[95] These movements (called *morphogenic movements*) determine the mediolateral and A-P orientation in the vertebrate embryo as well as in the fruit fly. These mechanisms regulate body axis orientation in all chordate species examined, including sea squirts, teleost (bony) fish, amphibians, birds, and mammals (reviewed in Reference 95).

Some of the genes involved in A-P axis formation and epithelial polarity have been identified. Two of these genes are serine/threonine kinases that have sequence homology between *C. elegans* and *Drosophila*, and with the human tumor suppressor gene *LKB1*. The kinase domain of the *Drosophila* LKB1 gene has 66% amino acid identity to human LKB1.[96] Interestingly, both the fruit fly and human LKB1 proteins have conserved prenylation and protein kinase A phosphorylation sites that are essential to the in vivo function of both proteins, strongly

suggesting that the two proteins are functional homologs as well as sequence homologs. Mutations in the human *lkb1* gene cause Purtz-Jeghers syndrome, which is characterized by the formation of intestinal polyps and a high incidence of adenocarcinomas, as well as a number of other epithelial cancers. Thus, *lkb1* has been classified as a tumor suppressor gene, whose function may relate to a role in cell cycle regulation or apoptosis. It is hypothesized that disruption of cellular polarity of epithelial tissues leads to polyp and tumor formation.[96] This is another example of how disruption of normal tissue developmental processes can lead to abnormal cellular behavior and potentially to cancer.

Mouse

Another morphogen gradient, one that involves mRNA decay, has been found in mouse embryos.[97] Head-to-tail patterning requires a strict time-dependent coordination and involves fibroblast growth factor (FGF) signaling. The investigators showed that transcription of FGF-8 mRNA was progressively degraded in the newly formed tissues, thus producing a gradient of FGF-8 mRNA in the posterior part of the embryo. The mRNA gradient correlates with a similar gradient of FGF-8 protein, which in turn correlates with a gradient of phosphorylation of Akt kinase, which is a downstream signal transduction effector of FGF-8. This provides another morphogenic mechanism, in addition to a pacemaker cell type that produces a growth factor such as FGF, transforming growth factor β (TGF-β), or other morphogen, creating a diffusion gradient for a regulatory protein.

Pathways: Getting to Know All the Players

The process of early development is a complicated one, and there are some similarities and some differences among various multicellular organisms. The biochemical signals and genes involved, as noted above, show a lot of evolutionary conservation. Various polypeptide growth factors have been shown to play a role in early morphogenesis.[98] For example, in early *Xenopus* development, there are a series of inductive events that involve growth factors, whose actions lead to differentiation of mesoderm at the interface between the animal and vegetal poles of the embryo. This induction is most efficiently achieved by a combination of members of the FGF and TGF-β families of growth factors. In *Xenopus, Drosophila*, and developing chick limb buds the role for members of the FGF and TGF-β families of polypeptide growth factors in early development appears to be regulation of expression of *hox* genes.[98–101] For example, growth factors regulate expression of a *hox* gene called *xhox3* in *Xenopus* that is required for A-P patterning. Similar observations have been made in *Drosophila*. Since *hox* genes themselves code for transcriptional regulators that can turn genes on or off, some of which may code for growth factor–like substances, one can visualize a cascade of events in which a local concentration of growth factor turns on a *hox* gene, which, in turn, activates another growth factor that turns on another *hox* gene in a responding cell. This process suggests a way that pattern formation could be transmitted from one cell region to another.

The activation of *hox* genes, however, does not clearly explain how, for example, within a given mesodermal area, different mesodermal cell types arise because *hox* genes are expressed, albeit perhaps at different times and levels, throughout the mesodermal layer. Thus, additional genes must be expressed in a carefully regulated way to lead to further "subspecialization" or differentiation events. One well-studied example of this is the expression of genes involved in the muscle differentiation pathway, such as the myogenic genes *myo D* and *myogenin*.

Another example is limb bud formation, studied in vertebrates.[102] This occurs in several stages. The first phase involves the establishment of signaling centers within the bud primordium. These signaling centers have positional determinants in the embryo: anterior-posterior, dorsal-ventral, and medial-lateral. The second phase is usually associated with increased cell proliferation mediated by various mitogens such as members of the FGF and Sonic hedgehog family of gene products. Ultimately, limb bud outgrowth ceases because of decreased release of mitogens and a balance between cell proliferation and programmed cell death. The way in

which the genes involved in this late phase are regulated is not totally clear, but it determines what regulates final organ size and the relationship of organ size to the overall size of the developing embryo.

The relationship of these processes to cancer is intriguing. Alterations of these events occur in malignancy: a turn-on of genes leading to cell proliferation, an alteration in the balance of cell proliferation and apoptosis, and a lack of feedback controls to limit organ size. Thus, an understanding of the regulation of these developmental events should go a long way toward understanding what goes wrong in the biochemistry of the cancer cell.

Some cellular developmental pathways have been closely linked to human cancers, e.g., the Hedgehog and Wnt pathways. Sonic hedgehog (Shh) is the mammalian version of the *Drosophila* morphogen and is known to mediate epithelial–mesenchymal interaction in lung development. Loss of Shh function results in severe lung defects as demonstrated by failure of bronchial tree branching. This information has led to the hypothesis that signaling via the Shh-mediated pathway is important in airway epithelium repair after toxic damage or carcinogenic insult.[103] Watkins et al.[103] have shown that the human hedgehog (Hh) pathway is extensively activated during repair of acute airway injury. Activation of this pathway was also seen in human small cell lung carcinoma (SCLC). It was shown that these cancers maintain their malignant phenotype via Hh pathway activation, thus suggesting that SCLC may develop by recapitulating an early Hh-mediated event in airway epithelial differentiation.

Members of the Hh family of molecules are secreted during early development and have essential roles in tissue patterning in organisms from *Drosophila* to humans. Although Hedgehog proteins are known to affect cell fate, they can also stimulate cell proliferation. Mutations that aberrantly activate Hh signaling have been observed in basal cell carcinomas and other human cancers.[104] In *Drosophila*, Hh acts on somatic ovarian stem cells to stimulate their proliferation.[105] Moreover, high production and signaling of Hh produces an abnormal number of ovarian stem cells. The authors of this study postulate that Hh is a stem cell proliferation factor and that its excessive signaling causes abnormal expansion of stem cell pools in human tissues, which may be part of the aberrant signaling during the malignant transformation process.

It has been shown that a wide range of digestive tract cancers, including carcinomas of the esophagus, stomach, biliary tract, and pancreas (but not colon), have increased activity of the Hh pathway. Interestingly, cyclopamine, an Hh pathway antagonist, suppresses cell proliferation in gastrointestinal (GI) tract cell lines and causes regression of GI tract cancer xenografts in vivo in athymic (nude) mice.[106] In addition, cell proliferation of digestive tract cancer cells in culture is enchanced by endogenous expression of Hh ligands (as determined by RT-PCR) and blocked by a Hh-neutralizing monoclonal antibody,[106] a finding suggesting that the Hh pathway would be a good target for anticancer drug development.

The Wnt pathway is another important developmental pathway involved in human cancers. The Wnt family of proteins are involved in cell–cell signaling and adhesion during many steps in animal development, including formation of the embryonic axes to end stage development of organs such as the kidney. This is the case both for invertebrates and humans. There are at least 19 genes encoding Wnt proteins in humans and ten receptors for these proteins.

There are three cellular signaling cascades activated by Wnt proteins that separately regulate cell differentiation, cell polarity, and cell adhesion (reviewed in Reference 107). These three cascades are (1) regulation of β-catenin intracellular location and function; (2) activation of a planar cell polarity pathway mediated via the *disheveled* gene, cdc 42, and Jnk signal transduction; and (3) activation of an increase in intracellular calcium levels and protein kinase C. All of these pathways are mediated by Wnt proteins interacting with G protein–coupled receptors.

Details of the Wnt/β-catenin pathway came from *Drosophila*, where this pathway functions in patterning body segments and appendages. Activation of this pathway by a Wnt ligand stabilizes β-catenin and facilitates its transfer to the nucleus, where in combination with chromatin-associated high-mobility group proteins, β-

catenin activates a number of genes that control cell fate. As noted above in the section on cell–ECM and cell–cell adhesion, dysregulation of the β-catenin pathway is observed in colon, breast, and other human cancers.

The *Notch* gene–regulated pathway also plays a central role in many developmental processes. Examples are peripheral neurogenesis in *Drosophila*, vulval development in *C. elegans*, and lymphoid development in mammals (reviewed in Reference 108). In zebrafish, activation of the Notch signaling pathway facilitates heart regeneration.[109] Zebrafish have a remarkable ability to regenerate various organs including heart and fins. These processes involve increased expression of the *notch lb* gene and subsequent activation of homeolox genes *msx*B and *msx*C. Notch activation is a key link in the decision-making process for proliferation and differentiation of stem cells in hematopoietic, neural, gastrointestinal, and skeletal muscle lineages of various organisms.

In a subset of acute lymphoblastic leukemias, there is a chromosomal translocation involving the human *notch* 1 gene and the T-cell receptor β locus. This rearrangement leads to constitutive expression of a truncated *notch* allele, which behaves like an oncogene in this situation (reviewed in Reference 108). Dysregulated Notch signaling has now been implicated in the pathogenesis of a broad range of human cancers. Somewhat surprising is the observation that Notch acts as a tumor suppressor in skin carcinogenesis, where it interacts with the Hedgehog and Wnt pathways. This demonstrates a principle that is common in cancer biology—i.e., that the consequences of activation or modification of a signal transduction pathway are dependent on cellular context.

There are, of course, a number of other tissue development and cell differentiation pathways that involve developmentally regulated genes and their encoded proteins. In fact, each differentiated tissue in an adult multicellular organism uses a combination of complex signaling pathways to achieve its final destination in the body and final functional state. Many of these pathways are ones that are disrupted or re-activated during the carcinogenic process. The take-home message here is that most of these pathways were originally discovered and characterized in lower organisms where the biology can be studied in a much less complicated way.

Other important parameters of morphogenesis include the ability of like cells to cluster together and "talk to each other" and the ability of cells to produce and interact with a specific tissue type ECM. Thus, the ability to regulate cell-cell and cell-ECM (cell-substratum) interactions is also key to normal development and cellular differentiation. Two families of adhesion molecules are involved: cell-cell adhesion molecules, or CAMs, and cell-substratum adhesion molecules, or SAMs.[110,111] CAMs produce cell-cell contact between like-minded cells that foster their interactions and cell sorting into homogeneous populations. As noted above, CAMs, or cadherins as they are also called, are large transmembrane proteins that interact through cytoplasmic connections called catenins that link cadherins to the cell cytoskeleton, thus providing an internal signaling process for CAMs that are in contact with the extracellular environment. These interactions are capable of modulating formation of actin cables in the cytoplasm, and, thus, of affecting cell migration and cell surface polarity.

Thus, a number of key interactions among growth factors, *hox* genes, CAMs, SAMs, the ECM, and specific genes involved in cell lineage–specific pathways occur during early development and early differentiation. Although mostly studied in lower organisms, all of these genes have homologous counterparts in mammalian, including human, cells.

Stimulation of Cancer Cell Differentiation

There are a number of examples of animal malignant tumors or human cancer cells in culture that can be induced to lose their malignant phenotype by treatment with certain differentiation-inducing agents. These include induction of differentiation of the Friend virus–induced murine erythroleukemia by dimethylsulfoxide (DMSO); differentiation of murine embryonal carcinoma cells by exposure to retinoic acid, cAMP analogues, hexamethylbisacetamide, or sodium butyrate; and differentiation of human acute promyelocytic (HL-60) cells in culture by a number of anticancer drugs, sodium butyrate, DMSO, vitamin D_3, phorbol esters, or retinoic acid analogues.[111]

Treatment of cancer through induction of cellular differentiation is an attractive idea because the therapy could be target-cell specific and most likely be much less toxic then standard chemotherapeutic agents. The best example of this is the treatment of acute promyelocytic leukemia in patients with all-transretinoic acid. A more recent example is induction of solid tumor differentiation by the peroxisome proliferator-activated receptor-γ (PPAR-γ) ligand troglitazone in patients with liposarcoma.[112] PPAR-γ is a nuclear receptor that forms a heterodimeric complex with the retinoid X receptor (RXR). This complex binds to specific recognition sequences on DNA and, after binding ligands for either receptor, enhances transcription differentiation–inducing genes, including those for the adipocyte-specific pathway. PPAR-γ appears to act as a tumor suppressor in the prostate and thyroid gland, but not in the colon, where its actions are more complex.[113] Nevertheless, agents that can exploit the proliferation-inhibiting effects of PPAR-γ in cancer tissue and have minimal metabolic side effects may be good targets for drug discovery.

Stem Cells

There has been tremendous excitement, not without some controversy of an ethical, political, and scientific nature, about stem cells. Until recently, it had always been thought that stem cells, those self-renewing, pluripotent cells that exist in an embryo, were only present in highly proliferative tissues such as the bone marrow. Even in tissues with a high cellular turnover rate, such as skin and gastrointestinal mucosa or in tissues such as liver that can regenerate, these cells were thought to be limited in their ability to generate cells of different lineages with various differentiated phenotypes. In some tissues such as the brain and the heart, self-renewing stem cells were thought to be nonexistent. It has been a true revolution in cell biology to find out that self-renewing, multipotent, and perhaps pluripotent stem cells exist in every organ in the body of mammals and most likely of humans. These cells presumably are called on to proliferate in response to tissue injury and are involved in tissue repair.

A "true" stem cell must satisfy the following criteria: (1) it must have unlimited self-renewal capacity; and (2) it must be able to divide into two types of daughter cells—one that replicates a self-renewing phenotype and one that attains the phenotype of a subsequent differentiated state.

Pluripotent stem cells can originate from the inner cell mass of an embryonic blastocyst or from the fetal gondal ridge of 8- to 10-week embryos. Although such cells, known as embryonic stem (ES) cells and embryonic germ (EG) cells, respectively, have been known about in mice for many years, it was only recently that ES cells[114] and EG cells[115] were isolated from human sources. Even more recently it has became known that most, and probably all, mammalian tissues also contain stem cells of amazing plasticity. These not only proliferate and renew damaged tissues but in some cases can even differentiate into cell types of another tissue. ES cells can be grown in cell culture and are capable of producing multiple cell types including vascular, neuronal, pancreatic, and cardiac muscle cells (reviewed in Reference 116). EG cells have also been shown to produce multiple lineages, including neurons, glial cells, vascular endothelium, hematopoietic cells, cardiomyocytes, and glucose-responsive, insulin-secreting cells.[115] In effect, ES and EG cells are capable of producing virtually any type of tissue, given the appropriate culture conditions.

The really astonishing thing is that some stem cells derived from adult tissues are also capable of generating other tissue cell types in addition to their own (Table 4–3). Although the mechanism for how this comes about is somewhat controversial (see below), there are extensive data showing that it can happen. For example, it has been observed that humans receiving nonautologous bone marrow transplants have cells of donor origin in nonhematopoietic tissues such as the liver.[117]

The ability of bone marrow–derived stem cells and stem cells from other tissue sources to generate mature functional cell types has led to the exciting possibility of tissue regeneration and repair for a variety of human diseases or conditions, including Parkinson's disease, diabetes, myocardial infarction, congestive heart failure, chronic limb ischemia, liver cirrhosis, and chronic lung disease.[118] While it is perhaps less surprising, given their greater plasticity that ES cells can differentiate into dopamine-producing neurons

Table 4–3. Potential Plasticity of Stem Cells

Location of Stem Cell	Type of Cells Generated
Brain	Neurons, oligodendrites, skeletal muscle, blood cells
Bone marrow	Endothelial cells, blood cells, cartilage, bone, adipocytes, cardiac muscle, skeletal muscle, neuronal cells, skin, oval cells, gastrointestinal tract cells, thymus, pulmonary epithelial cells
Skeletal muscle	Skeletal muscle, bone, cartilage, fat, smooth muscle
Myocardium	Myocytes, endothelial cells
Skin	Keratinocytes
Liver	Liver cells
Testis and ovaries	Gonads
Pancreatid ducts	Islet cells
Fatty tissue	Fat, muscle, cartilage, bone

From Rosenthal,[116] reprinted with permission from the Massachussetts Medical Society

that can regenerate function in an animal model of Parkinson's disease[119] or can produce insulin-secreting structures similar to pancreatic islets,[120] it is much less expected that bone marrow–derived cells can do the same thing. Although some reports indicate that hematopoietic stem cells (HSCs) from the bone marrow or cytokine mobilized peripheral HSCs are capable of multiple cell lineage differentiation, most likely bone marrow mesenchymal stem cells are the cell type that has this broad pluripotency. Nevertheless, cytokine mobilized HSCs from the bone marrow have been shown to repair myocardial infarctions in a mouse model,[121] a finding suggesting that HSCs themselves can transdifferentiate into multiple cell types and could be a relatively available source of cells for tissue regeneration. Bone marrow–derived cells of stromal origin have been shown to initiate pancreatic regeneration[122] and to produce neurons.[123]

Jiang et al.[124] reported that mesenchymal stem cells derived from the bone marrow of mice could differentiate into cells with visceral mesoderm, neuroectoderm, and endoderm characteristics in vitro. When single cells from this source were injected into early blastocysts, they contributed to most somatic cell types in the developed animal. In addition, when these cells were transplanted into a non-irradiated host, they engrafted and differentiated into hematopoietic lineage cells and the epithelium of the liver, lung, and gut. Thus bone marrow mesenchymal stem cells may be an ideal source for therapy of human inherited or degenerative diseases. This laboratory has also reported that co-culturing of mouse multipotent adult progenitor cells, derived from the bone marrow, with astrocytes facilitated neuronal differentiation, suggesting that astrocyte-derived factors are required for this process.[125]

A caveat to this work is that a number of laboratories have not been able to reproduce the extreme plasticity of bone marrow–derived cells,[126,127] and there is evidence to indicate that the appearance of differentiated cell types from either bone marrow stem cells or ES cells is due to fusion of these cells with adult differentiated cells in the organ in which these stem cells seed out. Regardless of the mechanism, however, it is exciting to think that stem cells from a donor could induce the production of differentiated cells that can regenerate or repair a tissue. Such cells could also be used to deliver genes as a form of gene therapy.[128]

A number of questions and caveats remain, however. For example, if adult stem cells are pluripotent and have unlimited self-renewal capacity, what prevents them from "taking off" on their own and escaping the body's feedback regulatory systems that stop them from becoming tumors? Since stem cells and cancer cells have a number of characteristics in common, are the stem cells present in various tissues the targets for carcinogenic agents? We already know that one of the drawbacks of ES cells as a source for tissue regeneration is that they can form teratomas, i.e., tumors that are made up of a wide variety of cell types without any organized organ structure. They are merely diffuse masses of cells. In addition, many of the cell surface receptors expressed on stem cells are also found on cancer cells[116,129] and the profile of expressed genes determined by DNA microarray for murine and human HSCs overlaps with genes expressed in cancer cells.[130] Also, a *ras*-like gene, *Eras*, is expressed in mouse ES cells, which may give these cells tumor-like properties.[131]

Since normal stem cells and cancer cells share a number of characteristics that facilitate the capacity for unlimited self-renewal, it seems likely that cancer cells acquire the machinery for cell proliferation that is expressed in normal tissue stem cells. What cancer cells lose, however, is the feedback systems to know when to stop proliferating and to start differentiating. Thus, it is not unexpected that a number of signal transduction pathways and their regulatory mechanisms are shared by stem cells and cancer cells. Such shared regulatory pathways include expression of genes involved in preventing apoptosis, e.g., *bcl*-2, and the developmentally regulated genes, Shh, Wnt, and Notch (Fig. 4–4; see color insert). Notch and Shh are involved in self-renewal of HSCs and also dysregulated in cancer cells. The Wnt pathway has been shown to regulate both self-renewal and oncogenesis in various organs. Data from transgenic mice suggest that activation of the Wnt signaling pathway in epidermal stem cells produces epidermal cancers.

These data lead to an obvious question: Are stem cells the targets for carcinogenic transformation? If so, what dysregulates the normal feedback-regulated self-renewal process in cells undergoing malignant transformation? One could further ask whether the steps leading to this dysregulation are likely to be the best targets for therapeutic intervention. Reya et al.[132] postulate that stem cells are the targets for transformation for two reasons: (1) stem cells already have the machinery for self-renewal turned on and it would require fewer genetic or epigenetic manipulations for a cell to become a cancer cell than if they had to turn all these genes on de novo; and (2) stem cells by their vary nature are set up to proliferate for several population doublings and thus have greater opportunity for carginogenic mutations to accumulate than in most mature cell types.

Both normal cells and cancer cells have the ability to generate heterogenous cell types. In the former case, these multiple cell types in a tissue assume some functional differentiated state peculiar to that tissue. In the latter case, a cancer becomes a heterogeneous mass of cells with little or no differentiated function. Although some of the cellular heterogenecity observed in malignant tumors is a result of genetic instability and the resultant continuing mutagenesis, it is likely that some of this heterogeneity is due to aberrant differentation of cancer cells.[132] Cancers often contain a mixture of cells, some of which have a partially differentiated phenotype, often reflective

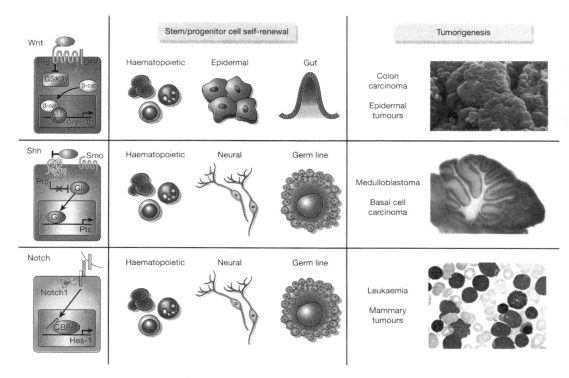

Figure 4–4. Signaling pathways that regulate self-renewal mechanisms during normal stem cell development and during transformation. Wnt and Notch pathways have been shown to contribute to the self-renewal of stem cells and/or progenitors in a variety of organs, including the hematopoietic and nervous systems. When dysregulated, these pathways can contribute to oncogenesis. Mutations of these pathways have been associated with a number of human tumors, including colon carcinoma and epidermal tumors for Wnt, mdeulloblastoma, basal cell carcinoma, and T-cell leukemias for Notch. (From Reya et al.,[132] reprinted with permission from Macmillan Publishers Ltd.)

of an early developmental stage, and as tumor progression occurs, cancers become less well differentiated. Also, there is variable expression of cellular differentiation markers in cancers. For example, there is variable expression of myeloid markers in chronic myeloid leukemia, of neuronal markers in neuroectodermal tumors, and of estrogen receptor in breast cancer (reviewed in Reference 132).

Another implication of the idea that tissue stem cells are the target for carcinogenic attack is that, in a given cancer the cancer stem cells maintain the proliferative capacity of the tumor and not the whole cancer cell mass. There is good evidence to support this concept. For example, when mouse myeloma cells were placed in an in vitro colony-forming assay, only one in several thousand cells was able to form colonies,

and when transplanted in vivo, only 1% to 4% formed spleen colonies.[133] For solid cancers, similar data have been obtained: only 1 in 1000 to 1 in 5000 lung cancer, ovarian cancer, or neuroblastoma cells were able to form colonies in soft agar,[134] a finding again suggesting that there is a subpopulation of cancer cells that proliferate to maintain progressive tumor growth. It has now been possible to distinguish the genetic and phenotypic characteristics of the subset of cells that are the more aggressive, self-renewing cells in a cancer. Al-Hajj et al.[135] found that when human breast cancer cells derived from breast cancer patients were grown in immunocompromised (SCID) mice, only a minority of breast cancer cells were able to form tumors. As few as 100 out of tens of thousands of cells were able to do this. The tumorigenic subpopulation was identified by their cell surface markers and identified as having a CD44$^+$/CD24$^-$ phenotype. When these cells were passaged into additional mice, tumors were generated that contained both CD44$^+$/CD24$^-$ cells and non-tumorigenic cells. These data demonstrate that only a few cells from human breast cancers have the ability to proliferate extensively, whereas the majority of cells from these tumors have only limited proliferative capacity in vivo. Similarly, a cancer stem cell population has been identified in human brain tumors.[136] These data are consistent with the concept that a cancer stem population lurking within a human cancer contains the cells responsible for the aggressive growth of cancers. Further, this may be the cell population for which biochemical markers need to be developed and implemented clinically to discern which breast neoplasms to treat aggressively and which may be more indolent and less dangerous. This would be a big help, for example, in discriminating which breast ductal carcinomas in situ (DCIS tumors) should undergo more extensive surgery and chemotherapy or hormonal therapy and which may be managed less aggressively. Similarly, such markers could be used to determine which prostate cancers should be excised, irradiated, or left for "watchful waiting."

In addition, it is the cancer stem cell population for which therapies should be targeted and developed. Currently available chemotherapeutic drugs were developed largely on the basis of their ability to shrink a tumor mass in an experimental model and in a human clinical trial. Since most cells in a cancerous tissue have limited proliferative potential, the ability of a drug to decrease a tumor mass largely reflects the ability of the drug to kill this less aggressive, potentially less dangerous type of cell, leaving behind the more proliferative clones. Thus, drugs more specifically targeted to the cancer stem cell population should result in more effective and durable responses.

Ways of determining the genetic and phenotypic markers for cancer cells are becoming more apparent. Gene expression arrays and proteomic analyses are beginning to tell us what these markers are (see Chapter 7). For example, Ma et al.[137] have shown that laser capture microdissection and DNA microarrays can be used to distinguish premalignant, preinvasive, and invasive stages of human breast cancer. Interestingly, genes highly expressed at the invasive stage were already expressed in preinvasive stages, suggesting that the cancer stem cell population may be present early in tumor development. In addition, the expression of a subset of genes was quantitatively correlated with the transition from preinvasive to invasive growth. It has been shown that the hedgehog and Bmi-1 signaling pathways are activated in human breast cancer stem cells.

CELL CYCLE REGULATION

Historical Perspectives

The development of our knowledge about cell cycle regulation is itself a fascinating story and takes us through a tale of fundamental discoveries in yeast, sea urchins, clams, fruit flies, frogs, mice, and humans. This story serves as a wonderful example of why fundamental, basic research should be supported for its own sake, even though its primary aim at the time may simply be the pursuit of knowledge.

The story of the factors involved in cell cycle regulation goes back many years. Definition of distinct phases of a division cycle, i.e., G$_1$, S, G$_2$, and M, became established in the mid- to late 1950s when tritum labeling and cell synchronization techniques became available to score mitoses and to measure the time between one mitotic wave and another in cycling cells (reviewed in Reference 138). This method consists of labeling cells with a pulse of tritiated thymidine, taking an aliquot of cells at various times after the labeling, fixing the cells, and counting the percentage of mitoses that are labeled in autoradiograms from each time point. The percentage of labeled mitoses will rise from zero to a peak as the cells that were in S phase at the time of the pulse go through mitosis. Following the peak, there will be a "trough" as the cells in G$_1$ at the time of label go through M. A second cycle will then show a similar peak of labeled M phase cells as the first "wave" comes back through the cell cycle, but the peak will be lower because of dilution of the [3H] thymidine as another round of DNA synthesis occurs and because of some spread of cell cycle times as synchrony is diminished (Fig. 4–5). In this way, the classic cell cycle of $G_1 \rightarrow S \rightarrow G_2 \rightarrow M$ was established.

It became clear from early studies of yeast mutants that certain genetically controlled factors played a key role in regulating the cell cycle. In the 1970s, Lee Hartwell and colleagues identified mutants of the budding yeast *Saccharomyces cerevisiae* that had defects at specific stages of the cell cycle. Temperature-sensitive mutants that were defective in initiation of DNA replication, DNA elongation, DNA ligation, tubulin assembly, spindle elongation, chromatin assembly, sister chromated separation, nuclear division, and cytokinesis were identified (reviewed in Reference 139). This led to the concept of "checkpoints" in the cell cycle that sense the completion of one event before allowing the cell to proceed to the next event. (This will be discussed in more detail below.) Hartwell et al. identified a series of genes called *cdc* genes, whose mutation produced defects in cell cycle progression. The mutated genes were postulated to be the ones involved in the checkpoints. One gene, called *cdc*28, appeared to control entry into mitosis. A similar gene, *cdc*2, was later discovered in the fusion yeast *S. pombe* by Nurse and Bisset.[140] That these genes had the same function was shown by the observation that the *cdc*28 gene could substitute for *cdc*2 mutants of *S. pombe* in allowing the cell cycle to proceed.[141]

Another piece fell into place when it was realized that a cytoplasmic factor called maturation promoting factor (MPF) in unfertilized frog (*Xenopus*) eggs, originally identified by Masui and Markert (reviewed in Reference 142), induced immature oocytes to undergo mitotic division and was conserved in oocytes of distantly related species such as starfish. MPF was found to be a protein complex containing a factor identical to the *cdc*2 gene product of *S. pombe* (reviewed in Reference 143). It was also later shown that a homologous human gene could substitute for a defective *cdc*2 gene in *S. pombe*,[144] indicating the evolutionary linkage for this key cell cycle gene.[145]

Another key to the puzzle was discovered when Tim Hunt and his students in the Physiology Course at Woods Hole Marine Biological Laboratory were looking for a simple experiment to study sea urchin egg development (reviewed in Reference 142 and 145). They labeled the eggs continuously with [^{35}S] methionine after fertilization and analyzed by SDS-gel electrophoresis the pattern of labeled proteins over time. They found that one protein about 55 kDa was strongly labeled after fertilization but that its presence seemed to oscillate with the division cycle in that it built up during interphase and was lost at about the time the egg divided. Hunt and colleagues made similar observations in a second species of sea urchins and in the clam *Spisula solidissima*, and they named this protein "cyclin."[146]

What remained was to link all these pieces together. The linkage of MPF to *cdc*2 and then to cyclin came from several approaches (reviewed in Reference 142). Purification of MPF from frog eggs showed that it contained CDC2, the gene product of *cdc*2, and a cyclin called cyclin B. CDC2 was shown to have a protein kinase activity that phosphorylated histone H1 and that oscillated in activity with cell cycle phase. Reconstitution experiments using frog egg cytoplasm depleted of endogenous transcripts showed that the presence of a cyclin whose synthesis varied with the cell cycle was tied to progression of cell division. Identification of cyclin homologs in the yeast *S. cerevisiae* (called CLN 1, 2, and 3), whose levels vary with various phases of the cell cycle and which bind to the CDC2 homolog CDC28 and regulate its kinase activity (as well as similar data from other

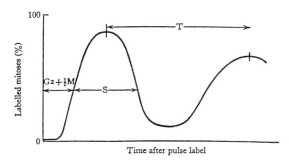

Figure 4–5. Diagram of labeled mitoses (metaphases) in successive cell samples after a pulse of tritiated thymidine. (From Mitchinson,[138] with permission.)

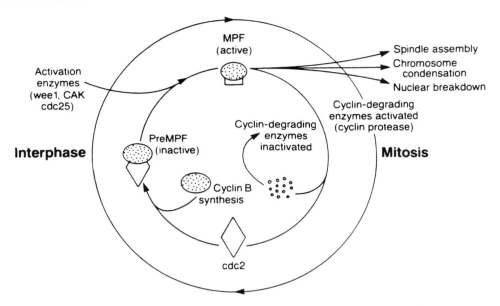

Figure 4–6. The cyclin–cdc2 cycle. During interphase, cyclin B accumulates and associates with cdc2 to form pre-MPF. This is then sequentially phosphorylated by wee1 (a cyclin-cdc2-specific protein tyrosine kinase) on Tyr15 of cdc2, then by CAK (cdc2-activating kinase) on *Thr*161: cdc25, a protein tyrosine phosphatase then dephosphorylates *Tyr*15, leaving active MPF. This triggers mitosis and activates cyclin protease. As cyclin is broken down, MPF disperses and the cyclin-degrading enzymes are inactivated. Thus, cyclin begins to accumulate once more. (From Kirschner,[142] with permission.)

organisms), provided an early picture of the cell cycle as shown in Figure 4–6.

Unfortunately or fortunately, depending on your point of view, the story has become much more complicated, even in yeast. Genetic studies in S. *pombe* have identified a negative regulator of CDC2, called wee 1, which is a tyrosine kinase that phosphoryates a specific tyrosine on CDC2, tyrosine 15, leading to inactivation of CDC2. CDC2 is re-activated by a phosphatase, called CDC25 in S. *pombe*, which removes the phosphate at tyrosine 15. A second phosphorylation step, phosphorylation of threonine 167, is required for activation of CDC2 kinase activity. Thus, the phosphorylation state of CDC2 is important for its regulation. Similar activation and inactivation events for CDC2-like kinases have been described in *Drosophila* and mammalian cells. Indeed, CDC2-like kinases are key cell cycle regulators in all cell types examined, making it the mother of all cell cycle kinases.

Lee Hartwell, Paul Nurse, and Tim Hunt were awarded the 2001 Nobel Prize for Physiology or Medicine in recognition of their pioneering work on cell cycle regulation. Hartwell was recognized for identification of yeast (S. *cerevisiae*) genes involved in cell cycle regulation, his lab identifying over 100 of them, including cdc28, also called *start*. He also introduced the concept of cell cycle checkpoints and identified five genes involved in those cell cycle stop signals.

Paul Nurse, working with another species of yeast, S. *pombe*, identified mutants that showed which genes were important in cell cycle regulation, and showed that one he identified as cdc2 was functionally equivalent to *start*. These two genes were later shown to encode enzymes belonging to the family of cyclin-dependent kinases (CDKs; see below). Homologs of the yeast CDKs were also found in human cells, indicating that these important cell cycle regulatory genes were conserved during evolution.

Tim Hunt's contribution was the discovery of cyclins and how their levels varied during phases of the cell cycle.

The Molecular Players

Cyclin-Dependent Protein Kinases

Cyclin-dependent protein kinases (CDKs), of which CDC2 is only one, are crucial regulators of the timing and coordination of eukaryotic cell cycle events. Transient activation of members of this family of serine/threonine kinases occurs at specific cell cycle phases. In budding yeast, G_1 cyclins encoded by the CLN genes (see above) interact with, and are necessary for the activation of, the CDC2 kinase (also called $p34^{cdc2}$), driving the cell cycle through a regulatory point called START (because it is regulated by the *cdc2* or *start* gene) and committing cells to enter S phase. START is analogous to the G_1 restriction point in mammalian cells.[147] The CDKs work by forming active heterodimeric complexes following binding to cyclins, their regulatory subunits (reviewed in Reference 148). CDK2, 4, and 6, and possibly CDK3 cooperate to push cells through G_1 into S phase. CDK4 and CDK6 form complexes with cyclins D1, D2, and D3, and these complexes are involved in completion of G_1. CyclinD–dependent kinases accumulate in response to mitogenic signals, and this leads to phosphorylation of the Rb protein. This process is completed by the cyclin E1- and E2-CDK2 complexes. Once cells enter S phase, cyclin E is degraded and A1 and A2 cyclins get involved by forming a complex with CDK2. There are a number of regulators of CDK activities; where they act in the cell cycle is depicted in Figure 4–7.

CDK Inhibitors

The inhibitors of CDKs include the Cip/Kip and INK4 family of polypeptides (reviewed in Reference 149). The Cip/Kip family includes $p21^{cip1}$, $p27^{kip1}$, and $p57^{kip2}$. The actions of these proteins are complex. Although the Cip/Kip proteins can inhibit CDK2, they are also involved in the sequestration of cyclin D-dependent kinases that facilitates cyclin E-CDK2 activation necessary for G_1/S transition.

The INK4 proteins target the CDK4 and CDK6 kinases, sequester them into binary CDK-INK4 complexes, and liberate bound Cip/Kip proteins. This indirectly inhibits cyclin E–CDK

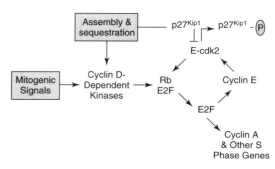

Figure 4–7. Restriction point control and the G_1-S transition. As cells enter the division cycle from quiescence, the assembly of cyclin D–dependent kinases in response to mitogenic signals requires Cip/Kip proteins, which are incorporated into catalytically active holoenzyme complexes. The cyclin D–dependent kinases initiate Rb phosphorylation, releasing E2F from negative constraints and facilitating activation of a series of E2F-responsive genes, the products of which are necessary for S-phase entry. Activation of cyclin E by E2F enables formation of the cyclin E–cdk2 complex. This is accelerated by the continued sequestration of Cip/Kip proteins into complexes with assembling cyclin D–cdk complexes. Cyclin E–cdk2 completes the phosphorylation of Rb, further enabling activation of E2F-responsive genes, including cyclin A. Cyclin E–cdk2 also phosphorylates $p27^{Kip1}$, targeting it for ubiquitination and proteasomal degradation. The initiation of the self-reinforcing E2F transcriptional program together with degradation of $p27^{Kip1}$ alleviates mitogen dependency at the restriction point and correlates with the commitment of cells to enter S phase. In subsequent cycles, cyclin D–dependent kinases remain active as long as mitogens are present, and levels of $p27^{Kip1}$ remain low. All $p27^{Kip1}$ in cycling cells is complexed with cyclin D–cdk complexes. Mitogen withdrawal results in cyclin D degradation, liberating $p27^{Kip1}$ from this latent pool. The resulting inhibition of cyclin D– and E–dependent kinases leads to cell cycle arrest, usually within a single cycle. (From Sherr,[149] reprinted with permission from the American Association for Cancer Research.)

and promotes cell cycle arrest. The INK4-directed arrest of the cell cycle in G_1 keeps Rb in a hypophosphorylated state and represses the expression of S-phase genes.[149]

Four INK4 proteins have been identified: $p16^{INK4a}$, $p15^{INK4b}$, $p18^{INK4c}$, and $p19^{INK4d}$. INKA4a loss of function occurs in a variety of cancers including pancreatic and small cell lung carcinomas and glioblastomas. INK4a fulfills the criteria of a tumor suppressor and appears to be the INK4 family member with the most active role in this regard.[149]

The INK4a gene encodes another tumor suppressor protein called ARF (p14^{ARF}). Mice with a disrupted ARF gene have a high propensity to develop tumors, including sarcomas, lymphomas, carcinomas, and CNS tumors. These animals frequently die at less than 15 months of age. ARF and p53 act in the same pathway to insure growth arrest and apoptosis in response to abnormal mitogenic signals such as *myc*-induced carcinogenesis (Fig. 4–8).

Cyclins

The originally discovered cyclins, cyclin A and B, identified in sea urchins, act at different phases of the cell cycle. Although both cyclins A and B interact with p34^{cdc2} to induce maturation and mitosis in *Xenopus* oocytes, the synthesis and destruction of cyclin A occurs earlier in the cell cycle than cyclin B. Cyclin A is first detected near the G$_1$/S transition and cyclin B is first synthesized during S phase and accumulates in complexes with p34^{cdc2} as cells approach the G$_2$-to-M transition. Cyclin B is then abruptly degraded during mitosis. Thus, cyclins A and B regulate S and M phase but do not appear to play a role in G$_1$ control points such as the restriction point (R point), which is the point where key factors have accumulated to commit cells to enter S phase.

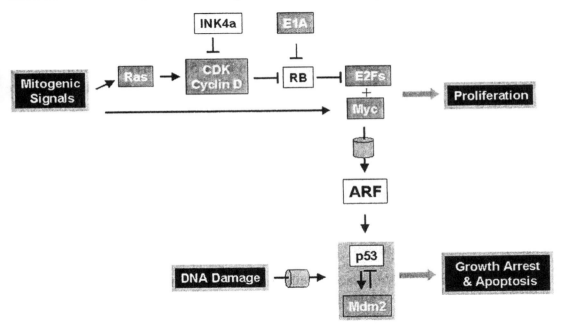

Figure 4–8. ARF tumor surveillance. When induced by inappropriate mitogenic signals, ARF antagonizes Mdm2 to activate p53. Hence, hyperproliferative signals are countered by ARF-dependent p53 induction, which diverts incipient cancer cells to undergo growth arrest and/or apoptosis. Loss of the ARF checkpoint (indicated by the vertical barrel) subverts this form of cell-autonomous tumor surveillance and allows proteins such as Ras, Myc, E1A, and E2F to function as "pure" proliferation enhancers. DNA damage signals engage various ARF-independent signaling pathways (shown collectively by the horizontal barrel) that stabilize p53, most commonly by inducing post-translational modifications in p53 and/or Mdm2 that prevent their interaction. Although ARF is not directly activated by ionizing radiation or various genotoxic drugs, it is still a potent modifier of the DNA damage response. ARF induction sensitizes cells to DNA damage signals; conversely, ARF loss increases the Mdm2 response and severely dampens the p53 response. All proteins enclosed by shaded boxes are potential oncogenes, whereas those illustrated by unfilled boxes are tumor suppressors. (From Sherr,[149] reprinted with permission from the American Association for Cancer Research.)

Three more recently discovered mammalian cyclins, C, D1, and E, are the cyclins that regulate the key G_1 and G_1/S transition points (reviewed in Reference 150). Unlike cyclins A and B, cyclins C, D1, and E are synthesized during the G_1 phase in mammalian cells. Cyclin C levels change only slightly during the cell cycle but peak in early G_1. Cyclin E peaks at the G_1–S transition, suggesting that it controls entry into S. Three distinct cyclin D forms, D1, 2, and 3, have been discovered and are differentially expressed in different mouse cell lineages. These D cyclins all have human counterparts. Cyclin D levels are growth factor dependent in mammalian cells: when resting cells are stimulated by growth factors, D-type cyclin levels rise earlier than cyclin E levels, implying that they act earlier in G_1 than E cyclins. Cyclin D levels drop rapidly when growth factors are removed from the medium of cultured cells. As noted above, all of these cyclins (C, D, and E) form complexes with, and regulate the activity of, various CDKs and these complexes control the various G_1, G_1–S_1, and G_2–M transition points.

A variety of viral oncogenic proteins also get into the act here. The adenovirus E1A protein binds the cyclin A-p34^{cdc2} complex, via its interaction with p107 and Rb[151] and may act to "strip" Rb or p107 from the cyclin–CDK complex, aiding in the CDK activation. Introduction of a constitutively acting c-*myc* gene into BALB/3T3 mouse fibroblasts activated cyclin A expression and produced a growth factor–independent association of cyclin A-CDK2 with the transcription factor E2F, which correlated with an increase in E2F transcriptional activity.[152] In this model system, *myc*-transformed cells reduced cyclin D1 expression in early G_1. In addition, both the *src* gene product p60^{c-src} [153] and SV 40 large T antigen[154] are phosphorylated by p34^{cdc2}, which suggests that this phosphorylation event is involved in the effects of these oncogenic proteins on DNA replication and cell proliferation.

Interestingly, negative growth regulators also interact with the cyclin-CDK system. For example, TGF-β1, which inhibits proliferation of epithelial cells by interfering with G_1-S transition, reduced the stable assembly of cyclin E-CDK2 complexes in mink lung epithelial cells, and prevented the activation of CDK2 kinase activity and the phosphorylation of Rb. This was one of

the first pieces of data suggesting that the mammalian G_1 cyclin-dependent kinases are targets for negative regulators of the cell cycle.[155]

Cell Cycle Checkpoints

The role of various CDKs, cyclins, and other gene products in regulating checkpoints at G_1 to S, G_2 to M, and mitotic spindle segregation have been described in detail elsewhere.[156–158] Alterations of one or more of these checkpoint controls occur in most, if not all, human cancers at some stage in their progression to invasive cancer. Examples of some of these alterations are given below.

A key player in the G_1–S checkpoint system is the retinoblastoma gene *rb*. Phosphorylation of the Rb protein by cyclin D–dependent kinase releases Rb from the transcriptional regulator E2F and activates E2F function. Inactivation of *rb* by genetic alterations occurs in retinoblastoma and is also observed in other human cancers, for example, small cell lung carcinomas and osteogenic sarcomas.

The *p53* gene product is an important cell cycle checkpoint regulator at both the G_1–S and G_2–M checkpoints but does not appear to be important at the mitotic spindle checkpoint because gene knockout of *p53* does not alter mitosis. The *p53* tumor suppressor gene is the most frequently mutated gene in human cancer, indicating its important role in conservation of normal cell cycle progression. One of *p53*'s essential roles is to arrest cells in G_1 after genotoxic damage, to allow for DNA repair prior to DNA replication and cell division. In response to massive DNA damage, *p53* triggers the apoptotic cell death pathway. Data from short-term cell-killing assays, using normal and minimally transformed cells, have led to the conclusion that mutated p53 protein confers resistance to genotoxic agents.

The spindle assembly checkpoint machinery involves genes called *bub* (budding uninhibited by benomyl) and *mad* (mitotic arrest deficient).[158] There are three *bub* genes and three *mad* genes involved in the formation of this checkpoint complex. A protein kinase called Mps1 also functions in this checkpoint function. The chromosomal instability, leading to aneuploidy in many human cancers, appears to be due to defective control of the spindle assembly checkpoint. Mutant alleles of the human *bub1* gene have been observed in colorectal tumors displaying aneuploidy. Mutations in these spindle checkpoint genes may also result in increased sensitivity to drugs that affect microtubule function because drug-treated cancer cells do not undergo mitotic arrest and go on to die.

Maintaining the integrity of the genome is a crucial task of the cell cycle checkpoints. Two checkpoint kinases, called Chk1 and Chk2

(also called Cds1), are involved in checkpoint controls that affect a number of genes involved in maintenance of genome integrity (Fig. 4–9; see color insert). Chk1 and Chk2 are activated by DNA damage and initiate a number of cellular defense mechanisms that modulate DNA repair pathways and slow down the cell division cycle to allow time for repair. If DNA is not successfully mended, the damaged cells usually undergo cell death via apoptosis (see below). This process prevents the defective

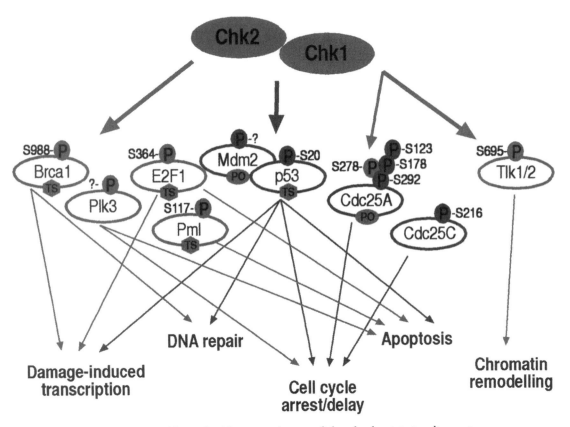

Figure 4–9. Chk1 and Chk2 as mediators of the checkpoint signaling network. Following their activation, Chk1 and Chk2 phosphorylate unique (green and red, respectively) and overlapping (blue) downstream effectors that further propagate the checkpoint signaling. Depending on the type of stress, velocity of DNA damage, and cellular context, this leads to (1) switch to the stress-induced transcription program (E2F1, Brca1, p53); (2) direct or indirect initiation of DNA repair (BRCA1, p53); (3) acute delay (degradation of Cdc25A) and/or sustained block (Cdc25C, p53, E2F1); (4) apoptosis (Pml[1], p53, E2F1); and (5) modulation of the chromatin remodeling pathways (Tlk 1/2). The known target sites of Chk1 (green), Chk2 (red), and both Chk1 and Chk2 (blue) on the individual substrates are shown. Some of the Chk1 and Chk2 downstream effectors are classified as proto-oncogenes (PO) or tumor suppressors (TS), as indicated. (From Bartek and Lucas,[159] reprinted with permission from Elsevier.)

genome from extending its paternity into daughter cells.

Upstream elements activating the checkpoint signaling pathways such as those turned on by irradiation or agents causing DNA double-strand breaks include the ATM kinase, a member of the phosphatidylinositol 3-kinase (PI3K) family, that activates Chk2 and its relative ATR kinase that activates Chk1. There is also cross talk between ATM and ATR that mediates these responses (reviewed in Reference 159). Chk1 and Chk2 phosphorylate CDC25A and C, which inactivate them. In its dephosporylated state CDC25A activates the CDK2-cyclin E complex that promotes progression through S phase. It should be noted that this is an example of dephosphorylation rather than phosphorylation activating a key biological function. This is in contrast to most signal transduction pathways, where the phosphorylated state of a protein (often a kinase) is the active state and the dephosphorylated state is the inactive one (see Signal Transduction Mechanisms, below). In addition, Chk1 renders CDC25A unstable, which also diminishes its activity (reviewed in Reference 160).

CDC25A also binds to and activates CDK1-cyclin B, which facilitates entry into mitosis. G_2 arrest induced by DNA damage induces CDC25A degradation and, in contrast, G_2 arrest is lost when CDC25A is overexpressed.

A number of proteins are now known to act as mediators of checkpoint responses by impinging on the Chk1 and 2 pathways. These include the BRCT domain–containing proteins 53BP1, BRCA1, and MDC1. These proteins are involved in activation of Chk1 and Chk2 by acting through protein–protein interactions that modulate the activity of these checkpoint kinases. In genereal, these modulators are thought to be tumor suppressors.

Chk1 and 2 have overlapping roles in cell cycle regulation, but different roles during development. Chk1 but not Chk2 is essential for mammalian development, as evidenced by the early embryonic lethality of Chk1 knockout mice. Chk2-deficient mice are viable and fertile and do not have a tumor-prone phenotype unless exposed to carcinogens, and this effect is more evident later in life (reviewed in Reference 159). Rare germline mutations of Chk2 have been observed in cancer-prone patients with the Li-Fraumeni syndrome (LFS), thus Chk2 mutations may be an alternative or overlapping genetic defect along with p53 mutations in these patients. Since LFS patients are susceptible to develop multiple types of tumors, including a predominant incidence of breast cancers and sarcomas, the Chk2 path may also be an important tumor suppressor for these tumors in non-LFS patients. Chk2 mutations have also been found in small subsets of "sporadic" human cancers, including carcinomas of the breast, lung, vulva, urinary bladder, colon, and ovary as well as in osteosarcomas and lymphomas.[159] In contrast, cancer-associated genetic defects in Chk1 are rare but have been observed in carcinomas of the colon, stomach, and endometrium.

As illustrated in Figure 4–9, there are interactions between the Chk kinases and the p53 pathway. Chk2 phosphorylates threonine-18 or serine-20 on p53, which attenuates p53's interaction with its inhibitor MDM2, thus contributing to p53 stabilization and activation. However, Chk2 and p53 only have partially overlapping roles in checkpoint regulation because not all DNA-damaging events activate both pathways. For example, some types of DNA damage that activate p53 do not activate Chk2 and vice versa. Thus, the two pathways are partly redundant and overlapping but not totally so, as evidenced by the fact that in Chk2-deficient cells, Chk1 can still phosphorylate and activate p53.

Cell Cycle Regulatory Factors as Targets for Anticancer Agents

The commonly observed defects in cell cycle regulatory pathways in cancer cells distinguishes them from normal cells and provides potential targets for therapeutic agents. One approach is to inhibit cell cycle checkpoints in combination with DNA-damaging drugs or irradiation. The rationale for this is that normal cells have a full complement of checkpoint controls, whereas tumor cells are defective in one or more of these and thus are more subject to undergoing apoptosis in response to excessive DNA damage. This has been accomplished by combining ATM/ATR inhibitors such as caffeine or Chk1 inhibitors in combination with DNA-damaging drugs. So far this approach hasn't been demonstrated clinically, and indeed is somewhat counterintuitive, since p53 mutant tumor cells are more resistant to many chemotherapeutic drugs. p53 is a key player in causing cell death in drug-treated, DNA-damaged cells (one exception to that is the microtubule inhibitor paclitaxel), and active, unmutated p53 is needed for this response.

Another approach is to target the cyclin-dependent kinases directly. Alteration of the G_1–S checkpoint occurs in many human cancers. Cyclin D1 gene amplification occurs in a subset of breast, esophageal, bladder, lung, and squamous cell carcinomas. Cyclins D2 and D3 are overexpressed in some colorectal carcinomas. In addition, the cyclin D–associated kinases CDK4 and CDK6 are overexpressed or mutated in some cancers. Mutations or deletions in the CDK4 and CDK6 inhibitor INK4 have been observed in familial melanomas, and in biliary tract, esophageal, pancreatic, head and neck, non–small cell lung, and ovarian carcinomas. Inactivating mutations of CDK4 inhibitory modulators p15, p16, and p18 have been observed in a wide variety of human cancers. Cyclin E is also amplified and overexpressed in some breast and colon carcinomas and leukemias.

Human cancers have a variety of mutations in cell cycle regulatory genes (reviewed in Reference 148). This includes overexpression of D1 and E1 cyclins and CDKs (mainly CDK4 and CDK6) as noted above. Loss of CDK inhibitory functions (mainly INK4a and 4b and Kip1) also occurs, as does loss of Rb, one of the first tumor suppressor genes identified (see Chapter 5). Loss of Kip1 function and overexpression of cyclin E1 occur frequently and are associated with poor prognosis in breast[161] and ovarian cancers.[162]

The mitogen-stimulated proliferation of cells is mediated via a retinoblastoma (Rb) pathway that involves phosphorylation of Rb, its dissociation from and activation of the E2F family of transcription factors, and subsequent turn-on of genes involved in G_1–S transition and DNA synthesis (reviewed in Reference 163). Disruption of this pathway by overexpression of cyclin D1, loss of the INK4 inhibitor p16, mutation of CDK4 to a p16-resistant form, or loss or mutation of Rb is frequently seen in cancer cells. The activation of CDK inhibitory factors such as p16[INK4] or p27[kip1] and inhibition of cyclin-dependent kinases are therefore potential ways to interdict the overactive cell proliferation pathways in cancer cells. Thus, inhibition of cyclins D1 and E and CDKs, especially CDK4 and CDK6, could be targets for inhibiting growth of cancers. As more knowledge of the complicated steps in cell cycle regulation is gained, more potential targets become available. For example, Bettencourt-Dias et al.[164] used RNA-mediated interference (RNAi) to carry out a genome-wide survey of protein kinases required for cell cycle progression in *Drosophila* and have found as many as 80 protein kinases, a number of them previously unknown, that regulate the cell division cycle. A number of these will no doubt have human orthologs.

APOPTOSIS

Apoptosis (sometimes called *programmed cell death*) is a cell suicide mechanism that enables multicellular organisms to regulate cell number in tissues and to eliminate unneeded or aging cells as an organism develops. The biochemistry of apoptosis has been well studied in recent years, and the mechanisms are now reasonably well understood.[165–167] The enzymatic machinery for this was first discovered in the nematode *C. elegans*, and later the homologues of these genes and their products were identified in mammalian cells, including human cells. The apoptosis pathway involves a series of positive and negative regulators of proteases called *caspases*, which cleave substrates, such as poly (ADP-ribose) polymerase, actin, fodrin, and lamin. In addition, apoptosis is accompanied by the intranucleosomal degradation of chromosomal DNA, producing the typical DNA ladder seen for chromatin isolated from cells undergoing apoptosis. The endonuclease responsible for this effect is called caspase-activated DNase, or CAD.

A number of "death receptors" have also been identified.[168] *Death receptors* are cell surface receptors that transmit apoptotic signals initiated by death ligands. The death receptors sense signals that tell the cell that it is in an uncompromising environment and needs to die. These receptors can activate the death caspases within seconds of ligand binding and induce apoptosis within hours. Death receptors belong to the tumor necrosis factor (TNF) receptor gene superfamily and have the typical cystine-rich extracellular domains and an additional cytoplasmic sequence termed the _death domain_. The best-characterized death receptors are CD95 (also called Fas or Apo1) and TNF receptor TNFR1 (also called p55 or CD120a).

The importance of the apoptotic pathway in cancer progression is seen when there are mutations that alter the ability of the cell to undergo apoptosis and allow transformed cells to keep proliferating rather than die. Such genetic alterations include the translocation of the _bcl-2_ gene in lymphomas that prevents apoptosis and promotes resistance to cytotoxic drugs. Other genes involved as players on the apoptosis stage include c-_myc_, p53, c-_fos_, and the gene for interleukin-1β-converting enzyme (ICE). Various oncogene products can suppress apoptosis. These include adenovirus protein E1b, _ras_, and _v-abl_.

Mitochondria play a pivotal role in the events of apoptosis by at least three mechanisms: (1) release of proteins, e.g., cytochrome c, that triggers activation of caspases, (2) alteration of cellular redox potential, and (3) production and release of reactive oxygen species after mitochondrial membrane damage.[169] Another mitochondrial link to apoptosis is implied by the fact that Bcl-2, the anti-apoptotic factor, is a mitochondrial membrane protein that appears to regulate mitochondrial ion channels and proton pumps.

Apoptosis occurs in most, if not all, solid cancers. Ischemia, infiltration of cytotoxic lymphocytes, and release of TNF may all play a role in this. It would be therapeutically advantageous to tip the balance in favor of apoptosis over mitosis in tumors, if that could be done. It is clear that a number of anticancer drugs induce apoptosis in cancer cells. The problem is that they usually do this in normal proliferating cells as well. Therefore, the goal should be to manipulate selectively the genes involved in inducing apoptosis in tumor cells. Understanding how those genes work may go a long way to achieving this goal.

Historical Perspectives

The study of cell death is a lively field, as evidenced by the tremendous spate of recent publications and scientific meetings covering the subject. But it wasn't always so. Evidence for the existence of two morphologically distinct types of cell death was obtained by Kerr[170] in 1965 from histochemical studies of ischemic injury to rat liver. Some cell death occurred with the typical changes seen in tissue necrosis: clumping of chromatin into ill-defined masses, swelling of organelles, flocculent densities in the matrix of mitochondria, membrane disintegration, and infiltration of inflammatory cells. Cells in some areas of the damaged liver, however, died a different death. They contained chromatin compacted into sharply delineated masses, condensation of the cytoplasm, and outcropping of cytoplasmic "blebs" or proturberances that became pinched off (apoptotic bodies) and released, to be devoured by tissue phagocytic cells. No inflammatory reaction, however, was noted around cells dying by this second mechanism. Further studies by Currie and Wyllie[171] showed that this second mechanism of cell death occurs as tissues undergo remodeling during development and in this sense is a "physiologic cell death." The original term for this phenomenon, _shrinkage necrosis_, didn't seem an appropriate one for this process, so Kerr et al. searched for another one. A colleague of theirs at the University of Aberdeen, professor James Carmack of the Department of Greek, suggested the term _apoptosis_ (pronounced apōtō, with the second _p_ silent), meaning "falling off" of petals from a flower or leaves from a tree. The term has stuck ever since. The concept of apoptosis however, was largely ignored until the mid- to late 1980s when the discovery of the _ced_ genes in the roundworm _Caenorhabditis elegans_ and of the _bcl-2_ gene in B lymphocytes (see below) put the field on a solid genetic basis.

A number of the cellular morphological and biochemical changes that occur during the apoptosis have been worked out. These include morphological changes that can be observed both at the light and electron microscopic levels (Figs. 4–10 and 4–11).[172] One of the most extensively studied biochemical events in apoptosis is the double-strand cleavage of nuclear DNA that occurs at linker strands between nucleosomes, producing fragments that are multiples of about 185 base pairs. These fragments can be observed as characteristic apoptotic DNA "ladders" by agarose gel electrophoresis.

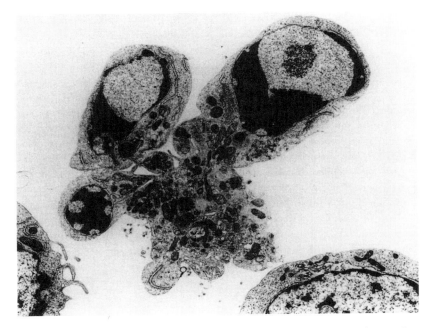

Figure 4–10. Apoptosis of murine NS-1 cell occurring spontaneously in culture. Note the discrete nuclear fragments with characteristic segregation of compacted chromatin, the crowding of organelles, and the marked convolution of the cellular surface. (From Kerr and Harmon,[172] with permission.)

Other biochemical changes that appear to correlate with the induction of the apoptotic cascade of events include CA^{2+} ion influx into cells, induction of transglutaminase that catalyzes the formation of ε (γ-glutamyl) lysine cross-lines between proteins, disruption of microtubules, induction of calmodulin expression, loss of cell membrane phospholipid asymmetry resulting in exposure of phosphatidylserine on the cell surface, activation of the cell surface receptor Fas/Apo1, activation of protein kinase C (some cells only), and induction of a neutral sphingomyelinase in the plasma membrane that releases phosphocholine and ceramide.[173] The biochemical mechanisms involved in apoptosis are described below.

Apoptosis can be triggered by a number of agents or stimuli, including events triggered in tissue differentiation during development; removal of growth factors or hormones required for cell survival; exposure to TGF-β, TNF, or glucorticoids in cell types sensitive to their negative regulation; and exposure to DNA-damaging anticancer drugs or environmental toxins. Sustained increases in intracellular free Ca^{2+} precede apoptosis induced by a number of agents or conditions, and apoptosis is delayed or inhibited when Ca^{2+} is depleted from the cellular growth medium. In addition, Ca^{2+} ionophores such as A23187, which carries Ca^{2+} into cells, induces apoptosis in some cell types.

Induction of transglutaminase accompanies apoptosis in several cell types. There is evidence that this enzyme causes protein-protein cross-linking to produce a protein net or meshwork that holds the cell intact and prevents leakage of cellular contents until the apoptotic bodies bud off and are consumed by phagocytes.[172]

Loss of microtubular structures and reorganization of the cytoskeleton occurs in apoptotic cells. Vinca alkaloids such as vincristine, which cause microtubule dissolution, can trigger apoptosis. However, cytochalasin B, an inhibitor of actin polymerization, inhibits formation of apoptotic bodies but not the other features of apoptosis.[174]

Biochemical Mechanism of Apoptosis

Multicellular organisms, from the lowest to the highest species, must have a way to get rid of

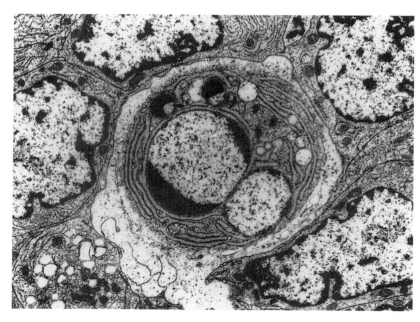

Figure 4–11. Apoptotic body containing well-preserved rough endoplasmic reticulum and four nuclear fragments, which has been phagocytosed by an intraepithelial macrophage in the rat ventral prostate 2 days after castration. (From Kerr and Harmon,[172] with permission.)

excess cells or cells that are damaged in order for the organism to survive. Apoptosis is the mechanism that they use to do this. It is the way that the organism controls cell numbers and tissue size and protects itself from "rogue" cells.

A simplified version of the apoptotic pathways can be visualized in Figure 4–12 (see color insert). The death receptor–mediated pathway is turned on by members of the death receptor superfamily of receptors including Fas receptor (CD95) and TNF receptor 1, which are activated by Fas ligand and TNF, respectively. Interaction of these ligands with their receptors induces receptor clustering, binding of the receptor clusters to Fas-associated death domain protein (FADD), and activation of caspase-8. This activation step is regulated by c-FLIP. Caspase-8, in turn, activates caspase-3 and other "executioner" caspases, which induce a number of apoptotic substrates (see below).

The DNA damage–induced pathway invokes a mitochondrial-mediated cell death pathway that involves pro-apoptotic factors like Bax (blocked by the anti-apoptotic protein Bcl-2). This results in cytochrome c release from the mitochondria and triggering of downstream

effects facilitating caspase-3 activation, which is where the two pathways intersect. There are both positive and negative regulators that also interact on these pathways. For example, the Smac/DIABLO protein blocks activation of inhibitor of apoptosis (IAP) proteins, which inhibit caspase-3. Cross talk between the two cell death pathways is mediated by Bid, a proapoptotic member of the Bcl-2 family. Bid is cleaved and activated by caspase-8. Bid then acts to promote cytochrome c release from mitochondria. Thus, one can see what a complex and tightly regulated pathway apoptosis is. Given the literal life-and-death importance of this pathway, it is easy to see why this is so.

Caspases

Caspases are a family of cysteine proteases that are activated specifically in apoptotic cells. This family of proteases is highly conserved through evolution all the way from hydra and nematodes up to humans. Over 12 caspases have been identified and although most of them appear to function during apoptosis, the function of all of them is not yet clear. The caspases are called

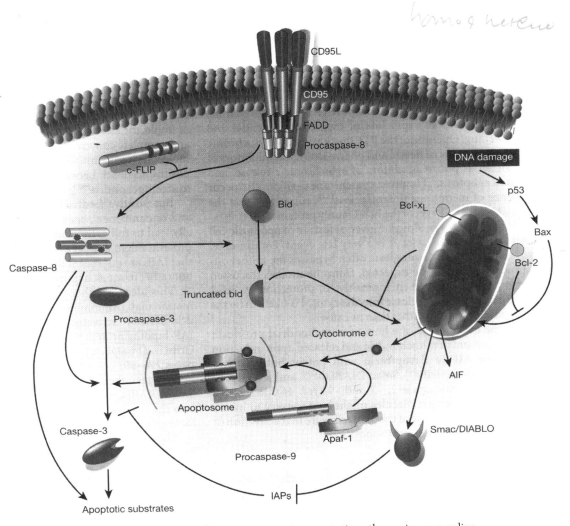

Figure 4–12. The roads to ruin: two major apoptotic pathways in mammalian cells. The death-receptor pathway is triggered by members of the death-receptor superfamily (such as CD95 and tumor necrosis factor receptor I). Binding of CD95 ligand to CD95 induces receptor clustering and formation of a death-inducing signaling complex. This complex recruits, via the adaptor molecule FADD (Fas-associated death domain protein), multiple procaspase-8 molecules, resulting in capase-8 activation through induced proximity. Caspase-8 activation can be blocked by recruitment of the degenerate caspase homologue c-FLIP. The mitochondrial pathway (right) is used extensively in response to extracellular cues and internal insults such as DNA damage. The death-receptor and mitochondrial pathways converge at the level of caspase-3 activation. Caspase-3 activation and activity is antagonized by the inhibitors of apoptosis (IAP) proteins, which themselves are antagonized by the Smac/DIABLO protein released from mitochondria. Downstream of caspase-3, the apoptotic program branches into a multitude of subprograms, the sum of which results in the ordered dismantling and removal of the cell. Cross-talk and integration between the death-receptor and mitochondrial pathways is provided by Bid, a pro-apoptotic Bcl-2 family member. Caspase-8-mediated cleavage of Bid greatly increases its pro-death activity, and results in its translocation to mitochondria, where it promotes cytochrome c exit. Note that under most conditions, this cross-talk is minimal, and the two pathways operate largely independently of each other. Clearly, additional death-inducing pathways must exist, as developmental apoptosis is by and large normal in mice defective in the caspase-8 and caspase-9 pathways. (From Hengartner,[166] reprinted by permission from Macmillan Publishers Ltd.)

cysteine-proteases because they have a cysteine in the active site that cleaves substrates after asparagines in a sequence of asp-X, with the four amino acids amino-terminal to the cleavage site determining a caspase's substrate specificity.

The importance of the caspases in apoptosis is demonstrated by the inhibitory effects of mutation or drugs that inhibit their activity. Caspases can either inactivate a protein substrate by cleaving it into an inactive form or activate a protein by cleaving a pro-enzyme negative regulatory domain. In addition, caspases themselves are synthesized as pro-enzymes and are activated by cleavage at asp-x sites. Thus, they can be activated by other caspases, producing elements of the "caspase cascade" shown in Figure 4–12. Also, as illustrated in Figure 4–12, caspases are activated in a number of steps by proteolytic cleavage by an upstream caspase or by protein–protein interactions, such as that seen for the activation of caspase-8 and the interaction of cytochrome c and Apaf-1 in the activation of caspase-9. A number of important substrates of caspases have been identified, including the caspase-activated DNase (CAD), noted above, which is the nuclease responsible for the DNA ladder of cells undergoing apoptosis. Activation of CAD is mediated by caspase-3 cleavage of the CAD-inhibitory subunit. Caspase-mediated cleavage of other specific substrates has been shown to be responsible for other typical changes seen in apoptotic cells, such as the cleavage of nuclear lamins required for nuclear shrinkage and budding, loss of overall cell shape by cleavage of cytoskeleton proteins, and cleavage of PAK2, a member of the p21-activated kinase family, that mediates the blebbing seen in dying cells (reviewed in Reference 166).

Bcl-2 Family

Mammalian Bcl-2 was first identified as anti-apoptotic protein in lymphomas cells.[175] It turned out to be a homolog of an anti-apoptotic protein called Ced-9 described in *C. elegans*.[176] In *C. elegans*, Ced-9 protects from cell death by binding to the pro-apoptotic factor Ced-4. Similarly, in mammalian cells, Bcl-2 binds to a number of pro-apoptotic factors such as Bax and Apaf-1. One concept is that pro- and anti- apoptotic members of the Bcl-2 family of proteins

form heterodimers, which can be looked on as reservoirs of plus and minus apoptotic factors waiting for the appropriate signals to be released. In this scenario, the one who wins depends on the relative amounts of the pro- or anti- apoptotic factors. This "war" of factors occurs at the mitochondrial membrane, where they compete to regulate release of cytochrome c (see below). However, in healthy cells the pro-apoptotic proteins Bax and Bak are not found in association with Bcl-2 family members. Rather, Bax and Bak appear to be directly activated under conditions of limiting survival signals.[167]

Role of Mitochondria in Apoptosis

When cells are functioning under normal homeostatic conditions, apoptosis is suppressed by strict compartmentalization of the cell death mediators. Mitochondria are key to this in that they contain and hold the cell death regulatory biochemical signals (reviewed in Reference 177). The major apoptotic pathway activator cytochrome c is confined to the mitochondrial intermembrane space. Upon activation of cell death signals leading to permeabilization or rupture of the outer mitochondrial membrane, cytochrome c is released and binds to the cytosolic factor Apaf-1, leading to the allosteric activation of caspase-9, which in turn activates caspase-3, as shown in Figure 4–12. In addition, Smac/DIABLO is released by mitochondrial permeabilization and neutralizes the cytosolic IAPs, thus releasing their IAP inhibition of caspases-3 and -9. The apoptosis-inducing factor (AIF) and the endonuclease CAD are also released from the mitochondrial intermembrane space, travel to the nucleus, and degrade chromatin to produce the typical DNA ladders.

Mitochondrial permeabilization can be induced by a number of pro-apoptotic second messengers such as CA^{2+}, reactive oxygen species, lipids such as ceramide and ganglioside GD3, and stress kinases. CA^{2+} release from the endoplasmic reticulum CA^{2+} "storage depot" appears to be a key pro-apoptotic event, and its uptake by mitochondria is mediated by the pro-apoptotic proteins Bax and Bak of the Bcl-2 family.[178] Conversely, cells are protected by appropriate levels of the anti-apoptotic proteins Bcl-2 or Bcl-X_L, which prevent activation of Bax and Bak.

The release of mitochondrial-associated cell death factors appears to be controlled by the amount of "openness" of mitochondrial membrane "pores" such as the voltage-dependent anion channels (VDACs) in the outer membrane and the adenine-nucleotide translocase (ANT) in the inner membrane. The openness of those pores is increased by activated Bax/Bak and inhibited by Bcl-2 and Bcl-X_L, which prevent activation of Bax and Bak.[179]

Anoikis

Anoikis is a form of apoptosis that occurs in normal cells that lose their adhesion to the substrate or extracellular matrix (ECM) on which they are growing. Adherence to a matrix is crucial for the survival of epithelial, endothelial, and muscle cells. Prevention of their adhesion usually results in rapid cell death, which occurs via apoptosis. Thus, anoikis is a specialized form of apoptosis caused by prevention of cell adhesion. The term *anoikis* means "homelessness" in Greek. Although the observation of this phenomen occurs only with cultured cells, it is likely to occur also in vivo because it is known that cell-cell and cell-ECM interactions are crucial to cell proliferation, organ development, and maintenance of a differentiated state (see Cell Proliferation versus Differentation, above). This may be a way that a multicellular organism protects itself from free-floating or wandering cells (such as occurs in tumor metastasis). The basic rule for epithelial and endothelial cells appears to be "attach or die."[180] Interestingly, cells that normally circulate in the body such as hematopoietic cells do not undergo anoikis.

As noted earlier in this chapter, cell attachment is mediated by integrins, and ECM-integrin interactions transduce intracellular signaling pathways that activate genes involved in cell proliferation and differentiation. Although the cell death pathways induced by disruption of these cell attachment processes aren't clearly worked out, cell detachment–induced anoikis does result in activation of caspases-8 and -3 and is inhibited by Bcl-2 and Bcl-XL, indicating some similarities to the typical apoptosis mechanisms. In addition, integrin-ECM interaction activates focal adhesion kinase (FAK) and attachment-mediated activation of PI3-kinase.

Both of these steps protect cells from anoikis, whereas inhibition of the PI3-kinase pathway induces anoikis (reviewed in Reference 180). Disruption of cell-matrix interactions also turns on the JNK/p38 pathway, a stress-activated protein kinase. The mitogen-activated kinase system may also be involved, since caspase-mediated cleavage of MEKK-1 occurs in cells undergoing anoikis.

As stated earlier, one of the hallmarks of malignantly transformed cells growing in culture is their ability to grow in an anchorage-independent manner, whereas normal cells do not. Thus, cancer cells may develop resistance to anoikis. This may be a way that metastatic cancer cells can survive in the bloodstream until they seed out in a metastatic site. Indeed, there is some evidence for this in that selection of a Cloudman melanoma cell line resistant to anoikis had enhanced metastatic potential.[180]

Resistance to Apoptosis in Cancer and Potential Targets for Therapy

It would be a mistake to portray apoptosis as only a mechanism to kill cells damaged by some exogenous insult such as DNA-damaging toxins, drugs, or irradiation. Apoptosis is, in fact, a normal mechanism used by all multicellular organisms to facilitate normal development, selection of differentiated cells that the organism needs, and control of tissue size. For example, studies of nematodes (*C. elegans*), fruit flies, and mice indicate that apoptotic-mediated mechanisms similar to those described here are intrinsic and required for normal development. Dysfunction of these pathways results in developmental abnormalities and disease states.[181]

In the human, development of the immune system is perhaps the best example of the role for apoptosis in normal development.[182] In the immune system, apoptosis is a fundamental process that regulates T- and B-cell proliferation and survival and is used to eliminate immune cells that would potentially recognize and destroy host tissues ("anti-self"). Mechanisms involving Apo-1/FAS (CD95)-mediated signaling of the caspase cascade as described above are employed in lymphocytic cell selection.

In the case of T lymphocytes, pre-T cells are produced in the bone marrow and circulate to the thymus where they differentiate and re-arrange their T-cell receptors (TCRs). Those cells that fail to rearrange appropriately their TCR genes, and thus cannot respond to self–major histocompatibility complex (MHC)–peptide complexes, die by "neglect." Those T cells that pass the TCR selection tests mature and leave the thymus to become the adult peripheral T-cell pool. The mature T-cell pool thus passes through a number of selection steps to ensure self-MHC restriction and self-tolerance. Apo-ptosis also is used to delete mature peripheral T cells that are insufficiently stimulated by positive growth signals, and this is a mechanism to down-regulate or terminate an immune response.

B lymphocytes undergo selection and matu-ration in the bone marrow and germinal centers of the spleen and other secondary lymphoid organs. Those with low antigen affinity or that are autoreactive are eliminated by apoptosis. Those that pass this test mature into memory B cells and long-lived plasma cells.

The ability of lymphoid progeny cells to avoid apoptosis may lead to lymphatic leukemias or lymphomas. For example, follicular lymphomas have a Bcl-2 translocation into the immunoglob-ulin heavy-chain locus that dysregulates Bcl-2 expression. As noted above, Bcl-2 overexpres-sion suppresses apoptosis and enhances cell proliferation. In addition, cancers develop mul-tiple mechanisms to evade destruction by the immune system such as decreased expression of MHC molecules on cancer cell surfaces and production of immunosuppressive cytokines (see Chapter 6).

Several cell proliferation–promoting events take place in cancer cells as they evolve over time into growth dysregulated, invasive, meta-static cell types. These events include activation of proliferation-promoting oncogenes such as *ras* and *myc* (see Chapter 5), overexpression of cell cycle regulatory factors such as cyclin D, increased telomerase to overcome cell senes-cence, and increased angiogenesis to enhance blood supply to tumor tissue. In addition, a number of mutations in apoptotic factors and up-regulation of anti-apoptotic factors occur in cancer cells during progression (reviewed in Reference 183). These include mutation or inactivation of p53 and overexpression of Bcl-2

and Bcl-X$_L$. Mutations or altered expression of p53 downstream effectors (PTEN, Bax, Bak, and Apaf-1) or upstream regulators (ATM, Chk2, Mdm2 and p19^{ARF}) also occur in human can-cers. Overexpression of inhibitors of apoptosis proteins (IAPs) and heat shock proteins (Hsps), which can inhibit caspase-9 activation, have also been observed in human cancers.

The above tumor-related disruptions occur in the so-called intrinsic, mitochondrial-mediated pathway. Tumorigenic disruptions also occur, though less frequently, in the death-receptor mediated pathway. For example, mutations in Fas (CD95) and TNF-related apoptosis-inducing ligand (TRAIL) receptors and downstream sig-naling pathways of these receptors have been seen in human cancers. Inactivation of the CD95 and TRAIL pathways may also allow tumors to escape from immune responses and thus pro-mote tumor expansion and metastasis.[183]

It is important that inhibition of apoptosis oc-curs at different steps in different tumor types, as this may be responsible for variability in drug responses in different cancers. A key to tumor-specific, effective therapy will be understanding which steps have gone awry in which tumors.

The cancer-related alterations in the apoptotic pathway provide a number of cancer chemo-therapeutic targets.[184,185] There have been no clinical breakthroughs yet, but there are number of Bcl-2 antagonists (both small molecules and anti-sense oligonucleotides) in preclinical devel-opment as well as TRAIL agonists and IAP antagonists such as Survivin. Survivin has been shown to be overexpressed in a wide variety of human cancers and provides an interesting target because it is not only anti-apoptotic but also up-regulated in angiogenically stimulated endothe-lium (reviewed in Reference 186). Thus, target-ing Survivin may also facilitate involution of new blood vessels in tumors.

GROWTH FACTORS

Historical Perspectives

Many "factors" that affect the proliferation of eukaryotic cells have been identified (Table 4–4). A *growth factor* is usually defined as a substance

Table 4–4. Characteristics of Some Representative Growth Factors

Factor	Original Source	Target Cell	Molecular Weight
Insulin	Beta cells of pancreas	General	6000
Insulin-like growth factors			
IGF-1	Human plasma	General	7650
IGF-2	Human plasma	General	7500
Basic fibroblast growth factor (FGF-2)	Bovine pituitary	Fibroblasts, myoblasts, smooth muscle, chondrocytes, glial cells, vascular endothelium	14,000
Nerve growth factor (NGF)	Mouse submaxillary gland, snake venoms, cultured cells	Sympathetic ganglia cells and sensory neurons	26,500
Epidermal growth factor (EGF)	Mouse submaxillary gland, human urine	Epidermal cells, various epithelial cells, vascular endothelial cells, chondrocytes, fibroblasts, glial cells	6000
Platelet-derived growth factor (PDGF) AA, AB, and BB	Human platelets	Fibroblasts, glial cells, arterial smooth muscle cells	24,000–31,000
Transforming growth factors			
TGF-α	Various virally transformed cell types and cancer cells	Similar to EGF	≃6000
TGF-β	Various transformed cell types, normal placenta, kidney, and platelets	Similar to EGF	25,000
Angiogenin	Human colon carcinoma cell line	Capillary endothelium	14,000
Colony-stimulating factors (human)			
GM-CSF	Placenta	Granulocyte/macrophage	22,000
G-CSF	Placenta	Granulocyte progenitor cells	20,000
M-CSF	Urine	Macrophage progenitor cells	45,000
Interleukins			
IL-1	Normal and malignant macrophages, keratinocytes, astrocytes	T lymphocytes, B lymphocytes, fibroblasts, hepatocytes, chondrocytes, hypothalamic cells (fever center)	15,000
IL-2 (TCGF)	T lymphocytes	T lymphocytes (that become T-helper cells or cytotoxic T cells)	23,000
IL-3 (Multi-CSF)	T lymphocytes	Eosinophil, mast cell, granulocyte, and macrophage progenitors; T lymphocytes	28,000
Mammary-derived growth factor (MDGF)	Human milk and mammary tumors	Normal mammary epithelial cells, epidermoid carcinoma cells	62,000
Uterine-derived growth factor (UDGF)	Pregnant sheep uterus	Rat mammary and uterine tumor cells	4000–6000

that stimulates cell proliferation and often also promotes cell differentiation of specific target cells. Excluded from this class of agents are substances that are simply nutrients, such as glucose, essential amino acids, vitamins, and key minerals. The impetus for studies of growth factors largely derives from earlier observations that most mammalian cells growing in culture require the presence of animal serum to grow and proliferate. Numerous attempts have been made to isolate growth factors from serum. The isolation and characterization of these factors

may help to elucidate the altered growth control of malignant cells because transformed cells usually have a lower growth requirement for serum and some transformed cells appear to produce their own growth factors.

Growth factors are now established as important regulators of embryonic growth and development, cellular proliferation and differentiation, and tumor cell proliferation. While the modern era of growth factors research dates to the late 1940s and early 1950s with the seminal work of Rita Levi-Montalcini, Stanley

Cohen, and Viktor Hamburger, research in this field actually started about 50 years earlier than that (reviewed in Reference 187) with the work of a young scientist named Thorburn Brailsford Robertson, who came from the University of Adelaide in 1905 to study with Jacob Loeb at the University of California. Loeb was studying the effects of chemicals and salt solutions on cellular motility of *Paramecium*. This work stimulated Robertson's interest in how biological functions can be modified by chemical substances and this in turn led to his interest in regulation of growth of developing organisms. In a series of papers published in 1916 in *The Journal of Biological Chemistry*, Robertson described the growth stimulatory effects of an anterior pituitary extract on the growth of juvenile mice from 20 to 60 weeks of age. The active component was extracted from desiccated pituitary glands with alcohol and precipitated in ether. He called this substance "tethelin," from the Greek word for "growing." Interest in tethelin as a pharmaceutical apparently stimulated production of it by a drug manufacturer in Philadelphia, H. L. Mulford's (the first biotech company!). There appears to be no record of its use in humans.

Other early work on "growth factors" included that of Alexis Carrel at the Rockefeller Institute and Eric Horning in Melbourne. In 1928, Carrel reported that an unstable protein extracted from embryonic tissue or lymphocytes stimulated growth of animal tissues in culture dishes. He called these "trephones," but did not characterize them further. About the same time, Horning, working with J. M. Byrne and K. C. Richardson, observed that cellular outgrowth from pieces of a mouse sarcoma was stimulated by extracts from embryonic or tumor tissue and that chick embryo intestinal fragments exhibited enhanced cellular proliferation in the presence of embryonic extract plus tethelin, which they obtained from Robertson. Unfortunately, the research trail on these growth-stimulating substances grew cold soon after these reports, with the death of Robertson in 1930 at age 46 and the subsequent lack of encouragement of Horning for this work. It was not until 20 years later that the work of Levi-Montalcini, Cohen, Hamburger, and their collaborators established this field on a firm footing, with the biochemical characterization of growth factors as definitive entities.

A large number of growth-promoting factors have now been found. In general, these are polypeptides of relatively low molecular weight (6000 to 30,000 Daltons). Specific receptors (high affinity, saturable) have been identified for many of them and these receptors are usually cell surface receptors, a number of which have been shown to possess endogenous protein kinase activity (e.g., insulin, IGFs, EGF, PDGF, FGFs, and TGF-α). Some of these receptors undergo receptor-mediated endocytosis, which may be involved in down-regulation of receptor-mediated action, transference of a receptor-mediated signal to the cell nucleus, or both. New growth factors continue to be discovered; only some of the more well-characterized ones will be discussed here. The almost ubiquitous presence of growth factors in a wide variety of tissues leads one to speculate that each tissue may have its own growth-modulating substances. During development and tissue differentiation, these substances probably act locally, either as paracrine- or autocrine-regulatory chemical messengers. They continue to act as needed during adult life for tissue renewal and wound repair. In cancerous tissue, the secretion of certain of these chemical messengers becomes unregulated and cellular proliferation continues unabated.

Some of the factors discussed here are primarily of historical interest, but because they shed light on how the whole field developed, they are included. For example, the discovery that insulin can be a growth-promoting substance for cultured cells and that there are insulin-like substances in human plasma that act similarly led to the discovery of the insulin-like growth factors (IGF-1 and IGF-2). The discovery in 1948 of a peptide factor (NGF) that stimulated nerve outgrowth in chick embryo limb buds was the initial finding that led investigators to look for more such factors and to the serendipitous discovery of epidermal growth factor (EGF). This discovery, in turn, led indirectly to the discovery of the transforming growth factors (TGF-α and -β). Similarly, biochemical characterization of the EGF receptor and elucidation of its amino acid sequence led to the discovery that certain cellular oncogenes code for growth factor receptors or parts thereof (see Chapter 5). Such is the wonderful and unpredictable nature of science.

Insulin

Soon after insulin was isolated it was found to help support growth of cells in culture.[188] The idea that insulin might be a growth-stimulatory factor was supported by the observation that many of the mitogenic peptides derived from blood have an insulin-like activity. Whether insulin has a physiologically important mitogenic activity in vivo, however, is questionable, because supraphysiologic concentrations of insulin are usually needed to stimulate cell proliferation and the mitogenic effect is usually small compared with that of total serum or other mitogenic peptides.[189] Insulin, however, frequently acts synergistically with other growth factors, probably because it is required for optimal uptake and utilization of needed nutrients.

Characterization of the cell surface receptor for insulin has led to a broader understanding of how insulin and insulinlike factors work. The insulin receptor is a tetrameric disulfide-linked complex containing two α subunits of 125,000 MW and two β subunits of 90,000 MW. It has been shown that the insulin receptor has intrinsic protein kinase activity that autophosphorylates its β subunit as well as certain other substrates on tyrosine residues.[190–192] Phosphorylation of the insulin receptor is activated by binding the ligand and it increases the kinase activity of the receptor for other substrates. As we will see, this kind of ligand–receptor interaction leading to receptor dimerization and activation of an intrinsic receptor kinase activity is a common feature of polypeptide growth factors (Fig. 4–13). Tyrosine kinase receptors interact with a number of signal transduction pathways. This will be discussed in more detail later in this chapter. The general structure of growth factor receptors and some of the ligands that activate them are shown in Figure 4–13. A general scheme for how a growth factor binding to its receptor can trigger a signal transduction pathway leading to gene activation is shown in Figure 4–14.[194]

Insulin-like Growth Factors

The discovery of IGFs came about from a variety of approaches but initially from the identification of insulin-like activities in plasma or serum. At first these activities had various names such as nonsuppressible insulin-like activities,

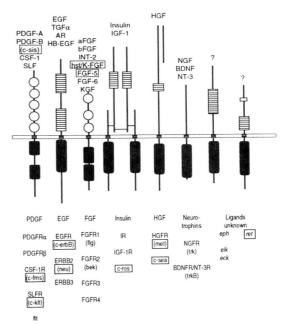

Figure 4–13. Transmembrane tyrosine kinases. Structural features of various receptor tyrosine kinase receptors are shown. Each receptor family is designated by a prototype ligand. Growth factors known to bind to receptors of a given family are listed above, and receptors that constitute each family are listed below. Boxes denote those growth factors or receptors whose genes were initially identified as activated oncogenes. The c-onc designation is used to specify cellular homologs of retroviral oncogenes. Open circles illustrate immunoglobulin-like repeats. Filled boxes indicate conserved tyrosine kinase domains. (From Aaronson,[193] with permission.)

somatomedins, and multiplication-stimulating activity.

A group of polypeptides present in plasma that have insulin-like activity but could not be neutralized by antibodies directed against insulin, called the nonsuppressible insulin-like activities (NSILA), were separated into two fractions: NSILA-P, which was precipitable by ethanol at low pH, and NSILA-S, which was soluble under these conditions. Two peptides of about 7000 MW with mitogenic activity were isolated from the NSILA-S fraction.[195] These have now been called insulin-like growth factors 1 and 2 (IGF-1 and IGF-2).

A number of years ago it was realized that the growth-promoting activity of pituitary growth hormone was mediated through factors present in the serum. These factors, termed *somatome-*

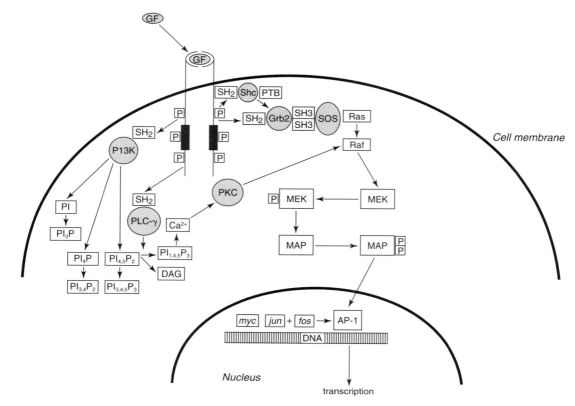

Figure 4–14. Schematic representation of the cell signaling pathway mediated by growth factor (GF) receptors with intrinsic tyrosine kinase activity. AP-1, activator protein 1; DAG, diacylglycerol; MAP, mitogen-activated protein kinase; MEK, MAP kinase kinase; P, phosphated region; PLC-γ, phospholipase c-γ; PI$_3$P, phosphatidylinositol(3)-phosphate; PI$_4$P, phosphatidylinositol(4)-phosphate; PI$_{3,4}$P$_2$, phosphatidylinositol(3,4,)-bisphosphate; PI$_{4,5}$P$_2$, phosphatidylinositol(4,5)-bisphosphate; PI$_{1,4,5}$P$_3$, phosphatidylinositol(1,4,5)-triphosphate; PI$_{3,4,5}$P$_3$, phosphatidylinositol(3,4,5)-triphosphate; PKC, protein kinase C; PLC-γ, phospholipase C-γ; PTB, PTB domain; SH2, SH2 domain; SH3, SH3 domain. (From Favori and DeCupis,[194] reprinted with permission from the American Society for Pharmacology and Experimental Therapeutics [ASPET].)

dins, which have a wide variety of stimulatory actions, including stimulation of sulfate uptake into cartilage and cartilage growth, insulin-like activity on muscle and adipose tissue, and proliferation-promoting activity for cultured cells, have been isolated. Somatomedin C was found to be a basic polypeptide, of about 7000 MW, that stimulates DNA synthesis in cartilage and cultured fibroblasts. Its activity circulates in plasma bound to carrier proteins, which may provide a "reservoir" for the polypeptide and slow its removal from the blood.

Dulak and Temin[196] isolated a small polypeptide from serum using the same initial steps as for NSILA purification, but avoiding the acid-

ethanol precipitation step. Further purification produced a fraction that was 6000 to 8000 times more active than whole calf serum in stimulating replication of chick embryo fibroblasts. The purification fraction, which they called *multiplication-stimulating activity* (MSA), had a minimal amount of insulin-like activity.

It is now known that the insulin-like growth-promoting activities just described are all related and belong to the IGF family of closely related mitogenic polypeptides. These polypeptides can be placed into two groups based on their isoelectric points. The basic group has isoelectric points above pH 7.5 and includes IGF-1, somatomedin C, and basic somatomedin. All of these are now known to be identical. Moreover, IGF-2 and MSA represent the same activity and belong to the neutral-acidic group of insulin-like activities with isoelectric points below pH 7.5. The circulating levels of IGF-1 depend on pituitary growth hormone and IGF-1 mediates its action, whereas plasma levels of IGF-2 do not appear to be so regulated. Because of the high levels of IGF-2 in fetal serum it is thought to be an important fetal growth factor.[197] IGFs are thought to be generated in the liver, but other cells such as cultured fibroblasts and fetal tissues can also produce them. In addition to growth hormone, such factors as nutritional status and circulating insulin levels can regulate IGF production.

The IGF-1 receptor is structurally similar to the insulin receptor, also having a tetrameric subunit structure with two α subunits of 125,000 MW and two β subunits of 90,000 MW and possessing tyrosine kinase activity, whereas the IGF-2 receptor is a single transmembrane chain with a small intracellular domain, lacking tyrosine kinase activity but possessing mannose 6-phosphate receptor activity (reviewed in Reference 198).

There is considerable cross-reactivity between the various ligands and receptors in the insulin and IGF family. For example, insulin can bind to both insulin and IGF-1 receptor, and IGF-1 and IGF-2 can bind to the insulin receptor.[199]

IGF-1 and IGF-2 as well as their receptors are present in a wide variety of human cancer types,[200] thus paracrine or autocrine growth factor stimulation of proliferation of tumor cells by IGFs may contribute to their malignant phenotype. IGF-1 and IGF-2 receptors are expressed on breast cancer cells, and IGF-1 and IGF-2 both have mitogenic effects on these cells.[201] The growth-promoting effect of estrogen on breast cancer cells has been postulated to be mediated by IGF-1. Interestingly, the antiestrogen drug tamoxifen has been reported to lower IGF-1 serum levels in breast cancer patients and to inhibit IGF-1 gene expression in the liver and lungs of rats bearing DMBA-induced mammary tumors,[202] a finding suggesting that part of tamoxifen antitumor action in breast cancer patients may be due to its ability to reduce production of the IGF-1. Moreover, the synthetic retinoid fenretinide has also been shown to lower plasma levels of IGF-1 in breast cancer patients, Providing a possible rationale for combination therapy with a retinoid and tamoxifen.[203]

IGF-1 and IGF-2 both have mitogenic and anti-apoptotic actions and they both have effects on the regulation of proliferation and differentiation in developing tissues. IGF-1 expression but not IGF-2 expression is turned on by growth hormone. Expression of IGFs is also affected by a variety of hormones including estrogens, adrenocorticotropic hormone (ACTH), thyrotropin (TSH), luteinizing hormone (LH), follicle-stimulating hormone (FSH), and human chorionic gonadotropin (hCG), as well as by growth factors EGF, FGF, and PDGF. Most of these hormones and growth factors stimulate expression of the receptor IGF-1R (reviewed in Reference 204). IGF has anabolic actions on protein and carbohydrate metabolism by increasing uptake of amino acids and glucose and by stimulating glycogen and protein synthesis. IGF-1's mitogenic actions include stimulation of cyclin D1 expression, acceleration of G_1 to S cell cycle progression, and inhibition of apoptosis by stimulating Bcl-2 expression and suppressing expression of Bax.

Binding of IGFs to IGF-1R activates the receptor's tyrosine kinase activity and this in turn activates two signal transduction pathways: the ras-Raf-Mek-Erk pathway and the phosphoinositol-3-kinase (PI3K) pathway. IGF-2R, by contrast, has no tyrosine kinase activity, binds only IGF-2, and appears to act like an antagonist by facilitating IGF-2 degradation.

Interestingly, IGF-1 levels appear to mediate the caloric restriction effect on tumor growth in that the effect of caloric restriction on growth of human prostate cancer xenografts in rats and growth of bladder cancer in mice correlates with a decrease in circulating IGF-1 and is reversed by IGF-1 (reviewed in Reference 204). Fasting has been shown to decrease circulating IGF-1 in human subjects. A 50% reduction in caloric intake or a 30% lowering of protein intake produces a decline in serum IGF-1 levels.

Seven IGF-binding proteins (IGFBPs) that modulate availability and function of the IGFs have been identified.[194] IGFBPs have complex actions and can either enhance or inhibit actions of IGFs, depending on the specific IGFBPs bound and the cellular context. The IGFBPs can act by (1) transporting IGFs in the bloodstream, (2) protecting them from degradation, and (3) decreasing the availability of IGFs for IGF-1R, since IGFBPs generally have higher affinity for IGFs than IGF-1R. Thus, for example, IGFBP-3 binds IGF-1 and inhibits its mitogenic and anti-apoptotic effects. More than 90% of the IGFs in the blood are bound to IGFBPs, predominantly to IGFBP-3.

IGF-1 acts synergistically with other mitogenic growth factors in stimulating cancer cell proliferation in culture and presumably in vivo (reviewed in Reference 204). For example, in breast cancer cells, estrogens induce expression of IGF-1 and IGF-1R. IGF-1 can also increase expression of the estrogen receptor (ER) in breast cancer cells. Interestingly, the ER antagonist–agonist tamoxifen inhibits the effects of IGF-1 in breast cells where it is antiproliferative and enhances IGF-1 effects in the uterine endometrium where it is mitogenic and carcinogenic.

Although the clinical and epidemiological data on the correlation of circulating IGFs and IGFBPs and on cellular IGF-1R levels are somewhat contradictory, a number of studies have reported an increased risk of solid tumors in association with circulating levels of IGF-1 and decreased risk with high circulating levels of IGFBP-3. However, clinical data do not consistently support an association between IGF-1 levels and age, menopausal status, tumor size, lymph node involvement, or tumor grade in breast cancer patients (reviewed in Reference 204). Nevertheless, converging results from epidemiological data and in vivo carcinogenesis models support the idea that high levels of circulating IGF-1 are associated with the risk of a number of common tumors such as colorectal, prostate, breast, and lung cancers.[205]

Nerve Growth Factor

Nerve growth factor (NGF) was discovered as a result of some experiments designed to test the effects of rapidly growing tissue on nerve development in the limb bud regions of chick embryos. In these experiments, two mouse sarcomas, S-37 and S-180, were transplanted to chick embryo limb bud regions. Subsequently, a pronounced sensory innervation of the tumor tissue was observed.[206] Further experiments showed that the nerve growth–promoting effects of the sarcomas were most pronounced for neurons of the sympathetic nervous system and the effector was a soluble protein.[207] Nerve growth factor, which has been purified from various snake venoms[208] and from mouse submaxillary gland,[209,210] consists of two subunits of about 13,000 MW, each with three intrachain disulfide bonds. When NGF is isolated from submaxillary gland, it is bound in a 7S protein complex of about 140,000 MW. It has several structural and functional similarities to insulin.[211] There is significant amino acid sequence homology with insulin, and most of the identical amino acid residues being clustered in regions of NGF that align with the insulin A and B chain segments of proinsulin but separated by the 35 residues needed for the C activation peptide of proinsulin. There are an additional 37 amino acid residues at the carboxyl-terminal end of NGF that extend beyond the sequences of its homology with proinsulin, but these residues are similar to the insulin B chain, suggesting a gene duplication event. Both insulin and NGF evoke similar biologic responses in their respective target tissues, including increased uptake of glucose and nucleosides and increased RNA, protein, and lipid synthesis. Thus, NGF appears to be a product of a gene that evolved from the same or a similar ancestral gene as proinsulin. In addition to its insulin-like actions, NGF increases nerve fiber outgrowth and induces specific enzymes involved in sympathetic nervous transmission, for example, tyrosine hydroxylase and dopamine β-hydroxylase, in developing nerve cells.

But these are only some of the cold, hard scientific facts. The story of how this all came about is much more interesting.[212] One of the co-discoverers of NGF, Rita Levi-Montalcini, was born in Turin, Italy, where she attended medical school and became board certified in neurology and psychiatry. She became interested in the regulation of neuronal development and began studying development of the nervous system in chick embryos in Giuseppe Levi's laboratory in Turin. Unfortunately, both she and Levi were barred from academic life in 1939 as a result of Mussolini's anti-Semitic "Manifesto delle razze." Determined to continue her work, she set up a primitive lab in her basement and continued to do research even as the bombs fell during World War II. In some of her early studies she found that fewer nerve cells grew into an area where the chick limb bud had been eliminated, suggesting that limb end cells were releasing some trophic factor that stimulates nerve cell growth. A paper published on this topic caught Viktor Hamburger's eye and he invited her to Washington University in St. Louis in 1946 to continue her studies. Intrigued by Bueker's observation[206] that mouse sarcomas transplanted to areas of the limb bud caused nerve cells to grow, she began to try to purify the trophic factor produced by the sarcoma cells. About that time, she began to collaborate with Stanley Cohen, a biochemist who was then at Washington University, on the purification. Initially, they found that the growth stimulatory material contained both protein and nucleic acid, so they added snake venom, which is known to contain a diesterase that breaks down nucleic acids, to see if they could determine which component contained the activity. What they found was that even their snake venom–only controls had nerve growth activity. This led to the idea that salivary glands might be a source of the active substance. This turned out to be the case, and the material we now know as NGF was ultimately purified from mouse salivary glands, which also proved to be a source for epidermal growth factor (EGF). In 1986, Levi-Montalcini and Cohen won the Nobel Prize for their pioneering studies on growth factors.

It is now known that there is a family of NGF-like factors (reviewed in Reference 213), which includes brain-derived neurotrophic factor (BDNF), and neurotrophins NT-3, NT-4, and NT-5. The members of this family have about 60% sequence homology and their activity appears to be limited to neuronal tissue. Various neurotrophins have growth-promoting activity for various nervous tissues—for example, NGF has positive cell survival and differentiation effects on sensory, sympathetic, and cholinergic neurons, as well as on PC12 pheochromocytoma cells; BDNF has positive effects on sensory, cholinergic, and dopaminergic neurons and on retinal ganglion cells, but not PC12 cells; and

NT-3, NT-4, and NT-5 are growth factors for sensory neurons, but have somewhat different effects on other nerve tissue cell types.

A high-affinity receptor for NGF has been identified as the 140 kDa glycoprotein gene product of the trk-A protooncogene.[214,215] This receptor, known as $gp140^{trk-A}$, has tyrosine kinase activity and appears to act through the Ras, Raf-1, MAP kinase signal transduction system.[216] Two Trk-A homologs, Trk-B and Trk-C, are receptors for other members of the neurotrophin family. Specificity for response of nerve tissue cells occurs because there is cell-type specificity for the Trk receptors. The transforming version of Trk was first observed in colon cancer cells as a truncated, chimeric protein fused to tropomyosin and having tyrosine kinase activity.[217] A second low-affinity receptor for NGF, called $gp75,^{NGFR}$ has also been identified and in some cell types it is involved in a crucial way with $gp140^{trk-A}$ in producing a response, but in other cells, $gp140^{trk-A}$ by itself is sufficient.[213]

In a study of tissue from 80 untreated neuroblastomas, the absence of $gp140^{trk-A}$ mRNA was associated with tumor progression.[218] Thus a lack of a response system for NGF may foster unregulated proliferation of malignant nerve cell tissue.

Epidermal Growth Factor

Epidermal growth factor (EGF) was discovered during the course of some experiments on NGF activity in submaxillary gland extracts.[219] When these extracts were injected into newborn animals, precocious eyelid opening and eruption of incisor teeth were observed. These phenomena were caused by stimulation of epidermal growth and by keratinization. The material responsible for these effects was isolated from mouse submaxillary gland and found to be a low-molecular-weight, heat-stable polypeptide,[219] which has been purified from mouse submaxillary gland and human urine. A single polypeptide chain of 53 amino acids, it has three intramolecular disulfide bonds that are required for biologic activity. The EGF activity isolated from submaxillary gland at neutral pH is bound to a carrier protein to form a high-molecular-weight (74,000) complex that consists of two molecules of EGF (6000 MW) and two molecules of binding protein

(29,300 MW). The binding of EGF to its carrier protein depends on the presence of the carboxyl-terminal arginine residue, and because the binding protein has arginine esteropeptidase activity, it has been suggested that biologically active EGF is generated from a precursor protein by the action of the carrier protein peptidase.[220] A similar but not identical carrier protein with arginine esteropeptidase activity has been discovered in association with NGF, and the carrier enzyme is also involved in formation of NGF from its precursor.[221] The EGF activity isolated from human urine is very similar in structure and function to mouse EGF and has significant amino acid sequence homology with it, suggesting that human and mouse EGF evolved from the same gene.[222] It has also been found that the gastrin antisecretory hormone urogastrone in human urine is, in fact, EGF; this finding suggests an additional biologic function of this factor.[223]

The mouse EGF gene has been cloned and is surprisingly large compared with the gene size required to code for the mature EGF polypeptide (reviewed in Reference 224). Mature EGF is only 53 amino acids long, yet the gene contains sufficient information to encode about 1200 amino acids. Contained within this 1200 amino acid sequence are eight EGF-like sequences, indicating that the gene codes for a large prepro-EGF molecule that is then processed to mature EGF and a family of EGF-like polypeptides in a manner similar to the production of ACTH from pro-opiomelanocortin.

A wide variety of cells, including epidermal cells, corneal cells, fibroblasts, lens cells, glial cells, granulosa cells, vascular endothelial cells, and a large variety of human cancer cells (see below), have specific cell surface receptors for EGF, indicating the ubiquitous nature of the target cells for this growth factor. Specific receptors for EGF have even been found in liver membrane fractions from certain fish. Thus, EGF is a phylogenetically old protein.

It is now realized that there is a family of EGF-like growth factors that includes TGF-α, amphiregulin (a growth factor first detected in phorbol ester–stimulated MCF-7 cultured breast carcinoma cells), and pox virus growth factors. All the members of the EGF family are 50–60 amino acids in length, have six half-cystines in the same register (indicating the structural importance of

the three intramolecular disulfide bonds), bind with high affinity to EGF receptors, and produce mitogenic responses in EGF-sensitive cells.[224] The role of the large, eight EGF sequence–containing prepro-EGF isn't clear, but in some cells it exists as a membrane protein and retains EGF-like biological activity. Several other membrane-associated, growth factor–containing proteins have also been discovered. A number of these have multiple EGF-like repeats (see below under TGF-α). In some cases, e.g., TGF-α, cleavage of the repeats releases soluble growth factors. In other cases, the membrane-anchored growth factor–containing polyproteins may be acting as "juxtacrine" growth factors in developmental processes that require cell–cell interaction. This function may not be brought about by secreted, diffusible growth factors.

The finding that a cultured human epidermal carcinoma cell line, A431, has a large number of specific EGF receptors has provided a model system in which to study the receptor and the effects of EGF–receptor interaction. The study of EGF receptors has produced some precedent-setting results that relate to a variety of growth factors.

The EGF receptor (EGFR) is an intrinsic membrane glycoprotein of about 170,000 MW.[225] It is monomeric, unlike the tetrameric insulin and IGF-1 receptors, and contains a core polypeptide of 1186 amino acids and N-linked oligosaccharides, the latter of which make up about 25% of the molecular weight of the receptor. The receptor binds EGF with high affinity ($K_D = 10^{-9}$ to 10^{-10} M) and high specificity. Occupation of the receptor by EGF induces an autophosphorylation of the receptor as well as phosphorylation of certain other cellular substrates (see below). The autophosphorylation site is a tyrosine[226] rather than a serine or threonine, which was a surprising finding at the time because the only other known tyrosine kinase was the pp60[src] product of the src oncogene (see Chapter 5). This observation for the EGF receptor came before similar findings for the insulin, IGF, and PDGF receptors and so was indeed precedent setting.

The extracellular domain of the EGFR binds EGF and EGF-like ligands with high affinity, contains 10 to 11 N-linked oligosaccharides, and has a high content of half-cystine residues that could form up to 25 disulfide bonds, providing a complicated tertiary structure. The cytoplasmic domain contains the tyrosine kinase activity, four sites for tyrosine autophosphorylation, and several sites for serine/threonine phosphorylation that are substrates for phosphorylation by non-receptor kinases such as protein kinase C, suggesting cross talk between kinase systems and additional control mechanisms for receptor activity. For example, PKC-mediated phosphorylation of threonine 654 attenuates EGF-activated autophosphosylation. The binding of EGF to its receptor induces a dimerization that is related to its functional activation. The Erb B oncoprotein is a truncated version of the EGF receptor and lacks most of the ligand-binding domain.

An important biological regulator in animals, EGF is capable of eliciting a wide variety of physiologic and cellular responses (Table 4–5). It is not clear, however, what its key biologic role is in vivo. In cell culture systems, EGF stimulates cell proliferation of a wide variety of cell types. A number of studies have shown that maximal stimulation of DNA synthesis in cultured cells occurs at about 25% saturation of EGF binding sites, indicating that only a fraction of available receptors need be occupied by EGF to trigger a mitogenic response. Moreover, cells grown in the presence of EGF continue to proliferate after the cultures become confluent, simulating the loss of density-dependent growth inhibition observed in transformed cells. This finding suggests that continued exposure to the growth-stimulating activity of EGF or EGF-like factors could be involved in the excessive proliferation of cells and the loss of feedback inhibition of growth seen in neoplasia. In this regard, it has been observed that certain transformed cell types produce their own EGF-like growth factor that appears to bind to EGF cell surface receptors (see below).

In addition to human epidermoid carcinoma cells, other human cancer cells have receptors for EGF or EGF-like growth factors. Among these are ovarian, cervical, renal, lung, bladder, and breast carcinomas, and glioblastomas (reviewed in Reference 227), but this is not a universal finding; human leukemia and lymphoma cells have low or absent EGF receptors. In human epidermoid A431 carcinoma cells as well as

Table 4–5. Biologic Effects of Epidermal Growth Factor

IN VIVO

Accelerated proliferation and differentiation
  Of skin tissue
  Of corneal epithelial tissue
  Of lung and trachea epithelia
Potentiation of methylcholanthrene carcinogenesis
Inhibition of gastric acid secretion
Increased activity of ornithine decarboxylase and
  accumulation of putrescine
Formation of fatty liver
Increase of disulfide group content in skin
Hepatic hypertrophy and hyperplasia
Potentiation of cleft palate

ORGAN CULTURES

Accelerated proliferation and differentiation
  Of skin tissue
  Of corneal epithelial tissue
  Of mammary gland epithelial tissue
Induction of ornithine decarboxylase and accumulation
  of putrescine
Enhanced protein synthesis, RNA synthesis
Inhibition of palate fusion

CELL CULTURES

Increased transport
  Of α-aminoisobutyrate
  Of deoxyglucose
  Of K^+
Activation of glycolysis
Stimulation of macromolecular synthesis
  Of hyaluronic acid
  Of RNA
  Of protein
  Of DNA
Enhanced cell multiplication
Alteration of membrane properties
Stimulated hCG secretion
Increased biogenesis of fibronectin
Enhanced prostaglandin biosynthesis
Alteration of viral growth
Increased squame production by keratinocytes

From Carpenter and Cohen[219]

in certain other human cancer cells (e.g., glioblastomas), the gene coding for the EGF receptor is amplified several-fold,[228] which explains why some cell types have high receptor content and why they may be so responsive to a proliferative signal provided by EGF-like growth factors. Truncation and deletion mutants of EGFR have also been observed in human cancers.[227]

Occupancy of EGF receptors by EGF triggers a mitogenic response. The molecular signals involved in this triggering mechanism have been defined and a number of cellular events associated with the EGF response have been identified. Stimulation of the mitogenic cascade involves activation of the Ras-Raf-MAP kinase system (see section on Signal Transduction, below). The cellular responses to EGF include both early and late events (reviewed in Reference 229 and 230). Within 5 minutes after EGF addition to cultures of responsive cells, increases in protein phosphorylation, ion fluxes, and membrane ruffling occur. Internalization of bound EGF, increased amino acid and glucose transport, onset of RNA synthesis, and increased protein synthesis occur, in that order, between 15 and 60 minutes after initiation of EGF binding to its cell surface receptor. In contrast, induction of DNA synthesis requires exposure of cells to EGF for about 12 hours, and onset of cell division follows within a few hours after that.

The relationship between the early and late cellular responses is not clear. Presumably the pleiotypic response to EGF, as well as to other growth factors, follows some closely regulated temporal cascade of events. Since increased phosphorylation of membrane and other cellular proteins is one of the earliest events observed after EGF binding, it is a likely candidate for the signaling event of the cascade.

Internalization of EGF occurs in cells by means of a process called receptor-mediated endocytosis, which is common to a number of polypeptide hormones and growth factors (Fig. 4–15).[231] After a polypeptide binds to its specific receptor, there is a clustering of ligand-bound receptor into patches on the cell surface. These clusters are then invaginated into organelles called clathrin-coated pits (clathrin is a protein lining the cytoplasmic face of these vesicle-like structures). The pits then pinch off from the cell membrane and form true vesicles. These vesicles become acidified and, under the acidic conditions, the ligand becomes dissociated from the receptor. These vesicles are then shuttled to lysosomes, where they fuse with the lysosomes, and the receptors are degraded in them. In some cases (e.g., transferrin receptors), the endocytotic vesicles appear to become associated with the Golgi complex and receptors may then be recycled to the cell surface. Because this process in effect removes receptors from the cell surface, the cells become refractory to further stimulation by exogenous growth factor. This refractory period, called receptor down-regulation, appears to be a common pathway for

regulation of cellular responses to a number of polypeptides, including EGF, insulin, transferrin, α_2-macroglobulin, immunoglobulins, and low-density lipoprotein (LDL).[232] In the case of EGF binding to cultured human fibroblasts, recovery of complete EGF binding capacity requires several hours and depends on RNA and protein synthesis,[233] indicating that resynthesis of EGF receptors is required.

A number of growth factors induce a rapid change in ion fluxes across cell membranes as an early event. These events include increased Na^+/H^+ exchange, increased levels of intracellular free Ca^{2+}, and a transient rise in intracellular pH.[234–236] These events are commonly induced changes in cells after exposure to a wide variety of mitogenic agents, and ultimately they trigger the events leading to DNA synthesis via a signal transduction cascade.

The role that EGF plays in vivo, particularly in humans, is not totally clear. EGF has been isolated from human urine, as noted, and thus presumably does have a physiologic effect in adult tissues, probably as a paracrine growth factor involved in tissue renewal and wound repair. EGF has been shown to be present in human milk, thus EGF may play a role in milk production and be a growth-promoting agent for the newborn.[237] Because EGF is acid stable and at least partially active after oral administration, it may conceivably have such a function. In addition, EGF has been reported to restore spermatogenesis in male mice whose submandibular glands (the major organ of EGF production in mice) have been removed, a finding suggesting a role for EGF in male reproductive function.[238] High-molecular-weight forms of EGF-like activities have been found in the urine of patients with disseminated cancer, and in a number of breast cancer cases there is a correlation with the presence of EGF-like material in the urine and advanced stage of the disease.[239]

The finding that the *erb* B oncogene encodes a protein that has significant sequence homology with the cytoplasmic, protein kinase–containing domain of the EGF receptor (see Chapter 5) suggests that cancer cells may elevate their ability to respond to endogenously or exogenously produced growth factors, which may be a key to their ability to grow autonomously with minimum regulatory control. Since this truncated receptor

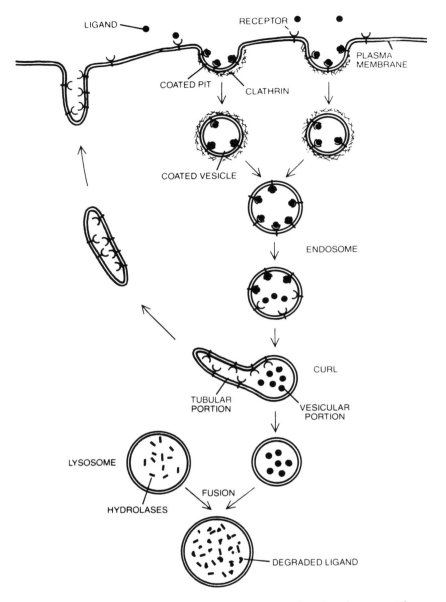

Figure 4–15. Mechanisms involved in receptor-mediated endocytosis. The pathway of receptors and ligands shown here was determined for galactose-terminal glycoproteins but is thought to apply in the case of other ligands and receptors as well. Ligand binds to receptors diffusely and then collects in coated pits, which invaginate and are internalized as coated vesicles, whose fusion gives rise to endosomes and then to a CURL (so designated because it is a *c*ompartment of *u*ncoupling of *r*eceptor and *l*igand). In the acidic CURL environment, ligand is dissociated from receptors. Ligand accumulates in the vesicular lumen of the CURL, and the receptors are concentrated in the membrane of an attached tubular structure, which then becomes separated from the CURL. The vesicular part moves deeper into the cell and fuses with a lysosome, to which it delivers the ligand for degradation. In some cases (e.g., transferrin receptors), the membranous tubular structure is thought to recycle receptors to the plasma membrane. This is not, however, the case for EGF receptors. (From Dautry-Varsat and Lodish,[231] with permission.)

protein lacks the ligand-binding domain, it may not be as subject to shut-off by the down-regulation mechanisms normally triggered by growth factor binding and thus may be constitutively active as a protein kinase.

EGF family members play an important role in normal development. As will be described in more detail below, the EGF family of growth factors activate one or more of four receptors: Erb B1 (EGF receptor or EGFR), Erb B2 (also called HER-2/neu), Erb B3, and Erb B4. Huotari et al.[240] showed that Erb B ligands EGF, TGF-α, heparin-binding EGF, beta-cellulin, and neuregulin-4 are detected in the developing pancreas of mouse embryos by day 13 of gestation. Some differential effects of the EGF family ligands were seen in organ cultures from mouse embryos, some of which were obtained from EGFR (−/−) gene knockout animals. Overall, their results "suggest that ligands of the Erb B1 and Erb B4 receptors regulate the lineage determination of islet cells during pancreatic development." These data as well as several other lines of evidence (reviewed in Reference 240) indicate a role for EGF family ligands receptors in pancreatic islet cell differentiation.

Binding of EGF family ligands to Erb B receptors' extracellular domain triggers formation of homo- and heterodimers, which activates the intrinsic-tyrosine kinases of these receptors. Erb B2 appears to be a co-receptor and the preferred partner for other ligand bound Erb B receptors (reviewed in Reference 241). The importance of Erb B2 heterodimers in development is demonstrated in Erb B2 knockout mice in which loss of Erb B2 produces defects in neuronal and cardiac development.

Erb B2 (Her-2/neu) overexpression (often by gene amplification) occurs in 25%–30% of breast cancers and correlates with lower survival. Inhibition of Erb B2 in cell culture blocks breast cancer cell proliferation. These and similar observations have led to the development of a successful treatment for Her-2/neu over-expressing breast cancers (see below). However, Erb B2 (Her-2/neu) action in stimulating cell proliferation depends on its interaction with Erb B3. Several lines of evidence point to this:[241] (1) co-overexpression of Erb B2 and B3 is seen in many human cancers, including breast, bladder,

and others; (2) in many Erb B2–overexpressing breast cancers, Erb B3 contains high levels of phosphotyrosine (indicating a high level of receptor activation); (3) Erb B2 inactivation decreases Erb B3 phosphorylation; and (4) inactivation of Erb B3 inhibits proliferation of breast cancer cells as efficiently as blockade of Erb B2. These results indicate that Erb B2/Erb B3 dimers act as an oncogenic "team" to stimulate cancer cell proliferation.

Mutant forms of the epidermal growth factor receptor also occur. EGFR (Erb B1) is amplified in human glioblastomas. In addition, the EGFR gene is mutated in about 50% of these cancers and is associated with a poorer prognosis (reviewed in Reference 242). The most common mutation is deletion of exons 2 through 7, producing a truncated EGFR, called EGFRvIII, lacking most of the extracellular domain. EGFRvIII doesn't bind EGF but has a ligand-independent tyrosine kinase activity that is constitutively turned on. It also appears to use a different set of downstream signaling pathways compared to unmutated EGFR (reviewed in Reference 242). In mouse models of glioblastoma, EGFRvIII carcinogenic activity requires mutations at the INK4a/ARF tumor repressor locus. Recall that INK4a is a repressor of cyclin-dependent kinase 4 (Cdk4) and that ARF regulates p53 (see Cell Cycle, above). These defects are also likely to be involved in human glioblastoma because the EGFRvIII and INK4a/ARF mutations are frequently seen in human glioblastomas.

EGFRvIII is also detected in cancers of the breast, ovary, and lung, and in medulloblastomas, but not in normal adult tissues. Tang et al.[243] have shown that overexpression of EGFRvIII transforms a nontumorigenic hematopoietic cell line into one with a highly tumorigenic phenotype in athymic mice and that this transforming ability of EGFRvIII is dose dependent. In addition, they showed that expression of a low level of EGFRvIII in the human breast cancer cell line MCF-7 significantly enhances the tumorigenicity of these cells in athymic mice, suggesting that the mutant EGFR gene could play a pivotal role in breast cancer progression to a more aggressive tumor.

The now well-known Herceptin treatment for breast cancer is the first and so far the best example of targeting the EGFR family in human cancers.[244] This is a great story in bench-to-bedside translational research. As noted above, 25%–30% of breast cancers here amplified Her-2/neu (Erb B2), and this is seen in the breast cancer cells themselves. Preclinical models showed that HER-2 amplification plays a direct role in transformation of cultured cells and in mammary carcinogenesis in mice and that the Her-2 pathway promotes hormone-independent growth of human breast cancer cells in culture (reviewed in Reference 244). It was also observed that women with breast cancers that overexpressed Her-2 had a more aggressive form of the disease and poorer overall survival. A humanized monoclonal antibody to Her-2 was shown to inhibit proliferation of Her-2 overexpressing human breast cancer cells in culture and in mouse xenografts, and to be synergistic with chemotherapeutic drugs in tumor inhibition in preclinical models. These results led to phase I and phase II clinical trials in which this antibody, called trastuzumab and later Herceptin, was safe and effective in women with Her-2-positive metastatic disease who had relapsed after chemotherapy. A phase III trial showed that "trastuzumab increased the clinical benefit of first-line chemotherapy in metastatic breast cancer that overexpresses Her-2."[244]

This success has led to other attempts to block one or more members of the EGFR family of receptors. These include monoclonal antibodies such as cetuximab (Erbitux) and small-molecule EGFR inhibitor such as gefitinib (Iressa). Clinical trials have shown some responses to cetuximab in colon cancer and to gefitinib in non–small cell lung carcinomas. Other monoclonal antibodies and small molecules that target EGFR are in development (reviewed in Reference 245).

Fibroblast Growth Factor

Using initiation of DNA synthesis in mouse 3T3 fibroblasts to monitor purification, Gospodarowicz[246] isolated from bovine pituitary a mitogenic polypeptide, which he called fibroblast growth factor (FGF). Distinct from the IGFs, FGF is a potent mitogen for 3T3 cells and other cultured fibroblasts; however, by itself it is only about 30% as potent as whole serum. A combination of FGF and glucocorticoids will replace whole serum as a growth promoter for some cell lines (e.g., 3T3) but not for others (e.g., human diploid fibroblasts). In the latter case, serum is still required for maximal cell proliferation. Although FGF is mitogenic for several cell types, it does not affect cells transformed with polyoma or SV40 virus,[189] presumably because these transformed cells have a lower growth factor requirement.

Originally, two FGF-like polypeptides were purified to homogeneity and characterized:[247] the FGF from bovine pituitary, called basic FGF, and an acidic FGF from bovine brain. Basic FGF has also been isolated from a variety of other tissues, including adrenal gland, corpus luteum, retina, and kidney. There is significant amino acid sequence homology between the two polypeptides, and they are potent mitogens for diverse cell types including capillary endothelial, vascular smooth muscle, adrenocortical, bone, and ovarian granulosa cells.

It is now clear that the FGFs constitute a large family of growth and differentiation factors for cells of mesenchymal and neuroectodermal origin. It includes the two original members acidic FGF (also called FGF1), basic FGF (FGF2), as well as at least seven other current members (reviewed in References 248 and 249). FGF3 (also known as Int-2) is a gene product first observed in mouse mammary tumor virus (MMTV)-transformed cells and is a 239–amino acid protein with 44% homology to FGF2. FGFs -4, -5, and -6 were found by screening tumors for oncogenes that could transform NIH 3T3 cells. FGF4 was also isolated from Kaposi's sarcoma and is called K-FGF. FGF7 (KGF) is a potent mitogen for keratinocytes and other epithelial cells and differs from the other FGFs in that it does not stimulate the proliferation of fibroblasts or endothelial cells. It is expressed in a restricted number of adult tissues, including kidney, colon, and ileum, but not in brain or lung as the others are. FGF8 is an androgen-induced growth factor cloned from mouse mammary carcinoma cells. FGF9, also called glia-activating factor (GAF), was purified from a human glioma cell line.[250]

The FGF family members have 30% to 70% amino acid sequence identity, with the greatest identity in a "core" region represented by most of the sequence of FGF2, which is considered the prototype FGF. All of the FGFs have the ability to bind heparan sulfate, which is important to their biological function (see below). Except for FGF1 and FGF2, all the FGFs have an N-terminal signal sequence and are secreted from cells. Yet FGF1 and FGF2 are abundant in the ECM, indicating that they must be exported from cells. Binding to the ECM appears to provide a reservoir for FGFs, which can then be mobilized in response to requirements for wound healing, angiogenesis, etc.

FGFs play an important role in embryonic development (reviewed in Reference 249). For example, FGF1 and FGF2 are produced in early mesoderm and appear to be involved in mesoderm induction. Members of the FGF family appear to regulate differentiation of a variety of cell types during development and they are expressed in a temporally and spatially regulated way. FGF4 and FGF5 are expressed in embryonic muscle cell precursors in the mouse. FGF1, -2, and -5 are expressed in nervous system tissue. FGF4 is the first FGF detected in early mouse development, as early as the four-cell stage, and continues at least through the early formation of tissue layers, i.e., mesoderm, endoderm, and ectoderm, and is later expressed in a tissue-specific manner. FGF3 and FGF5 are expressed prior to gastrulation and appear to be restricted to parietal and visceral endoderm, respectively. FGF3, -4, and -5 mRNAs are detectable during mesoderm formation, but in distinct spatial orientations, which suggests a specific role for each FGF in development of specific tissues. FGF1 and FGF2 are expressed as early as day 10 1/2 in mouse embryos and are detectable in a variety of tissues by day 13 1/2. FGF6 is observed during middle and late gestation, but peak levels are observed on day 15.

Five different FGF receptors (FGFRs) have been cloned and more may be found (reviewed in References 248 and 249). Isoforms of FGFRs are generated by alternate mRNA splicing. FGFRs belong to the tyrosine kinase receptor family and some of them were originally identified by cloning of tyrosine kinases, e.g., the hormone *flg*, chicken *cek*1, and mouse *bek* gene products. The receptors are now numbered FGFR 1 to 5. FGFR1 is the mammalian *flg* gene product (same as chicken *cek*1); FGFR2 is the murine and human *bek* product (also called K-Sam) and chicken *cek*3; FGFR3 is murine *flg*-2 and chicken *cek*2; FGFR4 and FGFR5 have no other names. As noted above, some FGFRs arise by alternate mRNA splicing. For example, FGFR2 and the receptor for KGF (KFGR) are both derived from the same gene (*bek*) by alternate splicing of one exon.

The FGFRs have several features in common: (1) an extracellular ligand-binding domain with three disulfide bonded loops and an immunoglobulin-like structure; (2) a single transmembrane domain, followed by a relatively long (80 amino acids) juxtamembrane domain; (3) two tyrosine kinase domains separated by a 14–amino acid insert; and (4) a carboxyl-terminal tail of about 50 amino acids, containing tyrosines that may be autophosphorylated upon ligand binding.

There is significant sequence homology among these receptors and overlap in their binding specificity for various FGFs. For instance, one isoform of FGFR1 binds FGF1 and FGF2 with slightly different affinities. An isoform of FGFR2 binds FGF1, -2, and -4 but not FGF5 or -7, whereas an FGFR2 splice variant binds FGF1 and -7, but not FGF2. FGFR3 can bind FGF1, -2, -4, and -5. FGFR4 binds FGF1 with high affinity and FGF4 and FGF5 with 10-fold lower affinity. This functional redundancy raises questions about what provides the specificity for response to the FGFs and why different receptors are needed to bind the same ligand. It seems likely that the tissue distribution of the FGFRs and the local production of specific arrays of FGFs provide some selectivity of response in various developing tissues. For example, FGFR1 is highly expressed in developing brain, skin, and growth plates of bones, whereas FGFR2 is highly expressed in choroid plexus, skin, lung, kidney, and brain temporal lobe. FGFR3 is abundant in the intestine, lung, kidney, and bone growth plates. FGFR4 is highly evident in adrenal gland, lung, kidney, and liver. Even though FGFR1 and FGFR2 have similar ligand-binding specificities, FGFR1 is seen predominantly in mesenchyme of limb buds and somites, whereas FGFR2 is in highest abundance in epithelial cells of skin and developing internal organs. These different sites suggest alternative and coordinate roles in these two tissue layers as organs develop. FGFR4 is found in muscle cell precursors and may have a prominent role in muscle development.

A feature shared by all FGFs is their high affinity for binding to heparan sulfate and heparan sulfate proteoglycans (HSPGs). A number of studies show an important role for heparan-sulfate-FGF complexes in binding of FGFs to their receptors. HSPG-like molecules are involved in presentation of FGF to the receptor and in stabilization of high affinity FGF-receptor complexes, which provides an "anchor" for the interaction with cells. Binding of FGF ligands to HSPGs may increase the half-life of FGFs by limiting proteolytic degradation. HSPGs also appear to act as a reservoir for FGFs by providing long-term storage sites. HSPGs can be divided into cell surface forms (syndecans and glypicans) and secreted extracellular matrix forms (e.g., perlecan). Specific HSPGs appear to act as co-receptors for FGF2 in stimulating tumor angiogenesis. For example, glypican-1 is up-regulated in gliomas and this enhances FGF2-induced angiogenesis in these tumors.[251]

FGFs have three possible signal transduction mechanisms: (1) ligand-initiated activation of receptor tyrosine kinase activity that leads to autophosphorylation and phosphorylation of other key cellular proteins such as Raf-1; (2) activation of phospholipase C-γ1 (PLC-γ1) and the phosphoinositol hydrolysis (PI) pathway leading to mobilization of intracellular calcium and activation of protein kinase C (see Signal Transduction Mechanisms below); and (3) internal localization of FGF in nuclei that could lead to DNA binding and direct activation of gene transcription. There is evidence for all three mechanisms. Interestingly, there are high-molecular-weight intracellular forms of FGF1 and FGF2 that don't appear to get secreted and have nuclear localization sequences. Forms of FGF1 and FGF2 have also been found in cell nuclei.

Various FGFs are produced by malignant tumors and constitutive expression of some FGFs induces a transformed phenotype in cultured cells. There are several examples of this. Basic FGF (bFGF, or FGF2) is not produced by normal melanocytes but is made constitutively by human metastatic melanoma cells grown in culture, and transfection of the bFGF gene into normal mouse melanocytes transforms them, although expression of bFGF by itself doesn't make them tumorigenic in vivo.[252] The ability of some FGFs to induce blood vessel growth has been observed in tumors (tumor angiogenesis).

An FGF (K-FGF or FGF4) isolated from Kaposi's sarcoma induces vascularization of this tumor. K-FGF's action is inhibited by a heparin analog called pentosan polysulfate, which blocks angiogenesis in a Kaposi's sarcoma growing in nude mice and induces tumor regression.[253] Basic FGF has been shown to be an autocrine growth factor for human and rat glioma cells, melanoma cells, and endometrial carcinoma cells (reviewed in Reference 254). Injection of an antibody to bFGF into tumor-bearing nude mice inhibited tumor growth, in part because the antibody blocked the potent angiogenic effects of bFGF.[254] Basic FGF is also found in areas of human squamous cell head and neck tumors with a high thymidine labeling index and significant endothelial cell proliferation, indicating a correlation between tumor vascularization, tumor cell proliferation, and bFGF production.[255] Through immunostaining for protein and in situ hybridization for mRNA, both acidic and basic FGFs have been detected in about 60% of surgical samples of human pancreatic cancer and their presence correlates with advanced tumor stage and poor prognosis.[256] Presence of bFGF in the cytoplasm of human renal cell carcinomas also correlates with shorter survival time.[257]

High concentrations of FGF2 have been detected in the urine and serum of cancer patients, and high serum levels are associated with a poor prognosis in small cell lung cancer.[258] Since secreted FGF2 is a mitogen for endothelial cells and a potent inducer of angiogenesis in vivo, the poor prognosis observed in patients with high circulating levels of FGF2 may reflect active angiogenesis and increased tumor growth and aggressiveness. Elevated FGF2 serum levels have also been observed in patients with head and neck cancer, colorectal carcinoma, non-Hodgkin's lymphoma, and chronic lymphocytic leukemia.[258]

Platelet-Derived Growth Factor

The discovery of platelet-derived growth factor (PDGF) arose from an observation by Balk,[259] who found that normal chicken embryo fibro-blasts proliferated better in culture medium supplemented with animal serum than in medium containing platelet-poor plasma. He also found that Rous sarcoma virus–transformed cells grew equally well in both, and he speculated that a "wound hormone" was released into serum during the process of clot formation. During the clotting of blood to form serum, a number of things happen, including the conversion of fibrinogen to fibrin and the clumping of platelets followed by platelet factor release. Thus, serum will contain some platelet-derived factors not present in plasma collected in the presence of anticoagulants and then centrifuged to remove the cellular elements and platelets. Other investigators have confirmed the observation that platelet-poor plasma is deficient in growth-promoting activity for cultured cells and have shown that platelet extracts could restore optimal growth to cells cultured in the presence of platelet-poor plasma. The growth-promoting factor from human platelets has been purified and characterized.[260–262]

Four PDGF polypeptide chains have been identified, and these constitute the five dimeric isoforms of PDGF: PDGF-AA, -AB, -BB, -CC, and -DD.[263] These isoforms act by binding three types of PDGF receptors: $\alpha\alpha$, $\alpha\beta$, and $\beta\beta$ (Fig. 4–16). These receptors are tyrosine kinases and have mitogenic effects on cancer cells, angiogenic effects in both normal tissues (e.g., wound healing), and cancer tissues, and paracrine effects in stromal fibroblasts and perivascular cells. As shown in Figure 4–16, all the PDGF isoforms, except PDGF-DD, induce PDGF α-receptor dimerization and activation. PDGF-BB and PDGF-DD activate receptors by dimerization. PDGF isoforms -AB, -BB, -CC, and -DD do the same for $\alpha\beta$ receptors. The PDGF ligand binding-mediated receptor subunit dimerization induces receptor auto-phosphorylation and triggers a downstream cascade of signal transduction involving Ras, PI-3 kinase, phospholipase C-γ, and protein kinase C as intermediaries.

PDGF receptor activation results in activation of a number of genes involved in cell proliferation, including *myc*, *fos*, c-*jun*, and *jun* B. As with EGF, the cellular actions of PDGF can be divided into early and late events. Early events (after 1 to 10 minutes) include tyrosine-specific phosphorylations, stimulation of phosphatidylinositol turnover, and reorganization of actin filaments.[264] Late events (after 30 to 180 minutes) include increased transcription of specific genes, stimulation of IGF binding, and increased amino acid transport. A clear difference between PDGF and EGF or IGF is that the stimulatory signal for cell division provided by PDGF is "locked in" after the cells are exposed to PDGF for only about 30 minutes, and PDGF can then be removed from the growth medium, whereas EGF and IGF must be present continually to stimulate cell division. Moreover, PDGF by itself does not induce a mitogenic response, but requires other plasma factors, among which are IGF and EGF. For example, when growth-arrested 3T3 fibroblasts are treated with PDGF for a short period of time and then transferred to platelet-poor plasma, the cells are observed to enter S phase 12 hours after the addition of plasma. This 12-hour lag before the onset of DNA synthesis occurs regardless of whether the plasma is added at the same time as PDGF or up to 13 hours after this. Thus, PDGF and plasma factors appear to control different events in the cell cycle. Scher et al.[265] have postulated that PDGF induces cells to become "competent" to enter the S phase of the cell cycle and that plasma factors allow only competent cells to undergo "progression" through the G_1 phase to enter S phase. Thus, PDGF is thought to prime cells to respond to other growth factors in plasma. As noted earlier, the PDGF-induced competent state must be stable for at least 13 hours after PDGF is removed, since the addition of plasma at any time up to 13 hours will permit progression into S phase.

Among the most potent factors in plasma that stimulate progression are the insulin-like growth factors. The evidence for this comes from experiments in which plasma from hypophysectomized rats was shown to be 20-fold less potent in permitting PDGF-treated competent 3T3 cells to enter S phase than normal rat plasma. The addition of a low concentration (10^{-9}M is 0.000000001 molar) of IGF-1 to cultures of PDGF-treated cells grown in plasma from

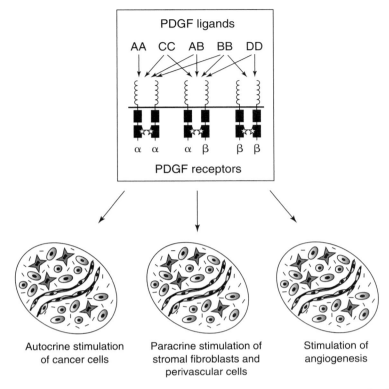

Figure 4–16. The platelet-derived growth factor (PDGF) system is involved in multiple tumor-associated processes. *Upper*: PDGF ligand binding specificity to PDGF receptors. *Lower*: PDGF receptors are expressed by many different cell types within tumors, and signaling from PDGF receptors can thus promote tumor progression in various ways. Tumor cells, purple; endothelial cells, red; fibroblasts, smooth muscle cells, and pericytes, green; extracellular matrix, dark solid lines. (From Pietras et al.,[263] reprinted with permission from Elsevier.)

hypophysectomized rats allowed the cells to enter S phase. Pure IGF-1 without plasma, however, did not do this, indicating that other materials in plasma are also needed for progression to occur. By using this assay system, the competence- and progression-stimulating activities of a variety of growth factors have been tested. Factors that have potent competence activity include PDGF, FGF, Ca^{2+}, and "wounding" (i.e., scraping a clear area through a sheet of confluent cells). Progression-inducing agents are IGF-1, IGF-2, insulin, and EGF. The mechanisms controlled by these two classes of growth factors are not clear, but they probably relate to the restriction point at which cells make a commitment to enter S phase. Once this commitment is made, cells enter S and com-

plete the cell cycle; if they are delayed somewhere past this restriction point, they frequently die. The competence-initiating factors may induce the synthesis of a critical initiator protein, whereas progression-inducing agents may promote cell division by stimulating the enzymes necessary for DNA synthesis. However, there appears to be more than one point at which PDGF-treated cells can be arrested before they enter into S phase,[265] which suggests a cascade of events, each of which might have different regulatory signals. The fact that a specific competence-inducing factor appears to be required for the proliferation of fibroblast-like cells suggests that other tissues may have similar requirements for competence activities. The less specific progression-stimulating activities, such

as those of the IGFs, may be general growth-promoting agents needed for the growth and development of many tissues in the body.

Of interest is the fact that transformation of fibroblasts with SV40 virus circumvents the need for both competence and progression activities. The SV40-transformed 3T3 cells grow to high density in either serum or platelet-poor plasma and require less serum than nontransformed cells. Thus the requirement for PDGF appears to be lost or greatly diminished during the process of transformation. In addition, transformed human and mouse cell lines that produce tumors in nude mice grow to high density in platelet-poor plasma, whereas cell lines that are not tumorigenic in nude mice grow poorly in platelet-poor plasma.[266] These data imply that transformed cells can produce a PDGF-like growth factor that occupies PDGF receptors and thus does not require exogenous PDGF. It has now been shown that a wide variety of murine and human cell lines transformed with oncogenic viruses or chemical carcinogens produce a PDGF-like substance that competes for binding to PDGF receptors and that antibody to PDGF can inhibit this PDGF-like activity.[267] Moreover, a number of human cancer cell lines have been shown to secrete PDGF-like factors; these include cells derived from osteosarcomas, glioblastomas, and fibrosarcomas. Some human tumor cells co-express both PDGF-like factors and PDGF receptors, setting the stage for self-stimulation of cell proliferation, given exposure to the right progression factors.[268]

When human melanoma cells are transfected with PDGF-BB cDNA, they produce in nude mice actively growing nests of tumor cells with a distinct stroma and abundant blood vessels, suggesting a role for PDGF as an inducer of a vascularized connective tissue stroma on which tumor cells can thrive.[269] Amplification and/or overexpression of PDGF receptors has been observed in human glioblastomas,[270] and activation of PDGF-B gene expression in concert with the PDGF-β receptor gene correlates with the conversion of human hydatidiform mole into choriocarcinoma.[271]

The first clear link was forged between growth factors and oncogene products when it was discovered that the PDGF B chain has a virtually identical amino acid sequence to the product of the sis oncogene first isolated from a sarcoma virus carried by the Woolly monkey (see Chapter 5). This cellular oncogene was apparently picked up during evolution by the simian sarcoma virus and is part of the transforming activity of the virus. The v-sis oncogene can activate α and β PDGF receptors and initiate cellular transformation when it binds intracellularly to PDGF receptors.[272]

The PDGF family of ligands and their receptors play an important role in normal development.[263] For example, activation of the PDGF-β receptor stimulates pericyte recruitment for blood capillary formation, development of vascular smooth muscle, and kidney development. Activation of PDGF-α receptors by PDGF-AA is needed for formation of lung alveoli, hair follicle development, villus formation in the gut, and oligodendrocyte production in the developing brain. In adults, PDGF actions foster wound healing by stimulation of fibroblast and smooth muscle cell proliferation. PDGF-β receptors also regulate interstitial fluid pressure by affecting fluid transport from the vasculature into the extracellular compartment of connective tissue.

PDGF-mediated activities can also be targets for chemotherapeutic attack. PDGF antagonists include antibodies, DNA aptamers, and small-molecule PDGFR-associated tyrosine kinase inhibitors. Clinically, the most effect anti-PDGF therapy has been with Gleevec, which inhibits receptor tyrosine kinase activities associated with the Bcr/Abl translocation in chronic myeloid leukemia, the Kit receptor in gastrointestinal stromal tumors (GIST), and the PDGF receptor isoforms in glioblastoma.

Transforming Growth Factors

The discovery of transforming growth factors (TGFs) came about as the result of experiments showing that mouse 3T3 cells transformed with murine or feline sarcoma viruses rapidly lost their ability to bind EGF, whereas cells infected with nontransforming RNA viruses maintained normal levels of cell surface EGF receptors.[273] These initial results suggested that the sarcoma virus genome produced something that altered EGF receptors. Later, however, it was found that murine sarcoma virus–transformed mouse fibroblasts produced a polypeptide growth factor that competed for binding with EGF on cell surfaces.[274] This factor was called *sarcoma growth factor* (SGF), a 6000 to 10,000 MW, heat-stable, trypsin-sensitive polypeptide that stimulated proliferation of transformed and untransformed fibroblasts. It competed with EGF for binding to EGF receptors, but it had a different molecular weight and was immunologically distinct from EGF. It also had the interesting property of being able to promote anchorage-independent growth in cultures of normal fibroblasts and thus to confer on normal cells properties associated with the transformed phenotype. This phenomenon was reversible, so that after the SGF was removed from the growth medium, the cells regained normal growth properties. Thus, SGF appeared to be a growth factor produced specifically by transformed cells and capable of stimulating their proliferation. This was the first observation suggesting that neoplastic cells are capable of autostimulation by producing their own growth factors. In this way, they could presumably escape the negative feedback systems of the normal host that control the production and release of endogenous hormones and growth factors.

Although the EGF-competing activity of SGF was contained in a 6000 to 10,000 MW fraction, the cellular transforming activity appeared to require, in addition, a fraction of higher molecular weight (20,000 to 25,000). The transforming activity of SGF isolated from murine or feline sarcoma virus–transformed cells was subsequently shown to be separable into two fractions: one of about 6000 MW, which competes with EGF for binding to EGF receptors and induces only small colonies of normal rat kidney (NRK) cells in soft agar, and one of about 25,000 MW, which does not compete for EGF binding but is required for production of large colonies of NRK cells in soft agar.[275] The former, EGF receptor-binding form has been termed transforming growth factor-α (TGF-α) and the latter is called TGF-β. Like EGF, TGF-α is a potent mitogen and appears to act through the same receptor, but by itself it is only a weak

inducer of anchorage-independent cell growth, as is EGF itself. TGF-β is also mitogenic for NRK cells and, to some extent, for mouse and human fibroblasts, but does not induce anchorage-independent growth if added alone to test cultures of NRK cells. However, it acts synergistically with either TGF-α or EGF to induce the transformed phenotype and anchorage-independent growth.

A wide variety of RNA and DNA tumor viruses stimulate production of TGF-like substances in cells transformed by them. Oncogenic RNA viruses such as Harvey, Kirsten, and Moloney murine sarcoma viruses as well as Abelson murine leukemia virus and DNA tumor viruses such as SV40 and polyoma have this property.[276] Not all the TGFs produced by cells transformed by these viruses are identical, and this finding as well as others led to the concept that there are families of TGF-α and TGF-β produced by different types of tumor cells. There is a close relationship between the release of TGF-α and cellular transformation by murine sarcoma viruses, as shown by the use of temperature-sensitive mutants of the transforming viruses.[276,277] When cells are grown at the temperature allowing cell transformation, TGF-α is produced, but when they are grown at the nonpermissive temperature, the factor is not produced.

TGF-α

As discussed previously, TGF-α belongs to a family of growth factors that includes EGF, amphiregulin, and vaccinia virus growth factor. TGF-α is produced as a 160–amino acid proTGF-α form that is cleaved to produce a 50 amino acid–soluble form of TGF-α (reviewed in Reference 278). Although TGF-α was first found in culture fluids of oncogenically transformed cells and is expressed by a wide variety of human cancer cells, its expression is not limited to neoplastic cells. During rodent embryogenesis it is expressed in maternal decidua and in developing kidney, pharynx, and otic vesicle. TGF-α mRNA and/or protein is also found in adult pituitary, brain, keratinocytes, ovarian theca cells, and macrophages, implying a role in the economy of normal adult tissues as well. TGF-α and EGF have a similar ability to promote proliferation and differentiation of mammalian mesenchymal and epithelial cells. This ability is not unexpected since they activate the same cell surface receptors.

In addition to the soluble form of TGF-α, there is a membrane-anchored form (the putative "juxtacrine" form). TGF-α shares this feature with several other membrane-anchored proteins bearing EGF-like repeats on their extracellular surface. Some of these membrane-bound glycoprotein forms are cleaved to yield soluble EGF-like growth factors, while others, such as the *Drosophila Notch*, *Delta*, and *Crumbs* gene products as well as the *C. elegans lin*-12 and *glp*-1, are not cleaved to release soluble factors. Although the role of these cell surface EGF-like repeats isn't clear, it seems probable that they interact with receptors on the surface of adjacent cells to sustain cell–cell adhesion and cell–cell regulation of proliferation and differentiation.[279]

While TGF-α plays a role in normal development, wound healing, ECM production, angiogenesis, and cellular adhesion, it is clear that its production at the wrong time or wrong place or its overproduction can favor neoplastic transformation and/or progression. Transfection of the TGF-α gene into cultured cells can be transforming. Overexpression or inappropriate production of TGF-α has been observed in human lung adenocarcinoma,[280] squamous cell carcinoma,[281] breast carcinoma,[282] endometrial adenocarcinoma,[283] and hepatocellular carcinoma.[284] TGF-α has been detected in the urine of patients with hepatocellular carcinoma[285] and in diffusion fluids of patients with a variety of cancers, including ovarian, breast, and lung cancers, often as a bad prognostic sign.[286] In addition, liver carcinomas have been shown to develop in transgenic mice that constitutively overexpress TGF-α.[284,287] This latter observation supports the notion that deregulated expression of TGF-α is a problem. It is expressed in developing liver, repressed in adult liver, and re-expressed in regenerating liver,[288] thus its expression may be coupled to cell proliferation and differentiation in a carefully regulated way.

TGF-β

There are three TGF-β polypeptides, but these are members of a much larger superfamily of TGF-β-related factors, including the bone morphogenetic proteins (BMPs), growth differentiation factors (GDFs), activins, inhibins, Müllerian inhibiting substance (MIS), Nodal, and Lefty 1 and 2 (Fig. 4–17).[289]

These TGF-β family members are translated as prepropeptide precursors with N-terminal signal peptides that put them on the secretory pathway. Typically, they have a number of intramolecular disulfide bonds that facilitate a conformational structure and intermolecular disulfide bonds that form covalent dimers. The TGF-βs are secreted as latent forms and are activated by proteolysis.

The TGF-β superfamily of ligands binds to and activates a family of transmembrane serine/threonine receptor kinases. These receptors are designated type I or type II, based on their structural and functional characteristics. The receptors have received a variety of names, but the one that seems to stick is activin-receptor-like kinase (ALK). There are seven ALK receptors (ALK-1 to -7) and a few others, including BMP-R2 and TGF-βR2.[289] In mammals, five type II and seven type I receptors have been identified. Type I and type II receptors exist as homodimers and binding of TGF-β type ligands facilitates formation of a type I–II tetramer, which then undergoes autophosphorylation and triggers a signal transduction pathway involving intermediates called Smads (Fig. 4–18).[290] The phosphorylated, activated Smads form complexes that translocate to the nucleus and via interaction with cell type–specific co-activators or co-repressors turn genes on or off, depending on the cellular context. The herculean effects of this family of growth factors is possible because of the multiple combinatorial interactions of ligand-bound receptor complexes and interaction with other signal transduction pathways (Fig. 4–19). Thus, a given TGF-β family ligand can induce different signaling pathways according to the composition of the receptor complexes. With all the effects that TGF-β family members have in development of various organ systems in the embryo, including the heart, skeleton, and craniofacial structures; left–right symmetry orientation; nervous system development; and effects in the adult organism, including reproductive function, wound healing, angiogenesis, extracellular matrix production, and modulation of the

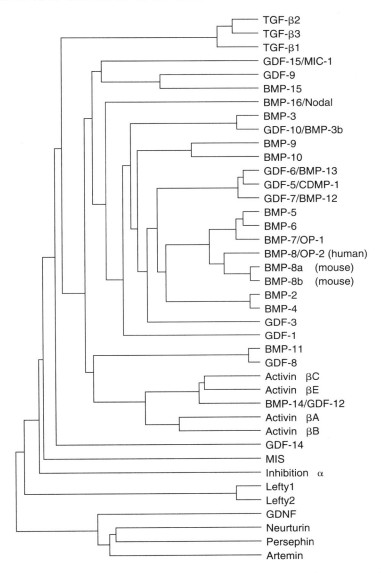

Figure 4–17. Transforming growth factor β (TGF-β) superfamily. Amino acid sequences of the carboxyl-terminal polypeptides of the mouse TGF-β super-family members (and human BMP-8) were aligned using the PILEUP program (Genetics Computer Group, Madison, WI). Mouse and human sequences are available for all sequences except BMP-8. In the mouse, there are two BMP-8 sequences (BMP-8a and BMP-8b), but only one in humans, because of a du-plication of the ancestral gene. (From Chang et al.,[289] with permission.)

immune system, this is indeed a herculean list of tasks.[289,290]

In carcinogenesis, TGF-β plays both a good-cop and a bad-cop role.[291] Originally, TGF-β received its name as a "transforming growth factor" because it assisted in inducing malig-nant transformation in cultured, nonmalignant fibroblasts. However, as time went on and more experiments were done, it became clear that TGF-β had an ubiquitous tissue distribution and a key role in normal development that was clearly at odds with its designation as a fac-tor responsible for inducing malignant transfor-mation. In fact, TGF-β was shown to be a

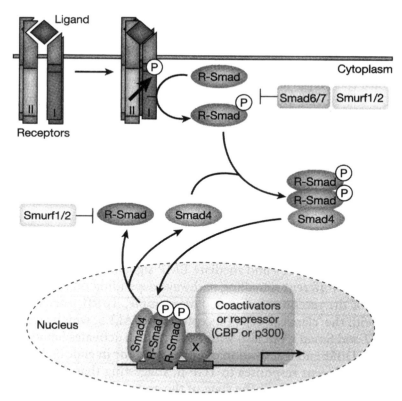

Figure 4–18. General mechanism of TGF-β receptor and Smad activation. At the cell surface, the ligand binds a complex of transmembrane receptor serine/threonine kinases (types I and II) and induces transphosphorylation of the GS segments (orange) in the type I receptor by the type II receptor kinases (blue). The consequently activated type I receptors phosphorylate selected Smads at C-terminal serines, and these receptor-activated Smads (R-Smads) then form a complex with a common Smad4. Activated Smad complexes translocate into the nucleus, where they regulate transcription of target genes, through physical interaction and functional cooperation with DNA-binding transcription factors (X) and CBP or p300 coactivators. Activation of R-Smads by type I receptor kinases is inhibited by Smad6 or Smad7. R-Smads and Smad4 shuttle between nucleus and cytoplasm. The E3 ubiquitin ligases Smurf 1 and Smurf 2 mediate ubiquitination and consequent degradation of R-Smads, yet can also interact with Smads6 and 7 and thereby ubiquitinate the type I receptors (not shown). (From Derynck and Zhang,[290] reprinted by permission from Macmillan Publishers Ltd.)

proliferation-suppressing factor in epithelial and lymphoid cells, so it was thought to be a tumor suppressor for these cell types. But "aha," the wheel turned again when it was realized that TGF-β actually facilitated invasiveness and metastasis of later, progressing tumors. The data indicate that activation of TGF-β signaling initially delays the appearance of mammary tumors in a Her2/neu-overexpressing mouse (reviewed in Reference 291). However, in mice expressing activated TGF-β receptor, an increased percentage of metastatic foci was observed. Some of these latter effects may be due to TGF-β's effects on extracellular matrix deposition and turnover, allowing cancer cells an escape route from tumors. In addition, TGF-β stimulates angiogenesis in vivo, and production and secretion of TGF-β by cancer cells suppress the activity of

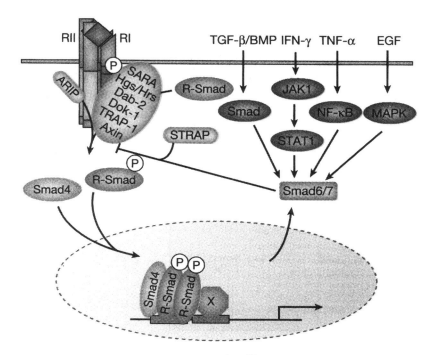

Figure 4–19. R-Smad activation is regulated by receptor-interacting proteins and Smad6 and Smad7. SARA, Hgs/Hrs, Dab3, Dok-1, TRAP-1 (TGF-β receptor–associated protein), Axin, and ARIP (activin receptor–interacting protein) (green) interact with type I or type II receptors and R-Smads. SARA or HRS and Dab2 stabilize the Smad2–Smad3 interactions with TGF-β type I receptors and function in internalization with the endocytic machinery in endosomes. Other proteins, such as the RasGAP-binding protein Dok-1, the PDZ-domain protein ARIP1, and axin, also probably control subcellular localization of receptors and link Smad2/Smad3 to the receptors. TRAP-1, a homologue of the yeast sorting protein Vam6p, interacts with TGF-β or activin type I receptors first, and then with Smad4 upon receptor activation, possibly facilitating Smad4 interaction with activated Smad2 or Smad3. Smad6 and/or Smad7 expression can be induced by several signaling pathways, including TGF-β/BMP signaling through Smads, and attenuates R-Smad activation. STRAP interacts with type I and type II receptors and with Smad7, thus stabilizing the interaction of Smad7 with the receptor complex. (From Derynck and Zhang,[290] reprinted by permission from Macmillan Publishers Ltd.)

infiltrating immune cells and hence allow tumors to escape immune surveillance.[291] High levels of circulating TGF-β are found in patients with invasive prostate cancer or colorectal cancer, and high urinary levels of TGF-β have been reported in patients with hepatocellular carcinoma.[291] The mechanism by which TGF-β can switch from a tumor suppressor function in early malignant transformation to a tumor progression-inducing factor in more advanced disease is illustrated in Figure 4–20. The large panoply of effects that the

TGF-β superfamily of ligands has on the pathogenesis of cancer makes this superfamily of factors and their signal transduction mechanisms a potentially large number of targets for cancer therapeutics.[292]

Hematopoietic Growth Factors

The hematopoietic growth factors (Table 4–6)[293] include erythropoietin, which stimulates red blood cell formation, the granulocyte, macrophage, and

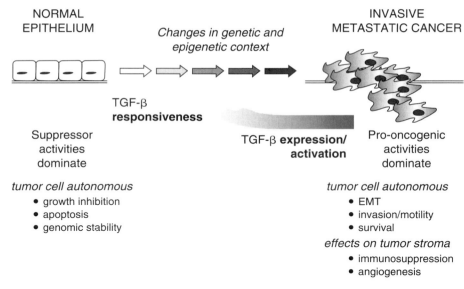

Figure 4–20. TGF-β switches from tumor suppressor in the premalignant stages of tumorigenesis to pro-oncogene at later stages of disease leading to metastasis. Progression to metastatic disease is generally accompanied by decreased or altered TGF-β responsiveness and increased expression or activation of the TGF-β ligand. These perturbations, along with other changes in genetic or epigenetic context of the tumor cell and its stromal environment, combine to alter the spectrum of biological responses to TGF-β. (From Roberts and Wakefield,[291] with permission.)

granulocyte-macrophage colony-stimulating factors (G-CSF, M-CSF and GM-CSF), the interleukins, of which there are at least 20,[294] which act on various stem cell populations in hematopoiesis, and various factors such as stem cell factor (SCF) and leukemia inhibitory factor (reviewed in Reference 293).

The first hematopoietic GF to be discovered was erythropoietin (EPO) in 1906. By the mid-1960s, semi-solid culture techniques that could support the growth of blood cell colonies from normal bone marrow became available, and it soon became clear that soluble substances released into the conditioned culture medium of such cell types were necessary to support the growth of these colonies. In the 1970s several of these "colony-stimulating factors" began to be purified and characterized, and by 1983, EPO, GM-CSF, G-CSF, M-CSF, and interleukins (IL) -1, -2, and -3 had all been purified. Between 1984 and 1986 human cDNAs for EPO, GM-CSF, and G-CSF became available, allowing their development for clinical use. These are considered by many observers to be the first clinically and commercially successful products of the new age of genetic engineering.

The colony-stimulating factors are a subset of regulatory polypeptides of the "cytokine" family that are involved in the proliferation and differentiation of granulocytes and monocyte or macrophages. The term *CSF* has stuck, and subsets of CSFs, based on their ability to stimulate particular pathways of hematopoietic cell differentiation, have been identified (Fig. 4–21).[295]

The CSFs are glycoproteins with 15,000 to 21,000 Da polypeptide chains and variable amounts of carbohydrate. They consist of a single polypeptide chain, except M-CSF, which is a homodimer. They are produced by multiple cell types, including fibroblasts, placenta, endothelial cells, lymphocytes, and bone marrow stromal cells. Blood levels of CSFs are normally low, but their production can be rapidly elevated in response to infection.

In vitro, some cell type specificity can be demonstrated: GM-GSF and IL-3 stimulate formation of granulocyte and macrophage colonies; G-CSF favors granulocyte colony formation;

Table 4–6. Some of the Hemopoietic Regulators

Regulator (Abbreviation)	Responding Hemopoietic Cells
Erythropoietin (Epo)	E, Meg
Granulocyte-macrophage colony stimulating factor (GM-CSF)	G, M, Eo, Meg, E
Granulocyte colony stimulating factor (G-CSF)	G, M
Macrophage colony stimulating factor (M-CSF)	M, G
Multipotential colony stimulating factor (Multi-CSF/IL-3)	G, M, Eo, Meg, Mast, E, Stem
Interleukin 1 (IL-1)	T, Stem
Interleukin 2 (IL-2)	T, B
Interleukin 4 (IL-4)	B, T, G, M, Mast
Interleukin 5 (IL-5)	Eo, B
Interleukin 6 (IL-6)	B, G, Stem, Meg
Interleukin 7 (IL-7)	B, T
Interleukin 9 (IL-9)	T, Meg, Mast
Interleukin 10 (IL-10)	T
Interleukin 11 (IL-11)	Meg, B
Interleukin 12 (IL-12)	NK
Megakaryocyte colony stimulating factor (Meg-CSF)	Meg
Stem cell factor (SCF)	Stem, G, E, Meg, Mast
Leukemia inhibitory factor (LIF)	Meg
Oncostatin M (OSM)	?
Macrophage inflammatory protein α (MIP-1α)	Stem

Abbreviations: B, B lymphocytes; G, granulocytes; E, erythroid cells; Eo, eosinophils; M, macrophages; Mast, mast cells; Meg; megakaryocytes; NK, natural killer cells; Stem, stem cells; T, T lymphocytes.
From Metcalf[293]

and M-CSF fosters macrophage colony growth. In addition to stimulating progenitor cell proliferation and cellular commitment to a particular differentiation pathway, the CSFs are also necessary to maintain functional activity of mature cells, for example, chemotaxis, phagocytosis, and production and release of cytotoxic factors.[296]

Interestingly, receptors for multiple CSFs are present on many hematopoietic progenitor cell types. There is a redundancy in the signaling process for hematopoietic cell proliferation, as if Nature has built in multiple mechanisms to protect the host from invading organisms and other stresses. For example, granulocyte-macrophage progenitor cells and their maturing progeny express membrane receptors for GM-, G-, M-, and multi-CSF. In addition, CSF-occupied receptors can initiate multiple functions in responding cells,

implying that broad signaling cascades are initiated by receptor occupancy. As one might expect from these observations, there are numerous potential interactions among CSFs. For instance, combinations of two CSFs can produce additive or synergistic responses. Because CSF receptor levels are low (a few hundred per cell),[296] occupancy of more than one type of CSF receptor may be required for an optimal proliferative response. Alternatively, progenitor cells may have multiple receptor types so that they are able to respond either to their normal, most appropriate ligand, and secondarily to another less optimal ligand that may be turned on by a different stress and/or a different CSF-producing cell type, so that the host can respond to any of a number of emergencies. In general, more mature, committed or single lineage cells can respond to stimulation by single growth factors, whereas less mature stem cells often require combined signalling from multiple factors.[293] It should be noted that CSFs were the first growth factors to show clearly that growth factors can both stimulate cell proliferation of developing stem cells and induce a differentiation pathway in cell-lineage progenitor cells. Some hematopoietic growth factors have a fairly limited range of target cells and others have a wide spectrum of target cells. For example, IL-3 and IL-6 act on hematopoietic precursor cell types, and leukemia inhibitory factor (LIF) acts on megakaryocytes, osteoblasts, neuronal tissue, hepatocytes, and adipocytes.[293]

Other ligand–receptor interactions have also been observed. Binding of GM-CSF to its receptor down-regulates expression of G-CSF receptors. GM-CSF and IL-3 compete for binding to the same receptor. Macrophages are induced by IL-3 and M-CSF to produce G-CSF and also by GM-CSF to produce M-CSF. Obviously, there is a great deal of cross talk between cells in the hematopoietic system, and it must require some finely tuned regulation, the mechanisms for which are only vaguely understood.

In spite of all the redundancies in the system, it is clear that in populations of bipotential progenitor cells, G-CSF fosters development of the cells in the granulocyte lineage and M-CSF fosters development of cells in the monocyte–macrophage lineage. This is borne out in vivo in that injection of G-CSF into mice induces a greater increase in peripheral blood granulocytes

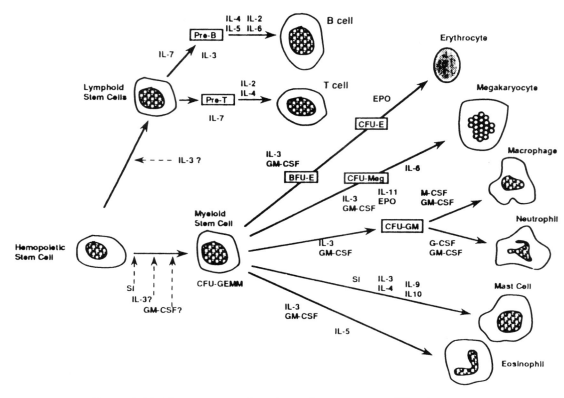

Figure 4–21. Hemopoiesis and cytokines. (From Miyajima,[295] with permission.)

than other blood cells, whereas GM-CSF induces a rise in both macrophages and granulocytes (reviewed in Reference 296). IL-3 administration elicits a rise in granulocytes, macrophages, eosinophils, and megakaryocytes, as might be expected from its broad target cell specificity.

Clinically, the CSFs have been used in AIDS, aplastic anemia, congenital or cyclic neutropenia, and in cancer. The latter use has been to restore bone marrow function after chemotherapy, often accompanied by bone marrow transplantation. Both G-CSF and GM-CSF have been shown to replenish peripheral blood neutrophils after high-dose chemotherapy followed by autologous bone marrow transplantation.[297,298] Positive benefits include decreasing the frequency of infections and shortening the stay in the hospital.

An intriguing sidelight of CSF therapy was the observation of a dramatic rise in the number of progenitor cells in the peripheral blood.[299] Usually, these cells are largely restricted to the bone marrow. This increase in the peripheral blood provides the capability of harvesting stem cells from the peripheral blood rather than the marrow, resulting in the recovery of many more stem cells and decreased trauma to the patient. A number of cytokines are currently being used to attain maximal mobilization of stem cells into the peripheral blood. Data from clinical trials indicate the use of peripheral stem cells is in many cases as effective as use of bone marrow cells in providing marrow reconsitution after chemotherapy.

The receptors for hematopoietic GFs have several common features (reviewed in Reference 293 and 300). They have highly related α chains with low-affinity binding sites, which when dimerized with β chains that provide some GF specificity produce a high-affinity receptor. Once receptor dimerization occurs, association with one of a family of cytoplasmic tyrosine kinases such as Tyk2 or JAK2 induces tyrosine phosphorylation on cellular substrates involved in the signal transduction cascade. Thus, although the CSF family of receptors are not themselves tyrosine kinases, once activated they become receptor tyrosine kinases in disguise. A model for these interactions is shown in Figure 4–22, where the α binding component of the cytokine binds to the α chain of the receptor, providing a stable "matrix" for sequestration of one β chain of the receptor, followed by dimerization of the receptor via a second β-chain binding site on the growth factor. This leads to activation of a cytoplasmic tyrosine kinase of the JAK family. Since different cytokine receptors are capable of generating qualitatively different signals in the same cells, it is likely that the receptors have some signalling specificity through their β subunits for either recruiting different sets of JAKs or attracting different substrates, or both.

Extracellular matrix components are known to up-regulate GM-CSF signaling in neutrophils and monocytes. There is evidence to suggest that this occurs via the ability of GM-CSF to release binding of the laminin receptors (LR) from the GM-CSF receptor, thus allowing binding of the neutrophils and monocytes to the ECM.[301] This ECM–cell attachment facilitates the signal transduction mediated by GM-CSF binding to its receptor.

Hepatocyte Growth Factor and Scatter Factor

Hepatocyte growth factor (HGF) and scatter factor (SF) were originally thought to be distinct cytokines that stimulated proliferation of cultured hepatocytes and that promoted motility of epithelial cells, respectively. HGF was first identified in the serum of partially hepatectomized rats as a potent mitogen for cultured rat hepatocytes and later also found in human plasma and serum, rat liver, and rat platelets (reviewed in Reference 302). HGF has been cloned and sequenced.[303] Scatter factor (SF) was originally found as a secretory product of fibroblasts that dissociates epithelial cell colonies into individual cells and stimulates migration of epithelial cells (reviewed in Reference 304). Purified SP also promotes invasiveness of cultured human carcinoma cells into collagen matrices, suggesting a role of SF in metastasis. Once both SF and HGF were cloned, it became clear that they were the same molecule.

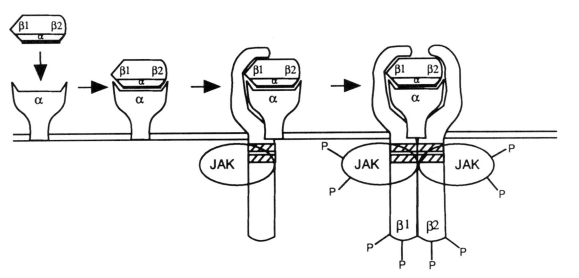

Figure 4–22. Stepwise formation of a generic cytokine receptor complex. Hatched boxes in β components represent conserved box 1 and box 2 sequences. (From Stahl and Yancopoulos,[300] with permission.)

HGF/SF is a disulfide-linked heterodimer of 55–65 kDa and 32–36 kDa subunits and is expressed in several tissues of mesodermal origin, including vascular smooth muscle cells, and has angiogenic properties[302] as well as the ability to induce morphological changes and induce anchorage-independent growth in HGF/SF transfected epithelial cells.[304]

The receptor for HGF/SF has been found to be identical to the c-*met* proto-oncogene, which is another member of the receptor tyrosine kinase family.[305] Activation of this receptor, which is found on a variety of cells including keratinocytes, melanocytes, endothelial cells, and other epithelial cells, triggers autophosphorylation and produces stimulation of the phosphatidyl inositol hydrolysis pathway and activation of Ras by shifting the equilibrium toward the active GTP-bound state.[306]

Miscellaneous Growth Factors

More modulators of cell proliferation and differentiation are being identified as they are looked for in normal and tumor tissues and cell lines derived from different tissues (reviewed in Reference 307). Some of these were called bone cell–derived growth factors (BDGF), uterine-derived growth factor (UDGF), mammary gland–derived growth factor (MDGF), melanocyte growth factor, lung cancer, ovarian cancer, and Wilms' tumor derived growth factors, as well as a family of estrogen-inducible growth factors called estromedins, found in a variety of tissues including uterus, kidney, and pituitary gland. Whether all of these growth factors are in fact distinct chemical entities or are in fact members of already identified growth factor families is yet to be determined.

The ubiquity of growth and differentiation factors leads one to predict that they will be found in all tissues in the body. These factors most likely act by paracrine or autocrine mechanisms to induce cell renewal or tissue repair of damage, but under some circumstances they may be released from the tissue and act on other organs through an endocrine mechanism like other known hormones. When cells undergo malignant transformation, they may continue to produce these factors, much as their normal proliferating stem cell counterparts do. Although the normal stem cells stop proliferating at some point and stop making these factors, tumor cells may continue to make them until they undergo enough genetic drift to become growth factor independent and capable of autonomous growth. The missing signal, or missing signal-receptor, that is necessary to convert the proliferating cancer cell type into a differentiating cell type is not known for most human cancers; clearly this is an important area for future research in cancer biology. Candidates for negative growth regulators have been found in a number of cell types, including lymphocytes, granulocytes, liver, mammary gland, epidermis, and fibroblasts. One of these growth regulatory substances, Oncostatin M, is 28–36 kDa polypeptide cytokine that is produced by activated T lymphocytes and phorbol ester–treated monocytes and inhibits the ability of melanoma cells and other cancer cell lines to grow in vitro. Paradoxically, it is also a potent mitogen for AIDS-derived Kaposi's sarcoma cells in culture.[308] These data indicate the complexity of growth factor–stimulated events and demonstrate the cellular "context" dependency of their actions, much as was discussed above for TGF-β.

A variety of growth regulatory factors for mammary gland epithelial cells have been reported (reviewed in Reference 309). Some of these factors have been detected in milk, some in conditioned medium of cultured mammary cells, some in mammary tissue extracts, and some even in neoplastic cells. Some appear to be produced constitutively and others are induced by anti-estrogens. One regulatory factor, called mammastatin, is produced by normal mammary cells[310] and can be detected in serum of women at the onset of menstruation and in rat mammary gland in late pregnancy. This finding suggests mammastatin or similar factors as candidates for differentiation-inducing agents that may provide the protective effect of early pregnancy for breast cancer.

SIGNAL TRANSDUCTION MECHANISMS

Some of the signal transduction pathways involved in cancer either as oncogenic or tumor suppressor functions are shown in Figure 4–23 (see color insert).[311] This is a simplified scheme and many other interconnecting pathway loops implicated as playing a role in cancer are still being discovered. In this section, some detail will be given for many of the key signaling pathways that play a role in the oncogenic process.

The first signal transduction pathway to be well defined is the cyclic AMP-dependent protein kinase system (reviewed in Reference 312). This system was the first of the so-called guanine nucleotide binding protein-coupled receptors, or GPCRs (reviewed in Reference 313), to be investigated. Study of the GPCR mechanism showed that hydrolysis of protein-bound GTP could act as a signaling switch, and this led to the discovery of the receptor as a seven-transmembrane domain protein. The first to be so identified was rhodopsin, but upwards of a thousand GPCRs have now been identified, making this the largest known receptor family. In addition, the largest number of marketed pharmaceutical agents interact with GPCRs.

Transmembrane signaling by phosphorylation and dephosphorylation mechanisms was defined during the 1980s and 1990s. Many of these turned out to be protein tyrosine kinase–coupled receptors (PTKRs) that phosphorylate tyrosine on substrates as opposed to serine or threonine kinases such as the cAMP-activated kinase PKA and the TGF-β receptors. The PTKRs are all transmembrane proteins with a cytoplasmic domain that has intrinsic kinase catalytic activity activated by ligand binding. A partial listing of the protein kinases identified in various organisms is shown in Table 4–7.

The tyrosine kinase–coupled receptors mentioned above are one potential target for carcinogenic alteration. Activation of these receptors can lead to phosphorylation of a number of key substrates. Many growth factor receptors mediate their cellular effects by intrinsic tyrosine kinase activity, which in turn may phosphorylate other substrates involved in mitogenesis. As noted in Chapter 5, a number of transforming oncogene products have growth factor or growth factor receptor–like activities that work via a tyrosine kinase–activating mechanism. For example, the v-*src* gene product is itself a cell membrane–associated tyrosine kinase. The v-*sis* oncogene product is virtually homologous to the B-chain of platelet-derived growth factor. The v-*erb* product is a truncated form of the EGF receptor. The *fms* gene product is analogous to

the receptor for colony stimulating factor CSF-1. The *met* and *tck* protooncogene products turn out to be receptors for hepatocyte growth factor and nerve growth factor, respectively.

Some of the key substrates for receptor–tyrosine kinase coupled activity include (1) phospholipase C (PLC-γ) which in turn activates phosphatidyl inositol hydrolysis, releasing the second messengers diacylglycerol (DAG) and inositol trisphosphate (IP_3) that activate protein kinase C (PKC) and mobilize intracellular calcium release (a number of tumor promoters also activate PKC); (2) the GTPase-activating protein GAP that modulates Ras proto-oncogene protein function; (3) Src-like tyrosine kinases; (4) phosphatidyl inositol kinase (PI3K) that associates with and may modulate the transforming activity of polyoma middle T antigen and the v-*src* and v-*abl* gene products; and (5) the *raf* proto-oncogene product that is itself a serine/threonine protein kinase.

Thus, activation of protein kinases is a key mechanism in regulating signals for cell proliferation. The substrates of these kinases include transcription regulatory factors such as those linked to mitogenic signaling pathways, e.g., proteins encoded by the *jun*, *fos*, *myc*, *myb*, *rel*, and *ets* proto-oncogenes.

The central role of tyrosine phosphorylation in cell proliferative signaling mechanisms also provides a target for chemotherapy. One should not forget, however, the catalysts for the other half of this reaction, the phosphatases. Although it has been known for a long time that protein phosphatases play a regulatory role in certain cellular metabolic functions, e.g., in the activation–inactivation steps for glycogen synthase and phosphorylase, it was more recently demonstrated that phosphatases play a role in the activity of various receptors and in the function of certain cell cycle–regulating genes (reviewed in References 315 and 316). For example, expression of a truncated, abnormal protein tyrosine phosphatase in baby hamster kidney (BHK) cells produces multinucleated cells, possibly by dephosphorylating the cyclin-dependent kinase p34^{cdc2}.[316] Activation of p34^{cdc2} requires dephosphorylation of a tyrosine residue, and this activation drives the cell from G_2 into M phase. The truncated phosphatase apparently interferes with the normal synchrony between nuclear formation and cell division.

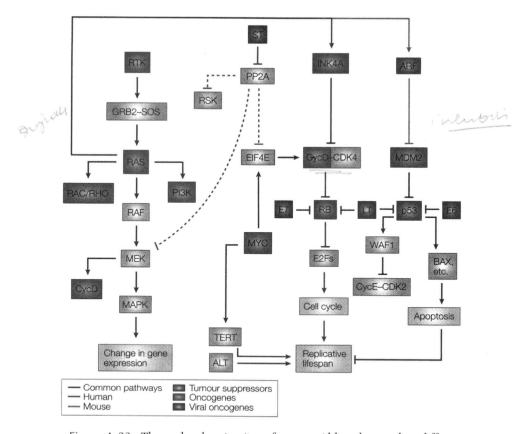

Figure 4–23. The molecular circuitry of cancer. Although countless differences between normal cells and cancer cells have been documented, much progress in identifying and connecting the fundamental pathways responsible for programming malignant cell growth has been made. Most cancer-associated mutations disrupt essential homeostatic mechanisms that regulate cell proliferation and survival. In many cases, particular mutations have been linked to specific biological phenotypes shown by cancer cells (yellow boxes). The cellular machinery responsible for controlling mammalian cell physiology is largely shared between human and mouse cells (black lines). Comparisons of human and mouse experimental cancer models identify several pathways that seem to have more prominent roles in human-cell transformation (red lines), as well as other molecular pathways that serve in dominant positions in mouse cancer models (blue lines). Perturbation of these five pathways (RB, p53, telomere maintenance, HRAS, and ST-PP2A) allows transformation of human cells. However, significant interactions among these pathways and other molecules that are implicated in the development of particular types of human cancer exist and remain to be characterized in detail. For the purposes of clarity, this representation is simplified and is illustrative rather than comprehensive. For example, clear evidence links RAS signaling to induction of cyclin D1 expression and the role of PP2A in most human cancers remains to be elucidated. An alternative method of displaying the molecular circuitry that programs the cancer phenotype can be found at http://www.nature.com/nrc/journal/v2/n5/wienberg_poster/. ALT, alternative lengthening of telomeres; CDK, cyclin-dependent kinases; Cyc, cyclin; E6, human papillomavirus E6 oncoprotein; E7, human papillomavirus E7 oncoprotein; LT, SV40 large T antigen; ST, SV40 small T antigen; TERT, telomerase reverse transcriptase; WAF1, also known as p21 (encoded by CDKN1A). (From Hahn and Weinberg,[311] reprinted by permission from Macmillan Publishers Ltd.)

Protein tyrosine phosphatases (PTPases) are a diverse family of enzymes that exist in cell membranes. Some of them are associated with receptors that have tyrosine kinase activity. Phosphatases are also in other intracellular locations. The aberrant phosphorylation state of tyrosine in certain key proteins such as c-Src or c-Raf that can lead to cellular transformation could theoretically come about through deregulation of a protein kinase or underexpression of a protein phosphatase.

Table 4–7. Mammalian, *Drosophila*, and Yeast Protein Kinases

Mammals	
Protein-Serine/Threonine Kinases	Protein-Tyrosine Kinases
CYCLIC NUCLEOTIDE REGULATED	*SRC* GENE FAMILY
cAMP-dependent protein kinases (C_α, C_β)	$pp60^{c\text{-arc}}$ (fibroblast, neuronal forms)
cGMP-dependent protein kinase	$pp62^{c\text{-yes}}$, $pp56^{ick}$
	fgr, hck, fyn, lyn proteins
CALMODULIN REGULATED	
Phosphorylase kinase (distinct liver and muscle forms?)	*ABL* GENE FAMILY
Myosin light chain kinases (skeletal, smooth muscle)	$p150^{c\text{-abl}}$ (type I and type II N-terminus)
Type II-calmodulin dependent protein kinase (brain α, β, β' subunits; liver α, α' subunits; muscle β, β' subunits)	*arg* protein
Calmodulin-dependent protein kinases I and III	*FPS* GENE FAMILY
	$p98^{c\text{-fps}}$
DIACYLGLYCEROL REGULATED	NCP94
Protein kinases Cs (α, β and β', γ, δ [RP14])	*c-fps*-related proteins (TKR11 and TKR16)
OTHERS	
Casein kinases I and II	GROWTH FACTOR RECEPTORS
Nuclear protein kinases N1 and N2	EGF receptor family
Protease-activated kinases I and II	EGF receptor (c-erbB protein)
Glycogen synthase kinases 3 and 4	*neu* protein (erbB2 protein)
Heme-regulated protein kinase	Insulin receptor family
Double-stranded RNA regulated protein kinase	Insulin receptor
Double-stranded DNA regulated protein kinase	IGF-1 receptor
S6 kinase	*c-ros, met, trk* proteins
β-adrenergic receptor kinase	PDGF receptor family
Rhodopsin kinase	PDGF receptor
Histone H1 kinase	CSF-1 receptor (*c-fms* protein)
Hydroxymethyglutaryl-CoA reductase kinase	*c-kit* protein
Pyruvate dehydrogenase kinase	*c-sea, ret* proteins
Branched charin ketoacid dehydrogenase kinase	
Polypeptide-dependent protein kinase	
Polyamine-stimulated protein kinase	OTHERS
c-mos, c-raf, A-raf, pks, pim-1 proteins	p75 (liver)
CDC-R (PSK-J3), CDC2Hs, PSK-H1, PSK-C3	p120 (brain)

Drosophila	
Protein-Serine/Threonine Kinases	Protein-Tyrosine Kinases
CYCLIC NUCLEOTIDE REGULATED	Dsrc64B protein
cAMP-dependent protein kinase-related (C0, C1, C2)	Dsrc28C protein
	Dash protein
cGMP-dependent protein kinase-related (G0 [2 genes], G1)	*fps*-related protein
	EGF receptor (types I, II, III N-terminus)
DIACYLGLYCEROL REGULATED	Insulin receptor
Protein kinase C	*sevenless* protein
OTHERS	
Casein kinase II	
raf protein	

(Continued)

Table 4–7. Mammalian, *Drosophila*, and Yeast Protein Kinases (*countinued*)

Yeast

CYCLIC NUCLEOTIDE REGULATED

cAMP-dependent protein kinase-related
TPK1, TPK2, TPK3, SRA3 (*S. cerevisiae*)

OTHERS

CDC28 (*S. cerevisiae* ($\approx cdc2^+$ in *S. pombe*)
CDC7 (*S. cerevisiae*)
KIN28 (*S. cerevisiae*)
wee1$^+$ (*S. pombe*)
nim1$^+$ (*S. pombe*)
STE7, STE11 (*S. cerevisiae*)
KIN1, KIN2 (*S. cerevisiae*)
SNF1 (*S. cerevisiae*)
ran1$^+$ (*S. pombe*)

Protein kinases are listed under protein-serine/threonine kinase and protein-tyrosine kinase in subfamily groups. Protein kinases included in this table have either been characterized as distinct by complete or partial protein purification, or have been identified as unique based on nucleotide sequencing. The reader should be aware, however, that many of the protein kinases whose existence is deduced from sequences of cDNA clones have not yet been proven to be protein kinases. Conversely, until the complete amino acid sequences are available for all of the protein kinases that have been identified on the basis of their enzymatic activity, one cannot be certain that they are distinct proteins. The primary references have been omitted to save space, but are available from the author,

(From Hunter,[314] with permission.)

For example, cells treated with vanidate, a PTPase inhibitor, had increased protein phosphotyrosine levels and a transformed phenotype.[317]

An interesting fallout from comparative genomic sequencing is the way in which kinase and phosphatase pathways have been conserved over evolution and how large a percentage of genes in the genomes of various organisms are devoted to protein phosphorylation–dephosphorylation regulation (reviewed in Reference 318). For example, in the yeast *S. cerevisiae*, there are 114 protein kinase genes (none of them protein tyrosine kinases) out of 6217 genes (1.8%). In *C. elegans* there are 400 protein kinase genes (92 are PTKs) out of 19,099 genes (total of 2.1%). In humans, there are predicted to be >1100 protein kinase genes (at least 150 of which are PTKs) out of a genome of about 30,000 genes (3.7%). In addition, in *C. elegans* there about 200 phosphatase genes and there are hints that there may be one phosphatase for each protein kinase.[318] Fine tuning indeed! It is also interesting that PTKs do not exist in yeast but do in *C. elegans*, one of the simplest multicellular organisms, a finding suggesting that PTKs evolved along with the need for intercellular communication. This makes sense because a majority of PTKs are transmembrane receptors that respond to secreted extracellular ligands, such as growth factors, that bind to the extracellular domains of these receptors.

An extension of the concept of ligand-activated transmembrane signaling was established by studies of the mechanism of activation of the Ras-Raf-Map kinase pathway. In this case, the signaling molecule Ras is anchored to the cell membrane via a linker, and its activation leads to GTP-GDP exchange and interaction with a Grb2/SH2/SH3 complex, which leads to activation of downstream effectors such as Raf, MAPK, and PI (see below). Interestingly, Ras was the first signaling protein identified as being conserved through evolution from yeast to humans, with homologs of many of the pathway components being functionally interchangeable among organisms.[318]

A large family of nuclear receptors has also been identified. These include the receptors for hormones such as estrogen, progesterone, and corticosteroids, as well as retinoic acid and other DNA-binding receptors. In this case, the ligand-activated receptor itself is the "second messenger." The unliganded receptors act as repressors because of their interaction with histone deacetylases. Ligand binding activates receptor-mediated gene transcription by releasing histone deacetylases and recruiting histone acetylases.

The phosphoinositol (PI)-mediated pathway was discovered in the 1960s. These studies led to the identification of a number of phospholipid-derived second messengers such as DAG and its role in regulating PKC, activation of which impacts a number of downstream effectors (see below). The idea that phospholipid metabolites constitute another class of second messengers produced in cells by a wide variety of hormones and growth factors came originally from the observation of Hokin and Hokin[319] of stimulated incorporation of ^{32}P into phospholipid in hormonally activated tissues. These investigators demonstrated increased phospholipid turnover in pancreatic tissue exposed to acetylcholine. Later it became clear that catabolism of inositol lipids is stimulated in many different tissues by many different external signals. The link of IP_3 generation to Ca^{2+} mobilization is now widely recognized and explains one pathway leading from receptor occupation to Ca^{2+} mobilization, PKC activation, and subsequent cellular responses.[320]

All the second-messenger signaling mediators described above involve small molecules such as nucleotides (e.g., cAMP, cGMP) or phospholipids (e.g., IP_3). The discovery that a gas, nitric oxide (NO), could also act as a second messenger was surprising. Originally described as endothelial cell–derived "relaxing factor," NO is now known to activate cytoplasmic guanylyl cyclase to elevate cGMP (reviewed in Reference 321), which in turn activates cGMP-dependent protein kinase and other cGMP-mediated events.

Some Key Signal Transduction Concepts

Transcriptional Regulation by Signal Transduction

After binding ligand, many transmembrane receptors activate downstream effector cascades that ultimately lead to a nuclear signal that turns genes on or off. The classic example of this is the protein kinase A (PKA)–cAMP-binding proteins (CREB) system in which cAMP activates PKA to release its catalytic subunit, which translocates to the nucleus and phosphorylates the CREB transcription factor. Phosphorylated CREB binds to co-activator factors, and this complex activates

transcription of cAMP-responsive genes. This sort of process is typical of many transcription factors, many of which are regulated by phosphorylation that enhances nuclear transport, DNA binding, or transactivation events. As will be seen under the description of individual signal transduction pathways below, many receptors (i.e., transmembrane, membrane associated, cytoplasmic, and nuclear) regulate gene transcription via downstream effectors.

Protein–Protein Interaction Domains

Protein–protein interaction domains determine the way that proteins talk to each other in forming functional complexes such as receptor multimeres, transcription factor–coactivator complexes, and DNA replication "machinery." In addition, protein conformational changes can be induced by phosphorylation or other chemical modifications such as acetylation or methylation, by ligand binding, or by interaction with extracellular matrix components to form functionally active (or in some cases inactive) proteins. Also, protein interaction domains are involved in association of proteins with phospholipids, small molecules such as drugs, and nucleic acids. Interaction domains are also involved in targeting proteins to specific subcellular locations, in providing recognition sites for post-translational modifications or second messengers, and in substrate binding to an enzyme. For example, phosphorylated tyrosines on PTK receptors bind phosphotyrosine recognition domains on Src homology peptides of the PTKR/Grb2/SOS complex involved in signal transduction from EGF receptors.

The folding of proteins into appropriate confirmations is key to achieving a functional structure.[322] Isolated interaction domains can be quite small and fold independently, with their amino- and carboxyl-termini juxtaposed in space in a way that leaves their ligand-binding domain available.[323] These domains recognize exposed interaction sites on their protein partners. The growth factor signal transduction pathway is a good example of this (Fig. 4–14). In most cases, a protein interaction domain recognizes a consensus recognition sequence and a specific structural conformation with flanking sequences that provide additional contacts and some binding selectivity.[323] Protein interaction domains involved in signaling are present in hundreds of copies in the human proteome, and they are used over and over again in different functional ways. For example, SH3 domains are used in different proteins to regulate signal transduction, protein trafficking, cytoskeletal organization, cell polarization, and organelle biosynthesis (reviewed in Reference 323). This re-use of interaction domains allows for re-shuffling of these domains for different functional purposes and this, over evolution, may have been one way to facilitate new cellular functions. This reshuffling can also have dire consequences because in cancer or hereditary disorders these interaction domains can be mutated, causing loss of important protein–protein interactions or creation of aberrant protein complexes. It should be noted that drugs can be designed to target protein interaction domains and either facilitate or inhibit protein–protein interactions.

Spatial and Temporal Regulation

Another important aspect of signal transduction is the spatial orientation of receptors and their downstream effectors as well as the timing of cellular responses following ligand–receptor interaction. For example, protein kinases and phosphatases as well as their substrates are frequently localized in different compartments in the cell. This allows for some specificity of interaction. Also, the orientation of signaling components may be organized differently in different cell types, providing some cell type selectivity. As noted above, some receptors are transmembrane, some are cytoplasmic, and some are nuclear and their downstream effectors may be similarly distributed. Although some second messengers such as cAMP are freely diffusible moieties, many second-messenger systems rely on protein–protein interactions at the cell membrane or in the cytoplasm and nucleus and have limited free diffusion. An example of this is the ligand–receptor interaction of glucocorticoids with their receptor in the cytoplasm and the assembly, disassembly, and nuclear transport of the corticosteroid receptor complex via heat shock proteins to the cell nucleus.[324] Protein localization is fundamental to the regulation of

signal transduction and provides a mechanism to prevent willy-nilly activation of signaling pathways. A good example is the translocation of transcription factors initiated by activation of cell surface receptors (Fig. 4–24). If these receptors were not separated from the site of their action or the enzymes regulating their phosphorylation state and stores of downstream effectors such as calcium, these transcription factors could be constitutively activated all the time, creating havoc in cell function. Take, for example, the activation of NFAT by the PLC-γ pathway, shown in Figure 4–24 (see color insert). NFAT nuclear localization is regulated by an activated receptor that stimulates PI turnover in the plasma membrane and that increases IP_3 production. IP_3 mediates the release of calcium ions from internal stores (e.g., in the endoplasmic reticulum). CA^{2+} ions stimulate the phosphatase calcineurin that dephosphorylates key sites on NFAT, exposing a nuclear import signal (reviewed in Reference 325).

Timing is also important in signal transduction. For example, the rates of effector activation in the various steps of a multistep cascade and the speed of movement of effector second messengers to their site of action regulate how cells respond to signals. Also, these rates may vary by cell type and provide some specificity of response as well as a well-regulated group of functions. The sequence and timing of this could be key to a normal versus an aberrant cellular response. A good example of this is activation of protein kinase C by CA^{2+}, DAG, or tumor promoters. The timing and duration of this activation differs depending on the type and half-life of the agonist (e.g., CA^{2+} vs. DAG vs. fatty acids; see Fig. 4–25).[326] For example, it is now apparent that activation of PKC may occur by a number of alternative routes and that some activations of PKC may be transient, leading to early or rapid cellular responses, while other activities may be more sustained, leading to late cellular responses such as cell proliferation and differentiation events.

Phospholipases C, A_2, and D all appear to be involved in PKC activation, with PLC being more involved in the early response pathway and PLA_2 and PLD in the late response pathway. The reaction products of phosphatidyl choline hydrolysis by PLA_2, i.e., the *cis*-unsaturated fatty

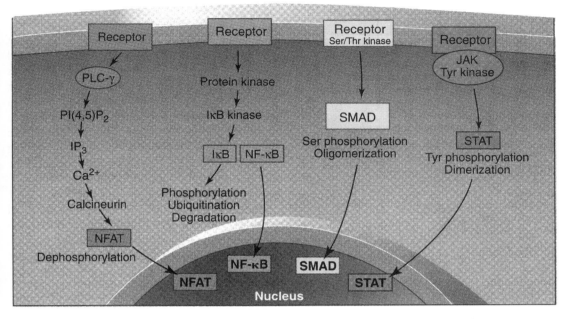

Figure 4–24. Getting to the nucleus. Signal transduction pathways initiated by activated receptors control translocation of transcription factors from the cytoplasm to the nucleus. Movement of the transcription factors NFAT, NF-κB, SMADs, and STATs is associated with the phosphorylation or dephosphorylation of signaling proteins by kinase and phosphatase enzymes. (From Cantley,[325] with permission from the American Association for the Advancement of Science.)

acids and lysophosphatidylcholine, can both enhance PKC activation. The tumor-promoting phorbol esters mimic DAG for activation of PKC and, being more stable than DAG, favor the sustained, late-response pathway involved in cell proliferation and differentiation.

Signaling Networks and Cross Talk

The old concept that metabolic and signaling pathways were sequential linear cascades has been shown to be incorrect in recent years. It is now known that both of these types of pathway have multiple linkages and form networks that were unpredicted when these pathways were initially discovered and delineated. This will be described in more detail in Chapter 5 under Systems Biology.

The linkage and networking of signal transduction pathways create two conceptual questions: (1) How can signaling specificity be maintained? (2) How can the redundancy implicit in overlapping or interconnected pathways be regulated in a way to select for the appropriate response only?

The specificity of signaling via different pathways or in different cells exposed to the same agonists (growth factors, hormones, drugs, etc.) can be at least partly explained by spatial and temporal separation of receptors and downstream effectors as noted above, but it is also because different cell types can have different repertoires of downstream effectors, different sets of transcription factors, and/or different sets or levels of co-stimulatory or co-repressor factors modulating gene expression.

Redundancy of signaling pathways is another challenge. Signal transduction pathway cross-talk can occur between pathways activated by a single receptor or among pathways activated by different receptors. Cross talk among signaling pathways can result in up- or down-regulation of one of them triggering coordinate responses in another one. Thus, inhibition of one component of a signal transduction pathway may be compensated for in the cell by up-regulation of

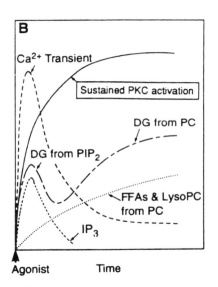

Figure 4–25. Schematic representation of agonist-induced membrane phospholipid degradation for sustained PKC activation (A). Time course of generation of various signaling molecules (B). DG, diacylglycerol; FFAs, free *cis* unsaturated fatty acids; IP$_3$, inositol 1,4,5-trisphosphate; LysoPC, lysophosphatidylcholine; PC, phosphatidylcholine; PIP$_2$, phosphatidylinositol 4,5-bisphosphate. (From Nishizuka,[326] with permission.)

another pathway. This has important therapeutic implications, because a drug that blocks an early or upstream component of a given pathway may be circumvented by activation of another parallel pathway. This phenomenon is seen, for example, in the development of resistance to some chemotherapeutic agents. A goal, then, is

to try to target the downstream events where transduction pathways converge in their ability to stimulate gene activation events.

An example of the cross talk among ligand–receptor triggered events is the binding of the growth factor beta platelet-derived growth factor (βPDGF) to its receptor βPDGFR (Fig. 4–26). This induces dimerization of the receptor, which in turn triggers signal transduction pathways. The βPDGF receptor becomes autophosphorylated on multiple tyrosines by activation of its receptor tyrosine kinase, which fosters binding to specific Src homology 2 domain (SH2)-containing proteins that are part of the Grb2-Sos-Ras-Raf-Mek-Erk pathway. In addition, there is cross talk with the phosphatidyl inositol kinase (PI3K) pathway. PI3K can also stimulate Rac GTPase, which can activate JAK/STAT signaling events. Activation of the SH2 domain protein PLC-γ1 can also potentially stimulate PKC signaling pathways. Thus, cytoplasmic signaling proteins form networks of interactions rather than simple, linear pathways.[327] These diverse signaling pathways, in turn, induce broadly overlapping sets of genes.[328]

GTP-binding protein (G-protein) signaling events are another ubiquitous pathway for gene activation, some of which are mediated by cyclic AMP that has protean effects on cellular processes (see below). Mutations in components of G protein–coupled pathways have been observed, some of which are involved in a number of human diseases, including cancer.

Overview of Some Signal Transduction Pathways Important in Cancer

G Protein–Linked Receptors

As noted above, guanine nucleotide binding protein–coupled receptors are a diverse set of ligand-activated receptors that regulate adenylate cyclase, ion channels, certain protein kinases, and other signal transduction mechanisms. They all share a common general structure (Fig. 4–3) with an external ligand-binding domain, membrane-spanning domains, and an intracellular domain that interacts with various G-protein complexes and contains sites for phosphorylation. The ligands that interact with such receptors include α- and β-adrenergic agonists

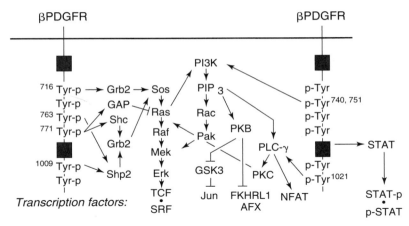

Figure 4–26. A signaling network from the activated βPDGFR. The activated receptor is a dimer. Each receptor chain becomes phosphorylated on multiple sites, some of which are depicted here, and binds specific SH2-containing proteins. The receptor itself has redundant interactions, for example, with Grb2. There are specific pathways leading from the receptor to the nucleus. There are also numerous potential cross-connections between distinct pathways, some of which are shown. See text for more detail. The figure is illustrative and by no means comprehensive; some binding partners and potential pathways, including Src family kinases, are not depicted. (From Pawson and Saxton,[327] reprinted with permission from Elsevier.)

and antagonists, angiotensin, serotonin, bombesin, bradykinin, acetylcholine (muscarinic type), vasopressin, and vasoactive intestinal polypeptide (VIP).

The first four G proteins discovered were designated G_s, G_t, G_i, and G_o. A large number of G proteins have subsequently been identified by cDNA cloning (reviewed in Reference 329). G proteins are heterotrimers composed of an α (39–46 kDa), β (37 kDa), and γ (8 kDa) subunit (reviewed in Reference 330). The β and γ subunits form a tightly associated complex ($G_{βγ}$) that functions as a unit and forms trimers with an α ($G_α$) subunit. The α subunit has a high-affinity binding site for GTP or GDP. The GDP-bound form of α binds tightly to the βγ complex and is inactive. When GTP is bound it displaces GDP, dissociates α from βγ, and induces the regulatory function of α. While it has been thought that the activated $G_α$ subunit alone regulates ion channels, adenylyl cyclase, phospholipase Cβ, and other enzymes, it is now apparent that the $G_{βγ}$ dimer also plays a role in modulating the activity of these effector systems. Currently, there are 20 $G_α$, 6 $G_β$, and 11 $G_γ$ subunits that have been identified.[329]

The $G_α$ subunits possess intrinsic GTPase activity. Modification of $G_α$ by cholera toxin activates $G_α$ proteins by inhibiting their GTPase activity, and binding of pertussis toxin blocks receptor-mediated activation of G proteins. Thus, these two toxins, which ADP ribosylate different sites on $G_α$ subunits, are often used as tools to investigate the role of G proteins in various physiological systems. G proteins are anchored in the plasma membrane of cells by lipid modifications of the subunits; γ subunits are prenylated and some γ subunits are myristolated.

The G_s and $G_γ$ families regulate the adenylyl cyclase and phospholipase C-beta (PLC-β) pathways, respectively. The G_i and G_o families' activities are more general and less well defined. The best understood is the G_i family regulation of the transducing pathway for light detection in the eye.[329] G_{12} and G_{13} pathways share downstream effectors, yet show some selectivity for certain ligands, e.g., LPA and thrombin. The four broad G-protein families transduce signals from a large number of diverse activating ligands and modulate a number of cellular functions such as homeostasis, embryonic development, metabolic regulation, gonadal development, and memory (Fig. 4–27). The G_s pathway was the original cell signaling pathway defined and is the one that regulates protein phosphorylation. A number of interconnecting linkages to other signaling pathways have been identified.

The G_i-mediated pathway was originally identified by its inhibiting activity on adenylyl cyclase; however, a number of hormones and neurotransmitters including epinephrine, acetylcholine, dopamine, and serotonin use the G_i and G_o pathway. Signal flow through this pathway is blocked by pertussin toxin via its ability to ADP-

ribosylate the G_α subunit. The G_q pathway is the one activated by calcium-mobilizing hormones and activates PLC-β to produce IP_3 and DAG.

The formation of cAMP activates protein kinase A (PKA), which can have a profound effect on cellular metabolism (Table 4–1). While many of the cellular actions of cAMP are due to activation of PKA, other actions are attributed to a direct action of gene transcription via binding to cAMP-binding proteins that act as transcription factors (see above).

Because cAMP appears to play a key role in cell proliferation and differentiation, a number

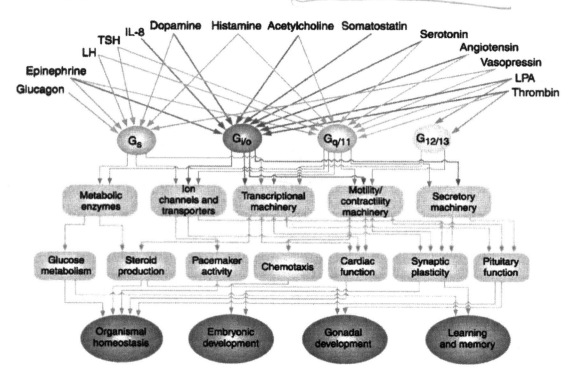

Figure 4–27. A schematic representation of how signaling through G-protein pathways can regulate systemic functions. Many extracellular agents, such as hormones (for example, glucagons, luteinizing hormone, and epinephrine), neurotransmitters (acetylcholine, dopamine, and serotonin), chemokines (IL-8), and local mediators (LPA), signal to the four main G-protein families to regulate such cellular machinery as metabolic enzymes, ion channels, and transcriptional regulators. Modulation of the activities of the cellular machines in turn gives rise to altered cellular functions, such as changes in glucose metabolism in liver and muscle or altered activities of pacemaker cells in the heart. These cellular activities contribute to the regulation of large-scale systems such as organismal homeostasis and learning and memory. Thus, G-protein pathways can propagate regulatory information through layers of increasing organizational complexity. At all levels, the examples shown here represent only a sample of extracellular agents that couple to the four G proteins, and the functions regulated by these pathways. (From Neves et al.,[329] reprinted with permission from the American Association for the Advancement of Science.)

of investigators have speculated that alterations in the cyclic nucleotide–generating or response systems may be altered during malignant transformation. A number of studies have shown that transformed fibroblasts regain a number of the characteristics of untransformed cells after treatment with cAMP analogues. These characteristics include a more flattened morphology, an increased adhesion to the substratum, a decreased agglutinability by lectins, and a decreased rate of cell proliferation. However, not all transformed cells respond to cAMP treatment in this way. The response is determined by the cell of origin of the transformed cell line. Fibroblastic cells, in general, tend to respond to cAMP in the previously described manner, whereas epithelial cells often do not. Experiments with various clones of rat kidney cells, for example, showed that in a fibroblastic clone, intracellular levels of cAMP rose as the cells reached confluency, but this was not the case in an epithelial clone.[331] Murine sarcoma virus—ransformed fibroblastic clones of rat kidney cells did not have elevated intracellular cAMP at confluency, and they responded to treatment with exogenous cAMP analogues by exhibiting a slower growth rate and a flattened cell morphology. Neither the growth nor the morphology of the epithelial clone was affected by exogenous cAMP.

It appears that certain parts of the response system for cAMP differ in certain kinds of cells. In support of this idea, experiments with the S49 lymphoma cell line showed that the proliferation of these cells was inhibited by cAMP but was not inhibited in a mutant S49 cell line defective in cAMP-dependent protein kinase.[332] Similar results have been obtained for mutant Chinese hamster ovary cells with a varient cAMP-dependent protein kinase. There may also be alterations in the cAMP-binding regulatory subunit of cAMP-dependent protein kinase or in the translocation step involved in the nuclear uptake of protein kinase, a step that appears to be required for response to cAMP in some cells. Decreased binding of cAMP and an altered cAMP-binding protein have been demonstrated in a cAMP-unresponsive line of Walter 256 carcinosarcoma cells compared with the responsive parent line.[333] The nuclear translocation of cAMP-binding proteins and pro-

tein kinase was also markedly diminished in the unresponsive tumor cells after treatment with dibutyryl cAMP.

It is clear that cAMP affects the proliferation rate of some normal and transformed cultured cells and that cAMP levels are lower in some transformed cell lines. It is not clear if changes in cAMP levels or the cAMP-response system are *responsible* for the appearance of the transformed phenotype and, more important, for the loss of normal growth control. In some cells, alterations in intracellular cAMP appear to be more closely related to the morphologic characteristics of the transformed phenotype than to growth control; in fact, the two events are clearly dissociable in certain cell types. Nevertheless, it is clear that induction of cAMP in several types of cultured neoplastic cells induces a more differentiated, less transformed phenotype.

Several lines of evidence implicate G protein–coupled receptors in malignant transformation.[334] Overexpression of acetylcholine or serotonin receptors in NIH 3T3 cells causes ligand-dependent transformation. Bombesin-like peptides are secreted by some small-cell lung carcinoma cells and stimulate their growth, and antibodies to bombesin inhibit tumor cell proliferation. Some pituitary, adrenal cortical, and ovarian tumors have point mutations in G proteins coupled to adenylyl cyclase that could lead to constitutive overproductions of cAMP.

The $\alpha_{1\beta}$-adrenergic receptor is a member of the G protein–coupled receptor superfamily and activates PI hydrolysis, a signaling pathway activated by a number of growth factors and that plays a crucial role in mitogenesis (see below). Mutation of three amino acid residues in the third intracellular loop increases the binding affinity of norepinephrine and its ability to stimulate PI hydrolysis by two to three orders of magnitude.[335] Moreover, this activating mutation renders the receptor constitutively active, stimulating PI turnover even in the absence of ligand. When the wild-type gene for the $\alpha_{1\beta}$ receptor is transfected into rat or NIH 3T3 fibroblasts, the cells express high levels of this receptor, become transformed in response to norepinephrine, and form tumors when injected into nude mice. When the mutated gene is transfected into fibroblasts, the cells spontaneously form transformed foci in the absence of ligand and have an enhanced ability to form tumors in nude mice. Thus, the $\alpha_{1\beta}$-adrenergic receptor gene acts like a proto-oncogene and when activated or overexpressed is a transforming oncogene. These data suggest that other G protein–coupled receptors of this type can act as oncogenes in certain cell types. This further suggests a host of strategies for chemotherapeutic interdiction of this system, for example, the design of specific antagonists of the G protein–coupled receptors that may be activated or overexpressed in tumor cells.

There is also evidence that alteration of G-protein subunits themselves can cause alterations in fibroblast growth characteristics. For example, transfection and overexpression of a mutated G-protein α_{i2}-subunit gene, a gene shown to be involved in proliferation of fibroblasts and differentiation of myeloid cells, in fibroblasts produces increased cell proliferation and anchorage-independent growth, indicating a role for this G-protein subunit in regulation of fibroblast cell proliferation and in transformation events.[336]

The Phosphoinositide 3-Kinase Pathway

The phosphorylated lipids phosphatidyl inositol and its additionally phosphorylated effectors phosphatidylinositol-4, -4,5, and -3,4,5 (PtdIns-3,4,5, or simply IP_3) are generated at cell membranes during signaling events and play a role in recruitment and activation of signaling components.[337] The kinase phophoinositide 3-kinase (PI3K) catalyzes the formation of IP_3, which in turn recruits a number of signaling proteins with pleckstrin homology (PH) domains to the cell membrane where they are activated (Fig. 4–28; see color insert). These signaling proteins include the serine/thrionine kinases Akt and PDK1, protein tyrosine kinases such as the Tec family, exchange factors for GTP-binding proteins (Grp1 and Rac), and adaptor proteins (GAB-1). These in turn modulate a number of cellular events including control of protein synthesis, actin polymerization, cell survival, and cell cycle entry.

Of particular interest to regulation of cancer cell proliferation and survival is the Akt-mediated

pathway and the central role of PDK1 in phosphorylation and activation of Akt.[338] Downstream protein targets of Akt include mTOR (see below) and p70S6 kinase that enhance protein synthesis; the apoptosis-facilitating protein Bad, whose phosphorylation inhibits its activity; and glycogen synthase kinase 3 (GSK3), whose phosphorylation releases GSK3's inhibition of c-Myc and cyclin D to promote cell cycle entry and cell proliferation. An important negative regulatory step in the Akt pathway is the dephosphorylation of IP_3, the primary product of PI3K, by PTEN (phosphatase and tensin homolog deleted from chromosome 10). PTEN is a tumor suppressor and its deletion has been observed in a number of human cancers.[339,340] PTEN works by preventing the activation of Akt by phosphorylation. Thus, with no Akt activation, the cancer cell proliferation and cell survival–promoting events stimulated by Akt are inhibited.

mTOR

The TOR (target of rapamycin) protein was originally identified in yeast mutants that were resistant to the growth inhibitory effects of the drug rapamycin (reviewed in Reference 341). TOR turned out to be a large protein (about 280 kDa), with protein kinase activity. Homologs to the yeast TOR have been found in Drosophelia, C. elegans, and mammals, including humans. The mammalian homolog is known as mTOR. As is the case for the other homologs, it binds rapamycin and the immunosuppressant agent FK506.

mTOR receives activating signals from a number of inputs and has a number of downstream effectors, including S6 kinases and 4EBP1, that modulate protein translation (Fig. 4–29). Signaling occurs mainly through the PI3K-Akt pathway, but mTOR can also be activated by growth factor receptors such as EGFR and IGF-1 receptor. Two additional intermediates in the PI3K-Akt pathway are the tuberous sclerosis complex (TSC1/2) and the Ras homolog enriched in brain (Rheb). Akt-mediated phosphorylation of TSC2 releases its inhibitory effect on mTOR, and Rheb also plays a role in activating mTOR.[342] Rheb is overexpressed in transformed malignant cells and appears to act like an oncogene. Cyclin D1 and c-Myc also appear to be downstream

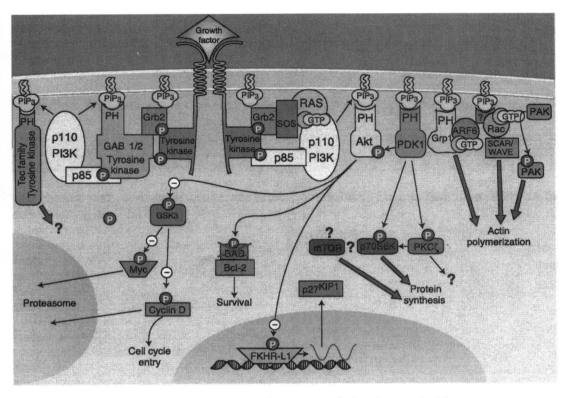

Figure 4–28. Signaling pathways downstream of phosphoinositide 3-kinase (PI3K) affect cell growth, cell survival, and cell movement. Activation of growth factor receptor protein tyrosine kinases results in autophosphorylation on tyrosine residues and transphosphorylation of adaptor proteins, such as GAB-1 on tyrosine. PI3K can also be stimulated by integrin-dependent cell adhesion and by G protein–coupled receptors (not shown). PI3K is brought to the membrane and activated by directly binding to phosphotyrosine residues of growth factor receptors or adaptors. The lipid product of PI3K, phosphatidylinositol-3,4,5-trisphosphate (PIP_3), recruits a subset of signaling proteins with pleckstrin homology (PH) domains to the membrane, where they are activated. These proteins include protein serine-threonine kinases (Akt and PDK1), protein tyrosine kinases (Tec family), exchange factors for GTP-binding proteins (Grp1 and Rac exchange factors), and adaptor proteins (GAB-1). Ultimately, these proteins initiate complex sets of events that control protein synthesis, actin polymerization, cell survival, and cell cycle entry. (From Cantley,[337] reprinted with permission from the American Association for the Advancement of Science.)

effectors of mTOR since levels of these proteins go down in cells treated with rapamycin.[342]

Several lines of evidence implicate mTOR in malignant transformation (reviewed in Reference 343). Among the pieces of evidence for this are the observations that (1) mTOR is essential for transformation events mediated by PI3K signaling and involves a number of mTOR downstream effectors, and (2) the anti-tumor effects of rapamycin are observed primarily in cancers that have overexpression of PI3K signaling.

The evidence that mTOR is also involved in human cancers includes the observations that mTOR is constitutively phosphorylated in prostate cancer cell lines lacking PTEN or overexpressing Akt and that Akt is overexpressed in several cancer types including gastric, breast, ovarian, pancreatic, and prostate cancers.[343] Another downstream mTOR effector, 4EBP, is overexpressed in lymphomas, cancers of the head and neck, and colon carcinomas. A number of mTOR inhibitors are being developed and

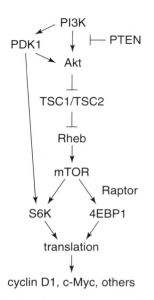

Figure 4–29. Signaling pathways involving mTOR. The diagram depicts the current view of mTOR regulation through the PI3K–Akt pathway based on biochemical and genetic studies. See text for more details. (From Sawyers,[342] reprinted with permission from Elsevier.)

are in clinical trial in an attempt to take advantage of the heightened activity of mTOR-mediated events in human cancer.

Tyrosine Kinase Pathways

Protein tyrosine kinases are a large family of signal transduction kinases that include many cell surface receptors for growth factors such as EGFR, insulin receptor, PDGFR, VEGFR, NGFR, and HGFR. There are also a large number of cytoplasmic tyrosine kinases such as Src, ABL, JAK, FAK, FES, and TEK (reviewed in Reference 344). These kinases activate a number of downstream effectors and mediate a herculean number of developmental, cell proliferation, and cell differentiation pathways. Their activity is usually under stringent regulatory control in cells, and mutations, overexpression, or other perturbations in their activity often cause malignant transformation of cells in culture. In addition, alternations of tyrosine kinase–mediated pathways are often observed in human cancers. Tyrosine kinase pathways network with several other signaling pathways, e.g., PI3K and STAT (see below) pathways, and ul-

timately induce or activate nuclear transcription factors that turn genes on or off (reviewed in Reference 345).

Protein Phosphatases

Although it has been known for a long time that protein phosphatases play a regulatory role in certain cellular metabolic functions, for example, in the activation–inactivation steps for glycogen synthase and phosphorylase, it was only later demonstrated that phosphatases play a role in the activity of various receptors and in the function of certain cell cycle–regulating genes. For example, expression of a truncated, abnormal protein tyrosine phosphatase in BHK cells produces multinucleated cells, possibly by dephosphorylating the cyclin-dependent kinase $p34^{cdc2}$. Activation of $p34^{cdc2}$ requires dephosphorylation of a tyrosine residue, and this activation drives the cell from the G_2 into the M phase. The truncated phosphatase apparently interferes with the normal synchrony between nuclear formation and cell division.

Protein tyrosine phosphatases (PTPases) are a diverse family of enzymes that exist in cell membranes. Some of them are associated with receptors that have tyrosine kinase activity. Phosphatases are also in other intracellular locations. The aberrant phosphorylation state of tyrosine in certain key proteins, such as c-Src or c-Raf, that can lead to cellular transformation could theoretically come about from deregulation of a protein kinase or underexpression of a protein phosphatase. For example as noted above, cells treated with vanidate, a PTPase inhibitor, have increased protein phosphotyrosine levels and a transformed phenotype.[346] Further evidence that PTPases are involved in cancer is the observation that receptor-linked PTPase γ (one of the PTPase isozymes) is located on chromosome 3, which has a deletion in renal cell and lung carcinomas, suggesting that the PTPase γ gene may act as a tumor suppressor gene. Thus, one could predict that a high level of expression of specific PTPases may be able to reverse the malignant phenotype, and one can think of strategies to transfect these genes into tumor cells or deliver inducers of the enzymes to tumor cells.

The protein tyrosine phosphatase PTEN has been found to be mutated in human brain, breast, and pancreatic cancers.[347] This finding was discovered by mapping homozygous deletions on human chromosome 10q23 that occur at high frequency in human cancers. Mutations of the *pten* gene were detected in 17% of primary glioblastomas as well as in human-derived cancer cell lines and xenografts of glioblastoma (31%), prostate cancer (100%), and breast cancer (6%). As discussed earlier, PTEN is a protein tyrosine phosphatase that dephosphorylates PIP3 in the phosphatidyl inositol pathway. Loss of PTEN activity increases PIP3 phosphorylation and leads to cellular transformation. Thus, PTEN is considered to have tumor suppressor function, and this protein and its substrates are potential targets for new therapeutic agents.

An effect of phosphatases opposite that of PTEN has been observed in metastatic human colon cancer. Saha et al.[348] have observed that the PRL-3 protein tyrosine phosphatase gene was overexpressed in each of 18 colon cancer metastases compared to nonmetastatic tumors and normal colorectal epithelium. This somewhat counterintuitive observation reminds us that the dysregulated state of phosphorylation events can have inhibitory or stimulatory effects on the cancer process, depending on the cell type and the microenvironment. Nevertheless, it does suggest that enzymes such as that encoded by PRL-3 can be targets for yet another approach to anticancer drug discovery.

A protein tyrosine phosphatase (PTP) gene superfamily has been identified. Eighty-three somatic mutations of this gene family have been found in human cancers.[349] These mutations are in six PTPs: PTPRF, PTPRG, PTPRT, PTPN3, PTPN13, and PTPN14. One or more of these mutations were observed in 26% of colorectal cancers and a smaller fraction of lung, breast, and gastric cancers. Fifteen of these mutations are nonsense, frameshift, or splice-site variants that lead to loss of phosphatase activity and five are missense mutations leading to reduced phosphatase activity. These data suggest that the mutated PTPs have lost tumor suppressor activity.

JAK-STAT Pathway

The Janus kinase (JAK) –signal transduction and activator of transcription (STAT) pathway is activated by a variety of extracellular signals transmitted via transmembrane receptors. The result of this activation is the targeting of gene promoters in the cell nucleus. This occurs without additional second messengers. The JAK-STAT pathway has been conserved from slime molds to humans.

The STATs are a family inactive cytoplasmic proteins activated by phosphorylation. They contain an Src homology 2 (SH2) phosphorylation-binding domain, a DNA interaction domain, and a number of protein–protein interaction domains for various receptors, transcription factors, and other components of the transcription machinery. Genetics experiments in mice, *Drosophila*, and *C. elegans* have defined a number of crucial functions for the STAT family of proteins.[350]

There are seven STAT genes in mammals: STAT -1, -2, -3, -4, -5A, -5B, and -6. Among these, there is enough diversity in their amino acid sequence and in their tissue-specific distribution to explain many of their diverse functions. The STAT proteins themselves, when activated by phosphorylation, act as transcription factors, which explains their lack of need for second messengers. The STATs in their inactive forms are located in the cytoplasm and are recruited to the SH2-phosphoprotein binding domain upon receptor activation by various ligands such as interferon, interleukin 6, and other cytokines (reviewed in Reference 351). The JAK kinases are required to phosphorylate STATs because STATs don't have their own intrinsic kinase activity. Four JAKs are known in mammalian cells: JAK -1, -2, and -3, and TYK2.

JAKs bind to the intracellular domains of cytokine receptors and catalyze ligand-stimulated autophosphorylation and phosphorylation of STAT docking sites on the receptor. This process leads to phosphoylation and dimerization of STATs and their translocation to the nucleus where the active STATs bind to specific DNA sequences consisting of an 8- to 10–base pair– inverted repeat with a consensus sequence of 5'-TT (N_{4-6}) AA-3'. The specificity of STAT-mediated gene expression is determined by the specific DNA sequence to which it binds. Activated STATs recruit nuclear co-activators (or in some cases co-repressors) to modulate gene transcription.[351]

Other protein kinases such as mitogen-activated kinases (MAPKs) can also activate STATs, indicating as we have seen so often, the existence of cross talk among signal transduction pathways (Fig. 4–30). STAT-activated signaling can be turned off by receptor endocytosis and degradation, or by dephosphorylation by receptor complex-associated or nuclear phosphatases. There is also a negative feedback loop for the JAKs, mediated by suppressor of cytokine signaling (SOCS) proteins.

Different STATs modulate different cellular functions. For example, the type I interferon (IFN α/β) receptor uses STAT1 and STAT2 and the type II interferon (IFN-γ) pathway uses STAT1. STAT3 is the one most studied in relation to cancer because it is used in the regulation of cell proliferation, inflammation, and embryonic development. STAT3 activation is observed in a number of hematologic and solid tumors such as leukemia, lymphoma, melanoma, head and neck cancers, multiple myeloma, and lung, prostate, breast, and ovarian carcinomas. STAT3 activation is also associated with malignant transformation in cell lines and animal models and blocking of STAT3 activity inhibits tumor growth in cultured cells and in vivo in tumor models. This is not surprising, since STAT3 promotes expression of cyclin D1, BCL-X$_L$, c-*myc*, and VEGF genes. STAT3 also down-regulates p53 expression and fosters immune system evasion by tumors (reviewed in Reference 352). Thus, STAT3 activation can foster tumor cell proliferation, increased survival (decreased apoptosis), angiogenesis, and immune system evasion. For all of these reasons, STAT3 and, to some extent STAT5, which also has tumor growth–promoting activities, are attractive ntargets for development of anticancer therapeutics.

They are also attractive because STAT3 and STAT5 are transcription factors and a number of upstream mitogenic stimuli converge on the STAT pathway. Potential inhibitors of STAT activity include antisense and decoy oligonucleotides, dominant-negative expression vectors, and small interfering RNA (siRNA) molecules.[352] However, these are probably long-range alternatives. More likely approaches to the development of inhibitors will focus on small molecules and inhibitory peptides that block

STAT3 phosphotyrosine–SH2 interactions, thus inhibiting STAT dimerization and DNA binding activity. Although the STATs are needed for normal embryonic development and maintenance of some functions in adult organisms, normal cells appear to be less sensitive than cancer cells to inhibitors of STAT3, perhaps because cancer cells have an increased dependence on STAT-mediated signaling.[352]

Estrogen Receptor Pathway

As mentioned earlier, estrogen receptor (ER) is one of the family of nuclear receptors that themselves, after hormone binding, become transcription factors in a manner similar to that of the STAT proteins. The ER status is, of course, one of the important parameters used to stratify breast cancer patients. As is the case for many receptor-mediated signaling events, the complexity of ER interactions has increased in direct proportion to knowledge of its protein–protein interactions. In addition, the discovery of a second estrogen receptor, ERβ, has complicated the issue even further. The ERα receptor is involved in most of the breast cell proliferation and differentiation effects of estrogen. Both ERs are widely distributed in human tissues, but they have distinct functions (reviewed in Reference 353).

ERα is the subtype mostly expressed in the uterus, liver, kidney, and heart. ERβ is the primary subtype expressed in ovary, prostate, lung, gastrointestinal tract, bladder, hemapoietic tissues, and the central nervous system.[353] Both are expressed in a number of tissues, including breast, bone, adrenal gland, thyroid gland, and some regions of the brain. Their transactivating functions generally work through different mechanisms. The estrogen antagonists–partial agonists tamoxifen and raloxifene are partial agonists for ERα but pure antagonists for ERβ. When both receptors are expressed, ERβ inhibits ERα-induced gene expression (see below). For example, estrogen-activated ERβ decreases the ability of estrogen binding to ERα and thus its ability to induce cyclin D1 expression, whose expression is part of the estrogen-induced cell proliferation response seen in breast cancer. Cyclin D1 is overexpressed in about 50% of human breast cancers.

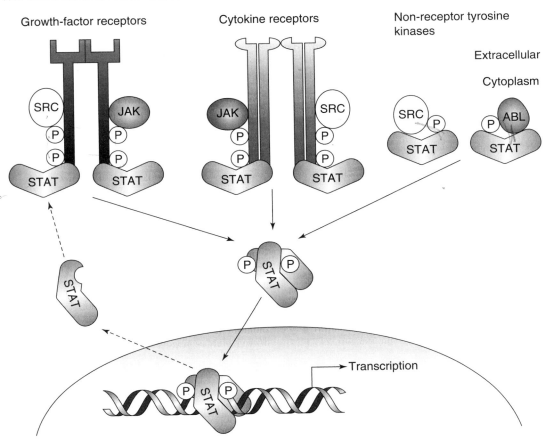

Figure 4–30. Signaling pathways that converge on STATs. STATs are an important point of convergence for many signaling pathways that are commonly activated in cancer cells. Binding of growth factors or cytokines to their receptors results in the activation of intrinsic receptor–tyrosine kinase activity or of receptor-associated kinases, such as the Janus kinase (JAK) or SRC tyrosine kinases. These tyrosine kinases subsequently phosphorylate the cytoplsmic tails of the receptor to provide docking sites for the recruitment of monomeric STATs. Once they have been recruited, STATs themselves become substrates for tyrosine phosphorylation. Non-receptor tyrosine kinases, such as the oncoproteins SRC and BCR-ABL (a fusion of the breakpoint-cluster region [BCR] and Abelson leukemia [ABL] proteins), can phosphorylate STATs independently of receptor engagement. Phosphorylated STATs dimerize and translocate to the nucleus, where the dimers directly regulate gene expression. Whereas STAT activation is tightly regulated in normal cells, the persistent activation of tyrosine kinases in cancer causes constitutive activation of STATs—in particular STAT3 and STAT5. This leads to permanent changes in the expression of genes that control fundamental cellular processes, which are subverted in cancer cells. Dashed arrows indicate the "recycling" of STAT proteins from the nucleus to the cytoplasm. (From Yu and Jove,[352] reprinted with permission from Macmillan Publishers Ltd.)

ERα is sequestered in target cells as part of a heat shock protein complex that keeps it inactive. Upon binding to estrogen, the receptor undergoes a conformational alteration that facilitates release from the heat shock protein complex and leads to formation of an ER dimer–cofactor complex that binds to DNA (reviewed in Reference 354). This interaction with target genes occurs either directly via estrogen response elements (EREs) or via interactions with other DNA-bound transcription factors. Depending on the ERα recruitment of co-activators or co-repressors, target gene transcription can either be up- or down-regulated. ERα has two transactivation domains that play a role in ERα–protein interactions and that also provide some cell-type specific estrogen responses. The number of these protein–protein interactions that have been identified has become immense and those that are key to estrogenic actions are still being worked out, but a number of these ER-associated proteins appear to play a co-activator or co-repressor role. The ERα–co-activator complexes involve recruitment of histone acetylase, which facilitates chromatin decondensation and gene expression (see Chapter 5).

Other regulation of ERα function occurs via ERβ, which can act as a dominant-negative inhibitor of ERα. ERβ levels can determine the type of response to estrogen observed in different tissues. Progesterone, acting through its receptor (PR), also negatively regulates estrogen action in some tissues, e.g., the uterine endometrium. In breast tissue, however, progesterone and estrogen are co-stimulatory. Other negative regulators of ER actions have also been identified (reviewed in Reference 354). ERα is the target for most endocrine therapy for cancer, and its level is used to predict response to estrogen agonists–antagonists such as tamoxifen.

Hypoxia-Inducible Factor

Hypoxia is a common event in human solid tumors and in rodent tumor models. It has been shown to correlate with poor survival in carcinomas of the cervix and head and neck and in soft tissue sarcomas. Hypoxia in tumors is also thought to be at least partly responsible for tumor cell resistance to chemotherapy and radiation therapy.[355] In tumors, this is due to the poorly regulated angiogenesis and the development of irregular blood vessels that have blind ends, arteriovenous shunts, irregular branching patterns, and an incomplete endothelial lining that makes them leaky (reviewed in Reference 356). This tumor neovasculature may also have endothelial cell surface components that vary from normal vasculature, allowing for selective targeting of therapeutic agents via integrin-binding peptide (see Angiogenesis, below).

The irregularities of blood vessel formation in tumors create oxygen gradients, resulting in significant areas within solid tumors becoming hypoxic and having diffusion-limited access to oxygen and other nutrients. The limiting diffusion distance for oxygen is 100–150 μmeters. Tumor hypoxia can either be "chronic," in the sense that certain areas of a tumor (often in the center of cores of tumors) stay beyond the reach of diffusible oxygen, or acute. The latter may occur as a result of the instability of tumor vasculature (forming, closing off, and reforming of new vessels). This acute-type hypoxia can create areas of solid tumors exposed to cycles of hypoxia and reoxygenation in time periods from 20 minutes to 2 hours, as determined in experimental animal models.[356] Acute hypoxia episodes in rodent tumor models have been shown to increase the number of lung metastases twofold, a finding suggesting that hypoxia can enhance the expression of genes that facilitate metastasis. For example, hypoxia induces a coordinated up-regulation of a number of genes involved in glucose transport, glycolysis, erythropoiesis, angiogenesis (VEGF and angiopoietin-related genes), insulin-like growth factor production, extracellular matrix remodeling, and cell cycle regulation.[357,358] The mechanism of turning these genes on is through a transcription factor called hypoxia-inducible factor 1 (HIF-1). The oxygen-regulated components of the HIF-1 transcription factor gene activation complex are the HIF-1α subunits. In tissue areas of normal oxygen tension, HIF-1α is rapidly degraded by binding to the von Hippel-Lindau (VHL) tumor suppressor protein that induces HIF-1α ubiquitination, targeting it for proteosomal destruction (reviewed in Reference 357). Mutations of the VHL gene are associated with renal cell carcinomas and other human cancers. In response to hypoxia, VHL is inactivated and HIF-1α levels are stabilized. HIF-1α activates Akt phosphorylation and stimulates the Akt pathway. Growth factors IGF-1, EGF, and heregulin (Erb B2), and TGF-α also increase HIF-1α levels and may act in concert with hypoxia, although the growth factor response appears to be slower in onset. In addition to VHL, PTEN decreases the response to HIF-1α, as would be expected for activation of an Akt-mediated pathway. MDM2 activation by Akt increases expression of HIF-1α protein synthesis and is the mechanism by which growth factors increase HIF-1α expression. Activation of Ras, Src, or Myc pathways also increases HIF-1α levels.

Tumor Necrosis Factor Receptor Signaling

Tumor necrosis factor (TNF) was described over a century ago but it took until 1982 for it to be purified and sequenced (see Chapter 6). TNF was originally thought to trigger primarily death of tumor cells, hence its name, but more recently it was found to enhance the malignancy of tumors under certain conditions, probably through its ability to enchance inflammation, which may facilitate tumor invasion. In its ability to mediate the inflammatory response and regulate immune function, TNF-α, the main TNF culprit, has been implicated in a wide variety of inflammatory and autoimmune diseases, including rheumatoid arthritis, inflammatory bowel disease, asthma, multiple sclerosis, diabetes, and osteoporosis. TNF-α inhibitors have found broad usefulness in some of these diseases.

TNF-α is a homotrimer of 157 amino acids, primarily produced by activated macrophages, and acts through two receptors, TNF-R1 and TNF-R2. TNF-R1 is the key receptor for initiating a majority of TNF-α's biological activities via binding of TNF-α to the extracellular domain of TNF-R1. This results in (1) release of an inhibitory protein silencer of death domains from the intracellular domains of the receptor, (2) the formation of a receptor–adaptor protein complex, and (3) recruitment of pathway specific enzymes (e.g., caspase-8 and IKKβ), which become activated to initiate a series of downstream events leading to NF-κβ and JNK-activation (Fig. 4–31).

TNF-α-induced activation of NF-κB occurs via phosphorylation-dependent ubiquitination and degradation of the inhibitor IκB (see Chapter 6). It is important to note here again the theme of signaling pathway cross talk, i.e., the cross talk between the NF-κB and c-JUN-NH₂-terminal kinase (JNK) pathways. In the absence of NF-κB activity, TNF-α fosters apoptosis, whereas activation of NF-κB by TNF-α protects against apoptosis.[359] In addition, TNF-α-induced activation of JNK is greater and more prolonged in cells lacking NF-κB, and NF-κB-activated gene products inhibit activation of JNK by TNF-α.

It is now known that the TNF family consists of 19 ligands and 29 receptors, including the recently discovered proliferation-inducing ligand APRIL and the B lymphocyte stimulator Bly5 (see Chapter 6). Thus, information about the TNF signaling pathway has become much more complicated.

Tumor Growth Factor-β Signal Transduction

The original TGF-β activity was discovered over 20 years ago. It is now known that there are over 30 members of the TGF-β family in vertebrates as well as a number of structural homologs in C. elegans and Drosophila (Fig. 4–17). TGF-βs regulate a large array of developmental and homeostatic functions. Mutations of TGF-β family members are involved in a number of human diseases, including cancer. TGF-β's function in cancer, as noted earlier, is double-edged: it can function both as an antiproliferative agent and as a tumor promoter, depending on the state of a tumor's progression.

Binding of TGF-β ligand to its dual receptors, type I and type II, induces them to associate and triggers phosphorylation of the type I receptor and its activation as a kinase (Fig. 4–18). This leads to phosphorylation of intermediate effectors called Smads, of which there are three subtypes: receptor Smads (R-Smads consisting of Smad-1, -2, -3, -5 and -8), co-Smads (Smad-4), and inhibitory Smads (Smad-6 and -7) (reviewed in Reference 360). Phosphorylation of R-Smads stimulates their translocation to the nucleus as heterodimeric complexes with co-Smad-4. In the nucleus, the Smads associate with transcriptional co-activators and co-repressors to positively or negatively regulate gene expression. The third class, the inhibitory Smads (Smad -6 and -7), counteract the R-Smads and antagonize TGF-β signaling.

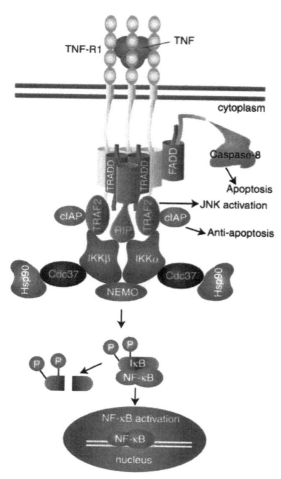

Figure 4–31. Tumor necrosis factor (TNF) signal transduction pathway. Engagement of TNF with its cognate receptor TNF-R1 results in the release of SODD and formation of a receptor–proximal complex containing the important adaptor proteins TRADD, TRAF2, RIP, and FADD. These adaptor proteins in turn recruit additional key pathway-specific enzymes (for example, caspase-8 and IKKβ) to the TNF-R1 complex, where they become activated and initiate downstream events leading to apoptosis, NF-κB activation, and JNK activation. (From Chen and Goeddel,[359] reprinted with permission from the American Association for the Advancement of Science.)

Cross talk of the TGF-β signaling pathway occurs with MAP kinases that can also phosphorylate Smads. In addition, TGF-β signaling can occur via Smad-independent pathways. For example, TGF-β activates the Rho family of GTPases, MAP kinases (ERK, p38, and JNK), and Akt. Mutations of TGF-β receptors and of Smads have been found in human cancers, particularly those of the colon.

Heat Shock Protein-Mediated Events

Heat shock proteins (hsps) were originally discovered in bacteria exposed to high temperatures, hence the name. Subsequent research over the years has shown that such proteins are found in cells ranging from bacteria to humans. Their function is to protect proteins from being denatured and degraded in cells undergoing any of a number of stress-related events, e.g., temperature extremes, or glucose or other nutrient limitation. It is now known that hsps are involved in a number of other cellular functions, including signal transduction mechanisms. For example, signal transduction from the steroid receptors, such as progesterone and glucocorticord receptors, involves a protein complex consisting of the steroid receptor; the heat shock proteins hsp90, hsp70, hsp40, Hop; and the immunophilin FK506-binding proteins 52, 51, and p23.[361] A multistep model for steroid receptor complex assembly and ligand-activated receptor nuclear translocation and gene activation has been determined.[361,362] The initiating event is binding of the steroid receptor to hsp40. Next, binding of hsp70 to the receptor occurs through an ATP-dependent mechanism. This receptor complex then associates with Hop and hsp90 to form the "intermediate complex." This complex is then converted to a "mature complex" containing hsp90, hsp70 and p23. At this stage, the ligand-binding domain of the receptor is folded into a native state that can bind steroid hormone. The activated hormone-bound receptor then disassembles from the hsp90 complex and forms a transcriptionally active, DNA-binding form that turns on gene expression.

Hsp90 is also a target for anticancer drug development. It makes up 1%–2% of total cellular protein under unstressed conditions, and it protects a number of "client" proteins from degradation and fosters their folding into a biologically active native conformation. Hsp90 exists in both ATP- and ADP-binding states. ATP hydrolysis to ADP triggers client protein release. Released proteins, particularly if they are in non-native or unfolded state, are then targeted to proteasomes for degradation.

Some hsp90 client proteins relevant to cancer are the protein kinases Raf-1, Akt, Cdk4, and

IKK; mutated signaling proteins p53, Kit, Flt-3, and B-Raf; chimeric proteins Bcr-Abl and NPM-Alk; transmembrane tyrosine kinases EGFR, Her2/neu, IGFR, and Met; HIF-1α; and oncogene proteins N-Ras and K-Ras and c-Src. These proteins are at some stage of their cellular life cycle dependent on hsp90 for stability.

Inhibitors of hsp90's ability to bind its clients have been discovered. The first effective one was geldanamycin, an ansamycin antibiotic analog that binds in the ATP binding pocket of hsp90 and inhibits its function.[363] Occupancy of the ATP/ADP binding pocket prevents ATP binding and the completion of client protein refolding, lending to proteasome-dependent degradation of proteins that require hsp90 for "conformational maturation." One of the key targets of this blockade is Akt, whose stability and ability to function as a kinase are inhibited by geldanamycin.[363,364]

Unfortunately, geldanamycin has limited clinical utility because of its high liver toxicity. Additional analogs, however, that have reduced liver effects have been synthesized. One of these is 17-allylamino-17-demethoxygeldanamycin (17-AAG), which is now in clinical trial.[365] Preclinical investigation has shown that 17-AAG depletes c-Raf-1, inhibits ERK-1/2 phosphorylation, and depletes N-Ras, K-Ras, and cAkt in human colon cancer cells.[365] The drug alone inhibits the growth of human ovarian carcinoma cells in culture and delays growth of human colon cancer xenografts in mice. A predicted synergy of 17-AAG would be with proteasome inhibitors such as Velcade. Indeed, an increase of 1-log in cell kill has been observed in cell culture systems when the two drugs are used together.

A key question is, why are tumor cells more sensitive than normal cells to 17-AAG? The answer isn't totally clear, but may be related to the fact that tumor cells have a higher amount of mutated proteins with non-native conformations, thus overloading cells with faulty proteins and "jamming" the protein degradation machinery of the cell when 17-AAG is present. In addition, it has been observed that hsp90 is present more as a complex in tumor cells and more as a free form in normal cells. The complex form has a higher binding affinity for geldanamycin and its 17-AAG analog.

ANGIOGENESIS

Development of a functional vasculature is a key event in normal embryonic development as well as in the adult for such things as wound healing, corpus luteum angiogenesis during the female reproductive cycle, and development of the placenta. The process of new blood vessel formation from mesodermal stem cells during embryonic development is called *vasculogenesis* (Fig. 4–32). *Angiogenesis*, by contrast, is the term used to describe development of new blood vessels from pre-existing ones. This is the process that takes place during wound healing, the reproductive cycle, and in tumors. In growing tumors, endothelial cells that will form the rudiments of new blood vessels may proliferate 20 to 2000 times faster than normal tissue endothelium in the adult (reviewed in Reference 367).

Initiation of the angiogenesis response is triggered by several factors. Among these are VEGF family members, basic FGF (bFGF or FGF-2), PDGF, angiopoietins, and factors that facilitate blood vessel formation by modulating extracellular matrix (ECM) production or differentiation of cell types involved in blood vessel formation. These latter factors include TGF-β, $\alpha_v\beta_3$ and $\alpha_v\beta_5$ integrins, ephrins, and plasminogen activators (Table 4–8).

It is of interest, and some therapeutic importance, that the endothelial cells in different tissues display organ-specific antigens on their surface (reviewed in Reference 366). This has been determined by a "biopanning" technique in which peptide libraries are used to screen for binding in vivo to endothelial cell (EC) surface molecules. For example, organ-specific EC surface markers have been observed in the lung, breast, prostate, brain, kidney, pancreas, and a number of other tissues. Of keen interest is that tumor EC cells also have sets of surface markers different from normal ECs. For example, aminopeptidase N was found to be a tumor EC-specific marker. In addition, gene expression arrays have shown that 46 transcripts are specifically elevated in the tumor ECs compared to normal adult endothelium (reviewed in Reference 366). Perhaps it's not surprising, however, that many of these transcripts are also found in developing embryonic vasculature and in

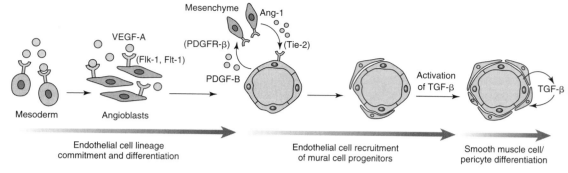

Figure 4–32. Vessel wall assembly. Angioblasts begin to differentiate into endothelial cells and assemble into tubes, most likely in response to VEGF signals
from surrounding tissues. Once endothelial cells form patent tubes, pericytes and
smooth muscle cells are recruited to form the vascular wall. In microvessels,
PDGF signals are involved in the recruitment of pericytes. In large vessels, the
Tie-2 and Ang-1 receptor–ligand pair is involved in the recruitment of smooth
muscle cells. (From Cleaver and Melton,[366] reprinted with permission from
Macmillan Publishers Ltd.)

remodeling of vasculature during wound healing
in adults. This phenomenon, once again, reiterates
the concept of the oncodevelopmental aspects
of malignant transformation (see Chapter 2).

The difference in tumor EC surface markers
can be taken advantage of therapeutically. For
example, Hoffman et al.[369] have shown by phage
display that peptides with the amino acid sequences CGKRK and CDTRL preferentially
bind to tumor neovasculature in skin carcinomas
compared to normal skin, and to some extent to
premalignant dysplastic skin lesions. Such differences in the molecular diversity of tumor
compared to normal ECs can be used to guide
anticancer agents selectively to cancer neovasculature and provide a novel mode of targeted
anticancer therapy.[370]

It has been known for more than 100 years
that solid tumors can become vascularized. It
was not appreciated until the 1950s, however,
that growing tumors elicit new capillary growth
from the host,[371] a process called *tumor angiogenesis*. The mechanism of this angiogenesis was
shown to involve release of some substance(s)
from growing tumors that stimulates outgrowth
of capillaries from the host's vasculature. This
was demonstrated by implanting tumors into the
cheek pouch of hamsters in such a way that the
normal stromal tissue of the host animal was
separated from the tumor tissue by a filter with
very small pores (0.45 μ in diameter) that would

not allow cells to migrate, but would allow large
molecules to diffuse between tumor and host
tissues.[372,373] In these experiments, the growing
tumors elicited the proliferation of new capillaries in the host tissue, indicating the release of
a diffusible substance by the tumor that stimulates capillary growth. This factor was called
tumor angiogenesis factor (TAF).[374] Folkman
and colleagues showed that tumor cells transplanted into the cornea of rabbits initially grew
slowly, but after about a week, small capillaries
began to grow outward from the iris toward the
tumor and when the capillaries reached the
tumor, it began to grow rapidly.[375] Corneal im-
plants of normal adult tissues or of rapidly dividing embryonic tissue did not induce capillary
growth. Injection of tissue extracts into the cornea and application of extracts directly onto the
chorioallantoic membrane of a fertile chicken egg
have been used to demonstrate the presence of
TAF. A wide variety of tumors have been examined for TAF activity, and many tumors have
been found to contain it. The ability to induce angiogenesis, however, is not restricted to
neoplastic cells. Angiogenesis can also be induced
by spleen lymphocytes, thymocytes, peritoneal
macrophages, and testicular grafts from newborn
mice and by leukocyte invasion of the cornea
(reviewed in Reference 376). It is now known that
the induction of capillary growth by tumors is, in
fact, the result of a combination of factors.

Table 4–8. Angiogenesis Activators and Inhibitors

Activators	Function	Inhibitors	Function
VEGF family members[†]	Stimulate angio- and vasculogenesis, permeability, leukocyte, adhesion	VEGFR-1, soluble VEGFR-1, soluble NRP-1	Sink for VEGF, VEGF-B, PlGF
VEGFR[†], NRP-1	Integrate angiogenic and survival signals	Ang2°	Antagonist of Ang1
Ang1 and Tie2[†°]	Stabilize vessels, inhibit permeability	TSP-1, -2	Inhibit endothelial migration, growth, adhesion, and survival
PDGF-BB and receptors	Recruit smooth muscle cells	Angiostatin and related plasminogen kringles	Suppress tumor angiogenesis
TGF-β1,° endoglin, TGF-β receptors	Stimulate extracellular matrix production	Endostatin (collagen XVIII fragment)	Inhibit endothelial survival and migration
FGF, HGF, MCP-1	Stimulate angio- and arteriogenesis	Vasostatin, calreticulin	Inhibit endothelial growth
Integrins αvβ₃, αvβ₅, α₅β₁	Receptors for matrix macromolecules and proteinases	Platelet factor-4	Inhibit binding of bFGF and VEGF
VE-cadherin; PECAM (CD31)	Endothelial junctional molecules	TIMPs, MMP inhibitors, PEX	Suppress pathological angiogenesis
Ephrins	Regulate arterial and venous specification	Meth-1, Meth-2	Inhibitors containing MMP, TSP, and disintegrin domains
Plasminogen activators, MMPs	Remodel matrix, release and activate growth factors	IFN-α, -β, -γ, IP-10, IL-4, IL-12, IL-18	Inhibit endothelial migration, down-regulate bFGF
PAI-1	Stabilize nascent vessels	Prothrombin kringle-2; antithrombin III fragment	Suppress endothelial growth
NOS, COX-2	Stimulate angiogenesis and vasodilation	Prolactin (Mₙ 16K)	Inhibit bFGF/VEGF
AC133	Regulate angioblast differentiation	VEGI	Modulate cell growth
Chemokines°	Pleiotropic role in angiogenesis	Fragment of SPARC	Inhibit endothelial binding and activity of VEGF
Id1/Id3	Determine endothelial plasticity	Osteopontin fragment	Interfere with integrin signaling
		Maspin	Protease inhibitor
		Canstatin, proliferin-related protein, restin	Mechanisms unknown

List of selected examples; further information and references are available at http://steele.mgh.harvard.edu. Abbreviations: bFGF, basic fibroblast growth factor; COX-2, cyclooxygenase-2; HGF, hepatocyte growth factor; Id1/Id3, inhibitors of differentiation 1/3; IP-10, inducible protein-10; MCP-1, monocyte chemoattractant protein 1; MMP, matrix metalloproteinase; NOS, nitric oxide synthase; NRP-1, neuropilin 1; PAI-1, plasminogen activator inhibitor-1; PDGF-BB, platelet-derived growth factor BB; PECAM, platelet/endothelial cell adhesion molecule-1; PEX, proteolytic fragment of MMP2; PlGF Pladienta growth factor; SPARC, inhibits endothelial binding and activity of VEGF; TGF, transforming growth factor; TIMPs, tissue inhibitors of MMP; VEGF, vascular endothelial growth factor; VEGFR, VEGF receptors; VEGI member of TNF family.

°Opposite effect in some contexts.

†Also present in or affecting non-endothelial cells

(From Carmeliet and Jain,[368] reprinted with permission from Macmillan Publishers Ltd.)

As noted above, angiogenesis is also a normal process by which new blood vessels are formed, for example, in development of the placenta, in vascularization of developing organs, and in wound healing. Under these conditions, however, angiogenesis is highly regulated, being turned on for specific periods of time and then shut off. It is an unregulated form of angiogenesis that occurs in tumors and in certain other diseases, such as arthritis, age-related macular degeneration (AMD), diabetic retinopathy (DR), and hemangiomas.

A number of steps are required for angiogenesis to occur: (1) local dissolution of the subendothelial basal lamina of the existing vessels; (2) proliferation of endothelial cells; (3) migration of endothelial cells toward the angiogenic stimulus; and (4) laying down of a basal lamina around the nascent capillary. Different angiogenesis factors modulate different parts of this cascade. For example, FGFs and VEGF are directly mitogenic for endothelial cells; TGF-β stimulates ECM deposition to help form a basal lamina; and angiogenin may help create new "tracks" for vessel formation by ribonucleolytic action.

The first purification of an angiogenesis factor was based on affinity of such factors for heparin and this led to the identification of basic and acidic FGFs as angiogenesis factors. Since then many others have been isolated and characterized, a number of such factors having been shown to be produced and secreted by human tissues. For example, VEGF is produced by human gliomas[377] and epidermoid carcinoma cells.[378] In some cases, angiogenesis factors are found in the urine[379] or effusion fluids[380] of cancer patients and their presence relates to conversion of hyperplasia to neoplasia[381] and to tumor progression.[382] Both tumor cells themselves and the surrounding stroma can produce angiogenic factors. Indeed, there is much evidence to suggest that neovascularization or conversion to the "angiogenic phenotype" is involved in tumor progression.

Most cancers in humans are of epithelial origin and may grow slowly and remain localized (in situ) for many years before they become invasive and metastatic (see Biology of Tumor Metastasis, below). Evidence suggests that part of this change from in situ carcinoma to invasive malignant cancer involves neovascularization of the tumor. There are data indicating that tumors of 1 to 2 mm in diamter can persist in tissue without a tumor-derived vasculature.[383] Epithelial cancers do not develop normal vascular beds like normal tissues and depend to a large extent on diffusion of oxygen and substrates for growth. When tumor cells are too far away from the capillary blood supply for diffusion to provide the needed nutrients the cells may die. This explains why the core of large solid tumors is often necrotic. As long as the tumors remain small, they can obtain sufficient nutrients by

diffusion; as they grow and progress to a more malignant cell type, however, this process becomes limiting. At that point, tumors may be stimulated to release angiogenic factors that induce capillary outgrowth from the host's surrounding normal tissues into the tumor. As noted above, tumor vascular beds are structurally and functionally abnormal. The vascular system in tumors is disorganized, tortuous, and dilated, leading to chaotic blood flow and variable regions of hypoxia.[368] Thus, although full vascularization of the tumor does not occur, it does provide nutrients for their growth. Since this process of angiogenesis is believed to be part of the process involved in converting in situ carcinomas to aggressive malignant tumors, blocking the process could inhibit or significantly slow this conversion. This concept led to a search for antiangiogenic agents, some of which are described below.

Vascular Endothelial Growth Factor

Vascular endothelial growth factor (VEGF) appears to play a critical rate-limiting role in physiological angiogenesis. It is also important in pathological angiogenesis, including that associated with tumor growth and invasion. There are a number of members of the VEGF family, including VEGFs A, B, C, and D, and placental growth factor (PLGF). VEGFA is a key regulator of blood vessel growth and development, whereas VEGFC and D regulate lymphatic angiogenesis (see below).

VEGFA is mitogenic for ECs derived from arteries and veins and acts as a survival factor for them in vitro and in vivo. It does so by activating the PI3K-Akt signal transduction pathway and by inducing the expression of the anti-apoptotic proteins Bcl-2 and A1 (reviewed in Reference 384). VEGF also acts as a vascular permeability factor and its unopposed action causes vessel leakiness, which is part of the pathophysiology of AMD and DR.

The VEGFA gene has eight exons, and alternative splicing produces four different isoforms: VEGF-121, -165, -189, and -206 (containing those numbers of amino acids). VEGF-165 is a heparin-binding form and plays a key role in EC mitogenesis, which is significantly decreased when the heparin-binding domains are deleted. VEGF-121 is a freely diffusible form and VEGF-189 and VEGF-206 are sequestered in the ECM. VEGF-165 is secreted by cells, but a significant amount remains bound to cell surfaces and the ECM. Hypoxia plays a critical role, via HIF-1α induction, in enhancing VEGF gene expression.

Several growth factors and oncogene proteins up-regulate VEGF gene expression. Stimulating growth factors include EGF, TGF-α, TGF-β, keratinocyte growth factor, IGF-1, FGF, and PDGF.[384] Inflammatory cytokines including IL-1α and IL-6 also induce expression of VEGF in synovial fibroblasts and some other cell types. Moreover, the Ras and Myc oncogenic pathways up-regulate VEGF gene expression. In this latter case of oncogene-mediated angiogenesis, the repression of the critical anti-angiogenic factor thrombospondin-1 (Tsp-1) is key.[385] Ras induces the sequential activation of PI3K, Rho, ROCK, and Myc. Myc in turn represses Tsp-1 gene expression. In addition, Ras can activate VEGF expression through activation of the Raf-Mek-Erk-AP1 pathway. These data support the concept that angiogenesis is under tight regulatory control in normal tissues through a baseline expression of angiogenesis inhibitors such as Tsp-1. Loss of this regulatory control is what occurs in cancers. The data also suggest that development of agents that mimic Tsp-1 could provide a new approach to anti-angiogenic therapy for cancer.

VEGFA signals through two related receptor tyrosine kinases, VEGFR-1 and VEGFR-2. A third receptor, VEGFR-3 (Flt-4) binds VEGFC and VEGFD (reviewed in Reference 384). VEGFR-1 (FLT-1) is up-regulated by H1F-1α and binds VEGFA, VEGFB, and PLGF. VEGFR-1 activation induces expression of matrix metalloproteinase-9 (MMP-9) in lung ECs and facilitates lung cancer metastasis. VEGFR-2 (KDR or Flk-1) is the major mediator of the mitogenic and permeability effects of VEGF. VEGFA, by its binding to VEGFR-2, induces EC proliferation via the Raf-Mek-Erk pathway and increases EC survival via the PI3K-Akt pathway.

VEGF mRNA expression is up-regulated in a wide array of human cancers, including, perhaps somewhat surprisingly, hematopoietic malignancies. Antibodies to VEGF and small-molecule VEGFR inhibitors block human tumor xenograft growth in nude mice. As noted above, cancer cells are the major source of VEGF production in tumors, but the tumor stroma also produces VEGF, thus there are at least two targets for anti-angiogenic therapy. A number of clinical trials are under way with anti-VEGF agents (discussed below).

Platelet-Derived Growth Factor

The platelet-derived growth factor (PDGF) family has angiogeneic effects in vitro and in vivo. The four PDGF polypeptides PDGF-A, -B, -C and -D can form homodimers and heterodimers upon ligand binding. Of these, PDGF-BB is one that plays a key role in angiogenesis and is expressed in a number of cell types including ECs and many tumors.[386] PDGF-BB acts via the PDGF-receptor β to enhance pericyte proliferation and migration. PDGF-BB also up-regulates VEGF expression in vascular smooth muscle cells, promoting EC proliferation and survival. Thus, anti-PDGF approaches to therapy may provide a way to do two things: (1) inhibit EC proliferation and survival, and (2) decrease formation and stabilization of an EC-friendly environment provided by pericytes and vascular smooth muscle cells.[386]

Angiopoietins

The angiopoietins (Ang-1 and Ang-2) were discovered as ligands for the Tie family of receptor tyrosine kinases that are selectively expressed in the vascular endothelium (reviewed in Reference 387). Ang-3 and Ang-4 have also been discovered but are less well characterized than Ang-1 and Ang-2.

Studies in gene knockout mice have defined many of the functions of the angiopoietins and their receptors. Mouse embryos lacking Ang-1 or Tie 2 develop a fairly normal vasculature; however, ECs in such embryos fail to associate properly with the underlying stroma, leading to defects in heart vasculature. Thus, Ang-1, acting with Tie 2 receptors, is thought to facilitate EC–stromal interactions. Overexpression of Ang-1 by transgene expression results in hypervascularization in skin, mostly due to increased vessel size.

This is in contrast to VEGF overexpression, which leads to increased vessel number. Combining the two in transgene overexpression experiments leads to profound hypervascularity. Another contrast between Ang-1 and VEGF is that VEGF expression by itself produces leaky vessels, but Ang-1 plus VEGF produces more mature, non-leaky vasculature.[387] Thus, both VEGF and Ang-1 appear to be required in normal angiogenesis.

Ang-2 was found on the basis of its homology to Ang-1 in cloning experiments. But Ang-2 has turned out to be a Tie 2 antagonist and is involved in vasculature remodeling. This concept is supported by experimental data from the remodeling vasculature in the ovary and in Ang-2 gene knockout experiments in mice. It is also supported by Ang-2-mediated vessel remodeling in tumors, where Ang-2 expression correlates with host vessel destabilization that allows tapping into the host's blood supply and facilitating VEGF-mediated endothelial proliferation. Ang-1 and Ang-2 are expressed in tumor cells and play a role in tumor angiogenesis. Ang-3, by contrast, inhibits tumor angiogenesis and blocks pulmonary metastasis in an experimental animal lung carcinoma model.[388]

Ephrins

The ephrin (Eph) family of growth factors is somewhat unique in that they must be tethered to the cell membrane to activate ephrin receptors.[387] The Eph-B2 growth factor binding to an Eph-B4 receptor mediates angiogenic remodeling, reminiscent of the data for Ang-1 and Tie 2. The localized distribution of Eph-B2 (arterial) and the EphB4 receptor (venous) suggest that they are involved in arterial–venous differentiation and junction formation. In tumors, the endothelium of new vessels re-expresses Eph-B2. Thus, the ephrins and their receptors are also potential targets for anti-angiogenic therapy.

Angiogenesis Inhibitors

A large number of potential therapeutic targets that could inhibit tumor angiogenesis have been identified. They can be divided into a number of subcategories: (1) inhibitors of proangiogenic factors (VEGF, Ang-1, bFGF, PDGF) or their receptors; (2) protease inhibitors (MMPs) that block vascular remodeling; (3) inhibitors of ECM production or cell–ECM adhesion needed for vessel stabilization (TGF-β, $\alpha_V\beta_3$ and $\alpha_V\beta_5$ integrins); (4) natural inhibitors (thrombospondin, angiostatin, endostatin); and (5) agents that block HIF-1α production.

Inhibitors of Proangiogenic Factors

The most common proangiogenic factor implicated in cancer growth is VEGF. It is mitogenic for endothelial cells and facilitates their survival. It is also a permeability factor, causing vessels to leak, and it is expressed in a high percentage of human tumors. Anti-VEGF agents inhibit in vivo tumor growth in a number of animal and xenograft tumor models. Inhibitors of VEGF action include antibodies to VEGF or its receptors, RNA aptamers, VEGF-Trap (a decoy receptor based on VEGFR-1 and VEGFR-2 fused to an Fc segment of IgG1), and small-molecule inhibitors of VEGF receptor-mediated signal transduction. Some tumors are more sensitive than others to anti-VEGF agents. For example, the Wilms' renal tumor is very sensitive to anti-VEGF antibody, whereas human neuroblastoma xenografts are only moderately sensitive and metastases are still formed.[389] The reason for this relative resistance is that neuroblastomas more tenaciously hang onto blood vasculature co-opted from surrounding tissues than do Wilms' tumors. Co-option of pre-existing host blood vessels occurs early in tumor development in a number of cancers. Later on, as tumors grow and become hypoxic, tumors express VEGF and other proangiogenic factors and neoangiogenesis is induced. Co-opted vessels then regress. While persistent existence of co-opted vasculature appears to be the resistance mechanism in experimental neuroblastomas, high doses of VEGF-Trap lead to tumor regression, suggesting that this agent also blocks tumor utilization of co-opted vessels.[389]

Inhibitors of other proangiogenic factors such as PDGF, FGF, and EGF are also under development and some of these are in clinical trial.[390] The angiopoietins Ang-1 and Ang-2 have also been shown to regulate tumor angiogenesis. As noted above, Ang-1 activates the receptor tyrosine kinase Tie-2, resulting in activation of the PI3K-Akt pathway and promoting endothelial cell survival.

Ang-2 is the naturally occurring antagonist of this Ang-1 effect. An effect of Ang-2 is to cause vessel destabilization, thus the ratio of Ang-2 levels to Ang-1 may initiate tumor angiogenesis. However, there is evidence that Ang-1 inhibits angiogenesis in human colon cancer xenografts in nude mice.[391] These effectors may have different effects in different cancers.

Metalloproteinases

Remodeling of the ECM by tissue proteases is an initiating event in vascular invasion and angiogenesis. The family of matrix metalloproteinases (MMPs) is key to this remodeling, as evidenced by the fact that mice deficient in MMP2 and MMP9 have reduced angiogenesis and decreased tumor progression in vivo (reviewed in Reference 392). There are also endogenous tissue inhibitors of metalloproteinases (TIMPs) that regulate the action of MMPs and have an anti-angiogenic mechanism. For example, TIMP3 has been shown to inhibit MMP action and to block the binding of VEGF to VEGFR-2, thus blocking VEGF's downstream signaling and angiogenesis in mouse tumor in vivo (reviewed in Reference 392). The MMPs and TIMPs are further discussed below in the section on tumor metastasis.

Integrins

Endothelial cell (EC) adhesion molecules are key to EC–extracellular matrix interactions required for capillary tube formation. The integrins $\alpha_V\beta_3$ and $\alpha_V\beta_5$ are adhesion factors involved in this. As such, they are attractive targets for angiogensis inhibitors. Neoangiogenic blood vessels in many species, including humans, express $\alpha_V\beta_3$, but normal quiescent vasculature does not express significant amounts (reviewed in Reference 393). Expression of both $\alpha_V\beta_3$ and $\alpha_V\beta_5$ is up-regulated in cancer cells. Antagonists to $\alpha_V\beta_3$ are potent angiogenesis inhibitors, and they include monoclonal antibodies, synthetic

peptides, small organic molecules, and antisense RNA to shut off $\alpha_V\beta_3$ expression.[393]

Endogenous Inhibitors

Thrombospondin is an endogenous factor, which when added in soluble form to a culture of ECs inhibits their proliferation. This effect may result from thrombospondin's ability to bind TGF-β and to modulate protease activity (reviewed in Reference 367). Low thrombospondin levels in patients with invasive urinary bladder cancer have been associated with increased recurrence rates, high microvessel density, and decreased overall survival.

Two other members of the endogenously produced anti-angiogenic proteins are angiostatin and endostatin.[394] Angiostatin is an internal polypeptide fragment of plasminogen, and endostatin is a proteolytic fragment of collagen XVIII. These two anti-angiogenic fragments were discovered in Judah Folkman's lab and have shown anti-angiogenic activity in a number of prelinical models. They have also been tested for activity in clinical trials with mixed results (see below). Their mechanism of action isn't totally clear, but endostatin appears to act by binding to α_V- and α_5-integrins on the surface of ECs.[395]

HIF-1α

As noted above in the section Signal Transduction Mechanisms, activation of HIF-1α by hypoxia or other stimulatory factors leads to enhanced expression of a number of genes, including VEGF. Ironically, at least for cancers at an early progressing stage, anti-angiogenic therapy for cancer may actually increase HIF-1α expression, leading to increased expression of a number of HIF-1α-activated genes that foster increased tumor cell proliferation, survival, invasion, and metastasis.[396] Increased metastatic dissemination of human melanoma xenografts has been observed after subcurative radiation treatment, most likely through a radiation-induced increase in hypoxic cells and hypoxia-induced up-regulation of urokinase-type plasminogen activator receptors.[397] This compensatory tumor response to lower blood flow and increased hypoxia may also facilitate the development of drug-resistant cancer cells. Thus, a combination of anti-endothelial agents plus anti-HIF-1α drugs is an attractive therapeutic approach. Anti-HIF-1α agents could prevent a compensatory turn on of genes favoring tumor progression and also prevent hypoxia-driven selection of resistant cells.

Miscellaneous Anti-angiogenic Agents

A number of large and small molecules with anti-angiogenic activity continue to be found, including the previously abandoned drug thalidomide. Other agents include pigment epithelium–derived factor (PEDF), which was first identified in the conditioned medium of cultured human retinal pigment epithelial cells;[398] peptides that selectively recognize tumor vasculature;[399] rapamycin, an immunosuppressive drug that also inhibits VEGF production and EC response to VEGF;[400] inhibitors of cyclooxygenase-1;[401] and my favorite of all, resveratrol, a natural compound found in red wine and grapes.[402]

Clinical Data

Although a number of anti-angiogenic agents have been efficacious in blocking tumor growth in preclinical animal models, clinical trials of these agents have had mixed and mostly disappointing results. Over 40 such agents are in clinical trial, and with the exception of a few glimmers of response, most haven't worked the way that was predicted.[386] There are several reasons why this might be the case.

1. Timing. In preclinical animal models, the anti-angiogenic agent is usually administered when tumors are very small and at the stage when new vasculature hasn't developed or is just developing. The way human clinical trials are carried out, the patients treated with new agents have later-stage, widespread disease. By that time, the neovasculature is developed and differentiated to the state where new vessels are less vulnerable, possibly because the surrounding pericytes and smooth muscle cells form a protective coat.
2. Site of the tumor. Animal studies are usually done by injecting tumor cells sub-cutaneously or in other sites that are foreign to the environment of the tumor cells. Thus, the type and role of the angiogenic factors and the supporting stroma may not provide a good index of how angiogenesis actually occurs in human cancers.
3. Multiple angiogenic factors. As noted in the discussions above, there are multiple angiogenic factors and endogenous angiogenesis blockers (e.g., thrombospondin and endostatin) in tissues. Thus, one set of angiogenic growth factors may be up-regulated in one tumor type but not be key for another type. The same applies for the endogenous blockers. Angiogenesis is, after all, a balance between stimulators and inhibitors in normal tissues as well as in tumors.
4. Tumor heterogeneity. Tumors have a variety of cell types, particularly as they progress and undergo genetic drift. Some sites in a tumor may have adapted to hypoxia more than others and hence be more resistant to anti-angiogenic agents.

One of the brighter spots in anti-angiogenic therapy is the anti-VEGF antibody bevacizumab (Avastin). This agent in combination with carboplatin and taxol produced an increased time-to-progression (TTP) in patients with non–small cell lung cancer and also showed an improved response rate and TTP in combination with 5-fluorouracil and leucovorin in colon cancer patients (reviewed in Reference 386). Bevacizumab also has shown an increased TTP in patients with metastatic renal cancer.[403] The drug didn't work as well in breast cancer trials, perhaps because angiogenesis in colon cancers are more VEGF dependent than in breast cancers.[390] Since angiogenesis inhibitors appear to be mostly cytostatic in the sense that they usually cause tumor growth inhibition rather than killing of established tumors, it makes sense to combine these agents with cytotoxic drugs. This indeed seems to be the most logical approach, as evidenced by the early clinical trials.

Having a surrogate marker to measure the effectiveness of anti-angiogenic therapy would provide a key element in clinical trials. One such marker may be circulating peripheral blood endothelial cells (CECs) or their progeniotr cells (CEPs). Kerbel and colleagues[404] have shown a correlation between bFGF- or VEGF-induced angiogenesis and CEC or CEP levels among eight inbred mouse strains. In addition, they showed that treatment of mice bearing the Lewis lung carcinoma with an antibody to the VEGF receptor VEGFR-2 caused a dose-dependent reduction in circulating endothelial cell precursors that correlated with the antitumor response.

Lymphangiogenesis

It has been known for a long time that carcinomas spread initially through lymphatic channels. Eventually, cancer cells find their way into the bloodstream and bone marrow. There are numerous connections between lymph channels and the vascular system that allow this. Questions that have recently arisen about these connections are: Do tumors also have ways to foster lymphangiogenesis like they do angiogenesis? Could this be involved in tumor spread and metastasis? The answer to both questions is "yes," at least in animal tumor or xenograft models. Both VEGF-C and VEGF-D induce tumor lymphangiogenesis in such model systems.[405–407] This is also likely to be the case for human cancers growing in patients. However, even though VEGF-C levels are elevated in a tumor doesn't necessarily mean that *new* lymphatic vessels are needed to cause metastasis. For example, Podera et al.[408] have shown that in a mouse tumor growing in vivo, VEGF-C increased lymphatic channel surface area in the tumor margin and in lymphatic metastasis but did not produce an increase in the number of functional lymphatic vessels. These data suggest that lymphatics at the tumor margin are sufficient for tumor cells to spread and metastasize. They also suggest that VEGF-C-induced lymphatic surface area growth could be a target for therapy. There are, however, some caveats to that approach: over half of human cancers will already have lymphatic spread by the time of diagnosis and a potential disruption of lymphatic drainage by anti-VEGF-C agents could cause lymphedema in patients.[409]

Another approach is to use antibodies or selected peptides that target tumor lymphatics to treat tumors known to metastasize via lymphatic channels. Rouslahte and colleagues[410] have used such an approach. They treated mice bearing human breast cancer xenografts with a homing peptide that targets tumor lymphatics and showed that the treated tumors had increased apoptosis and reduced numbers of lymphatic vessels.

Tumor Dormancy

A long-range goal of cancer therapy is to prevent the progression, invasiveness, and metastasis of cancer cells. Although it has proven difficult to kill or remove every last cancer cell from the body therapeutically, it is conceivable that with the right combination of drugs a state of dormancy could be induced, which in effect could turn cancer into a chronic, but controlled, disease, like rheumatoid arthritis. Thus, without totally eliminating a cancer, it might be possible to let patients live out a normal life span and have a reasonably good quality of life. Experimentally, this can be done. For example, Folkman, O'Reilly, and colleagues have shown that with the appropriate regimen of the anti-angiogenic agents angiostatin and endostatin administered to mice bearing the Lewis lung carcinoma or to nude mice bearing human tumor xenografts, a state of tumor dormancy could be induced (reviewed in Reference 411). Interestingly, metastases from the Lewis lung carcinoma were also held in the dormant state. Similarly, when the right scheduling of angiogenesis inhibition and cyclophosphamide was employed to treat cyclophosphamide-resistant Lewis lung carcinomas, apoptosis of endothelial cells within the tumors was induced and the drug-resistant tumors eradicated.[412] Somewhat surprising, perhaps, is that the induction of tumor dormancy wasn't due to total eradication of the tumor because if the residual small module left at the tumor site was transplanted into another mouse, it regrew.

Tumor dormancy is also a clinically relevant phenomenan. Breast cancer, for example, may recur in patients 10 to 15 years after apparent eradication of the primary tumor. Such recurrence of tumors could be due to a quiescent tumor stem cell population that gets reactivated, loss of immune surveillance, angiogenesis, or all of the above.[413] Surprisingly, tumors themselves can be sources of anti-angiogenic factors that promote dormancy. For example, in experimental systems, presence of the primary tumor was observed to prevent development of micrometastases and when the primary tumor was removed, micrometastasis quickly developed.[414] Tumor progression clinically has been observed to be associated with a decrease in expression of the endostatin precursor collagen XVIII. Expression of collagen XVIII mRNA was measured in five hepatocellular carcinomas (HCCs) from 57 patients.[415] Tumors expressing the highest levels of collage XVIII were smaller and had lower microvessel density than tumors expressing low levels. Moreover, patients whose cancer recurred within 2 years of primary tumor resection had 2.2-fold lower collagen XVIII mRNA in their tumors than tumors in patients whose cancer recurred. These data suggest that production of endogenous angiogenesis inhibitors by tumors can regulate angiogenesis at primary and metastatic sites. This obviously has profound clinical implications and poses a question about the wisdom of surgical removal of carcinomas in situ that would most likely remain dormant and not progress to invasive, metastatic tumors in a patient's normal lifetime. This question is, of course, a heretical thought and will only be proved or disproved by the use of genomic and proteomic techniques to discriminate among tumors of various invasive and metastatic potential.

BIOLOGY OF TUMOR METASTASIS

The "Classic" Theory of Tumor Metastasis

In humans, the earliest detectable malignant lesions are often referred to as in situ cancers (Fig. 4–33). These are small tumors (usually only a few millimeters in diameter) that are localized in tissues. They are usually detected only if they can be endoscopically or directly visualized, for instance, as in the case of carcinoma in situ of the uterine cervix, urinary bladder, or skin, or by examination of biopsy material, as for ductal carcinoma in situ (DCIS) of the breast. At this stage, the tumor is usually avascular, lacking its own network of blood vessels to supply oxygen

and nutrients. The latter are provided primarily by diffusion, and this limitation results in slow growth of the tumor. As a result, these lesions may remain dormant for several years. The critical events that trigger the conversion of a dormant tumor into a more rapidly growing invasive neoplasm are not well understood, but this conversion is associated with the vascularization of tumors, stimulated by tumor angiogenesis factors (see Angiogenesis, above). The vascularized tumor begins to grow more rapidly. It compresses surrounding tissue, invades through basement membranes, and metastasizes. Metastasis occurs early for some tumors (e.g., melanoma, small cell carcinoma of the lung) and late for others (e.g., some thyroid carcinomas). Metastatic potential is related to the invasiveness of a subpopulation of cells in a given tumor; however, the establishment of a metastatic tumor site requires the expression of additional genes. Historically, it had been thought that metastasis reflects the size of the primary tumor and the duration of tumor progression (number of population doublings); however, it is now clear that there are exceptions to this.

Vascularization appears to contribute to tumor progression and invasion, since the increased supply of nutrients and the resulting increased number of proliferating cells favor the propagation of more aggressively growing cells and the appearance of new subclones of cells with a more malignant phenotype.

The progression of growth of a human solid tumor is shown schematically in Figure 4–34. Because the ordinate (number of cancer cells) is a log scale, the magnitude of the changes that occur after vascularization of the tumor is somewhat deceiving. The dormant phase of growth, which may go on for several years, achieves a diameter of only a few millimeters (10^6 cells, 1 mg mass). Once vascularization occurs, the growth becomes more rapid, and a clinically detectable tumor (10^9 cells, 1 g mass) may be achieved within a few months or years, depending on the cell type. A tumor of this size would be about 1 cm in diameter and just within the realm of detection by sensitive diagnostic methods such as CT and MRI scans, but the newer methods of molecular imaging are allowing detection of smaller tumours (see Chapter 7). In other words, the patient will have a tumor burden of about one billion cells before

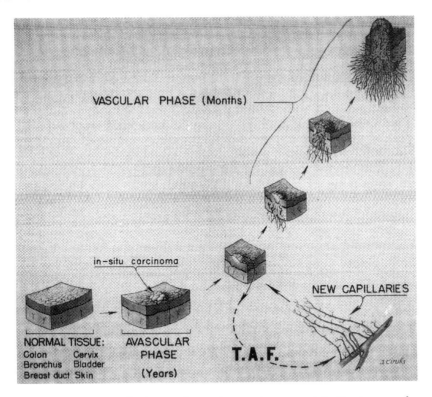

Figure 4–33. Growth phases of a carcinoma. A tumor originating in normal epithelium is separated from the vascular system by the basement membrane separating the epithelium from the underlying connective tissue stroma. Such in situ carcinomas can remain dormant for years. During tumor progression, more aggressively growing cancer cells evolve. These cells may evoke a response from the host that involves invasion of the tumor by immune lymphocytes, macrophages, and polymorphonuclear leukocytes. Vascularization of the tumor can occur by direct invasion through the basement membrane and into small blood vessels and by release of tumor angiogenesis factors by the tumor and/or by lymphocytes and macrophages in the tumor bed. Once the tumor becomes vascularized, it grows more rapidly and can metastasize. (From Folkman,[416] with permission.)

it can be diagnosed clinically. Within another few years, a tumor that is not treated could theoretically approach a tumor burden (10^{12} cells, 1 kg mass). Although unthinkable, a few patients do have huge tumors by the time they come to see a physician. If one thinks of tumor growth as represented in Figure 4–33 in terms of the number of cancer cell doublings, by the time a cancer is detected clinically it will have already gone through approximately two-thirds of its lifetime, with about 30 cell population doublings.[417] If tumor growth is unchecked, five more population doublings would produce a cancer of about 32 g that would be about 4 cm in diameter if it were all in one solid sphere. By five more doublings, a tumor of 1 kg mass (10^{12} cells) would be reached. The growth curve depicted in Figure 4–33 resembles a typical Gompertzian growth curve for cells growing in culture, in that it has a lag phase of relatively slow cell proliferation, a logarithmic growth phase of rapid cell doubling, and a phase of slow growth, eventually reaching a steady state of cell proliferation and cell loss. Similar growth kinetics are seen for some tumors transplanted into experimental animals.[418] For many human cancers, however, the growth kinetics depicted in Figure 4–33 are only an approximation. For this kind of growth

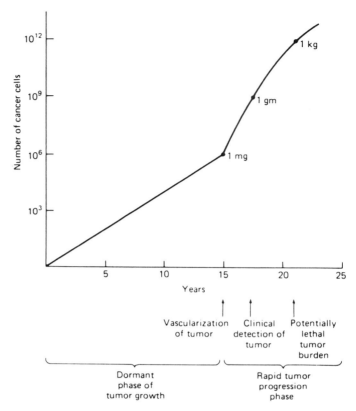

Figure 4–34. Theoretical growth curve of a human carcinoma. The dormant phase of growth may occur over many years, leading eventually to a more aggressive, invasive tumor. This is followed by a more rapid growth phase. Tumors are generally not clinically detectable until they reach a diameter of about 1 cm ($\cong 10^9$ cells). A tumor burden of 10^{12} cells is approaching lethality. (Used with permission.)

kinetics to occur, there would have to be virtually no cell loss as a tumor progresses through its logarithmic phase of growth. This is not true for most human tumors, because a significant amount of cell loss due to cell death and/or exfoliation from the tumor usually occurs during growth. Limitation of nutrients in the central, less well-vascularized areas of tumors and host defense mechanisms also play a role in tumor cell loss. In addition, some human tumors have short mass doubling times (e.g., small cell carcinoma of the lung, embryonal carcinoma of the testis), whereas others have much longer mass doubling times (e.g., carcinomas of the breast and colon). Some tumors are better vascularized than others. Some will become lethal more rapidly because they invade surrounding tissues and metastasize very early in their development.

Some will grow in areas limited by fixed anatomic boundaries and be compressed or contained, and others will grow in critical areas and be lethal before they get very large because of their early compromising of critical functions (e.g., brain tumors). Nor do these kinetics hold for most leukemias or a number of lymphomas. For example, the onset and course of acute lymphocytic leukemia in children is much more rapid, occurring frequently in early childhood and becoming lethal within 6 months to 1 year if untreated. Similarly, the growth kinetics of Burkitt's lymphoma reflect a much more rapid growth. The mass doubling time of Burkitt's lymphoma is 1 to 2 days, as opposed to about 50 days for a typical breast carcinoma.[419] One point is clear, however, which is that most human tumors are relatively far advanced by the time of diagnosis, and about 50% of patients have metastatic spread by the time their cancer is clinically detected.[417]

Alternate Theory of Tumor Metastasis

The classic hypothesis about how tumor metastasis occurs is the tumor progression model championed by Vogelstein and colleagues.[420] Indeed, there is significant evidence supporting this concept, including the following observations: (1) cancer may take many years to develop after initiation (e.g., the 20-pack year observation for lung cancer); (2) biopsies frequently show early lesions that over time, if not totally removed, lead to cancer in later years (e.g., colonic adenomatous polyps preceding colon carcinoma or actinic keratoses foreshadowing carcinomas of the skin); and (3) a progressive enhanced expression of oncogenes and loss of tumor suppressor gene function occurs over a time frame that corresponds to the progression of early precancerious lesions to malignant cancers.

More recent data, however, indicate that even small and apparently noninvasive tumors can contain within them the cells that are going to lead to development of life-threatening metastases. Such cells appear to be present almost from the outset, suggesting that some cancers are pre-ordained to be bad actors from the beginning.

Microarray data now indicate that patients with breast cancer who are lymph node negative and should have a good prognosis can be stratified into distinct groups: one with a good prognosis and one with a poor prognosis, based on the gene expression profile.[421] Similarly, Ramaswamy et al.[422] have compared gene expression profiles of metstases and primary tumors from several types of tumors and identified a specific genetic fingerprint that correlated with metastasis and poor survival. They found that a subset of primary tumors resembled metastatic tumors, based on their gene expression signature, and that these were the primary tumors most likely to metastasize. These data suggest that the ability to metastasize is an inherent quality of the tumor from the get-go. The identity of these metastasis-prone cells and the means of identifying them are not yet totally clear, but a genetic profile has been found for breast cancer stem cells (see Stem Cells, above) that identifies the cells that possess the seeding and proliferation phenotype leading to tumor survival, progression, and metastasis.[135] For example, 100 to 200 cells with the cancer stem cell genotype exist in a population of ten thousand or more cells obtained from human breast cancer tissues. It is these 100 to 200 cells that can form tumors in mouse xenograft models, whereas the bulk of the 10,000 plus cells do not. Schmidt-Kittler et al.[423] have also reported that disseminated cells from early stage breast cancers at stage MO (no metastasis), obtained from bone marrow aspirates, contain fewer chromosomal aberrations than the primary tumors. In addition, the chromosomal aberrations in disseminated cells from M1 (one mestastatic site) stage cancers were different from the primary tumors in these patients. These authors concluded that "human breast cancer cells disseminate much earlier in genomic development than expected from a sequential model of cancer progression." Their data also suggest that since a number of MO-stage patients relapse after complete resection, the "seed cells" of distant mestastasis must have spread before surgery and first diagnosis.

Invasion and Metastasis: The Hallmarks of Malignant Neoplasia

The steps involved in the invasion and metastatic spread of cancer cells are illustrated in Figure 4–35. Tumor cells can spread by direct extension into a body cavity, such as the pleural or peritoneal space, or the cerebrospinal fluid. In these cases, tumor cells released into the body space can seed out onto tissue surfaces and develop new growths where they become embedded. Examples of cancers that spread in this way are lung cancers that enter the pleural cavity, ovarian cancers that shed cells into the peritoneal cavity, and brain tumors that shed cells into the cerebrospinal fluid. Tumor cells metastasize by invading blood vessels or lymphatic channels. Although it has frequently been said that carcinomas metastasize primarily through the lymphatic system and sarcomas through the blood vessels, this distinction is somewhat arbitrary, since the blood and lymph systems communicate freely, and it has been shown that cancer cells that invade lymphatic

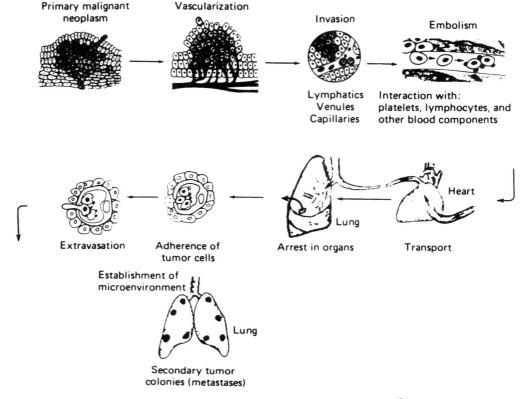

Figure 4–35. Steps involved in cancer metastasis. (From Fidler,[424] with permission.)

channels enter the bloodstream and vice versa.[425] Capillaries, venules, and lymph vessels offer little resistance to penetration by tumor cells because of their thin walls and relatively "loose" intercellular junctions. Arteries and arterioles, by contrast, are surrounded by dense connective tissue sheaths made up of collagen and elastic fibers, and hence are rarely invaded by tumor cells.

The mechanisms for invasion of tumor cells through tissue barriers and into blood and lymphatic vessels are not well understood, but they appear to involve both mechanical and enzymatic processes. As a tumor grows, the pressure exerted on surrounding tissue tends to force tumor cells between intercellular spaces. It is unlikely that this process, in itself, could explain the penetration of cancer cells through tissue barriers such as basement membranes. For this to occur, the release of certain degradative enzymes appears to be necessary. Indeed, as dis-

cussed below, tumors are known to contain and secrete a variety of proteolytic enzymes that may be involved in this step. Some of the lytic enzymes found in high concentration in tumor fluids are listed in Table 4–9. The enzyme activities released by growing tumors destroy surrounding cells and degrade tissue barriers, allowing tumor cells to penetrate.

After tumor cells invade the lymphatic or vascular vessels, they may form a local embolus by interaction with other tumor cells and blood cells and by stimulating fibrin deposition. Individual cells or clumps of cells are then shed from these sites and spread to distant organs by the lymph or blood vessels. Tumor cells that enter the lymphatic system travel to regional lymph nodes in which some tumor cells may be trapped and produce a metastatic growth. However, all the tumor cells are not necessarily trapped or "filtered out" in the first few lymph nodes draining an area of tissue containing a

Table 4–9. Relative Activities of Some Lytic Enzymes in Tumor Fluids*

Fluid Assayed	Relative Activities					
	Dipeptidase	Arginase	Acid Phosphatase	β-glucuronidase	Aryl Sulfatase	Cathespin
Normal mouse plasma	0.2	0.1	6	1	2	5
Normal mouse intraperitoneal fluid	1.3	0.2	1	1	1	12
Peripheral tumor fluid	20	2	21	3	5	15
Central tumor fluid	24	2	43	9	11	63

*Activities are based on rates of hydrolysis of substrates in a standardized assay. Tumor data include interstitial fluid from unicentric transplants of sarcoma 37, MCIM rhabdomyosarcoma, methylcholanthrene-induced sarcoma, the Harding-Passey melanoma, and mouse mammary carcinomas.

(Data from Sylven[426])

tumor. Fisher and Fisher[427] found that some tumor cells are not trapped in the first groups of lymph nodes and can thus reach the circulatory system early in the course of tumor cell dissemination. A technique called *sentinel-node biopsy* has been demonstrated as a way to detect the lymph node(s) where the first metastatic tumor cells are trapped. From these sites, further metastatic spread can occur. This technique has been employed for melanoma and for breast cancer (reviewed in Reference 428). Recently, this technique is also being tested as a way to predict cancer spread in colon and other cancers.

In this technique, a blue dye with or without a radioactive tracer (or tracer alone) such as technetium-[99m]-labeled particles of colloidal human albumin that are injected into or around the tumor. The dye or tracer then follows lymphatic drainage from the tumor site to detect the lymph nodes where cancer cells have seeded out. The advantage of this technique is that it pinpoints the most likely site for early metastatic spread and permits a more thorough pathological evaluation of a smaller, more manageable number of lymph nodes without the need for the morbidity associated with a complete axillary- or other nodal-region dissection. Key questions associated with this technique are the following: (1) Is the technique sensitive enough to find the nodes of first metastatic spread? (2) How does it compare to full axillary-node dissection for breast cancer? (3) Is overall survival of patients who had sentinal-node assay different from that for those who had axillary-node dissection?

Veronesi et al.[428] have data indicating that in breast cancer patients, sentinel-node biopsy is a safe and accurate method of screening axillary-node involvement and overall survival is identical between patients who had this procedure and those who underwent axillary-node dissection in a randomized clinical trial. Two large clinical trials, one sponsored by the National Surgical Adjuvant Breast and Bowel Project (NSABP) and another by the American College of Surgeons Oncology Groups (ACOSOG), are under way to answer definitively the question of whether there is a difference between sentinal-node biopsy and axillary dissection.[429]

The presence of tumor cells in blood does not invariably mean that distant metastases will form. The vast majority of circulating tumor cells shed from solid tumors do not survive in the blood, and only about 0.1% live long enough to form a distant metastasis.[430] During circulation in the vascular system, tumor cells can undergo a variety of interactions, including aggregation with platelets, lymphocytes, and neutrophils, which lead to the formation of emboli that can become lodged in the capillary bed of a distant organ. These clumps of cells adhere to the capillary endothelium and elicit the formation of a fibrin matrix that appears to favor the survival of the cancer cells. A number of years ago, Wood[431] showed that adherence of cancer cells to capillary endothelium and subsequent thrombus formation are involved in metastasis. He injected V₂ carcinoma cells (a type of rabbit carcinoma) stained with the dye Trypan blue into small arteries in the rabbit's ear in which a chamber had been inserted to visualize the capillary circulation. The carcinoma cells quickly stuck to the capillary endothelium, even though the blood was flowing briskly. Within 30 minutes, a thrombus formed around the cancer cells. By 24 hours, the cancer cells began to divide. By 48 hours, invasion of the endothelium was apparent, and by 72 hours a tumor metastasis was established. Subsequent experiments indicated that fibrinolytic agents that lysed tumor cell–containing thrombi reduced metastasis formation.

It has been experimentally demonstrated that both the size of a circulating tumor cell–containing embolus and its "deformability" during passage through capillary beds are related to the formation of metastatic foci in a given tissue;[432,433] however, these factors do not solely determine the site of localization of a metastasis. A substantial amount of evidence indicates that there is a tissue "tropism" for some circulating cancer cells. For example, the distribution of metastasis after an injection of certain kinds of tumor cells reflects a preference for certain tissue sites rather than just the first capillary system encountered by the tumor cells. This suggests that there are specific adhesive interactions between circulating tumor cells and cells of given host tissues.

The adhesion of tumor cells to capillary endothelium in susceptible organs appears to damage the vessel walls and to lead to the accumulation of neutrophils that may penetrate the spaces between endothelial cells and open up a channel through which tumor cells can also penetrate.[434] Moreover, platelets that aggregate at the site of the thrombus release mediators, such as histamine, which promote capillary permeability, allowing the migration of tumor cells through the endothelium. The role of platelets in this process is implied from several lines of evidence.[435,436] Many murine tumors aggregate platelets in vitro and in vivo. Addition of fibroblasts to a tumor cell inoculum enhances platelet aggregation and the number of metastases, whereas induction of thrombocytopenia in the host animal or treatment with aspirin, at doses that decrease platelet aggregation, decrease tumor metastases. Aggregation of platelets and release of their contents can be induced

by a number of factors, including collagen, thrombin, and arachidonic acid. Platelets accumulate in areas of endothelial cell regeneration following trauma, and platelet-released factors have a mitogenic effect on a number of different cell types, including endothelial cells. Elastase and collagenase are released from platelets, thus altering the connective tissue of the vessel wall. Platelet aggregation also produces an increase in serum thrombin, which in turn increases the amount of fibrin deposited on the endothelial wall. This deposition of fibrin stimulates the release of plasminogen activator from neutrophils, macrophages, and other cells to induce fibrinolysis through plasmin, thus generating more proteolytic activity in the area of the tumor thrombus. Once tumor cells migrate through the vascular wall, they quickly establish themselves in the new environment and begin to proliferate. This is fostered by the release of angiogenesis factors from tumor cells or host lymphocytes and macrophages that promotes vascularization of the nidus of tumor cells. In the presence of platelets or platelet-released factors, the mitogenic activity of angiogenesis factors for endothelial cells growing on a collagen substratum is greatly enhanced.[436] Thus, the local aggregation of platelets in the area of a tumor cell–containing thrombus activates a whole cascade of events that can promote the extravasation and new growth of tumor cells at a metastatic site.

Somewhat paradoxically, the presence of immune lymphocytes that recognize tumor cells may enhance the colonization of metastatic sites. For example, the metastatic potential of a mixture of intravenously injected murine melanoma cells and immune lymphocytes is enhanced if the ratio of immune lymphocytes to melanoma cells is about 1000:1, but if the ratio is 5000:1, a reduction in metastases is observed.[437] The lower number of sensitized lymphocytes may favor the formation of tumor thrombi necessary for extravasation; the higher number leads to killing of the majority of tumor cells. In other words, as far as the tumor is concerned, "a little immunity is a good thing." The ability of cancer cells to take advantage of the host's own inflammatory response mechanisms, and at the same time avoid destruction by the host's immunologic defense system, gives the cancer cells a tremendous selective advantage.

Metastasis Is at Least Partly a Selective Process

Clinical observations on the pathogenesis of cancer metastases have shown that there is a tendency for primary tumors arising in a given organ to metastasize to particular distant sites (Table 4–10). Some of this tendency is explained by the nature of the venous and lymphatic drainage of a given tissue and by the presence of natural channels created by fascial planes and nerve sheaths. In addition, as already mentioned, tumors impinging on body cavities, such as the pleural or peritoneal space, can spread by shedding tumor cells directly into the cavity. For example, many carcinomas, such as those of the breast, stomach, colon, and lung, metastasize most frequently to regional lymph nodes, but they also metastasize to certain distant organs more frequently than to others. In general, the lungs and the liver are the most common sites of visceral metastases because of their large bulk and abundant blood supply. However, a tendency of some carcinomas to metastasize to certain other tissues is also seen. Adenocarcinomas of the breast often metastasize to lungs, liver, bones, adrenals, and ovaries. Stomach carcinomas spread to liver, lungs, and bone, and, by direct extension, into the peritoneal cavity. Lung cancers frequently metastasize to brain. Carcinomas of the prostate have a predilection for metastasis to bones of the spine. In this latter case, the fact that venules draining the vertebral column anastomose with those draining the prostate explains how some tumor cells could directly reach the bones of the spine. Such clinical observations led Paget[440] to propose the "seed and soil" hypothesis, which states that certain cancer cells with metastatic potential ("seeds") will grow readily in certain tissues that provide a growth advantage to the metastatic cells ("soil").

Studies in experimental animals have also suggested a propensity for certain types of cancer cells to metastasize to certain organs. Murine thymomas, plasmacytomas, melanomas, fibrosarcomas, and histiocytomas have been

Table 4–10. Most Frequent Sites of Metastasis for Some Human Carcinomas

Site of Primary Tumor	Most Common Sites of Metastasis
Breast	Axillary lymph nodes, opposite breast through lymphatics, lung, pleura, liver, bone, brain, adrenal, spleen, and ovary
Colon	Regional lymph nodes, liver, lung, by direct extension to urinary bladder or stomach
Kidney	Lung, liver, and bone
Lung	Regional lymph nodes, pleura, diaphragm (by direct extension), liver, bone, brain, kidney, adrenal, thyroid, and spleen
Ovary	Peritoneum, regional lymph nodes, lung, and liver
Prostate	Bones of spine and pelvis, regional lymph nodes
Stomach	Regional lymph nodes, liver, lung, and bone
Testis	Regional lymph nodes, lung, and liver
Urinary bladder	By direct extension to rectum, colon, prostate, ureter, vagina, bone; regional lymph nodes; bone; lung; peritoneum; pleura; liver; and brain
Uterine endometrium	Regional lymph nodes, lung, liver, and ovary

Data from Anderson[438] and Rubin[439]

demonstrated to have some organ specificity to the pattern of their metastatic spread (reviewed in Reference 424). Thus, both clinical and experimental animal studies indicate that the metastatic spread of cancer cells is not a random event, but reflects properties of individual host tissues and, possibly, of the circulating malignant cells themselves.

Fidler[441] has developed a model system with which to determine whether tumor cells themselves can choose the site of metastatic spread and whether cells of a given malignant neoplasm have the same or different metastatic potential (Fig. 4–36). He injected a B16 mouse melanoma line (B16-F0) intravenously into syngeneic C57 BL/6 mice and 2 to 3 weeks later removed colonies of melanoma cells growing in the lungs. The melanoma cells were dissected free of lung tissue

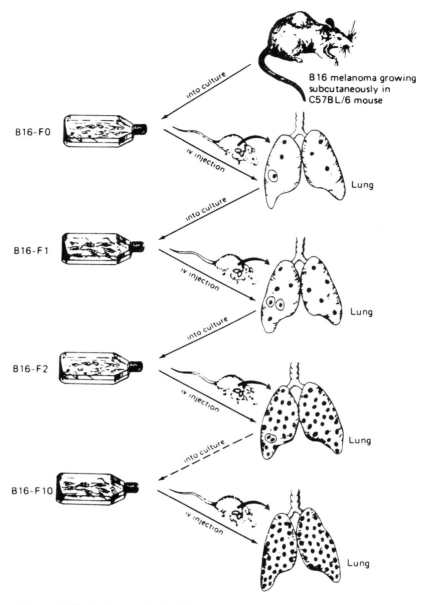

Figure 4–36. Selection of a highly metastatic line of mouse B16 melanoma. C57 BL/6 mice were injected intravenously with the cultured parent line (FO) of B16 melanoma cells. About 2 weeks later, lung colonies were isolated and placed into cell culture dishes (F1 line). When these cells grew to a sufficient number, they were again injected intravenously into mice, and after a few weeks, lung colonies were again isolated and placed in culture. This procedure was carried out several times until a line of highly metastatic, "lung-seeking," B16 cells (F10 line) was isolated. (From Fidler,[442] with permission.)

and established in tissue culture as a line called B16-F1. Cells from this cell culture line were then injected into C57 BL/6 mice, and 3 weeks later the lung colonies produced by these cells were again removed and cultured to yield B16-F2. This procedure was continued until a line of cells, designated B16-F10, that produced significantly more lung tumors per input cell than the

B16-F1 line after either intravenous or intracardiac (left ventricle) injection was obtained.[443] The increased number of lung colonies per input cell for the F10 line was not due to nonspecific trapping in the first capillary bed encountered, since injection into the left ventricle of the same number of cells (radioactively labeled with [125 I]iododeoxyuridine) of the F1 and F10 cell lines resulted in equivalent numbers of cells localized in the lung 2 minutes later, but by 2 weeks postinjection, eightfold more metastatic colonies were found in the lungs of animals injected with F10 cells than in those that received F1 cells. Since cells injected into the left ventricle had to pass through the capillary beds of other tissues before they reached the lung, the ability of F10 cells to seed out and grow in the lung selectively must be due to something other than nonspecific trapping. Moreover, the observation that B16-F10 formed primarily lung colonies after left ventricular injection, whereas the F1 line produced both pulmonary and extrapulmonary colonies, indicated that the invasion and growth of the F10 cells were organ-site specific. Nicolson and colleagues[444,445] have also selected B16 melanoma lines that selectively colonize brain, adrenal, or ovary. Taken together, these data suggest that at least part of the organ-site specificity of cancer metastases is determined by the cancer cells themselves.

In experiments in which the parent B16-F1 line was cloned in cell culture prior to injection into syngeneic mice, Fidler and Kripke[446] determined that individual cells of a given cancer have different metastatic potentials. The original uncloned F1 line produced similar numbers of metastases in different animals, but the cloned sublines differed markedly, with some clones producing a very high number of lung metastases and some very few. This finding indicates that a highly metastatic population of cells pre-exists in the parent melanoma line. Similar marked heterogeneity in metastatic potential has been found in cloned lines of an ultraviolet-induced fibrosarcoma of the C3H mouse, a murine mammary tumor, a methylcholanthrene-induced fibrosarcoma, a murine sarcoma virus–induced fibrosarcoma, and a transformed rat epithelial cell line (reviewed in Reference 447). Thus, the data indicate that primary malignant tumors contain subpopulations of cells with vastly differing metastatic potential, adding another important parameter to the list of phenotypic characteristics for which malignant neoplasms are heterogeneous. Both human and animal neoplasms also display heterogeneity in their growth rate, metabolic characteristics, immunogenicity, and sensitivity to irradiation and cytotoxic drugs. The development of resistance to therapy by human cancers probably reflects this heterogeneity. Moreover, because it is the metastatic cells that ultimately prove fatal to the patient, directing studies toward defining the specific biochemical characteristics of metastatic cells would seem to be one of the more promising ways to design anticancer therapies that will cure patients of their disease.

Biochemical Characteristics of Metastatic Tumor Cells

Metastatic cells share certain biochemical characteristics, among which are the ability to invade through basement membranes, evade the host's immune defenses, attach to endothelial surfaces, extravasate into the tissue parenchyma at a distant site (probably by means of similar mechanisms involved in initial invasion), and elicit development of a vascular network in their new home. Thus, determination of the biochemical parameters associated with the ability to do these things could still provide a common point for therapeutic attack. Some of the candidates for this commonality are discussed below.

Relationship of Cancer Metastasis to Normal Tissue Invasion Events

It is important to keep in mind that malignant cells use some of the same tissue degradative and invasive mechanisms that are used by normal processes, such as cell migration and tissue remodeling in embryonic development, wound healing, and trophoblast invasion of the uterine wall during normal pregnancy. The latter event can be used as an example.

In order for a successful implantation of the blastocyst into the uterine wall to occur in pregnancy, trophoblast cells must cross the basement membranes of the uterine epithelium and vasculature. Several lytic enzymes, proteolytic enzyme regulators, and growth factors are involved in this process (reviewed in Reference 448). For example, the production of protease of the urokinase-type plasminogen activator (u-PA) by murine trophoblasts coincides with invasion of the mouse blastocyst into the uterine wall. Similarly, human trophoblast cells express u-PA receptors and bind u-PA. Metalloproteinases such as stromelysin and the 92 kDa form of type IV collagenase are also produced by trophoblast cells, as are tissue inhibitors of metalloproteases (TIMPS).

Trophoblast adhesion to the uterine wall is key to implantation and placenta formation. This adhesion step is mediated by L-selectin, a molecule that enables circulating leukocytes to bind to blood vessel endothelium.[449] L-selectin is an oligosacchride-binding protein that recognizes specific oligosaccharide structures on epithelial surfaces. The trophoblast cells take advantage of this, either by "capturing" selectin or by up-regulating its expression. At the same time, the hormonal-induced changes in the uterine lining up-regulate expression of L-selectin-binding molecules, fostering adherence of the trophoblast to the uterine wall. The adherence and implantation steps are integrin-dependent processes and mimic the mechanism of leukocyte adherence and extravasation. This process simulates in many respects the adherence and cell transmigration process of cancer cells.

It has been observed that human trophoblast invasion of the uterine wall is stimulated by human placental growth hormone (hPGH), which differs from pituitary growth hormone by 13 amino acids and an N-linked glycosylation site.[450] The extravillous cytotrophoblast cells secrete hPGH and express its receptor, hPGHR. The invasion of the uterine wall by the cytotrophoblast is mediated by a JAK-2 signal transduction mechanism. These results suggest an autocrine–paracrine role of hPGH in trophoblast invasion, and if such a pathway is reactivated in tumors (another potential example of oncodevelopmental reversion), it could provide another target for therapy.

The difference, however, between normal trophoblast invasion of the uterine wall and cancer cell metastasis is that the former is a tightly regulated process with stringent termination signals. For example, production of TIMP at the time of termination of implantation shuts off metalloprotease activity. Expression of TGF-β is activated at this time, and it induces differention of cytotrophoblasts into syncytiotrophoblasts, which are noninvasive. TGF-β also induces expression of TIMP and an inhibitor of u-PA. An intriguing possibility is that there is a feedback loop here, in that production of u-PA by the invading trophoblast could release TGF-β from the extracellular matrix of the uterine epithelium, which in turn activates the production of u-PA inhibitors and TIMP to terminate the invasive process. Interestingly, TGF-β actually appears to stimulate growth of advanced, metastatic melanoma cells instead of inhibiting their proliferation as is seen for normal melanocytes or early melanoma lesions, an action suggesting that metastatic cells inappropriately respond to negative signals.

Role of Lytic Enzymes in the Metastasis Cascade

In order for cancer cells to carry out a successful metastasis, a group of cells within the primary tumor must invade through the host tissue cells and ECM, enter the circulation, arrest at a distant vascular bed, extravasate into the target organ's ECM and interstitium, proliferate as a new colony, and induce a new blood supply (reviewed in Reference 451). A number of these steps require the release of lytic enzymes.

The invasive and metastatic potential of cancer cells has been correlated in a number of studies with the activity of various protease activities, including serine proteases such as plasmin (activated by plasminogen activator), thiol proteases such as the cathepsins, and metalloproteases such as type IV collagenase. These proteolytic activities don't go unabated in tissues, even tumor tissues, because there are a number of tissue protease inhibitors that keep them in check under normal conditions. Proteases, as noted above, are needed for a number of natural processes such as normal tissue repair, tissue remodeling during development, and implantation of the blastocyst and growth of the placenta during normal pregnancy. In these instances, in contrast to highly malignant tumors, the proteases and anti-proteases are kept in a tightly regulated balance, the mechanisms for which aren't entirely clear, but probably involve the local release of growth factors and feedback from the ECM. Thus, normally there is a stringently regulated process that controls the release of proteases and then inactivates them once they have done their job. Tumor cells of the metastatic variety have lost or don't respond to this control mechanism.

It is also clear that individual proteases don't act alone in the metastatic process; they act as part of a cascade of lytic activity. For instance, plasminogen activator activates plasmin, which in turn can activate type IV collagenase. Of the plasminogen activators, the urokinase type (u-PA) has been most closely linked to the metastatic phenotype.[451] Several studies also support an important role for type IV collagenase in tumor metastasis. Moreover, benign proliferative lesions of the breast, benign polyps of the colon, as well as normal colon and gastric mucosa have low levels of a 72 kDa form of type IV collagenase, but their invasive counterparts express high levels of this enzyme. Also, type IV collagenolytic activity can be inhibited by retinoic acid and this correlates with loss of the invasive phenotype in cultured human melanoma cells.[452]

Another important concept for understanding the biology of tumor metastasis is the interaction of cancer cells with the surrounding stroma on which they grow. Cross talk among the cancer cells, the ECM, and the supporting stroma occurs. As an epithelial tumor grows and breeches the ECM, the tumor cells come into contact with the fibroblasts and other mesenchymal cells in the supporting stroma. Via production and secretion of various growth factors and cytokines and interaction among tumor cells, stromal cells, and ECM components, the process of invasion and metastasis goes on. This is also part of the process by which tumors become vascularized. As noted above in the angiogenesis section, tumors secrete angiogenesis factors that induce the growth of vascular endothelial channels through the stroma and ECM to reach the tumor. This appears to be the time in the life cycle of a malignant neoplasm when it undergoes a spurt a growth and becomes more aggressive.

Plasminogen activators (PAs) of either the u-PA or tissue (t-PA) type are neutral serine proteases whose primary proteolytic activity is to convert the zymogen plasminogen into plasmin, which is a "nondiscriminate" protease that degrades a number of ECM components including fibronectin, laminin, and type IV collagen and activates other matrix metalloproteases. Thus the PAs may have a pivotal role in activating a hydrolytic cascade capable of attacking the ECM (reviewed in Reference 453). The u-PA type of PA has been most closely linked to the metastatic phenotype, and antibodies to u-PA can block human hepatoma cell invasion in the chick chorioallantoic membrane assay as well as mouse melanoma B16-F10 metastasis following tail vein injection.[451] High levels of u-PA and its inhibitor PAI-1 have been found in cytosolic extracts of human breast carcinomas and to correlate with poor prognosis.[454]

The cathepsins are a family of cysteine proteases that also appear to be involved in the metastatic process. The cathepsin family includes serine, cysteine, and aspartyl type proteases. There are 11 family members in the human genome (cathepsins B, C, H, F, K, L, O, S, V, W, and X/Z) (reviewed in Reference 455). The cathepsins are primarily intracellular proteases involved in lysosomal protein degradation and protein processing in other organelles such as secretory granules. They are also involved in bone remodeling, epidermal remodeling, and antigen presentation in antigen-presenting cells (see Chapter 6). Cathepsin activity has been shown to be up-regulated in a number of human tumors, and cathepsin activity is increased during experimental tumorigenesis.[455] Cathepsin B activity is elevated in a variety of human and animal tumors and is found at higher levels in metastatic than in nonmetastatic B16 melanoma cells.[456] Cathepsin L is expressed at higher levels in a wide variety of human cancers than in their normal counterpart tissues.[457] An inhibition of tumor formation and metastasis by human melanoma xenografts has been observed by transfecting an anti-cathepsin L single-chain variable fragment into the cells before transplantation.[458]

Cathepsin B mRNA levels are elevated in human colorectal carcinoma tissue in a tumor stage–specific way.[459] The increased cathepsin B gene expression was found at a time when colorectal tumors were in the process of invading the bowel wall. Cathepsin levels in breast cancer tissue have been correlated with poor prognosis,[460] but examination of breast cancer cell lines didn't show a correlation with their in vitro metastatic potential, suggesting that the high tumor tissue levels of cathepsin are due to the stromal components of the tumor such as infiltrating inflammatory cells rather than the cancer cells themselves.[461]

The matrix metalloproteinases (MMPs) comprise a family of over 20 proteins involved in ECM breakdown and remodeling. A number of them have been implicated in invasion and metastasis, and some MMPs are overexpressed in human cancers. MMP family members are products of different genes and each one has some substrate specificity for various components of the ECM. However, they all show some functional and structural characteristics, with a Zn^{2+} metal binding catalytic domain and a pro-domain that keeps the enzyme inactive until cleaved (reviewed in Reference 462).

The activity of MMPs is tightly regulated in vivo. They are generally expressed at low levels in normal tissues, but expression is up-regulated during times of tissue remodeling (e.g., ovulation, menstruation, and trophoblast implantation), wound healing, and during cancer progression. MMP-2, -3, -7, and -9 have all been implicated in tumor progression. MMP-9 has been shown to induce VEGF receptor-1 activation.[463] The MMPs are secreted by most cell types that make them, except in macrophages and neutrophils, where they are packaged in secretory granules. At least one form is membrane bound: MT1-MMP.[464] The secreted forms are released as pro-enzymes that must be cleaved to be active. The tissue-specific inhibitors of MMPs (TIMPs) are endogenous regulators and inhibit MMP activity. Normally, there is a balance between these two activities, but in a cancerous lesion, the up-regulation of MMP activity can predominate.

MMPs can be expressed by tumor cells and by their surrounding stroma. Their expression is induced by a number of oncogenes and growth factors involved in tumor progression. Some MMPs have elevated expression in early stages of tumor progression, but their levels may rise dramatically in invasive and metastatic tumors. Their role in metastasis, though thought originally to be primarily via degrading the ECM and

creating passageways for cancer cells, is now known to include a role in angiogenesis and tumor growth.[462] This quality has made them attractive targets for drug design, and a number of MMP inhibitors have advanced to clinical trial. However, their effectiveness has been disappointing, probably because they would work best only in early-stage cancers. They also have a disturbing side effect of multiple joint tendonitis.

There are a lot of data linking type IV collagenase to tumor progression and metastasis. Type IV collagenase comes in two varieties: a 72 kDa form originally purified from a metastatic mouse tumor, and a 92 kDa form identified originally in neutrophils (reviewed in Reference 465). Both enzymes cleave type IV collagen at a single site, and they also degrade elastin and gelatin. As it turns out, the 72 and 92 kDa type IV collagenases are two of the members of the MMP family that also includes interstitial collagenase (MMP-1), stromelysin-1 (MMP-3), stromelysin-3 (MMP-11), as well as the 72 kDa collagenase (MMP-2) and the 92 kDa collagenase (MMP-9). The 72 kDa and/or the 92 kDa collagenase have been found at elevated levels in a wide variety of human cancers, including melanoma and other skin cancers, and colon, breast, lung, prostate, and bladder cancers. In most of these cases, the elevated levels correlated with a higher tumor grade and invasion. A factor produced by mouse and human cancer cell lines stimulates production of MMP-1, MMP-2, and MMP-3 by human fibroblasts, indicating a way in which human cancer cells could induce these lytic enzymes in stromal tissue with which they come into contact.[466] MMP-2 levels are significantly elevated in the serum of patients with metastatic lung cancer, and in those patients with high levels response to chemotherapy was poor.[467]

Collagenases are not unique to tumor cells, being produced also by inflammatory cells and by normal involuting epithelial duct cells. However, there is evidence to indicate that transplanted animal tumors can produce collagenases in vivo as well as in vitro. Thus, both tumor cells themselves and infiltrating inflammatory cells may contribute to the destruction of the basement membrane in tumor tissue. Collagenase IV activity is consistently higher in malignant tumors than that of corresponding benign tumors, and within a series of cell lines of variable metastatic potential, there is a quantitative relationship between the amount of type IV collagenase activity and the degree of metastasis.[468]

Tissues in the body have mechanisms to limit the amount of basement membrane turnover. During wounding, tissue repair, and various inflammatory conditions, lytic enzymes similar to those produced by invasive tumor cells are released. Various kinds of protease and collagenase inhibitors circulate in the blood and are found in normal tissues. For example, α_2-macroglobulin is a potent, circulating inhibitor of collagenase. The extracellular matrix also contains a family of cationic proteins that inhibit collagenase activities. Treatment with collagenase inhibitors has been shown to retard invasion of human tumors in nude mice[469] and tumor cell invasiveness in an in vitro assay.[470] It is likely, therefore, that in the area of an invasive tumor, the normal balance between collagenases and anticollagenase activity is lost, perhaps because the tumor locally overwhelms the available anticollagenase activity, because invasive tumor cells secrete an anticollagenase inhibitor, or because in the area where tumor cells breach the basement membrane, extracellular matrix components are not laid down in a normal fashion.

Heparanase is another lytic enzyme that has been implicated in tumor metastasis. Heparanase is an endoglycosidase that degrades heparan sulfate, a key polysaccharide component of the extracellular matrix proteoglycans. The heparanase gene is expressed in a wide variety of human cancers and malignant cell lines. Its expression has been shown to play a role in tumor invasiveness, angiogenesis, and metastasis. Silencing of its expression by anti-heparanase gene ribozymes or siRNA gene-silencing vectors decreases tumorigenic properties of cultured cells, suggesting a target for antimetastatic therapy.[471]

As more is learned about the human genome and its protease genes, more protease and antiprotease activities that appear to have a role in cancer are being identified. There appear to be 500 to 600 protease genes in the human genome—the human "degradome." For example, an emerging role of human tissue kallikreins (hKs) is being demonstrated (reviewed in Reference 472). The hKs are a family of 15 homologous single-chain serine endopeptidases of 25 to 30 kDa. They are ubiquitously present in mammalian species and were first identified in pancreas tissue in the 1930s. The term comes from the Greek word for pancreas, "kallikreas." Accumulating experimental results indicate that kallikrein expression is dysregulated in human cancers and is often associated with poor prognosis. Interestingly, a clue to their role is the fact that one of the most well-characterized hKs is hK3, otherwise known as prostate-specific antigen (PSA). A number of hK family members activate the uPA-uPAR-MMP proteoloytic cascade, and this is most likely the mechanism by which they degrade extracellular matrix and other tissue barrier components.

Role of Plasma Membrane Components in Metastasis

A role of plasma membrane components in organ colonization by metastatic cells has been demonstrated by experiments in which plasma membrane vesicles shed from B16-F10 cells were fused with B16-F1 cells. The results were that the B16-F1 cells with fused membranes from F10 cells were converted from a low to a high lung-colony-forming cell type.[473] An indication of the role of cell surface glycoproteins in tissue-specific arrest of cells in vivo was obtained by incubating F16-F1 or B16-F10 cells in culture with tunicamycin or swainsonine, drugs that block formation and maturation of asparagine-linked oligosaccharides, respectively, before injecting them intravenously into mice.[474] The drug-treated cells remained viable but formed significantly fewer lung colonies. Furthermore, the tunicamycin-treated cells did not adhere to endothelial cell monolayers in culture, a finding suggesting that cell surface glycoproteins containing asparagine-linked (N-linked) oligosaccharides are required for the tumor–host cell interactions involved in tissue arrest and metastatic colony formation.

Several lines of evidence indicate a relationship between cell surface sialic acid content and metastatic potential of tumor cells. For 10 cell lines derived from a polyoma virus–induced rat renal sarcoma, the ability of the cell lines to metastasize spontaneously from subcutaneous sties in syngeneic hosts correlated with the degree of sialylation of cell surface glycoconjugates and the platelet-aggregating activity of these glycoconjugates.[475] Moreover, the ability of a wide variety of rodent tumor cell lines, including rat renal sarcoma, rat mammary adenocarcinomas, chemically and virally transformed mouse lines, and B16 melanoma lines, to metastasize from subcutaneous sites correlated with the total sialic acid content and the degree of sialylation of galactose and N-acetylgalactosamine residues of cell surface glycoconjugates.[476] The fact that these latter two sugars are found on O-linked (serine or threonine) oligosaccharides of cell surface glycoproteins and on the O-linked GAG chains of proteoglycans suggests that these molecules are also important for tumor cell attachment and metastasis.

Another example of the importance of cell surface oligosaccharides in tumor cell attachment and invasion is the finding that the expression of aberrant tri- and tetra-antennary β1, 6-N-acetylglucosamine-bearing N-glycans, whose formation is mediated by up-regulation of glucosaminyl transferase V, is increased in invasive but not in noninvasive human gliomas.[477]

Differences in lectin-binding characteristics between low and high metastatic cell lines also suggest an important role for cell surface carbohydrates contained in glycoproteins or glycolipids. In a study of DBA/2 mouse T cell lymphoma sublines with variable metastatic behavior, it was shown that low metastatic tumor lines expressed receptor sites for soybean agglutinin and *Vicia villosa* lectin, whereas in metastatic lines the respective lectin-binding sites were blocked by sialic acid (shown by treating the cells with neuraminidase). This result strongly implies that differences in sialylation patterns on the cell surface, which are involved in the masking or unmasking of terminal sugars, influence metastatic potential of tumor cells. Supporting this concept are the findings of Dennis et al.,[478] who demonstrated that a wheat germ agglutinin (WGA)-resistant mouse tumor cell line (MDW40), derived from a highly metastatic line called MDAY-D2 by exposure to cell-killing concentrations of WGA, loses its metastatic phenotype. Although the cell glycoproteins of both MDW40 and MDAY-D2 have as a major class of their N-linked oligosac-charides Man$_{5-9}$-containing forms, the nonmetastatic mutant cell type contains a unique triantennary class of N-linked oligosaccharides that lacks sialic acid and galactose, indicating the presence of incompletely processed N-linked oligosaccharide chains. Interestingly, when WGA-resistant cells re-acquire the metastatic phenotype, they regain the oligosaccharide composition of the metastatic parent cell type.

Alterations in oligosaccharide composition of another key cell surface molecule, MUC1, also appears to be involved in the metastatic phenotype. As noted earlier in this chapter, MUC1 is a polymorphic, highly glycosylated transmembrane protein expressed on the surface of epithelial cells in pancreas, breast, gastrointestinal tract, and lung. MUC1 is overexpressed and has altered glycosylation patterns in adenocarcinomas that arise in these tissues (reviewed in Reference 479). The altered glycosylation pattern of tumor-associated MUC1 exposes additional sequences of the amino acid tandem repeats of the protein core. This alters the cell–cell and cell–ECM functions of MUC1 and facilitates detachment of tumor cells from their stroma, thus fostering their invasive properties.[479]

Role of Extracellular Matrix Components and the Basement Membranes in Tumor Metastasis

As noted in the section Growth Regulation of Malignant Cells above, the basement membrane of epithelial tissues consists of type IV collagen, fibronectin, laminin, entactin, heparan-sulfate-containing proteoglycans, and, in some tissues, type V collagen. Invading tumor cells encounter basement membranes in a variety of ways. They have to breach this barrier to invade the underlying stroma. To invade adjacent tissues, the tumor cells must also locally disrupt the basement membrane of that tissue. To gain access to the blood vasculature, they must invade through the basement membrane of capillary endothelium. Finally, when an embolus containing tumor cells lodges in a distant capillary bed, the tumor cells have to attach to endothelial basement membranes and invade it once again. Disruption of the basement membrane, on which an in situ carcinoma develops (step 1 in metastasis), could occur through release of degradative enzymes, as discussed previously, or abnormalities in the quantity or quality of basement membrane components laid down by tumor cells, or through a combination of these mechanisms. The latter mechanism could result from decreased sythesis or decreased assembly of extracellular matrix components.

Laminin is one of the key ECM components of the basement membranes and one that tumor cells are capable of producing. Four categories of laminin production by human cancer cells in culture can be defined[480]: (1) "laminin producers and secretors," which produce a considerable amount of complete laminin (a 950 kDa trimer made up of one A and two B subunits)[481] and shed about 25% of that produced into the culture medium; (2) "high laminin secretors," which shed over 60% of the synthesized 950 kDa laminin molecule from their surface; (3) "laminin A subunit–deficient cells," characterized by cancer cells that produce the B but not the A subunits; and (4) "low laminin producers," which produce only trace quantities of the laminin A and B subunits. Seven of 10 human cell lines tested were either unable to biosynthesize one or both laminin subunits or to retain laminin in a cell-associated matrix. This situation contrasts with most nonmalignant human cell types tested in that normal cells produced complete laminin and shed very little into the culture medium, a finding suggesting that normal epithelial cells deposit the laminin they produce into a more stable ECM than their cancerous counterparts do. Moreover, because epithelial cells are a principal biosynthetic source of the basal lamina to which they attach, the aberrant basement membranes frequently observed in human carcinomas may arise at least in part as a result of the impaired ability of malignant cells to synthesize or to deposit basal lamina components.

The 67 kDa high-affinity laminin receptor has been associated with a cancer cell's metastatic capability, in that highly metastatic cells express higher levels of laminin receptors of their surface than do less metastatic or benign tumor cells of the same tissue type. A number of examples can be cited: the number of laminin receptors on breast carcinoma cells correlates with the extent of lymph node metastases and poor prognosis in patients; and the number of 67 kDa laminin receptors correlates with the degree of invasiveness and metastasis of colon carcinoma cells in patients with that disease (reviewed in Reference 482).

A number of studies indicate that metastatic spread of tumors in experimental animal models can be influenced by substances that interfere with tumor cell binding to adhesion molecules in basment membranes. For example, co-injection of tumor cells with high metastatic potential together with antibodies against laminin reduces metastases.[483] Proteolytic fragments of laminin that bind to the 67 kDa receptor in the surface of tumor cells have been shown to inhibit metastasis of melanoma cells (reviewed in Reference 484). Pretreatment of murine melanoma cells with the synthetic peptide Arg-Gly-Asp-Ser (RGDS), which binds the laminin receptor, inhibits their metastasis after injection into syngeneic mice and results in an increased rate of clearance of melanoma cells from the pulmonary microcirculation.[485] However, a 20–amino acid synthetic polypeptide that represents the laminin-binding domain of the 67 kDa receptor[486] or a 19–amino acid polypeptide representing a sequence of the lamina A chain,[487] when injected into mice prior to injection of melanoma cells, increases attachment of the melanoma cells to subendothelial matrix and enhances lung metastasis. In this latter experiment, laminin-like peptides apparently foster tumor cell attachment by, in effect, acting like laminin itself. In the case of the RGDS peptide, the peptide binds to the tumor cell surface and prevents efficient attachment to laminin, whereas the laminin peptide fosters formation of the cell–receptor–laminin complex.

Fibronectin is another important ECM attachment factor for some cell types, such as fibroblasts, and it has been postulated that inability of transformed cells to produce or deposit fibronectin on their cell surface contributes to their invasive and metastatic potential. However, no simple, direct relationship between fibronectin production and metastatic potential can be made (see below).

A comment should be made here on the differences between tumor cell invasion into capillary endothelial basement membrances and that into lymphatic channels. Lymphatic capillaries lack a "tight" basement membrane containing type IV collagen and laminin.[488] Thus a tumor cell that has already invaded into the stromal interstitial space does not need to cross another basement membrane to enter the lymphatic circulation. Infiltration into lymphatic channels, then, is a path of lesser resistance for invading epithelial cancer cells and probably accounts for the fact that lymphatic spread is usually the first type of metastasis observed clinically in patients with carcinomas, the most common kind of human cancer.

At the other end of the circuit, namely, the place where metastatic cells lodge in the capillary beds of other organs and invade, several distinct steps involving the basement membrane have been postulated: (1) adhesion to endothelium, (2) retraction of endothelial cells, (3) migration onto the endothelial basement membrane, (4) breaching of the basement membrane, and (5) locomotion or invasion into the interstitial space of the target tissue.[489] Some investigators favor the idea that tumor cells themselves can attach directly to the endothelial basement membrane by means of attachment factors of their own or receptors for host–tissue attachment factors; indeed, there is evidence to support this idea. Presumably this could occur in open spaces between endothelial cell–cell contact or after distortion of endothelial cell–cell boundaries induced by the tumor cell embolus. Attachment of normal cells to underlying matrix occurs through the cell surface integrins that bind the glycoproteins fibronectin or laminin, and tumor cells also take advantage of these attachment mechanisms. These attachment factors may be synthesized by the tumor cells themselves, or the cells may use factors already present in the matrix.

Fibronectin can mediate attachment of certain kinds of tumor cells to collagen, endothelial-derived matrix, or plastic culture dishes in vitro, and this can be blocked with antifibronectin antibodies. Thus, some investigators have postulated that fibronectin is an important attachment factor for metastatic tumor cells.[490] Other data, however, suggest that fibronectin production or attachment is not crucial to metastasis. For example, when parental mammary adenocarcinoma cells and their metastasis-derived clones were examined for ability to produce or release fibronectin, no difference in these characteristics was found between these different types of tumor cells.[491]

Tissue Adhesion Properties of Metastatic Cells

The organ site–specific localization of metastatic tumor cells has been suggested to result from specific adhesion of metastatic cells to the endothelial cells or endothelial basal lamina of the tissues for which they have a tropism. This specific adhesive quality appears to relate to the composition of cell surface glycoproteins, which form the attachments between cells as well as between cells and the basal lamina.

Organ-specific adhesion of metastatic tumor cells to cryostat sections of specific tissues has been shown. B16-F10 melanoma cells adhere much more to lung than to liver, brain, heart, or testis, and a murine reticulum cell sarcoma (M5076), which metastasizes specifically to liver in vivo, adheres to liver cryostat sections much more than to lung, brain, heart, or testis.[492] This specific adhesion process was inhibited by first treating the tumor cells with neuraminidase plus β-galactosidase or with tunicamycin to block addition of asparagine-linked oligosaccharides to glycoproteins, indicating the importance of glycoproteins for the observed tissue-specific adhesion. Cryostat sections have also been used to select organ-specific metastatic cells: unselected B16 cells with high metastatic potential for lung over other tissues in vivo have been obtained by repeated incubations with mouse lung tissue sections.[493] In contrast, B16 cells selected on cryostat sections of mouse brain did not show a significant change in their metastatic organ site tropism in vivo. This indicates either a difference in the adhesive qualities of the tissues or a difference in the mechanisms by which metastatic cells attach.

Adhesion studies have also been carried out between specific types of endothelial cells and metastatic tumor cells. Teratoma cells with ovary-seeking properties in vivo have been shown to adhere preferentially to mouse ovary endothelial cells over brain endothelial cells, whereas glioma cells adhered preferentially to brain endothelial cells.[494]

Other attachment factors are also important in the metastatic process. Some of them act to inhibit the metastatic potential of tumor cells by favoring *intercellular* adhesiveness and limiting cell detachment. Although cell–cell adhesion is a complex process involving at least four families of adhesion molecules (integrins, immunoglobulins, selectins, and cadherins), a significant amount of data implicate the Ca^{2+}-dependent E-cadherin as a critically important adhesion factor to maintain epithelial integrity (reviewed in References 495 and 496). E-cadherin plays a key role in the normal development of epithelial tissues, and antibodies to it disturb developmental processes in the early embryo. Loss or aberrant expression of E-cadherin has been implicated in the invasive and metastatic potential of tumor cells.[497] Oncogene *ras*-transformed, invasive Madin-Darby canine kidney (MDCK) cells lack E-cadherin expression, but if the E-cadherin cDNA is transfected into these cells, they lose their invasiveness.[498] Similarly, noninvasive clones of *ras*-transformed MDCK cells were rendered invasive by transfection of a plasmid encoding E-cadherin-specific antisense RNA. Moreover, human cancer cell lines from bladder, breast, lung, and pancreas carcinomas were noninvasive by an in vitro assay if they expressed E-cadherin and invasive if they did not.[499] The former could be rendered invasive if treated with monoclonal antibodies to E-cadherin, and the latter could be made noninvasive by transfection with E-cadherin cDNA.

In human bladder cancers, decreased expression of E-cadherin was observed in only 5 of 24 superficial tumors, but in 19 of 25 invasive cancers, a correlation of low expression with increased stage, grade, and poor survival was found.[495] Another cadherin, N-cadherin, has an opposite effect to that of E-cadherin. N-cadherin has been implicated in fostering tumor metastasis (reviewed in Reference 500). It is up-regulated in invasive cancer cell lines from human melanomas. It induces an invasive phenotype in squamous tumor cells and stimulates migration, invasion, and metastasis of breast cancer cells. N-cadherin acts synergistically with bFGF to do these things. In the presence of N-cadherin, bFGF causes activation of the MAPK-ERK signal transduction pathway, leading to enhanced transcription of MMP-9 and cellular invasion by MCF-7 human breast cancer cells.[500] The mechanism for this effect appears to be the protection

of the FGF receptor-1 from ligand-induced internalization into cells.

Other members of the cell–cell adhesion molecule (CAM) family are also be involved in tumor cell metastasis. For example, expression of NCAM-B has been found to down-regulate MMP-1 and MMP-9, indicating that expression of NCAM-B on the cell surface can regulate the turnover of the surrounding extracellular matrix.[501]

Some CAMS expressed in tissues may foster, however, the ability of tumor cells to seed out in vascular beds in a tissue-specific way. For instance, a 90 kDa lung-specific, melanoma cell–binding molecule called Lu-ECAM is expressed on the endothelia of pleural and subpleural capillaries and venules and fosters the attachment of "lung-seeking" melanoma cells.[502] Pretreatment with antibodies to Lu-ECAM inhibited colonization of lungs in mice by lung-seeking B16-F10 cells but had no effect on liver metastatic colonies produced by liver-seeking B16-F10 cells or on lung metastases produced by other lung-metastatic cell lines.

Weinberg and colleagues[502a] have found that the transcription factor Twist, which is a "master regulator" of embryonic morphogenesis first identified in *Drosophila* and later in mammalian cells, plays an important role in metastasis. Blocking Twist expression in highly metastatic mammary carcinoma cells in murine models inhibits their ability to metastasize to lung. Thus, Twist is another oncodevelopmental gene product whose reexpression in cancer cells favors their metastatic spread. Twist reexpression, as well as that of other genes whose expression is normally limited to early development such as Fox C2, induces an epithelial-mesenchymal transition (EMT) that occurs during expression of the metastatic phenotype. EMT induction results in loss of E-cadherin-mediated cell–cell adhesion, activation of mesenchymal markers such as vimentin and fibronectin, and increased cell motility. Ectopic expression of Twist is correlated with invasive lobular carcinoma in patients, and reexpression EMT markers such as Twist and Fox C2 may turn out to be early diagnostic markers for aggressive cancer.

Taken together, these data indicate the importance of cell adhesion to the ECM and of

CAMs in the expression of the metastatic phenotype. Strategies to increase the expression of normal CAMS in tumor tissue might be thought of as ways to modulate this phenotype.

Ability of Metastatic Tumor Cells to Escape the Host's Immune Response

Cell surface antigens representing the major histocompatibility complex (MHC) antigens of the mouse (H-2 genes) and human (HLA genes) play a role in immune surveillance, tumorigenicity, and metastatic potential in both mouse and human cancer (see Chapter 6). Cytolytic lymphocytes recognize cell surface alterations of neoplastic cells associated with MHC antigens. Experiments in mice have demonstrated that metastatic properties of certain mouse tumors are correlated with the expression of class I MHC antigens. Using cloned cell lines of differing metastatic capability derived from the 3-methylcholanthrene-induced fibrocarcoma T10 of mice, a correlation was observed between the in vivo metastatic potential and expression of the H-2D and H-2K antigens.[503] Metastatic clones express only H-2D^b and H-2D^k MHC antigens, but lack H-2K^b and H-2K^k expression. Nonmetastatic clones have H-2D^b on their cell surface but not H-2D^k, suggesting that the D^k antigen contributes to the metastatic potential of these clones. Furthermore, when genes coding for the H-2K region are transfected into the metastatic cloned T10 cells, these cells express K^b and K^k antigens on their surface and lose their metastatic ability in vivo, even though they remain locally tumorigenic. These results stongly imply that the MHC system is involved in immune surveillance that limits the viability of circulating, metastatically potent tumor cells. This contention is supported by the fact that the H-2K gene–transfected cells are more immunogenic and more susceptible to killing by cytolytic lymphocytes than their H-2K-negative counterparts.[503]

Chemotactic Factors in Cancer Cell Migration

Cellular migration occurs normally throughout the life of multicellular organisms. In early embryonic development, migration of neural crest cells, hematopoietic cells, and germ cells occurs, enabling the embryonic progenitor cells to reach their destination for organogenesis. Examples of migratory cells in adult life are the motility of spermatozoa during fertilization, movement of cells during wound healing and tissue repair, and migration of leukocytes and macrophages in the inflammatory process. In these instances, movement of cells is under the influence of several regulatory signals, including cell-to-cell contact, the nature of the extracellular matrix, and chemotactic factors that regulate cellular motility and directionality of cell movement. In cancer cell metastasis, similar kinds of mechanisms come into play, but in an unregulated fashion.

Chemotactic factors for both leukocytes and nonleukocytic cells have been identified (reviewed in Reference 504). Sources of chemotactic factors include native types I, II, and III collagen; collagenolytic breakdown products; lymphocyte-derived chemotactic factors; and complement-derived peptides. Chemotactic factors generated from the fifth complement component (C5) are chemotactic for both leukocytic and nonleukocytic cell types. A C5-derived fibroblast chemotactic factor has a molecular weight of about 80 kDa and is clearly distinguishable from smaller leukocyte and tumor-cell chemotactic factors generated from C5.

In the early 1970s, Hayashi and coworkers described chemotactic responses in several tumor cell lines.[505] They demonstrated that metastatic tumors developed at skin sites injected with chemotactic factors and proposed that this mechanism was similar to that of leukocyte migration and localization at sites of inflammation. Ward and colleagues as well as others (reviewed in Reference 504) observed chemotactic responses for a number of tumor-cell types. Potent tumor-cell chemotactic factors were generated from intact C5 as well as from the C5a fragment of C5, the latter of which is chemotatic for leukocytes. The active component was shown to be generated from C5 or C5a by proteolytic cleavage to a 6000 MW peptide. Since both C5a and lysosomal proteases that can generate tumor-cell chemotactic factor from C5a are present in inflammatory exudates, it was predicted that inflammatory sites would favor the generation of tumor-cell chemotactic factor in vivo, and this was subsequently shown to be correct in animal models. Interestingly, when inflammatory reactions were generated in vivo, or when preformed tumor-cell chemotactic factors were injected intraperitoneally, an increased number of metastases formed in tumor-bearing treated animals compared with tumor-bearing control animals.

C5-derived chemotactic factor has also been observed in human neoplastic effusion fluids (peritoneal and pleural effusions as well as cerebrospinal fluids). Human tumor cells were also found to undergo chemotactic factor–stimulated motility when incubated with effusion fluid from the same patient or with C5-derived factor generated in vitro.[504] Thus, these findings suggest that tumor-cell chemotactic factors can modulate the metastatic potential of tumor cells in vivo and may explain the tendency of tumors to metastasize to areas of inflammation. The role of inflammation in cancer invasion and metastasis is discussed in more detail in Chapter 6.

Chemoattractants called chemokines and their receptors may also play a role in tumor metastasis. Tumor cell migration, invasion, and metastasis share a number of similarities with leukocyte migration and extravasation. Both of these cell types use chemokines and their receptors to carry out these steps. Muller et al.[506] have reported that the chemokine receptors CXCR4 and CCR7 are highly expressed in human breast carcinomas and their metastases. The ligands for these receptors, CXCL12/SDF-1α and CCL21/6Ckine, are highly expressed in organs that are primary sites for breast cancer metastases (thus providing the "soil" for the "seeds"). The selective expression of CXCL12 in lung, liver, lymph nodes, and bone marrow suggests that this chemokine-mediated event plays a role in the metastasis of breast cancer cells to these sites. A similar phenomenom was observed in malignant melanoma. These findings indicate that chemokines and their receptors play a critical role in determining the sites of metastatic spread for various human cancers.

These data also suggest a third alternative for cancer metastasis.[507] The first theory holds that invasive tumor cells leave the organ site of primary tumors and seed out only in tissues with appropriate growth factors and stroma. A second theory holds that endothelial cells that line the blood vessels in certain organs have the right adhesion molecules to bind circulating tumor cells and hold them in these target organs. A third hypothesis is exemplified by the cytokine–chemoattractant findings described above. In this theory, organ-specific chemoattractant molecules are present in the organs to which tumor cells traffick, and the receptors for these chemoattractants on the circulating tumor cells provide the "stop here" signal.

This is another example of how tumor cells take advantage of a normal homing mechanism. For example, chemokines induce leukocytes to migrate and adhere to receptors on endothelial cells of blood vessels. In addition, stem cells use a similar mechanism to home to various organs during embryonic development. Macrophages also play a role in facilitating tumor cell migration and invasiveness. The presence of macrophages in primary tumors is associated with increased metastatic activity. Using a chemotaxis-based in vivo invasion assay, Wyckoff et al.[508] showed that an interaction between macrophages and tumor cells enhances tumor cell movement. This interaction involves a paracrine effect of reciprocal signaling events between tumor cells that have EGF receptors and macrophages that have colony stimulating-1 (CSF-1) receptors. CSF-1 is secreted by tumor cells and can activate macrophages to secrete EGF receptor ligands. This increases the migration of both macrophages and tumor cells, and inhibition of either CSF-1 or EGF signaling inhibits migration of both cell types.

It should be pointed out that the three theories of metastasis are not mutually exclusive; most likely, all three mechanisms play a role.

Role of Oncogenes in Tumor Metastasis

In Chapter 5, the role of oncogene activation in tumorigenesis and tumor progression will be discussed. Suffice it to say here that activation of a number of oncogenes has been associated with the invasive, metastatic phenotype in different tumor types, but whether this is a direct cause of the induction of metastasis or is a reflection of the increased survival potential that goes along with oncogene activation in cells that have a selective advantage to survive in a new tissue environment isn't clear. Another possibility is that activation of cellular oncogenes is involved in producing the genetic instability that leads to metastasis. Such an observation was made by tranfection of the v-H-*ras* gene into rat mammary carcinoma cells induced by DMBA.[509] In this case, the v-H-*ras*-transfected cells became genetically unstable, as demonstrated by the acquisition of additional chromosomal abnormalities, and developed more distant metastases than mammary tumor cells transfected with a control plasmid. Co-expression of v-*fos* in a *src*-transformed rat cell line induced cells with a greater invasive capacity in in vitro assays and a higher metastatic potential in vivo.[510] Transfection of a c-*erbB*-2 gene, activated by mutation, into low-metastatic potential mouse colon carcinoma cells significantly enhanced lung metastasis.[511] In humans, the detection of ErbB2-positive cells in the bone marrow correlated with the incidence of metastasis.[512] In patients with overt metastases, the incidence of metastatic ErbB2-positive cells in the bone marrow was 68% in breast cancer patients and 28% in colorectal cancer patients and correlated with clinical stage of tumor progression. These data suggest that Erb expression is a marker for cells that exhibit the metastatic phenotype and that have a selective advantage for survival during the metastatic process. Overexpression of c-*myc*, c-*erb*, c-K-*ras*, and *hst* oncogenes has been observed in metastatic gastric cancers,[513] and *mdm*2 gene amplification has been seen in metastatic osteosarcomas.[514] This activation or overexpression of cellular oncogenes is a common phenomenon in metastatic cancers. There may also be tissue-associated oncogene-related growth factors that differentially stimulate the growth of tumor cells in specific tissues, as was shown for a lung-derived growth factor that stimulates the proliferation of lung-seeking metastatic cells in a mouse model system.[515]

Identification of the "Metastatic Genes" and "Metastasis Suppressor Genes"

Clearly, it is of utmost importance to know which cellular genes are involved in the expression of the metastatic phenotype and to learn how they are regulated. Theoretically, an approach similar to the one used to isolate transforming genes from human and animal cancer cells could be used here (see Chapter 5). Therefore, if one selected for metastatic clones of cancer cells, created a library of genes from such cells, and tested for metastatic potential of the transfected cells, one should be able to determine if specific genes are associated with the metastatic phenotype. Individual genes could then be sequenced and the amino acid sequence of putative protein products predicted. By then comparing the amino acid sequences with that of known proteins in the computerized databases, one could determine if the "metastasis proteins" are related to known cellular proteins. In addition, gene expression microarrays could be used to compare and contrast metastatic and nonmetastatic tumor cells. Also, synthetic peptides could be made from the putative amino acid sequence, monoclonal antibodies prepared, and the presence of such proteins looked for in metastatic versus nonmetastatic cells.

Kang et al.[516] have identified a number of genes that contribute to a gene expression signature of MDA-MB-231 human breast cancer cells that metastasize to bone. Metastatic subpopulations of cells were selected by in vivo passage in nude mice, expansion in culture, and re-injection into mice. Breast cancer cells with the gene expression signature of metastatic cells were found in the parental cells, indicating that they were expressed in preselected cells. Genes in this group included genes involved in cell homing (CXCR4), pericellular proteolyses and invasion (MMP-1, ADAMTS-1), angiogenesis (FGF5 and CTGF), osteoblastogenesis (IL-11, osteopontin), growth factor regulation (follistatin), and extracellular matrix alteration (proteoglycan-1). A number of these gene products act together to promote bone metastasis, since it took combined expression of at least three of these genes (IL-11 and osteopontin together with CXCR4 or CTGF) to see a metastatic phenotype. Furthermore, TGF-β plays a role in activating some of these genes, e.g., IL-11 and osteopontin. The functions of the CXCR4, CTGF, IL-11, and osteopontin proteins could be fulfilled by different mediators that provide the appropriate seed and soil components for other tumor types, since the homing, invasion, angiogenesis, and ECM remodeling steps seen here are common steps in the metastatic process of most if not all cancers.

Kang et al. suggest that their data bridges the gap between the early-expression and progressive-expression theories of metastasis, in that the parental MDA-MB-231 cell population possesses a "poor prognosis gene expression signature." However, the additional expression of genes that confer the aggressive metastatic phenotype occurs during the selection step(s) and is superimposed on the poor prognosis phenotype already present.

Serial analysis of gene expression (SAGE) has been used by Walter-Yohrling et al.[517] in combination with tumor cell cluster stromal invasion assay to identify metastasis-related gene expression profiles in eight human cancer cell lines, including breast, colon, renal cell, lung, hepatocellular, ovarian, pancreatic, and prostate cancers. Genes commonly expressed in the metastatic subtypes of these cancers included bone marrow stromal antigen 2 (BST2) protein overexpressed in multiple myeloma cells; stathmin-like 3, a microtubule-destablizing phosphoprotein; tumor necrosis factor receptor 5 (TNFR5), which induces MMP9 expression; and hepatocyte growth factor–regulated tyrosine kinase substrate. A number of these expressed genes appear to fulfill parts of the metastatic cascade: cell adhesion, ECM remodeling, and growth factor signaling.

Metastasis *suppressor* genes have been looked for in cell fusion experiments examining the metastatic potential of hybrid cells prepared by fusion of high-metastatic and low-metastatic tumor cells. It was then asked which chromosomes were found in hybrid cells that were nonmetastatic. Ichikawa et al.,[518] for example, fused highly metastatic rat mammary carcinoma cells, transfected with v-H-*ras*, with nonmetastatic parent mammary carcinoma cells. Several hybrid clones of cells were isolated that grew as primary tumors but were nonmestatic. Interestingly, these cells continued to express v-H-*ras*. These data strongly suggest the presence of a metastatic suppressor gene that could overcome the ability of high levels of v-H-*ras* to foster metastasis. With continued serial passage in vivo of a rat prostate cancer, some animals developed distant metastasis that was correlated with deletion of chromosome 2, suggesting that a metastasis suppressor gene had been lost.[519]

It is important to note that metastasis suppressor genes are not the same as tumor suppressor genes (for a description of the latter, see Chapter 5). By definition, *metastasis suppressor genes* inhibit the formation of metastatic foci *without* affecting tumor formation, latency period, or growth rate. Through microcell-mediated chromosomal transfer experiments, metastasis suppressor activity has been found associated with human chromosomes 1, 6, 7, 8, 10, 11, 12, 16, and 17, and a number of metastasis suppressor genes have been identified, including nm23, KA11, KiSS1, BrMS1, and MKK4 (reviewed in Reference 520).

The suppressor gene called nm23 was identified by mRNA subtraction experiments comparing the content of mRNA found in metastatic vs. nonmetastatic murine melanoma cells.[521] The levels of nm23 mRNA were 10-fold lower in melanoma cell lines of high metastatic potential compared to those with low potential. Subsequently, a similar gene has been found in human cells and low levels of its expression have been correlated with metastasis and poor patient prognoses in breast, hepatocellular, and ovarian carcinomas and malignant melanoma (reviewed in Reference 482). However, in human colon tissue, nm23 mRNA levels were increased in colon carcinoma cells compared to normal colonic mucosa, a finding suggesting that nm23 gene expression is controlled differently in different tissues. The nm23 gene codes for a nucleoside diphosphate (NDP) kinase.

NDP kinases are an ubiquitous family of enzymes that catalyze the transfer of the terminal phosphate group of 5′-triphosphate nucleotide donors to diphosphate nucleotide acceptors, e.g., GDP to GTP via ATP. These kinases participate in functions that could affect tumor cell proliferation and metastasis by an action on G protein–coupled signal transduction mechanisms that regulate microtubule assembly, since GTP is required for this function. The NDP kinase coded for by the homologous *awd* gene in *Drosophila* larvae is associated with microtubules.[522] The role that this might have in tumor metastasis is speculative at this point, but because microtubules are important for cell locomotion and for response to external signals mediated by the ECM, loss of regulatory mechanisms mediated by NDP kinases could result in loss of normal matrix–cell interactions. This idea is supported by evidence that *nm23* gene transfection into murine melanoma cells or human breast carcinoma cells inhibits their motility in response to serum, PDGF, or IGF-1.[523] Another interesting observation is that the human purine-binding transcription factor gene PuF has an identical sequence to the *nm23*-H2 gene, a member of the *nm23* gene family.[524] Since PuF encodes a transcription factor that regulates c-*myc* expression, this suggests a direct link between *nm23* and expression of c-*myc*.

The KA11 metastasis suppressor gene was discovered by probing for the suppressor activity associated with chromosome 11. This activity was demonstrated by transfection of KA11 into Dunning rat prostate cancer cells and assaying the metastatic potential of these cells in SCID mice (reviewed in Reference 520). KA11-coded proteins decrease invasiveness and motility of cells in vitro and alter cell–cell interactions. Lowered expression of KA11 has been observed in pancreatic, hepatocellular, bladder, breast, non–small cell lung, and esophageal carcinomas and in lymphomas (reviewed in Reference 520).

Another putative metastasis suppressor gene has been identified in human mammary epithelial cells. The product of this gene, called *maspin*, is related to the serpin family of protease inhibitors (serine protease inhibitors) (reviewed in Reference 525). Maspin is expressed in normal mammary epithelial cells but not in most mammary carcinoma cell lines. Transfection of the *maspin* gene into a human mammary carcinoma cell line did not alter the cells growth properties in vitro, but reduced the ability of the transfected cells to induce tumors and metastasize in nude mice. These cells also had a reduced ability to invade through a basement membrane matrix in vitro. One of Maspin's activities is to inhibit angiogenesis. Maspin expression was also reduced or lost in advanced breast cancer specimens from patients, which suggests that Maspin is a tumor metastasis suppressor in vivo.

Another candidate metastasis suppressor gene is differentiation-related gene-1 (*Drg-1*), which was originally found to be induced in differentiating colon epithelial cells in vitro and to be down-regulated in colorectal cancers (reviewed in Reference 526). Additional genes of this family have been found: *Drg*-2, -3, and -4. The protein encoded by *Drg*-1 has a molecular weight of 43,000 and is phosphorylated by protein kinase A. *Drg*-1 is expressed in most normal organs, but is especially high in prostate, ovary, intestine, and kidney. Its expression is repressed by c-Myc and the N-Myc–Max complex and enhanced by PTEN and p53 in cell culture systems. Expression of other metastasis suppressor genes such as *maspin* and KA11 is also increased by p53, suggesting that the tumor suppressor function of p53 and PTEN could be part of their cell protection–linked function.[525]

In a study of human prostate cancer, *Drg*-1 expression was lower in patients with a high Gleason score and significantly reduced in patients with lymph node or bone metastasis compared with those patients with localized prostate cancer, findings suggesting a tumor suppressor function for this gene.[526] These investigators' work with a SCID mouse–human prostate cancer xenograft model supported this concept, in that *Drg*-1 expression almost completely inhibited lung metastasis but not growth of the primary tumor transplant.

As the human genome is further explored by studies of "functional genomics," it is likely that additional metastasis suppressor genes will be found. Their expression may be identified by gene expression arrays and proteomic analysis (see Chapter 7).

References

1. A. Balmain: Cancer genetics: from Boveri and Mendel to microarrays. *Nat Rev Cancer* 1:77, 2001.
2. O. T. Avery, C. M. McLeod, and M. McCarty: Studies on the chemical nature of the substance inducing transformation of pneumococcal types. Induction of transformation by a desoxyribonucleic acid fraction isolated from *Pneumococcus* type III. *J Exp Med* 79:137, 1944.
3. J. D. Watson and F. H. C. Crick: Molecular structure of nucleic acids. A structure for deoxyribose nucleic acid. *Nature* 171:964, 1953.

4. W. R. Earle: Production of malignancy in vitro. IV. The mouse fibroblast cultures and changes in the living cells. *J Natl Cancer Inst* 4:165, 1943.
5. D. Stehelin, R. V. Guntaka, H. E. Varmus, and J. M. Bishop: Purification of DNA complementary to nucleotide sequences required for neoplastic transformation of fibroblasts by avian sarcoma viruses. *J Mol Biol* 101:349, 1976.
6. W. K. Cavenee, T. P. Dryja, R. A. Phillips, et al.: Expression of recessive alleles by chromosomal mechanisms in retinoblastoma. *Nature* 305:779, 1983.
7. A. G. Knudson: Two genetic hits (more or less) to cancer. *Nat Rev Cancer* 1:157, 2001.
8. H. T. Lynch, R. E. Harris, H. A. Guirgis, et al.: Familial association of breast/ovarian carcinomas. *Cancer Res* 41:1543, 1978.
9. P. C. Nowell and D. A. Hungerford: A minute chromosome in human chronic granulocytic leukemia. *Science* 132:1497, 1960.
10. J. D. Rowley. A new consistent chromosomal abnormality in chronic myelogenous leukemia identified by quinacrine fluorescence and Giemsa staining. *Nature* 243:290, 1973.
11. L. Hartwell and T. Weinert: Checkpoints: Controls that ensure the order of cell cycle events. *Science* 246:629, 1989.
12. M. O. Hengartner and H. R. Horvitz: Programmed cell death in *C. elegans*. *Curr Opin Genet Dev* 4:581, 1994.
13. C. Norburg and P. Nurse: Animal cell cycles and their control. *Annu Rev Biochem* 61:441, 1992.
14. E. W. Sutherland. Studies on the mechanism of hormone action. *Science* 177:401, 1972.
15. O. Warburg: *The Metabolism of Tumors*. London: Arnold Constable, 1930.
16. H. Lu, R. A. Forbes, and A. Verma: Hypoxia-induced factor 1 activation by aerobic glycolysis implicates the Warburg effect in carcinogenesis. *J Biol Chem* 277:23111, 2002.
17. R. A. Gatenby and R. J. Gillies: Why do cancers have high aerobic glycolysis? *Nat Rev Cancer* 4:891, 2004.
18. J. L. Griffin and J. P.l Shockcor: Metabolic profiles of cancer cells. *Nat Rev Cancer* 4:551, 2004.
19. V. W. S. Liu, H. H. Shi, A. N. Y. Cheung, et al.: High incidence of somatic mitochondrial DNA mutations in human ovarian carcinomas. *Cancer Res* 61:5998, 2001.
20. J. P. Greenstein: *Biochemistry of Cancer*. New York: Academic Press, 1954.
21. E. C. Miller and J. A. Miller: The presence and significance of bound aminoazo dyes in the livers of rats fed *p*-dimethylaminoazobenzene. *Cancer Res* 7:468, 1947.
22. V. R. Potter: The biochemical approach to the cancer problem. *Fed Proc* 17:691, 1958.
23. V. R. Potter: Biochemical perspectives in cancer research. *Cancer Res* 24:1085, 1964.
24. H. P. Morris: Studies in the development, biochemistry, and biology of experimental hepatomas. *Adv Cancer Res* 9:227, 1965.
25. S. Weinhouse: Glycolysis, respiration, and anomalous gene expression in experimental hepatomas: G. H. A. Clowes Memorial Lecture. *Cancer Res* 32:2007, 1972.
26. G. Weber: Enzymology of cancer cells (part one). *N Engl J Med* 296:486, 1977.
27. W. E. Knox: *Enzyme Patterns in Fetal, Adult, and Neoplastic Rat Tissues*, 2nd ed. Basel: S. Karger, 1976.
28. R. D. Klausner: The fabric of cancer cell biology—Weaving together the strands. *Cancer Cell* 1:3, 2002.
29. J. A. Thomson, J. Itskovitz-Eldor, S. S. Shapiro, et al.: Embryonic stem cell lines derived from human blastocysts. *Science* 282:1145, 1998.
30. W. C. Hahn, C. M. Counter, A. S. Lundberg, et al.: Creation of human tumor cells with defined genetic elements. *Nature* 400:464, 1999.
31. M. Abercrombie and J. E. M. Heaysman: Social behavior of cells in tissue culture II. Monolayering of fibroblasts. *Exp Cell Res* 6:293, 1954.
32. R. Dulbecco: Topoinhibition and serum requirement of transformed and untransformed cells. *Nature* 227:802, 1970.
33. R. W. Ruddon: In Chapter 4. *Cancer Biology*, 3rd ed. New York: Oxford University Press, 1995.
34. T. G. Graeber and D. Eisenberg: Bioinformatic identification of potential autocrine signaling loops in cancers from gene expression profiles. *Nature Genetics* 29:295, 2001.
35. F. K. Sanders and B. O. Burford: Ascites tumours from BHK 21 cells transformed in vitro by polyoma virus. *Nature* 201:786, 1964.
36. M. C. Alley, C. M. Pacula-Cox, M. L. Hursey, et al.: Morphometric and colorimetric analyses of human tumor cell line growth and drug sensitivity in soft agar culture. *Cancer Res* 51:1247, 1991.
37. S.-I. Hakomori: Biochemical basis of tumor-associated carbohydrate antigens. Current trends, future perspectives, and clinical applications. *Immunol Allergy Clin North Am* 10:781, 1990.
38. H. C. Wu, E. Meezan, P. H. Black, et al.: Comparative studies on the carbohydrate-containing membrane components of normal and virus-transformed mouse fibroblasts. I. Glucosamine-labeling patterns in 3T3, spontaneously transformed 3T3, and SV-40-transformed 3T3 cells. *Biochemistry* 8:2509, 1969.
39. L. Warren, J. P. Fuhrer, and C. A. Buck: Surface glycoproteins of normal and transformed cells: a difference determined by sialic acid and

a growth-dependent sialyl transferase. *Proc Natl Acad Sci USA* 69:1838, 1972.

40. S. Ogata, T. Muramatsu, and A. Kobata: New structural characteristic of the large glycopeptides from transformed cells. *Nature* 259:580, 1976.

41. S.-I. Hakomori: Aberrant glycosylation in cancer cell membranes as focused on glycolipids: Overview and perspectives. *Cancer Res* 45:2405, 1985.

42. S.-I. Hakomori: The glycosynapse. *Proc Natl Acad Sci USA* 99:225, 2002.

43. M. M. Fuster and J. D. Esko: The sweet and sour of cancer: Glycans as novel therapeutic targets. *Nature* 5:526, 2005.

44. K. Yamashita, Y. Tachibana, T. Ohkura, and A. Kobata: Enzymatic basis for the structural changes of asparagine-linked sugar chains of membrane glycoproteins of baby hamster kidney cells induced by polyoma transformation. *J Biol Chem* 260:3963, 1985.

45. T. Mizuochi, R. Nishimura, C. Derappe, et al.: Structures of the asparagine-linked sugar chains of human chorionic gonadotropin produced in choriocarcinoma: Appearance of triantennary sugar chains and unique biantennary sugar chains. *J Biol Chem* 258:14126, 1983.

46. S. Takamatsu, S. Oguri, M. T. Minowa, et al.: Unusually high expression of N-acetylglucosaminyltranferase-IVa in human choriocarcinoma cell lines: A possible enzymatic basis of the formation of abnormal biantennary sugar chain. *Cancer Res* 59:3949, 1999.

47. E. H. Holmes and S-I. Hakomori: Enzymatic basis for changes in fucoganglioside during chemical carcinogenesis: induction of specific α-fucosyltransferase and status of an α-galactosyltransferase in precancerous rat liver and hepatoma. *J Biol Chem* 258:3706, 1983.

48. R. Salvini, A. Bardoni, M. Valli, and M. Trinchera: β1,3-galactosyltransferase β3Gal-T5 acts on the GlcNAcβ1→3Galβ1→4GlcNAcβ1→R sugar chains of carcinoembryonic antigen and other N-linked glycoproteins and is down-regulated in colon adenocarcinomas. *J Biol Chem* 276:3564, 2001.

49. J. G. M. Bolscher, D. C. C. Schaller, H. von Rooy, et al.: Modification of cell surface carbohydrates and invasive behavior by an alkyl lysophospholipid. *Cancer Res* 48:977, 1988.

50. I. Kijima-Suda, Y. Miyamoto, S. Toyoshima, et al.: Inhibition of experimental pulmonary metastasis of mouse colon adenocarcinoma 26 sublines by a sialic-nucleoside conjugate having sialyltransferase inhibiting activity. *Cancer Res* 46:858, 1986.

51. H. E. Wagner, P. Thomas, B. C. Wolf, et al.: Inhibition of sialic acid incorporation prevents hepatic metastases. *Arch Surg* 125:351, 1990.

52. J. W. Dennis, K. Koch, S. Yousefi, and I. VanderElst. Growth inhibition of human melanoma tumor xenografts in athymic nude mice by swainsonine. *Cancer Res* 50:1867, 1990.

53. G. K. Ostrander, N. K. Scribner, and L. R. Rohrschneider. Inhibition of v-*fms*-induced tumor growth in nude mice by castanospermine. *Cancer Res* 48:1091, 1988.

54. S. J. Gendler, A. P. Spicer, E-N. Lalani, et al.: Structure and biology of a carcinoma-associated mucin, MUC. *Am Rev Respir Dis* 144:542, 1991.

55. S. B. Ho, G. A. Nichens, C. Lyftogt, et al.: Heterogeneity of mucin gene expression in normal and neoplastic tissues. *Cancer Res* 53:641, 1993.

56. K. R. Jerome, D. L. Barnd, K. M. Bendt, et al.: Cytotoxic T-lymphocytes derived from patients with breast adenocarcinoma recognize an epitope present on the protein core of a mucin molecule preferentially expressed by malignant cells. *Cancer Res* 51:2908, 1991.

57. B. Agrawal, S. J. Gendler, and B. M. Longenecker: The biological role of mucins in cellular interactions and immune regulation: prospects for cancer immunotherapy. *Mol Med Today* 9:397, 1998.

58. M. A. Hollingsworth and B. J. Swanson: Mucins in cancer: Protection and control of the cell surface. *Nat Rev Cancer* 4:45, 2004.

59. T. E. Hardingham and A. J. Sosang: Proteoglycans: Many forms and many functions. *FASEB J* 6:861, 1992.

60. K. Sugahara, I. Yamashina, P. De Waard, et al.: Structural studies on sulfated glycopeptides from the carbohydrate-protein linkage region of chondroitin 4-sulfate proteoglycans of swamp rat chondrosarcoma. *J Biol Chem* 263:10168, 1988.

61. E. Ruoslahti: Proteoglycans in cell regulation. *J Biol Chem* 264:13369, 1989.

62. T. Lind, F. Tufaro, C. McCormick, et al.: The putative tumor suppressors EXT1 and EXT2 are glycosyltransferases required for the biosynthesis of heparin sulfate. *J Biol Chem* 273:26265, 1998.

63. P. D. Yurchenco and J. C. Schittny: Molecular architecture of basement membranes. *FASEB J* 4:1577, 1990.

64. C. H. Streuli and M. J. Bissell: Expression of extracellular matrix components is regulated by substratum. *J Cell Biol* 110:1405, 1990.

65. N. A. Bhowmick, E. G. Neilson, and H. L. Moses: Stromal fibroblasts in cancer initiation and progression. *Nature* 432:332, 2004.

66. M. M. Mueller and N. E. Fusenig: Friends or foes—Bipolar effects of the tumour stroma in cancer. *Nat Rev Cancer* 4:839, 2004.

67. J. A. Joyce: Therapeutic targeting of the tumor microenvironment. *Cancer Cell* 7:513, 2005.

68. E. Ruoslahti and M. D. Pierschbacher: New perspectives in cell adhesion: RGD and integrins. *Science* 238:491, 1987.
69. F. G. Giancotti and E. Ruoslahti: Integrin signaling. *Science* 285:1028, 1999.
70. R. W. Ruddon: Chapter 5. In *Cancer Biology*, 3rd ed. New York: Oxford University Press, 1995.
71. D. B. Stewart and W. J. Nelson: Identification of four distinct pools of catenins in mammalian cells and transformation-dependent changes in catenin distributions among these pools. *J Biol Chem* 272:29652, 1997.
72. T. Uemura. The cadherin superfamily at the synapse: more members, more missions. *Cell* 93:1095, 1998.
73. S. J. Vermeulen, J. Bruyneel, E. A. Bracke, et al.: Cell–cell interactions. *Cancer Res* 55:4722, 1995.
74. P. Guilford: E-cadherin downregulation in cancer: fuel on the fire? *Mol Med Today* 5: 172, 1999.
75. P. Guilford: E-cadherin germline mutations in familial gastric cancer. *Nature* 392:402, 1998.
76. X. P. Hao, T. G. Pretlow, J. S. Rao, and T. P. Pretlow: β-catenin expression is altered in human colonic aberrant crypt foci. *Cancer Res* 61:8085, 2001.
77. T. Brabletz, A. Jung, S. Reu, et al.: Variable β-catenin expression in colorectal cancers indicates tumor progression driven by the tumor environment. *Proc Natl Acad Sci USA* 98: 10356, 2001.
78. W. J. Rutter, R. L. Pictet, and P. W. Morris: Toward molecular mechanisms of developmental processes. *Annu Rev Biochem* 42:601, 1973.
79. S. A. Chervitz, L. Aravind, G. Sherlock, C. A. Ball, E. V. Koonin, et al.: Comparison of the complete protein sets of worm and yeast: orthology and divergence. *Science* 282:2022, 1998.
80. J. M. Stuart, E. Segal, D. Koller, and S. K. Kim: A gene–coexpression network for global discovery of conserved genetic modules. *Science* 302:249, 2003.
81. R. H. Kessin: Making streams. *Nature* 422:481, 2003.
82. E. W. Sutherland: Studies on the mechanism of hormone action. *Science* 177:401, 1972.
83. I. Pastan and M. Willingham: Cellular transformation and the "morphologic phenotype" of transformed cells. *Nature* 274:645, 1978.
84. J. Hochman, P. A. Insel, H. R. Bourne, P. Coffino, and G. M. Tomkins: A structural gene mutation affecting the regulatory subunit of cyclic AMP-dependent protein kinase in mouse lymphoma cells. *Proc Natl Acad Sci USA* 72: 5051, 1975.
85. Y. S. Cho-Chung, T. Clair, P. N. Yi, and C. Parkinson: Comparative studies on cyclic AMP binding and protein kinase in cyclic AMP–responsive and –unresponsiveness Walker 256 mammary carcinomas. *J Biol Chem* 252:6335, 1977.
86. Y. S. Cho-Chung, T. Clair, and P. Huffman: Loss of nuclear cyclic AMP binding in cyclic AMP-unresponsive Walker 256 mammary carcinoma. *J Biol Chem* 252:6349, 1977.
87. S. A. Aaronson: Growth factors and cancer. *Science* 254:1146, 1991.
88. P. Taylor and P. A. Insel: Molecular basis of drug action, In W. B. Pratt and P. Taylor, eds.: *Principles of Drug Action*. New York: Churchill Livingstone, 1990, pp. 103–220.
89. L. F. Allen, R. J. Lefkowitz, M. G. Caron, and S. Cotecchia: G-protein-coupled receptor genes as protooncogenes: Constitutively activating mutation of the α_{1b}-adrenergic receptor enhances mitogenesis and tumorigenicity. *Proc Natl Acad Sci USA* 88:11354, 1991.
90. S. Hermouet, J. J. Merendino, Jr., J. S. Gutkin, and A. M. Spiegel: Activating and inactivating mutations of the α subunit of G_{12} protein have opposite effects on proliferation of NIH 3T3 cells. *Proc Natl Acad Sci USA* 88:10455, 1991.
91. H. B. Fraser, A. E. Hirsh, L. M. Steinmetz, C. Scharfe, and M. W. Feldman: Evolutionary rate in the protein interaction network. *Science* 296:750, 2002.
92. E. H. Davidson, D. R. McClay and L. Hood: Regulatory gene networks and the properties of the developmental process. *Proc Natl Acad Sci USA* 100:1475–1480, 2003.
93. A. A. Teleman, M. Strigini, and S. M. Cohen: Shaping morphogen gradients. *Cell* 105:559, 2001.
94. E. H. Davidson, J. P. Rast, P. Oliveri, A. Ransick, C. Calestani, et al.: A genomic regulatory network for development. *Science* 295:1669, 2002.
95. R. Keller: Shaping the vertebrate body plan by polarized embryonic cell movements. *Science* 298:1950, 2002.
96. S. G. Martin and D. St. Johnston: A role for *Drosophila* LKB1 in anterior-posterior axis formation and epithelial polarity. *Nature* 421:379, 2003.
97. J. Dubrulle and O. Pourquié: *fgf8* mRNA decay establishes a gradient that couples axial elongation to patterning in the vertebrate embryo. *Nature* 427:419, 2004.
98. C. J. Tabin: Retinoids, homeoboxes, and growth factors: toward molecular models for limb development. *Cell* 66:199, 1991.
99. D. A. Melton: Pattern formation during animal development. *Science* 252:234, 1991.
100. L. Reid: From gradients to axes, from morphogenesis to differentiation. *Cell* 63:875, 1990.
101. G. M. Edelman: Morphoregulatory molecules. *Biochemistry* 27:3533, 1988.

102. B. L. M. Hogan: Morphogenesis. *Cell* 96:225, 1999.

103. D. N. Watkins, D. M. Berman, S. G. Burkholder, B. Wang, P. A. Beachy and S. B. Baylin: Hedgehog signaling within airway epithelial progenitors and in small-cell lung cancer. *Nature* 422:313, 2003.

104. C. Wicking, I. Smyth, and A. Bale: A hedgehog signaling pathway in tumorigenesis and development. *Oncogene* 18:7844, 1999.

105. Y. Zhang and D. Kalderon: Hedgehog acts as a somatic stem cell factor in the *Drosophila* ovary. *Nature* 410:599, 2001.

106. D. M. Berman, S. S. Karhadkar, A. Maitra, R. Montes de Oca, M. R. Gerstenblith, et al.: Widespread requirement for Hedgehog ligand stimulation in growth of digestive tract tumours. *Nature* 425:846, 2003.

107. C. Niehrs: Solving a sticky problem. *Nature* 413:787, 2001.

108. I. Maillard and W. S. Pear: Notch and cancer: Best to avoid the ups and down. *Cancer Cell* 3:203, 2003.

109. A. Raya, C. M. Koth, D. Büscher, Y. Kawakami, T. Itoh, et al.: Activation of notch signaling pathway precedes heart regeneration in zebrafish. *Proc Natl Acad Sci USA* 100:11889, 2003.

110. M. Takeichi: Cadherins: A molecular family important in selective cell–cell adhesion. *Annu Rev Biochem* 59:237, 1990.

111. R. W. Ruddon. Chapter 5. In *Cancer Biology*, 3rd ed. New York: Oxford University Press; 1995.

112. G. D. Demetri, C. D. M. Fletcher, E. Mueller, et al.: Induction of solid tumor differentiation by the peroxisome proliferators-activated receptor-γ ligand troglitazone in patients with liposarcoma. *Proc Natl Acad Sci USA* 96:3951, 1999.

113. E. D. Rosen and B. M. Spiegelman: PPARγ: a nuclear regulator of metabolism, differentiation, and cell growth. *J Biol Chem* 276:37731, 2001.

114. J. A. Thomson, J. Itskovitz-Eldor, S. S. Shapiro, M. A. Waknitz, J. J. Swiergiel, et al.: Embryonic stem cell lines derived from human blastocysts. *Science* 282:1145, 1998.

115. M. J. Shamblott, J. Axelman, J. W. Littlefield, P. D. Blumenthal, G. R. Huggins, et al.: Human embryonic germ cell derivatives express a broad range of developmentally distinct markers and proliferate extensively in vitro. *Proc Natl Acad Sci USA* 98:113, 2001.

116. N. Rosenthal: Prometheus's vulture and the stem-cell promise. *N Engl J Med* 349:267, 2003.

117. M. R. Alison, R. Poulsom, R. Jeffery, A. P. Dhillon, A. Quaglia, et al.: Hepatocytes from non-hepatic adult stem cells. *Nature* 406:257, 2000.

118. M. Körbling and Z. Estrov: Adult stem cells for tissue repair—A new therapeutic concept? *N Engl J Med* 349:570, 2003.

119. J-H. Kim, J. M. Auerbach, J. A. Rodriguez-Gómez, I. Velasco, D. Gavin, et al.: Dopamine neurons derived from embryonic stem cells function in an animal model of Parkinson's disease. *Nature* 418:50, 2002.

120. N. Lumelsky, O. Blondel, P. Laeng, I. Velasco, R. Ravin, and R. McKay: Differentiation of embryonic stem cells to insulin-secreting structures similar to pancreatic islets. *Science* 292:1389, 2001.

121. D. Orlic, J. Kajstura, S. Chimenti, F. Limana, I. Jakoniuk, et al.: Mobilized bone marrow cells repair the infracted heart, improving function and survival. *Proc Natl Acad Sci USA* 98:10344, 2001.

122. D. Hess, L. Li, M. Martin, S. Sakano, D. Hill, et al.: Bone marrow-derived stem cells initiate pancreatic regeneration. *Nature Biol* 21:763, 2003.

123. D. Woodbury, E. J. Schwarz, D. J. Prockop, and I. B. Black: Adult rat and human bone marrow stromal cells differentiate into neurons. *J Neurol Res* 61:364, 2000.

124. Y. Jiang, B. N. Jahagirdar, R. L. Reinhardt, R. E. Schwartz, C. D. Keene, et al.: Pluripotency of mesenchymal stem cells derived from adult marrow. *Nature* 418:41, 2002.

125. Y. Jiang, D. Henderson, M. Blackstad, A. Chen, R. F. Miller, and C. M. Verfaillie: Neuroectodermal differentiation from mouse multipotent adult progenitor cells. *Proc Natl Acad Sci USA* 100:11854, 2003.

126. C. Holden and G. Vogel: Plasticity: time for a reappraisal? *Science* 296:2126, 2002.

127. Y. Kanazawa and I. M. Verma: Little evidence of bone marrow–derived hepatocytes in the replacement of injured liver. *Proc Natl Acad Sci USA* 100:11850, 2003.

128. E. H. Kaji and J. M. Leiden: Gene and stem cell therapies. *JAMA* 285:545, 2001.

129. M. Schuldiner, O. Yanuka, J. Itskovitz-Eldor, D. A. Melton, and N. Benvenisty: Effects of eight growth factors on the differentiation of cells derived from human embryonic stem cells. *Proc Natl Acad Sci USA* 97:11307, 2000.

130. N. B. Ivanova, J. T. Dimos, C. Schaniel, J. A. Hackney, K. A. Moore, and I. R. Lemischka: A stem cell molecular signature. *Science* 298:601, 2002.

131. K. Takahashi, K. Mitsui, and S. Yamanaka: Role of Eras in promoting tumour-like properties in mouse embryonic stem cells. *Nature* 423:541, 2003.

132. T. Reya, S. J. Morrison, M. F. Clarke, and I. L. Weissman: Stem cells, cancer, and cancer stem cells. *Nature* 414:105, 2001.

133. D. E. Bergsagel and F. A. Valeriote: Growth characterizatics of a mouse plasma cell tumor. *Cancer Res* 28:2187, 1968.

134. A. W. Hamburger and S. E. Salmon: Primary bioassay of human tumor stem cells. *Science* 197:461, 1977.

135. M. Al-Hajj, M. S. Wicha, A. Benito-Hernandez, S. J. Morrison, and M. F. Clarke: Prospective identification of tumorigenic breast cancer cells. *Proc Natl Acad Sci USA* 100:3983, 2003.

136. S. K. Singh, I. D. Clarke, M. Terasaki, V. E. Bonn, C. Hawkins, et al.: Identification of a cancer stem cell in human brain tumors. *Proc Natl Acad Sci USA* 63:5821, 2003.

137. X-J. Ma, R. Salunga, J. T. Tuggle, J. Gaudet, E. Enright, et al.: Gene expression profiles of human breast cancer progression. *Proc Natl Acad Sci USA* 100:5974, 2003.

138. J. M. Mitchison: *The Biology of Cell Cycle*. New York: Cambridge University Press, 1971.

139. L. H. Hartwell and T. A. Weinert: Checkpoints: Controls that ensure the order of cell cycle events. *Science* 246:629, 1989.

140. P. Nurse and Y. Bisset: Gene required in G1 for commitment to cell cycle and in G2 for control of mitosis in fission yeast. *Nature* 292:558, 1981.

141. D. H. Beach, B. Durkacz, and P. Nurse: Functionally homologous cell cycle control genes in budding and fission yeast. *Nature* 300:706, 1982.

142. M. Kirschner: The cell cycle then and now. *Trends Biochem Sci* 281, 1992.

143. A. W. Murray and M. W. Kirschner: Dominoes and clocks: The union of two views of the cell cycle. *Science* 246:614, 1989.

144. M. G. Lee and P. Nurse: Complementation used to clone a human homologue of the fission yeast cell cycle control gene cdc2. *Nature* 327:31, 1987.

145. J. Pines: Cyclins: Wheels within wheels. *Cell Growth Differ* 2:305, 1991.

146. T. Evans, E. T. Rosenthal, J. Youngblom, D. Distel, and T. Hunt: Cyclin: A protein specified by maternal mRNA in sea urchin eggs that is destroyed at each cleavage division. *Cell* 33:389, 1983.

147. A. B. Pardee: G_1 events and regulation of cell proliferation. *Science* 246:603, 1989.

148. M. Malumbres and M. Barbacid: To cycle or not to cycle: A critical decision in cancer. *Nat Rev Cancer* 1:222, 2001.

149. C. J. Sherr: The pezcoller lecture: cancer cell cycles revisited. *Cancer Res* 60:3689, 2000.

150. C. J. Sherr: Mammalian G1 cyclins. *Cell* 73:1059, 1993.

151. B. Faha, M. E. Ewen, L.-H. Tsai, D. M. Livingston, and E. Harlow: Interaction between human cyclin A and adenovirus E1A-associated p107 protein. *Science* 255:87, 1992.

152. P. Jansen-Dürr, A. Meichle, P. Steiner, M. Pagano, K. Finke, J. Botz, J. Wessbecher, G. Draetta, and M. Eilers: Differential modulation of cyclin gene expression by MYC. *Proc Natl Acad Sci USA* 90:3685, 1993.

153. D. O. Morgan, J. M. Kaplan, J. M. Bishop, and H. E. Varmus: Mitosis-specific phosphorylation of $p60^{c-src}$ by $p34^{cdc2}-$ associated protein kinase. *Cell* 57:775, 1989.

154. D. McVey, L. Brizuela, I. Mohr, D. R. Marshak, Y. Gluzman, and D. Beach: Phosphorylation of large tumour antigen by cdc2 stimulates SV40 DNA replication. *Nature* 341: 503, 1989.

155. A. Koff, M. Ohtsuki, K. Polyak, J. M. Roberts, and J. Massagué: Negative regulation of G1 in mammalian cells: Inhibition of cyclin E-dependent kinase by TGF-β. *Science* 260:536, 1993.

156. D. G. Johnson and C. L. Walker: Cyclins and cell cycle checkpoints. *Annu Rev Pharmacol Toxicol* 39:295, 1999.

157. C. J. Sherr: Cancer cell cycles. *Science* 274: 1672, 1996.

158. T. L. Orr-Weaver and R. A. Weinberg: A checkpoint on the road to cancer. *Nature* 392:223, 1998.

159. J. Bartek and J. Lukas: Chk1 and Chk2 kinases in checkpoint control and cancer. *Cancer Cell* 3:421, 2003.

160. N. Sagata: Untangling checkpoints. *Science* 298:1905, 2002.

161. K. Keyomarsi, S. L. Tucker, T. A. Buchholz, M. Callister, Y. Ding, et al.: Cyclin E and survival in patients with breast cancer. *N Engl J Med* 347:1566, 2002.

162. J. Farley, L. M. Smith, K. M. Darcy, E. Sobel, D. O'Connor, et al.: Cyclin E expression is a significant predictor of survival in advanced, suboptimally debulked ovarian epithelial cancers: A gynecologic oncology group study. *Cancer Res* 63:1235, 2003.

163. H. Cam and B. D. Dynlacht: Emerging roles for E2F: Beyond the G1/S transition and DNA replication. *Cancer Cell* 3:311, 2003.

164. M. Bettencour-Dias, R. Giet, R. Sinka, A. Mazumdar, W. G. Lock, et al.: Genome-wide survey of protein kinases required for cell cycle progression. *Nature* 432:980, 2004.

165. D. L. Vaux and S. J. Korsmeyer: Cell death in development. *Cell* 96:245, 1999.

166. M. O. Hengartner: The biochemistry of apoptosis. *Nature* 407:770, 2000.

167. D. R. Green: Apoptotic pathways: ten minutes to dead. *Cell* 121:671, 2005.

168. A. Ashkenazi and V. M. Dixit: Death receptors: signaling and modulation. *Science* 281:1305, 1998.

169. D. R. Green and J. C. Reed: Mitochondria and apoptosis. *Science* 281:1309, 1998.

170. J. F. R. Kerr: A histochemical study of hypertrophy and ischaemic injury of rat liver with special reference to changes in lysosomes. *J Pathol Bacteriol* 90:419, 1965.

171. J. F. R. Kerr, A. H. Wyllie, and A. R. Currie: Apoptosis: A basic biological phenomenon with wide-ranging implications in tissue kinetics. *Br J Cancer* 26:239, 1972.

172. J. F. R. Kerr and B. V. Harmon: In L. D. Tomei and F. O. Cope, eds.: *Apoptosis: The Molecular Basis of Cell Death*. Cold Spring Harbor, NY: Cold Spring Harbor Laboratory Press, 1991, p. 5.

173. W. D. Jarvis, R. N. Kolesnick, F. A. Fornari, R. S. Traylor, D. A. Gewirtz, and S. Grant: Induction of apoptotic DNA damage and cell death by activation of the sphingomyelin pathway. *Proc Natl Acad Sci USA* 91:73, 1994.

174. S. J. Martin, D. R. Green, and T. G. Cotter: Dicing with death: Dissecting the components of the apoptosis machinery. *Trends Biochem Sci* 19:26, 1994.

175. S. J. Korsmeyer: Bcl-2: A repressor of lymphocyte death. *Immunol Today* 13:285, 1992.

176. M. O. Hengartner, R. E. Ellis, and H. R. Horvitz: Caenorhabditis elegans gene ced-9 protects cells from programmed cell death. *Nature* 356:494, 1992.

177. M. P. Mattson and G. Kroemer: Mitochondria in cell death: Novel targets for neuroprotection and cardioprotection. *Trends Mol Med* 9:196, 2003.

178. L. Scorrano, S. A. Oakes, J. T. Opferman, E. H. Cheng, M. D. Sorcinelli, et al.: BAX and BAK regulation of endoplasmic reticulum Ca^{2+}: A control point for apoptosis. *Science* 300:135, 2003.

179. E. H.-Y. Cheng, T. V. Sheiko, J. K. Fisher, W. J. Craigen, and S. J. Korsmeyer: VDAC2 inhibits BAK activation and mitochondrial apoptosis. *Science* 301:513, 2003.

180. Z. Zhu, O. Sanchez-Sweatman, X. Huang, R. Wiltrout, R. Khokha, et al.: Anoikis and metastatic potential of cloudman S91 melanoma cells. *Cancer Res* 61:1707, 2001.

181. P. Meier, A. Finch, and G. Evan: Apoptosis in development. *Nature* 407:796, 2000.

182. P. H. Krammer: CD95's deadly mission in the immune system. *Nature* 407:789, 2000.

183. R. W. Johnstone, A. A. Ruefli, and S. W. Lowe: Apoptosis: A link between cancer genetics and chemotherapy. *Cell* 108:153, 2002.

184. D. W. Nicholson: From bench to clinic with apoptosis-based therapeutic agents. *Nature* 407:810, 2000.

185. J. C. Reed: Apoptosis-targeted therapies for cancer. *Cancer Cell* 3:17, 2003.

186. D. C. Altieri: The molecular basis and potential role of survivin in cancer diagnosis and therapy. *Trends Mol Med* 7:542, 2001.

187. A. W. Burgess: Reflections on biochemistry. *Trends Biochem Sci* 14:117, 1989.

188. G. O. Gey and W. Thalhimer: Observations on the effects of insulin introduced into the medium of tissue cultures. *JAMA* 82:1609, 1924.

189. D. Gospodarowicz and J. S. Moran: Growth factors in mammalian cell culture. *Annu Rev Biochem* 45:531, 1976.

190. M. Kasuga, F. A. Karlsson, and C. R. Kahn: Insulin stimulates the phosphorylation of the 95,000-Dalton subunit of its own receptor. *Science* 215:185, 1982.

191. L. Petruzzelli, R. Herrera, and O. M. Rosen: Insulin receptor is an insulin-dependent tyrosine kinase: Copurification of insulin-binding activity and protein kinase activity to homogeneity from human placenta. *Proc Natl Acad Sci USA* 81:3327, 1984.

192. K-T. Yu and M. P. Czech: Tyrosine phosphorylation of the insulin receptor β subunit activates the receptor-associated tyrosine kinase activity. *J Biol Chem* 259:5277, 1984.

193. S. A. Aaronson: Growth factors and cancer. *Science* 254:1146, 1991.

194. R. E. Favoni and A. De Cupis: The role of polypeptide growth factors in human carcinomas: new targets for a novel pharmacological approach. *Pharmacol Rev* 52:179, 2000.

195. E. Rinderknecht and R. E. Humbel: The amino acid sequence of human insulin-like growth factor I and its structural homology with proinsulin. *J Biol Chem* 253:2769, 1978.

196. N. C. Dulak and H. M. Temin: A partially purified polypeptide fraction from rat liver cell conditioned medium with multiplication-stimulating activity for embryo fibroblasts. *J Cell Physiol* 81:153, 1973.

197. J. Zapf and V. R. Froesch: Insulin-like growth factors/somatomedins: structure, secretion, biological actions and physiological roles. *Hormone Res* 24:121, 1986.

198. M. P. Czech: Signal transmission by the insulin-like growth factors. *Cell* 59:235, 1989.

199. J. E. Fradkin, R. C. Eastman, M. A. Lesniak, and J. Roth: Specificity spillover at the hormone receptor-exploring its role in human disease. *N Engl J Med* 320:640, 1989.

200. W. H. Daughaday: Editorial: The possible autocrine/paracrine and endocrine roles of insulin-like growth factors of human tumors. *Endocrinology* 127:1, 1990.

201. K. J. Cullen, D. Yee, W. S. Sly, J. Perdue, B. Hampton, M. E. Lippman, and N. Rosen: Insulin-like growth factor receptor expression and function in human breast cancer. *Cancer Res* 50:48, 1990.

202. H. T. Huynh, E. Tetenes, L. Wallace, and M. Pollak: In vivo inhibition of insulin-like growth factor I gene expression by tamoxifen. *Cancer Res* 53:1727, 1993.

203. R. Torrisi, F. Pensa, M. A. Orengo, E. Catsafados, P. Ponzani, F. Boccardo, A. Costa, and A. Decensi: The synthetic retinoid fenretinide lowers plasma insulin-like growth factor I levels in breast cancer patients. *Cancer Res* 53:4769, 1993.

204. H. Yu and T. Rohan: Role of the insulin-like growth factor family in cancer development

and progression. *J Natl Cancer Inst* 92:1472, 2000.

205. M. N. Pollak, E. S. Schernhammer, and S. E. Hankinson: Insulin-like growth factors and neoplasia. *Nat Rev Cancer* 4:505, 2004.

206. E. D. Bueker: Implantation of tumours in the hind limb field of the embryonic chick and the developmental response of the lumbosacral nervous system. *Anat Rec* 102:369, 1948.

207. S. R. Cohen, R. Levi-Montalcini, and V. Hamburger: A nerve growth stimulating factor isolated from sarcomas 37 and 180. *Proc Natl Acad Sci USA* 40:1014, 1954.

208. R. A. Hogue-Angeletti, W. A. Frazier, J. W. Jacobs, H. D. Niall, and R. A. Bradshaw: Purification, characterization, and partial amino acid sequence of nerve growth factor from cobra venom. *Biochemistry* 15:26, 1976.

209. S. Varon, J. Nomura, and E. M. Shooter: The isolation of the mouse nerve growth factor protein in a high molecular weight form. *Biochemistry* 6:2202, 1967.

210. V. Bocchini and P. U. Angeletti: The nerve growth factor: Purification as a 30,000-molecular-weight protein. *Proc Natl Acad Sci USA* 64:787, 1969.

211. W. A. Frazier, R. H. Angeletti, and R. A. Bradshaw: Nerve growth factor and insulin. *Science* 176:482, 1972.

212. M. Holloway: Finding the good in the bad. *Sci Am* 1:31, 1993.

213. R. A. Bradshaw, T. L. Blundell, R. Lapatto, N. Q. McDonald, and J. Murray-Rust: Nerve growth factor revisited. *Trends Biochem Sci* 2:48, 1993.

214. D. R. Kaplan, B. L. Hempstead, D. Martin-Zanca, M. V. Chao, and L. F. Parada: The trk proto-oncogene product: A signal transducing receptor for nerve growth factor. *Science* 252:554, 1991.

215. C. Cordon-Cardo, P. Tapley, S. Jing, V. Nanduri, E. O'Rourke, et al.: The *trk* tyrosine protein kinase mediates the mitogenic properties of nerve growth factor and neurotrophin-3. *Cell* 66:173, 1991.

216. K. W. Wood, C. Sarnecki, T. M. Roberts, and J. Bienis: ras mediates nerve growth factor receptor modulation of three signal-transducing protein kinases: MAP kinase, Raf-1, and RSK. *Cell* 68:1041, 1992.

217. D. Martin-Zanca, S. H. Hughes, and M. Barbacid: A human oncogene formed by the fusion of truncated tropomyosin and protein tyrosine kinase sequences. *Nature* 319:743, 1986.

218. T. Suzuki, E. Bogenmann, H. Shimada, D. Stram, and R. C. Seeger: Lack of high-affinity nerve growth factor receptors in aggressive neuroblastomas. *J Natl Cancer Inst* 85:377, 1993.

219. G. Carpenter and S. Cohen: Epidermal growth factor. *Annu Rev Biochem* 48:193, 1979.

220. J. M. Taylor, W. M. Mitchell, and S. Cohen: Characterization of the binding protein for epidermal growth factor. *J Biol Chem* 249:2188, 1974.

221. P. Frey, R. Forand, T. Maciag, and E. M. Shooter: The biosynthetic precursor of epidermal growth factor and the mechanism of its processing. *Proc Natl Acad Sci USA* 76:6294, 1979.

222. R. H. Starkey, S. Cohen, and D. N. Orth: Epidermal growth factor: Identification of a new hormone in human urine. *Science* 189:800, 1975.

223. H. Gregory: Isolation and structure of urogastrone and its relationship to epidermal growth factor. *Nature* 257:325, 1975.

224. G. Carpenter and S. Cohen: Epidermal growth factor. *J Biol Chem* 265:7709, 1990.

225. S. Cohen, H. Ushiro, C. Stosbeck, and M. Chinkers: A native 170,000 epidermal growth factor receptor–kinase complex from shed plasma membrane vesides. *J Biol Chem* 257:1523, 1982.

226. H. Ushiro and S. Cohen: Identification of phosphotyrosine as a product of epidermal growth factor-activated protein kinase in A-431 cell membranes. *J Biol Chem* 255:8363, 1980.

227. I. E. Garcia de Palazzo, G. P. Adams, P. Sundareshan, A. J. Wong, J. R. Testa, et al.: Expression of mutated epidermal growth factor receptor by non-small cell lung carcinomas. *Cancer Res* 53:3217, 1993.

228. T. A. Libermann, H. R. Nusbaum, N. Razon, R. Kris, I. Lax, et al.: Amplification, enhanced expression and possible rearrangement of EGF receptor gene in primary brain tumours of glial origin. *Nature* 313:144, 1985.

229. J. E. Dancey: Predictive factors for epidermal growth factor receptor inhibitors—the bull's-eye hits the arrow. *Cancer Cell* 5:411, 2004.

230. C. F. Fox, P. S. Linsley, and M. Wrann: Receptor remodeling and regulation in the action of epidermal growth factor. *Fed Proc* 41:2988, 1982.

231. A. Dautry-Varsat and H. F. Lodish: How receptors bring proteins and particles into cells. *Sci Am* 250:52, 1984.

232. J. L. Goldstein, R. G. W. Anderson, and M. S. Brown: Coated pits, coated vesicles, and receptor-mediated endocytosis. *Nature* 279:679, 1979.

233. G. Carpenter and S. Cohen: ^{125}I-labeled human epidermal growth factor: Binding, internalization, and degradation in human fibroblasts. *J Cell Biol* 71:159, 1976.

234. W. H. Moolenaar, R. Y. Tsien, P. T. van der Saag, and S. W. de Laat: Na$^+$/H$^+$ exchange and cytoplasmic pH in the action of growth factors in human fibroblasts. *Nature* 304:645, 1983.

235. P. Rothenberg, L. Glaser, P. Schlesinger, and D. Cassel: Activation of Na$^+$/H$^+$ exchange by

epidermal growth factor elevates intracellular pH in A431 cells. *J Biol Chem* 258:12644, 1983.

236. C. P. Burns and E. Rozengurt: Extracellular Na$^+$ and initiation of DNA synthesis: Role of intracellular pH and K$^+$. *J Cell Biol* 98:1082, 1984.

237. G. Carpenter: Epidermal growth factor is a major growth-promoting agent in human milk. *Science* 210:198, 1980.

238. O. Tsutsumi, H. Kurachi, and T. Oka: A physiological role of epidermal growth factor in male reproductive function. *Science* 233:975, 1986.

239. K. Eckert, A. Granetzny, J. Fisher, E. Nexo, and R. Grosse: A M_r 43,000 epidermal growth factor–related protein purified from the urine of breast cancer patients. *Cancer Res* 50:642, 1990.

240. M-A. Huotari, P. J. Miettinen, J. Palgi, T. Koivisto, J. Ustinov, et al.: ErbB signaling regulates lineage determination of developing pancreatic islet cells in embryonic organ culture. *Endocrinology* 143:4437–4446, 2002.

241. T. Holbro, R. R. Beerli, F. Maurer, M. Koziczak, C. F. Barbas, and N. E. Hynes: The ErbB2/ErbB3 heterodimer functions as an oncogenic unit: ErbB2 requires ErbB3 to drive breast tumor cell proliferation. *Proc Natl Acad Sci USA* 100:8933, 2003.

242. S. J. Lavictoire, D. A. E. Parolin, A. C. Klimowicz, J. F. Kelly, and I. A. J. Lorimer: Interaction of Hsp90 with the nascent form of the mutant epidermal growth factor receptor EGFRvIII. *J Biol Chem* 278:5292, 2003.

243. C. K. Tang, X-Q. Gong, D. K. Moscatello, A. J. Wong, and M. E. Lippman: Epidermal growth factor receptor vIII enhances tumorigenicity in human breast cancer. *Cancer Res* 60:3081, 2000.

244. D. J. Slamon, B. Leyland-Jones, S. Shak, H. Fuchs, V. Paton, et al.: Use of chemotherapy plus a monoclonal antibody against HER2 for metastatic breast cancer that overexpresses HER2. *N Engl J Med* 344:783, 2001.

245. V. Grunwald and M. Hidalgo: Developing inhibitors of the epidermal growth factor receptor for cancer treatment. *J Natl Cancer Inst* 95:851, 2003.

246. D. Gospodarowicz: Purification of a fibroblast growth factor from bovine pituitary. *J Biol Chem* 250:2515, 1975.

247. F. Esch, A. Baird, N. Ling, N. Ueno, F. Hill, et al.: Primary structure of bovine pituitary basic fibroblast growth factor (FGF) and comparison with amino-terminal sequence of bovine brain acidic FGF. *Proc Natl Acad Sci USA* 82:6507, 1985.

248. D. Givol and A. Yayon: Complexity of FGF receptors: genetic basis for structural diversity and functional specificity. *FASEB J* 6:3362, 1992.

249. K. Miller and A. Rizzino: Developmental regulation and signal transduction pathways of fibroblast growth factors and their receptors. In M. Nielsin-Hamilton, ed.: *Growth Factors and Signal Transduction Pathways in Development*. New York: Wiley-Liss, 19–49, 1994.

250. M. Miyamoto, K-I. Naruo, C. Seko, S. Matsumoto, T. Kondo, and T. Kurokawa: Molecular cloning of a novel cytokine cDNA encoding the ninth member of the fibroblast growth factor family, which has a unique secretion property. *Mol Cell Biol* 13:4251, 1993.

251. D. Qiao, K. Meyer, C. Mundhenke, S. A. Drew, and A. Friedl: Heparan sulfate proteoglycans as regulators of fibroblast growth factor-2 signaling in brain endothelial cells. *J Biol Chem* 278:16045, 2003.

252. G. P. Dotto, G. Moellmann, S. Ghosh, M. Edwards, and R. Halaban: Transformation of murine melanocytes by basic fibroblast growth factor cDNA and oncogenes and selective suppression of the transformed phenotype in a reconstituted cutaneous environment. *J Cell Biol* 109:3115, 1989.

253. A. Wellstein, G. Zugmaier, J. A. Califano III, F. Kern, S. Paik, and M. E. Lippman: Tumor growth dependent on Kaposi's sarcoma–derived fibroblast growth factor inhibited by pentosan polysulfate. *J Natl Cancer Inst* 83:716, 1991.

254. J. L. Gross, W. F. Herblin, B. A. Dusak, P. Czerniak, M. D. Diamond, et al.: Effects of modulation of basic fibroblast growth factor on tumor growth in vivo. *J Natl Cancer Inst* 85:121, 1993.

255. S. Schultz-Hector and S. Haghayegh: β-fibroblast growth factor expression in human and murine squamous cell carcinomas and its relationship to regional endothelial cell proliferation. *Cancer Res* 53:1444, 1993.

256. Y. Yamanaka, H. Friess, M. Buchler, H. G. Beger, E. Uchida, et al.: Overexpression of acidic and basic fibroblast growth factors in human pancreatic cancer correlates with advanced tumor stage. *Cancer Res* 53:5289, 1993.

257. D. M. Nanus, B. J. Schmitz-Dräger, R. J. Motzer, A. C. Lee, V. Vlamis, et al.: Expression of basic fibroblast growth factor in primary human renal tumors: Correlation with poor survival. *J Natl Cancer Inst* 85:1597, 1993.

258. T. Ruotsalainen, H. Joensuu, K. Mattson, and P. Salven: High pretreatment serum concentration of basic fibroblast growth factor is a predictor of poor prognosis in small cell lung cancer. *Cancer Epidiol Biomarkers Prev* 11:1492, 2002.

259. S. D. Balk: Calcium as a regulator of the proliferation of normal, but not of transformed, chicken fibroblasts in a plasma-containing medium. *Proc Natl Acad Sci USA* 68:271, 1971.

260. H. N. Antoniades, C. D. Scher, and C. D. Stiles: Purification of human platelet-derived growth factor. *Proc Natl Acad Sci USA* 76:1809, 1979.

261. C.-H. Heldin, B. Westermark, and A. Wasteson: Platelet-derived growth factor: Purification and partial characterization. *Proc Natl Acad Sci USA* 76:3722, 1979.

262. E. W. Raines and R. Ross: Platelet-derived growth factor. I. High yield purification and evidence for multiple forms. *J Biol Chem* 257:5154, 1982.

263. K. Pietras, T. Sjöblom, K. Rubin, C-H. Heldin, and A. Östman: PDGF receptors as cancer drug targets. *Cancer Cell* 3:439, 2003.

264. C. D. Stiles: The molecular biology of platelet-derived growth factor. *Cell* 33:653, 1983.

265. C. D. Scher, R. C. Shepard, H. N. Antoniades, and C. D. Stiles: Platelet-derived growth factor and the regulation of the mammalian fibroblast cell cycle. *Biochim Biophys Acta* 560:217, 1979.

266. C. D. Scher, W. J. Pledger, P. Martin, H. N. Antoniades, and C. D. Stiles: Transforming viruses directly reduce the cellular growth requirement for a platelet derived growth factor. *J Cell Physiol* 97:371, 1978.

267. D. F. Bowen-Pope, A. Vogel, and R. Ross: Production of platelet-derived growth factor–like molecules and reduced expression of platelet-derived growth factor receptors accompany transformation by a wide spectrum of agents. *Proc Natl Acad Sci USA* 81:2396, 1984.

268. C. Betsholtz, B. Westermark, B. Ek, and C-H. Heldin: Coexpression of a PDGF-like growth factor and PDGF receptors in a human osteosarcoma cell line: Implications for autocrine receptor activation. *Cell* 39:447, 1984.

269. K. Forsberg, I. Valyi-Nagy, C-H. Heldin, M. Herlyn, and B. Westermark: Platelet-derived growth factor (PDGF) in oncogenesis: Development of a vascular connective tissue stroma in xenotransplanted human melanoma producing PDGF-BB. *Proc Natl Acad Sci USA* 90:393, 1993.

270. T. P. Fleming, A. Saxena, W. C. Clark, J. T. Robertson, E. H. Oldfield, et al.: Amplification and/or overexpression of platelet-derived growth factor receptors and epidermal growth factor receptor in human glial tumors. *Cancer Res* 52:4550, 1992.

271. L. Holmgren, F. Flam, E. Larsson, and R. Ohlsson: Successive activation of the platelet-derived growth factor β receptor and platelet-derived growth factor B genes correlates with the genesis of human choriocarcinoma. *Cancer Res* 53:2927, 1993.

272. B. E. Bejcek, R. M. Hoffman, D. Lipps, D. Y. Li, C. A. Mitchell, et al.: The vis-sis oncogene product but not platelet-derived growth factor (PDGF) a homodimer activate PDGF α and β receptors intracellularly and initiate cellular transformation. *J Biol Chem* 267:3289, 1992.

273. G. J. Todaro, J. E. DeLarco, and S. Cohen: Transformation by murine and feline sarcoma viruses specifically blocks binding of epidermal growth factor to cells. *Nature* 264:26, 1976.

274. A. B. Roberts, L. C. Lamb, D. L. Newton, M. B. Sporn, J. E. DeLarco, and G. J. Todaro: Transforming growth factors: Isolation of polypeptides from virally and chemically transformed cells by acid/ethanol extraction. *Proc Natl Acad Sci USA* 77:3494, 1980.

275. J. Massagué: Type β transforming growth factor from feline sarcoma virus–transformed rat cells. Isolation and biological properties. *J Biol Chem* 259:9756, 1984.

276. B. Ozanne, T. Wheeler, and P. L. Kaplan: Cells transformed by RNA and DNA tumor viruses produce transforming factors. *Fed Proc* 41:3004, 1982.

277. M. B. Sporn and A. B. Roberts: Autocrine growth factors and cancer. *Nature* 313:745, 1985.

278. J. Massagué: A model for membrane-anchored growth factors. *J Biol Chem* 265:21393, 1990.

279. S. T. Wong, L. F. Winchell, B. K. McCune, H. S. Earp, J. Teixido, et al.: The TGF-α precursor expressed on the cell surface binds to the EGF receptor on adjacent cells, leading to signal transdution. *Cell* 56:495, 1989.

280. K.-I. Imanishi, K. Yamaguchi, M. Kuranami, E. Kyo, T. Hozumi, and K. Abe: Inhibition of growth of human lung adenocarcinoma cell lines by anti-transforming growth factor-α monoclonal antibody. *J Natl Cancer Inst* 81:220, 1989.

281. M. Reiss, E. b. Stash, V. F. Vellucci, and Z-L. Zhou: Activation of the autocrine transforming growth factor α pathway in human squamous carcinoma cells. *Cancer Res* 51:6254, 1991.

282. S. E. Bates, N. E. Davidson, E. M. Valverius, C. E. Freter, R. B. Dickson, et al.: Expression of transforming growth factor α and its messenger ribonucleic acid by estrogen and its possible functional significance. *Mol Endocrinol* 2:543, 1988.

283. Y. Gong, G. Ballejo, L. C. Murphy, and L. J. Murphy: Differential effects of estrogen and antiestrogen on transforming growth factor gene expression in endometrial adenocarcinoma cells. *Cancer Res* 52:1704, 1992.

284. H. Takagi, R. Sharp, C. Hammermeister, T. Goodrow, M. O. Bradley, et al.: Molecular and genetic analysis of liver oncogenesis in transforming growth factor α transgenic mice. *Cancer Res* 52:5171, 1992.

285. Y. C. Yeh, J. F. Tsai, L. Y. Chuang, H. W. Yeh, J. H. Tsai, et al.: Elevation of transforming growth factor α and α-fetoprotein levels in patients with hepatocellularcarcinoma. *Cancer Res* 47:896, 1987.

286. C. L. Arteaga, A. R. Hanauske, G. M. Clark, C. K. Osborne, P. Hazarika, et al.: Immunoreactive α-transforming growth factor activity in effusions from cancer patients as a marker of tumor burden and patient prognosis. *Cancer Res* 48:5023, 1988.

287. G. H. Lee, G. Merlino, and N. Fausto: Development of liver tumors in transforming growth factor α transgenic mice. *Cancer Res* 52:5162, 1992.

288. J. E. Mead and N. Fausto: Transforming growth factor α may be a physiological regulator of liver regeneration by means of an autocrine mechanism. *Proc Natl Acad Sci USA* 86:1558, 1989.

289. H. Chang, C. W. Brown, and M. M. Matzuk: Genetic analysis of the mammalian transforming growth factor-β superfamily. *Endocrine Rev* 23:787, 2002.

290. R. Derynck and Y-E. Zhang: Smad-dependent and smad-independent pathways in TGF-β family signaling. *Nature* 425:577, 2003.

291. A. B. Roberts and L. M. Wakefield: The two faces of transforming growth factor β in carcinogenesis. *Proc Natl Acad Sci USA* 100:8621, 2003.

292. J. M. Yingling, K. L. Blanchard, and J. S. Sawyer: Development of TGF-β signaling inhibitors for cancer therapy. *Nature Rev Drug Disc* 3:1011, 2004.

293. D. Metcalf: Hemopoietic regulators. *Trends Biochem Sci* 286, 1992.

294. H. Blumberg, D. Conklin, W. Xu, A. Grossmann, T. Brender, et al.: Interleukin 20: Discovery, receptor, identification, and role in epidermal function. *Cell* 104:9, 2001.

295. E. Passegué, C. H. M. Jamieson, L. E. Ailles, and I. L. Weissman: Normal and leukemic hematopoiesis: are leukemias a stem cell disorder or a reacquisition of stem cell characteristics? *Proc Natl Acad Sci USA* 100:11842, 2003.

296. D. Metcalf: Control of granulocytes and macrophages: Molecular, cellular, and clinical aspects. *Science* 254:529, 1991.

297. W. P. Hammond IV, T.H. Price, L.M. Souza, and D.C. Dale: Treatment of cyclic neutropenia with granulocyte colony-stimulating factor. *N Engl J Med* 320:1306, 1989.

298. J. Vose, P. Bierman, A. Kessinger, P. Coccia, J. Anderson, et al.: The use of recombinant human granulocyte-macrophage colony stimulating factor for the treatment of delayed engraftment following high dose therapy and autologous hematopoietic stem cell transplantation for lymphoid malignancies. *Bone Marrow Transp* 7:139, 1991.

299. U. Duhrsen, J-L. Vileval, J. Boyd, G. Kannourakis, G. Morstyn, and D. Metcalf: Effects of recombinant human granulocyte colony-stimulating factor on hematopoietic progenitor cells in cancer patients. *Blood* 72:2074, 1988.

300. N. Stahl and G. D. Yancopoulos: The alphas, betas, and kinases of cytokine receptor complexes. *Cell* 74:587, 1993.

301. J. Chen, J. M. Cárcamo, O. Bórquez-Ojeda, H. Erdjument-Bromage, P. Tempst, and D.W. Golde: The laminin receptor modulates granulocyte-macrophage colony-stimulating factor receptor complex formation and modulates its signaling. *Proc Natl Acad Sci USA* 100:14000, 2003.

302. F. Bussolino, M. F. DiRenzo, M. Ziche, E. Bocchietto, M. Olivero, et al.: Hepatocyte growth factor is a potent angiogenic factor which stimulates endothelial cell motility and growth. *J Cell Biol* 119:629, 1992.

303. T. Nakamura, T. Nishizawa, M. Hagiya, T. Keki, M. Shimonishi, et al.: Molecular cloning and expression of human hepatocyte growth factor. *Nature* 342:440, 1989.

304. Y. Uehara and N. Kitamura: Expression of a human hepatocyte growth factor/scatter factor cDNA in MDCK epithelial cells influences cell morphology, motility, and anchorage-independent growth. *J Cell Biol* 117:889, 1992.

305. D. P. Bottaro, J. S. Rubin, D. L. Faletto, A. M.-L. Chan, T. E. Kmiecik, et al.: Identification of the hepatocyte growth factor receptor as the c-met proto-oncogene product. *Science* 251:802, 1991.

306. A. Graziani, D. Gramaglia, P. dalla Zonca, and P. M. Comoglio: Hepatocyte growth factor/scatter factor stimulates the ras-guanine nucleotide exchange. *J Biol Chem* 268:9165, 1993.

307. R. W. Ruddon: Chapter 9. In *Cancer Biology*, 3rd edition. New York: Oxford University Press, 1995.

308. S. A. Miles, O. Martinez-Maza, A. Rezai, L. Magpantay, T. Kishimoto, et al.: Oncostatin M as a potent mitogen for AIDS-Kaposi's sarcoma–derived cells. *Science* 255:1432, 1992.

309. F. R. Miller, D. McEachern, and B. E. Miller: Growth regulation of mouse mammary tumor cells in collagen gel cultures by diffusible factors produced by normal mammary gland epithelium and stromal fibroblasts. *Cancer Res* 49:6091, 1989.

310. P. R. Ervin, Jr., M. S. Kaminski, R. L. Cody, and M. S. Wicha: Production of mammastatin, a tissue-specific growth inhibitor, by normal human mammary cells. *Science* 244:1585, 1989.

311. W. C. Hahn and R. A. Weinberg: Modelling the molecular circuitry of cancer. *Nat Rev Cancer* 2:331, 2002.

312. E. G. Krebs and J. A. Beavo: Phosphorylation–dephosphorylation of enzymes. *Annu Rev Biochem* 48:923, 1979.

313. A. G. Gilman: G proteins, transducers of receptor-generated signals. *Annu Rev Biochem* 56:615, 1987.

314. T. Hunter: A thousand and one protein kinases. *Cell* 50:823, 1987.

315. T. Hunter: Protein-tyrosine phosphatases: The other side of the coin. *Cell* 58:1013, 1989.

316. H. Fischer, H. Charbonneau, and N. K. Tonks: Protein tyrosine phosphatases: A diverse family of intracellular and transmembrane enzymes. *Science* 253:401, 1991.

317. J. K. Klarlund: Transformation of cells by an inhibitor of phosphatases acting on phosphotyrosine in proteins. *Cell* 41:707, 1985.

318. T. Hunter: Signaling-2000 and beyond. *Cell* 100:113, 2000.

319. M. R. Hokin and L. E. Hokin: Enzyme secretion and the incorporation of P^{32} into phospholipides of pancreas slices. *J Biol Chem* 203:967, 1953.

320. M. J. Berridge and R. F. Irvine: Inositol trisphosphate, a novel second messenger in cellular signal transduction. *Nature* 312:315, 1984.

321. F. Murad: Regulation of cytosolic guanylyl cyclase by nitric oxide: The NO-cyclic GMP signal transduction system. *Adv Pharmacol* 26:19, 1994.

322. R. W. Ruddon, S. A. Sherman, and E. Bedows: Protein folding in the endoplasmic reticulum: lessons from the human chorionic gonadotropin β subunit. *Prot Sci* 5:1443, 1996.

323. T. Pawson and P. Nash: Assembly of cell regulatory systems through protein interaction domains. *Science* 300:445, 2003.

324. W. B. Pratt and D. O. Toft: Regulation of signaling protein function and trafficking by the hsp90/hsp70-based chaperone machinery. *Exp Biol Med* 228:111, 2003.

325. L. C. Cantley: Translocating tubby. *Science* 292:2019, 2001.

326. Y. Nishizuka: Intracellular signaling by hydrolysis of phospholipids and activation of protein kinase C. *Science* 258:607, 1992.

327. T. Pawson and T. M. Saxton: Signaling networks—Do all roads lead to the same genes? *Cell* 97:675, 1999.

328. D. Fambrough, K. McClure, A. Kazlauskas, and E. S. Lander: Diverse signaling pathways activated by growth factor receptors induce broadly overlapping, rather than independent, sets of genes. *Cell* 97:727, 1999.

329. S. R. Neves, P. T. Ram, and R. Iyengar: G protein pathways. *Science* 296:1636, 2002.

330. J. R. Hepler and A. G. Gilman: G proteins. *Trends Biochem Sci* 17:383, 1992.

331. I. Pastan and M. Willingham: Cellular transformation and the "morphologic phenotype" of transformed cells. *Nature* 274:645, 1978.

332. J. Hochman, P. A. Insel, H. R. Bourne, P. Coffino, and G. M. Tomkins: A structural gene mutation affecting the regulatory subunit of cyclic AMP-dependent protein kinase in mouse lymphoma cells. *Proc Natl Acad Sci USA* 72:5051, 1975.

333. Y. S. Cho-Chung, T. Clair, and P. Huffman: Loss of nuclear cyclic AMP binding in cyclic AMP-unresponsive Walker 256 mammary carcinoma. *J Biol Chem* 252:6349, 1977.

334. S. A. Aaronson: Growth factors and cancer. *Science* 254:1146, 1991.

335. L. F. Allen, R. J. Lefkowitz, M. G. Caron, and S. Cotecchia: G-protein-coupled receptor genes as protooncogenes: Constitutively activating mutation of the α_{1b}-adrenergic receptor enhances mitogenesis and tumorigenicity. *Proc Natl Acad Sci USA* 88:11354, 1991.

336. S. Hermouet, J.J. Merendino, Jr., J.S. Gutkin, and A.M. Spiegel: Activating and inactivating mutations of the α subunit of G_{i2} protein have opposite effects on proliferation of NIH 3T3 cells. *Proc Natl Acad Sci USA* 88:10455, 1991.

337. L. C. Cantley: The phosphoinositide 3-kinase pathway. *Science* 296:1655, 2002.

338. A. Toker and A. G. Newton: Cellular signaling: Pivoting around PDK-1. *Cell* 103:185, 2000.

339. J.-O. Lee, H. Yang, M-M. Georgescu, A. Di Cristofano, T. Maehama, et al.: Crystal structure of the PTEN tumor suppressor: Implications for its phosphoinositide phosphatase activity and membrane association. *Cell* 99:323, 1999.

340. S. A. Weaver and S. G. Ward: Phosphoinositide 3-kinases in the gut: a link between inflammation and cancer? *Trends Mol Med* 7:455, 2001.

341. T. Schmelzle and M. N. Hall: TOR, a central controller of cell growth. *Cell* 103:253, 2000.

342. C. L. Sawyers: Will mTOR inhibitors make it as cancer drugs? *Cancer Cell* 4:343, 2003.

343. P. K. Vogt: PI 3-kinase, mTOR, protein synthesis and cancer. *Trends Mol Med* 7:482, 2001.

344. P. Blume-Jensen and T. Hunter: Oncogenic kinase signalling. *Nature* 411:355, 2001.

345. J. Schlessinger: Cell signaling by receptor tyrosine kinases. *Cell* 103:211, 2000.

346. J. K. Klarlund: Transformation of cells by an inhibitor of phosphatases acting on phosphotyrosine in proteins. *Cell* 41:707, 1985.

347. J. Li, C. Yen, D. Liaw, et al.: PTEN, a putative protein tyrosine phosphatase gene mutated in human brain, breast, and prostate cancer. *Science* 275:1943, 1997.

348. S. Saha, A. Bardelli, P. Buckhaults, et al.: A phosphatase associated with metastasis of colorectal cancer. *Science* 294:1343, 2001.

349. Z. Wang, D. Shen, D. Williams Parsons, A. Bardelli, J. Sager, et al.: Mutational analysis of the tyrosine phosphatome in colorectal cancers. *Science* 304:1164, 2004.

350. J. E. Darnell, Jr.: STATs and gene regulation. *Science* 277:1630, 1997.

351. D. S. Aaronson and C. M. Horvath: A road map for those who don't know JAK-STAT. *Science* 296:1653, 2002.

352. H. Yu and R. Jove: The stats of cancer—New molecular targets come of age. *Nat Rev Cancer* 4:97, 2004.

353. J. Matthews and J.-A. Gustafsson: Estrogen signaling: a subtle balance between ERα and ERβ. *Mol Interv* 3:281, 2003.

354. D. P. McDonnell and J. D. Norris: Connections and regulation of the human estrogen receptor. *Science* 296:1642, 2002.

355. J. M. Brown: The hypoxic cell: A target for selective cancer therapy-eighteenth Bruce F. Cain memorial award lecture. *Cancer Res* 59:5863, 1999.

356. R. A. Cairns, T. Kalliomaki, and R. P. Hill: Acute (cyclic) hypoxia enhances spontaneous

metastasis of KHT murine tumors. *Cancer Res* 61:8093, 2001.

357. T. Seagraves and R. S. Johnson: Two HIFs may be better than one. HIF-2α overexpression directly contributes to renal clear cell tumorigenesis: Evidence for HIF as a tumor promoter. *Cancer Cell* 1:211, 2002.

358. A. Lal, H. Peters, B. St. Croix, Z. A. Haroon, M. W. Dewhirst, et al.: Transcriptional response to hypoxia in human tumors. *J Natl Cancer Inst* 93:1337, 2001.

359. G. Chen and D. V. Goeddel: TNR-R1 Signaling: A beautiful path. *Science* 296:1634, 2002.

360. L. Attisano and J. L. Wrana: Signal transduction by the TGF-β superfamily. *Science* 296:1646, 2002.

361. M. P. Hernandez, W. P. Sullivan, and D. O. Toft: The assembly and intermolecular properties of the hsp70-Hop-hsp90 molecular chaperone complex. *J Biol Chem* 277:38294, 2002.

362. Y. Morishima, K. C. Kanelakis, P. J. M. Murphy, E. R. Lowe, G. J. Jenkins, et al.: The Hsp90 cochaperone p23 is the limiting component of the multiprotein Hsp90/Hsp70-based chaperone system in vivo where it acts to stabilize the client protein-Hsp90 complex. *J Biol Chem* 278:48754, 2003.

363. A. D. Basso, D. B. Solit, G. Chiosis, B. Giri, P. Tsichlis, and N. Rosen: Akt forms an intracellular complex with heat shock protein 90 (Hsp90) and Cdc37 and is destabilized by inhibitors of Hsp90 function. *J Biol Chem* 277:39858, 2002.

364. W. Xu, X. Yuan, Y. J. Jung, Y. Yang, A. Basso, et al.: The heat shock protein 90 inhibitor geldanamycin and the ErbB inhibitor ZD1839 promote rapid PP1 phosphatase-dependent inactivation of AKT in ErbB2 overexpressing breast cancer cells. *Cancer Res* 63:7777, 2003.

365. I. Hostein, D. Robertson, F. DiStefano, P. Workman, and P. A. Clarke: Inhibition of signal transduction by the Hsp90 inhibitor 17-allylamino-17-demethoxygeldananmycin results in cytostasis and apoptosis. *Cancer Res* 61:4003, 2001.

366. O. Cleaver and D. A. Melton: Endothelial signaling during development. *Nat Med* 9:661, 2003.

367. A. W. Griffioen and G. Molema: Angiogenesis: Potentials for pharmacologic intervention in the treatment of cancer, cardiovascular diseases, and chronic inflammation. *Pharmacol Rev* 52:237, 2000.

368. P. Carmeliet and R. K. Jain: Angiogenesis in cancer and other diseases. *Nature* 407:249, 2000.

369. J. A. Hoffman, E. Giraudo, M. Singh, L. Zhang, M. Inoue, et al.: Progressive vascular changes in a transgenic mouse model of squamous cell carcinoma. *Cancer Cell* 4:383, 2003.

370. R. Pasqualini, W. Arap, and D. M. McDonald: Probing the structural and molecular diversity of tumor vasculature. *Trends Mol Med* 8:563, 2002.

371. R. W. Merwin and G. H. Algire: The role of graft and host vessels in the vascularization of grafts of normal and neoplastic tissue. *J Natl Cancer Inst* 17:23, 1956.

372. R. L. Ehrmann and M. Knoth: Choriocarcinoma: Transfilter stimulation of vasoproliferation in the hamster cheek pouch-studied by light and electron microscopy. *J Natl Cancer Inst* 41:1229, 1968.

373. M. Greenblatt and P. Shubik: Tumor angiogenesis: Transfilter diffusion studies in the hamster by the transparent chamber technique. *J Natl Cancer Inst* 41:111, 1968.

374. J. Folkman, E. Merler, C. Abernathy, and G. Williams: Isolation of a tumor factor responsible for angiogenesis. *J Exp Med* 113:275, 1971.

375. M. A. Gimbrone, Jr., R. S. Cotran, S. B. Leapman, and J. Folkman: Tumor growth and neovascularizaton: An experimental model using the rabbit cornea. *J Natl Cancer Inst* 52:413, 1974.

376. P. M. Gullino: Angiogenesis and oncogenesis. *J Natl Cancer Inst* 61:639, 1978.

377. K. H. Plate, G. Breier, H. A. Weich, and W. Risau: Vascular endothelial growth factor is a potential tumour angiogenesis factor in human gliomas in vivo. *Nature* 359:845, 1992.

378. Y. Myoken, Y. Kayada, T. Okamoto, M. Kan, G. H. Sato, and J. D. Sato: Vascular endothelial cell growth factor (VEGF) produced by A-431 human epidermoid carcinoma cells and identification of VEGF membrane binding sites. *Proc Natl Acad Sci USA* 88:5819, 1991.

379. M. Nguyen, H. Watanabe, A. E. Budson, J. P. Richie, and J. Folkman: Elevated levels of the angiogenic peptide basic fibroblast growth factor in urine of bladder cancer patients. *J Natl Cancer Inst* 85:241, 1993.

380. K.-T. Yeo, H. H. Wang, J. A. Nagy, T. M. Sioussat, S. R. Ledbetter, et al.: Vascular permeability factor (vascular endothelial growth factor) in guinea pig and human tumor and inflammatory effusions. *Cancer Res* 53:2912, 1993.

381. J. Folkman, K. Watson, D. Ingber, and D. Hanahan: Induction of angiogenesis during the transition from hyperplasia to neoplasia. *Nature* 339:58, 1989.

382. M. Maxwell, S. P. Naber, H. J. Wolfe, E. T. Hedley-Whyte, T. Galanopoulos, et al.: Expression of angiogenic growth factor genes in primary human astrocytomas may contribute to their growth and progression. *Cancer Res* 51:1345, 1991.

383. D. Hanahan and J. Folkman: Patterns and emerging mechanisms of the angiogenic switch during tumorigenesis. *Cell* 86:353, 1996.

384. N. Ferrara, H.-P. Gerber and J. LeCouter: The biology of VEGF and its receptors. *Nat Med* 9:669, 2003.

385. R. S. Watnick, Y-N. Cheng, A. Rangarajan, T. A. Ince, and R. A. Weinberg: Ras modulates Myc activity to repress thrombospondin-1 expression and increase tumor angiogenesis. *Cancer Cell* 3:219, 2003.

386. M. F. McCarty, W. Liu, F. Fan, A. Parikh, N. Reimuth, et al.: Promises and pitfalls of anti-angiogenic therapy in clinical trials. *Trends Mol Med* 9:53, 2003.

387. G. D. Yancopoulos, S. Davis, N. W. Gale, J. S. Rudge, S. J. Wiegand, and J. Holash: Vascular-specific growth factors and blood vessel formation. *Nature* 407:242, 2000.

388. Y. Xu, Y.-J. Liu, and Q. Yu: Angiopoietin-3 inhibits pulmonary metastasis by inhibiting tumor angiogenesis. *Cancer Res* 64:6119, 2004.

389. E. S. Kim, A. Serur, J. Huang, C. A. Manley, K. W. McCrudden, et al.: Potent VEGF blockade causes regression of coopted vessels in a model of neuroblastoma. *Proc Natl Acad Sci USA* 99:11399, 2002.

390. J. Marx: A boost for tumor starvation. *Science* 301:452, 2003.

391. O. Stoeltzing, S. A. Shmad, W. Liu, M. F. McCarty, J. S. Wey, et al.: Angiopoietin-1 inhibits vascular permeability, angiogenesis, and growth of hepatic colon cancer tumors. *Cancer Res* 63:3370, 2003.

392. J. H. Qi, Q. Ebrahem, N. Moore, G. Murphy, L. Claesson-Welsh, et al.: A novel function for tissue inhibitor of metalloproteinases-3 (TIMP3): Inhibition of angiogenesis by blockage of VEGF binding to VEGF receptor-2. *Nat Med* 9:407, 2003.

393. S. M. Dallabrida, M. A. De Sousa, and D. H. Farrell: Expression of antisense to integrin subunit β_3 inhibits microvascular endothelial cell capillary tube formation in fibrin. *J Biol Chem* 275:32281, 2000.

394. J. Folkman and R. Kalluri: Cancer without diseases. *Nature* 427:787, 2004.

395. M. Rehn, T. Veikkola, E. Kukk-Valdre, H. Nakamura, M. Ilmonen, et al.: Interaction of endostatin with integrins implicated in angiogenesis. *Proc Natl Acad Sci USA* 98:1024, 2001.

396. M. V. Blagosklonny: Antiangiogenic therapy and tumor progression. *Cancer Cell* 5:13, 2004.

397. E. K. Rofstad, R. Mathiesen, and K. Galappathi: Increased metastatic dissemination in human melanoma xenografts after subcurative radiation treatment: Radiation-induced increase in fraction of hypoxic cells and hypoxia-induced up-regulation of urokinase-type plasminogen activator receptor. *Cancer Res* 64:13, 2004.

398. J. Tombran-Tink and C. J. Barnstable: Therapeutic prospects for PEDF: more than a promising angiogenesis inhibitor. *Trends Mol Med* 9:244, 2003.

399. W. Arap, W. Haedicke, M. Bernasconi, R. Kain, D. Rajotte, et al.: Targeting the prostate for destruction through a vascular address. *Proc Natl Acad Sci USA* 99:1527, 2002.

400. M. Guba, P. von Breitenbuch, M. Steinbauer, G. Koehl, S. Flegel, et al.: Rapamycin inhibits primary and metastatic tumor growth by antiangiogenesis: involvement of vascular endothelial growth factor. *Nat Med* 8:128, 2002.

401. R. A. Gupta, L. V. Tejada, B. J. Tong, S. K. Das, J. D. Morrow, et al.: Cyclooxygenase-1 is overexpressed and promotes angiogenic growth factor production in ovarian cancer. *Cancer Res* 63:906, 2003.

402. E. Bråkenhielm, R. Cao, and Y. Cao: Suppression of angiogenesis, tumor growth, and wound healing by resveratrol, a natural compound in red wine and grapes. *FASEB J* 15:1798, 2001.

403. J. C. Yang, L. Haworth, R. M. Sherry, P. Hwu, D. J. Schwartzentruber, et al.: A randomized trial of bevacizumab, an anti-vascular endothelial growth factor antibody, for metastatic renal cancer. *N Engl J Med* 349:427, 2003.

404. Y. Shaked, F. Bertolini, S. Man, M. S. Rogers, D. Cervi, et al.: Genetic heterogeneity of the vasculogenic phenotype parallels angiogenesis: Implications for cellular surrogate marker analysis of antiangiogenesis. *Cancer Cell* 7:101, 2005.

405. M. Skobe, T. Hawighorst, D. G. Jackson, R. Prevo, L. Janes, et al.: Induction of tumor lymphangiogenesis by VEGF-C promotes breast cancer metastasis. *Nat Med* 7:192, 2001.

406. T. Karpanen, M. Egeblad, M. J. Karkkainen, H. Kubo, S. Ylä-Herttuala, et al.: Vascular endothelial growth factor C promotes tumor lymphangiogenesis and intralymphatic tumor growth. *Cancer Res* 61:1786, 2001.

407. S. A. Stacker, C. Caesar, M. E. Baldwin, G. E. Thornton, R. A. Williams, et al.: VEGF-D promotes the metastatic spread of tumor cells via the lymphatics. *Nat Med* 7:186, 2001.

408. T. P. Padera, A. Kadambi, E. di Tomaso, C. Mouta Carreira, E. B. Brown, et al.: Lymphatic metastasis in the absence of functional intratumor lymphatics. *Science* 296:1883, 2002.

409. J. E. Gershenwald and I. J. Fidler: Targeting lymphatic metastasis. *Science* 296:1811, 2002.

410. P. Laakkonen, M. E. Akerman, H. Biliran, M. Yang, F. Ferrer, et al.: Antitumor activity of a homing peptide that targets tumor lymphatics and tumor cells. *Proc Natl Acad Sci USA* 101:9381, 2004.

411. D. Hanahan and J. Folkman: Patterns and emerging mechanisms of the angiogenic switch during tumorigenesis. *Cell* 86:353, 1996.

412. T. Browder, C. E. Butterfield, B. M. Kräling, B. Shi, B. Marshall, et al.: Antiangiogenic scheduling of chemotherapy improves efficacy against experimental drug-resistant cancer. *Cancer Res* 60:1878, 2000.

413. G. N. Naumov, I. C. MacDonald, P. M. Weinmeister, N. Kerkvliet, K. V. Nadkarni,

et al.: Persistence of solitary mammary carcinoma cells in a secondary site: A possible contributor to dormancy. *Cancer Res* 62:2162, 2002.

414. M. Guba, G. Cernaianu, G. Koehl, E. K. Geissler, K-W. Jauch, et al.: A primary tumor promotes dormancy of solitary tumor cells before inhibiting angiogenesis. *Cancer Res* 61:5575, 2001.

415. O. Musso, M. Rehn, N. Théret, B. Turlin, P. Bioulac-Sage, et al.: Tumor progression is associated with a significant decrease in the expression of the endostatin precursor collagen XVIII in human hepatocellular carcinomas. *Cancer Res* 61:45, 2001.

416. J. Folkman: The vascularization of tumors. *Sci Am* 234:58, 1976.

417. V. T. DeVita, Jr., R. C. Young, and G. P. Canellos: Combination versus single agent chemotherapy: A review of the basis for selection of drug treatment of cancer. *Cancer* 35:98, 1975.

418. F. M. Schabel, Jr.: The use of tumor growth kinetics in planning "curative" chemotherapy of advanced solid tumors. *Cancer Res* 29:2384, 1969.

419. G. G. Steel: Cytokinetics of neoplasia. In J. F. Holland and E. Frei III, eds.: *Cancer Medicine*. Philadelphia: Lea & Febiger, 1973, pp. 125–140.

420. E. R. Fearon and B. Vogelstein: A genetic model for colorectal tumorigenesis. *Cell* 61:759, 1990.

421. M. J. van de Vijver, Y. D. He, L. J.l van 't Veer, H. Dai, A. A. M. Hart, et al.: A gene-expression signature as a predictor of survival in breast cancer. *N Engl J Med* 347:1999, 2002.

422. S. Ramaswamy, K. N. Ross, E. S. Lander, and T. R. Golub: A molecular signature of metastasis in primary solid tumors. *Nat Genet* 33:49, 2003.

423. O. Schmidt-Kittler, T. Ragg, A. Daskalakis, M. Granzow, A. Ahr, et al.: From latent disseminated cells to overt metastasis: Genetic analysis of systemic breast cancer progression. *Proc Natl Acad Sci USA* 100:7737, 2003.

424. I. J. Fidler: Tumor heterogeneity and the biology of cancer invasion and metastasis. *Cancer Res* 38:2651, 1978.

425. B. Fisher and E. R. Fisher: The interrelationship of hematogenous and lymphatic tumor cell dissemination. *Surg Gynecol Obstet* 122:791, 1966.

426. B. Sylven: Biochemical factors accompanying growth and invasion. In R. W. Wissler, T. L. Dao, and S. Wood, Jr., eds.: *Endogenous Factors Influencing Host-Tumor Balance*. Chicago: University of Chicago Press, 1967, pp. 267–276.

427. E. R. Fisher and B. Fisher: Recent observations on the concept of metastasis. *Arch Pathol* 83:321, 1967.

428. U. Veronesi, G. Paganelli, G. Viale, A. Luini, S. Zurrida, et al.: A randomized comparison of sentinel-node biopsy with routine axillary dissection in breast cancer. *N Engl J Med* 349:546, 2003.

429. D. Krag and T. Ashikaga: The design of trials comparing sentinel-node surgery and axillary resection. *N Engl J Med* 349:6, 2003.

430. I. J. Fidler: Metastasis: Quantitative analysis of distribution and fate of tumor emboli labeled with ^{125}I-5-iodo-2′-deoxyuridine. *J Natl Cancer Inst* 45:775, 1970.

431. S. Wood, Jr., R. R. Robinson, and B. Marzocchi: Factors influencing the spread of cancer: Locomotion of normal and malignant cells in vivo. In R. W. Wissler, T. L. Dao, and S. Wood, Jr., eds.: *Endogenous Factors Influencing Host–Tumor Balance*. Chicago: University of Chicago Press, 1967, pp. 223–237.

432. I. J. Fidler: The relationship of embolic homogeneity, number, size, and viability to the incidence of experimental metastasis. *Eur J Cancer* 9:223, 1973.

433. I. Zeidman: The fate of circulating tumor cells. I. Passage of cells through capillaries. *Cancer Res* 21:38, 1961.

434. B. A. Warren: Environment of the blood-borne tumor embolus adherent to vessel wall. *J Med* 4:150, 1973.

435. G. J. Gasic, T. B. Gasic, N. Galanti, T. Johnson, and S. Murphy: Platelet–tumor cell interactions in mice: the role of platelets in the spread of malignant disease. *Int J Cancer* 11:704, 1973.

436. A. M. Schor, S. L. Schor, and S. Kumar: Importance of a collagen substratum for stimulation of capillary endothelial cell proliferation by tumor angiogenesis factor. *Int J Cancer* 24:225, 1979.

437. I. J. Fidler: Immune stimulation-inhibition of experimental cancer metastasis. *Cancer Res* 34:491, 1974.

438. W. A. D. Anderson, ed.: *Pathology*, 4th ed. St. Louis: C. V. Mosby, 1961.

439. P. Rubin, ed.: *Clinical Oncology*, 4th ed. Rochester: American Cancer Society, 1974.

440. S. Paget: The distribution of secondary growth in cancer of the breast. *Lancet* 1:571, 1889.

441. I. J. Fidler: Selection of successive tumor lines for metastasis. *Nat New Biol* 242:148, 1973.

442. I. J. Fidler: General considerations for studies of experimental cancer metastasis. In H. Busch, ed.: *Methods in Cancer Research, Vol. XV*. New York: Academic Press, 1978, pp. 399–439.

443. I. J. Fidler and G. L. Nicolson: Organ selectivity for survival and growth of B16 melanoma variant tumor lines. *J Natl Cancer Inst* 57:1199, 1976.

444. K. W. Brunson, G. Beattie, and G. L. Nicolson: Selection and altered properties of brain-colonizing metastatic melanoma. *Nature* 272:543, 1978.

445. G. L. Nicolson and K. W. Brunson: Organ specificity of B16 melanomas: In vivo selection for organ preference of blood-borne metastasis. *Gann Monogr Cancer Res* 20:15, 1977.

446. I. J. Fidler and M. L. Kripke: Metastasis results from preexisting variant cells within a malignant tumor. *Science* 197:893, 1977.

447. G. Poste and I. J. Fidler: The pathogenesis of cancer metatheses. *Nature* 283:139, 1980.

448. S. Strickland and W. G. Richards: Invasion of the trophoblasts. *Cell* 71:355, 1992.

449. O. D. Genbacev, A. Prakobphol, R. A. Foulk, A. R. Krtolica, D. Ilic, et al.: Trophoblast L-selectin-mediated adhesion at the maternal-fetal interface. *Science* 299:405, 2003.

450. M. C. Lacroix, J. Guibourdenche, T. Fournier, I. Laurendeau, A. Igout, et al.: Stimulation of human trophoblast invasion by placental growth hormone. *Endocrinology* 146:2434, 2005.

451. L. A. Liotta, P. S. Steeg, and W. G. Stetler-Stevenson: Cancer metastasis and angiogenesis: An imbalance of positive and negative regulation. *Cell* 64:327, 1991.

452. M. Nakajima, D. Lotan, M. M. Baig, R. M. Carralero, W. R. Wood, M. J. C. Hendrix, and R. Lotan: Inhibition of retinoic acid of type IV collagenolysis and invasion through reconstituted basement membrane by metastatic rat mammary adenocarcinoma cells. *Cancer Res* 49:1698, 1989.

453. A. M. P. Montgomery, Y. A. DeClerck, K. E. Langley, R. A. Reisfeld, and B. M. Mueller: Melanoma-mediated dissolution of extracellular matrix: Contribution of urokinase-dependent and metalloproteinase-dependent proteolytic pathways. *Cancer Res* 53:693, 1993.

454. J. Grondahl-Hansen, I. J. Christensen, C. Rosenquist, N. Brummer, H. T. Mouridsen, et al.: High levels of urokinase-type plasminogen activator and its inhibitor PAI-1 in cytosolic extracts of breast carcinomas are associated with poor prognosis. *Cancer Res* 53:2513, 1993.

455. J. A. Joyce, A. Baruch, K. Chehade, N. Meyer-Morse, E. Giraudo, et al.: Cathepsin cysteine proteases are effectors of invasive growth and angiogenesis during multistage tumorigenesis. *Cancer Cell* 5:443, 2004.

456. J. Rozhin, A. P. Gomez, G. H. Ziegler, K. K. Nelson, Y. S. Chang, et al.: Cathepsin B to cysteine proteinase inhibitor balance in metastatic cell subpopulations isolated from murine tumors. *Cancer Res* 50:6278, 1990.

457. S. S. Chauhan, L. J. Goldstein, and M. M. Gottesman: Expression of cathepsin L in human tumors. *Cancer Res* 51:1478, 1991.

458. N. Rousselet, L. Mills, D. Jean, C. Tellez, M. Bar-Eli, and R. Frade: Inhibition of tumorigenicity and metastasis of human melanoma cells by anti-cathespin L single chain variable fragment. *Cancer Res* 64:146, 2004.

459. M. J. Murnane, K. Sheahan, M. Ozdemirli, and S. Shuja: Stage-specific increases in cathepsin B messenger RNA content in human colorectal carcinoma. *Cancer Res* 51:1137, 1991.

460. S. M. Thorpe, H. Rochefort, M. Garcia, G. Freiss, I. J. Christensen, et al.: Association between high concentrations of $M_r52,000$ cathepsin D and poor prognosis in primary human breast cancer. *Cancer Res* 49:6008, 1989.

461. M. D. Johnson, J. A. Torri, M. E. Lippman, and R. B. Dickson: The role of cathepsin D in the invasiveness of human breast cancer cells. *Cancer Res* 53:873, 1993.

462. L. J. McCawley and L. M. Matrisian: Matrix metalloproteinases: multifunctional contributors to tumor progression. *Mol Med Today* 6:149, 2000.

463. S. Hiratsuka, K. Nakamura, S. Iwai, M. Murakami, T. Itoh, et al.: MMP9 induction by vascular endothelial growth factor receptor-1 is involved in lung-specific metastasis. *Cancer Cell* 2:289, 2002.

464. K. Hohnbeck, P. Bianco, and H. Birkedal-Hansen: MT1-MMP: A collagenase essential for tumor cell invasive growth. *Cancer Cell* 4:83, 2003.

465. R. E. B. Seftor, E. A. Seftor, W. G. Stetler-Stevenson, and M. J. C. Hendrix: The 72 kDa type IV collagenase is modulated via differential expression of $\alpha_v\beta_3$ and $\alpha_5\beta_1$ integrins during human melanoma cell invasion. *Cancer Res* 53:3411, 1993.

466. H. Kataoka, R. DeCastro, S. Zucker, and C. Biswas: Tumor cell–derived collagenase-stimulatory factor increases expression of interstitial collagenase, stromelysin, and 72-kDa gelatinase. *Cancer Res* 53:3154, 1993.

467. S. Garbisa, G. Scagliotti, L. Masiero, C. DiFrancesco, C. Caenozzo, et al.: Correlation of serum metalloproteinase levels with lung cancer metastasis and response to therapy. *Cancer Res* 52:4548, 1992.

468. L. A. Liotta, C. N. Rao, and S. H. Barsky: Tumor invasion and the extracellular matrix. *Lab Invest* 49:636, 1983.

469. D. P. DeVore, D. P. Houchens, A. A. Ovejera, G. S. Dill, and T. B. Hutson: Collagenase inhibitors retarding invasion of a human tumor in nude mice. *Exp Cell Biol* 48:367, 1980.

470. U. P. Torgeirsson, L. A. Liotta, T. Kalebric, and I. M. K. Margulies: Effect of natural protease inhibitors and a chemoattractant on tumor cell invasion in vitro. *J Natl Cancer Inst* 69:1049, 1982.

471. E. Edovitsky, M. Elkin, E. Zcharia, T. Peretz, and I. Vlodavsky: Heparanase gene silencing, tumor invasiveness, angiogenesis, and metastasis. *J Natl Cancer Inst* 96:1219, 2004.

472. C. A. Borgoño and E. P. Diamandis: The emerging roles of human tissue kallikreins in cancer. *Nat Rev Cancer* 4:876, 2004.

473. G. Poste and G. L. Nicholson: Arrest and metastasis of blood-borne tumor cells are modified by fusion of plasma membrane vesicles from highly metastatic cells. *Proc Natl Acad Sci USA* 77:399, 1980.

474. M. J. Humphries, K. Matsumoto, S. L. White, and K. Olden: Oligosaccharide modification by swainsonine treatment inhibits pulmonary colonization by B16-F10 murine melanoma cells. *Proc Natl Acad Sci USA* 83:1752, 1986.

475. E. Pearlstein, P. L. Salk, G. Yogeeswaran, and S. Karpatkin: Correlation between spontaneous metastatic potential, platelet-aggregating activity of cell surface extracts, and cell surface sialylation in 10 metastatic-variant derivatives of a rat renal sarcoma cell line. *Proc Natl Acad Sci USA* 77:4336, 1980.

476. G. Yogeeswaran and P. L. Salk: Metastatic potential is positively correlated with cell surface sialylation of culture murine tumor cell lines. *Science* 212:1514, 1981.

477. H. Yamamoto, J. Swoger, S. Greene, T. Saito, J. Hurh, et al.: β1,6-N-Acetylglucosamine-bearing N-glycans in human gliomas: Implications for a role in regulating invasivity. *Cancer Res* 60:134–142, 2000.

478. J. W. Dennis, J. P. Carver, and H. Schachter: Asparagine-linked oligosaccharides in murine tumor cells: Comparison of a WGA-resistant (WGA^r) nonmetastatic mutant and a related WGA-sensitive (WGA^s) metastatic line. *J Cell Biol* 99:1034, 1984.

479. K. G. Kohlgraf, A. J. Gawron, M. Higashi, J. L. Meza, M. D. Burdick, et al.: Contribution of the MUC1 tandem repeat and cytoplasmic tail to invasive and metastatic properties of a pancreatic cancer cell line. *Cancer Res* 63:5011, 2003.

480. G. P. Frenette, T. E. Carey, J. Varani, D. R. Schwartz, S. E. G. Fligiel, et al.: Biosynthesis and secretion of laminin and laminin-associated glycoproteins by nonmalignant and malignant human keratinocytes: Comparison of cell lines from primary and secondary tumors in the same patient. *Cancer Res* 48:5193, 1988.

481. B. P. Peters, R. J. Hartle, R. F. Krzesicki, T. G. Kroll, F. Perini, et al.: The biosynthesis, processing, and secretion of laminin by human choriocarcinoma cells. *J Biol Chem* 260:14732, 1985.

482. R. W. Ruddon: *Cancer Biology*, 3rd edition. Chapter 11, New York: Oxford University Press, 1995.

483. V. P. Terranova, L. A. Liotta, R. G. Russo, et al.: Role of laminin in the attachment and metastasis of murine tumor cells. *Cancer Res* 42:2265, 1982.

484. J. B. McCarthy, A. P. N. Skubitz, S. L. Palm, and L. T. Furcht: Metastasis inhibition of different tumor types by purified laminin fragments and a heparin-binding fragment of fibronectin. *J Natl Cancer Inst* 80:108, 1988.

485. M. J. Humphries, K. Olden, and K. M. Yamada: A synthetic peptide from fibronectin inhibits experimental metastasis of murine melanoma cells. *Science* 233:467, 1986.

486. G. Taraboletti, D. Belotti, R. Giavazzi, M. E. Sobel, and V. Castronovo: Enhancement of metastatic potential of murine and human melanoma cells by laminin receptor peptide G: Attachment of cancer cells to subendothelial matrix as a pathway for hematogenous metastasis. *J Natl Cancer Inst* 85:235, 1993.

487. T. Kanemoto, R. Reich, L. Royce, D. Greatorex, S. H. Adler, N. Shiraishi, G. R. Martin, Y. Yamada, and H. K. Kleinman: Identification of an amino acid sequence from the laminin A chain that stimulates metastasis and collagenase IV production. *Proc Natl Acad Sci USA* 87:2279, 1990.

488. S. H. Barsky, A. Baker, G. P. Siegel, S. Togo, and L. A. Liotta: Use of anti-basement membrane antibodies to distinguish blood vessel capillaries from lymphatic capillaries. *Am J Surg Pathol* 7:667, 1983.

489. G. L. Nicolson, T. Irimura, M. Nakajima, and J. Estrada: Metastatic cell attachment to and invasion of vascular endothelium and its underlying basal lamina using endothelial cell monolayers. In G. L. Nicolson and L. Milas, eds.: *Cancer Invasion and Metastasis: Biologic and Therapeutic Aspects*. New York: Raven Press, 1984, pp. 145–167.

490. R. H. Kramer, R. Gonzalz, and G. L. Nicolson: Metastatic tumor cells adhere preferentially to the extracellular matrix underlying vascular endothelial cells. *Int J Cancer* 26:639, 1980.

491. A. Neri, E. Rouslahti, and G. L. Nicolson: Distribution of fibronectin on clonal cell lines of a rat mammary adenocarcinoma growing in vitro and in vivo at primary and metastatic sites. *Cancer Res* 41:5082, 1981.

492. P. A. Netland and B. R. Zetter: Organ-specific adhesion of metastatic tumor cells in vitro. *Science* 224:1113, 1984.

493. P. A. Netland and B. R. Zetter: Metastatic potential of B16 melanoma cells after in vitro selection for organ-specific adherence. *J Cell Biol* 101:720, 1985.

494. G. Poste and G. L. Nicolson: Arrest and metastasis of blood-borne tumor cells are modified by fusion of plasma membrane vesicles from highly metastatic cells. *Proc Natl Acad Sci USA* 77:399, 1980.

495. P. P. Bringuier, R. Umbas, H. E. Schaafsma, H. F. M. Karthaus, F. M. J. Debruyne, and J. A. Schalken: Decreased E-cadherin immunoactivity correlates with poor survival in patients with bladder tumors. *Cancer Res* 53:3241, 1993.

496. Y. Koki, H. Shiozaki, H. Tahara, M. Inoue, H. Oka, K. Iihara, T. Kadowaski, M. Takeichi, and T. Mori: Correlation between E-cadherin expression and invasiveness in vitro in a human esophageal cancer cell line. *Cancer Res* 53:3421, 1993.

497. U. Cavallaro and G. Christofori: Cell adhesion and signaling by cadherins and ig-cams in cancer. *Nat Rev Cancer* 4:118, 2004.

498. K. Vleminckx, L. Vakaet, Jr., M. Mareel, W. Fiers, and F.V. Roy: Genetic manipulation of E-cadherin expression by epithelial tumor cells reveals an invasion suppressor role. *Cell* 66:107, 1991.

499. U.H. Frixen, J. Behrens, M. Sachs, G. Eberle, B. Voss, A. Warda, D. Lochner, and W. Birchmeier: E-cadherin-mediated cell–cell adhesion prevents invasiveness of human carcinoma cells. *J Cell Biol* 113:173, 1991.

500. K. Suyama, I. Shapiro, M. Guttman, and R. B. Hazan: A signaling pathway leading to metastasis is controlled by N-cadherin and the FGF receptor. *Cancer Cell* 2:301, 2002.

501. K. Edvardsen, W. Chen, G. Rucklidge, F. S. Walsh, B. Obrink, and E. Bock: Transmembrane neural cell–adhesion molecule (NCAM), but not glycosyl-phosphatidylinositol-anchored NCAM, down-regulates secretion of matrix metalloproteinases. *Proc Natl Acad Sci USA* 90:11463, 1993.

502. R. C. Johnson, D. Zhu, H. G. Augustin-Voss, and B. U. Pauli: Lung endothelial dipeptidyl peptidase IV is an adhesion molecule for lung-metastatic rat breast and prostate carcinoma cells. *J Cell Biol* 121:1423, 1993.

502a. J. Yang, S. A. Mani, J. L. Donaher, S. Ramaswamy, R. A. Itzykson, et al.: Twist, a master regulator of morphogenesis, plays an essential role in tumor metastasis. *Cell* 117:927, 2004.

503. R. Wallich, N. Bulbuc, G. J. Hammerling, S. Katzav, S. Segal, and M. Feldman: Abrogation of metastatic properties of tumour cells by de novo expression of H-2K antigens following H-2 gene transfection. *Nature* 315:301, 1985.

504. J. Varani and F. W. Orr: Chemotaxis in nonleukocytic cells. In P. A. Ward, ed.: *Handbook of Inflammation, Vol. 4.* Amsterdam: Elsevier, 1983, pp. 211–244.

505. K. Ushijima, H. Nishi, A. Ishikura, and H. Hayashi: Characterization of two different factors chemotactic for cancer cells from tumor tissue. *Virchows Arch* B21:119, 1976.

506. A. Muller, B. Homey, H. Soto, N. Ge, D. Catron, et al.: Involvement of chemokine receptors in breast cancer metastasis. *Nature* 410:50, 2001.

507. L. A. Liotta: An attractive force in metastasis. *Nature* 410:24, 2001.

508. J. Wyckoff, W. Wang, E. Y. Lin, Y. Wang, F. Pixley, et al.: A paracrine loop between tumor cells and macrophages is required for tumor cell migration in mammary tumors. *Cancer Res* 64:7022, 2004.

509. T. Ichikawa, N. Kyprianou, and J. T. Isaacs: Genetic instability and the acquisition of metastatic ability by rat mammary cancer cells following v-H-*ras* oncogene transfection. *Cancer Res* 50:6349, 1990.

510. S. Taniguchi, M. Tatsuka, K. Nakamatsu, M. Inoue, H. Sadano, H. Okazaki, H. Iwamoto, and T. Baba: High invasiveness associated with augmentation of motility in a *fos*-transferred highly metastatic rat 3Y1 cell line. *Cancer Res* 49:6738, 1989.

511. K. Yusa, Y. Sugimoto, T. Yamri, T. Yamamoto, K. Toyoshima, and T. Tsuruo: Low metastatic potential of clone from murine colon adenocarcinoma 26 increased by transfection of activated c-*erb*B-2 gene. *J Natl Cancer Inst* 82:1633, 1990.

512. K. Pantel, G. Schlimok, S. Braun, D. Kutter, F. Lindermann, et al.: Differential expression of proliferation-associated molecules in individual micrometastatic carcinoma cells. *J Natl Cancer Inst* 85:1419, 1993.

513. G. N. Ranzani, N. S. Pellegata, C. Previdere, A. Saragoni, A. Vio, M. Maltoni, and D. Amadori: Heterogeneous protooncogene amplification correlates with tumor progression and presence of metastases in gastric cancer patients. *Cancer Res* 50:7811, 1990.

514. M. Ladanyi, C. Cha, R. Lewis, S.s C. Jhanwar, A. G. Huvos, and J. H. Healey: MDM2 gene amplification in matastatic osteosarcoma. *Cancer Res* 53:16, 1993.

515. P. G. Cavanaugh and G. L. Nicolson: Purification and some properties of a lung-derived growth factor that differentially stimulates the growth of tumor cells metastatic to the lung. *Cancer Res* 49:3928, 1989.

516. Y. Kang, P. M. Siegel, W. Shu, M. Drobnjak, S. M. Kakonen, et al.: A multigenic program mediating breast cancer metastasis to bone. *Cancer Cell* 3:537, 2003.

517. J. Walter-Yohrling, X. Cao, M. Callahan, W. Weber, S. Morgenbesser, et al.: Identification of genes expressed in malignant cells that promote invasion. *Cancer Res* 63:8939, 2003.

518. T. Ichikawa, Y. Ichikawa, and J. T. Isaacs: Genetic factors and suppression of metastatic ability of v-Ha-ras-transfected rat mammary cancer cells. *Proc Natl Acad Sci USA* 89:1607, 1992.

519. T. Ichikawa, Y. Ichikawa, and J. T. Isaacs: Genetic factors and suppression of metastatic ability of prostatic cancer. *Cancer Res* 51:3788, 1991.

520. B. A. Yoshida, M. M. Sokoloff, D. R. Welch, C. W. Rinker-Schaeffer: Metastasis-suppressor genes: A review and perspective on an emerging field. *J Natl Cancer Inst* 92:1717, 2000.

521. P. S. Steeg, G. Bevilacqua, L. Kooper, U. P. Thorgeirsson, J. E. Talmadge, L. A. Liotta, and M. E. Sobel: Evidence for a novel gene associated with low tumor metastatic potential. *J Natl Cancer Inst* 80:200, 1988.

522. J. Biggs, E. Hersperger, P. S. Steeg, L. A. Liotta, and A. Shearn: A *Drosophila* gene that is homologous to a mammalian gene associated with tumor metastasis codes for a nucleoside diphosphate kinase. *Cell* 63:933, 1990.

523. J. D. Kantor, B. McCormick, P. S. Steeg, and B. R. Zetter: Inhibition of cell motility after

nm23 transfection of human and murine tumor cells. *Cancer Res* 53:1971, 1993.

524. E. H. Postel, S. J. Berberich, S. J. Flint, and C. A. Ferrone: Human c-myc transcription factor PuF identified as nm23-H2 nucleoside diphosphate kinase, a candidate suppressor of tumor metastasis. *Science* 261:478, 1993.

525. M. J. C. Hendrix: De-mystifying the mechanism(s) of maspin. *Nat Med* 6:374, 2000.

526. S. Bandyopadhyay, S. K. Pai, S. C. Gross, S. Hirota, S. Hosobe, et al.: The *Drg-1* gene suppresses tumor metastasis in prostate cancer. *Cancer Res* 63:1731, 2003.

5

Molecular Genetics of Cancer

In 1914, Boveri[1] formulated the somatic mutation hypothesis of the origin of cancer. He thought that the origin of the cancer cell was due to a "wrongly combined chromosome complex," occurring in a somatic cell (rather than a germ cell) and that this caused abnormal cell proliferation. He believed further that this defect was passed on to all cellular descendants of the original cancer cell. He also thought that a single abnormal chromosome combination could account for the malignant character of a cancer cell. It is now well established that some human cancers have a familial distribution (see Chapter 3) and that certain chromosomal rearrangements are associated with human malignant neoplasia.

One of the first chromosomal abnormalities definitively associated with human cancer is the so-called Philadelphia chromosome (Ph[1]), described by Nowell and Hungerford[2] in patients with chronic myelocytic leukemia (CML). The Ph[1] chromosome was at first thought to result from a deletion of part of the long arm of chromosome 22 in the leukemic cells. Rowley[3] later found that the lesion was really a translocation of a piece of chromosome 22 to chromosome 9. More than 90% of patients with CML have the Ph[1] chromosome in their leukemic cells, and the presence of this chromosome is perhaps the strongest argument that chromosomal aberrations are causally related to cancer. This hypothesis is strengthened by the fact that the Ph[1] chromosome is an acquired characteristic of leukemic cells; only the CML-affected individual in a set of identical twins has the Ph[1] chromosome

in his or her bone marrow cells.[4] The development of chromosomal-banding techniques provided a great advance in the identification of chromosome rearrangements in cancer cells. The interaction of certain alkylating fluorochrome (e.g., quinacrine mustard) and histochemical stains (e.g., Giemsa stain) with specific regions of chromosomes produces "bands" along the chromosomes that can be used to fingerprint each chromosome pair.[5,6] A number of definitive assignments of chromosomal changes in various cancers have now been made (see below).

The average chromosome band observed with standard banding techniques contains 5×10^6 nucleotide pairs; deletions or duplications of 2×10^6 nucleotide pairs or less are difficult to detect by this technique.[7] Since the gene size needed to code for an average protein of 50,000 MW is about 1200 nucleotide pairs, approximately 1000 genes could be duplicated or deleted without being detected. Furthermore, it is likely that the entire gene would not have to be duplicated or deleted for its function to be altered. For example, the alteration of gene cis-regulatory elements or transcription termination sequences could have profound effects on gene function. Thus, the chromosome derangements detected in cancer cells by banding underestimated the actual number of cancers that altered gene function. Other, more sensitive methods for detecting genetic alterations are discussed below and in Chapter 7. To think about the potential impact of molecular genetic changes at the cellular level, it is important to first understand the structure and function of chromatin, the DNA–protein complex in cells that determines how genes are packaged and expressed. Most human malignancies are associated with somatic alterations of the human genome that lead to oncogene activation, tumor suppressor gene inactivation, or both.[8]

CHROMATIN STRUCTURE AND FUNCTION

If one stretched out the total amount of DNA in a single human cell, it would be a strand about 2 meters long. Thus, the genetic material of a cell has to be compressed and packaged in a way to fit into the nucleus with a diameter of only 10 μm. This is not just some random process like spooling a long piece of string, but involves a very specific, ordered packaging that allows chromatin to be unwound during DNA replication and cell division and then rewound in daughter cells. Chromatin must also be packaged in a way that allows specific genes to be active or silent during various phases of embryonic development and in response to various internal and external signals, for example, hormones, growth factors, contact with neighboring cells and the extracellular matrix, and exogenous chemical substances. This sequence of activation and silencing of genes goes on in cells throughout life and is a very dynamic process.

The basic unit of chromatin is the nucleosome (described below). It is composed of 146 base pairs of double-helical DNA wrapped around two copies of each of the histone proteins H2A, H2B, H3, and H4. In mammalian cells, a variant form of H2, called H2AX, accounts for about 10%–15% of the H2A composition of chromatin. H2AX has an extended carboxyl-terminal tail compared to that of H2A, and it becomes phosphorylated when DNA is damaged. When one or both copies of the H2AX gene were inactivated in combination with p53 inactivation, a dramatic increase in tumor formation in mice was observed (reviewed in Reference 9). Moreover, evidence of genome instability was observed in mice that lacked both H2AX and p53. Thus, H2AX appears to be one of the "genome caretaker" proteins that acts with p53 as a tumor suppressor.

Components of Chromatin

Chromatin proteins are divided into two classes: histones and nonhistones. There are five major types of histones in eukaryotic cells: the very lysine-rich histone class, H1 ($M_r = 22,000$), the lysine-rich histones H2A ($M_r = 14,000$) and H2B ($M_r = 13,700$), and the arginine-rich histones H3 ($M_r = 15,000$) and H4 ($M_r = 11,300$). These proteins have basic isoelectric points and are highly conserved on an evolutionary scale; there is little variation in amino acid sequence, particularly for the arginine-rich histones, between organisms widely separated on the phylogenetic tree. The estimated mutation rate of histone H4 is 0.06 per 100 amino acid residues per 100 million years, which makes this the most highly conserved protein known.[10] H3 is also highly conserved, but H2a and H2b have undergone more evolutionary changes.

From an evolutionary point of view, H1 is the most divergent histone and also has several detectable subfractions in various organisms and even in different tissues of the same organism. In addition, H1 and the other histones can be modified posttranslationally by the addition of methyl, acetyl, or phosphate groups (see below) that modify their function. There may be a difference of 15% to 20% in the amino acid sequence of histone H1 between species.[11] The amino acid substitutions frequently involve interchanges of lysine, alanine, proline, and serine, and some of these substitutions have potentially important functional results. For example, an alanine is substituted for a serine in some species at position 37, a site at which cAMP-dependent phosphorylation of serine is known to occur in hormone-stimulated rat liver cells.[12]

The nucleosome core histones H2A, H2B, H3, and H4 have multiple domain structures consisting of randomly coiled, highly basic amino-terminal regions and globular carboxy-terminal domains.[13] The flexible, randomly coiled amino-terminal "tails" constitute about 25% of the mass of the core histones. H2A and H3 have, in addition, short carboxyl-terminal tails.

The histone amino-terminal tails are enriched in basic amino acids lysine and arginine, which are highly positively charged at physiologic pH and thus able to bind tightly to the negatively charged phosphates on DNA. The binding of histone tails to DNA is modulated by acetylation of lysine residues in the tails, reducing the charge–charge interaction, and creating potentially more open regions of DNA. Other important post-translational modifications of histones such as phosphorylation, methylation, ubiquitination, and ADP-ribosylation also occur on the tail regions. These post-translational modifications of histones modulate chromatin function (see below). For example, the amino-terminal tail of histone H4 contains four lysine residues, at positions 5, 8, 12, and 16, that are sites for in vivo acetylation, which affects chromatin transcriptional activityl.[13] The cell cycle and gene activation relationships with histone modifications have led to the concept of a "histone code," which postulates that the "language" encoded in the histone tail domains, with their modifiable sequences, is read by other proteins.[14] The histone code concept is that combinations of histone alterations at specific times in the cell cycle "mark" histone tails by these chemical modifications in a way that enables them to recruit or "de-recruit" other chromatin-modifying proteins such as transcriptional co-activators or co-repressors.[15] The lysine-rich histone H1 and its variants H1° and H5 (the H1 variant specific to nucleated erythrocytes) are called *linker* histones because they interact with the linker DNA between nucleosomes, sealing two turns of DNA around the nucleosome core.[16]

The chromatin-associated nonhistone proteins (NHP) are a tremendously diverse group of polypeptides. Several hundred polypeptides have been seen on two-dimensional polyacrylamide gels after electrophoresis of NHP extracted from nuclei. They have a ratio of acidic-to-basic amino acid residues of 1.2 to 1.6, isoelectric points from less than 4 to 9, and a range in molecular weight from about 10,000 to more than 200,000. There is some evidence for a tissue-specific distribution of NHP, but one of the key questions is how many of them are actually involved as structural components of chromatin. This class of proteins contains a number of enzymes, including DNA and RNA polymerases, nucleases, DNA ligase, phosphoprotein kinases, proteases, histone acetyltransferase and methylase, terminal deoxynucleotidyl transferase, and topoisomerases I and II. In addition, the DNA-binding proteins that modulate gene transcription, i.e., transcription factors and

transcriptional co-activators and co-repressors, belong to this group of proteins (see below).

One abundant subset of proteins in the NHP class are the high-mobility group (HMG) proteins, so called because their electrophoretic mobility at low pH is greater than that of most of the other NHP. The HMG proteins were first isolated by Goodwin et al.[17] from calf thymus nuclei. Four distinct proteins in this group, HMG-1, -2, -14, and -17, occur in all animal tissues studied. They are among the more abundant NHP, being present at about 10^6 molecules per nucleus,[18] and they appear to play an important role in the structure and function of chromatin.

The histones are generally stable proteins and are synthesized primarily during the S phase of the cell cycle in conjunction with DNA synthesis. The NHP fraction, however, contains a number of proteins with relatively short intracellular half-lives. Alterations in the pattern of synthesis and turnover of NHP have been observed when cells are induced by hormones or drugs to produce new proteins or are stimulated to progress from a resting to a proliferative state. Regulation of expression of HMG proteins has been correlated with events in cellular differentiation. For example, during muscle development (myogenesis) the levels of HMG-14 and -17 mRNA and protein are down-regulated, and reinduction of HMG-14 expression in differentiating myoblasts prevents normal myogenesis.[19]

Chemical Modifications of Chromatin-Associated Proteins

Important postsynthetic chemical modifications occur for both histone and nonhistone chromatin proteins; they affect the binding of these proteins to DNA and play a role in the control of DNA replication and transcription. A greater variety of postsynthetic modifications occurs for the histones. Two such modifications, acetylation of ε-amino groups of histone lysyl residues and phosphorylation of histone seryl, threonyl, lysyl, and histidyl residues, tend to neutralize positive charges at specific sites in histone molecules. A number of methylations also occur in histones to yield mono-, di-, and trimethyl lysine derivatives, methyl guanidino arginine, and N-methyl histidine. These latter changes tend to increase positive charges at specific sites in histones. These histone modifications occur not only at specific sites in the molecules but also at specific times during the cell cycle, further suggesting that they have some important function in the control of gene expression. For example, in synchronized cultures of Chinese hamster ovary cells, the methyllysine content of histones H2A and H3 begins to rise in the early S phase, peaks after termination of DNA and histone synthesis, coincident with the beginning of the mitotic phase, and begins to fall by mid-M phase.[20] The methyllysine content of histone H2B, by contrast, peaks in the early S phase, coincident with initiation of DNA synthesis, and rapidly falls to its original unmethylated state by late S phase. Acetylation of histones also plays a role in the function of chromatin; for instance, it has been shown that acetylation of arginine-rich histones occurs before the large increase in RNA synthesis observed in phytohemagglutinin-stimulated human lymphocytes and in posthepatectomized rat liver.[21] In both instances, the acetylation occurs before the increased cell division resulting from the mitogenic stimulation. Histone methylation has also been linked to increased expression of certain genes, and recently a specific demethylase has been identified that represses gene expression by maintaining unmethylated histones (reviewed in Reference 22).

As noted above, the positively charged core histone tails bind to the negatively charged phosphodiester backbone of DNA. This interaction is thought to prevent access of transcription factors to DNA promoter–enhancer regions and limit RNA transcription. In support of this idea, it has been shown that the H3 and H4 core histones prevent binding of the transcription factor TFIIIA (see below) to a specific DNA sequence (the gene for 5S RNA) and that acetylation of these two histones allows TFIIIA binding.[23] Moreover, acetylation of histone H4 has been correlated with induction of loop formation in the transcriptionally active regions of the so-called lampbrush chromosomes of oocytes from the amphibian *Triturus cristatus*,[24] a classic model for chromatin activation.

Some of the histone modifications appear to involve reciprocal alterations that affect DNA transcription. For example, Nakayama et al.[25] have shown in fission yeast that histone H3 methylation of lysine-9 is linked to H3 deacet-

ylation on lysine-14, both of which events are necessary for formation of heterochromatin, the form that is inactive in transcription. In contrast, the "on" state of chromatin active in gene transcription is related to acetylation of lysine-14 and phosphorytion of serine-10.[15]

Other similar reciprocal chromatin activating and deactivating events have been observed in mammalian (HeLa) cells, in which methylation of arginine-3 of histone H4 facilitates subsequent acetylation of H4 amino acid "tails," leading to transcriptional activation of a nuclear hormone receptor.[26]

Some gene-silencing events require both histone deacetylation and DNA methylation. Methylation of DNA by DNA methyltransferase recruits methyl-binding proteins and histone deacetylases. This coupling of DNA methylation and histone deacetylation correlates with silent transcriptional regions in chromatin.[27] These processes of controlling chromatin structure and function are key to understanding cell differentiation and the altered gene expression that occurs in malignant transformation.

Some of the genes involved in the acetylation and deacetylation of histones have been identified.[28] The acetylation genes are of two categories: *hat1* and *hat2*. Acetylation of histone H4, for example, reduces the affinity of the histone amino-terminal tail for DNA and allows a reduction of DNA wrapping around the histone octamer and a subsequent decrease in the tightness of nucleosome packaging. This makes more DNA sequences available for transcription. Mutations in yeast deacetylases have been identified that allow H3 and H4 acetylation to be maintained. This would be expected to result in constitutively unfolded regions of chromatin and increased gene transcription. Disruption of deacetylase activity that alters expression of many genes in yeast as well as mammalian cells has been observed.[29] Mutations in histone acetylases, deacetylases, and components of these complexes have significant effects in yeast cells, and similar mutations may have implications for human disease, including cancer. Recent data have shown that members of the histone deacetylase-1 and -2 (HDAC1 and HDAC2) family of genes belong to a network of genes coordinately regulated and involved in chromatin remodeling during cell differentiation.[29]

In addition to acetylation, phosphorylation of histones is also important for chromatin structure and function.[30] H1 interacts with DNA, links adjacent nucleosome cores, and further condenses chromatin structure. Phosphorylation of H1 is thought to play a role in increased gene transcription. Phosphorylation of histone H3, by contrast, is required for proper chromosome condensation and segregation during mitosis.[30] In addition, during the immediate-early response of mammalian cells to mitogens, histone H3 is rapidly and transiently phosphorylated by a kinase called Rsk-2.[31] This suggests that chromatin remodeling is part of the cascade involved in mitogen-activated protein kinase–regulated gene expression.

A "cancer–chromatin connection" is implicated by the observations relating to the role of the tumor-suppressor gene *rb* in the regulation of the histone deacetylase HDAC1.[32] Rb acts as a strong transcriptional repressor by forming a complex with the transcriptional activating factor E2F and HDAC1, tethering these activities to E2F-responsive promoters, including the cyclin E promoter region. Repression of E2F-bound promoters by Rb is released by mitogenic signals that activate cyclin-dependent kinase phosphorylation of Rb, thereby releasing Rb from the complex and allowing histone acetylation to occur. This increases accessibility of gene promoter sequences to transcriptional activators. Point mutations of the *rb* gene observed in some tumors abolish Rb-induced repression and Rb-associated deacetylase activity, allowing increased E2F-mediated gene expression. Viral oncoproteins can disrupt the interaction between Rb and HDAC1. In addition, nonliganded retinoic acid receptors (RARs) have been shown to repress transcription of target genes by recruiting the histone deacetylase complex to these genes.[33] Mutant forms of RAR-α result from chromosomal translocations seen in human acute promyelocytic leukemia (APL). These mutant forms prevent appropriate deacetylase activity and result in dysregulated gene activation. This dysregulation can be diminished by all-*trans*-retinoic acid, at doses that induce APL cell differentiation. These findings suggest that oncogenic alterations in RARs mediate leukemogenesis via aberrant regulation of the histone acetylation state.

The phosphorylation of histones, particularly H1, has also been associated with various phases of the cell cycle. An increase in H1 phosphorylation occurs at specific sites during S phase and additional sites are phosphorylated in M phase.[34] The first phosphorylations may play a role in DNA replication, and the latter in chromosome condensation. At least 50% of H1 molecules are phosphorylated in rapidly dividing cells, and a rapid dephosphorylation of H1 occurs as the cells move into early G_1 phase,[35] indicating specific cell cycle control of these events. Phosphorylation of H1 also occurs following hormone stimulation of certain tissues, but in this case only about 1% of the total H1 is phosphorylated. In glucagon-stimulated rat liver, a cascade of events occurs, leading to the elevation of intracellular cAMP, the activation of cAMP-dependent histone phosphokinase, and the phosphorylation of lysine 37 in H1.[36] Because the phosphorylation of H1 that occurs during S phase is not cAMP-dependent, there are clearly two different kinds of events, one that presumably affects DNA synthesis while the other affects transcription of DNA into RNA. Animal tissues also contain specific phosphatases that can remove phosphate groups and thereby regulate the phosphorylation state of histones. Thus, specific phosphorylation–dephosphorylation of histones appears to play a key role in gene activation events.

Phosphorylation is the most important post-synthetic modification of the nonhistone proteins. Phosphorylated NHP are highly heterogeneous, but they have a tissue-specific distribution and a subfraction appears to bind specifically to the DNA of the tissue of origin.[37] Increased phosphorylation of NHP occurs at times of gene activation—for example, after stimulation of lymphocytes with phytohemagglutinin, following hormone treatment of various tissues, and in the actively transcribing chromosome "puffs" of insect chromosomes. The rates of NHP phosphorylation also vary during the cell cycle, being most rapid during periods of high RNA synthesis (G_1 and early S) and decreasing when RNA synthesis is suppressed (G_2 and M).[38] One of the ways in which NHP phosphorylation could be involved in chromatin activation is by destabilizing the charge–charge interactions between positively charged histones and negatively charged phosphates of DNA.[37] This could open up gene sites for active transcription. Similar to the histones, specific phosphorylation and dephosphorylation ezymes regulate the phosphorylation state of NHP. Both cAMP-dependent and cAMP-independent phosphoprotein kinases have been shown to be involved in the phosphorylation of NHP.[37,39] Among the NHP whose functions are regulated by their phosphorylation state are those involved in cell cycle regulation.

The HMG chromosomal proteins are also subject to a number of postsynthetic modifications, including acetylation, methylation, glycosylation, phosphorylation, and ADP ribosylation. The latter reaction (reviewed in Reference 40) involves the addition of adenosine diphosphoribose groups to carboxyl groups of the proteins, a reaction catalyzed by a chromosomal enzyme, ADP-ribose synthetase. Several ADP-ribosyl groups can be added—up to 50 on some proteins. Histone H1 is also subject to this modification, which is of interest because the nature of the association of H1 histone and HMG proteins with DNA may determine gene activity (see below). Such post-translational modifications most likely affect the binding of these proteins to DNA. HMG proteins 1, 2, 14, and 17 have been shown to contain ADP-ribosyl groups in intact cells.[41] HMG proteins 14 and 17 are also phosphorylated in intact cells, and the extent of phosphorylation varies with different physiologic conditions;[40] for example, the phosphorylation of HMG-14 is higher in metaphase chromosomes than in interphase chromatin and in the G_2 phase of the cell cycle than in the G_1 phase. Phosphorylation of HMG-17 is increased in the early S phase of the cell cycle and in the log phase of growth over that in stationary phase cell cultures. Phosphorylated HMG-14 and -17 have also been shown to induce correct spacing in chromatin nucleosomes assembled in vitro, supporting the concept that they are important in determining chromatin structure in intact cells.[42]

It should be noted here that various chemical carcinogens and oncogenic viruses can alter modification of chromatin proteins by directly reacting with them or by affecting the modifying enzymes. Thus, one way oncogenic agents can bring about the alterations of gene readout observed in malignantly transformed cells is by changing the function of chromatin-associated proteins.

Packaging of Chromatin

Chromatin is a complicated structure, and the way it is "packaged" in cells affects its function (reviewed in Reference 43). The basic organization of chromatin was derived from the results of three kinds of studies: direct visualization of electron microscopy, isolation of chromatin units after nuclease digestion, and neutron and X-ray diffraction analyses. In 1974, Olins and Olins[44] reported the electron microscopic visualization of chromatin isolated from rat thymus, rat liver, and chicken erythrocytes. They observed linear arrays of spherical chromatin particles, which they called v bodies. These are spherical particles about 70 Å in diameter, separated by connecting strands of about 15 Å. This structural arrangement has now been observed for several eukaryotic cells and has come to be known as the "beads-on-a-string" structure of chromatin. The repeating particles of chromatin, the v bodies, or nucleosomes, as they are now called, can be obtained by mild nuclease digestion that clips that connecting DNA links between the "beads" to produce subunit monomers, dimmers, trimers, or higher oligomers (Fig. 5–1). Endogenous endonuclease, staphylococcal nuclease, pancreatic DNase I, and spleen acid DNase II are all preferentially cut between nucleosomes to produce the typical repeating subunit pattern. The nucleosome core may also be cleaved internally by continued digestion with these nucleases to produce pieces of DNA separated by integral multiples of 10, which suggests that the DNA as it is coiled in the nucleosome has a kind of periodicity exposing every tenth base pair to nuclease attack. The size of the DNA in nucleosome monomers varies from about 200 to 240 base pairs, depending on the organism and the tissue from which the nucleosomes were isolated. However, after further nuclease digestion, a nucleosome core that contains 146 (±2) base pairs is produced. This number is invariant for all species and tissues studied so far. Thus, the variation in size of DNA in the nucleosome oligomers is due entirely to the variation in length of the spacer DNA between the cores.[45] The DNA in the nucleosome is wrapped around a core of histones, representing an octamer of two each of histones H2A, H2B, H3 and H4.[9] Histone H1 is not part of the core but binds to the outside of the nucleosomes, and removal of H1 exposes the

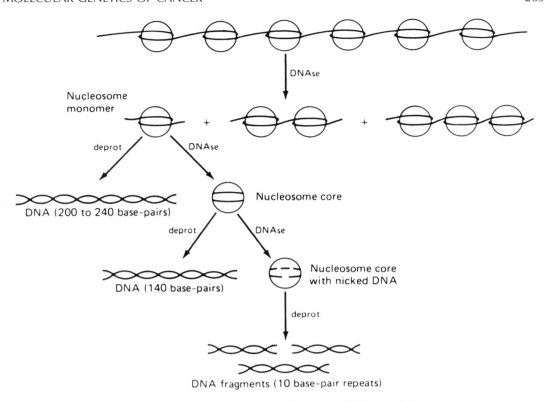

Figure 5–1. Schematic representation of digestion of chromatin by DNase. Chromatin is digested by mild DNase treatment to yield nucleosomes of monomer, dimer, trimer, or higher oligomer size. The DNA content of a nucleosome monomer varies from about 200 to 240 base pairs, depending on the cell type. This can be seen by removing protein from the nucleosome ("Deprot"). Further digestion of nucleosomes by DNase yields the nucleosome core structure, which has a DNA content of 146 base pairs. Still further digestion with DNase yields a core structure with DNA nicked at 10- (or multiples of 10) base-pair intervals.

DNA spacer region between cores to nuclease attack; this arrangement suggests that histone H1 is involved in the tight packing of nucleosomes that occurs in native chromatin.[43] If histone H1 is removed from chromatin by selective extraction, the condensed structure of chromatin is converted to a looser, more open configuration. The HMG proteins appear to provide some selectivity to sites of nuclease attack and to be involved in regulation of gene transcription.

The orders of chromatin packaging are illustrated schematically in Figure 5–2. Data from neutron and X-ray scattering behavior indicate that the nucleosomes are actually cylindrical rather than spherical and have a diameter of 110 Å and a height of about 55 Å.[45] The DNA makes almost two turns around each core, corresponding to about 70 base pairs per turn. A higher order of nucleosome packing is assumed by additional coiling of the nucleosome structure in which the H1 histones appear to play a major role by forming bridges between superhelical turns of DNA in adjacent nucleosomes to form 30 nm–thick chromatin fibers (order II). The HMG proteins are bound to chromatin in a way that provides selectivity of digestion of DNase 1, and this effect appears to require a ratio of only 1 HMG protein molecule for every 10 or 20 nucleosome monomers.[46]

The supercoiled nucleosome is further packed into a solenoid-like structure (order III), with a diameter of about 300 Å and a pitch of 110 Å per

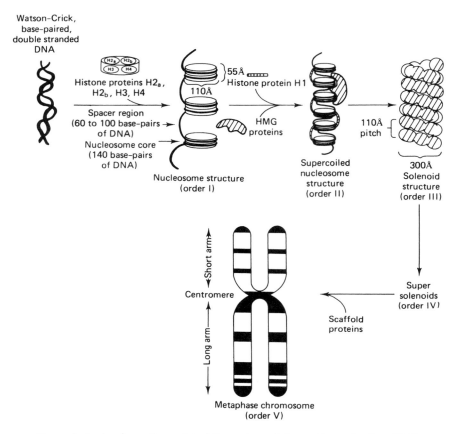

Figure 5–2. A schematic model of the orders of chromatin packaging. DNA and histones H2a, H2b, H3, and H4 interact to form the basic structure of chromatin, the nucleosome (order I). Through additional coiling and bridging of nucleosomes by H1, a supercoiled nucleosome is achieved (order II). This can be further wound into a solenoid-like structure (order III). Two or three further orders of packaging probably occur before the state achieved by the metaphase chromosome (order V) during mitosis can occur.

turn of six nucleosomes.[47] More condensed packing can be achieved by coiling the solenoidal form into yet another order of helix or by winding a number of solenoids around each other. This can be accomplished by kinks in the DNA double helix, in addition to the supercoiling that occurs in achieving the lower orders of packaging.[47] It is likely that two or three higher orders of packaging are achieved before the fully condensed metaphase chromosome is attained. Proteins other than histones appear to be involved in this highest order of packaging because even after the histones (and most of the nonhistones) are removed by extraction from metaphase chromosomes, chromosomal DNA is

still highly organized and relatively compact.[48] The structure of histone-depleted chromosomes is due to the presence of a subset of nonhistone proteins that form a central DNA-containing "scaffold" resembling intact chromosomes in size and shape. Even when the DNA is removed from these structures by nuclease digestion, the scaffold remains intact.

Structure and Function of Interphase Chromosomes

When cells are preparing for cell division and enter the mitotic (M) phase of the cycle, chromosomes condense and become more identifiable as discrete entities. During metaphase they line up along an axis in preparation for cell division and it is the metaphase chromosomes that are used in cytogenetic studies to identify chromosomal defects. However, most of a cell's life is spent in interphase. It is here that the cell spends its time making proteins specific to its differentiated function. Even in interphase, the structure of chromatin, though more diffuse and less visible, has a distinct structure that determines how it functions and which genes are transcribed. Orderly, cell-specific transcription involves highly folded chromosomal domains consisting of large sequences of DNA, up to hundreds of kilobases, and this three-dimensional structure contributes to the gene readout that distinguishes the phenotype of different cells.[49]

The term *chromatin* usually refers to the lower orders of packaging as observed at the level of nucleosomes or nucleosomes wrapped into solenoid-like structures alternatively called *chromonema fibers*.[50] Each full turn around a solenoid contains about 1200 bases of DNA, less than the length of many transcriptional units.[49] Some single genes may be as large as 30,000 bases or up to one million bases. Thus, a transcriptional unit must be able to "read through" several nucleosomes. Superimposed on this solenoid structure are areas of highly condensed or packed solenoids known as heterochromatin. These areas of chromatin are generally transcriptionally inactive. More open regions of solenoidal chromatin are called *euchromatin* and these are generally more active in gene transcription.

Noncoding DNA, i.e., "silent" DNA not transcribed into mRNA, makes up more than 90% of the mammalian genome. This DNA includes satellite DNA, long, interspersed repeated elements, and smaller nontranscribed DNA sequences. This apparently inactive DNA likely plays an important structural role in determining which genes of chromatin are silent and which are transcribed in a given tissue, as supported by evidence that regions of noncoding DNA are nonrandomly dispersed and organized in chromosomes of different mammalian species.

Actively transcribed genes are more accessible to the transcription machinery, e.g., transcription factors and RNA polymerase, than highly condensed regions of chromatin. One index of this accessibility, or "openness," is the sensitivity to nucleases. The domains of chromatin that are open for gene transcription vary from cell type to cell type and even within the same cell type during the differentiation process. For example, during red blood cell differentiation (erythropoiesis), there is a progressive change of chromatin to a more heterochromatic, less transcriptionally active state as the cells become more terminally differentiated. The opposite occurs in neuronal cells. As early neuroblasts differentiate into neurons, chromatin becomes more euchromatic, consistent with the high transcriptional activity of brain cells. A number of events that damage DNA during oncogenesis may also depend on the packaging of chromatin. For example, more open regions of chromatin may be more subject to insertion of viral genes, damage by chemical carcinogens, chromosomal breakage, and translocation.

One question in chromatin research that has yet to be answered is how chromatin packaged in nucleosomes can be transcribed. In other words, how does the transcriptional machinery get access to DNA wrapped around a core of histones and linked together by H1 histones? One thing is agreed upon, which is that H1 histone is present at lower concentrations in active chromatin than it is in inactive chromatin, allowing some loosening of the tightly packed array of nucleosomes. But the way in which transcription proceeds from there is unclear. There are several theories to explain this:[51–54] (1) the displacement of core histones H2A and H2B, allowing a partial release of DNA from the core such that it can more easily complex with RNA polymerase; (2) a "rolling stone" effect such that RNA polymerase induces a progressive, transient displacement of histone H2A-H2B dimers as it passes through the nucleosome; and (3) the transfer of histone octamers to sites behind the transcribing polymerase once it has passed through a region of DNA. Some type of opening of the tight nucleosome packing is probably required for efficient gene transcription. As noted above, there is evidence that acetylation of histones, particularly H4, and ubiquitination (the addition of ubiquitin to a lysine side chain) of H2A and H2B occur in transcriptionally active regions of chromatin, suggesting that these modifications may open up sites for initiation of transcription (i.e., binding of transcription factors).

In the yeast *S. cerevisiae* most transcription factor–binding motifs were devoid of nucleosomes.[51] Whether these transcription factor–promoter binding sites are nucleosome-free all the time or whether transcription factor complexes "push" nucleosomes out of the way is an open question. Currently, we know of no mechanism that clearly explains how transcription occurs on chromatin arranged in nucleosome arrays. There is even in vitro evidence that no histone octamer dissociation is required for T7 RNA polymerase to transcribe DNA present in nucleosomes.[53]

It is clear that histone modifications have a lot to do with "marking" gene activation sites in nucleosomes (reviewed in reference 54a). For example, a triple methylation mark on lysine 4 of histone H3 (designated H3K4 me3) is a hallmark for all active genes. It has been shown that the nucleosome remodeling factor CHD1 binds preferentially through a domain called the chromodomain to nucleosomes containing the H3K4 me3 mark.[54a] Nucleosome remodeling factors are enzymes that disrupt histone-DNA interactions, leading to local opening of the chromatin, recruitment of histone acetyltransferase, and downstream chromatin modifications that open repressed (inactive in transcription) chromatin.

Another nucleosome remodeling factor called NURF can also be recruited to H3K4 me3-marked chromatin, which plays a role in developmentally regulated gene activation.[54b] The BPTF subunit of NURF interacts with H3K4me3-marked sites via a protein domain called the PHD finger, a protein structural fold coordinated by two zinc atoms affinity for methylated histone tails, which is a general feature of PHD fingers.[54b,c]

One might also ask how DNA bundled into nucleosomes replicates itself during S phase in the cell cycle. Taken together, the data indicate that DNA replication in higher eukaryotes occurs just as it does in simple eukaryotes and bacteria—that is, replication occurs in a bidirectional manner from a specific origin of replication and proceeds via a replication fork producing one continuous daughter strand and one discontinuous strand (Okasaki fragments) (Fig. 5–3). Newly synthesized DNA is rapidly assembled into nucleosomes. Newly synthesized and assembled histone octamers are randomly distributed to both arms as the replication fork passes through each nucleosome.[55]

Nuclear Organization

The cell nucleus has a structural framework that consists of the nuclear envelope, nuclear lamina, and elements of the nucleolus as well as interactive contact points with chromatin. The nuclear envelope mediates bidirectional molecular traffic between the cytoplasm and nucleus via nuclear pores. The nuclear lamina is a fibrillar meshwork that lines the nucleoplasmic surface of the nuclear envelope and interacts with the envelope via the lamin B receptor (LBR). There are eight transmembrane domains of the LBR in the inner nuclear membrane (reviewed in Reference 56). The LBR also forms a link between the nuclear lamina and chromatin. This linker involves a protein called *heterochromatin protein 1* (HP1), which, as the name implies, is involved in formation of transcriptionally inactive heterochromatin.

The nuclear lamina is a network of polymeric filaments consisting of lamin proteins and associated lamin binding proteins. Lamins have affinity for chromatin structures and are involved in the spatial organization of chromatin during its various functional activities: DNA replication, chromosome condensation and nuclear breakdown during mitosis, and positioning of active DNA sequences for gene transcription. The lamins are 60 to 78 kDa intermediate filament proteins.

There are three types of lamins: closely homologous lamins A and C and a distinct lamin B. Human cells have three lamin genes: *lmna*, which encodes four isoforms including lamins A and C, and *lmnb*-1 and -2, which encode B-type lamins.[57] The lamins are essential for eukaryotic cell function, as evidenced by their key role in DNA replication, chromosomal segregation during mitosis, gene transcription, and nuclear integrity. How they do all this isn't clear, but they are players in all these events.

There are at least two human diseases associated with hereditary mutations in the *lmna* gene: Emery-Dreifuss muscular dystrophy, a muscle-wasting disease, and Dunnigan-type familial partial lipodystrophy. These mutations affect the charge and hydrophobicity of lamins, which disrupts the stability of lamin dimers and multimers and thus the integrity of the nuclear lamina.[58] Lamin dysfunction could also cause impaired interactions between mutant lamins and chromatin, nuclear envelope proteins, transcription factors,

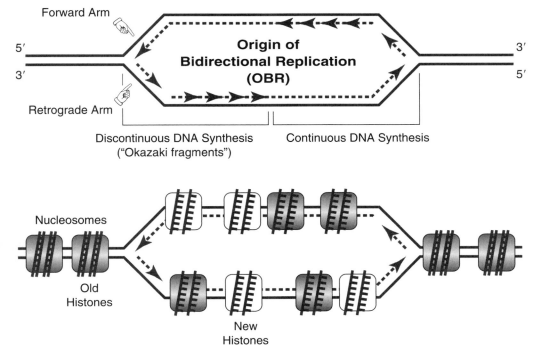

Figure 5–3. Replication fork model with random segregation of prefork histone octamers. (From DePamphilis,[55] with permission.)

or other components of the genetic machinery. Why only two rather rare human diseases have been found associated with lamin dysfunction isn't clear, but it is likely that more will be discovered. Since there is differential expression of various lamins and lamin-associated proteins in different tissues, the effects of lamin-gene mutations might vary, depending on which tissues are primarily affected.[58]

The observation that sites of active gene transcription appear to be at contact points of chromatin with the nuclear lamina (sometimes called the *nuclear matrix*) suggests that these lamina (matrix) association regions may provide transcription start sites.[59]

It has also been shown that the protein composition of the nuclear lamina depends on the differentiation state.[60] The lamina protein lamin B is expressed throughout differentiation in a number of cell types, and lamins A and C are absent in certain undifferentiated cell types. Some differences have been observed between normal cells and cancer cells in the composition of their nuclear matrix proteins. For example, it has been reported that prostate adenocarcinoma cells have a different array of nuclear matrix proteins from that of normal prostate.[61] Another study showed that nuclear matrices derived from normal human breast tissue and from breast cancer tissue share some common nuclear matrix proteins but also have some specific differences in their protein composition.[62] Colon cancer tissue has also been shown to contain an array of nuclear matrix proteins different from those in normal colon.[63] Nuclear matrix proteins, most likely derived from cell death and release of nuclear contents, have also been detected in the sera of patients with breast, colon, uterine, lung, ovarian, and prostate cancer at higher levels than in normal sera.[64] A number of characteristic differences in the nuclear matrix architecture between normal and cancer cells have also been observed.[65]

Nuclease Sensitivity

The fact that only about 10% to 20% of the total DNA contained in the chromatin of eukaryotic cells is transcribed suggests that there must be something structurally different about the way active sequences are packaged. Much of the evidence indicating that actively transcribed genes are packaged differently from the bulk of nontranscribed DNA has been obtained by the use of endonucleases that can cleave between base pairs within a DNA sequence. Weintraub and Groudine[66] have shown that the globin gene in chick erythrocyte nuclei is preferentially sensitive to digestion with pancreatic DNase (DNase I) but not to digestion with micrococcal nuclease. The resistance of globin genes to micrococcal nuclease suggests that these genes are contained in nucleosomelike particles, whereas the sensitivity to DNase I indicates that the chromatin regions containing these genes are conformationally different from most nucleosomes.

Active transcriptional units from some species can actually be observed in the electron microscope, for example, the Balbiani rings on the large chromosome IV in the salivary glands of *Chironomus tentans* and the ribosomal RNA genes in *Oncopelius fusciatus* embryos.[67] Inactive chromatin isolated from these organisms has a uniformly beaded appearance, whereas the areas of active transcription have "streamers" of growing, nascent RNA chains attached, and the nucleosome beads, although present in the area of intense transcription, are decreased in frequency along the chromatin fiber. In addition, a smooth unbeaded segment is visible in the electron microscope about 500 base pairs 5′ to the region of the growing RNA chains. This area may correspond to the "hypersensitive" DNase sites (see below). A similar change in the frequency of nucleosome packing is also seen in early development of *Oncopeltus*. At 32 hours after fertilization (early blastula), only tightly beaded chromatin is seen, but 6 hours later nonbeaded stretches of chromatin appear in the region of ribosomal RNA synthesis before actual RNA chains can be detected biochemically.

Francis Collins and colleagues[68] have developed a way to identify gene regulatory elements by genome-wide recovery of DNase hypersensitive sites. They generated a library of DNase I–hypersensitive sites from quiescent primary CD4[+] T lymphocytes and cloned the sequences. Sequences from these cloned DNase I–digested segments were frequently mapped to regions upstream of genes or within CpG islands (fre-

quently found in regulatory elements; see below). These cloned sequences also tended to map near genes that are actively transcribed in CD4[+] T cells, showing that these were transcriptionally active regions of the genome in T cells. This method should be useful to fish out active genes and their regulatory elements in any cell type and perhaps delineate how these regulatory elements actively transcribe genes change during embryonic development and tissue differentiation and during disease initiation and progression.

Transcriptional Activation and the Cancer Connection

As can be discerned from the previous discussion, for gene transcription to occur, some degree of chromatin remodeling must occur. Some of this remodeling is mediated by a multiprotein complex called *SWI/SNF*, a nine (or more) protein complex that is evolutionarily conserved from yeast to humans. There is an invariant core complex and variable subunits that contribute to transcriptional activation or repression. The complex uses the energy of ATP hydrolysis to mobilize nucleosomes and remodel chromatin. The function of this complex is involved in cancer development, and a number of its subunits have tumor-suppressor activity or are required for the activity of other tumor suppressor genes.[69]

The subunits of SWI (faulty mating-type switching) and SNF (sucrose nonfermenting factor) were originally discovered in yeast mutant screens (reviewed in Reference 69). A number of genes were isolated from these mutants and were designated SWI or SNF according to the function with which they were linked. Later it was found that the products of these genes were members of a linked multiprotein complex, hence the term *SWI/SNF complex*. In yeast, transcription of about 5% of all genes are regulated by SWI/SNF. In actuality, SWI/SNF represses more genes than the complex activates, and interestingly, the repressed genes are interspersed throughout the genome.

During screens for suppressors of SWI/SNF mutations, it was found that SWI/SNF function was linked to chromatin structure and function (reviewed in Reference 69). SWI/SNF is recruited to chromatin, where it hydrolyzes ATP and uses this energy to remodel nculeosomes, as evidenced by alteration of DNAse I–sensitive sequences, extrusion of DNA from nucleosomes, and movement of nucleosomes along the DNA helix. The function of some of the SWI/SNF subunits has been defined: BRG1 or BRM (human homologs of the yeast ATPase subunits SWI2/SNF2), SNF5 (INII), BAF155, and BAF170 are required for in vitro nucleosome remodeling. The function of the other SWI/SNF subunits is still being defined. An important point to note here is that there is cell-to-cell heterogeneity in the composition of the SWI/SNF complex.

In addition to regulation of gene transcription, the SWI/SNF complex has been linked to DNA synthesis, viral gene integration and expression, and mitosis, all of which could be related to SWI/SNF's role in tumor suppression (reviewed in Reference 69). SNF5 is a core subunit of the complex, is present in all known variants of the complex, and is required for both in vitro and in vivo chromatin remodeling. Inactivating mutations or deletions of SNF5 have been observed in a number of human cancers, including malignant rhabdoid tumors, pediatric choroid plexus carcinomas, meningiomas, medulloblastomas, and primary neuroectodermal tumors.[69] Children suffering from the rhabdoid tumors were found to be carriers of a germline mutation in one allele of the snf5 gene and the remaining allele was deleted in the tumors (a classic example of the Knudson two-hit hypothesis). The tumor suppression function of SNF5 isn't clear, but it appears to be due to its ability to induce cell cycle arrest by increasing expression of the cell cycle regulators INK4A, CDK4, and/or Rb. Thus SNF5 may act upstream of Rb in control of cell cycle progression.[69]

In addition, mutation or loss of the ATPase subunits BRG1 and BRM have been observed in a variety of human cancers: BRG1 was found to be mutated in pancreatic, breast, lung, and prostate cancer cell lines, and 10% of non–small cell lung cancer primary tumors lack expression of both the brg1 and brm genes.

The SWI/SNF complex also has been shown to have interactions with tumor suppressor genes rb and brca1 and with the c-myc and mll (mixed lineage leukemia) oncogenes. The ability of the Rb protein to prevent cell cycle progression is dependent on the BRG1 and BRM subunits, since cells lacking these subunits are unable to undergo Rb-mediated cell cycle arrest. The Rb repressor complex contains histone deacetylase (HDAC) and SWI/SNF. This complex blocks transcription of cyclins E and A, producing G_1 cell-cycle arrest.

BRCA1 forms a multiprotein complex with SWI/SNF. A dominant-negative mutant form of BRG1 blocks the BRCA1-mediated stimulation of p53-regulated genes, activity suggesting an additional tumor-suppressing effect of SWI/SNF subunits. On the other side of the oncogenesis coin, the SNF5, BRG1, and BAF53 subunits of SWI/SNF are involved in activation of the c-myc gene, and several lines of evidence indicate that the MLL oncogenic fusion proteins function by interaction with the SWI/SNF complex.

Control of Gene Expression During Embryonic Stem Cell Differentiation

Gene expression that occurs during embryonic stem cell differentiation must be carefully regulated in order for ES cells to proceed from the totopotent to pleuripotent to a specific differentiated state. Some of the key steps in this process have been defined, and they involve the way chromatin is packaged and nucleosomes are arranged to alternatively repress and express differentiation genes. This regulatory pathway involves polycomb repressive complexes (PRCs), specific methylation sites on histone H3, and binding of transcription factors OCT4, SOX2, and NANOG (reviewed in Reference 69a).

PRCs are composed of polycomb groups proteins (PcGs), which can form multiple PRCs. The PRCs have a function conserved from Drosophila to humans. The PRCs bind to chromatin and function to repress gene expression by epigenetic modification of chromatin structure. One of the PRCs, PRC2, catalyzes methylation of lysine-27 in histone H3. This methylation event provides a binding site for PRC1, which causes chromatin oligomerization, condensation, and inhibition of remodeling that maintains silencing of gene expression. PRC2 contains a subunit called SUZ12, and PRCs containing this subunit bind to and repress a broad array of developmental genes.[69a] The sites occupied by PRC2-SUZ12 are on nucleosomes containing trimethylated lysine-27 on H3. The PRC2-inhibited target genes include the ES cell regulatory transcription factors OCT4, SOX2, and NANOG, which are required for propagation of ES cells in culture. Thus, co-occupation of nucleosomes with PRC-2 and these transcription factors appears to block emtry into a developmental pathway and keep ES cells in the pluripotent state,

A puzzling thing about these data is that both repressors and activators of the genes involved in ES differentiation are bound to the same nucleosome sites. Thus, it is as if these sequences of the genome that are inhibited are at the same time poised to trigger the differentiation pathway. This conundrum is at least partly explained by how histone H3 is methylated. Bernstein et al.[69b] have found that specific H3 methylation patterns create "bivalent domains" that silence developmental genes in ES cells while keeping them poised for activation. H3 methylated lysine-27 is a "repressive mark" and H3 methylated at lysine-4 is a "permissive mark" for gene expression. Gene domains in chromatin containing both marks are repressed. Those containing only the lysine-4 mark are activated. During differentiation of ES cells. Some DNA sequence domains retain the lysine-27 mark and these genes remain repressed. Other domains retain only the lysine-4 mark, allowing differentiation to proceed. The direct signals for lysine-27 methylation and the segregation of domains containing the two histone H3 marks are not yet clear. Interestingly, lysine-4 methylated H3 bound domains are associated with CpG island-rich DNA, suggesting an additional way that expression of these CpG-rich sequences (which are often in promoter sequences) can be regulated.

SPLIT GENES AND RNA PROCESSING

Another important aspect of the transcriptional capability of chromatin is the combined question of how genes are put together and how the transcriptional products of genes are processed. The isolation and characterization of the *restriction endonucleases*,[70] enzymes that cleave DNA at specific sequences, have revolutionized our understanding of the organization of genes and led to the ability to clip out specific genes, insert them into bacteria, and produce millions of copies of a specific gene and the product of that gene. This ability has brought about the revolution in molecular biology known as "genetic engineering." One of the startling discoveries from these studies

is the finding that many genes of eukaryotic cells are split into a number of pieces. This means that there is yet another order of control interposed in the flow of information from DNA to RNA to protein, another control point that could malfunction in the generation of neoplastic cells. The sequences of DNA converted into functional mRNA are called *exons*; the sequences of DNA intervening between these are called *introns*.[71] RNA polymerase transcribes a whole gene complex with its introns and exons. The RNA transcript is then "processed" to remove the intron segments and "spliced" to put the pieces of mRNA together that are to be translated into protein.

The mechanisms of the splicing and processing reactions are not entirely understood, but a number of the key components have been recognized. The nucleic acid–base sequences at the points of the exon–intron junctions have a specific arrangement, known as "Chambon's rule," because it was first clearly demonstrated for the ovalbumin gene in Chambon's laboratory, that introns begin with a GT and end with an AG base sequence.[72] This same sequence arrangement has been found in most exon–intron junctions that have been sequenced in eukaryotic genes. Additional "consensus" sequences have been found around the junctions of many genes. The generality of the GT-AG rule and the presence of other consensus sequences strongly suggest that the splicing enzymes (i.e., the specific endonucleases and RNA ligases) involved in the processing of precursor messenger RNA (mRNA) molecules into mature mRNA are phylogenetically very old enzymes that have evolved from a single ancestral enzyme system. For example, when the precursor mRNA of chicken ovalbumin is introduced into mouse cells in culture, the pre-mRNA is cut and spliced correctly by the mouse RNA-splicing enzymes.

The primary transcript is further modified by the addition of a methylated guanine "cap" at the 5' end and of a polyadenylate "tail" at the 3' end. These modifications play a role in correctly initiating translation of the mRNA into protein and in protecting the mRNA from nuclease digestion. The ligated, capped mature mRNA is transferred by means of a protein carrier[73] through nuclear pores to the cytoplasm, where it binds to polyribosomes and is translated into protein.

The recognition signals for the cutting and splicing points in primary RNA transcripts involve the binding of "small nuclear RNA" (snRNA) present in ribonucleoprotein complexes called *small nuclear ribonucleoprotein particles* (U-snRNPs), so named because they contain a family of uracil-rich snRNAs found in eukaryotic cells. It is now clear that the snRNPs are key factors for RNA sequence recognition during mRNA splicing.[74] Changes in these factors could regulate mRNA splicing patterns as cells differentiate or as cells are exposed to different physiological stimuli. A splicing reaction involves the formation of a "spliceosome" containing at least three snRNPs: U2, U5, and U4 + U6. A U2 snRNP complex is a likely intermediate in the formation of the spliceosome. The U1 and U5 snRNPs appear to be required for recognition of the 5′ and 3′ splice sites, respectively, on RNA. Mutations in the highly conserved AG-containing sequences at the 3′ splice site and deletions in the polypyrimidine sequence that binds U2 snRNP prevent normal splicing[75] and may explain how mutations in these regions could produce an abnormal mRNA.

There are three different types of introns: group I, group II, and nuclear pre-mRNA introns.[76] Group I and group II introns have been found in *Tetrahymena*, yeast, *Neurospora*, *Physarum*, and various other fungi, bacteria, and plants. However, similar elements may exist in higher organisms. Group I and group II introns have distinguishing structural features. For example, group II intorns have a highly conserved secondary structure with a core sequence and six looped-out helixes that define different sequence domains. Nuclear pre-mRNA introns have the conserved 5′ and 3′ sequence motifs described above but do not have the conserved structural sequence motifs of group I and II introns. The group I and II introns are mobile—i.e., they can move around in the genome, and they contain open reading frames that encode proteins. Unlike most other mobile DNA transpositions that involve nonhomologous donor and recipient sequences, group I intron mobility is site-specific, resulting in intron insertion at specific alleles and sometimes involving repeated rounds of insertion.[76] This intron mobility requires a site-specific double-stranded DNA endonuclease encoded by the intron itself. Other proteins encoded by group I introns include RNA splicing enzymes. Some group II–encoded sequences have homology with

the enzyme reverse transcriptase, which would allow for production of reverse-transcribed cDNA from processed RNA and could be a mechanism for preserving intron sequences and providing mobility and reinsertion substrates.

Group I and group II introns can be viewed as "ribozymes," in that they produce RNA products that encode products that catalyze their own splicing and play a role in their mobility. In one sense, they can be viewed as "infectious introns" because they can move around within a genome and possibly even between species.[77] Group I introns are said to be "homing" introns because they are site-specific and characteristically unidirectional, i.e., they produce nonreciprocal insertions into allelic related genes that involve intron-encoded site-specific endonuclease. Group II intron "homing" is less well characterized and most likely involves the intron-encoded reverse transcriptase-like activity. The cDNA copies of excised introns' mRNA may be the actual "vectors" for transposition to other sites in the genome and in that sense resemble the transposable elements (transposons) described originally in corn by McClintock (see below). However, both group I intron endonucleases and group II intron reverse transcriptases function in RNA splicing.[77]

Self-splicing group I and group II introns are thought to be vestiges of primordial evolution. For example, group I introns of a cyanobacteria tRNA gene are found in exactly the same location in the analogous tRNA gene in plant chloroplasts. This similarity suggests that this intron was present prior to the endosymbiotic incorporation of cyanobacteria into plant progenitor cells, thought to have occurred about one billion years ago.[77] This intron is absent, however, in the analogous tRNA gene in mitochondria, which suggests that it was sporadically distributed after that time.

Group II introns have only been found in cellular organelles such as chloroplasts and mitochondria, and are considered to be the evolutionary precursors of the nuclear pre-mRNA introns that code for snRNAs involved in spliceosome formation and RNA splicing in higher organisms. In any case, the presence of introns in genes has provided a mechanism for "exon shuffling," alternative mRNA splicing, and gene regulation that may well have conferred selective advantages to the host organisms over evolutionary time.

A question arising from the discovery of the split-gene arrangement of eukaryotic genes is whether multiple protein products could result from one primary transcript, depending on the stop–start points of the cutting and splicing mechanisms. The answer is clearly "yes." An example is the generation in the rat of three different mRNAs coding for different types of the cellular matrix protein fibronectin;[78] two of these mRNAs arise from the use of different 3′ splice sites contained within a single exon. Similarly, rat muscle myosin light chains 1 and 3 are produced from a single gene by differential RNA splicing of a single primary gene transcript.[79] In addition, a single mouse α-amylase gene can transcribe two different α-amylase mRNAs in different tissues of the mouse.[80] In all these cases, the splice site sequences follow Chambon's rule of GT at the 5′ end and AG at the 3′ end; however, the sequences around these splice sites vary. This implies tissue-specific and, in some instances, differentiation-specific recognition of these splice sites. Candidates for providing such specific recognition are (1) different splice enzymes activated at different times during differentiation or in different tissues; (2) tissue-specific or differentiation-specific snRNAs that bind to these sites; and (3) protein signals that differentially mark the splice sites. The first alternative seems unlikely in view of the commonality of splicing enzymes noted earlier. The other two candidates, or a combination of them, seem likely possibilities.

The importance of carefully maintaining the correct splicing steps in mRNA processing is indicated by what happens when processing goes awry. For example, in a type of β$^+$-thalassemia, characterized by anemia with reduced β-globin synthesis, decreased production of normally functional β-globin mRNA is caused by abnormal splicing of the β-globin gene primary transcript.[81] The β-globin gene from such patients contains a single base substitution 22 base pairs upstream from the 3′ junction between intron 1 and exon 2 of the gene, creating an alternative splice site within intron 1 and resulting in the retention of 19 extra bases from the 3′ end of intron 1 in the mRNA. This abnormally spliced mRNA appears to be less stable and poorly transported to the cytoplasm.

One can understand how such a delicate and complicated series of events as the correct transcription and splicing of mRNA sequences could be interrupted or upset by agents that interact with DNA or chromatin proteins. There are a number of guanines around the key exon–intron junction sites, guanine bases in DNA being one of the primary targets of alkylating-type agents, which frequently are chemical carcinogens (see Chapter 2). Irradiation damage at these key DNA sequences could also disrupt the correct transcription or splicing of mRNA. Oncogenic viruses are known to insert their DNA into the host's DNA at various points. If these alterations occurred in a key functional gene sequence, such as one involved in coding for an enzyme involved in a crucial metabolic step, the result would probably be fatal to the cell. But if such alterations or insertions were at a key control point, such as one involved in switching genes on or off or in the correct processing of exon–intron sequences in mRNA, the cell might survive, although in a way that would allow it to transcribe, process, and translate mRNA abnormally. It has been shown, for example, that there are "intron mutants" in yeast for the mitochondrial gene of cytochrome b, an enzyme in the respiratory chain.[82] These mutations result in an abnormality of RNA splicing that produces an altered but in many cases still functional protein.

Clearly, alternative pre-mRNA splicing can play an important role in embryonic development, tissue differentiation, and human disease. It is somewhat surprising that more than half of all human genes have alternatively spliced transcription products. Johnson et al.[83] used DNA microarrays to monitor splicing at every exon–exon junction in more than 10,000 mutli-exon human genes in 52 tissues and cell lines. Their results showed that 74% of human multi-exon genes are alternatively spliced. The frequency of this phenomenon may explain the estimation that 15% of point mutations causing human disease are those that affect pre-mRNA splicing. While the samples with the highest frequencies of alternative splicing events were cell lines, many such events were observed in human tissue samples. Of the genes affected, the highest frequencies were found for genes involved in cell communication, receptor tyrosine kinases, and enzyme regulation.[83] Similar tissues had similar patterns of alternative splicing—e.g., fetal liver was like adult liver; stomach was similar to small intestine; heart similar to skeletal muscle; and neuronal tissues were similar to each other.

Aberrant alternative splicing occurs in human cancers. In a genome-wide screen of 11,014 genes in human cancers, Wang et al.[84] identified 26,258 alternative splicing isoforms, of which 845 were associated with human cancers (liver, brain, lung, kidney, and prostate cancers), and 54 were specifically associated with liver cancers. Interestingly, canonical GT-AG base splice junctions were used significantly less frequently in the alternative splicing isoforms generated in tumors.

Another interesting result of research on splice variants is the finding that the so-called junk DNA that exists in introns is really useful stuff. In simple organisms like yeast, algae, and bacteria, introns actually code for some proteins called *inteins*. Some of these inteins are enzymes that can insert DNA sequences into precise loci in the genome, providing a unique way to do protein engineering. This technique has been used to create new proteins not ordinarily made in bacteria. In yeast, one of the inteins codes for an endonuclease that clips DNA in a recipient cell during yeast cell mating and introduces a new sequence. By genetic recombination the intron sequence is then inserted into the donor cells' own DNA.[85] Hundreds of other group I intron-coded enzymes have now been found in yeast, algae, viruses, and the mitochondria and chloroplast genomes of higher plants. Group II introns have also been found to encode inteins. Using plasmids carrying the enzyme encoding introns, some investigators have designed new protein "manufacturing" systems in bacteria (reviewed in Reference 85). For example, intron technology has been used to add viral resistance to cheese-making lactic acid bacteria. In the future, inteins could be used to mass-produce purified proteins, produce hard-to-make proteins, and link proteins to small molecules such as drugs.

In primate, including human, cells there are short (about 300 nucleotides in length) repeat sequences called *Alu* elements. These elements have a copy number of 1.4 million, making up 10% of the total human genome. These sequences are interspersed in the genome and were originally thought to be nonfunctional junk DNA. It is now known that parts of these *Alu* elements can be inserted into mature mRNA by splicing, a process called *exonization*. This process presumably occurs via recognition by the splicing machinery of splice motif sequences in *Alu* elements. These movable elements have most likely contributed to the unique features of primates that have been achieved through evolution. Over 5% of human alternatively spliced exons are *Alu* derived, and most if not all *Alu*-containing exons are alternatively spliced.[86] These splice variants have a good news–bad news connotation. The good news is that unique helpful phenotypes have been achieved over evolutionary time. The bad news is that some *Alu* insertions can cause genetic disorders and disease.

GENETIC RECOMBINATION

Genetic recombination is the process by which new combinations of nucleic acid sequences are generated by shifting around genetic elements in the genome. While most organisms judiciously guard their genetic material, there are times in the life cycle of an organism, during early development in a multicellular organism, for example, when it is advantageous to move genetic information around to achieve a new set of genetic alterations. One obvious example is the ability of microorganisms to mutate and adapt to new environments rapidly, e.g., to gain resistance to antibiotics or to an immunological challenge from the invaded host. Indeed, a lot of what is known about genetic recombination has been gleaned from bacteria and parasitic organisms such as trypanosomes.[87,88]

Because gene rearrangements are potentially dangerous to the viability of an organism, they are well regulated in higher organisms, unlike in an organism like the pathogenic yeast *Candida albicans*, which can switch among at least seven phenotypes with a frequency of 10^{-4} per cell division.[87] In humans, willy-nilly, inopportune gene rearrangements can lead to inherited disease or cancer. "Incidental" or unprogrammed rearrangements can arise from errors in DNA replication, repair, or recombination, from the movement of mobile elements such as transposons (see below) or from the insertion of viral DNA. Most of these events are deleterious. Programmed gene rearrangements, by contrast, are part of the normal developmental process and are developmentally regulated.

Genetic recombination is classified into three categories: (1) general recombination between homologous DNA regions; (2) transpositional recombination carried out by transposable elements (transposons) that jump from one chromosomal location to another; and (3) conservative site-specific recombination that produces rearrangements occuring in specific gene sequences within chromosomes (e.g., the gene rearrangements that occur in V(D)J antigen receptor genes of lymphocytes; see below).[88]

The first inklings of genetic recombination came from studies of Bateson and colleagues in 1905, when they found some non-Mendelian inheritance traits in the sweet pea (*Lathyrus odoratus*) (reviewed in Reference 89). Certain combinations of traits occurred more frequently than expected and others less frequently. We now know that this was due to genetic recombination.

General recombination was discovered in *Drosophila* in 1911 by Morgan, who coined the term *crossing over*.[90] This term was used to describe the exchange that gave rise to new combinations of linked genetic traits in fruit flies. Genetic recombination involves exchanges of genetic information at equivalent positions along the length of two chromosomes with significant sequence homology. A number of models theorizing how this occurs have been proposed, but the mechanism remains unclear.[89]

Transpositional recombination was discovered by McClintock in maize in the 1940s, but her findings were ignored for many years. She was finally recognized for her contribution with the awarding of the Nobel Prize in Physiology and Medicine in 1983.[91] At the time of Dr. McClintock's experiments, maize plants provided the best material for locating known genetic traits along a chromosome and for determing the breakpoints in chromosomes that had undergone various types of gene rearrangements. The crucial experiment began with the growing of about 450 plants, each of which started its development with a zygote having received a damaged chromosome, with a broken end, from each parent. Such ruptured chromosomes could be produced by X-ray exposure of germinating plants. The seedlings developed from these matings produced totally unexpected phenotypic variants. Variegated seedlings had startlingly different pro-

duction of chlorophyll, this altered expression being confined to clearly defined sectors in a given leaf. McClintock concluded that the modified expression of the genes regulating chlorophyll production had to be related to an event in the precursor cell that gave rise to that sector of the leaf. For this differential gene expression to occur in a localized area of the leaf, all of whose cells had arisen from the same parent cell, some cell component had apparently been unequally segregated at mitosis. Dr. McClintock originally called these "controlling elements," but we now know, based on her pioneering work, that these are transposable genetic elements, or *transposons*. The mechanism for the heritable segregation of traits in maize was not simply a result of the passage of unrepaired broken chromosomes, because it was clearly demonstrated that broken ends of ruptured chromosomes found each other and fused in the telophase nucleus. Rather, the selective appearance of different genetic traits was due to the activation of transposable elements carried normally in the silent state in the maize genome. Once these elements are activated, their mobility allows them to enter different gene loci and "take over control of action of the gene wherever one may enter."[91]

Movable genetic elements, or transposons, are now known to be present in many organisms, including bacteria, yeast, and the fruit fly *Drosophila*. In *Drosophila*, these elements are present as repeated gene sequences—as many as 30 copies per cell. These repeated sequences are dispersed among several chromosomes and appear to be nomadic. Similar sequences exist in higher organisms, including mammals. In humans, the *Alu* family of sequences are likely candidates for mobile DNA elements.[92] The bacterial and maize elements turn genes on and off as they move around in their respective genomes. Moreover, they promote chromosomal rearrangements, thus giving rise to mutations.

Developmental processes involving the breakage and rejoining of DNA at defined sites are now widely known in nature.[93] They lead to the programmed elimination of DNA in some daughter cells and the realignment of rejoined sequences, producing the readout of entirely new patterns of genes. The eliminated DNA may comprise a significant portion of the genome present in the germline—as much as 10% to 20% in the ciliated protozoa *Tetrahymena*, for example[94]— and is removed from somatic cell nuclei at specific stages in development. In *Tetrahymena* there appear to be more than 5000 rearrangement sites in the genome.[93]

Site-specific recombination, resulting in rearrangements that occur in highly sequence-specific loci in chromosomes, was first described by Campbell[95] for the integration of bacteriophage λ chromosome into its bacterial host's chromosome. In site-specific recombination, specific DNA sequences are bridged by protein–protein interactions between DNA-binding proteins linked to the recombinatorial DNA strands.[89] Key enzymes involved in site-directed recombination are called *recombinases*. These promote breakage and rejoining of DNA via covalent DNA-protein recognition sites surrounding the sites of cleavage and strand exchange.[88] The recombinase family of enzymes includes the integrases and resolvase-invertases, whose activities have been studied in detail in bacterial systems and yeast. Similar enzyme activities exist in higher organisms (see below).

It should be pointed out that the three classes of recombination events noted above share some characteristics. For example, general recombination occurs between homologous DNA molecules, but sequence homology is also important in site-specific recombination.[88] While the specific class of recombinase enzymes is involved in site-specific recombination, sequence-specific recognition of DNA by cleavage and rejoining enzymes also occurs in general and transpositional recombination. Moreover, although transposons can jump to multiple sites in a target DNA, there are examples of site-specific sequence recognition.[88]

An important example of site-specific genetic recombination in mammalian organisms is that which occurs during the rearrangement of immunoglobulin (Ig) genes in B-lymphocyte differentiation. Early in the differentiation pathway for B-type lymphocytes—that is, the cells that produce antibodies as their specific differentiated function—there is a commitment to express only one subset of Ig molecules, containing one light-chain and one heavy-chain gene product out of the large numbers of available genes that code for such proteins. The expression of Ig genes is achieved through specific genetic recombinations that occur during the ontogeny of B cells. This specific gene expression results from the splicing of one variable (V) region gene, out of a large pool of V genes, with a constant (C) region gene.

Each Ig molecule contains a heavy (H) and a light (L) (κ or λ type) chain, both of which have variable and constant regions that are so designated according to the amount of variation in the amino acid sequence. Both heavy- and light-chain genes undergo splicing to produce a specific type of Ig molecule (Fig. 5–4).[96]

There are several steps in the production of mature B lymphocytes that produce specific types of immunoglobulins. One of the first steps is the chromosomal rearrangement of one V_H, D_H and J_H gene with a Cμ gene and the transcription of μ mRNA from this spliced gene. A cell at this stage, called a *pre-B cell*, does not secrete Ig. The next stage of differentiation occurs when one set of V_L and J_L genes is rearranged with its appropriate constant-region κ- or λ-type light-chain gene to produce an Ig light chain, which combines with the heavy chain to produce a complete Ig molecule of the IgM type. This is expressed on the cell surface, but is not secreted, and the cell is now an immature B lymphocyte.

The next stages in B-cell differentiation involve "isotype" switching, in which a V_H-D_H-J_H set becomes associated with different constant region genes. The intermediate stages in B-cell differentiation are also signaled by the appearance of a variety of cell surface proteins, or "markers," involved in regulating B-cell migration, proliferation, and cell recognition. Immature B lymphocytes express membrane-bound IgM and later coexpress IgD molecules that share the same V_H-D_H-J_H and light chains as IgM but have the heavy-chain IgD determinant. Some B cells undergo a further switch (from IgM and IgD) to IgG, IgE, or IgA production by splicing of the V-D-J gene set with appropriate Cγ, Cε, or Cα gene, a process that involves deleting intervening genes up to the 5′ end of the one that is expressed. The fully mature, terminally differentiated B lymphocyte, or plasma cell, produces and secretes one subset of antibody molecules.

The changes in gene readout of B cells described in the preceding paragraphs take place in a specific time frame during development. For example, in mice the first heavy chain produced by B-lymphocyte precursor cells is the Cμ chain, which is detected in the cytoplasm of fetal liver

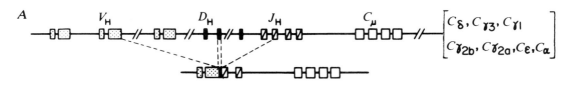

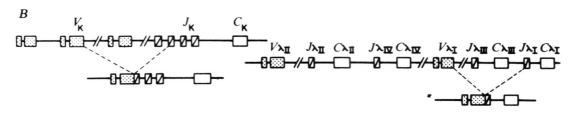

Figure 5–4. Schematic diagram of immunoglobulin gene rearrangement in the mouse: (A) at the mouse heavy (H) chain loci; (B) at the mouse light (κ and λ) chain loci. The three loci indicated are on three different chromosomes. There are probably a few hundred V_κ elements in the κ locus, about the same number of V_H elements in the H locus, with perhaps one-tenth that number of D_H gene segments. None of these numbers is known precisely, however. The mouse λ locus contains only two V genes and four J-C regions, of which one ($J_{\lambda 4}$-$C_{\lambda 4}$) may be inactive. Immunoglobulin gene expression requires precise fusion of V, D, and J segments in the H locus and precise fusion of V and J in L loci, as shown. The exon shown upstream of each of the V segments encodes the signal sequence for a secreted protein. J-C fusion and removal of other introns occur during RNA processing. The λ-gene rearrangement shown, fusion of $V_{\lambda I}$ with $J_{\lambda I}$, is the one most frequently encountered of the locus. With the exception of Cδ, stable expression of heavy-chain constant-region genes other than the Cμ requires additional DNA rearrangements that move the fused V-D-J segment close to one of the downstream C_H genes. (From Coleclough,[96] with permission.)

cells obtained from 11- to 12-day embryos (reviewed in Reference 97). Membrane-associated IgM is found on lymphocytes obtained from livers of 15- to 17-day mouse embryos, and 3 to 5 days after birth IgD molecules appear on the surface of these cells. By 7 weeks of age, more than 90% of the precursor B cells express both IgM and IgD. Further maturation occurs upon antigen stimulation. When B cells are exposed to an antigen and the appropriate growth factors (lymphokines), membrane IgM and IgD are lost and a secretory form of IgM is synthesized or IgM production is replaced by IgG, IgA, or IgE synthesis and secretion occurs by the mechanisms indicated earlier. In the adult organism, multiple clones of functional B lymphocytes continue to be produced, each synthesizing its own specific type of antibody. Even during adult life, continual exposure to various antigenic substances in the environment can call

forth the expansion of a clone of B cells ready to produce antibody against that antigen. Thus, the tremendous antibody diversity available to the adult organism is developed through a series of differentiation events resulting from a programmed rearrangement of the genes coding for these proteins. This is a clear example of how cells can change their differentiation phenotype by the mechanism of genetic recombination.

V(D)J recombination is a complex gene rearrangement event requiring a number of regulatory activities. These include (1) recognition of conserved DNA sequences (called RS sequences) that flank germline V, D, or J segments; (2) introduction of site-specific double-strand breaks between the RS sequences and the piece of DNA to be rearranged; (3) in some cases deletion or addition of nucleotides at coding junctions; (4) polymerization; and (5) ligation.[98]

Two recombinase-like genes, RAG-1 and RAG-2, are involved in V(D)J rearrangements. The RAG gene products are crucial for V(D)J recombination, as shown by the fact that cell lines lacking the RAG genes cannot carry out this event, whereas cells transfected with the RAG genes can.[98] At least two other genes, both also involved in DNA repair processes, are also involved. Interestingly, cells from immune-deficient SCID mice are impaired in their ability to complete normal V(D)J recombination and also have a defect in double-strand break DNA repair.

Another form of rearrangement involving reverse-transcribed DNA→RNA→DNA events has been observed in human cells. Retrotransposable elements called *L1 elements* are highly repetitive sequences found in all mammals, including humans, in whom L1 elements make up about 5% of the genome. There appears to be about 3500 six-kilobase L1 elements, of which a subset is actively expressed, reverse transcribed, and transposed to other regions of the genome (reviewed in Reference 99). The L1 family appears to consist of a small number of "master genes" controlling a larger number of retrotransposably active L1 elements. The L1 elements have some features in common with the repetitive *Alu* sequences noted above and may be involved in the retrotransposition of *Alu* sequences, which themselves lack the encoded reverse transcriptase activity to carry out their own transposable function. If these retrotransposable elements are reinserted into the human genome at an inappropriate place, disease can result. For example, two retrotranspositions of truncated L1 elements into an exon of the factor VIII gene on the X chromosome have been observed in patients with hemophilia A, and L1 insertions have been found in the *dystrophin* gene involved in Duchenne's muscular dystrophy and in the *APC* gene, a tumor suppressor gene altered in colorectal cancer (reviewed in Reference 99).

Another rearrangement that can have dire consequences is activation of the *myc* proto-oncogene in lymphocytes by a translocation that juxtaposes the *myc* and immunoglobulin genes. This results in the deregulation of the *myc* gene, which thus loses its own regulatory sequences and comes under the influence of the Ig gene regulatory sequences. The *myc* gene, normally not expressed in adult lymphocytes, may be expressed and produce a type of malignant lymphoma. Whether this translocation is the cause or the result of the carcinogenic process is not yet clear, but it appears to be involved in this process (see Chapter 7).

GENE AMPLIFICATION

Increasing the number of gene copies (*gene amplification*) is a mechanism by which cells meet the demand for increased amounts of certain gene products (e.g., enzymes, structural proteins, ribosomal RNA). Cells also appear to use this mechanism during the process of differentiation to produce high levels of cellular components called for at different developmental stages. That developing organisms use gene amplification as a differentiation mechanism has been known for several years from studies in lower animals. During the maturation of oocytes in amphibians, for example, there is a large amplification of rRNA genes, which disappear into the cytoplasm after the oocyte matures and are no longer active.[100] In *Drosophila* oogenesis, gene amplification occurs on two different chromosomes in response to the need for large amounts of chorion proteins.[101] By the time the ovarian follicle cells degenerate and oogenesis is completed, the number of chorion protein genes has been amplified 16- to 60-fold.

It is clear that gene amplification is not restricted to lower animals. Selective gene amplification in mammalian cells was first documented in 1978 as a mechanism for acquisition of resistance to the anticancer drug methotrexate (MTX).[102] The target enzyme for MTX is dihydrofolate reductase (DHFR), a key enzyme in nucleic acid biosynthesis. This enzyme is inactivated by MTX by forming a very stable drug–enzyme complex. Cells can circumvent this inhibition by producing increased amounts of enzyme, through amplification of the DHFR gene. Drug-resistant cell variants can be obtained with as many as 100 to 1000 DHFR gene copies.

Many other examples of gene amplification in mammalian cells, including human cells, have been reported. These include the genes for metallothionein, hypoxanthine-guanine phosphoribosyl transferase, hydroxymethylglutaryl coenzyme A reductase, adenosine deaminase, glutamine synthetase, ornithine decarboxylase, and uridylate synthetase (reviewed in Reference 103). Most of the reports of gene amplification relate to the development of drug resistance in somatic cells undergoing a strong selective pressure provided by a cytotoxic drug. In fact, development of drug resistance and multidrug cross-resistance to anticancer drugs by this mechanism appears to be a widespread and common phenomenon.[104] It is also clear the gene amplification can occur spontaneously in the absence of selection pressure, and in cultured mammalian cells at least, a two-fold increase in gene copy number occurs in 1 out of every 1000 cell replications in the absence of drug.[103] This kind of variation in gene copy number is remarkable in view of the dogma of stringent genomic replication during cell mitosis. However, such amplifications appear to be relatively unstable in the absence of a sustained selection pressure. The extra genes may be extrachromosomal and appear in the cell as minichromosomes known as *double minutes* (DM) or incorporated into chromosomes, in which case they often show up as homogeneously staining regions (HSRs). The DMs are usually unstable and may disappear within as few as 20 cell population doublings in the absence of selection pressure. Not being part of the chromosomal apparatus and not having a centromere, DMs may be unequally proportioned between daughter cells at mitosis, and thus could provide a way to set up a different differentiation pathway in one of two daughter cells.

Amplification of certain genes also appears to be related to the carcinogenic process in certain types of cancer cells. Agents that damage DNA, such as ultraviolet light and the chemical carcinogen N-acetoxy-N-acetoaminofluorene, enhance amplification of the DHFR gene,[104] and carcinogenic agents can also induce amplification of simian virus 40 (SV40) DNA sequences in cultured Chinese hamster ovary (CHO) cells.[105] Moreover, exposure of cells to the tumor-promoting phorbol ester TPA, either at the time of initial exposure to methotrexate or during subsequent cloning of resistant sublines, can enhance DHFR gene amplification 100-fold.[106] During tumor progression, tumor cells gain an increased ability to amplify genes as they lose cell cycle control and tumor suppressor gene activity.

Cells derived from cancer tissue often have amplified oncogene sequences, including the *myc* and HER-2/*neu* gene sequences. Certain experiments suggest that gene amplification may be directly involved in the unregulated growth potential of cancer cells. For example, Levan and Levan[107] have shown that cells taken directly from a mouse tumor (SEWA) contain multiple DMs, suggesting extensive gene amplification. When these cells are grown in culture, they progressively lose DMs with continuous passage. If, however, the cultured SEWA tumor cells are reinjected into a mouse, the cells of the growing tumor again contain multiple DMs. This result suggests that growth constraints (immune mechanisms, etc.) indigenous to the mouse induce a stress on the tumor cells such that they produce multiple copies of certain genes to ensure their continued unregulated growth, whereas such constraints are not present in cell culture, and no continued selection pressure exists to maintain the amplified genes.

Gene amplification in somatic cells can occur by means of a least two mechanisms: (1) unequal crossing over of sister chromatids and (2) multiple replication of individual gene sequences during the S (DNA synthetic) phase preceding a cell division. The former mechanism has been shown to occur in bacteria and results in a gain of genes by one sister chromatid and a loss by its pair, followed by multiple mitoses of the cells containing the extra gene copies. The other mechanism, so-called disproportionate replication,[103] is the most likely mechanism for gene amplification in higher organisms, although both mechanisms may occur. A schematic model for this mechanism is shown in Figure 5–5.[108] An example of this is the amplification of the chorion genes in *Drosophila* that results in the generation of multiple gene copies of variable lengths, starting from multiple initiations of replication at specific sites in DNA, producing replication "bubbles" visible by electron microscopy.[109]

It is now known that gene copy member also varies among humans and contributes to human genetic variation and diversity. Indeed, large-scale copy number polymorphisms (CNPs) of 100 kilobases or more have been observed in human populations.[110] Using oligonucleotide microarrays of 20 individuals, Sebat et al.[110] found 221 copy number differences representing 76 unique CNPs. On average, people differed by 11 CNPs. Copy number variation of 70 different genes was observed, including genes

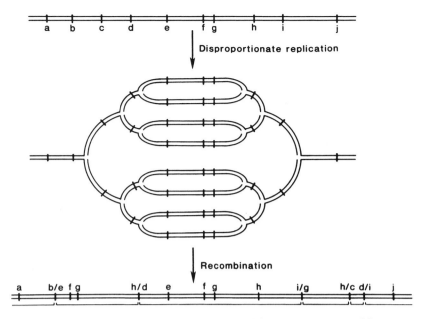

Figure 5–5. Schematic model of gene amplification. Gene amplification proceeds through two steps. Disproportionate replication of a restricted chromosomal region may occur many times in one cell cycle. Three rounds of replication are depicted here. The resulting structure resolves into an irregular linear array by means of homologous recombination between repeated DNA sequences. (From Roberts et al.,[108] with permission.)

involved in neurological function and regulation of cell growth and metabolism. In addition, CNPs include genes involved in regulation of food intake and body weight and genes implicated in leukemia and drug resistance in breast cancer. Thus, CNPs may be related to general health problems such as obesity as well as neurological diseases and cancer.

Many of the points raised earlier about chromatin packaging, transcript splicing to create different gene products, and recombination events, as well as the increasing information about the role of RNA in gene regulation (see RNAi later), raise the question: "What is a gene?"[110a] The genome contains many overlapping gene sequences. A transcript can start at a DNA sequence coding for one protein and run through into a gene for a different protein, leading to a fused transcript. This may account for up to 4–5% of transcribed genes (reviewed in reference 110a). Depending on how these transcripts are processed, a cell may be able to generate a greater variety of proteins from a limited number of exons—sort of a "DNA without borders." Some estimates suggest that as

much as 63% of the mouse genome is transcribed, yet only 3% of the genome is thought to contain "normal" exons that are transcribed into mRNA. What in the world is all this other RNA doing? We now have a glimmer. The evidence that a fair amount of this "extra" RNA has a regulatory function is now emerging. Some of it is processed into microRNA that regulates gene transcript function. Other RNA molecules also contribute to the regulation of many cellular processes, leading to the view that RNA, other than mRNA, is actively involved in carrying out instructions in the genome. This somewhat muddles the concept of just what a *gene* is, at least in the context of how information is transduced from a DNA sequence into a cellular function.

CIS-ACTING REGULATORY ELEMENTS: PROMOTERS AND ENHANCERS

The presence of promoter sequences in DNA for RNA transcription has been known for a long time and was discovered first in bacteria. It is now known that similar regulatory regions are also present in the DNA of eukaryotic cells. The regulation of gene transcription in eukaryotes, however, is more complicated than in prokaryotes, and includes higher-level orders of chromatin packaging, methylation, binding of nuclear proteins, and the other mechanisms detailed in the preceding sections of this chapter. There are also regulatory elements on DNA that affect the function of promoters in directing RNA transcription. They act in a *cis* manner in doing so—that is, they directly affect the function of gene sequences on the same DNA strand, even though they may be several hundred base pairs upstream or downstream from the initiation site for RNA transcription. Other mechanisms of gene regulation are said to be *trans*, the term for a factor, a regulatory protein, for example, coded for by a distant gene, that modulates transcription of genes not associated directly with DNA sequences in the same strand.

There are two broad categories of *cis*-acting regulatory elements: (1) sequences near the transcription-initiation site, termed *promoters* by analogy with their prokaryotic counterparts that define the start site and "loading efficiency" for RNA polymerase II, the enzyme that transcribes messenger RNA; and (2) *enhancers* (or *activator* genes) that are more remotely located from the gene and that increase transcription efficiency by mechanisms not yet perfectly defined but which appear to affect DNA topology in a way that facilitates access of RNA polymerase to initiation sites. A number of things are remarkable about these enhancer sequences. They can exert influence over genes that may be as far as 10,000 base pairs away; they can be effective whether they are in front of (at the 5′ end) or behind (at the 3′ end) the transcribable gene; and they function regardless of which way the sequence is oriented (i.e., which direction the sequence reads). Because enhancer elements can act over such large distances, they are candidates for a more regional or global type of gene regulation than individual promoter sequences—the kind of programmatic regulation, in other words, that might be involved in cellular differentiation or neoplastic transformation.

Promoters are organized as a group of "control modules" clustered around the initiation site for RNA polymerase II.[111] Early work on the structural organization of promoters for the herpes simplex virus (HSV) thymidine kinase (tk) gene and simian virus 40 (SV40) early transcription genes provided the background for how we think about promoters.[112] They are made up of discrete sequences of 7 to 20 base pairs (bp) of DNA and have recognition sites for transcriptional regulatory proteins (transcription factors; see below). One of the functions of promoters is to position the start site for RNA transcription. For many, but not all, gene promoters a consensus TATA base-containing sequence, called the *TATA box*, is this positioning element. Other promoters, typically 30 to 110 bp upstream (toward the 5′ end) from the transcription start site, regulate the frequency of transcription initiation.[111]

Enhancer sequences were first discovered in 1981 by two laboratories studying the regulation of SV40 virus gene transcription. Benoist and Chambon[113] and Gruss et al.[114] described the presence of a *cis*-acting sequence located more than 100 nucleotides upstream (i.e., before the 5′ end) of the cap site of the so-called early SV40 genes, or those genes transcribed shortly after infection. Deletion of this sequence reduced early gene expression 100-fold and abolished viral viability. This prototype enhancer is a 72-bp tandem repeat located upstream from three 21-bp, GC-rich repeats and the TATA box.

Subsequently, it was found that SV40 enhancer as well as similar elements from other animal viruses could function when linked not only to their natural genes but also to other genes (so-called enhancer swap experiments). Many other viruses have now been shown to contain enhancer elements in their genomes; these include polyoma, BK virus, adenovirus, Moloney sarcoma virus (MSV), bovine papilloma virus, and Rous sarcoma virus (RSV) (reviewed in Reference 115). In the case of some DNA viruses (e.g., polyoma, BK, adenovirus, and bovine papilloma virus), the enhancer sequences, as in SV40, are present as short tandem repeats of 50 to 100 nucleotides. The enhancer sequences of the RNA retroviruses (e.g., MSV and RSV) have been identified within the *long terminal repeat* (LTR) regions, the portions of the retroviral genomes known to contain transcriptional regulatory sequences. These sequences can augment transcriptional activity of heterologous genes, and when viral LTRs are integrated into a eukaryotic cell's genome, they can activate cellular genes that come under their influence.[116]

Such a process can lead to activation of cellular proto-oncogenes (see below) and may lead to neoplastic cellular transformation.

In addition to viruses, cellular genes have been found to contain enhancer elements. Originally, they were found to be associated with insulin, chymotrypsin, and Ig genes. Enhancers, like promoters, may contain several modules, sometimes called *enhansons*,[111] and also can bind to positive or negative transcription factors. Enhancers and promoters have to "talk to each other" if a cell is to coordinate its developmental program as well as its response to environmental cues. Both enhancers and promoters regulate transcription, bind transcription factors, and have similar modular organization. Three different (but not mutually exclusive) mechanisms have been proposed for enhancer–promoter cross talk. In the first mechanism, enhancer and promoter elements may be brought together by binding of distant DNA sequences to bring enhancers and promoters into contact. This binding would be facilitated by transcription factors. In the second one, bridging between enhancer and promoter sequences could be brought about by protein–protein interactions between transcription factors binding the two DNA domains. Finally, a DNA-tracking mechanism could accomplish the interaction by allowing transcription factors or the transcriptional complex to "slide" along DNA, thus conveying the transcriptional signal from one domain to another. Evidence for the latter mechanism has been obtained for regulation of bacteriophage T4 late gene transcription.[117] The DNA-tracking mechanism would predict that the sliding of a transcriptional activator complex is a DNA structural organization rather than a specific DNA sequence. Otherwise, it is difficult to see how passage of such a complex over long sequences of differing base composition could be accomplished. Alternatively, tracking could be accomplished by the DNA-binding or -looping mechanism of the first mechanism described.

Evidence for the protein–protein bridging model comes from experiments of Cullen et al.,[118] who studied the estrogen-induced interaction between enhancer and promoter regions of the prolactin (PRL) gene. Estrogen induces the transcription of the PRL gene by binding to the estrogen receptor (ER), which binds to the estrogen response element (ERE). The ERE is at the 3' end of a distal enhancer region that is between −1550 and −1578 bp away from the transcription start site. Thus, for the ER complex to function in PRL gene transcriptional regulation, it somehow has to bridge this distance. What Cullen et al. found, by using a unique chromatin ligation assay, was that the distal enhancer and proximal promoter regions of the rat PRL gene are juxtaposed and that estrogen enhanced bridging between these domains two- to threefold over non-estrogen-treated chromatin. Previous findings showed that although the chromatin surrounding the ERE and the promoter became nuclease sensitive, the intervening regions between the ERE and the promoter remained nuclease insensitive. These data favor the chromatin-binding, protein–protein bridging models of enhancer–promoter interaction.

Another point to bear in mind is that the modular organization of enhancers and promoters helps explain a way in which cells can modulate their response to external signals and why there are "strong" and "weak" promoters and enhancers. For example, if a promoter or enhancer contains two different types of modules, recognized by different transcription factors, promoter or enhancer activity might be seen only if both factors are present at the same time in the responding cell, or activity might only be "half-maximal" if only one of the two factors were present. Similarly, an enhancer or promoter with three or more modules might be a "stronger" transcriptional regulator than those with only one or two such modules.

Some promoters and enhancers bind transcription factors that are constitutively expressed, i.e., they are continually made in cells and not only when induced by exogenous stimuli like the estrogen-induced ERE response described above. It isn't clear why similar regulatory mechanisms would be needed for constitutively expressed genes, which may include general "housekeeping" genes, but it may be that low or baseline levels of expression of some genes needs to go on continually so that they are "primed" and ready to go when the cell is stimulated or stressed by exposure to an environmental factor. Examples of such genes include the metallothionein gene, which is involved in protection against heavy-metal toxicity and whose promoter contains binding sites for the transcription factor SP-l (see below), and the

Ig genes, whose enhancers have binding sites for constitutively expressed transcription factors.[111]

The elegant and complex gene regulation imposed by enhancers and promoters apparently evolved over eons of time. Based on studies of simple prokaryotic and viral genes, it seems likely that single promoter or enhancer regions arose in scattered regions of a genome, perhaps randomly at first, and then gradually through genetic recombination and duplication events found their way into proximity with structural genes on whom they imposed their control. A selective advantage to the organism may have thus been gained by its imposing a regulated rather than a random response to the environment once the promoter–enhancer elements took on the ability to bind specific proteins.

TRANSCRIPTION FACTORS

As noted above, the regulation and initiation of transcription in eukaryotic cells is an intricate and complex process. It involves interaction of DNA-binding transcription factors (TFs) with short consensus DNA sequence motifs in enhancer and promoter regions. These interactions produce, by cooperative binding reactions, the formation of a transcription complex (see below) and changes in chromatin structure that foster the binding of RNA polymerase II and initiation of transcription.

On the basis of experiments initially performed primarily in yeast and later in other eukaryotes, one can formulate the following characteristics of eukaryotic TFs:[119–122]

1. Transcriptional factors are multi-domain proteins that have DNA-binding, activation, dimerization (most but not all TFs), nuclear localization, and ligand-binding domains.
2. DNA-binding and activation domains are often interchangeable among TFs.
3. DNA-binding domains have specific structures and recognize specific DNA sequences, whereas activation domains have less precisely defined structures characterized by an acidic amino acid–rich (i.e., rich in aspartic and glutamic acids), glutamine-rich, or proline-rich sequences.

4. Transcription factors that work in yeast cells will, in general, work in plant, insect, and mammalian cells for genes with the required promoter–enhancer sequences.
5. Transcription factors undergo cooperativity in binding with other components of the transcription complex. This cooperative binding ensures specificity and reversibility through multiple, low-affinity interactions.
6. Transcription factors with more potent activation domains (i.e., more activating sites) can act at greater distances on the DNA.
7. Transcription factors have DNA-binding domains with typical structural characteristics such as helix-turn-helix, zinc-finger, and basic region domains (see below). A number of TFs also have a characteristic helix-loop-helix (HLH) structure (e.g., Myo D). Helix-turn-helix domains form structures that make contact with DNA. Helix-loop-helix domains are involved in protein–protein interactions.
8. Many TFs form dimers through leucine-rich "zipper" domains. (e.g., Jun-Fos).
9. Transcription factors have nuclear localization motifs that enable them to move into the nucleus after synthesis.
10. The ligand-binding domains enable a number of TFs to be activated by hormones, growth factors, developmental morphogens (e.g., retinoic acid), and other exogenous stimulatory agents.

Structural Motifs of Regulatory DNA-Binding Proteins

DNA-binding proteins that have a regulatory function, i.e., transcriptional activators and repressors, display a number of common structural motifs. These include the helix-turn-helix, zinc finger, leucine zipper, and acidic domains, mentioned above (Fig. 5–6).[121] The helix-turn-helix motif was the first studied and most well-characterized structural motif and is found in prokaryotic activator and repressor proteins as well as those of higher organisms. As implied by the name, this class of proteins contains two α helices separated by a β turn. This geometry is common to all types of TFs with this motif, even though the primary amino acid sequence can

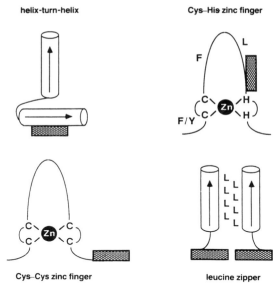

helix-turn-helix

Cys–His zinc finger

Cys–Cys zinc finger   **leucine zipper**

Figure 5–6. Schematic representation of the four structural motifs described in the text. α-Helices are represented by cylinders with arrows indicating directionality; conserved amino acid residues are shown in one-letter code, and zinc ions are shown as black circles. The shaded boxes indicate the regions of the proteins proposed to be involved in specific contacts to DNA. (From Struhl,[121] with permission.)

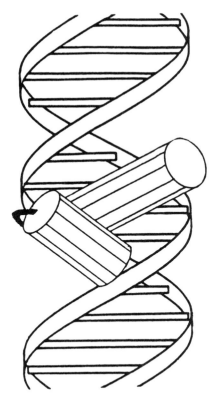

Figure 5–7. The helix-turn-helix with the recognition helix of the major groove of DNA. (Reprinted with permission from Schleif,[123] and the American Association for the Advancement of Science.)

vary. One of the α helices, the DNA-recognition helix, contacts bases in the major groove of DNA (Fig. 5–7).[123] The other helix lies across the major groove, in eukaryotic cells; the helix-turn-helix motif was first described for the homeobox-encoded proteins of *Drosophila*. The homeobox proteins contain conserved polypeptide sequences called *homeodomains*. Homeodomains have now been identified in eukaryotic organisms from yeast to human.

The zinc finger motif was first proposed for the transcription factor TFIIIA, a "general" TF required for transcription of the 5S RNA genes by RNA polymerase III.[121] TFIIIA has 7 to 11 zinc atoms per molecule and contains 9 repeating units of about 30 amino acids each. Two kinds of zinc fingers have been proposed (Fig. 5–6), one in which each "finger" contains two cysteines and two histidines arranged in a tetrahedral coordination complex that binds zinc, and a second type in which the coordination complex is made up of four cysteines. Similar zinc finger domains have been found by DNA sequence analysis of organisms

from yeast to human, as seen, for example, in the GAL4 transcriptional activator of yeast and the steroid hormone receptor of mammals. Some of these genes code for transcriptional activators. Both the zinc finger region and zinc itself appear to be necessary for DNA binding, and contact with DNA is thought to occur via the outstretched fingers. "Finger swapping" experiments have shown that, in the case of four cysteine-containing zinc fingers, the finger motif is essential for DNA binding but that direct contact with DNA involves additional amino acid sequences.[121]

The leucine zipper motif is contained in DNA-binding proteins such as the yeast GCN4 transcriptional activator and the Jun, Fos, and Myc oncogene proteins. This structural motif is produced by runs of four or five leucine residues spaced exactly seven residues apart, forming interdigitating α helices and leading to protein dimer formation important for the transcriptional activity of some TFs, e.g., Jun-Fos dimers (see below). Dimerization appears to involve a "coiled-coil" structure consisting of two parallel α helices with hydrophobic contact sites that bring into close proximity a region of each subunit rich in basic amino acids. This interaction provides a bimolecular contact with DNA. Leucine-zipper motifs have also been found in some proteins that contain zinc finger or helix-turn-helix domains, which suggests that more than one mechanism for DNA contact may be used by some TFs.

It has become increasingly clear that transcriptional regulatory mechanisms have been amazingly conserved over evolution. For example, some yeast TFs can function in a variety of other eukaryotic organisms, including vertebrates and vice versa; yeast and mammalian TATA-box TFs are functionally interchangeable Interestingly, although eukaryotic cells have some related TFs that recognize similar DNA sequences, each organism's TFs may regulate a different set of functions. For example, in yeast the GCN4 TF activates amino acid biosynthesis and oxygen utilization, while its homolog in mammalian cells, c-Jun, activates a different series of events involved in so-called immediate-early events in response to external signals such as growth factors. Thus, during evolution the structural genes responding to TFs have diverged even though the transcriptional regulatory sequences upstream or downstream from them have been remarkably conserved.

Acidic domain motifs, as noted above, are contained in the transcriptional activation regions of the TFs. Again, these domains have several features in common among different organisms of the phylogenetic tree. The negative charge of these domains must be important because if it is removed, activating function is lost. Yet additional characteristics are also clearly important. These appear to be repeating α-helical structures, amphipathic in character, that favor interactions with other proteins in the transcription machinery.[121] As will be discussed below, the acidic regions of TFs appear to be involved in the interactions that bring an enhancer sequence into proximity with the RNA transcription start-sites via binding to TATA-box binding proteins. Other types of activation domains contain glutamine- or proline-rich sequences; the mechanisms of these different types of TFs has not been clearly defined.[122]

The transcriptional machinery can go awry in cancer cells. For example, oncogenic conversion of normal cells into cancer cells involves changes in transcription factors. Such changes are exemplified by conversions of the genes coding for TFs, e.g., c-*jun* to v-*jun*, c-*fos* to v-*fos*, c-*myb* to v-*myb*, and c-*erbA* to v-*erbA*.

Repressors

Precise regulation of gene expression requires both positive and negative control. During evolution two kinds of gene regulation mechanisms have evolved. The negative factors are sometimes called *transcriptional down-regulators* or *repressors*. Some of these have DNA-binding domains but lack functional activation domains and they compete with transcriptional activators for binding to regulatory elements in DNA.[124] Some can form heterodimers with activators and then block the ability of TFs to bind to DNA or to activate transcription. Others may bind to TFs already attached to DNA and prevent the interaction of the acidic domain of TFs with the TATA-box binding proteins. Another type of repressor can sequester activators into inactive multimeric complexes.

A somewhat surprising finding is that transcriptional activators and repressors can be encoded by the same gene.[124] One way this can be accomplished is by alternate mRNA splicing such that an activator becomes a repressor. Examples of this include the alternate splicing of the *erbA* gene mRNA, which modifies the carboxyl-terminal domain and prevents binding to its ligand, and the alternate splicing of the *fos* B gene mRNA to create a defective activation domain (reviewed in Reference 124).

Another perhaps surprising finding is that activators and repressors can be co-expressed in cells, and during development of an organism the balance of expression of activators and repressors can control gene expression in a temporal and tissue-specific developmental pattern. The function of activators and repressors is often regulated by post-transcriptional modifications such as phosphorylation. For example, the ability of the retinoblastoma gene product Rb to repress gene expression is lost when it becomes phosphorylated. Also, some TFs can activate one gene or set of genes, and at the same time repress another gene or set of genes. Such activity has, for example, been shown for some homeobox proteins in *Drosophila* and for steroid hormone receptors in mammals (reviewed in Reference 125).

Another important concept to bear in mind is that there is "cross talk" among TFs. In some cases, they may work together to activate a gene or genes. In other cases, one TF may turn on the gene for a second TF. In still other instances, one TF may inhibit the action of a second. Some examples will illustrate these points.

The virus-inducible enhancer of the human interferon β (IFN-β) gene has overlapping regulatory elements recognized by the TFs NF-κB, IRF-1, and ATF-c-Jun (reviewed in Reference 122). None of these function on their own, and two or more of the TF-activated enhancers are required to turn on the IFN-β gene. This apparently results from protein–protein interactions between the TFs to form a stereospecific complex that brings the enhancer and promoter regions together.

Interleukin-1 (IL-1) is a cytokine that mediates a variety of cell proliferative responses by activating c-*myc* gene expression, which produces a TF involved in a variety of cell proliferative mechanisms. The action of IL-1 in turning on c-*myc* is accomplished by activating the TF NF-κB, which in turn activates c-*myc* via NF-κB response elements in the c-*myc* gene.[126]

Yin-Yang-1 (YY1) is a zinc finger DNA-binding protein that, depending on the context, can function as an activator, a repressor, or an initiator of transcription (reviewed in Reference 127). YY1 repressor actions include repression of the adeno-associated virus P5 promoter, the c-*fos* promoter, the human papillomavirus 18 promoter, the LTR of Moloney murine leukemia virus, and the N-*ras* promoter. In contrast, YY1 activates the c-*myc* promoter and the promoter of ribosomal proteins L30 and L32. As a feedback mechanism, c-Myc protein binds to YY1 and inhibits both its repressor and activator functions.[127]

Thus, the interaction of TFs is both complex and context dependent. Depending on the cell type, the environmental signals, and the state of differentiation, the amount and type of TFs expressed may vary. The requirement for them to form dimers and protein–protein complexes allows them to act in specific ways in different cell types according to who their partners are. One

way to look at this scenario is that there are multiple competing TFs for the same DNA-binding sites and whoever gets there "the firstest with the mostest" wins.

General (Basal) Transcription Factors

The regulatory factors involved in initiation of transcription by RNA polymerases, first discovered by Roeder in sea urchins (reviewed in Reference 128), are usually divided into two classes: general transcription factors (GTFs) and promotor (or enhancer)-specific transcription factors (STFs).[129–132]

In eukaryotic organisms, three different RNA polymerases are involved in gene transcription. RNA polymerase I (RNA pol I) transcribes ribosomal RNA; RNA pol II transcribes protein coding mRNAs and many small nuclear (sn) RNAs; and RNA pol III carries out synthesis of tRNA and 5S RNA. The transcriptional machinery for each of these types of genes has features in common. All three RNA polymerases require formation of a transcription complex containing a TATA box–like binding protein (TBP) and a series of TBP-associated factors (TAFs) that are somewhat different for each of the polymerase complexes.

The factors for RNA polymerase II have been studied in the most detail, thus we know the most about them. Many but not all protein-encoding genes that use RNA pol II have a TATA-containing DNA sequence near the transcription site (e.g., many housekeeping genes lack a TATA element). It is now known that class I (using RNA pol I) and class III (using RNA pol III) also contain TBPs in their transcription complexes. In the case of RNA pol I, a TBF-TAF complex called SL1 makes up a key part of the transcription complex. In the case of RNA pol III, transcription requires at least two TAFs as well as the TBP.[133] These various TBP-TAF complexes are specific for each type of gene (class I, II, or III), ensuring that each type of gene forms a complex only with its appropriate RNA polymerase.[130,133]

As noted above, the most attention has been focused on mRNA-transcribing class II genes transcribed by RNA pol II. The general transcription factors required for accurate transcription of class II genes include the TATA-binding proteins TFIIA, TFIIB, TFIID, TFIIE, TFIIF, and

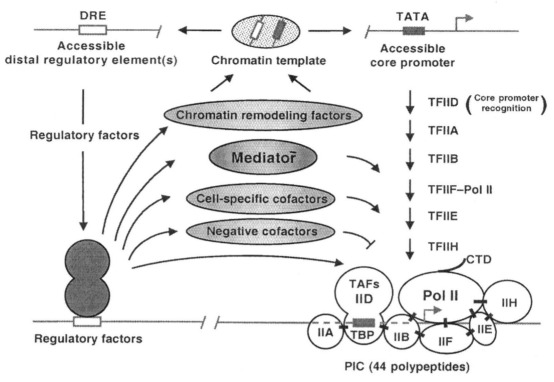

Figure 5–8. General initiation factors and PIC assembly pathway for class II
genes with a TATA-containing core promoter, and regulation by gene-specific
factors and interacting cofactors. Assembly of a PIC containing Pol II and
general initiation factors (yellow) is nucleated by binding of TFIID to the
TATA element of the core promoter. A model for the regulation of PIC
assembly and function involves, sequentially: (1) binding of regulatory factors
to distal control elements; (2) regulatory factor interactions with cofactors
that modify chromatin structure to facilitate additional factor interactions;
and (3) regulatory factor interactions with cofactors that act after chromatin
remodeling to facilitate, through direct interactions, recruitment or function
of the general transcription machinery. TAFs, TBP-associated factors. (From
Roeder,[128] reprinted with permission from Macmillan Publishers Ltd.)

TFIIH. These are thought to assemble in an ordered fashion on a promotor to form a preinitiation complex with RNA pol II.[132] The first step
is binding of TFIID to the TATA box, a step
facilitated by TFIIA. TFIIB binding to this
DNA–protein complex fosters recruitment of
RNA pol II at the promoter site. This step requires the RAP30 subunit of TFIIF. TFIIE also
appears to play a role in formation of an active
complex. In the case of class II genes, the TATA
box–binding protein TBP is a subunit of TFIID,
called TFIIDτ.[134] The composition of the preinitiation complex (PIC) for class II genes is
shown in Figure 5–8.

In order for efficient transcription to occur,
specific transcription factors must come into
play. The regulatory DNA sequences to which
the STFs bind are often a long distance away
from the transcription start site, and it isn't clear
how they make contact with the transcription
complex. One way this could happen is shown in
Figure 5–9.[135] In this model, the activation region of an STF (promotor-specific activator),
bound to its enhancer, induces binding of the
transcription factor–enhancer complex to approximate it to the transcription initiation complex.
This complex then binds a target site on an already
formed transcription complex to stimulate initi-

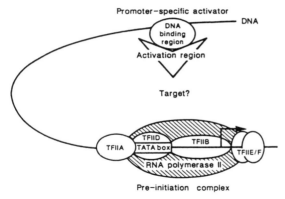

Figure 5–9. A simplified mechanism for transcription stimulation by promoter-specific activators. The activator recruits one or more of the general transcription factors to facilitate assembly of a pre-initiation complex and then enhances a step following assembly of the general factors into a pre-initiation complex. The promoter-specific activator is shown bound to its site on the DNA loop. The site of transcription initiation is indicated by the arrow. (Adapted from Lillie and Green,[135] with permission.)

ation of transcription. In the absence of an activator, nonproductive or inefficient preinitiation complexes could form, producing a low baseline level of transcription. In the presence of activator, TFIID and TFIIB would assemble in highly productive manner and transcription would be increased. An alternative model is one in which the activator interacts with TFIIB-TFIID and helps them assemble into a preinitiation complex. In any case, it is clear that interaction between the activation domain–containing TF and its DNA-binding region with TFIIB is crucial for transcriptional activation.[136]

Promoter- and Enhancer-Specific Transcription Factors

Eukaryotic promoters and enhancers contain a unique array of DNA sequences that bind STFs. The STFs include AP1, ATF/CREB, SP1, octamer-binding transcription factors Oct-1 and Oct-3, YY1, LEF-1, E2F, and a number of others.[137] These are proteins that have DNA-binding domains and activation domains as noted above. Many of them also contain specific ligand-binding domains that enable them to recognize and bind to external signals such as hormones and growth factors. In addition, there are families of STFs

that induce the expression of tissue-specific genes and play a key role in cellular differentiation. These include the MyoD factors involved in muscle differentiation, Pit-1 involved in gene expression in pituitary cells, and HNF-1 involved in liver cell differentiation. Some promoter-specific transcription factors, their DNA-binding sites and size, and the promoters and enhancers that they activate are shown in Table 5–1.[138] Some of these factors are discussed below.

AP-1/Fos/Jun

The AP-1 promoter-specific factor was identified as a *trans*-acting factor that binds to the SV40 virus enhancer element. The SV40 enhancer was the first such element described and is often used as a prototype because it contains a number of prototypical STF-binding sequences, (e.g., SP-1 and AP-2, -3, -4, and -5).[138] Subsequently, it was found that AP-1 binding sites could bind a variety of transcriptional activators of the Fos/Jun cellular oncogene family as well as the tumor promoter phorbol ester (TPA) binding factor and the glucocorticoid receptor. Thus, AP-1 DNA binding sites are also known as TPA-responsive element (TRE) and the glucocorticoid receptor element (GRE). This family of transcriptional activators forms dimers via leucine zippers as noted above and is encoded by a family of genes including *fos, fra-1, fra-2, fos* B, c-*jun, jun* B, and *jun* D.[139] For these factors to be active, they must dimerize. Fos/Jun heterodimers are the most active; Jun-Jun homodimers are weakly active; and Fos-Fos homodimers are difficult to form and are not active. The activity of Fos and Jun appears to be regulated by their phosphorylation state.[140] The AP-1 binding proteins are induced by mitogenic, differentiation-inducing, and neuronal-specific stimuli.[139]

ATF/CREB

The ATF/CRE enhancer sequence (TGAC GTCA) was identified as the activating transcription factor (ATF) binding site or the cyclic-AMP response element (CRE) that bound the cAMP response element binding protein (CREB). The ATF/CREB subfamily of transcription factors includes CREB, CRE-BP1, ATF-3, and ATF-4.[139] This family of STFs has

Table 5–1. Characteristics of Some Specific Transcription Factors

Factor	Binding Site	Size	Promoter and Enhancer	Comments
AP1	T(T/G)AGTCA	47 kDa	SV40/Py enhancers BLE of hMTIIA Collangenase Stromolysin α_1-anti-trypsin Transthyretin MHC-H2 AdE3	Binding site is the TPA-responsive element (TRE). API sites are bound by Jun/Fos heterodimer.
CREB ATF	TGACGTCA	43 kDa	Somatostatin (CREB) E1A/E2A/E3E4 Ad early genes (ATF), c-*fos*, hsp70, tyrosine hydroxylase, α-gonadotropin, VIP, fibronectin, HTLV-II LTR, HTLV-1 LTR, BLV LTR	ATF/CREB family includes CREB, CREBP1, ATF-3, and ATF-4; binding site is the cAMP-responsive element (CRE).
MLTF/USF	GGCCACGTGACC	46 kDa	MLP of adenovirus α-fibrinogen Mouse Mt1	
CTF/NFI	TGGCT(N_3) AGCCAA	52–66 kDa	$\alpha\alpha$-, β-globin, hsp70, HSV *tk, ras,* Ad2/5 origin AdE3 c-*myc*, albumin	A family of factors, required for transcriptional stimulation and stimulation of adenovirus DNA replication in vitro. Gene has been cloned and recognizes multiple mRNA species. A half-site is sufficient for binding.
SP1	GGGCGG	105 kDa	SV40 early promoter, hMTIIA human ADA, type II procollagen, E1B, HSVIE-3, DHFR, HIV LTR AdlTR	Human Spl cDNA cloned
Octamer binding proteins	ATTTGCAT		heavy and light Ig, histone H2B snRNA genes, SV40 enhancer	
Ubiquitous OTF1 OBP100 NF111/octB1A IgNFA1 octB3		90 kDa 100 kDa		
B-cel Specific OTF11 1gNFA2/octB2 octB1B		62–58.5 kDa		
E2F	TTTCGCGC	54 kDa	adenovirus E2A E1A enhancer	Binding activity detected in infected but not uninfected HeLa cells; also detected in undifferentiated F9 cells
AP3	GGGTGTGGAAAG°		SV40 enhancer Py enhancer	Overlaps the core motif of SV40. Induced by TPA
EBP20	TGTGG(A/T)(A/T) (A/T)G CCAAT		MSV enhancer SV40 enhancer Py enhancer	This protein was originally purified by virtue of its ability to bind to the SV40 enhancer; if also binds to the CCAAT sequence. Both EBP20 (enhancer binding activity) and CBP (CCAAT binding activity) reside on one polypeptide encoded by a single gene.
TFIID	TATA box		Many genes	

Factor	Binding Site	Size	Promoter and Enhancer	Comments
AP2	CCCCAGGC	52 kDa	SV40 promoter and enhancer Py enhancer/ origin Pre-proenkephalin collagenase, mouse H2K Ad MLP, human hsp70 hMT11A	
AP4	CAGCTGTGG		SV40 enhancer Py enhancer	
GT11-1B			Pre-proenkephalin	
AP5	CTGTGGAATG		SV40 enhancer	
EF,E			F441 Py enhancer	
GT11-C				
PEA2	GACCGCA		Py enhancer	
EF,C	GTTGCN₂GGCAAC		Py enhancer Hepatitis B enhancer	
E2aE-Cβ	TGGGAATT		E2A (E2aE-Cβ)	
E4EF2			E4 (E4F2)	
E4TFI	GGAAGTG		E4	

*This is the SV40 binding site and not a consensus.
(From Jones et al.[138]).

been implicated in cAMP-, calcium-, and virus-induced alterations in gene transcription.

While Fos/Jun and ATF/CREB protein families were originally thought to be distinct sets of STFs that share the basic region/leucine-zipper motif but have different DNA-binding specificities, it is now known that members of each family can cross-dimerize to form active STFs.[139] The three Jun protein family members bind to DNA as homodimers or as heterodimers among themselves or with members of the Fos or ATF/CREB families of proteins.[140] The four members of the Fos family bind to DNA as heterodimers with members of the Jun family or the ATF/CREB family. More than 50 different complexes among the Fos/Jun and ATF/CREB families of proteins have been identified, many of them in intact cells.[140]

Some fine-tuning of gene regulation occurs via the interactions of the AP-1 and ATF/CREB families of STFs. For example, while the glucocorticoid receptor (GR) and AP1 (Jun/Fos) are primary regulators of two different signal transduction pathways, triggered by glucocorticoids and by mitogens such as TPA, respectively, they can modulate each other's activity. It has been observed that a "composite" GRE could bind either GR or a Jun/Fos heterodimer. In the presence of c-Jun, the GRE was active, but it was inactive when a Jun/Fos complex with high Fos content was added.[141] Moreover, there is evidence that GR and Jun/Fos can reciprocally repress one another's transcriptional activation.[142] Overexpression of c-Jun prevented GR-induced activation of genes with a GRE promoter and, conversely, GR could repress genes carrying an AP-1 binding element. These data suggest that members of these two classes of STFs (i.e., GR and Jun/Fos) can modulate one another's activity in either a positive or negative direction, depending on the environment of the cell. In addition, members of the Jun family of STFs can function in opposing ways. For example, when NIH 3T3 cells become growth inhibited, the level of c-Jun falls and JunD accumulates.[143] When resting cells are stimulated by the addition of serum, JunD is degraded and c-Jun synthesis is increased, followed by resynthesis of JunD later in G_1. Overexpression of JunD results in cells accumulating in G_0/G_1, whereas c-Jun overexpression drives cells into S/G_2 and M phase. Also, JunD partially suppresses cell transformation by an activated *ras* oncogene, but c-Jun cooperates with *ras* to transform cells.[143]

AP-1 (Jun/Fos) activity has also been shown to be required for TPA- or epidermal growth factor (EGF)-induced transformation of JB6 mouse epidermal cells, and a block of AP-1 activity by introducing a dominant-negative mutant of c-Jun inhibits tumor promoter-induced transformation.[144]

SP1

SP1 is a member of a family of transcription factors that includes SP2, SP3, and SP4. These bind to GC-rich sequences found in the promoters of many genes. SP1 was first detected as a protein in HeLa cells that could bind to and activate the SV40 early promoter.[145] It was later shown that SP1 binds selectively to GC-rich sequence elements in a wide variety of viral and cellular promoters. SP1, a single polypeptide with a molecular weight of 105 kDa, contains three zinc fingers near the C-terminal end of the protein, which provide the DNA binding domains of the protein. The activation domains are glutamine-rich sequences in the N-terminal half of the protein.[146] Although SP1 is generally considered to be a "proximal promoter factor" in that its functional binding sites are usually found within a few hundred base pairs of the transcription start site, there is evidence that SP1 can also act at distal promoter sites by recruiting distally bound activators to a position more proximal to the basal transcription complex.[146] There is also evidence that SP1 can bind the retinoblastoma gene product (Rb) control element (RCE) within the c-fos, c-myc, and TGF-β1 genes and that SP1 stimulates RCE-dependent transcription in vivo, suggesting that Rb may regulate transcription by an interaction with SP1.[147] Cooperative interactions between SP1 and other DNA-bound transcription factors has been observed and can contribute to context-dependent transcriptional regulation.[137]

Oct-3

A number of transcription factors active in early embryonic development, including that of mammals, have been identified. These include proteins that bind particular DNA octamer base sequence motifs. One of these, Oct-3, first detected in mouse embryonal carcinoma (EC) cells and in mouse embryonic stem (ES) cells, specifically binds the DNA octamer motif ATTTG-CAT.[148] Oct-3 is one of a family of transcription factors known as the POU family (see below).

Oct-3 is expressed during early mouse development, from the early one- and two-cell embryo stage up to the point that the inner cell mass and primitive ectoderm differentiate into primitive endoderm and early ectoderm, mesoderm, and endoderm. At this point Oct-3 expression is down-regulated.[148] Expression of Oct-3 in EC and ES cells is also down-regulated when they are induced to differentiate by addition of retinoic acid. In contrast, homeobox genes are expressed at low levels in undifferentiated EC and ES cells and activated when differentiation is induced, thus a reciprocal relationship between the regulation of expression of these families of genes may be inferred. The decrease of Oct-3 expression as early embryonic cells lose pluripotency suggests that its expression is important for the proliferative, highly undifferentiated state of totipotent and pluripotent stem cells. Other members of the POU family, namely Oct-1 and Oct-2, can stimulate DNA replication of adenovirus DNA in vitro. Other octamer-motif binding proteins can stimulate SV40 DNA replication in cells, supporting this notion.[148] Oct-3 is also overexpressed in a number of human germ cell tumors.[149]

The Superfamily of Hormone Receptors

Receptors for steroid hormones (including glucocorticoids, estrogen, and progesterone), thyroid hormones, retinoic acid, and vitamin D_3 belong to a superfamily of ligand-binding proteins that can bind to DNA and activate or repress gene transcription. They most likely arose from a common ancestral gene and share a number of common structural features.[150,151] After binding ligand, these intracellular receptors act as dimeric transcription factors to activate or repress target genes by binding to specific DNA promoter–enhancer sequences called hormone response elements (HREs).

This family of receptors all have DNA-binding, ligand-binding, dimerization, nuclear localization, and activation domains. The DNA-linking domains are highly conserved and consist of two zinc finger–like structures that recognize HREs consisting of an inverted TGTTCT palindromic repeat (the glucocorticoid, mineralocorticoid, androgen, and progesterone receptors) or HREs consisting of an inverted repeat of TGACCT or a closely related sequence (estrogen, thyroid hormone, retinoic acid, and vitamin D_3 receptors).[151] The ligand-binding domains also share a number of common features and homology.

Since the DNA-binding and ligand-binding domains are rather well conserved among this family of receptors, one might ask: how functional diversity is achieved. This is an important question, because these hormones have different roles during development and in the adult organism. One explanation is that their ligand-binding domains are sufficiently different as to recognize different hormones, even though their DNA-binding domain may be quite conserved. A second point is that different target cells have a different relative abundance of receptors for a given hormone and/or different, cell-specific post-translational modifications (e.g., phosphorylation) of receptors. It should also be noted that some hormone receptors such as the thyroid hormone receptor may function as a gene repressor in the absence of ligand and as a gene activator in the presence of ligand.[150] It is also interesting that truncated or mutated forms of numbers of this superfamily of receptors can act as oncogenic proteins.

YYI

YYI is a transcription factor expressed in multiple mammalian cell types. It can act as an activator or repressor, depending on the promoter context. YYI contains four C-terminal zinc finger DNA-binding motifs and binds a specific DNA sequence: CCATNTT. It interacts with a variety of basal and specific transcription factors, including TFIIB, RNA pol II, SP1, c-Myc, and ATF/CREB. The ability of YYI to induce DNA-bending toward a basal transcription complex appears to be critical for its promoter activation function; however, this bending process can also repress transcription from the c-*fos* and AP-1 promoters.[137] YYI can also serve as a transcription initiator binding protein via its ability to recruit TFIIB and RNA pol II, but this process still requires upstream activators for full activity. In some cell environments in which c-Myc levels are high, YYI is recruited into a c-Myc–YYI complex that inhibits YYI's ability to act as an activator or repressor of transcription (reviewed in Reference 137).

LEF-1

LEF-1 is a member of the T-lymphocyte factor family of transcription factors that in mammals includes TCF-1, -3, and -4. These factors contain an 85–amino acid DNA-binding domain that is homologous to the HMG protein DNA-binding domains. LEF-1 is another of the specific or context-dependent transcription factors that binds DNA after it binds to enhancer sequences. In doing so, it promotes interactions between ATF/CREB and Ets promoter–bound complexes.[137] LEF-1 also interacts with the β-catenin co-activator, which enhances LEF-1's ability to activate LEF-1-regulated genes   The oncogene c-*myc* is activated by LEF-1 and its family members in tumor cells containing high levels of β-catenin.

E2F

The E2F family of transcription factors, containing six or more members, forms heterodimers with the DNA-binding proteins DP1 or DP2, and these complexes bind TTSSCGC (S = C or G) consensus DNA sequences. E2F binds Rb and its family members p107 and p130. This interaction with Rb is critical to E2F's ability to regulate expression of its target genes. In some cellular contexts, E2F acts as a repressor for some genes, e.g., B-*myb*, via its binding to hypophosphorylated Rb and associated histone deacetylase. In tumor cells where Rb is either lost or hyperphosphorylated, E2F becomes a transcriptional activator (reviewed in Reference 137).

Tissue-Specific Transcription Factors

Tissue-specific gene expression during early organ development and tissue differentiation is regulated to a great extent at the level of gene transcription. Several transcription factors have been identified that carry out this regulation. These include MyoD, involved in muscle differentiation; HNF-1, involved in hepatocyte differentiation; and Pit-1, a pituitary activator of growth hormone and prolactin gene expression during normal ontogeny and necessary for the differentiation of lactotroph, somatotroph, and thyrotroph cells in the anterior pituitary gland. Some of these transcription factors will be discussed here.

MyoD

The ability to induce muscle cell differentiation in undifferentiated cultured mouse fibroblasts has provided a unique tool to look at lineage-specific events that regulate commitment and terminal differentiation. A family of myogenic-inducing transcription factors have been isolated and cloned, and the way in which they work has provided a model for cellular differentiation in multiple organ systems.

The first of these to be identified, MyoD, was initially cloned by subtractive hybridization with cDNAs prepared from mRNA transcripts expressed in myoblasts, but not in undifferentiated 10T 1/2 mouse cells used as the model cell line in which the myogenic pathway can be induced (reviewed in Reference 152 and 153). MyoD is a nuclear protein of 318 amino acids that forms heterodimers of an HLH motif and binds to muscle-specific enhancers. It is now known that the *myo*D gene, encoding the MyoD protein, is one of a family of myogenic genes that includes *myogenin*, *myf*-5, and *mrf*-4. Each of these factors is expressed only in skeletal muscle; gene knockout experiments indicate that mice lacking MyoD are viable and have normal muscle-specific gene activation mechanisms.[153]

Members of the MyoD family are about 80% homologous in a region that includes a basic amino acid region followed by an HLH motif in which two amphipathic α-helices are separated by an intervening unstructured loop. HLH regulatory proteins have been found in mammals, a wide variety of other vertebrates, the fruit fly *Drosophila*, the worm *C. elegans*, and the sea urchin.[153] Of interest is the fact that myogenic factors from *C. elegans* and sea urchin can activate the myogenic differentiation pathway in mouse 10T 1/2 cells, indicating the highly conserved nature of these factors. The HLH motif is also typical of the *myc* family of oncogene products as well as their dimer-forming partner Max protein and certain other "E-proteins" such as E12, E47, and HEB.

The HLH motif provides an interface for heterodimerization between MyoD and other E-proteins, forming a dimer that recognizes a DNA sequence CANNTG (where N can be any base) called the E-box.[152,153] This base sequence motif is in the regulatory region of muscle-specific genes such as muscle creatine kinase and of other cell type–specific genes regulated by HLH-type proteins. Functional activity of myogenic HLH proteins requires heterodimer formation with E-proteins such as E12 and E47.[154] As with other transcriptional factors, MyoD and the HLH family have DNA-binding domains, dimerization domains, and activation domains. In addition, their activity is modulated by their phosphorylation state.

Muscle cell determination and differentiation are dictated by a balance of factors that determine the proliferation potential and the shutdown of proliferation that accompanies differentiation. For example, once myoblasts enter the terminal differentiation pathway, they stop proliferating and fuse with neighboring myoblasts to form multinucleated myotubes. This process is inhibited by growth factors such as fibroblast growth factor (FGF), transforming groth factor β (TGF-β), and serum.[153] Moreover, expression of a variety of oncogenes including *src*, *ras*, *fos*, *jun*, *fps*, *erb*A, *myc*, and E1A as well as mitogens such as phorbol esters inhibit myogenic differentiation.[152] Differentiation also cannot be induced in a number of tumor cell types even though MyoD is expressed. Rhabdomyosarcoma cells derived from a muscle cell–type malignant tumor differentiate only poorly, although they express *myo*D. These data suggest that during the process of oncogenesis, some factors required for normal differentiation are lost or their function inhibited.

Some HLH proteins lack the functional basic regions but can still form dimers with transcriptional factors of the HLH type, thus forming nonfunctional dimers. These proteins can act as negative regulators of E-box-dependent transcription. One of these proteins is the Id (inhibitor of differentiation) protein, which can dimerize with MyoD and other E-proteins and render them inactive as inducers of myogenesis and other tissue-specific differentiation processes in which E-proteins are involved.[155]

The four members of the MyoD family are first expressed in early embryos when myogenic precursor cells appear in the somites and they are present in the limb bud during muscle terminal differentiation. Yet each gene is activated at a slightly different time, which suggests that they are expressed in response to somewhat different signals. Alternatively, activation of one of the early myogenic genes could lead to a cascade of timed activation of subsequent genes in the pathway. There is evidence that members of the *myo*D gene family can positively autoregulate one anothers' expression.[153] The timed expression of these genes produces the right mix of transcription factors that in turn induce the expression of the genes responsible for making a muscle cell.

Factors that inhibit muscle cell differentiation may act in different ways. For example, activated Ras and c-Fos proteins block MyoD transcription; TGF-β inhibits MyoD activity but not its transcription; and c-Jun can block MyoD function through a protein–protein interaction between the leucine zipper domain of c-Jun and the HLH region of MyoD.[156]

LIVER-SPECIFIC TRANSCRIPTION FACTORS

At least six liver-specific TFs are functional in development and terminal differentiation of the liver: HNF-1, C/EBP, DBP, HNF-3, HNF-4, and LF-A1 (reviewed in Reference 157). These act together in development of the hepatocyte phenotype, yet none of these appears to have expression limited to the liver, suggesting that interplay with other environmental or endogenous signals is important.

HNF-1 can bind to the promoters of a variety of liver-specific genes including α fibrinogen, α fetoprotein, α_1 antitrypsin, albumin, and transthyretin. HNF-1 binds as a homodimer to an inverted palindrome of the sequence GTTA ATNATTAAC. Optimal transcription of liver-specific genes depends on interaction among transcription factors. For example, the albumin gene contains six *cis*-regulatory elements A–F. Basal expression of the albumin gene can be achieved by binding of an ubiquitous TF, known as NF-1, to the C element, but fuller activity is achieved by binding of HNF-1 to the B element and DBP or C/EBP to the D element.[157] Augmented expression is achieved by binding of C/EBP to the A and F elements.

Since the liver expresses over 1000 liver specific genes, it seems unlikely that the whole panoply of liver gene expression would be controlled by only six transcription factors. Thus, many more likely await discovery. Some additional regulatory diversity may be achieved by

molecules similar in structure to HNF-1 or other liver TFs that can form dimers having different DNA sequence specificity. By mixing and matching various dimer motifs, a much wider variety of genes could possibly be regulated.

PIT-1

As noted above, Pit-1 is a pituitary gland–specific TF that regulates the development of hormone-producing cells in the anterior pituitary. Pit-1 is a POU domain TF, and Pit-1-activated promoters respond to EGF, cAMP, and phorbol esters. Pit-1 is phosphorylated in pituitary cells at two different sites in response to cAMP and phorbol esters, and phosphorylation causes conformational changes in Pit-1 that alter its DNA-binding specificity, with increased binding at some promoter sites and decreased binding of others.[158]

E2A

The E2A gene codes for the E-box transcription factors E12 and E47. These E-box elements are present in the immunoglobin heavy-chain enhancer and in genes involved in muscle, pancreas, and B-lymphocyte differentiation. As noted above, they are HLH proteins and bind to DNA after forming heterodimers with other E-type proteins. E2A genes have been found to be mutated in B-cell leukemias and are involved in chromosomal translocations that result in chimeric proteins being formed between E2A-encoded proteins and other DNA-binding proteins. One example is a chimera formed by a t(1;19) translocation that brings the N-terminal coding region into proximity with the C-terminal coding region of the transcription factor Pbx1, which is normally not expressed in B cells.[159] This converts a nonactivating DNA-binding protein into a potent transcriptional activator, and may be involved in the leukemogenic process.

NF-κB

NF-κB was first detected as a protein that could form a complex with a 10-bp site in the κ light-chain enhancer called κB (reviewed in Reference 160). Since it is constitutively expressed in B cells undergoing κ light-chain expression and is crucial for κ gene enhancer function, it was thought to be a tissue-specific transcription factor. Later work revealed, however, that it is ubiquitously expressed and involved in regulation of gene expression in multiple cell types, leading to the idea that it may be a "pleiotropic mediator of inducible and tissue-specific gene control."[160]

NF-κB is induced by several T-cell mitogens and by antibodies that activate T cells and appears to be important for T-cell activation through a variety of mechanisms. NF-κB is involved in transcriptional regulation of a variety of genes including interleukin-2α receptor (IL-2αR), β-interferon, interleukin-6 (IL-6), tumor necrosis factor-α (TNF-α), and lymphotoxin.[160] Several agents induce NF-κB expression, including the *tax* gene product of human T-lymphotropic virus HTLV-1 and the viral *trans*-activators of cytomegalovirus (CMV) and hepatitis B virus. In addition, CMV, SV40, and human immunodeficiency virus (HIV-1) enhancer regions all have NF-κB binding sites, suggesting an important role of NF-κB in viral replication.

POU-DOMAIN BINDING PROTEINS

Several mammalian transcription factors and developmental TFs in *C. elegans* have a unique structural motif called *POU*. POU proteins have a highly conserved N-terminal region of 76 amino acids (the POU-specific domain), a variable linker region of 15 to 27 amino acids, and a 60–amino acid homeodomain (the POU homeodomain) that diverges with species of organism (reviewed in Reference 161). The entire POU domain is required for high affinity and sequence-specific DNA binding. POU proteins appear to be able to bind DNA as either a monomer or a dimer and they bind particular octamer (8 bp) sequences. POU-family proteins include the octamer-binding TFs OCT-3 and Pit-1 noted above, as well as other developmental regulators that generate specific cell phenotypes.

ETS1 AND ETS2

The Ets proteins are transcription factors that bind to a GGAA purine-rich core DNA sequence seen in promoters or enhancers of various cellular and viral genes. Examples of genes under Ets protein transcriptional regulation include murine sarcoma virus LTR, stromelysin, urokinase-type plasminogen activator, and IL-2 (reviewed in Reference 162). The *ets* 1 and *ets* 2 genes can act like oncogenes when overexpressed or expressed at an inappropriate time. For example,

overexpression of *ets* 1 and *ets* 2 can transform murine fibroblasts, and *ets* 1 is expressed in embryonal neuroectodermal tumors such as neuroblastoma.[162] This theme occurs again and again: a transcription factor normally expressed during a certain stage or stages of development and sometimes only in specific embryonic cell types becomes a transforming oncoprotein when activated, mutated, or expressed at the wrong time or in the wrong cell type later in life.

In the case of the *ets* genes, *ets* 1 expression is limited to certain cell types during fetal development of the mouse. Its expression is seen predominantly in lymphoid tissues, brain, and organs such as the lung, kidney, and salivary gland when they are undergoing branching morphogenesis.[162] In neonatal development, *ets* 1 is expressed in lymphoid tissues and brain, but in adult mice it is only expressed in lymphoid tissue. It is also expressed in bone during bone formation or remodeling, a role that may be mediated by the Ets 1 transcriptional regulation of metalloproteinases such as stromelysin and of plasminogen activator. Activation of these proteinases most likely is important for the degradation and remodeling of extracellular matrix components during branching morphogenesis and bone deposition.

Ets 2 expression, by contrast, is widespread in all organs of embryonic, neonatal, and adult mice, consistent with a more fundamental role in cell growth control such as regulation of cell mitosis.[162]

HOMEOBOX PROTEINS

Among the many complex things that multicellular organisms must accomplish during development is to develop an anterior–posterior axis. This is done by turning on, in a spatial and temporal pattern, a series of genes called *homeotic selector genes* or *homeobox genes*. These genes were originally discovered in the fruit fly *Drosophila*; when they were mutated abnormal structural development was seen. It is now known that these genes code for a series of transcription factors and they are present in organisms as diverse as acorn worms and humans.[163] The homeobox family of genes has been detected in simple hydrozoans (*Sarsia* species), thus that they are at least 500 million years old.[164,165]

Although the first homeotic mutation in *Drosophila* was reported in 1915, it wasn't until the advent of DNA technology that responsible genes were identified (reviewed in Reference 166). These mutations cause dramatic alterations in the architecture of the fruit fly, such as a second pair of wings (a mutation called *bithorax*) or the growth of legs instead of antennae on the head (a mutation called *Antennopedia*, or *Antp*). Identification and sequencing of the genes involved in these and other structural defects in *Drosophila* revealed that there was a lot of cross-homology among them. An 180-bp DNA segment, the so-called homeobox region, was highly conserved in all of them. These genes encode the homeobox proteins that have several structural features in common: they share a 60–amino acid motif with a helix-turn-helix structure. A typical structure is that of the *Antp* protein with a flexible amino-terminal arm, three well-defined α helices, and a fourth, more flexible helix.[166] A recognition helix binds to specific base pairs in the major groove of DNA and a flexible amino-terminal domain contacts DNA bases in the minor groove, while two helices make contact with the DNA backbone.

One interesting feature of the homeobox genes is that they are clustered in a 3'-to-5' orientation that is exactly in the same order as the anterior–posterior segments of the body whose formation they regulate—that is, the genes located most 3' in the cluster are expressed in the most anterior segments, and those located toward the 5' end are expressed more posteriorly (Fig. 5–10). Furthermore, the *Antp* class homeobox clusters, called HOM-C in *Drosophila*, have remarkably similar organization and expression in the mouse and human genomes, where they are called *Hox* genes. Several loss-of-function and gain-of-function mutations of these genes have been identified in mice and frogs (Table 5–2).

In *Drosophila*, there is a single cluster of homeobox genes, whereas in mice and humans there are four clusters on four different chromosomes; however, the principles of organization are conserved.[166] Thus, the homeobox genes are a highly conserved class of "master control genes" that regulate the structural orientation of the "body plan" during development.

The genes or functions that the homeobox proteins in turn regulate are not entirely clear, but they have been shown to regulate expression of cell adhesion proteins such as the neural cell

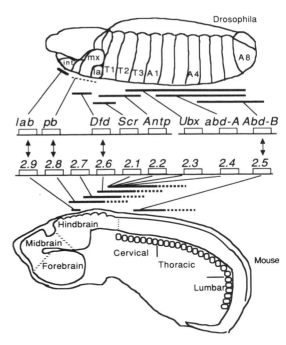

Figure 5–10. Summary of HOM-C and *Hox*-2 expression patterns. The *upper half* of the figure contains a diagram of a 10-hour *Drosophilia* embryo, with the approximate extents of the epidermal expression domains of the *HOM-C* genes *lab* through *Abd-B* indicated by the horizontal bars. The expression domains of these genes also approximately correspond to the indicated limits within the embryonic central nervous system (CNS). The *pb* expression pattern is represented by a thin dotted bar, as *pb* has no detectable function in the embryonic head. int, mx, and la designate intercalary, maxillary, and labial segments, respectively. T1, T2, and T3 indicate thoracic segments 1–3. A1–A8 indicate abdominal segments 1–8. The *lower half* of the figure contains a schematic diagram of a 12-day mouse embryo, with the approximate extents of *Hox*-2 expression domains in the CNS indicated by the horizontal bars. The dotted extensions of the *Hox*-2.8 through *Hox*-2.5 patterns indicate that these expression domains extend in overlapping fashion to posterior regions of the CNS. *Hox*-2.1, *Hox*-2.2, *Hox*-2.3, and *Hox*-2.4 have subtly different boundaries in the posterior regions of the hindbrain; for simplicity, their expression domains are represented together. (From McGinnis and Krumlauf,[165] with permission.)

adhesion molecule N-CAM, which is important in nervous system development.[167] Cell adhesion molecules are crucial for embryonic development, and regulating their expression may be one way that *Hox* gene proteins control tissue growth and development in a spatiotemporal manner.

Table 5–2. Alterations to *Hox* Expression and Resulting Phenotypes in Vertebrate Embryos

Gene (Species) (Homolog)	Type of Mutation	Phenotype
Hox-1.6 (mouse)	Loss-of-function targeted disruption in ES cells	Recessive. Neonatal lethal. Defects concentrated at the level of rhombomeres 4–7 in structures derived from paraxial and head mesoderm, neural crest, placodal, and neuroectoderm: e.g., missing motor nucleus of facial (VII) nerve; alterations to basioccipital and exoccipital bones, the inner ear, and cranial sensory ganglia; in neural crest mostly neurogenic components abnormal.
Hox-1.5 (mouse)	Loss-of-function targeted disruption in ES cells	Recessive. Neonatal lethal. Defects focused in head and thorax in structures and organs derived from mesoderm, pharyngeal endoderm, neural crest: missing thymus, parathyroids, lesser horn of hyoid bone; altered thyroid, heart, maxilla, mandible, third and fourth branchial arch, circulatory system; in neural crest mostly mesenchymal components abnormal.
Hox-1.1 (mouse)	Gain-of-function ectopic expression of a *Hox*-1.1 transgene from a β-actin promoter	Dominant lethal. Several craniofacial abnormalities, including secondary cleft palate. In axial skeleton, normal vertebrae up to C3, and variations in C1 and C2 consistent with transformations, new proatlas and a vertebral body associated with the atlas.
Hox-1.4 (mouse)	Gain-of-function ectopic expression of a *Hox*-1.4 transgene from its own promoter	Dominant lethal. Highly elevated levels of expression in the gut are associated with hyperproliferation of the colon, which leads to compaction and death in the adult. Decreases density of innervation by enteric ganglia in gut. Resembles megacolon phenotype associated with Hirschsprung's disease.
XlHbox6 (*Xenopus*) (mouse *Hox*-2.5)	Gain-of-function injection of *XlHbox*6 mRNA in two-cell *Xenopus* embryos and Einsteck grafts	Dominant alteration to axial properties of mesoderm. Animal cap region exposed to injected mRNA and grafted into blastocoel resulted in secondary axis with tail-like structures; the injected homeodomain protein also respecifies animals caps exposed to growth factors.
XlHbox1 (*Xenopus*) (mouse *Hox*-3.3)	Loss-of-function injection of antibodies to long form of XlHbox1 protein	Dominant alteration to anterior spinal cord. Hindbrain appears enlarged and extends to more posterior regions relative to pronephros. This is accompanied by XlHbox1 expression in the abnormal region. There are dorsal fin defects in neural crest cells that express the protein.
XlHbox1 (*Xenopus*) (mouse *Hox*-3.3)	Gain-of-function injection of mRNA for long and short XlHbox1 proteins	Dominant alteration to somitic segmentation and myotome markers when long form of protein used. Dominant alteration to anterior spinal cord, giving appearance of expanded hindbrain, similar to antibody injections, but phenotype extends more posterior. Also localized neuronal asymmetry.
Xhox-1A (*Xenopus*) (mouse *Hox*-2.6)	Gain-of-function injection of mRNA for *Xhox-1A*	Complex dominant phenotypes. One major alteration was the perturbation to paraxial segmentation. Regional variation in the myotome component of somites.

From McGinnis and Krumlauf[165]

Moreover, homeodomain proteins have been shown to enhance the DNA-binding activity of growth-stimulatory signals such as serum response factor.[168] This may be another way in which *Hox* gene transcription factors control segmental growth, i.e., by being turned on at a specific time and place when growth in a particular part of the body is called for.

What happens when expression of these genes goes awry? Since they are a class of master growth control genes, one might predict dire consequences if they were expressed at the wrong time or in the wrong place. One clue to answering this question is the role that *Hox* genes play in hematopoiesis. The *Hox* 3.3 gene has been shown to be involved in an early step in proliferation of the erythroid colony-forming (CFU-E) precursor cells in red blood cell formation, and this gene is also expressed in erythroleukemia cells.[169] Moreover, a subset of T-cell acute lymphocytic leukemias possess a t(10;14) or t(7;10) chromosomal translocation that involves a *Hox* gene (*Hox* 11) present on chromosome 10, suggesting that its activation is involved in the leukemogenic process.[170,171]

DNA METHYLATION

Every individual in a population of cells, organisms, or people is unique. This is true even for cells, organisms, or people who have the same genotype, such as cloned cats or identical twins.[172] How can this be, since genes are the master control elements of a cell? Obviously, there are other influences affecting how and when genes are expressed that provide additional diversity. The study of these other influences is called the science of *epigenetics*.

In mammals, two of the principle epigenetic events that contribute to diversity are DNA methylation and post-translational modifications of histones. The DNA methylation machinery is composed of DNA methyltransferases (DNMTs), which establish and maintain DNA methylation patterns, and the methylated DNA sequence binding proteins, which are involved in recognizing and binding to methylated sequences.[173] The role of epigenetic events in cancer (primarily DNA methylation events) has been thoroughly studied in recent years (reviewed in Reference 174 and see below).

The presence of the methylated base 5-methylcytosine was first reported in 1948 (reviewed in Reference 175). The presence of this minor base was first detected in calf thymus DNA, but since then it has been found in all vertebrate and plant species examined and in a wide variety of other organisms. In mammalian cells, between 2% and 7% (depending on the species and tissue) of all cytosine residues present in the genome are methylated on position 5

in the pyrimidine ring. Methylation of newly formed DNA occurs shortly after replication in proliferating cells and is an enzymatic process catalyzed by DNA methyltransferases utilizing S-adenosylmethionine as the active methyl donor. More than 90% of the 5-methylcytosine residues occur on CpG dinucleotide sequences in DNA (i.e., where C is 5′ in position to G), and they are present in a tissue- and species-specific pattern. The CpG "islands" are found in the 5′-regulatory regions of about half of all human genes. The pattern of methylation can change during differentiation from one cell type into another or during carcinogenesis, as will be discussed below.

The nonrandom and tissue-specific distribution of 5-methylcytosine residues (m^5C) in tissues has suggested a role for methylation in gene function, supported by evidence accumulating over the past several years. The basis of this evidence comes from five different kinds of studies: (1) the relationship of the amount of m^5C residues in a gene to its transcriptional activity; (2) changes in DNA methylation patterns during differentiation of specific cell lineages; (3) correlation of transcription of transfected genes into heterologous cell types with the methylation state of those genes; (4) the effects of the drug 5-azacytidine, which blocks DNA methylation, on gene activity; and (5) the close relationship between DNA methylation and genetic imprinting during gametogenesis.

Detection of m^5C residues in DNA is usually based on the sensitivity of DNA to clipping by specific endonucleases that recognize CpG sites, depending on whether they are methylated or not. For example, in a typical experimental approach, the restriction endonuclease Hpa II recognizes a CCGG sequence only when the middle CpG dinucleotide is unmethylated, whereas the endonuclease Msp I recognizes the CCGG sequence and cuts DNA at the middle CpG regardless of the methylation state of the internal cytosine.[175] Using this specific "cutting" technique, total genomic DNA is digested with Hpa II or with Msp I. The DNA fragments are then separated according to size by agarose gel electrophoresis and transferred to nitrocellulose sheets by blotting the gel on the sheets.

The fragments containing the gene of interest can be detected on the nitrocellulose sheets through hybridization with a radioactively labeled mRNA or cDNA probe. This is the well-known blotting technique developed by Southern. When the blotting patterns for the Msp I and Hpa II digestions of genomic DNA are compared, one can discern whether a gene sequence is cut differentiatlly by the two enzymes. If the patterns are different in different cell types or in cells at different stages of differentiation, it can be deduced in which cell type the gene is methylated (i.e., it is not cut by Hpa II but by Msp I). Additional evidence for the presence of m^{5}C can be obtained by sequence analysis of the DNA fragments by means of the Maxam-Gilbert DNA sequencing technique.

The relative amounts of m^{5}C in active versus inactive genes have been examined using the differential sensitivity to restriction endonucleases as a probe of DNA methylation. McGhee and Ginder[176] first reported that specific CCGG sequences in the region of the chicken β-globin gene are less methylated in erythrocytes and reticulocytes, which produce β globin, than in oviduct cells, which do not. Mandel and Chambon[177] found a correlation between the expression of the ovalbumin gene in tissues of the chicken and its undermethylation. Similar conclusions have been reached for a wide variety of genes, including human and rabbit globin genes, the α-fetoprotein, immunoglobulin, and metallothionein genes of the mouse, and various virus genes.

Although scientists were first able to analyze DNA methylation at specific gene loci by digestion with restriction endonucleases in conjunction with Southern blot analysis, this approach proved to be cumbersome. It has now been replaced by sodium bisulfite treatment of DNA, which converts unmethylated cystosines to uracil residues. This is followed by a methylation-specific polymerase chain reaction (PCR).[178] This technique has revolutionized the detection of methylated DNA sequences.

Changes in DNA methylation patterns have been observed during differentiation of various cell lineages. The fact that the pattern of DNA methylation is not identical in all tissues of a given organism strongly suggests that changes in gene methylation occur during tissue differentiation. Razin et al.[179] found that mouse teratocarcinoma cells induced to differentiate in culture by exposure to retinoic acid undergo a high degree of demthylation of their DNA. Up to 30% of the teratocarcinoma cells DNA methyl groups are lost after prolonged exposure to the differentiation-inducing agent, and this demethylation can be observed at specific sites in representative genes such as dihydrofolate reductase, β globin, and histocompatibility genes. This phenomenon appears to mimic developmental processes in vivo because the extent of DNA methylation in mouse embryo yolk sac and placenta is significantly lower than that found in mouse sperm. Adult tissue DNA, however, has a high methylation content similar to that of sperm DNA. This similarity suggests that as genes are turned on in the embryo after fertilization, a demethylation process occurs, and as tissues subsequently differentiate, methylation of the genome increases again, most likely as specific genes are shut off in those tissues by the process of genomic restriction. Thus, although large changes in DNA methylation patterns appear to occur during early embryogenesis, the methylation patterns of adult somatic cells seem to be relatively stable.

Mechanisms that have been invoked to explain the role of methylation in gene activity include (1) inhibition of binding of transcription factors by methylated DNA sites; (2) inhibition of binding of transcription factors by DNA-binding proteins that specificially bind to methylated DNA sequences; (3) methylation-induced change in chromatin structure that renders it less open for gene transcription;[180,181] and (4) co-recruitment of DNA methyltransferase and histone deacetylases to DNA methylation sites.[27] There is some evidence for each of these scenarios. Levine et al.[180] found that inhibition of promoter activity correlated with the density of methyl CpG sites at the preinitiation domain of the promoter (TATA) box but was not effected by methyl CpG sequences distant from this domain. Their evidence also suggested that a methyl CpG binding protein was involved in this inhibition. Interestingly, some DNA templates were able to establish functional preinitiation complexes even in their methylated state.

DNA Methyltransferases

There are four known DNA methyltransferases: DNMT1, 2, 3a and 3b and a number of proteins that bind methylated DNA; these include MBD1, 2, 3, and 4 and MeCP1 and MeCP2.[182] Functionally, the DNMTases are of two classes. DNMT1 is a maintenance methylase and, if knocked out in mice, causes genome-wide demethylation and developmental arrest. This finding supports the observation that DNMT1's role is in propagating parental DNA methylation during DNA replication (reviewed in Reference 183). DNMT1 can methylate CpG sites packaged into nucleosomes; however, there are significant differences in accessibility among DNA sequences.[183] For example, DNA sequences located in the central region of a nucleosome core where histones H3 and H4 contact the DNA are refractory to methylation. Since histone-DNA interactions occur exclusively in the minor groove, recognition of CpG sites that protrude from the major groove facilitates their recognition by DNMTs.

The other functional family of DNMTs includes DNMT3a and DNMT3b, which are the DNA methyltransferases involved in establishing new (de novo) DNA methylation patterns during development; it has been shown that DNMT3a and 3b are essential for development.[184] Inactivation of both genes blocks DNA methylation in early mouse embryos, but does not affect maintenance methylation involved in gene imprinting. Mutations of human DNMT3b are found in a rare autosomal recessive condition called the ICF syndrome (immunodeficiency, centromeric instability, facial anomalies), further indicating an important role for this methyltransferase in human disease.[184] DNMT1, 3a, and 3b are all reported to be overexpressed in human cancer cells.[185]

Methyl DNA Binding Proteins

The methyl DNA binding proteins play a role in transcriptional silencing. The original member of this family, MeCP2, is a single polypeptide chain that contains a methyl-CpG-binding domain (MBD) and a transcriptional repression domain (TRD).[182] MeCP2 recognizes a single CpG dinucleotide methylated on both strands in either naked DNA or a sequence situated in packaged chromatin. However, not all MeCP2 binding sites are available in nucleosomes, perhaps because MeCP2 needs to contact up to five base pairs on either side of the methylated CpG nucleotide.[183] MeCP1 exists in many cell types and can also inhibit transcription.

Four additional methylated DNA binding proteins, related to MeCP2, have been found. These are MBD1 to MBD4 that all share a consensus methyl CpG binding domain with MeCP2.[182] MBD1 binds selectively to methylated DNA and inhibits transcription in in vitro assays and represses transcription in intact Drosophila and mammalian cells. MBD1, 2, and 3 have all been implicated as transcriptional repressors and at least part of this repression appears to involve recruitment of histone deacetylases to these methylated sites. MBD2b has demethylase activity and may be involved in reversing the transcriptional inhibition of DNA methylation.[186] MBD4 is a thymine DNA glycosylase involved in DNA repair.

Methylation of DNA is also capable of inducing structural changes in chromatin. For example, as noted above, transcriptionally active chromatin is DNase I hypersensitive, whereas the same chromatin when methylated becomes transcriptionally inactive and DNase I resistant.[187] Moreover, methylated DNA has been shown to be resistant to the restriction enzyme Msp1 when it is complexed to H1 histone, whereas "naked" methylated DNA is digested by Msp1, and unmethylated DNA is digested by Msp1, regardless of whether it is complexed to histone H1.[188] These data support the concept that histone H1 binding to methylated DNA sequences renders them cryptic so that these regions would be inaccessible to binding proteins such as transcription factors.

A crucial determinant of the effectiveness of methylation to regulate gene expression is the density of methyl CpGs near the promoter sequence. Weak promoters can be fully repressed by a low level of methylation, but activation of transcription can be restored by interaction with an enhancer sequence.[181] Even strong promoters or promoter–enhancer interactions can be repressed by heavy methylation at a promoter site. The degree of repression appears to be related to the level of MeCP1 binding to the methylated sequence. It should be noted that some CpG-rich regions of the genome are typically not methylated, do not bind MeCP1, and are active in transcription.[181] CpG island methylation plays a role in inactivation of one of the X-chromosomes in females.

Additional evidence for the role of DNA methylation in gene expression has come from experiments in which the pyrimidine antimetabolite 5-azacytidine (5-azaCR) was shown to block DNA methylation and activate previously silent genes. Early studies of this came from Taylor and Jones,[189] who showed that 5-azaCR could induce differentiation of 10T 1/2 mouse fibroblasts into muscle and other differentiated cell types, suggesting that demethylation of genes leads to activation of a differentiation program. Later evidence indicated, however, that hypermethylation of certain promoter regions occurs as cells are carried in culture and is not their de novo state in actual tissues.

It is clear that maintenance of methylation patterns in the genome is somehow crucial for normal development. Evidence for this comes from study of transgenic mice in which a homozygous deletion of a methyltransferase that methylates DNA is introduced.[190] Homozygous mutant mouse embryos resulting from these transgenic manipulations are stunted in developmental growth and die at mid-gestation. One way that this may occur is through genetic imprinting (see below).

DNA Methylation and Cancer

Although genetic abnormalities such as chromosomal aberrations (e.g., translocations, aneuploidy) and base mutations have long been implicated in cancer initiation and progression, it is only more recently that epigenetic changes have been found to play a key role in these events. The term *epigenetic* is used to define a change in the pattern of gene expression that occurs by mechanisms other than mutations in the primary nucleotide sequence or chromosomal abnormalities such as deletions, translocations, and amplifications. DNA methylation, because of its role in regulating chromatin packaging and gene expression, is a key mechanism that can go awry in cancer and frequently does. Aberration of DNA methylation is one of the epigenetic mechanisms through which malignant changes in cells are produced. Interestingly, there is little or no DNA methylation in lower organisms such as yeast or *Drosophila*. This is one of the evolutionary changes that has

allowed additional fine-tuning of gene expression. In higher organisms such as mammals that contain large amounts of transcriptionally silent DNA, DNA methylation is one of the critical mechanisms for maintaining transcriptional silence. Heavily methylated DNA is replicated later in S phase. This late replication event helps maintain the formation of inactive chromatin and protect cells from activation of potentially harmful inserted and methylated genes such as viral sequences or transposons.[191]

In addition to CpG islands that are the targets for methylation in gene promoter sequences, other CpG sites exist in other parts of the genome. The CpG islands in promoters of active genes are usually unmethylated, whereas CpG sites outside the CpG island–containing promoter regions are usually methylated and may play a role in packaging chromatin into an inactive conformation. Curiously, this pattern is often reversed in cancer—that is, unmethylated CpG islands in promoters become methylated and methylated CpG sites in silent chromatin become unmethylated. This reversal does two things: it causes actively transcribed tumor suppressor genes to be silenced and normally unexpressed genes such as proto-oncogenes to be transcribed. The effects of hypomethylation have, for example, been shown in the overexpression of putative oncogenes in pancreatic ductal carcinoma.[192] Genome-wide hypomethylation has been observed in human colon cancer[193] and various other human cancers (reviewed in Reference 192). In such cases, the extent of hypomethylation appears to correlate with tumor grade and prognosis. Gaudet et al.[194] have shown a direct link between DNMT1 deletion and tumor induction in mice carrying a hypomorphic *dnmt1* allele, which reduced DNMT1 expression to 10% of wild type and resulted in significant genome-wide hypomethylation. These mice developed aggressive T-cell lymphomas and had a high incidence of chromosome 15 trisomy, findings suggesting that genomic hypomethylation promotes carcinogenesis by causing chromosomal instability.

Methylation of CpG islands in gene promoter sequences is a mechanism for inactivation of tumor suppressor genes, which, of course, can also be caused by loss of heterozygosity and base sequence mutations. However, the number of tumor suppressor genes inactivated by epigenetic inactivation at least equals if not exceeds those inactivated by mutation.[191] There are also other mechanisms for epigenetic inactivation, including chromatin remodeling and histone deacetylation. DNA methylation and histone deacetylases, as pointed out above, act synergistically in this regard. Furthermore, CpG island methylation by DNMTs does not by itself cause transcriptional silencing. Such an effect requires the formation of complexes containing methylated DNA binding proteins. Also, when CpG islands in promoter regions are densely methylated, nucleosomes are tightly packed, inhibiting access to promoter regions by the transcriptional machinery.

In the interaction between DNA methylation and histone deacetylation, DNA methylation appears to be dominant. Indeed, binding of DNMT and its associated methyl DNA binding proteins recruits HDAC to these sites. A drug that only inhibits HDAC can increase the expression of genes without methylated promoters, but not those with promoters that are hypermethylated. Therapeutically, then, it makes sense to use combinations of DNMT and HDAC inhibitors. This synergy has been shown, for example, by the observation that inhibition of DNA methylation by 5-aza-2-deoxycytidine plus HDAC inhibition by sodium phenylbutyrate prevented tobacco-induced lung cancer in mice.[195]

The pathways disrupted by gene promoter hypermethylation are legion and include gene silencing of a number of tumor suppressor or other cancer-related genes, such as those that regulate cell cycle control (Rb, p16, p15, p14, p73), genes involved in DNA repair (MLH1, 0^6-MGMT, GST-Pi, and BRCA1) and apoptosis (DAP kinase, caspase 8, TMS-1), and inhibitors of tumor invasion (E-cadherin, VHL, APC, LKB1, TIMP-3, THBS1) (reviewed in Reference 191). The function of many of these genes has already been described in Chapter 4, but two that are worth mentioning here are GST-Pi and 0^6-MGMT.

The gene coding for GST-Pi (glutathione S-transferase Pi) is hypermethylated in about 90% of prostate cancers. This is an interesting finding because this enzyme is involved in detoxification of a number of carcinogens. Hypermethylation of this gene has also been detected in prostate intraepithelial neoplasia (PIN), which may be an early biochemical lesion. The gene was not hypermethylated in normal prostate tissue. When used to diagnose prostate cancer in patients undergoing prostatectomy, GST-Pi was 70% accurate by examination of methylated DNA sequences in plasma or serum (reviewed in Reference 178).

When cells lose the function of 0^6-MGMT (0^6-methylguanine DNA methyltransferase), they have a diminished capacity to repair DNA damaged by alkylating agents, making cells more susceptible to the cytotoxic effects of anticancer drugs that alkylate DNA and to some environmental carcinogens. Inability to remove alkylated guanisine leads to guanine-to-adenine mutations. However, inability to repair DNA alkylated by anticancer drugs may actually enhance drug response. For example, early evidence indicates that brain tumors with hypermethylated 0^6-MGMT respond better to alkylating agent therapy than those that do not contain the hypermethylation (reviewed in Reference 191).

Clinically, there is an important difference between gene silencing caused by mutation and that caused by epigenetic mechanisms. Mutations are essentially irreversible, whereas epigenetic events are potentially reversible. Thus, reversal of epigenetic events provides a target for cancer treatment and for cancer prevention. The latter would be the case, for example, if one could reverse the progression of early intraepithelial neoplasms by inducing re-expression of a silenced tumor suppressor gene. Such might be the case in the example of methylation of the GST-Pi gene in PIN described above. Reactivation of the hypermethylated cyclin-dependent kinase inhibitor p15 gene in patients with myelodysplastic syndrome, a precursor to leukemia, has been observed in patients treated with 5-aza-deoxycytidine (reviewed in Reference 192). Such observations hold out hope that reversal of the silencing of hypermethylated genes could be an important chemopreventative approach. Unfortunately, this treatment would probably have to continue for months or years, because aberrant promoter methylation and gene silencing return once the treatment with DNMT inhibitors is stopped.

Abnormal DNA methylation patterns have been detected in tumor tissue and/or body fluids from patients with a wide variety of human cancers, indicating that aberrant DNA methylation is a correlate if not a cause of cancer. Altered

DNA methylation has been observed in breast,[196] colorectal,[197] non–small cell lung,[198,199] bladder,[200] pancreatic,[201] kidney,[202] and gastric[203] carcinomas.

Methylated DNA sequences can also be used as tumor markers. DNA-based markers are stable, and methylated DNA sequences can be detected in serum, urine, sputum, and other body fluids. Furthermore, hypermethylation of gene promoters is common in over 70% of tumors from the major cancer types, and methylated sequences, even though they may be fragments, have enough sequence information to determine which genes they came from. All of these factors contribute to the importance of methylated DNA as a universal tumor marker.[191]

Although the association of promoter DNA methylation and resultant tumor suppressor gene silencing with cancer is strong, a cause-and-effect relationship is difficult to prove and remains an open question.[204] For example, silencing of a gene that has a potential methylation site in a specific gene promoter may reflect loss of gene expression due to a mutational event in a signal transduction cascade or a transcription factor network, or a mutational event that affects chromatin packaging, rather than the methylation event. Furthermore, a number of tumor cell–related genomic methylation events have been observed in cultured cell lines and may not be the same in primary tumors. On the other hand, data showing that genes mutated in certain familial cancers are the same genes that are frequently hypermethylated in their promoters is a strong argument that hypermethylation can be a key event. Such is the case for the *VHL* gene in renal cancers, which can be both mutated and hypermethylated.[204] A question could be which came first, but these data indicate that epigenetic silencing can hit genes involved in the cancer process.

GENOMIC IMPRINTING

Genomic imprinting is the process by which the expression of one of the two parental genes is shut off in the embryo. Mammals inherit two complete sets of chromosomes, one from each parent, and thus two copies of every autosomal gene. Both copies of parental genes may be expressed, but sometimes only one of the two parental genes is expressed. (The other allele is said to be "imprinted"). Most of the evidence obtained so far indicates that the mechanism for genetic imprinting involves DNA methylation.[205] Methylation of DNA appears to have developed over evolutionary time as a way for an organism to protect itself from foreign DNA, i.e., "if it ain't us, methylate it." This prevents foreign DNA from being expressed. About 50 mammalian genes are known to be imprinted.[206] Genes imprinted in the mouse include insulin-like growth factor-2 gene (*IGF-2*); *IGF-2* receptor gene; *H19*, a gene coding for a regulator of *IGF-2* expression; and *Snrpn*, a gene that encodes a ribonucleoprotein that catalyzes RNA splicing (reviewed in Reference 205). In the mouse, the *IGF-2* gene and *Snrpn* are exclusively paternal in expression and the *IGF-2* receptor gene and *H19* are maternal in expression. Since the repressed locus does not express any mRNA, the gene must be switched off. The *IGF-2* receptor gene contains a region that is methylated in a developmentally regulated way, as is another transgene, *TG-A*, artificially introduced into transgenic mice. Somewhat counterintuitively, the maternal locus of the *IGF-2* receptor gene is methylated and yet it is the allele that is expressed.[205] Thus, methylation is not the only factor involved in the turning off and on of genes during genetic imprinting. Other factors such as histone methylation, which appears to have a role in establishment of CpG methylation patterns,[207] and histone acetylation, which affects chromatin conformation,[208] are most likely involved. As noted above, methylation patterns change during development, even though methylation patterns in the gamete may direct how later parent-specific methylation occurs in the embryo. The methylation pattern is heritable and becomes specific for given tissue types in differentiated adult cells via "maintenance methylases" that ensure the heritability of the methylation profile.[175] Gamete DNA is highly methylated, representing a highly repressed genome. During early development, demethylation of multiple genes needed for cell proliferation, cell migration, invasion into the uterine wall, and many "housekeeping" functions (e.g., substrate transport, protein and carbohydrate metabolism, nucleic acid synthesis) are demethylated in correlation with their increased transcription. As tissues differentiate, some of these genes become remethylated and turned off while new genes become operational, producing a different specific methylation in each adult somatic tissue. A number of the phenotypic characteristics of early developing embryos, which are turned off as the embryonic tissues differentiate, are similar to those of malignant cancer cells. These include, for example, invasiveness, metastasis (i.e., migration through tissues), and rapid cell proliferation. In cancer cells, the genes controlling these functions are somehow turned back on inappropriately.

Another potential mechanism for maintenance of methylation patterns as parent cells divide into daughter cells is the regeneration of a discrete chromatin structure that allows or disallows methylation at various genetic loci. Such a structure could be determined by the complement of sequence-specific DNA binding proteins and/or the coiling of DNA. A model has been proposed by Selker[209] in which a specific subset of proteins capable of holding chromatin in a given form can associate with DNA and reassociate with it after DNA replication, thereby recapitulating a form that is available (or not available) for methylation.

A peculiarity of DNA methylation is that it creates sites of high mutability.[181,209] Deamination of 5-methylcytidines occurs spontaneously in DNA at a fairly high frequency, producing thymidine, and when DNA replicates, this introduces a CpG→TpG base transition mutation if not repaired. The importance of this process is suggested by the fact that CpG→TpG base transitions are thought to produce point mutations involved in about one-third of all human genetic diseases.[181] For Nature to tolerate this potential for mutation, DNA methylation must have been preserved for a very important reason. It likely has to do with the importance of assuring the fidelity of parental inheritance and preventing foreign DNA from replicating in an organism's cell, these aims outweighing the danger of mutation. Although this is a teleological argument, it has some attraction.

In a number of human cancers, loss of imprinting (LOI) occurs, allowing both the maternal and paternal alleles to be expressed. If this occurs for a growth factor, such as IGF-2, cells get a double dose of a growth stimulatory signal. LOI of IGF2 has been observed in about 45% of a series of patients with colorectal cancer.[210] Interest-

ingly, this LOI could also be detected in patients' circulating leukocytes, thus this may be an alteration that precedes the onset of neoplasia and could be used as a screening test for cancer susceptibility. Somewhat paradoxically, LOI can be reversed by drugs that are DNA methyltransferase inhibitors, such as 5-aza-2-deoxycytidine, so an aberrant DNA methylation event may induce LOI.[211] LOI of the IGF-2 gene appears to be involved in tumor progression, leading to a more invasive phenotype.[212]

Loss of imprinting of IGF-2 was first observed in Wilms' tumor, a kidney cancer that is the most common solid tumor in children, and subsequently found in other embryonal tumors of childhood and in a variety of adult cancers, including uterine, cervical, esophageal, prostate, lung, and germ cell tumors (reviewed in Reference 213). LOI of IGF-2 is correlated with biallelic hypermethylation of five CpG sites in the "insulator" CTCF-binding element of the H19 gene locus in both tumor and normal tissue of patients with microsatellite instability-positive colorectal cancers.[214] Methylation of the H19 locus in turn regulates the silencing of the IGF-2 gene on the same chromosome. CTCF is a multivalent transcription factor that acts as a chromatin insulator by binding to the H19 gene differentially methylated region when it is unmethylated, thus separating ("insulating") IGF-2 from its enhancer and allowing monoallelic expression of IGF-2 (reviewed in Reference 215). Although CTCF gene mutations or CTCF gene silencing have been found in some of the same tumors that have IGF-2 LOI, aberrant methylation of CTCF binding sites appears to be necessary but not sufficient for IGF-2 gene LOI, at least in Wilms' tumor.[216] However, loss of methylation rather than gain of methylation appears to cause IGF-2 LOI in colorectal cancer,[217] which suggests that the mechanism of LOI may be different in different tumor types. Nevertheless, the explanation for DNA methyltransferase inhibitors such 5-aza-2-deoxycytidine restoring LOI may be that when the imprinted gene is demethylated, CTCF can no longer bind and silence expression of the imprinted allele.

LOI observed in human cancers produces a twofold to threefold in increase in the tumor as well as in other tissues. Interestingly, in colon cancer patients, there is also an LOI-induced expansion of progenitor stem cells in their colons. Since deregulated Myc expressionb is enhanced by IGF2 (as well as IGF1 and PDGF), LOI of IGF2 may increase cancer risk by expanding the stem cell population and by augmenting expression of Myc.[217a]

Using the Min mouse model that has a propensity to develop colon cancer, Feinberg and colleagues have shown that deletion of the H19 gene and its upstream differentially methylated region (DMR) reactivates the normally silenced allele of *IGF-2*. In mice containing the *H19* deletion, there is a twofold increase of *IGF-2* expression and an increase in number of colon tumors (reviewed in 103b and in Reference 217a) as well as a shift to a more undifferentiated phenotype in the intestinal epithelium. These alterations appear to primarily effect tumor initiation rather than progression, leading to the concept that LOI is an epigenetic alteration that increases the abnormal pool of progenitor cells, which in turn facilitates a mutation of "gate keeper" genes and the induction of a primary tumor. Later events occur in this background of altered progenitor cells.

The relative roles of epigenetic alterations such as methylation-related silencing of tumor suppressor or DNA repair genes and genetic mutations in the causation of cancer are under vigorous discussion. One point of view is that epigenetic changes such as LOI have a causal role in many human cancers,[218] and the other is that genetic instability caused by the mutator phenotype dominates over the gradual accumulation of DNA hypermethylation that occurs in cancer.[219] In fact, both of these types of events occur in human cancer, thus the two hypotheses are not mutually exclusive.

LOSS OF HETEROZYGOSITY

Deletion of genetic material is a very common event in human cancer. Indeed, it is the most frequently observed genetic abnormality in solid tumors. These deletion events often involve loss of heterozygosity (LOH) of the expression of either the maternal or paternal alleles of a gene. If this is accompanied by mutation of the remaining allele, as is sometimes the case for a tumor suppressor gene such as *p53*, an important mechanism to regulate cell proliferation and differ-

entiation is lost. An early observation of LOH in human cancer was by Solomon et al.,[220] who showed that about 20% of human colorectal cancers had undergone allelic loss on chromosome 5q. Vogelstein and colleagues subsequently reported how a series of genetic alterations, including LOH of alleles at chromosomal regions 5q (*apc* gene), 17p (*p53* gene), and 18q, are involved in progression of colorectal cancer.[221]

It is now recognized that LOH occurs in most if not all human solid tumors and may involve up to 20% of the genome. In some cancers, including lung, ovarian, and colorectal cancers, LOH is an early event and may occur at the stage of dysplasia or carcinoma in situ. The prevalence of LOH differs at different positions within the genome and is more prevalent at certain "hot spots." Frequently involved allelic loss occurs in cancer cells on chromosomes 3p, 5q, 7q, 8q, 9p, 13q, 17p, and 18q. These losses often involve regions containing tumor suppressor genes. The tumor suppressor gene functions contained in these regions include *p53*, *brca1*, *rb*, *brca2*, *apc*, *vhl*, and *p16*. LOH is detected by using molecular genetic techniques such as restriction fragment length polymorphism (RFLP) or PCR. Frequently the same genes that have undergone LOH in hereditary cancers also undergo LOH in "spontaneous" cancers.

TELOMERES AND TELOMERASE

Normal human cells undergo a finite number of cell divisions when grown in culture and ultimately stop dividing and undergo what is called *replicative senescence*. For human cells, the number of cell divisions attained before senescence ensues is about 50.[222] One difference between young, replicating cells and their senescent counterparts is the length of specialized "tails" at the end of chromosomes, called *telomeres*. In human cells, telomeres are made up of an average of 5000 to 15,000 base-pair repeats containing the sequence (TTAGGG)n together with telomere-binding proteins.[223] Younger cells have the longer telomeres. Every time a cell divides, 50 to 100 base pairs are lost, and a cellular signal is eventually triggered to stop cell division.

Cells of higher eukaryotic organisms maintain telomere length by the activity of an enzyme complex called *telomerase*. This is a ribonucleo-protein complex that contains several proteins and RNA. The catalytic component of this complex is a reverse-transcriptase, *human telomerase reverse transcriptase* (hTERT), that uses the RNA contained in the complex as a template for reverse transcription to replicate the DNA sequences in the telomere. Germ cells and pluripotent tissue stem cells have telomerase activity, although telomerase is turned off in cells from most tissues as they differentiate. Most human cancers appear to be able to reactivate telomerase activity, thus rejuvenating their proliferative capacity;[224] however, 10%–15% of human cancers do not express telomerase and apparently maintain telomere length by a different mechanism.[225] Telomerase has been a hot target for both diagnostic and therapeutic approaches to cancer. A problem with the use of telomerase inhibitors for cancer therapy is the slow onset of action of such agents because tumor cells can continue to proliferate until telomere length reaches a critical length. Moreover, normal stem cells such as those involved in hematopoiesis and wound healing are negatively affected by telomerase inhibition.[225] There are also data indicating that restoration of telomerase in human cells extends their life span,[226] suggesting that senescence can be overcome and perhaps provide a way to maintain human stem cells for replacement of aging or damaged tissues.

POST-TRANSCRIPTIONAL REGULATION

After genes are transcribed into mRNA, a whole series of events regulate how an mRNA gets translated into a functional protein. These events include (1) splicing of the high-molecular-weight precursor mRNA (pre-RNA) transcripts into mRNA; (2) capping, polyadenylation, and editing of the mRNA; (3) nuclear-cytoplasmic transport; (4) initiation of translation; (5) alternate translation from overlapping reading frames; (6) turnover of the mRNA; (7) protein folding and processing; (8) post-translational modifications of the protein; and (9) intracellular translocation of the mature protein, leading to secretion or sequestration into its functional compartment. Some of these events have already been discussed. Others are discussed briefly below.

RNA editing is a post-transcriptional process that produces an mRNA with a nucleotide sequence that differs from that of the transcribed DNA.[227,228] It is another mechanism for modulating gene expression. RNA editing was first described as a mechanism for mitochondrial gene expression in protozoa, where the insertion or deletion of uridine in an mRNA was observed. Other examples include the conversion of a cytidine to uridine in mammalian apolipoprotein-B mRNA, insertion of two guanosine residues in a paramyxovirus transcript, and conversion of a cytidine to uridine at multiple positions in the mRNA for subunit II of cytochrome-*c* oxidase in wheat mitochondria.[227,228] In these instances, mRNA editing either changes a nontranslated message into a translated one or modifies a translatable message into one that generates a protein with a different amino acid sequence. The mechanism for such mRNA editing isn't clear, but it appears to involve an error-prone base-pairing mechanism with a "guide RNA."[229] In some organisms such as trypanosomes, the edited proteins accumulate mutations about twice as fast as unedited proteins,[229] a finding suggesting that pre-mRNA editing plays a role in the process of evolution.

Nuclear-cytoplasmic transport of mRNA is required to get the message to the polyribosomes where they are translated. This is also a regulated event and requires RNA-binding proteins.[230]

In higher eukaryotic cells, mRNA translation is regulated by structural features of the mRNA as well as by the translation–initiation machinery, including a finely tuned series of initiation factors (reviewed in Reference 231). The translational machinery is shown in Figure 5–11.[232] Structural features of the mRNA for modulating translation include (1) the m7G cap at the 3′-end; (2) the primary base sequence around the AUG initiation codon; (3) the position of the AUG codon (whether it is first or in a place where a second initiation can occur; see below); (4) secondary structure upstream and downstream from the AUG codon; and (5) length of the leader sequence. The sequence of events, briefly, is as follows. The 40S ribosomal subunit, bearing a methionine transfer RNA (tRNA) for the AUG codon and the appropriate set of initiation factors, attaches to the 5′ end of the mRNA and migrates along the mRNA until it finds the first AUG codon. At this

(a) Formation of the cap-binding complex

(b) Formation of the ternary translation initiation complex

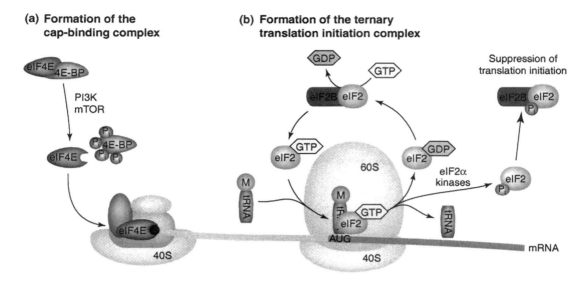

Figure 5–11. Rate-limiting steps in the initiation of translation. *a*. The eukaryotic translation initiation factor 4E(eIF4E) is retained in an inactive form by 4E-binding proteins (4E-BPs), but is released after phosphorylation of the 4E-BPs by phosphoinositol 3-kinase (PI3K) and mTOR (mammalian target of rapamycin). Free eIF4E then binds to the mRNA cap (black circle) as part of the eIF4F complex that recruits the small 40S ribosomal subunit. *b*. A ternary complex, consisting of eIF2, Met-tRNA$_i^{Met}$, and GTP, facilitates AUG-codon recognition and initiation of protein synthesis. eIF2 is recycled through the exchange of GDP for GTP by the associated guanine-nucleotide-exchange factor, eIF2B. However, phosphorylation of the eIF2α subunit by an eIF2α kinase prevents dissociation of eIF2 from eIF2B, leading to inhibition of translation initiation. M, methionine; P$_i$, phosphate. (From Calkoven,[232] with permission.)

time a 60S ribosomal subunit joins the 40S subunit and the first peptide bond is formed.

Certain characteristics of the mRNA or of the initiation factors can alter the way in which these translational events occur. For example, in some cases, two different proteins can result from different AUG initiation codons within the same mRNA. This can occur when the first AUG site is, at a particular time or cellular environment, in a less favorable conformation than that of a second one downstream. Thus, different proteins can be produced from an overlapping reading frame, depending on which start site is used.[233] Such alternate production of two proteins by initiation at the first or second AUG codon has been observed for a variety of viral mRNAs and some human oncogenes.[233]

Alterations in levels of initiation factors can also modulate mRNA translation. For example, it has been shown that overexpression of the translation initiation factor eIF-4E in NIH 3T3 or rat-2 fibroblasts causes their tumorigenic transformation,[234] apparently as a result of loss of regulation of initiation of protein synthesis.

The turnover rate of various mRNAs is also an important variable in modulating mRNA translation. Some hormones and external factors that induce gene expression may do so by stabilizing an mRNA with a relatively short cellular half-life. Half-lives of eukaryotic mRNAs vary from a few minutes for highly regulated mRNAs such as cellular oncogenes and rate-limiting enzymes to more than 100 hours for very stable mRNAs, such as certain housekeeping and structural proteins.[235] The average half-time for turnover for mRNA in eukaryotic cells is 10–20 hours, whereas the average t 1/2 for proteins is about 48–72 hours, although many turn over faster than that. One notable example is the tumor suppressor protein p53, which in the normal, wild-type form has a half-life of about 1–2 hours, whereas its mutated form has a t 1/2 of about 6–8 hours (see below). In addition, rates of degradation of mRNA and proteins may change during the cell cycle (e.g., the cell cycle regulatory cyclins), in response to stress (e.g., heat-shock proteins), availability of nutrients, or during various stages of differentiation (e.g., oocyte mRNAs after early stages of embryogenesis).

Post-transcriptional regulation can be thought of as two separate but linked "buckets," the first one holding the machinery involved in quality control (QC) of mRNA function and the second one containing the translational machinery. Quality control of mRNA involves several steps through which mRNA must pass successfully or it is degraded or otherwise prevented from being translated.[236] These steps involve mRNA splicing, capping, transport out of the nucleus, and correct interaction with the ribosomal apparatus. Some of these steps were listed above. It should be noted that defects in capping, splicing, and 5′ and 3′ end formation inhibit mRNA export from the nucleus. In addition, efficient mRNA export requires binding to "shuttle" ribonucleoprotein (hnRNP).

Quality control is also invoked during mRNA translation (reviewed in Reference 236). The mRNA caps and poly(A) tail protect mRNA from degradation in the cytoplasm and help initiate formation of the translational complex that recruits the 40S ribosome (Fig. 5–11). Inappropriately processed mRNAs are prevented from being translated by a mechanism called *nonsense-mediated mRNA decay* (NMD), which triggers mRNA degradation if the QC system detects premature amino acid chain termination. This process occurs in every organism from yeast to humans.

Two crucial QC events occur during initiation of translation: recruitment of mRNA to the ribosomal complex and selection of the AUG initiation codon. Both of these events are directed by formation of multiprotein complexes and regulated by phosphorylation (reviewed in Reference 232). The eIF4E protein complex binds the 5′ cap of mRNA and initiates recruitment of the 40S ribosome, which in turn initiates scanning of the ribosomal complex for the AUG initiation codon. eIF4E is kept in an inactive form by binding 4E-binding proteins, whose phosphorylation by kinases is activated by mitogen or growth factor signaling. A second rate-limiting event (Fig. 5–11) is formation of a ternary complex between a G protein called eIF2, the initiator for the first codon (Met-tRNA$_i$ ^{met}), and GTP. This process leads to AUG recognition and initiation of protein synthesis.[232]

Mutations that affect the regulation of eIF2 have been observed in inherited diseases such as the Wolcott-Rallison syndrome and leukoencephalopathy. Mutations in other components of the translational machinery also occur (reviewed in Reference 232). Fragile-X mental retardation syndrome is caused by a loss of function of fragile-X mental retardation protein (FMRP) by transcriptional silencing or single amino acid change. In chronic myelogenous leukemia (CML), the hnRNPE2 protein is overexpressed, causing inhibition of translation of the C/EBPα transcription factor that induces myeloid precursor cell differentiation. Mutation of the *c/ebpα* gene occurs in acute myelogenous leukemia (AML).

A number of oncogenic-related signal transduction pathways also involve dysregulation of translational control mechanisms. For example, overexpression of eIF4E has been observed in malignant transformation of rodent fibroblasts and may involve cooperation with v-Myc or adenovirus E1A oncoproteins. Increased levels of eIF4E levels have also been found in human colon, breast, and bladder carcinomas as well as in non-Hodgkin's lymphoma.

Even though only about 10% of vertebrate proteins have their expression regulated at the translational level, a number of these proteins have a significant impact on cell proliferation, apoptosis, and differentiation. These proteins include cell cycle regulators p27 and cyclin D1, thrombopoietin, Bcl-2, C/EBP-α and -β, and the transcription factor ATF-4 (reviewed in Reference 232).

MOLECULAR GENETIC ALTERATIONS IN CANCER CELLS

Cancer is essentially a genetic disease, in that all cancer cells have some alteration of gene expression or function. These genetic alterations include chromosomal translocations, the Philadelphia chromosome in CML described earlier being an example of this, inversions, deletions,

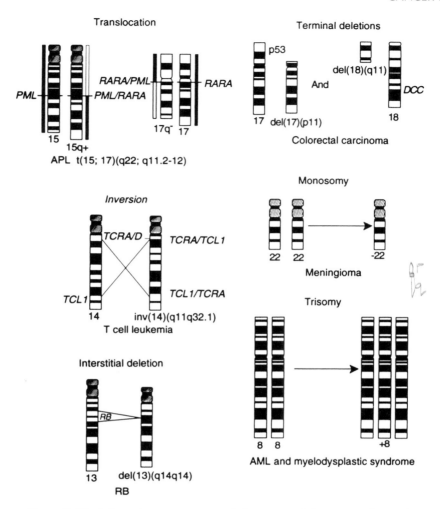

Figure 5–12. Schematic representation of chromosomal aberrations observed in tumors. Shown are the t(15;17)(q22;q11.2–12). seen in APL; the inv(14) (q11q32.1) observed in T-cell leukemia; the cel(13)(q14q14) associated with RB; the terminal deletions of chromosomes 17p and 18p seen colorectal carcinoma; monosomy 22 associated with meningioma; and trisomy 8 seen in AML and myelodysplastic syndrome. (Reprinted from Solomon et al.,[237] with permission from the American Association for the Advancement of Science.)

amplifications, point mutations, and duplications or losses of whole chromosomes (aneuploidy) (Fig. 5–12).[237] Most of the original information about genetic alterations in cancer came from studies of leukemias and lymphomas[237,238] because it is easier to obtain relatively pure, single-cell dispersions of populations of these types of cells from peripheral blood or bone marrow samples than from pure-cell populations from solid tumors such as colon, lung, or breast. Nevertheless, a significant amount of information

has been obtained about the genetic alterations in solid tumors. Some of these are described in subsequent sections of this chapter.

Translocations and Inversions

Reciprocal translocations are typical of leukemias, lymphomas, and sarcomas. Although chromosomal reciprocal translocations are less common in solid tumors, they do occur.[239] More than 100 commonly occurring translocations have been observed.[237] The fact that many of these occur consistently in certain specific cancer types argues strongly that they are involved in a key way in generating the malignant phenotype. The gene rearrangements caused by translocations have two principal effects: (1) they cause activation of proto-oncogenes by relocation to the site of active gene regulatory elements, and (2) they generate fusion gene products resulting from breakpoints within introns of two genes on two different chromosomes.

As noted above, the first constant translocation observed was the reciprocal translocation between the long arm (called q) of chromosome 9, band 34 (the number indicates the location of the band on each arm; the short arm p is above the centromere and long arm q is below the centromers, and they are divided numerically) and band 11 of the q arm of chromosome 22. The shorthand used by cytogeneticists to describe this is t(9;22)(q34;q11).

Later it became apparent that the t(9;22) translocation in CML involved a breakpoint near the Abelson (abl) proto-oncogene. Indeed, as it turned out, this was just one of many such translocations involving proto-oncogenes in leukemia and lymphoma (Table 5–3). In fact, the common involvement of proto-oncogenes in these breakpoints is strong evidence for the involvement of these genes in the malignant process of leukemia and lymphoma. The first translocation junction involving a proto-oncogene to be analyzed was actually the t(14;18) translocation seen in Burkitt's lymphoma[240,241] (Fig. 5–13). This rearrangement results in the translocation of the myc cellular proto-oncogene from chromosome 8 to chromosome 14 near the immunoglobulin heavy-chain Cμ (Ig-Cμ) gene, resulting in the activation of the myc gene.

The genes involved in breakpoint junction of CML were the next to be identified. In 1982, Hagemeijier et al.[242] showed that the c-abl gene was translocated from chromosome 9 to the Philadelphia chromosome. This was identified because of its homology to the viral oncogene v-abl isolated from a mouse pre-B-cell leukemia. Using a probe derived from the v-abl gene, Heisterkamp et al.[243] identified, by chromosomal "walking" across the translocation junction, sequences derived from chromosome 22, thus proving a reciprocal translocation event. The

breakpoints on chromosome 22 in 17 of 17 CML patients examined occurred within a 5.8 kilobase segment, which they called the breakpoint cluster region, or bcr. The breakpoints in abl on chromosome 9 occur at variable sites, but always in introns. As a result of the translocation, the abl gene piece containing exon II through to its 3' terminus is moved to the midpoint of the bcr gene, which encodes a GTPase activating protein (GAP), forming a fusion gene that codes for a chimeric Bcr-Abl protein. This protein has high tyrosine kinase activity and a signal transduction mechanism often deregulated in cancer cells. It has been the target for the drug Gleevec, one of the few examples of a drug that targets specifically a cancer molecular defect.

A conundrum arose when similar t(9;22) translocations were found in a significant number of patients with adult acute lymphocytic leukemia (ALL). Since ALL is a very different disease from CML, it was difficult to reconcile this difference with a similar cause–effect relationship for these two diseases. It was later found, however, that the Bcr-Abl fusion protein from CML cells results from a somewhat different breakpoint than that seen in most ALL patients. The CML fusion protein is 210 kDa in size, whereas that seen in ALL is 190 kDa.[244] Both have tyrosine kinase activity, but the Bcr-Abl fusion protein from ALL has higher activity, which may relate to ALL being a more aggressive disease.[245]

The conundrum described above will arise again and again because similar genetic changes occur in very different kinds of cancer (see below). For example, mutations in the ras proto-oncogene occur in several different cancer types, as do mutations or deletions of the p53 tumor suppressor gene. The explanation seems to be that different patterns of gene alterations can produce common phenotypic changes in cells that lead to misregulated cell proliferation, invasion, and metastasis. Another possibility is that similar patterns of altered gene expression can lead to different end points in different cell types. Evidence for the former comes from the multiplicity of genetic changes seen in human cancer cells (Tables 5–3 and 5–4). Evidence for the latter comes from the observation of common genetic translocations in CML and B-cell ALL, and in Burkitt's lymphoma (BL) and T-cell ALL, for example. In any case, "all roads lead to Rome" in

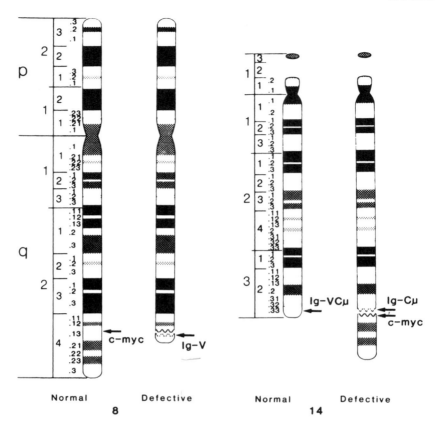

Figure 5–13. Location of c-*myc* oncogene and heavy-chain immunoglobulin variable (V) and constant μ (C$_μ$) genes on normal and defective chromosomes 8 and 14 in Burkitt's lymphoma, represented at the 1200-Giesma band stage. The defective chromosome 8 loses the c-*myc* and gains V genes. The defective chromosome 14 gains c-*myc* from chromosome 8, becoming contiguous or near to C$_μ$. Arrows point to the normal and rearranged location of these genes. Broken ends of defective chromosomes indicate breakpoint sites. (From Yunis,[253] with permission.)

the sense that cancer cells share several common features.

Another well-defined translocation involves the *bcl1* gene, originally defined by its rearrangement with immunoglobin heavy-chain locus (IgH) in B-cell chronic lymphocytic leukemia (B-CLL), diffuse B-cell lymphoma, and multiple myeloma.[246] Another oncogene identified by translocation is *bcl2*, which is observed by the t(14;18) translocation in follicular lymphoma.[247] *bcl-2* is involved in regulation of lymphocyte proliferation and differentiation and acts to prolong cell survival by blocking programmed cell death (apoptosis).[248]

The mixed-lineage leukemia (*MLL*) gene on chromosome 11 (band 23) is a gene frequently translocated in human leukemias. MLL chromosomal translocations produce an in-frame gene fusion and the production of a chimeric mRNA and fusion protein (reviewed in Reference 249). The N-terminal portion of the MLL protein is fused to the C-terminal portions of various fusion partners. Several MLL fusion partners have been identified and different MLL fusion proteins have been found in different leukemias.

MLL-associated fusion proteins have been found in about 10% of ALL and AML patients and are associated with a poor prognosis. MLL is

Table 5–3. Molecularly Characterized Neoplastic Rearrangements

Part	Disease	Rearrangement	Gene	Protein type
A	BL	t(8;14)(q24;q32)	*MYC*	HLH domain
		t(2;8)°(p11;q24)		
		t(8;22)(q24;q11)		
	B-CLL	t(11;14)(q13;q32)	*BCL1 (PRAD1?)*	PRAD1 is a G1 cyclin
	Follicular lymphoma	t(14;18)(q32;q21)	*BCL2*	Inner mitochondrial membrane
	B-CLL	t(14;19)(q32;q13)	*BCL3*	CDC10 motif
	Pre-B ALL	t(5;14)(q31;q32)	*IL-3*	Growth factor
B	T-ALL	t(8;14)(q24;q11)	*MYC*	HLH domain
	T-ALL	t(7;19)(q35;p13)	*LYL1*	HLH domain
	T-ALL	t(1;14)(p32;q11)	*TCL5 (TAL1, SCL)*	HLH domain
	T-ALL	t(11;14)(p15;q11)	*RBNT1*	LIM domain
	T-ALL	t(11;14)(p13;q11)	*RBNT2*	LIM domain
	T-ALL	t(7;9)(q35;q34)	*TAN1 (TCL3)*	Notch homolog
	T-ALL	t(10;14)(q24;q11)	*HOX11 (TCL3)*	Homeodomain
C	Parthyroid adenoma	inv(11)(p15;q13)?	*PTH* deregs *PRAD1*	PRAD1 is a G1 cyclin
	B-CLL	t(8;12)(q24;q22)	*BTG1* deregs *MYC*	MYC has an HLH
D	CML, B-ALL	t(9;22)(q34;q11)	*BCR-ABL*	BCR, GAP for p21^ras ABL, tyrosine kinase
	APL	t(15;17)(q22;q11.2-12)	*PML-RARA*	PML, Zn finger RARA, Zn finger
	AML-M2, AML-M4	t(6;9)(p23;q34)	*DEK-CAN*	DEK, nuclear CAN, cytoplasmic
	Pre-B ALL	t(1;19)(q23;p13)	*E2A-PBX*	E2A, HLH PBX, homeodomain
	NHL	ins(2;2)(p13;p11.2-14)	*REL-NRG*	REL, NF-κB family NRG, no homology

Key: A, oncogenes juxtaposed to Ig loci; B, oncogenes juxtaposed to TCR; C, oncogenes juxtaposed to other loci; D, fusion oncoproteins. AML-M2, acute myeloblastic leukemia; AML-M4, acute monomyelocytic leukemia; APL, acute promyelocytic leukemia; B-ALL, B-cell acute lympocytic leukemia; B-CLL, B-cell chronic lymphocytic leukemia; BL, Burkitt's lymphoma; CML, chronic myeloganous leukemia; deregs, deregulates; HLH, helix-loop-helix; inv, inversion; NHL, Non-Hodgkin's lymphoma; Pre-B ALL, precursor B–cell ALL; T-ALL, T-cell ALL.
Source: Reprinted with permission from Solomon et al.,[237] and the American Association for the Advancement of Science.

a nuclear protein, and while its function isn't clear, it appears to involve aberrant expression of homeobox (*Hox*) genes. Evidence for this involvement comes from the observation that homozygous loss of the *mll* gene in mice is embryonically lethal at day 11, and these embryos lack expression of major *Hox* genes.

Genes encoding transcriptional regulatory factors are frequently involved in translocation breakpoints seen in hematologic malignancies. For example, two related helix-loop-helix (HLH)-type transcriptional regulators, LYL1 and TCL5, are rearranged in T-cell ALL.[237] Myc is an HLH protein that is translocated and deregulated in both B- and T-cell neoplasms. The TCL3 locus identified in the t(10;14) translocation of some T-cell ALLs codes for a homeobox protein, HOX11, also a transcriptional regulator.

The multiplicity of translocation events in various cancers strongly suggests that they have a causal relationship in inducing the cancer phenotype. However, some of them may occur as secondary events in the evolution of more aggressive phenotypic changes. The inherent genetic instability of malignant cells leads to further karyotypic abnormalities as the disease progresses, reflecting additional genetic alterations that increase growth potential. Evidence that malignant transformation of cells doesn't usually result from single translocation events comes from patients with ataxic telangiectasia, who have an increased likelihood of developing leukemia. These patients may have T lymphocytes with a translocation present for several years before leukemia develops.[250] Similarly, some patients with benign follicular hyperplasia have *bcl2* gene rearrangements.[251]

Table 5–4. Translocations in Solid Tumors

Tumor	Translocation
Breast adenocarcinoma	t(1)(q21-23)
Glioma	t(19)(q13)
Ewing's sarcoma	t(11;22)(q24;q12)
Leiomyoma (uterus)	t(12;14)(q13-15;q23-24)
Lipoma	t(3;12)(q27-28;q13-15)
	t(6)(p22-23)
	t(12)(q13-15)
Liposarcoma (myxoid)	t(12;16)(q13;p11)
Melanoma	t(1)(q11-q12)
	t(1;6)(q11-12;q15-21)
	t(1;19)(q12;p13)
	t(6)(p11-q11)
	t(7)(q11)
Myxoid chondrosarcoma	t(9;22)(q22;q11.2)
Malignant histiocytosis	t(2;5)(p23;q35)
Ovarian adenocarcinoma	t(6;14)(q21;q24)
Pleomorphic adenoma	t(3;8)(p21;q12)
	t(9;12)(p13-22;q13-15)
	t(12)(q13-15)
Renal cell carcinoma	t(3;8)(p21;q24)
Rhabdomyosarcoma (alveolar)	t(2;13)(q35-37;q14)
Synovial sarcoma	t(X;18)(p11;q11)

Source: Reprinted with permission from Solomon et al.,[237] and the American Association for the Advancement of Science.

Thus, additional genetic mutation events seem to be needed to trigger the development of the full-blown malignant phenotype.

One argument against the multiple-hit theory, at least for leukemia, is that a single initiating target appears to be what drives the malignant phenotype in CML. The drug Gleevec targets the *bcr/abl* translocation fusion protein Bcr/Abl, which is the dysregulated tyrosine kinase activity present in most CML cells.

A similar event has been observed in human prostate cancer by Chinnaiyan and colleagues.[252] Using a unique DNA microarray analysis called the *cancer outlier profile analysis* (COPA) to identify overexpressed genes against a large background of gene expression "noise" and subjecting this information to an Oncomine database developed by Chinnaiyan's group, a translocation event present in about 80% of prostate cancers was found. What's intriguing about this translocation is that it produces a fusion gene coupling an androgen-responsive promoter element to an on-cogene (either *ERG* or *ETV1*) of the *ETS* family of oncogenes. Since most prostate cancers are androgen hormone driven, at least in the early stages, and since this translocation was identified in such a high proportion of the prostate cancers examined, it is likely that this is a key carcinogenic event, if not *the* key event, in driving the progression of prostate cancer. If this scenario is indeed true, it opens the door for the development of highly specific, targeted drugs for the treatment of prostate cancer, just as Gleevec is for CML. Additionally interesting is the idea that similar previously unidentified translocation events may be found in other solid tumors, including breast, colon, and lung. This would be a game-changer for the way that most common cancers are diagnosed and treated.

Chromosomal Deletions

The history of the study of chromosomal abnormalities in cancer closely parallels advances in chromosomal banding techniques, as well as the development of molecular biology. Before the discovery of banding techniques, only one cancer, chronic myelocytic leukemia, had been clearly associated with a consistent chromosomal defect, the Philadephia chromosome. This was reported in 1960.[2] After the introduction of banding techniques in 1970,[5] more than 30 neoplastic conditions were shown to have consistent chromosomal anomalies. Whereas before 1970 chromosomal defects were generally thought to be secondary or late changes in neoplasia, by the early 1980s it was widely believed that they are found in most cancers and that each cancer is associated with specific chromosomal lesions.[253] We are now in a third phase of evolution in this field, its driving force being the application of techniques of molecular biology, such as gene cloning, in situ hybridization, PCR analysis of gene transcription, and DNA microarrays. As noted above for prostate cancer, the use of these techniques has led to the conclusion that a given chromosomal abnormality may be associated with a variety of neoplasms and that a given oncogene can be activated in a variety of human cancers.

Certain general statements can be made about the kinds of chromosomal abnormalities seen. The most common defects usually observed in solid tumors have been deletions in specific gene sequences, sometimes observed as loss of a part of a banding region or the loss of heterozygosity of a specific genetic allele.[254] As described earlier, gene amplifications are sometimes observed as homogeneously staining regions on chromosome banding patterns or as small, chromosome-like fragments in cells called *double-minute chromosomes*. Single base substitutions or point mutations also occur in a variety of cancers (see below). As in the case of the translocations discussed above, many of the genetic changes seen in solid tumors result in activation of a cellular oncogene. In tumors with genetic deletions, a tumor suppressor gene may be lost.

Deletion of genetic material in a cancer cell suggests loss of function that regulates cell proliferation or differentiation. More than 20 human solid tumors have been shown to have some type of chromosomal deletion (Table 5–5). Some chromosome deletions appear to be specific for certain tumor types. These include deletion del(13)(q14q14) seen in retinoblastoma that results in loss of the *rb* tumor suppressor gene, the 11p13 deletion in Wilms' tumor, and deletion of the *dcc* (deleted in colon cancer) gene in colon carcinoma. Deletions in the long arm of chromosome 5 (del 5q) are seen in a number of hematologic diseases, including acute nonlymphcytic leukemia and chronic myeloproliferative disorders. These deletions commonly involve the 5q21-31 region that contains genes encoding growth factors and growth factor receptors involved in myeloid cell differentiation.[255]

Other chromosome deletions are observed in multiple kinds of cancer. These include deletions in the chromosome 3p13-23 region in small cell carcinoma and adenocarcinoma of the lung, renal cell carcinoma, and ovarian adenocarcinoma; deletion in the 1p32-36 region in neuroblastoma and glioma; and 1p11-22 deletions in melanoma, breast adenocarcinoma, intestinal leiomyosarcoma, mesothelioma, and malignant fibrous histiocytoma (Table 5–5). The 1q21-23 region is often subject to deletions in uterine and bladder adenocarcinomas and is involved in translocations in breast adenocarcinomas. Deletions of 6q11-27 have been reported in melanoma, glioma, and ovarian carcinoma. Portions of the 7q21-34 region are lost in uterine leiomyoma, prostate carcinoma, glioma, and acute

Table 5–5. Deletion and Loss of Heterozygosity in Solid Tumors

Tumor	Chromosomal Deletion in Tumor	Allele Loss
CLONED		
RB	13q14	13q
Colorectal carcinoma	17p 18q	5q; 17p; 18q
WT	11p13	11p
NOTED		
Bladder adenocarcinoma	1q21-23 Monosomy 9	9q; 11p; 17
Breast adenocarcinoma	1p11-13 3p11-13 3q11-13	1p; 1q; 3p; 11p; 13q; 16q; 17p; 17q; 18q
Glioma	1p32-36 6p15-q27 7q22-q34 8p21-23 9p24-p13	17
Leiomyosarcoma (intestine)	1p12-12	NT
Leiomyoma (uterus)	6p21 7q21-31	NT
Lipoma	13q12-13	NT
Lung adenocarcinoma	3p13-23	3p; 13q; 17p
Lung small cell carcinoma	3p13-23	3p; 13q; 17p
Mesothelioma	3p21-25	NT
Mesothelioma (pleura)	1p11-13	NT
Malignant fibrons histiocytoma	1q11	NT
Melanoma	1p11-22 6Q11-27	1p
Meningioma	Monosomy 22 22q12-13	22q12-qter
Neuroblastoma	1p32-36	1p
Ovarian adenocarcinoma	3p13-21 6q15-23	3p; 6q; 11p; 17q
Prostatic adenocarcinoma	7q22 10q24	10; 16
Renal cell carcinoma	3p13-21	3p
Uterine adenocarcinoma	1q21-23	3p

NT, not tested.

Reprinted with permission from Solomon et al.,[237] and the American Association for the Advancement of Science.

myeloid leukemia. The *p53* tumor suppressor gene–containing region of chromosome 17p is deleted or mutated in a wide variety of human cancers.

The fact that there is such commonality among cancer cell types in the loss of chromosomal material strongly suggests that these regions contain genes coding for regulatory factors involved in cell proliferation and/or differentiation of a wide variety of cell types. Many of these regions contain genes involved in cell cycle regulation through interaction with cyclins or signal transduction pathways that regulate response elements of particular growth regulatory genes.

Induction of the malignant neoplastic process is thought to involve at least two genetic "hits," as described in Chapter 2. In the case of genetically predisposed tumors, the first genetic alteration may be inherited through the germline, with the second alteration occurring after birth. In genetically predisposed cells, the remaining single normal allele may be sufficient to maintain normal growth regulation, and a second deletion or mutation is required to inactivate the remaining normal allele. In the case of a tumor suppressor gene, both alleles are then in effect lost or inactivated. Loss of heterozygosity at a genetic locus has frequently been the mechanism for detecting deletion or mutation of genes involved in cancer causation.

Gene Amplification

The mechanisms involved in gene amplification were described earlier in this chapter. It is a relatively common event in cancers. Amplification of genes observed in human cancers include amplification of the N-*myc* gene in stage III and IV neuroblastoma, of the epidermal growth factor receptor–regulated gene *her-2/neu* in advanced breast and ovarian carcinomas, and of the *int-2*, *hst-1*, and *prad*1 oncogenes in breast and squamous cell carcinomas and in melanoma.

Trisomy of chromosome 8 has been observed in AML, ALL, and myeloproliferative disease. Trisomy of chromosome 9 has been seen in myeloproliferative disorders and of chromosome 12 in malignant lymphoma and lymphoproliferative disorders. Some malignant lymphomas have a trisomy of 3, and a trisomy of 7 has been found in some carcinomas and neurogenic tumors (reviewed in Reference 237).

Point Mutations

Point mutations that lead to single base changes in a DNA sequence will be discussed in more detail in the discussion of oncogenes and tumor suppressor genes below. These mechanisms are involved in chemical carcinogenesis, activation of proto-oncogenes, and loss of function of some tumor suppressor genes. Suffice it to say here that reaction of DNA with carcinogenic chemicals or as a result of spontaneous mutations due to oxidative damage can lead to formation of base adducts that can cause base mispairing during DNA replication or loss of an adducted nucleic acid base producing an abasic site in the DNA chain. Such abasic sites may then be filled with an inappropriate base during DNA repair or replication, leading to a point mutation. If this mutation is in a regulatory element of a gene, loss or alteration of regulation of gene expression can occur. If the mutation is in a coding region of a gene, an altered protein may be formed.

Aneuploidy

The genetic instability manifested during tumor progression is characterized by a variety of aberrations in the genome, including point mutations; gene deletions, rearrangements, and amplifications; chromosome translocations; and abnormal chromosome number, known as *aneuploidy*. Although the more subtle changes in the genome—namely, point mutations, gene deletions, and gene rearrangements—may be associated with initiation of the malignant transformation process, gross changes in the number of chromosomes usually occur as tumors progress in malignancy. As noted earlier, certain chromosomal deletions, translocations, and trisomies are characteristically associated with a particular form of cancer; these are called *nonrandom chromosomal alterations*. Changes in cell ploidy, however, are associated with a variety of tumor types in their advanced stages and may be random in the sense that no definitive pattern of chromosome number is associated with a given tumor type. In advanced cancers, both random and nonrandom chromosomal alterations may be found. These continuing genomic changes bring about tumor heterogeneity and the natural selection of more highly invasive and metastatic cancers. Thus, tumor progression may be viewed as a highly accelerated evolutionary process.[256]

Evidence for changes in ploidy during tumor progression comes from both human and experimental animal cancers. For example, Frankfurt et al.[257] examined chromosomal ploidy in 45 human prostate carcinomas by staining DNA with a fluorochrome and scanning cells for DNA content by flow cytometry. They found that localized tumors that had not metastasized had a much lower incidence of aneuploidy than more poorly differentiated tumors and tumors that had spread beyond the pelvis. In nearly two-thirds of patients with aneuploid tumors, pelvic or distant metastases were found. In general, the frequency of aneuploidy increased with progressive stages of the disease. Human urinary bladder carcinomas demonstrate the same association between degree of malignancy and degree of aneuploidy: most aneuploid bladder tumors have a high histologic grade and are invasive.[258,259] Primary human breast cancers are mostly diploid, as determined by flow cytometry and karyotype analysis, whereas cells taken from metastatic sites are often aneuploid.[260] "Dedifferentiation," as evidenced by loss of estrogen receptors, and poor prognosis, has also been associated with aneuploidy. However, the diploid breast carcinoma cells had the ability to invade human amnion basement membrane in culture,[260] so they may have achieved an invasive phenotype before they became aneuploid.

In a series of Dunning rat prostatic tumors of different stages in malignant progression, Wake et al.[261] found that the original parent tumor had a normal karyotype, whereas cytogenetic analysis of the Dunning sublines of later progression indicated a correlation between degree of aneuploidy and more advanced malignant phenotype. During the spontaneous evolution of Chinese hamster cells in culture to highly tumorigenic cells, there is a multistep progression of karyotypic changes, as determined by Giemsa banding and flow cytometry.[262] Four stages of neoplastic progression were identified: trisomy of chromosome 5, a change in banding pattern of chromosome 8, an insertion in chromosome 3(3q$^+$), and a trisomy of chromosome 8. Trisomy of chromosome 5 preceded the acquisition of the cells' ability to be tumorigenic in nude mice.

There has been a lot of debate about the relative roles of gene mutations and aneuploidy in the oncogenic process.[263,264] The sequential, progressive gene mutation theory has been championed by Vogelstein and colleagues.[265] Although leukemias and lymphomas, as noted above, often contain reciprocal translocations and point mutations, they generally remain diploid or near diploid. This is not the case for carcinomas. In the latter, dramatic gains and losses of chromosomal material (aneuploidy) frequently occur. So the chicken-and-egg question pertains here: Which came first and which is the most important causal event?

"Gate-keeper" genes that control cell proliferation and cell death and "caretaker" genes that protect the genome are often those found to be mutated in human carcinomas. These include gain-of-function mutations of oncogenes such as ras, flt-3, and c-kit and loss-of-function mutations of tumor suppressor genes such as p53, rb, and apc. The question is whether these mutations are enough to cause a full-blown human cancer. The cancer "mutator phenotype" postulated by Loeb contends that tumors gain mutations over time, and this is what causes cancer progression. While there is some evidence for this scenario, direct measurements of mutations and gene chip analyses of colon and other human cancers do not show the high number of mutations that would be predicted from the Loeb model.[263,264] Instead, what such data show is that hundreds, if not thousands, of genes have different expression levels in normal compared to tumor tissue.[263,264] This finding is more consistent with chromosomal imbalance than with sequential mutation of a few genes. Other data also support the concept that a main difference between normal and malignantly transformed cells is the gene copy number and the number of genes altered, rather than the type of genes differentially expressed (reviewed in Reference 263 and 264).

Another point is that aneuploidy can help explain the genetic drift of cancer cells, because aneuploidy produces an imbalance, through effects on gene dosage, of large numbers of genes, including those involved in mitosis, which can result in chromosomal instability. Interestingly, this imbalance could also explain the high propensity for carcinomas to become resistant to multiple and chemically unrelated chemotherapeutic agents.[266]

A number of experiments in animal models support the aneuploidy theory. For example, it has been shown that telomerase-null mice, which developed carcinomas with gains and losses of chromosomes very similar to human carcinomas, had aneuploidy and extensive chromosomal instability involving similar genes gained or lost in orthologous human carcinomas (reviewed in Reference 264).

The genome scrambling typical of carcinomas in mice and humans most likely results from ineffectual repair of double-strand DNA (dsDNA) breaks or from eroded telomere ends that are sensed and processed by cells similar to a dsDNA break (reviewed in Reference 264). The importance of this mechanism in human cancer is supported by the increased incidence of cancers in patients with inherited DNA repair defects, such as those with ataxia telangiectasia, who exhibit unrepaired dsDNA breaks.

The above arguments and data are convincing in the sense that aneuploidy clearly can produce many of the changes seen in the genotype and phenotype of human cancers. And it occurs in a high percentage of human carcinomas—some estimates are as high as 99%.[263] The data, however, don't exclude the mutation theory of oncogenesis; in fact, these two kinds of genetic events almost certainly go hand in hand. Indeed, some genetic mutations are permissive in a critical way to aneuploidy and genetic instability, for example, mutations in the *p53* gene and other cell cycle checkpoint genes.

Disomy

An unusual type of inheritance pattern has been observed in patients with a genetically determined, large-fetus syndrome called Beckwith-Wiedemann syndrome (BWS). In these patients, there is a propensity to develop malignant neoplasms, particularly Wilms' tumor of the kidney, but hepatoblastomas and rhabdomyosarcomas also occur. These patients have a uniparental paternal disomy for the 11p15.5 region of chromosome 15 and a loss of the maternal allele of this locus in the tumors that they develop.[267] In an analogous fetal overgrowth syndrome in the mouse, a region of chromosome 7 homologous

to human chromosome 11p15.5 also contains a paternal disomy.[268] Interestingly, this locus contains the gene for insulin-like growth factor-2 (*IGF-2*), and an increased level of IGF-2 mRNA is seen in the tumors of BWS patients. Since the maternal allele of this locus is lost in these tumors, it suggests that overexpression of growth promoting genes (IGF-2) and loss of a tumor suppressor function on the maternal chromosome locus 11p15 combine to cause the malignant tumors in these individuals.

Trinucleotide Expansion

In the human genome there are interspersed repeated DNA sequences widely dispersed throughout the genome. These interspersed repeats are frequently close to or even within structural genes. While structural genes in general have a low mutation rate (e.g., about 1 amino acid out of 400 per 200,000 years),[269] repeated sequences have much higher mutation frequencies.[270] Because repeated sequences are usually in noncoding regions of the genome, these mutations are tolerated by the organism and may even be beneficial by allowing genetic recombination and alternate splicing events to produce new gene arrangements that help an organism adapt to new environments.

The interspersed repeated DNA sequences can undergo a unique form of mutation, namely variation in copy number. This form of mutation, sometimes called *dynamic mutation*,[270] results from an increase in copy number of repeated trinucleotide sequences, hence the term *trinucleotide expansion*. This form of mutation has now been linked to a number of genetic diseases, including the fragile X syndrome, in which a CGG trinucleotide is amplified;[271,272] myotonic dystrophy and spinal bulbar muscular atrophy, in which the amplified repeat is trinucleotide CAG (reviewed in Reference 270); and Huntington's chorea, in which the amplified repeat is also CAG.[273] Whereas normal individuals may have 6–60 copies of these repeats, unaffected transmitting individuals may have 60–200 copies, and severely affected persons may have more than 1000 copies of a trinucleotide repeat. An unusual feature of this type of mutation is that the copy number increases with succeeding generations, explaining the phenomenon of "genetic anticipation" in which asymptomatic carriers in earlier generations pass on the mutant chromosome to their offspring such that in successive generations the repeat length and the severity of the disease increase.

Sequencing studies have revealed that the trinucleotide repeats occur in the 3' untranslated regions of certain genes. In the case of the CAG repeat in myotonic dystrophy, for example, the repeat occurs near a region with a cyclic AMP–dependent protein kinase–like sequence. It isn't apparent why these amplified trinucleotide repeats in uncoded regions near a gene would so dramatically affect function, but the data suggest that these sequences have some regulatory action. In the case of the fragile X syndrome, the amplification blocks transcription of a gene called FMR-1.[274] If one scans the human GenBank, more than 30 sequences with five or more copies of trinucleotide repeats can be found. For example, at least 10 human genes contain p(CCG)n repeats of five or more copies (Table 5–6).

While no such mutations have been reported in human cancer, it seems likely that similar genetic changes will be identified in individuals with a susceptibility to develop cancer, especially since some proto-oncogenes and other regulatory genes contain such trinucleotide repeats (Table 5–6).

Microsatellite Instability

DNA sequences termed *microsatellites* are one to six nucleotide motifs randomly repeated numerous times in the human genome. For example, about 100,000 (CA)n dinucleotide repeats are found scattered throughout the human genome and many of these exhibit genetic polymorphisms in the length of the repeats. In colorectal cancer, a number of studies have shown differences in the repeat (CA)n length between tumor and normal DNA from colon specimens from the same patient (reviewed in Reference 275). This microsatellite instability (MSI) correlated with tumor location in the ascending colon, with patient survival, and inversely, with loss of heterozygosity for chromosomes 5q, 17p, and 18q. MSI has been found in both "sporadic" and familial colorectal cancers. Similarly, MSI has been reported in breast cancer,[275] small cell lung cancer,[276] non–small cell lung cancer,[277] urinary bladder cancer,[278] and gastric cancer.[279] Thus, the data suggest that MSI is a common genetic alteration in human cancer.

Mismatch DNA Repair Defects

The frequent occurrence of MSI and other genetic alterations suggests a generalized defect in human cancer. For instance, patients with hereditary non-polyposis colorectal cancer (HNPCC) syndrome contain frequent alterations within (CA)n and other simple repeated sequences. This syndrome, which affects as many as 1 of 200 individuals in the Western world, predisposes affected persons to cancers of the colon, endometrium, ovary, and other organs, often before age 50.[280] Such alterations as those seen in the instability of (CA)n sequences indicate a DNA replication error called the *RER phenotype*. RER tumor cells display a biochemical defect in mismatch DNA repair analogous to a similar defect

Table 5–6. (CCG)n Repeats in Human Genes

Gene or Encoded Protein	GenBank Symbol	Copy Number	Location
Znf6 (zinc finger transcription factor)	HUMZNF	8, 3, 3	5' Untranslated region
CENP-B (centromere autoantigen)	HUMCENPB	5	5' Untranslated region
c-cbl (proto-oncogene)	HUMCCBL	11	5' Untranslated region
Small subunit of calcium-activated neutral protease	HUMCANPO2	10, 6	Coding region (N-terminal)
CAMIII (calmodulin)	HUMCAM3X1	6	5' Untranslated region
BCR (breakpoint cluster region)	HUMBCRD	7	5' Untranslated region
Ferritin H chain	HUMFERH	5	5' Untranslated region
Transcription elongation factor SII	HUMTEFSII	7	5' Untranslated region
Early growth response 2 protein	HUMEGR2A	5	Coding region (central)
Androgen receptor	HUMAR	17	Coding region (central)

From Richards and Sutherland[270]

in bacteria and yeast. A gene containing this DNA repair defect has been found and maps to chromosome 2P. This gene, *hMSH2* (human *mutS* homologue 2), is homologous to the bacterial gene *mutS*, which is responsible for strand-specific mismatch repair.[281,282]

Gene Derepression in Cancer Cells

The fact that many malignant neoplasms produce polypeptides, oligosaccharides, and lipids that are inappropriate for the cell types of their tissue of origin indicates a derangement in the flow of genetic information in the transformed malignant cell. This derangement could occur by means of an alteration of gene expression, resulting from gene amplification, rearrangement, translocation, or point mutations, as discussed above. Any of these mechanisms could result in the so-called derepression of genes normally present in cells, but not at all or only minimally expressed in normal adult cells. The inappropriate production by cancer cells of certain proteins and other cellular products has been called *ectopic production*, and it has been observed that the pattern of ectopically produced proteins and hormones often resembles more closely that of the embryonic or fetal state than of the adult state of differentiation. This observation has led to the concept that the expression of these genes in cancer cells results from the derepression of "oncodevelopmental genes," that is, genes expressed normally during embryonic development and that are usually shut off or only minimally transcribed by differentiated adult cells.

Ectopic Hormone Production by Human Cancers

The first examples of ectopic polypeptide production by human tumors came from the observations of hormonally related syndromes in patients with nonendocrine tumors. In 1928, Brown[283] reported that a patient with small-cell carcinoma of the lung had a clinical syndrome manifested by diabetes, hirsutism, hypertension, and adrenal hyperplasia. In other words, the patient had the symptoms of Cushing's disease, which is caused by excess adrenocorticotropic hormone (ACTH) release. At that time it was not appreciated that tumors could produce ACTH. By 1959, about 40 well-documented reports of Cushing's syndrome in patients with nonendocrine tumors had been reported. It was not until 1961, however, that human nonendocrine gland tumors were observed to produce an ACTH-like substance, and the term *ectopic ACTH* was coined by Liddle and colleagues[284] to describe this phenomenon. Although the clinical syndrome associated with ectopic ACTH production is observed in about 2% to 3% of patients with carcinomas of the lung,[285] it has been reported that a high percentage of lung cancers contain detectable amounts of ACTH by radioimmunoassay.[286] The reason for this discrepancy may be that most of the immunoreactive ACTH present in tumor tissue and plasma of cancer patients is present as a precursor form called "big" ACTH, which has only about 4% of the biologic activity of ACTH secreted by the normal pituitary gland.[285] It has also been observed that 33% of patients with chronic obstructive pulmonary disease, mostly due to heavy smoking, have elevated plasma immunoreactive ACTH.[286] This finding, coupled with the observation that the lungs of dogs with atypical hyperplasia, resulting from forced inhalation of cigarette smoke, contained "big" ACTH, whereas lung tissue from dogs with no significant histologic changes did not, suggests that ACTH production may be stimulated in lung tissue undergoing preneoplastic changes.

The ectopic production of parathyroid hormone (PTH) by tumors has also been observed. The first clue to this occurrence was a report in 1936 by Gutman et al.[287] of a patient who had hypercalcemia and hypophosphatemia in association with a nonendocrine tumor not involving bony tissue. Later it was shown that removal of nonosseous, nonparathyroid tumors in patients with this syndrome corrected the ionic imbalance.[288] Tashjian et al.[289] and Sherwood et al.[290] demonstrated that nonparathyroid tumor extracts from several hypercalcemic patients contained a material that was immunologically similar to PTH. A variety of tumors containing PTH-like material have now been identified; these include certain carcinomas of the lung, kidney, pancreas, colon, adrenal, and parotid gland.

The clue to a possible connection between cancer and the ectopic production of antidiuretic hormone (ADH) came in a report in 1938 by Winkler and Crankshaw,[291] who observed that some patients with lung cancer excreted a very concentrated urine high in salt content, but the authors did not suggest a possible hormonal explanation for their observation. In 1957, Schwartz et al.[292] described two patients with lung cancer who excreted a hypertonic urine containing high concentrations of sodium, and these investigators attributed their findings to an inappropriate secretion of ADH. Ectopic ADH has been found in several lung tumors (mostly small cell carcinomas) as well as in duodenal, pancreatic, and other carcinomas.

Another example of ectopic hormone production by human tumors is the secretion of the placental hormone human chorionic gonadotropin (hCG). In 1959, Reeves et al.[293] demonstrated by bioassay the presence of an hCG-like substance in a hepatic carcinoma from a boy with precocious puberty. Since that time, hCG has been shown to be produced by a wide variety of human cancer cells both in vivo and in vitro.[294–296]

A summary of the ectopic hormones produced by various human cancers is given in Table 5–7.

Possible Mechanisms of Ectopic Protein Production

The inappropriate expression of polypeptides by tumor cells could result from a rearrangement or mutation in a regulatory gene that leads to the increased transcription of structural genes coding for oncodevelopmental proteins and hormones. The evidence accumulated to date indicates that the amino acid composition of ectopic hormones produced by tumors is the same as that made by the normal hormone-producing cell, indicating that the ectopic product does not simply result from "chaotic" protein synthesis coded for by a scrambled or mutated structural gene.

A second means by which ectopic polypeptide production could occur is by an increased abundance of the mRNA coding for a particular protein. This could occur as a result of an increased "gene dosage" (resulting from gene amplification or gain of chromosomal material, as noted earlier) or of an increased rate of gene transcription, an elevated rate of mRNA processing,

Table 5–7. Ectopic Hormones Produced by Various Human Cancers

Hormone	Tumors Producing Hormone Ectopically	Associated Clinical Syndrome
ACTH	Carcinomas of lung, colon, pancreas, thyroid, prostate, ovary, cervix; thymoma; pheochromocytoma; carcinoid tumors	Cushings' syndrome
ADH	Carcinomas of lung, duodenum, pancreas, ureter, prostate; thymoma; lymphoma; Ewing's sarcoma	Inappropriate antidiuresis; hyponatremia
Calcitonin	Carcinomas of lung, breast, prostate, bladder, pancreas, liver, esophagus, stomach, colon, larynx, testis; carcinoid tumors; insulinoma; pheochromocytoma; melanoma	No apparent syndrome
Erythropoietin	Hemangioblastoma; uterine myofibroma; pheochromocytoma; carcinoma of liver, ovary	Polycythemia (erthrocytosis)
Gastrin	Carcinoma of pancreas	Zollinger-Ellison syndrome (gastric hypersecretion with intractable peptic ulceration)
Glucagon	Carcinoma of kidney	Hyperglycemia, malabsorption, gastrointestinal stasis
Growth hormone	Carcinomas of lung, stomach, ovary, breast	Hypertrophic pulmonary osteoarthropathy, acromegaly
HCG	Carcinomas of breast, stomach, small intestine, pancreas, parotid, ovary, testis, spleen, breast	Gynecomastia, precocious puberty
Prolactin	Carcinomas of lung, kidney	Galactorrhea, gynecomastia
PTH	Carcinomas of kidney, lung, liver, adrenal, pancreas, parotid, ovary, testis, spleen, breast	Hypercalcemia
TSH	Carcinomas of lung, breast	Hyperthyroidism

or a decreased degradation of mRNA. Evidence is accumulating that mammalian cells contain "leaky" genes, in the sense that there is a very low rate of transcription of many genes in cells that do not make functional amounts of the protein encoded by those genes. For example, normal tissues other than placenta have been found to produce small amounts of hCG,[297] and a small amount of hemoglobin transcription has been found in nonerythroid cells.[298] These results suggest that a low-level transcription of certain oncodevelopmental genes continues in adult differentiated tisues and that ectopic protein production is an expansion of production of these proteins, some of which may provide a selective growth advantage for transformed cells.

Pearse[299] suggested a third means by which ectopic protein production could occur in neoplasms: the clonal expansion of certain cells that produce the ectopic product continuously and are normally present in only very small numbers in adult tissues. He developed the idea that certain cellular derivatives of embryonic neuroectoderm tissue are present in normal adult tissues. These cells retain the ability to synthesize and secrete certain hormones, and they may proliferate after carcinogenic alteration of normal tissues. The increased proliferation of these clones of cells after malignant transformation could then lead to an elevated ectopic production of certain proteins and hormones. Pearse has coined the term *APUD-oma* (amine precursor uptake and decarboxylation) to describe certain features of this class of cells. This definition relates to the ability of these cells to take up and decarboxylate amine precursors involved in the synthesis of certain hormones and neurotransmitters (e.g., epinephrine, norepinephrine, and serotonin). Although this hypothesis could account for ectopic hormone production by tumors arising in certain tissues known to contain these type of cells, it does not explain the general phenomenon of the re-expression of oncodevelopmental products by tumor cells, since many of the latter are not products characteristic of neuroectoderm cells, for example, carcinoembryonic antigen, hCG, α-fetoprotein, and placental alkaline phosphatase.

The "leakiness" of gene expression increases with aging. Ono and Cutler[298] found that the amount of globin mRNA present in mouse brain and liver cells increased in aging mice. Thus the controls regulating gene expression may become less stringent as an organism ages. Since cancer is primarily a disease of aging organisms, it could be that a lack of stringent gene control contributes to the emergence and clonal expansion of transformed malignant cells.

Chromosomal Abnormalities in Leukemic Patients Exposed to Genotoxic Agents

A question that often gets asked is what environmental agents might cause or facilitate the genetic alterations described above.[300] In this case, the occurrence of leukemia in some patients exposed to various genotoxic agents is instructive. A number of studies have shown a relationship between exposure to know mutagenic and carcinogenic agents and hematologic malignancies, particularly for acute nonlymphocytic leukemia (ANLL). Mittelman et al.[301] found that 32% of 162 patients with ANLL had occupational exposure to insecticides, solvents, or petroleum products, and 75% of the exposed ANLL patients, as opposed to 32% of ANLL patients with no history of such exposure, had chromosomal abnormalities in their bone marrow cells. Chromosomal abnormalities have also been observed in leukemias and the myelodysplastic syndrome that have arisen after previous treatment with antineoplastic drugs, a number of which are themselves mutagenic, for other cancers.[302]

The Fourth International Workshop on Chromosomes in Leukemia[303] summarized data from 716 patients with ANLL and reported that chromosomal abnormalities were observed in about 55% of patients with no history of previous anticancer therapy, whereas 75% who had previous exposure to anticancer drugs had abnormalities, usually involving chromosomes 5 and/or 7. Karyotypic abnormalities involving chromosomes 5 and 7 are associated with poor prognosis in ANLL patients. In only 13% of "spontaneous" ANLL cases were these chromosomes involved.

Exposure to benzene is also associated with ANLL, and in these patients chromosomal abnormalities are frequently observed.[304] Chromosomal abnormalities have also been observed in myelodysplastic syndrome (MDS), a preleukemic condition that progresses to ANLL. However, no significant difference was found between MDS patients who had a history of previous exposure to genotoxic agents and those who didn't in the type of chromosomal abnormalities seen.[305]

Cancer Genetic Changes Summed Up

By now the reader is getting the point that multiple genetic lesions are associated with individual human cancers and that many of these defects show up consistently in cancers of very different tissue types. This phenomenon leads to the conclusion that there are families of tumor suppressor genes, or perhaps what are more appropriately called growth regulatory genes, located on different chromosomes and probably activated to regulate cell proliferation in different cell types at different stages of their embryonic development and/or in their tissue renewal stem cells at different stages of their growth and differentiation phases. Thus, in one tissue type a growth suppressor gene may be important in an early step of stem cell proliferation and differentiation, whereas in other tissues it may be more important at a later step. This may explain why some genes seem to undergo allelic loss early in the tumorigenesis of one type of cancer but later in another type of cancer. The large number of allelic losses of tumor suppressor genes that show up consistently in cancer cells as tumors progress supports this idea.

The other side of the coin is the activation or mutation of cellular oncogenes. Oncogene activation is also a frequent phenomenon in human cancer and provides the second edge of the two-edged sword of uncontrolled cell proliferation and loss of ability of cells to differentiate. The potency of these two events occurring simultaneously or sequentially in cells has been shown by experiments in which co-transfection of the *ras* oncogene and a mutated *p53* gene into rat cells induced their malignant transformation even if the wild-type *p53* gene was still expressed.[306]

We have also seen how aneuploidy is a key event in the oncogenic process. Which of these events are the cause and which are the effect of the carcinogenic process? Presumably, those genetic effects that occur early in the carcinogenic process and consistently in a high percentage of cases are associated with the underlying causes of the disease, and those that occur later are associated with progression of a tumor cell

into a more invasive, metastatic cancer. Because of the proliferative advantage and higher rate of cell division of cancer cells compared to their normal counterparts and because of the genetic instability of transformed cells, additional genetic defects are likely to accumulate in malignant cells as they evolve into more aggressive cancers. Some of the genetic changes that occur may just "be along for the ride" and not be involved in a crucial way in the carcinogenic process. If such genes are located in chromosomal regions that are translocated, deleted, or amplified in cancer cells, their expression could be increased or decreased. This scenario might partly explain the production of certain "ectopic" proteins by cancer cells. These gene products may have nothing to do directly with the cancer process—for example, the production of hCG-β subunit by a variety of nongonodal tumors.[307]

To all of this discussion one must add the findings from studies of hereditary susceptibility genes. A number of these genes have been found, e.g., the germline mutations of *p53* in Li-Fraumeni syndrome and the *apc* gene in familial polyposis. Others are yet to be identified. One point that is becoming clear is that the ability to repair DNA is crucial to protecting the genome from carcinogenic damage. Several genes show up in this category, including *p53*, which stops cells in the cell cycle until DNA is repaired or targets cells for death if it can't, and the DNA mismatch repair genes *hMSH2* and *HMLH1*. It is likely that a concatenation of events involving several gene types can lead to the loss of cellular control that produces cancer.

ONCOGENES

Historical Perspectives

The Provirus, Protovirus, and Oncogene Hypotheses

Much of our understanding of the molecular mechanisms involved in cellular transformation of RNA oncogenic viruses is an outgrowth of the seminal work of Temin and Baltimore and their colleagues (reviewed in References 308 and 309). In the early 1960s, Temin demonstrated that mutations in the Rous sarcoma virus (RSV) genome of RSV-infected chicken cells could be induced at a high rate, that mutation of an RSV gene present in an infected cell often changes the morphology of the infected cell, and that the virus genome was stably inherited by subsequent progeny cells. These findings led to the idea that virus genetic information was contained in a regularly inherited structure of the host cell as a "provirus" and that this provirus was integrated into the host cell's genome. The problem with the provirus hypothesis was that there was no known way for the RNA of the tumor virus to be converted into a form of DNA that could be integrated into the host's DNA. The central dogma of molecular biology at the time was that genetic information was transferred *only* from DNA to RNA to protein. When actinomycin D, a drug that specifically blocks DNA-directed RNA synthesis, was added to RSV-producing cells, virus production was blocked, indicating to Temin that the flow of genetic information for RSV could go from RNA to DNA to RNA to protein. Further experiments showed that new DNA synthesis was, in fact, required for RSV production to occur and that new RSV-specific DNA was present in infected cells. On the basis of these results, Temin proposed the DNA provirus hypothesis in 1964. The basis of the hypothesis is that the RNA of the infecting RSV acts as a template for the synthesis of viral DNA, which is then integrated as a provirus into host cell DNA, where it can subsequently act as a template for the synthesis of progeny RSV RNA. He then set about trying to obtain evidence for an RNA-directed DNA polymerase, which would have to be present if his idea were correct.

The DNA provirus hypothesis was largely ignored for about 6 years until Temin and colleagues[310] and Baltimore and coworkers,[311] working independently, demonstrated the presence of a virus-contained RNA-directed DNA polymerase activity, which came to be known as *reverse transcriptase*. Although the discovery of reverse transcriptase explained how the DNA–provirus mechanism could work, formal proof of the hypothesis was not obtained until it was demonstrated that radioactively labeled RSV RNA hybridized to the DNA of infected chicken cells to a much greater extent than to the DNA of uninfected cells[312] and that DNA obtained from RSV-infected cells could transfect uninfected cells, leading to the production of complete RSV.[313]

In an extension of the DNA provirus hypothesis, Temin later proposed the protovirus theory, in which he postulated that the genome of oncogenic viruses arose during evolution, in part from normal cellular DNA that had perhaps been altered by some exogeneous carcinogen. This theory would help explain the known hybridization of oncogenic viral nucleic acid sequences with normal cellular DNA (see discussion of *src* gene, below).

The normal cellular homologues of viral oncogenes have come to be known as cellular *proto-oncogenes* rather than protovirus genes because it is now clearly established that their origin is cellular and that they have been present in cells over a vast range of evolution. Such conservation implies a central role for these genes in normal cellular function, and it is likely that their oncogenicity derives from a rare event, such as translocation, amplification, or mutation of a key nucleotide sequence. Although the term *proto-oncogene* has found wide acceptance, it is somewhat misleading because it is not reflective of the role of these genes in normal cell differentiation and function.

The highly oncogenic viruses presumably arose from genetic recombination events between viruses of low oncogenicity and an evolutionarily stable set of nucleotide sequences of cellular origin, the combination of which has produced a highly transforming viral genome. Since many of these viruses are replication defective, they do not form complete virus unless the cells are co-infected with a "helper" virus. Thus, recombination between these replication-competent helper viruses and cellular genes, some of which may be involved in regulating cell proliferation, may have produced the highly oncogenic virus strains.

Certain essential features of these theories—namely, that normal cells contain sequences (cellular proto-oncogenes) homologous to those of oncogenic viruses and that these sequences can be activated during carcinogenesis (perhaps by a mutation or chromosomal rearrangement)—are now well established. Whether oncogenic viruses pick up by transduction "cancer genes" mutated long ago in evolution by some carcinogen, as originally proposed in the protovirus hypothesis, is not clear. What is clear is that cellular homologues of genes carried by oncogenic viruses are present in untransformed cells spanning the evolutionary scale from yeast to humans. The means by which these sequences came to reside in viruses appear to involve recombination events at the DNA rather than the RNA level, followed by transcription and splicing of mRNA coded by these genes, and packaging into the retrovirus. This process will be described in more detail below for the *src* gene.

The src Gene

The cellular origins of oncogenes (*onc* genes) were first clearly established for the RSV oncogene v-*src*, which is derived from its cellular progenitor (c-*src*). The identification of this gene sequence is an elegant story in molecular biology, and the methods used to detect it have been used as a precedent in the discovery of other *onc* gene sequences. We will describe these experiments here in some detail because of their prototypical significance.

If one assumes that the transforming oncogenic sequences are not necessary for virus replication, as in the case of RSV, but are "extra" genes that provide some selective advantage for cell proliferation, then cells transformed with oncogenic viruses should contain some nucleic acid sequences not present in nontransforming viruses of the same class. Moreover, this idea predicts that a transforming virus should code for a transformation-specific protein that is not needed for viral replication, but has a particular function in transformed host cells.

It has been known for some time that the transforming avian sarcoma viruses (ASV; of which RSV is an example) contain more genetic information than transformation-defective strains of these viruses.[314] The transformation-defective variants of ASVs do not induce sarcomas in animals or transform fibroblasts in culture, and they lack 10% to 20% of the genetic information (RNA) contained in the parent transforming viruses; yet these "defective" viruses can infect cells and replicate perfectly well. Hence, their genetic deletion does not appear to affect virus-replicative functions. Taking advantage of these transformation-defective strains, Stehelin et al.[315] designed an experiment to isolate the portion of the genome related to the transforming activity of these viruses. They used RNA isolated from a transforming strain of RSV called Prague strain Subgroup C as a template to synthesize radio-

actively labeled DNA pieces complementary to the transforming virus RNA, in vitro, by means of reverse transcriptase. The complementary DNA pieces synthesized in vitro were then hybridized to the RNA of a transformation-defective strain of the same virus. Since those pieces of DNA that contained the transformation-specific sequences would not be represented in the defective strain's RNA, they would not hybridize to it (hybridization requires a significant amount of homology between nucleic acid base pairs). Thus, these transforming DNA pieces did not form double-stranded duplexes with the RNA and were separated from the nontransforming DNA pieces that did by chromatography on hydroxylapatite, which binds only the double-stranded nucleic acid hybrids. The transforming pieces were finally purified by hybridization with RNA from the transforming virus and separated from the RNA by hydrolysis in alkali, leaving radioactively labeled, transforming DNA. This DNA, originally designated "cDNA sarc" for complementary DNA-bearing sarcoma-producing gene sequences, was then used as a probe to determine whether these sequences are present only in transformed cells or in normal cells as well.

The answer to this question was surprising. When the cDNA sarc was used as a probe, it was found that sequences homologous to, though not completely identical with, cDNA sarc were present in uninfected avian cells as well as in normal salmon, mouse, calf, and human cells. DNA sequences homologous to v-src* were also subsequently found in *Drosophila* and in other lower organisms.

Because the evolutionary separation of birds from teleosts and mammals occurred 400 million years ago, this finding suggests that a portion of the *src* gene has been conserved in cells of higher organisms for several evolutionary epochs. Moreover, RNA sequences corresponding to cDNA *scr* were also found in the cellular RNA of normal and neoplastic avian cells, indicating that the gene is

---

*In current terminology, the specific transforming gene of avian sarcoma viruses has been designated v-*src*; the DNA sequences of normal cells that correspond to v-*src* are called c-*src*. A similar designation has been used for other oncogenes as well, for example, v-*ras* and c-*ras*, v-*myc*, and c-*myc*, and so on.

not only present but also transcribed in normal and in transformed cells.[316] The amount of sarcoma-specific RNA and its intracellular location have also found to be essentially identical in RSV-transformed cells and RSV-transformed revertant (transformed phenotype lost) cells, supporting the conclusion that neither the presence of the src gene nor its transcription is specific to transformed cells.[317] Thus, at this point in these studies, it was concluded that if the src gene is related to transformation, it must be at the level of translation of src gene RNA into protein (i.e., the amount of Src protein present in the cell) or the specific effect of the src gene protein on cellular targets.

Detection of the src gene product was accomplished by using serum from rabbits bearing ASV-induced tumors to immunoprecipitate proteins from uninfected and ASV-transformed chicken and hamster cells grown in culture.[318] A 60,000 MW protein, designated p60^{src}, was precipitated by this serum from transformed chicken and hamster cells, but was not observed in uninfected cells or in cells infected with nontransforming viruses. Also, p60^{src} was synthesized in cell-free systems programmed by the addition of the 3' one-third of the ASV viral RNA, the region that contains the src gene.[319] This p60^{src} protein has turned out to be a phosphoprotein (thus, its phosphorylated form is called pp60^{src}) that has protein kinase activity of the cAMP-independent type.[320–322] The site of phosphylation by this kinase activity is on tyrosine residues rather than serine or threonine residues, the usual phosphorylation sites of previously discovered cellular protein kinases.[323] This finding raised the possibility that phosphorylation of specific cellular proteins was involved in host cell transformation by ASV. However, as discussed in Chapter 4, it is now known that receptors for a number of normal growth factors also have tyrosine kinase activity.

Oncogene Families

A number of oncogene sequences have now been identified and characterized (Table 5–8). New sequences continue to be found in various eukaryotic organisms, including humans. This might give one the impression that there are innumerable oncogene sequences in nature.

Table 5–8. Some Viral Oncogenes*

Oncogene	Animal Retrovirus	Species Of Origin
abl	Abelson murine leukemia virus	Mouse
fos	FBJ osteosarcoma virus	Mouse
int-1	Mouse mammary tumor virus	Mouse
int-2	Mouse mammary tumor virus	Mouse
mos	Moloney murine sarcoma virus	Mouse
raf†	3611 Murine sarcoma virus	Mouse
fes†	ST feline sarcoma virus	Cat
fgr	Gardner-Rasheed feline sarcoma virus	Cat
fms	McDonough feline sarcoma virus	Cat
kit	H-Z feline leukemia virus	Cat
fps†	Fujinami sarcoma virus	Chicken
erb-A	Avian erythroblastosis virus	Chicken
erb-β	Avian erythroblastosis virus	Chicken
ets	E26 virus	Chicken
mil(mht)†	MH2 virus	Chicken
myb	Avian myeloblastosis virus	Chicken
myc	MC29 myelocytomatosis virus	Chicken
ros	UR II avian sarcoma virus	Chicken
ski	Avian SKV770 virus	Chicken
src	Rous sarcoma virus	Chicken
yes	Y73 sarcoma virus	Chicken
sis	Simian sarcoma virus	Woolly monkey
H-ras	Harvey murine sarcoma virus	Rat
K-ras	Kirsten murine sarcoma virus	Rat
neu/erb-B (HER-2/neu)	None	Rat
rel	Reticuloendotheliosis virus	Turkey
hst	None	Human
met	None	Human
N-ras	None	Human
N-myc	None	Human
L-myc	None	Human
trk	None	Human

*The names of the viral oncogenes are loosely derived from the names of the viruses in which they were identified or from the types of cancers they cause (src from Rous sarcoma virus or ras from rat sarcoma, for example). A half-dozen or so additional transforming genes, some related to the viral oncogenes and some not, have been identified. In addition, the early region genes EIA and EIB of adenoviruses, the T antigens of SV 40 and polyoma viruses, and E6 and E7 proteins of papilloma viruses are considered oncogenes.

†fes and fps are feline and avian versions of the same oncogene; raf and mil(mht) are murine and avian oncogene counterparts.

However, there is now clear evidence that these genes exist as families or even "superfamilies" of related sequences that have been derived from a much smaller number of ancestral genes. This conclusion is drawn from the repeated appearance of the same *onc* genes in a variety of independent viral isolates. For example, the *myc* gene has been found in the genomes of four isolates of avian myelocytomatosis virus, and *fes*, which was originally discovered in feline sarcoma virus, has also been found in a chicken sarcoma virus and termed *fps*. Both *fes* and *fps* turn out to be members of the *src* gene family, which also includes the *onc* genes, *mos*, *raf*, *erb*-B, *yes*, and *abl*. A common ancestor for all these genes has been postulated on the basis of relatedness of their conserved sequences. These are very old genes indeed. *fps* and *fes*, for example, appear to have diverged about 200 million years ago.

Similarly, a large family of *ras*-related oncogenes is rampant in nature. This family includes the *ras* gene group: Harvey *ras* (H-*ras*), Kirsten *ras* (K-*ras*), neuroblastoma-derived *ras* (N-*ras*), and the *rho* gene family. Thus, the *ras*-like genes may be members of a superfamily of genes that diverged from a common ancestral gene hundreds of millions of years ago.[324] Again, the "ancientness" of these genes and their highly conserved sequences suggest that they are very important in the economy of eukaryotic cells. Indeed, as will be discussed later, several functional groupings can be devised according to the products of these genes. For example, oncogene products can be functionally grouped into those that have tyrosine kinase activity (e.g., the *src* family), growth factor–like activity (e.g., *sis*), growth factor receptor domains (e.g., *erb*-B), that function as DNA-binding (e.g., *myc* and *myb*), RNA-binding (e.g., *mil*), or guanine-nucleotide-binding (e.g., the *ras* family) proteins.

It seems highly likely that not all oncogene-related sequences have been discovered yet. Because their discovery has most often occurred by their appearance in viral isolates, it could be that a much larger pool of proto-*onc* genes exists but that they are not readily available for transduction by retroviruses. Readily transducible proto-*onc* genes are probably those that are active in many cell types or organisms and at many stages of growth, and thus are readily available

for retroviral integration, such as genes required for cell proliferation or basic common metabolic functions. Moreover, the close homology between viral *onc* genes and cellular proto-*onc* genes does not necessarily mean that all cellular proto-*onc* genes *cause* malignant transformation of cells, even though the term implies that they do.

It should be noted that two types of sequence conservation have occurred during evolution of the *onc* genes: conservation of nucleic acid sequence, as measured by nucleic acid hybridization, and conservation of the amino acid sequence despite many base changes in nucleotide sequence—that is, conservative base changes that do not alter the sequence of the protein significantly. For example, of 15 *onc* genes originally defined as distinct, based on lack of homology in nucleic acid hybridization experiments, several were later found to be related when the amino acid sequence of the gene product was used as a criterion of homology. A case in point is that of the H-*ras* and K-*ras* oncogenes; these appeared to be only weakly related by DNA-DNA hybridization, and yet the amino acid sequence of the p21 gene product encoded by these genes is 80% homologous.[325]

The finding of various *onc* genes, particularly the *ras* genes, in lower organisms such as *Xenopus laevis*, *Drosophila melanogastor*, and the yeast *Saccharomyces cerevisiaei*[326] provides a powerful tool to look at the function of *onc* genes and how mutations affect their function. *S. cerevisiaei* contains two genes (*RAS*-1 and *RAS*-2) that are closely related to mammalian *ras* and that code for proteins 90% homologous to the Ras p21 protein at their amino termini, but the yeast and mammalian proteins are dissimilar at their carboxy termini. When both *RAS* genes are deleted or made nonfunctional, it is lethal to yeast cell survival, but disruption of only one of them is not. Notably, a high percentage of yeast cells remain viable if the yeast *RAS* genes are replaced by a human *ras* gene.[327] Yeast cells survive and grow well if a chimeric *ras* gene, made up of the first part of the human gene and the second part of the yeast *RAS*-2 gene, is transfected into the yeast cells. However, when the *RAS*-2 yeast gene is mutated by a single base substitution at a specific site, the yeast do not sporulate, they fail to accumulate carbohydrates, and they have poor viability. However, they do have a high content of cAMP, which reflects the increased GTP-binding activity of the p21 protein and thus its increased ability to stimulate adenylate cyclase.[328] Interestingly, this mutation is equivalent to a mutation in a human *ras* gene found in human bladder cancer (see below). Moreover, when a deletion mutant of the yeast *RAS*-1 gene is further altered by a point mutation analogous to the one that increases the transforming activity of mammalian *ras* genes, the yeast *RAS* gene so altered induces transformation in cultured mouse 3T3 cells.[329] The yeast *RAS*-1 gene is larger than the mammalian *ras* gene, primarily in the region coding for the carboxyl terminus of the protein product, and deletion of this part of the gene enhances transforming activity. This is perhaps due to augmentation of the GTP-binding activity, since the amino-terminal region of the p21 gene product is where the GTP-binding and GTPase activity reside.[330] These results indicate that mammalian and yeast *ras*-like genes have similar biologic functions. The ability of human *ras* genes to function in yeast allows the biologic effects of mutations in human *ras* genes to be tested and enables determination of the mutations likely to be important for the altered function of these genes in cancer cells.

Cell Transforming Ability of *onc* Genes

Another approach for discovering oncogenes involved the procedure of gene transfer (DNA transfection). Several laboratories have reported that DNA segments from a variety of animal and human tumors can cause transformation of cultured NIH-3T3 mouse fibroblasts (reviewed References 331 and 332). Through use of probes developed to the oncogenes of retroviruses, it has been found that these transforming DNA segments contain sequences homologous to known v-*onc* genes. This startling discovery led to the concept that activation of cellular *onc* genes can occur either by recombination with retroviral genomes, as described earlier, or by some sort of somatic mutational event leading to activation or abnormal expression of cellular proto-*onc* genes. There is now experimental evidence to show that point mutations, gene amplification, and chromosomal translocation events can lead to activation or increased transcription of cellular proto-*onc* genes.

The rationale for attempting to find transforming or "cancer" genes in the cellular DNA of malignantly transformed cultured cells and tumors goes back to experiments performed in the late 1940s by Avery et al.,[333] who were the first to demonstrate that DNA isolated from a virulent strain of pneumococci could transform a nonvirulent strain into a virulent one with the cellular markers characteristic of the latter bacterial type. These findings, coupled with experiments such as those of Hill and Hillova,[313] who showed that DNA from RSV-infected cells could transform cells as well as produce complete RSV, led to the idea that DNA from cells transformed by chemical carcinogens or DNA from cancerous cells themselves might be able to transform normal cells into malignant cells. In the late 1970s, a number of laboratories began experimenting with this idea. This first demonstration, that DNA from cells transformed with chemical carcinogens could transform other cells, came from the work of Weinberg and colleagues, who showed that DNA from 3-methylcholanthrene (3-MC)-transformed mouse fibroblasts could morphologically transform a line of "normal" 3T3 mouse fibroblasts known as the NIH/3T3 line,[334] which has become the gold standard for testing for transforming DNA (but is not without its problems, as will be discussed later). These experiments were made possible by the development of procedures to transfer intact DNA into whole cells. The transformed cells are visible because the original transformants multiply to give little colonies or foci of transformed cells that pile up on one another instead of growing as flat monolayers of cells as normal fibroblasts do. In addition to this morphologically detectable transformation, the altered cells, when plucked from the culture dishes, will grow in suspension in soft agar, a typical characteristic of transformed cells (see Chapter 4), and they will form tumors when inoculated into mice. When DNA from untransformed NIH/3T3 cells is used in the transfection assay, the recipient cells are not morphologically transformed, nor are they tumorigenic. Thus, treatment of "normal" fibroblasts with the chemical carcinogen 3-MC somehow changes the cells' DNA so that it now carries the genetic information to induce a malignant phenotype in cells into which it is transfected. Subsequently, other laboratories repeated and confirmed these results, thus adding chemical carcinogen-alteration of DNA to retroviral DNA as a means to induce, after integration into a cell's genome, malignant transformation. Other chemically activated transforming DNAs include those extracted from ethylnitrosourea-induced rat neuroblastomas, 7,12-dimethylbenz(a)anthracene [benzanthracene (DMBA)-induced mouse bladder carcinomas, benzo(a)pyrene (BP)-induced rabbit bladder carcinoma,[335] and N-methyl-N'-nitro-N-nitrosoguanidine (MNNG)-transformed human cells.[336] As will be discussed in more detail later, DNA from a wide variety of animal and human cancers has now also been shown to contain segments that will transform NIH/3T3 cells, thus establishing a link between chemically induced transformation in vitro and spontaneously arising cancers in vivo, and showing clearly that alteration of DNA can induce a malignant phenotype in cells.

Identification of the transforming DNA segments contained in chemically transformed cells, or in cancer cells, involved a series of elegant experiments by laboratories that took different but related fundamental approaches to DNA transfection.[337–339] These procedures produced a tremendous enrichment of transforming DNA sequences. For example, whereas it took on average 2 μg of the original total bladder cancer DNA to produce one colony of transformed NIH/3T3 cells, the same amount of isolated, cloned transforming DNA induced about 50,000 transformed foci.[331]

These cloning experiments strongly suggested that a single gene sequence was responsible for the transforming activity in each case. Further evidence for this was obtained by the use of specific endonucleases that cut DNA at specific base sequences. If, for example, a specific endonuclease destroyed transforming activities obtained from a number of clones, whereas another endonuclease did not, this would suggest that the transforming activity from the different clones was the same. Such evidence has been obtained for the transforming activities of DNA isolated from four different chemically transformed mouse fibroblast lines.[340] Specific patterns of restriction-endonuclease sensitivity were subsequently also found for a wide variety of animal and human tumor cells. All of these data point to the conclusion that transforming DNA isolated in each tumor type is carried in a single or small number of genes. Moreover, similar patterns of susceptibility to restriction-endonuclease cleavage have been observed among certain types of human cancers, suggesting that the same or similar onc genes may be activated in these cancers. DNA sequence analysis of the isolated transforming DNA sequences has now confirmed this, and it appears that the same transforming genes are activated in neoplasms of the same differentiated cell type, regardless of whether the neoplasm was virally or chemically induced or occurred spontaneously.

The questions that followed related to the identity of these transforming genes and whether they corresponded to any known proto-onc gene or retroviral onc genes. The answers to these questions came quickly, the answer to the second question being a resounding affirmative. Given the probes developed to the c-onc and v-onc genes, the experiment to test their sequence homology against the cloned transforming genes isolated from various neoplasms and transformed cell lines was straightforward. Initially, probes developed to the v-onc sequences src, myc, fes, ras, erb, mos, myb, and sis were used to test, by nucleic acid hybridization, sequence homology to the isolated transforming sequences (reviewed in Reference 332). These experiments had some surprising results: the transforming genes of human bladder and lung carcinoma detected by DNA transfection in the NIH/3T3 transformation assay were homologous to the ras genes of v-H-ras and v-K-ras, respectively.[341–343] Other human carcinomas and human tumor cell lines also possess the K-ras gene, including carcinomas of the lung, pancreas, colon, gallbladder, and urinary bladder, as well as a rhabdomyosarcoma.[344] In addition, a third ras-like gene was found in the transforming sequences from a human neuroblastoma weakly homologous to both v-H-ras and v-K-ras.[345] This transforming gene represents a third member of ras gene family and has been designated N-ras. The involvement of different ras genes in different types of human cancers suggests that members of the ras gene family may be involved in some general way in regulating the phenotypic characteristics of a variety of human malignant neoplasms. Whether this is a cause or effect of the malignant transformation events is still a matter of some debate (see below). Nevertheless, the activation of cellular *ras* genes in human cancers provides the first direct link between the transforming genes of retroviruses and human cancer.

That human cancer–transforming genes are indeed induced by the activation of cellular proto-*onc* genes has been shown by the following types of experiments. Hybridization analysis of restriction endonuclease–digested cellular DNAs from human bladder and lung carcinomas and from normal human cells with cloned probes of v-H-*ras* and v-K-*ras* sequences and with cloned probes of the biologically active transforming gene from human bladder cancer has shown that the activated transforming genes of bladder and lung carcinomas are homologous to the *ras* proto-*onc* genes of normal cells (reviewed in Reference 332). Furthermore, when viral transcriptional promoter LTR sequences from murine or feline retroviruses are linked to the *ras* proto-*onc* gene isolated from normal human cells, oncogenic transformation of NIH/3T3 mouse fibroblasts is achieved and an increased expression of the p21 gene product of the proto-*onc ras* gene is observed in the transformed cells,[346] a finding suggesting that elevated expression of a "normal" proto-*onc* gene can induce oncogenic transformation. However, as we shall see later, it is also possible to activate proto-*onc* (c-*onc*) genes through other mechanisms, including somatic mutation and gene amplification.

A wide variety of human tumors has now been examined for expression of cellular c-*onc* genes by DNA-RNA hybridization using v-*onc* gene cDNA probes. Expression (transcription into RNA) of genes homologous to v-*onc* genes in human tumors occurs in a variety of leukemias and lymphomas, carcinomas, various sarcomas, neuroblastoma, teratocarcinoma, and choriocarcinoma.

The DNA transfection experiments suggest that transforming ability is a dominant trait; in other words, if a transforming *onc* gene is transfected into or activated in a normal cell, it captures the cell's genetic machinery and turns it into a cancer cell. This conclusion is most likely wrong for the following reasons. The NIH/3T3 cell is already two-thirds a cancer cell. Indeed, subpopulations of "untransformed" NIH/3T3 cell cultures are tumorigenic and metastatic under the right conditions,[347] although transformation with a *ras* gene markedly increases the malignant potential of these cells. Moreover, as

will be discussed later, transfection with at least two *onc* genes is needed to transform normal diploid fibroblasts in culture, supporting the idea that malignant transformation is a multistep process. Finally, cell hybridization between malignant and normal cells indicates that the hybrid cells formed are more likely than not to be nontumorigenic. Thus, expression of the complete malignant phenotype is not likely to be due to insertion or activation of a single "cancer gene," and in most cases appears to involve the loss of tumor suppressor genes.

Activation of c-*onc* genes can take place by means of many of the mechanisms described above for activation of genes during cell transformation or tissue differentiation. These mechanisms include point mutations, gene rearrangement, gene amplification, and increased transcription due to alterations in chromatin packaging. In addition, insertion of retrovirus enhancer regions (LTRs) next to c-*onc* genes or mutation in c-*onc* gene coding sequences can alter their function. Some of these mechanisms have been identified only by test tube or animal experiments, but all could potentially be involved in c-*onc* gene activation during carcinogenesis in humans. Proto-oncogenes are present in all human cells, just as they are in animal cells, and they apparently have to be activated by some endogenous (e.g., faulty repair of oxidative damage from normal cellular processes) or exogenous agent (e.g., ultraviolet light, chemical carcinogens) to trigger the cancer process.

A number of genetic lesions have been observed in human tumors,[348] and these most likely are part and parcel of the carcinogenic process. Direct proof that they are cause rather than effect can only be deduced, however, from cell culture and transgenic animal experiments, where an altered gene can be introduced and its effect on cell transformation or tumor development directly observed.

Functional Classes of Oncogenes

Oncogenes and their normal cellular counterparts, the proto-oncogenes, can be classified by their function into several different categories (Table 5–9).[349] A number of these genes encode growth factors, e.g., *sis* (PDGF B-chain), *int-2*, and *hst* (FGF-like factor). These oncogene growth

Table 5–9. Functions of Cell-Derived Oncogene Products°

CLASS 1—GROWTH FACTORS			*N-ras*	Membrane-associated GTP-binding/GTPase
sis	PDFG B-chain growth factor		gsp	Mutant activated form of G_s α
int-2	FGF-related growth factor		gip	Mutant activated form of G_i α
hst (KS3)	FGF-related growth factor			
FGF-5	FGF-related growth factor		**CLASS 5—CYTOPLASMIC PROTEIN-SERINE KINASES**	
int-1	Growth factor?		*raf/mil*	Cytoplasmic protein-serine kinase

CLASS 2—RECEPTOR AND NONRECEPTOR PROTEIN-TYROSINE KINASES

src	Membrane-associated nonreceptor protein-tyrosine kinase
yes	Membrane-associated nonreceptor protein-tyrosine kinase
fgr	Membrane-associated nonreceptor protein-tyrosine kinase
lck	Membrane-associated nonreceptor protein-tyrosine kinase
fps/fes	Nonreceptor protein-tyrosine kinase
abl/bcr-abl	Nonreceptor protein-tyrosine kinase
ros	Membrane-associated receptor-like protein-tyrosine kinase
erbB	Truncated EGF receptor protein-tyrosine kinase
neu	Receptor-like protein-tyrosine kinase
fms	Mutant CSF-1 receptor protein-tyrosine kinase
met	Soluble truncated receptor-like protein-tyrosine kinase
trk	Soluble truncated receptor-like protein-tyrosine kinase
kit (W locus)	Truncated stem-cell receptor protein-tyrosine kinase
sea	Membrane-associated truncated receptor-like protein-tyrosine kinase
ret	Truncated receptor-like protein-tyrosine kinase

CLASS 3—RECEPTORS LACKING PROTEIN KINASE ACTIVITY

mas	Angiotensin receptor
α1β	Angiotensin receptor

CLASS 4—MEMBRANE-ASSOCIATED G PROTEINS

H-*ras*	Membrane-associated GTP-binding/GTPase
K-*ras*	Membrane-associated GTP-binding/GTPase

CLASS 5—CYTOPLASMIC PROTEIN-SERINE KINASES

raf/mil	Cytoplasmic protein-serine kinase
pim-1	Cytoplasmic protein-serine kinase
mos	Cytoplasmic protein-serine kinase (cytostatic factor)
cot	Cytoplasmic protein-serine kinase?

CLASS 6—CYTOPLASMIC REGULATORS

crk	SH-2/3 protein that binds to (and regulates?) phosphotyrosine-containing proteins

CLASS 7—NUCLEAR TRANSCRIPTION FACTORS

myc	*Sequence-specific DNA-binding protein*
N-*myc*	Sequence-specific DNA-binding protein
L-*myc*	Sequence-specific DNA-binding protein
myb	Sequence-specific DNA-binding protein
lyl-1	Sequence-specific DNA-binding protein?
p53	Mutant form may sequester wild-type p53 growth suppressor
fos	Combines with c-jun product to form AP-1 transcription factor
jun	Sequence-specific DNA-binding protein; part of AP-1
erbA	Dominant negative mutant thyroxine (T_3) receptor
rel	Dominant negative mutant NF-κB-related protein
vav	Transcription factor?
ets	Sequence-specific DNA-binding protein
ski	Transcription factor
evi-1	Transcription factor
gli-1	Transcription factor
maf	Transcription factor
pbx	Chimeric E2A-homeobox transcription factor
Hex2.4	Transcription factor?

OTHER

dbl	Cytoplasmic truncated cytoskeletal protein?
bcl-2	Inhibits apoptosis

°The table is somewhat selective and obviously incomplete. These oncogenes were originally detected as retroviral oncogenes or tumor oncogenes. Others were identified at the boundaries of chromosomal translocations and at sites of retroviral insertions in tumors, or were found as amplified genes in tumors and shown to have transforming activity. EGF, epidermal growth factor; FGF, fibroblast growth factor; PDCGF, platelet-derived growth factor.

(From Hunter[349])

factors can stimulate tumor cell proliferation by paracrine or autocrine mechanisms, but by themselves may not be sufficient to sustain the transformed phenotype.

A second type of oncogene codes for altered growth factor receptors, many of which have associated tyrosine kinase activity. These include the *src* family of oncogenes, *erb* B (EGF receptor), and *fms* (CSF-1 receptor). For some of these receptor-like, tyrosine kinase–associated membrane proteins, the actual ligand is not known (e.g., *trk*, *met*, and *ros*).

A third receptor class that doesn't have associated tyrosine kinase activity is the *mas* gene product (angiotensin receptor) and the α1β-adrenergic receptor.

A fourth class of oncogene products is membrane-associated, guanine nucleotide–

binding proteins such as the Ras family of proteins. These proteins bind GTP, have associated GTPases, and act as signal transducers for cell surface growth factor receptors. The transforming *ras* oncogenes have been mutated in such a way as to render them consitutively active by maintaining them in a GTP binding state, most likely because of a defect in the associated GTPase activity.

A fifth category is the cytoplasmic oncoproteins with serine/threonine protein kinase activity. These include the products of the *raf*, *pim*-1, *mos*, and *cot* genes. A prototype of this class is the c-Raf protein, activated by a variety of tyrosine kinase–associated receptors. There is clear evidence that c-Raf acts as an intermediate in the signaling pathway between Ras and the cell nucleus by activating the mitogen activated protein (MAP) kinase cascade (see below). The oncogenic form of Raf has lost part of its regulatory amino-terminal sequence and appears to be constitutively active. c-Crk is also a cytoplasmic protein, and it appears to act by stabilizing tyrosine kinases associated with the Src family of oncoproteins.

A sixth type is cytoplasmic regulators like *crk*, which affect phosphotyrosine-containing proteins.

A seventh, large class of oncogenes are those that code for nuclear transcription factors such as *myc*, *myb*, *fos*, *jun*, *erb* A, and *rel*. For a number of these, the oncogenic alteration that makes them transforming oncoproteins is a mutation that leads to loss of negative regulatory elements (e.g., for *jun*, *fos*, and *myb*), and in other cases (e.g., *erb-A* and *rel*) the activating mutations cause the loss of their active domains, producing a mutant protein that prevents the activity of the normal gene product—a so-called dominant-negative mutation. Mutations of the tumor suppressor gene *p53*, in sort of a "reverse twist," produce a dominant-negative effect by producing a protein that in this case prevents the action of a tumor suppressor function.

Characteristics of Individual Oncogenes

ras

The most frequently detected alterations in oncogenes in both animal tumor model systems and in human cancers are mutations in the *ras* family of oncogenes.[350,351] The three most commonly involved oncogenes in human cancer are Harvey (H), Kirsten (K), and neuroblastoma (N) *ras*. The 21 kDa proteins encoded by these genes include the transforming proteins of the murine sarcoma viruses v-H-Ras and v-K-Ras, which are oncogenic mutants of normal cellular c-H-Ras and c-K-Ras proteins. The K-*ras* gene mRNA is alternatively spliced to produce two protein isoforms, K-RasA and K-RasB. Other members of the Ras family include M-Ras, R-Ras, Rap 1 and 2, and Ral, all of which share at least 50% sequence identity to other members of the family.[350]

Ras proteins are membrane-bound GTP/GDP binding proteins of about 190 amino acids that are highly conserved in the N- and C-termini. The difference between the proteins lies primarily in the C-terminal hypervariable domain of 25 amino acids. Activation of Ras requires covalent addition of a lipid linker moiety by a prenylation step involving addition of either farnesyl (15-carbon) or geranylgeranyl (20-carbon) groups to a conserved C-terminal cysteine-containing sequence CAAX (C = cysteine, A = aliphatic amino acid [leucine, isoleusine, or valine], X = methionine, serine, leucine, or glutamine). These steps are catalyzed by farnesyl transferase or geranylgeranyl transferase. This is followed by a proteolytic cleavage step and a methylation step (Fig. 5–14). The signal transduction steps were described in Chapter 4 and will be outlined here. Binding of a growth factor such as epidermal groth factor (EGF), insulin-like growth factor (IGF), or platelet-derived growth factor (PDGF) triggers receptor dimerization and receptor tyrosine kinase activation, receptor autophosphorylation, and binding of adapter proteins Grb 2 and SOS. This leads to release of GDP and binding of GTP. (Ras must be in its membrane-bound form for this to happen.) Activated Ras can turn on a number of downstream effectors, including the Raf-MEK-ERK, phospholipase C-DAG-PKC, and PI3K-Akt, and Ral-Rac-Rhu pathways (Fig. 5–15). This ultimately leads to activation of transcription factors such as C-Jun, Fos, and Myc that stimulate gene expression and facilitate (in most cell types) cell proliferation. In normal cells, these Ras signaling cascades are only transiently activated because an intrinsic GTPase activity hydrolyzes GTP, converting Ras-GTP to

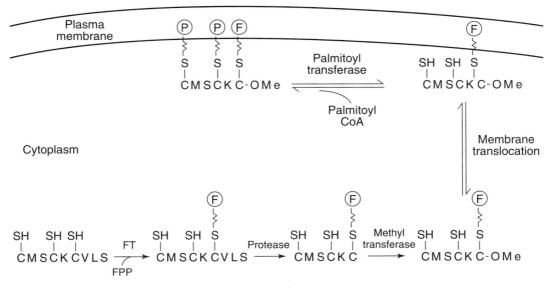

F : Farnesyl group
P : Palmitoyl group

Figure 5–14. Simplified scheme of the post-translational processing of H-ras. Farnesyl transferase (FT) transfers a farnesyl group (F) from faresyl-pyrophosphate (FPP) to the thiol group of the cysteine residue of the CAAX motif (CVLS in the case of H-ras). The terminal tripeptide is cleaved by a specific endoprotease in the endoplasmic reticulum. The methyl donor for the reaction catalyzed by a prenyl protein-specific methyl transferase is S-adenosylmethionine. Palmitoylation of C-terminal cysteine residues occurs before membrane localization. CoA, coenzyme A. (From Adjei,[350] with permission.)

its inactive Ras-GDP form. Oncogenic mutations of Ras prevent or greatly slow this GTPase-mediated step leading to constitutively active Ras. There is also evidence that Ras can be activated on Golgi membranes by phospholipase Cγ, and this activation may have additional cellular consequences.[352]

Mutations in the *ras* gene usually involve codons for amino acids 12, 13, 59, and 61, all of which prevent GTPase-activating protein (GAP)-induced GTP hydrolysis. One of the most well-studied animal systems for *ras* oncogene mutation is the mouse skin model. A number of mouse skin carcinomas initiated by application of carcinogens such as *N*-methyl-*N*-nitro-*N*-nitrosoguanidine (MNNG), methylnitrosourea (MNU), 3-methylcholanthrene (3-MC), and 7,12-dimethylbenz(a)anthracene (DMBA) contain mutated H-*ras* genes. Interestingly, these agents cause somewhat different mutations, and the mutations seen in skin papillomas are not the

same as those seen in skin carcinomas,[353] thus distinct mutagenic events may be involved in the initiation phases and in the promotion–progression phases of skin tumorgenesis. For example, in one study the alkylating agents MNNG and MNU caused G→A transitions at codon 12, whereas 3-MC caused G→T transversions at codon 13 and A→T transversions at codon 61 in papillomas.[353] Only the G→T mutation was seen in carcinomas, a finding suggesting that the cells bearing that mutation were the ones destined to progress to carcinoma. In the case of DMBA-induced skin tumors, a codon 61 A→T transversion could be seen well before the appearance of papillomas, indicating that this mutation is an early event in initiation.[354] Mutations of *ras* gene family members have also been observed in chemically induced liver, pancreatic, and mammary carcinomas and in ultraviolet-induced skin cancers in rodents (reviewed in Reference 355). There is a lot of evidence to

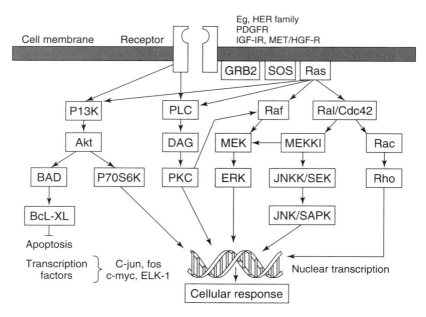

Figure 5–15. Simplified drawing of ras signaling and its effector pathway. P13K, phosphoinositide 3′-kinase; PLC, phospholipase C; PKC, protein kinase C; MEK, mitogen-activated protein kinase kinase; JNK, Jun amino-terminal kinase; SAPK, stress-activated protein kinase. The well-characterized Ras/Raf/mitogen-activated protein (MAP) kinase pathway illustrates a typical MAP kinase-signaling module. Raf is an MAPKKK (MEKK). MEK is an MAPKK. Activated MAPK (ERK, i.e., extraceullular signal-regulated kinase) phosphorylates and activates various transcription factors in the nucleus, which control cellular responses. Although simplified, note the cross-talk and redundancy of the signaling pathways. BAD, pro-apoptotic protein of the Bcl-2 family; DAG, diacyl glycerol; SEK, stress-activated protein (SAP)/Erk-Kinase; Eg, for example; PDGFR, platelet-derived growth factor receptor; IGF-IR, insulin-like growth factor receptor type 1; MET/HGR-R, hepatocyte growth factor receptor (a product of the c-met proto-oncogene). (From Adjei,[350] with permission.)

indicate that the action of a single oncogene is insufficient to cause neoplastic transformation of cells, particularly normal diploid cells. Instead, the action of at least two oncogenes appears to be required. In the case of ras, cooperativity with the myc oncogene, for example, has been shown to be required to induce neoplastic transformation of primary embryo fibroblasts or prostate tissue organ cultures.[356] In addition, v-H-ras and v-fos co-infected keratinocytes produced squamous cell carcinomas in nude mice, whereas v-fos alone produced only skin hyperplasia, and v-H-ras alone produced only papillomas.[357] In human cells, malignant transformation is even more tightly controlled and requires an immortalization step (activation of telomerase) combined with two oncogenes (SV40 large T antigen and activated Ras; reviewed in Reference 358).

Mutations in members of the ras gene family are found in a wide variety of human cancers, including colon, pancreas, lung, breast, skin, thyroid, bladder, liver, and kidney carcinomas, as well as in seminomas, melanomas, and some forms of leukemia (reviewed in Reference 359). Most of these mutations are in codons 12, 13, or 61. While ras gene mutations can be found in a variety of human tumor types, the incidence of such mutations varies greatly. For example, the highest incidences are found in carcinomas of the pancreas (90%), colon (50%), lung (30%), thyroid (50%), and myeloid leukemia (30%), while a much lower incidence is found in urinary bladder (6%)[360] and ovarian (15%)[361] cancers. Ras mutations are also infrequent in breast, stomach, esophagous, and prostate carcinomas.[350] Some diagnostic advantage may be taken from the fact that some cancers have a high incidence of *ras* gene mutations. For example, eight of nine cases of colorectal cancer have K-*ras* mutations at codons 12 or 13 in biopsy samples of the tumor, and have detectable mutations in sloughed cells obtained in the feces.[362] Similarly, K-*ras* gene mutations were found in codon 12 in the pancreatic juice of seven of seven patients with pancreatic carcinomas.[363] This finding suggests a way to differentiate between chronic pancreatitis and pancreatic carcinoma for this difficult to diagnose cancer.

Genomic searches have found a mutation in another Ras-associated protein called BRAF.[364] The *braf* gene encodes a serine/threonic kinase that appears to cause constitutive activation of the Raf-Mek-Erk pathway primarily in melanomas. Fifty-nine percent of human melanoma cell lines have this mutation, which suggests that BRAF is part of a melanocyte-specific Raf-Mek-Erk pathway. BRAF mutations are more common in melanomas occurring in the skin of individuals with intermittent sun exposure.[364] Thus BRAF may be a target for skin cancer chemoprevention.

As one might predict, various parts of the activated Ras signal transduction pathway are potential targets for anticancer drug development. Indeed, both farnesyl and geranylgeranyl transferase inhibitors have shown activity in clinical trials and a number of Raf and Mek kinase inhibitors have been developed (reviewed in Reference 350). A point to consider in development of such drugs is whether upstream or downstream pathway inhibition would be most efficacious. Inhibition of Ras downstream effectors such as Mek could inhibit a number of signaling events, because it is an intersection point for a number of pathways (Fig. 5–15). While this may increase the number of tumor types against which such drugs could be effective and decrease the chance for a redundant pathway to go around the blockage point, it could also increase the chance for toxicity to normal cells. In contrast, inhibiting an upstream component such as an EGF, IGF, or PDGF receptor could limit the anti-tumor spectrum and allow escape via a redundant pathway.

myc

The *myc* gene was discovered by looking for the cell-transforming sequence of the avian retrovirus

MC29,[365] later identified in vertebrate genomes. A number of studies have shown that *myc* plays a key role in cell proliferation and differentiation events. Deregulation of c-*myc* can occur via either gene rearrangement or amplification, and both have been observed in human cancers.

The other members of the *myc* gene family are N-*myc*, cloned in 1983 and found to be amplified in human neuroblastoma; L-*myc*, identified in 1985 and highly expressed in small cell lung cancer; B-*myc*, encoding a truncated version of a Myc protein and primarily expressed in brain; and s-*myc*, a putative suppressor of neoplastic transformation (reviewed in Reference 366).

Expression of c-Myc protein is higher in proliferating cells and falls as terminal differentiation proceeds. In model systems for differentiation such as murine erythroleukemia cells, 3T3-L1 pre-adipocytes, or F9 teratocarcinoma cells, continued expression of c-*myc* from a transfected gene blocks differentiation (reviewed in Reference 366). Thus down-regulation of c-Myc may be necessary for cessation of cell division during terminal differentiation.

Rearrangement of *myc* is one of the classic examples of *onc* gene activation by chromosomal translocation. One of the clear examples of this is the B-cell lymphoma known as Burkitt's lymphoma (BL), a primarily pediatric disease of high incidence in equatorial Africa. A number of types of translocation events have been found in various cell lines derived from patients with this disease, including translocations between chromosomes 8 and 14 (the most common), 8 and 2, and 8 and 22. These findings led Klein to postulate that the intrachromosomal breakages and rearrangements in normal lymphoid cells that are involved in V-region and C-region joining of immunoglobulin genes (see Chapter 6) would in transformed lymphoid cells bring an *onc* gene and the Ig gene segments together in such a way that the *onc* gene would be derepressed by coming under the influence of the normal Ig gene promoters.[367] This idea turned out to be particularly perspicacious, for it was soon found that the *myc* oncogene was located on chromosome 8 in humans and was in a region translocated to chromosome 2, 14, or 22 in BL cells.[368,369] Similar translocations between chromosome 15 (the chromosome bearing the *myc* gene in the mouse) and chromosome 12 (location of Ig genes in the mouse) have been observed in murine plasmacytoma cells,[370] the mouse cell type equivalent to Ig-producing B-cell malignancies of humans.

Normally, expression of the c-*myc* gene is carefully regulated. Expression is low in resting normal lymphoid cells and is turned on in proliferating lymphoid cells, for example, in cells stimulated by lectins or antigens. Thus the c-*myc* gene product may play a critical role in DNA synthesis and mitosis. In BL cells, however, c-*myc* expression is not regulated and becomes "constitutive" (i.e., not turned on or off in response to normal metabolic stimuli), perhaps because of the loss or alteration of parts of exon 1, which appear to be involved in regulation of c-*myc* expression. In summary, there may be a variety of mechanisms, involving different types of gene rearrangements of c-*myc*, that lead to c-*myc* gene derepression and unregulated expression so that the cell has a continued transcription of c-*myc* and cannot return to the resting, nonproliferating state.

Members of the *myc* gene family are also deregulated by gene amplification. In fact, the first reported *onc* gene amplification was for the c-*myc* gene in a human promyelocytic leukemia cell line, HL-60, and in primary leukemic cells taken from the patient from whom the cell line has been derived.[371] The gene was amplified about 20 times in these cells. The c-*myc* gene was later found to be amplified in human cell lines derived from colon carcinoma[372] and small-cell lung carcinoma (SCLC).[373] In the latter case, the highest degree of c-*myc* amplification (20- to 76-fold) was found in the SCLC cell types with the least differentiated and most highly malignant phenotype. Amplification, rearrangements, and deregulated, enhanced expression of c-*myc* have also been observed in chemically induced rodent tumors.[374]

One or more of three members of the *myc* gene family, c-*myc*, N-*myc*, and L-*myc*, are amplified in various human cancers and in cell lines derived from them. In one study of SLSCs, all three *myc* gene family members were found to be amplified, this amplification occurring more commonly in tumors from patients after treatment with chemotherapy.[375] Myc protein levels are also elevated in colon carcinomas.[376] Amplification of c-*myc* occurs in breast carcinomas and

appears to be an independent prognostic marker of overall survival.[377] Treatment with hydroxyurea of human cancer cell lines that have amplified *myc* genes, present as extrachromosomal double-minute elements, decreases the *myc* copy number and the tumorigenicity of these cells in nude mice.[378] These results suggest a way to eliminate amplified *myc* genes in vivo and perhaps improve patient survival.

The c-Myc protein product of the c-*myc* gene is a DNA-binding, nuclear phosphoprotein that has all the characteristics of a transcription factor. It has a transcriptional activation domain, a DNA-binding domain, a nuclear localization signal, a site for phosphorylation by a nuclear protein kinase, an HLH motif, and a leucine zipper motif typical of transcription factors that have to form dimmers to be active (reviewed in Reference 366).

For a long time, the dimerization partner of c-Myc was a missing link in our understanding of its action as a transcription factor. At first it was thought that c-Myc formed homodimers with itself, but this process did not seem to occur inside cells. This problem was solved when Blackwood and Eisenman[379] cloned a human gene coding for a protein they called Max, which dimerized with c-Myc. A mouse homolog of Max called Myn was cloned a short time later.[380] Max can dimerize with c-Myc, N-Myc, or L-Myc, but not other basic HLH–leucine zipper proteins.[379] Max can bind DNA as a homodimer, but doesn't seem to be able to activate transcription, which suggests that it lacks a transcriptional activation domain. This ability to bind DNA and not activate genes most likely explains the ability of Max-Max homodimers to antagonize the ability of Myc-Max heterodimers to stimulate gene transcription. Myc exists as a heterodimer and its activity is determined by the partner it is bound to. Myc-Max dimers activate transcription; Mad-Max dimers block transcription. Mad is another member of the Myc-associated family and its interaction with Max appears to cause repression by chromatin remodeling. Myc-Max heterodimers bind to the consensus sequence CACGTG, to which Max homodimers also bind.[381] Myc alone doesn't form homodimers efficiently and doesn't bind to DNA except at high concentrations in vitro. Thus, Myc acts as a transcriptional activator that requires dimerization with Max, and Max homodimers act as a repressor of Myc-Max action. It has also been found that Max over-expression in cells represses transcription of reporter genes bearing the CACGTG regulatory sequence, and Max-induced repression is relieved by overexpression of c-Myc.[381,382]

Moreover, the oncogenic transforming ability of c-Myc requires dimerization with Max.[383] Although c-Myc-induced cellular transformation is associated with gene amplification and high expression of a normal coding sequence, it can also involve mutations of the translocated *myc* alleles.[384]

Myc's action as a transcription factor includes induction of ornithine decarboxylase, cyclin A, and cyclin E, all of which are involved in cell proliferation. Somewhat paradoxically, increased c-Myc production in some cell types (e.g., B lymphocytes) is associated with programmed cell death (apoptosis). How can c-Myc promote both cell proliferation and apoptosis? The answer to this problem involves the finding that mutant *myc* alleles, derived, for example, from human Burkett's lymphoma, uncouple proliferation from apoptosis and, as a result, are even more effective than overexpressed wild-type Myc in promoting lymphomagenesis in mice.[384] Mutant Myc proteins retain their ability to stimulate proliferation and to activate the p53-driven apoptotic pathway, but despite the latter ability, mutant Myc is defective in promoting apoptosis because it doesn't induce Bim, a protein also needed to induce apoptosis. Wild-type Myc is as efficient as mutant Myc in producing lymphomas if either Bim or p53 function is inactive.[384]

The transcription factor NF-κB regulates c-*myc* expression in a number of cell types. Another player here is Mga, which can interact with Max to form a complex with the E2F inhibitor E2F-6.[385] Normally, this shuts off E2F and Myc-responsive genes, leading to cell quiescence and a G_0 state, but in some malignant cells E2F-6 appears to be inactive, thus preventing cells from entering the quiescent state.

src

As noted above, *src* was the first transforming oncogene discovered. It exemplifies yet again another way in which a cellular proto-oncogene is altered to become a transforming oncogene. In this case, data comparing the normal cellular gene product of c-*src* to its viral, transforming counterpart from v-*src* showed that the normal protein pp60[c-src] differed from a pp60[v-src] in that various isolates of the latter contained a number of scattered single amino acid differences between residues 1 and 514 (but only one common one: Thr 338→Ile) as well as truncations and alterations at the carboxyl terminus (reviewed in Reference 386). This finding implied that the C-terminus of pp60[src] plays an important role in regulating its transforming ability. Because pp60[src] is a non-receptor tyrosine kinase (as opposed to a receptor tyrosine kinase such as EGFR and PDGFR) and regulates cell function by its ability to phosphorylate key cellular substrates, it was thought that the C-terminus somehow regulates the kinase activity of the Src protein. A clue that this idea was correct lay in the observation that Src tyrosine kinase activity was enhanced by phosphatases.[387] Subsequently it was found that phosphorylation of tyrosine-527 in the C-terminus of Src was inhibitory, and dephosphorylation of this residue stimulated Src kinase activity (reviewed in Reference 388). In most transforming Src mutants, Tyr-527 is either missing or underphosphorylated, compared to wild-type Src. In addition, phosphorylation of Tyr-416 in the catalytic domain of the kinase activity was required for full activity. Mutations that activate Src have been mapped to the kinase domain itself, to SH2 and SH3 domains, and to the C-terminus,[388] all of which appear to produce constitutive activation of Src kinase activity. A tyrosine kinase called Csk (C-terminal Src kinase) has been identified that can phosphorylate the C-termini of Src and all its family members; Csk, when overexpressed, inhibits the cell-transforming ability of high levels of Src (reviewed in Reference 388).

There are at least eight members of the *src* gene family in vertebrates: *yes, fgr, lyn, lck, fyn, hck, blk,* and *src* itself.[389] Morever, alternate translational initiation codons and tissue-specific mRNA splicing results in more than 14 different Src-type proteins being expressed selectively in various cell types.[388] They all have tyrosine kinase activity, an N-terminal myristorylation signal presumably required for their association with cell membranes, SH2 and SH3 domains, a kinase domain, and a C-terminal regulatory "tail." The structural relationships of

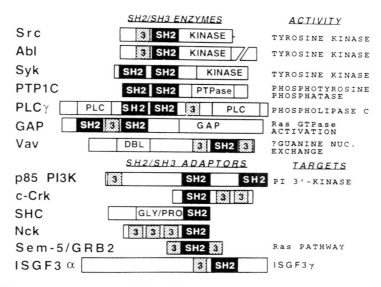

Figure 5–16. Structures of selected SH2-containing proteins. These polypeptides are divided into those with intrinsic enzymatic activity and those without
known catalytic domains, which may act as adaptors to couple tyrosine kinases to
downstream targets. No catalytic activity has been shown for Vav, but it contains
a region of homology to Dbl (indicated as DBL), which has guanine nucleotide
exchange activity. ISGF3α is three closely related proteins of 84, 91, and 113
kDa. 3, SH3 domain; PTPase, phosphototyrosine phosphatase domain; GLY/
PRO, glycine-proline-rich region. (From Pawson and Gish,[392] with permission.)

all these proteins strongly suggest that they are
all subject to the same regulatory mechanisms,
yet they appear to have very cell-specific functions (see below).

c-Src is present at low levels in most cell
types, but high levels are found in neural tissue,
platelets, lymphocytes, monocytes, and chro-
maffin cells (reviewed in Reference 390). High
levels are also found in human neuroblastoma,
small-cell lung, colon, and breast carcinomas,
and rhabdomyosarcoma.[390,391]

Mention should be made here of the Src homology domains SH2 and SH3. Even though these
domains are found in several proteins that interact with tyrosine kinase receptors and other proteins in signal transduction pathways, they are part
of the sequence of the *src* gene–encoded protein.

The SH2 domain is a conserved motif of about
100 amino acids found in a diverse array of proteins involved in signal transduction (Fig. 5–16).
A number of these proteins also contain SH3
domains, which are sequences about 50
amino acids long. These domains are found in a
variety of *onc* gene tyrosine kinases, phospholi-

pase c-γ (PLC-γ), intermediates in the guanine-
nucleotide exchange pathway involving Ras (see
above), GAP, phosphatidylinositol (PI)3′-kinase,
and an ever-increasing list of other proteins involved
in various phosphorylation–dephosphorylation
cascades.

A typical scenario is as follows: an external ligand such as a growth factor or hormone binds to
its receptor, inducing dimerization and autophosphorylation of the receptor. These steps
create binding sites for SH2 domains that recognize phosphotyrosine (or phosphoserine–
phosphothreonine) adaptor proteins that are the
linkers to the next step in a signal transduction
pathway. In the case of the Ras pathway, as noted
above, this linker is the Sem5/GRB2 protein.
Specificity is provided in that high-affinity binding of an SH2 domain requires recognition of
a phosphotyrosine within a specific amino acid
sequence.[392] For example, PI3′-kinase, Ras-GAP,
and PLC-γ each bind to different autophosphorylated sites on the PDGF-β receptor. Thus,
the binding of SH2-containing adaptor proteins
to a given receptor depends on the amino acid

sequence (and presumably the peptide conformation) at the autophosphorylation site.

The SH3 domains of adaptor proteins recognize, in a context- and conformation-dependent way, guanine nucleotide exchange factors such as Sos in the Ras pathway and GTPase-activating proteins. Both SH2 and SH3 domains appear to be involved in the regulation of *onc* gene product tyrosine kinase activities. For example, deletion of the SH3 domain from the c-Abl protooncogene protein activates its tyrosine kinase activity and renders it transforming; moreover, alterations of the amino acid sequence of the SH2 domain of c-Abl also render it transforming.[393] These results indicate that both SH2 and SH3 domains play a role in regulating the activity of oncogene protein tyrosine kinases.

The Src protein is protean in its ability to modulate cellular functions (Fig. 5–17). Among its many actions are its interactions with several signal transduction pathways that facilitate or modulate cell proliferation, cell survival, metastasis, intracellular trafficking, and cell adhesion (reviewed in Reference 389). A prominent effect of activated Src (v-Src) kinase is to decrease cell adhesion. It appears to do this by phosphorylating R-Ras.[394] In this respect, activated Src and Ras proteins, working together, reduce cell adhesion and fibronectin production for the extracellular matrix. Phosphorylated R-Ras inhibits integrin activity involved in cell adhesion and in so doing most likely contributes to tumor invasive properties of Src-transformed cells.

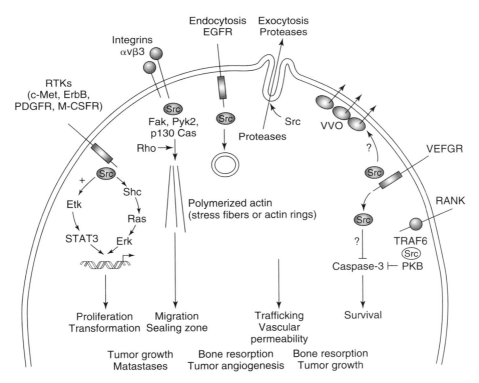

Figure 5–17. The known and putative functions of Src in various cell types. Abbreviations: EGFR, epidermal growth factor receptor; Erk, extracellular regulated kinase; M-CSFR, macrophage colony-stimulating factor receptor; PDGFR, platelet-derived growth factor receptor; PKB, protein kinase B; RANK, receptor activating nuclear factor κB; RTKs, receptor tyrosine kinases; STAT3, signal transducer and activator of transcription 3; TRAF6, tumor necrosis factor receptor–associated factor 6; VEGFR, vascular endothelial growth factor receptor; VVO, vesiculo-vacuolar organelles. (From Susa et al.,[389] with permission.)

jun and fos

The role of c-*jun* and c-*fos* as transcription factors was covered earlier in this chapter. Here, we will focus on the transforming ability of the Jun and Fos proteins when they are expressed in an inappropriate way or at inappropriate times in the life of a cell.

The oncogene v-*jun* was originally discovered as the transforming gene of avian sarcoma virus 17 that can induce fibrosarcomas in chickens via its p65 *gag-jun* gene–derived fusion protein.[395] Both v-Jun and c-Jun form dimers with c-Fos and act as transcription factors that bind to AP-1 regulatory sequences. Overexpression of c-Jun in rat embryo fibroblasts co-transfected with c-H-*ras* gives rise to immortalized cell lines that grow in soft agar and produce tumors in nude mice.[396] However, v-Jun has even greater transforming ability and significantly greater transcriptional activity than c-Jun. Sequence analysis of c-Jun and v-Jun has revealed that v-Jun has three amino acid substitutions in the C-terminus and deletion of a 27–amino acid run, called the δ region, from the N-terminus.[397] The δ region contains a negative regulatory domain, the lack of which renders v-Jun more active as a transcription factor, apparently because it lacks a site for binding of a cellular factor that modulates c-Jun activity. Since the leucine zipper motif of Fos is required for transformation by overexpression of c-Jun or expression of v-Jun, it can be concluded that transcriptional activation of AP-1-regulated genes is involved in the cellular transformation events.

Other members of the *jun* gene family include *jun*B and *jun*D, which share significant sequence homology and which can all bind to AP-1 sites in the presence of c-Fos. However, members of the *jun* family are expressed at different levels in different cell types and their production responds differently to extracellular signals. Moreover, they differ in their transforming ability. For example, JunB is less potent than c-Jun in transforming rat embryo cells co-transfected with *ras*, and co-transfection of *jun*B with c-*jun* into *ras*-activated cells decreases transformation compared to c-*jun*/*ras*–transfected cells.[398]

The *fos* oncogene was detected in two independent isolates of mouse osteosarcoma viruses: FBJ-MUSV, isolated from a spontaneous osteosarcoma, and FBR-MuSV, isolated from a radiation-induced osteosarcoma.[399] Both viruses can induce chondrosarcomas or osteosarcomas when inoculated into newborn mice and are able to induce transformation in mouse fibroblast cell lines. The FBJ-encoded v-Fos protein is similar to its cellular homolog c-Fos except for a frameshift mutation at its C-terminus, whereas the FBR v-Fos has N-terminal and C-terminal truncations, internal deletions, and several single–amino acid changes (reviewed in Reference 400). Sequences in the C-terminal half of c-Fos have a regulatory role in its activity, and alteration of this domain in v-Fos correlates with its transforming activity. An amino acid substitution at residue 138 (Glu→Val) activates the cell-immortalizing activity of v-Fos.

ets

The v-*ets* oncogene was first identified in the E26 acutely transforming retrovirus of the chicken. The E26 virus is unique in that it produces a transforming fusion protein containing v-Ets and v-Myb (see below). It is now clear that there is a large family of *ets*-related oncogenes present in vertebrate and lower organisms (Table 5–10). Ets proteins have a conserved DNA-binding domain, but the DNA-binding motif differs from other DNA-binding proteins in that the typical zinc fingers, leucine zipper, and helix-turn-helix motifs appear to be absent. The Ets-family proteins are transcription factors, and some of them (e.g., Ets-1 and Ets-2) cooperate with the Jun/Fos (Ap-1) transcription factor, whereas others (e.g., Elk-1 and SAP-1) act by forming complexes with the serum response factor (SRF) (reviewed in Reference 401). An Ets binding site (EBS) has been identified in a number of regulatory elements of genes, and all of these contain a consensus GGAA or GGAT sequence. EBS sequences have been found in regulatory elements for interleukin-2, SV40, HTLV-1, stromelysin-1, c-*fos*, T-cell receptor α genes, and a number of other genes. The v-Ets protein differs from c-Ets in that v-Ets has a different C-terminus that appears to alter its DNA-binding affinity and make it a transforming protein. Since Ets-1 and Ets-2 bind to the promoters of the stromelysin-1,

Table 5–10. The *ets* Gene Family

Protein	Source	Molecular Mass (kDa)	Amino-Acid Homology to ETS Domain of Ets-1 (%)	Human Chromosomal Location	Expression and Features
Ets-1	Human Mouse Chicken	39-52 63 54/68	100	11q23	Elevated expression in thymus and endothelial cells; phosphorylated; alternatively spliced; positiviely autoregulates transcription
Ets-2	Human Chicken	58/62	90	21q22	Expression was induced following macrophage differentiation and T-cell activitation; alternatively spliced; phosphorylated
Erg	Human	41/52	70	21q22	Alternatively spliced; 98% homologous to Fli-1
Fli-1	Human Mouse	51	68	11q23	Activated by proviral insertion of Friend MuLV: 98% homologous to Erg
Elk-1	Human	60	76	Xp11.2	ETS domain located in the amino terminus of the protein; forms ternary complex with SRF; shows three regions of homology with SAP-1
SAP-1 a/b	Human	58/52	75	ND	SRF accessory protein 1, which, like Elk-1, forms a ternary complex with SRF over the c-fos SRE; contains three regions of homology to Elk-1, including the ETS domain, which is located in the amino terminus of the protein; the two isoforms, SAP-1a and SAP-1b, differ in their carboxyl termini
Spi-1/PU1	Human Mouse	30	38	11p11.22	Activated in Friend erythroleukemia by proviral insertion of the SFFV; normal expression of the PU-1 transcription factor is restricted to B cells and macrophages
E74A/B	*Drosophila*	110/120	50	*Drosophila* chromosome 3L74EF	E74A is induced by ecdysone and regulates the expression of E74B, which is also *ets* related
Elf-1	Human	68	50	ND	The ETS domain is the human homologue of the E74A protein of *Drosophila*; binds to the NF-AT and NFIL-2B sites in the interleukin-2 promoter and the human immunodeficiency virus 2 LTR
GABP-α	Rat	51	82	ND	High-level expression in rat thymus; complexes with GABP-β, which contains ankyrin repeats, and is related to the Notch protein
D-Elg	*Drosophila*	15	64	*Drosophila* chromosome 3R97D	Contains only a DNA-binding domain; maternally expressed message and also expressed throughout embryogenesis
PEA3	Mouse	68	63	ND	Expressed in mouse brain and epididymis and in fibroblast and epithelial cell lines; down-regulated in embryonic cell lines in response to retinoic acid–induced differentiation
TCF1-α	Human	55	ND	ND	Very limited homology to ETS domain exists within the HMG box of this factor; expression is restricted to the thymus and is induced following T-cell activation; regulates activity of the TCRα enhancer

Abbreviations: GABP-β, GA-binding protein; HMG, high mobility group; LTR, long terminal repeat; MuLV, murine leukemia virus; ND, not determined; NF-AT, nuclear factor of activated T cells; NFIL-2B, nuclear factor of interleukin 2B; PEA3, polyomavirus enhancer activator 3; SFFV, spleen focus forming virus; SRF, serum response factor; TCR, T-cell receptor.

(From McLeod et al.[401])

collagenase, and urokinase plasminogen activator genes, it has been postulated that Ets proteins play an important role in modulating degradation of the extracellular matrix and that this may play a role in cancer metastasis.[401]

As for other oncogenes, the transforming ability of v-*ets* is complemented by other oncogenes. For example, v-*ets* and v-*myb* co-expression results in an increased transformation of erythroid cells compared to either gene individually.[402] The human *ets*-1 and *ets*-2 genes are translocated in some forms of acute leukemia; both t(11:4) and t(21;8) translocations have been observed, findings suggesting that these genes may be deregulated by these gene rearrangements.

As noted earlier, in a high percentage of human prostate cancers there is a translocation that produces a fusion gene containing androgen-responsive promoter elements and the Ets oncogenes *ERG* or *ETV1*.[252]

bcr/abl

The v-*abl* oncogene was identified as the transforming gene of the Abelson murine leukemia virus and was shown to have tyrosine kinase activity.[403] In comparison to its cellular homologue c-*abl*, the v-*abl* kinase activity is a deregulated chimeric protein.

An intriguing discovery led to our understanding of the role of the c-*abl* tyrosine kinase in human cancer: the c-*abl* gene was translocated from its normal position on chromosome 9 into a sequence called *bcr* (breakpoint cluster region) on chromosome 22, producing the Philadelphia chromosome seen in chronic myeloganous leukemia and some forms of acute lymphocytic leukemia in humans (reviewed in Reference 404 and 405). The hybrid *bcr/abl* gene of the Philadelphia chromosome produces a hybrid 210 kDa phosphoprotein in CML cells and a 185 kDa phosphoprotein in ALL cells. Both of these proteins have the same c-Abl component, in which exon 1 of the c-*abl* gene has been lost, but differ in the Bcr component because the 185 kDa form results from a breakpoint further downstream, resulting in the loss of certain exons of *bcr*. Both Bcr/Abl forms have deregulated tyrosine kinase activity and both forms cause hematopoietic malignancies when placed as transgenes into transgenic mice, although the 185 kDa form

tends to produce lymphoid malignancies[406] and the 210 kDa form favors production of myeloid leukemias[405] in the recipient mice.

Because both the p210 and p185 Bcr/Abl proteins as well as the p160 transforming v-Abl protein lose the exon 1 component of the c-*abl* gene product, it was assumed that nothing more was needed for their transforming activity. But this assumption turned out not to be the case. Additional key changes in the transforming proteins are the substitution of the Bcr sequence in place of the exon 1 component and mutation or deletion of parts of the SH3 domain of the Abl protein, which has been shown to up-regulate the Abl tyrosine kinase activity.[404] Overexpression of c-Abl itself can also lead to excess tyrosine kinase activity in cells, but this doesn't seem to be a factor in human cancer. Overexpression of c-*abl* in NIH/3T3 cells inhibits growth by causing cell cycle arrest, which suggests that c-*abl* itself has a tumor suppressor function like Rb and p53.[407]

The deletion or mutation of regulatory components of oncogene products is by now becoming familiar to the reader, and Bcr/Abl provides yet another example. There is a cellular inhibitor that interacts with normal c-Abl to regulate its activity. The ability of this normal regulator to bind to c-Abl is inhibited by the substitution of the Bcr component for the exon 1 component of c-Abl. Some of the details of this loss of a regulatory mechanism have been worked out.[404,408] Amino acid sequences within the first exon of Bcr activate the transforming potential of Bcr/Abl. Bcr/Abl forms complexes with our old friend GRB2, which binds to Bcr/Abl via its SH2 domain by interacting with a sequence containing a phosphorylated tyrosine at position 177 in the Bcr first exon. This Bcr/Abl-GRB2 interaction activates the Ras signal transduction pathway described above. If tyrosine-177 of Bcr is mutated to phenylalanine, the binding of Bcr/Abl to GRB2 is blocked, the Ras pathway is not activated, and the ability of Bcr/Abl to transform primary bone marrow cultures is abrogated.[408] Thus, it appears that the normal regulatory mechanism for the c-Abl tyrosine kinase, and presumably its protein substrates, is substituted by one that activates the Ras system to an inappropriate degree, leading to stimulation of the kinase cascade favoring cell proliferation.

There is a large body of experimental and clinical evidence that the Bcr/Abl fusion protein is both an initiating event and one involved in maintenance of the malignant state in CML. This situation is unusual in oncology. Most cancers, particularly solid tumors, have multiple genetic changes, as noted above. It is not that CML cells don't have other genetic abnormalities; they do. However, the expression of the Bcr/Abl kinase is the key event, providing a unique target for an anticancer therapy. The development of the drug ST1-571 (Gleevec) is unique in that it is the first highly successful, small-molecule "molecular-targeted" therapy. In CML patients, it induces a high rate of remissions, by selectively inhibiting the Abl kinase. "Selectively" is used here to mean that ST1-571 only inhibits a few kinases: Abl, PDGF receptor kinase, and c-kit (and maybe a few others). This effect is somewhat surpising because the drug binds to the ATP-binding domain, which all kinases have. The specificity occurs apparently because ST1-571 binds to Abl when the kinase is in an inactive, or "off," conformation, which is structurally different from the "off" conformation of other kinases such as Src (reviewed in Reference 409). Unfortunately, resistance develops even to this targeted therapeutic, likely because of mutations in the kinase domain of Abl that diminish binding of ST1-571 and/or amplification of the bcr/abl gene.

myb

As noted above, the E26 chicken acute transforming virus contains a fusion protein of two transforming oncogenes, v-ets and v-myb. v-myb is also found in avian myeloblastosis virus (AMV). Both of these retroviruses block monocyte–macrophage differentiation of infected cells. These viruses encode a truncated version of c-Myb, the normal proto-oncogene, that is a highly conserved nuclear phosphoprotein involved in the differentiation of hematopoietic cells (reviewed in Reference 410). c-Myb is expressed in immature hematopoietic cell lineages and down-regulated during terminal differentiation of blood-forming cells. Constitutive expression of c-Myb blocks this differentiation event. Both c-Myb and v-Myb act as transcription factors, but v-Myb has lost a phosphorylation site for the nuclear kinase CK-II, which site when phos-

phorylated in c-Myb inhibits its binding to DNA.[410] Expression of c-Myb is important for the response of T lymphocytes to PHA and antigenic stimulation, and it appears to be deregulated by overexpression in human T-cell leukemia, since down-regulation of its expression by antisense oligonucleotides blocks DNA synthesis in T leukemia cells taken from patients.[411]

bcl-2

The bcl-2 proto-oncogene is activated by a common chromosomal translocation observed in non-Hodgkin's B-cell lymphomas, namely the t(14;18)(q32:q21) translocation. This event juxtaposes the bcl-2 gene (so called because it was identified in B-cell lymphomas) from chromosomal locus 18q21 next to the immunoglobin heavy-chain (IGH) locus at 14q32, resulting in increased expression of the bcl-2 gene.[412] Thus, activation of this gene is similar to that of myc gene activation in Burkitt's lymphoma in that the abnormally high levels of expression result from the placement of the gene under the influence of the IGH enhancer. Among oncogenes, bcl-2 has a unique action in that it enhances lymphoid cell survival by inhibiting programmed cell death (apoptosis) rather than stimulating cell proliferation. Thus, its primary action appears to allow B cells to accumulate by prolonging their survival.

The bcl-2 gene undergoes translocation in approximately 85% of follicular lymphomas, 20% of diffuse large-cell lymphomas, and 10% of B-cell chronic lymphocytic leukemias.[412] Experiments with transgenic mice suggest that B cells are the primary target for abnormal expression of this gene because the most pronounced effect in the mice receiving the bcl-2 transgene was clonal expansion of B cells, with a lesser expansion of the T-cell pool.[413] However, a survey of other tissues indicates that Bcl-2 is expressed in a variety of cell lineages such as gastrointestinal epithelium, skin, and developing nervous system tissue (primarily in the stem cell or proliferating compartments of these tissues), suggesting a broader role for Bcl-2 in sustaining the progenitor cells of various cell lineages (reviewed in Reference 414).

The bcl-2 gene encodes two proteins, one of 26 kDa called Bcl-2α, and one of 22 kDa called

Bcl-2β, which result from alternative splicing of the mRNA.[412] Other *bcl*-2-related genes have also been identified, *bcl*-X$_L$ and *bax*, which can render cells resistant to apoptosis and prevent *bcl*-2 overexpression from preventing apoptosis, respectively (see Apoptosis section, Chapter 4). Thus there are both positive and negative regulators of Bcl-2 function. Not all mechanisms involved in induction of apoptosis can be prevented by Bcl-2. For example, Bcl-2 blocks apoptosis in hematopoietic cell lines deprived of IL-3, IL-4, or GM-CSF as growth factors, but not in cell lines dependent on IL-2 or IL-6,[413] and Bcl-2 can prevent apoptosis in embryonic neurons deprived of nerve growth factor (NGF) but not those dependent on another growth factor.[414] These data suggest that there are multiple mechanisms for inducing and/or preventing apoptosis. One way to explain this is that cells may have a variety of mechanisms to regulate *bcl*-2 gene expression or Bcl-2 protein function.

Oncogenic tyrosine kinases increase expression of the anti-apoptotic protein Bcl-X$_L$, and this causes resistance to DNA-damaging anticancer drugs that induce apoptosis (reviewed in Reference 415). This is true for both receptor tyrosine kinases such as up-regulated EFG receptor (EFGR) and non-receptor tyrosine kinases such as Src and Bcr/Abl. Part of the action of ST1-571 is due to its ability to inhibit the Bcr/Abl-induced increase in Bcl-X$_L$.[415] Another potential way to circumvent drug resistance may be to block oncogenic tyrosine kinase stimulation of Bcl-X$_L$.

NF-κB/rel

NF-κB was originally identified as a nuclear protein that is bound to the κB site in the immunoglobin enhancer in B-lymphoid cells.[416] It was later found that NF-κB is involved in the regulation of a large number of genes in different cell types (reviewed in Reference 417). NF-κB was the first oncogene transcription factor whose functional regulation was found to depend on its cellular localization rather than its level of transcription. NF-κB is held in an inactive form in the cell cytosol in a complex with an inhibitor protein IκB. After treatment of cells with a variety of agents that induce NF-κB activation, including phorbol ester, lipopolysaccharide,

tumor necrosis factor α, double-stranded DNA, and IL-1, the IκB protein is phosphorylated by a cytosolic kinase. This phosphorylation disociates IκB from NF-κB, releasing the latter, which is then translocated to the nucleus via a now-exposed translocation signal. Once in the nucleus, NF-κB can bind to its regulatory DNA sequences and induce the transcription of several genes including cytokines, cytokine receptors, MHC antigens, serum amyloid A protein, and viral gene expression of HIV-1, cytomegalovirus, and SV40.[417]

NF-κB consists of two heterodimeric proteins of 50 kDa (p50) and 65 kDa (p65). The p50 subunit contains the DNA-binding site, and p65 is required for binding to IκB. Interestingly, it was later found that NF-κB has high sequence homology with the proto-oncogene c-*rel* and its viral counterpart v-*rel* as well as with the gene *dorsal* in *Drosophila*.[417,418] Part of this homology includes DNA-binding and dimerization domains, and the dimerization of p50/p65 is apparently required for gene activation.[419]

The oncogene v-*rel* is carried by the reticuloendotheliosis virus strain T, which causes acute leukemia in turkeys. The transforming and immortalizing properties of v-*rel* are related to some small deletions and 14 amino acid substitutions that distinguish it from c-*rel*. The proto-oncogene c-*rel* has been cloned from turkey, mouse, and human cells, and it appears to require cytosol-to-nucleus translocation to be active, just as NF-κB does.

It is now known that the NF-κB/Rel proteins are a family of transcription factors whose members share homologous DNA-binding and dimerization domains. In lymphoid tissues of transgenic mice, the NF-κB p50/Rel B protein heterodimer constitutively activates a reporter gene, whereas the NF-κB p50/p65 heterodimer activates the reporter gene in mouse embryo fibroblasts, suggesting that different members of the NF-κB/Rel family of transcriptional activators are involved in gene activation in a tissue-specific manner.[419]

erbA

Yet another functional type of oncogene is represented by the *erb*A gene, originally identified as the v-*erb*A gene in avian erthryoblastosis virus

(AEV), which induces erythroleukemias and fibrosarcomas in chickens.[420] AEV carries two oncogenes, v-*erb*A and v-*erb*B, that cooperate in the transforming action of AEV. The v-*erb*B gene product is a truncated version of the EGF receptor with tyrosine kinase activity (see below). In contrast, the v-*erb*A gene product is an altered version of the thyroid hormone receptor family of DNA-binding proteins.[421] Other members of this family include the steroid hormone, retinoic acid, and vitamin D$_3$ receptors, all of which have ligand-binding, DNA-binding, and transactivating domains. Although there is some sequence homology in the DNA-binding domains and some overlap in their consensus sequence binding sites, all these receptors bind to specific response elements, called T3 response element (T3RE), glucocorticoid response element (GRE), retinoic acid response element (RARE), and vitamin D$_3$ response element. These hormone receptors can either stimulate or inhibit gene expression, depending on the cell type and the response element involved. Thus, they are "content dependent" in their action.

The v-*erb* oncogene is a highly mutated version of its cellular homolog c-*erb*A, having truncations at both the N- and C-terminis and several scattered point mutations. This alteration results in loss of binding of thyroid hormone T3, but it retains its DNA-binding capacity.

Two mechanisms of transforming action have been proposed for v-*erb*A. One is that it acts as a dominant-negative repressor of ligand hormone receptors such as T3 receptor or retinoic acid receptor (RAR), which can act as growth-slowing, differentiating agents in certain cell types. Evidence for this comes from the observation that when T3 receptor and v-ErbA are co-expressed in the same cells, v-ErbA functions as an antagonist of T3 action.[421] A second proposed action for the transforming ability of v-*erb*A is based on data showing that the v-ErbA oncoprotein blocks an RAR-induced function in slowing cell proliferation.[422,423] This slowing of proliferation appears to occur via repression of AP-1 transcriptional activation. Recall that AP-1 is the Fos/Jun transcriptional activation complex that is turned on by a number of mitogenic signals via the Ras-Raf-Mek-MAP kinase pathway. Thus, RA, by binding with its receptor, shuts this mechanism off by interaction with the AP-1 complex, inhibiting cell proliferation, and v-ErbA reverses this by blocking the ability of RAR to carry this out.[422]

sis

The *sis* oncogene encodes an oncoprotein that mimics a growth factor. The history behind this statement starts in July 1983, when two startling articles, one in *Science*[424] and the other in *Nature*,[425] proposed a direct link between growth-regulating factors and oncogene products. These reports indicated for the first time how an oncogene product could directly stimulate cell proliferation.

As described in Chapter 4, platelet-derived growth factor is made up of dimers of two distinct chains, PDGF-A and PDGF-B, with subunits of about 14,000 to 17,000 Daltons. PDGF is derived from platelets, and is a potent mitogen for connective tissue and glial cells in culture. The amino acid sequence of PDGF was reported in May 1983 by Antoniades and Hunkapiller.[426] Russell Doolittle, who was establishing a computer bank of known protein sequences, plugged in the data on PDGF soon after reading the report of its sequence. What fell out was the sequence of the simian sarcoma virus (*sis*) oncogene, which had previously been sequenced by Aaronson and colleagues.[427] The homology was strong: 87% of 70 amino acids in the sequence of PDGF-B were homologous to the *sis* oncogene product, and what discrepancies there were could be explained by expected species variation, since PDGF was isolated from outdated human platelet preparations, and simian sarcoma virus (SSV) was isolated from a fibrosarcoma of a woolly monkey. The cloned transforming gene of SSV, v-*sis*, is known to produce a 28,000 Dalton gene product in transformed cells.[426] Similarly, Waterfield et al.,[425] who were also working on the sequence of PDGF, found a region of 104 contiguous amino acids virtually identical to the predicted sequence of p28[sis], the 28,000 Dalton protein isolated from cells transformed by the cloned transforming gene, v-*sis* of SSV. This finding led to the speculation that continued production of this growth factor by v-*sis*-transformed cells could account for the malignant phenotype induced by this gene.

Since the cellular homologue (c-*sis*) of the viral *sis* oncogene is present in the human genome as a single gene, in all likelihood SSV or one of its ancestors picked up the normal cellular gene coding for PDGF. Because PDGF primarily stimulates the proliferation of cells of connective tissue origin, such as fibroblasts, smooth muscle cells, and glial cells, it was logical to look for expression of *sis* in cell lines derived from cancers of connective tissue cells. Investigators found that a PDGF-like product is often produced by tumor cells of connective tissue origin, whereas cancers derived from epithelial cells usually do not make it. Furthermore, cell lysates and conditioned medium of SSV-transformed cells growing in culture contain a PDGF-like mitogenic factor[428,429] that can be partially neutralized by anti-PDGF antibodies. The PDGF-like material produced by SSV-transformed cells binds to cells with PDGF receptors in a manner competitive with PDGF, and the ability of SSV-transformed cells to grow in nude mice correlates with the production of p28[sis] by the cells growing in culture.[429] Anti-PDGF antibodies also inhibit the growth of high-PDGF-producing SSV-transformed cell lines. Interestingly, SSV-induced tumors appear to be restricted to the cell types that have PDGF receptors (e.g., gliomas, fibrosarcomas).[430]

The v-*sis* gene actually encodes a 271–amino acid protein whose N-terminal 51 amino acids are derived from the viral envelope protein; the remainder is derived from c-*sis*, the cellular homolog of v-*sis*. It is the c-*sis* gene that encodes a polypeptide precursor of the B chain of PDGF. The production of the v-*sis* gene product is now known to be more complicated than originally thought (reviewed in Reference 431). In SSV-transformed cells, the v-*sis* gene product is synthesized as a 36 kDa glycoprotein with one N-linked oligosaccharide chain. It then forms a 72 kDa dimer that is proteolytically processed sequentially into p68, p58, and p44 forms, the latter of which is secreted but most of which remains bound to the cell surface. Part of the p44 form is cleaved into a 27 kDa form, which most likely accounts for the earlier observation of a product of about 28 kDa being secreted by SSV infected cells. The high affinity of the secreted v-*sis* gene product for the cell surface suggests a way in which autocrine stimulation of v-*sis* transformed

cells could occur, i.e., by release of the v-*sis* protein and immediate binding to cell surface PDGF receptors. However, other data suggest that the "autocrine loop" is not extracellular but intracellular, in that a p27 form is also generated intracellularly. It has been postulated that this form binds nascent PDGF receptors, creating a signal transducing signal without ever having to exit the cell.[431]

erbB

A second link between oncogenes and growth factors came from studies of the structure of the EGF receptor. Downward et al.[432] reported that the amino acid sequence of six peptides derived from the EGF receptor isolated by immunoaffinity purification from cultured human epidermoid carcinoma A431 cells and from human placenta was identical, at 74 of 83 residues sequenced, to the transforming protein of the v-*erb*B oncogene of avian erythroblastosis virus (AEV). However, the *erb*B sequence was missing a large segment of the amino-terminal end of the EGF receptor. It is now known that the v-*erb*B oncogene encodes a truncated EGF receptor, containing only the transmembrane hydrophobic region and the cytoplasmic tyrosine kinase domain, but not the cell surface domain associated with ligand binding. These results suggest that the v-*erb*B oncogene could transform cells through an uncontrolled receptor function in which, even in the absence of ligand binding, a constitutive expression of receptor function could occur. It has been difficult, however, to demonstrate a chronic, constitutive activation of EGF receptors in AEV-transformed cells, partly because the tyrosine kinase activity of the *erb*B gene product appears to be quite low.[430] It is possible that a very specific subset of substrates are phosphorylated by ErbB kinase activity and that they are difficult to detect; it is also possible that the transforming activity of v-*erb*B is pronounced only in the presence of other activated oncogenes.

erbB-2 (HER-2/neu)

The *neu* oncogene was initially identified in rat neuroblastomas, and a human homolog called c-*erb*B-2 (also called Her-2) was later found to be amplified in some human adenocarcinomas (reviewed in Reference 433). ErbB-2 is similar to the ErbB oncoprotein discussed above in that it is an altered, truncated version of the EGF receptor and has intrinsic tyrosine kinase activity that can carry out autophosphorylation as well as other phosphorylation steps. A difference is that ErbB is a 170 kDa protein, whereas ErbB-2 is a 185 kDa protein. A single amino acid change (val^{664}→ Glu^{664}) in the transforming rat *neu* gene (compared to its normal c-*neu* counterpart) significantly increases its autophosphorylating protein kinase activity and turns it into a potent transforming gene.[434] The transforming potential of the *neu* oncogene can also be activated by overexpression due to gene amplification or deletion of part of the N-terminal extracellular domain.[433–435] It is noteworthy that overexpression of c-*neu* and the EGF receptor in rodent fibroblast lines act synergistically to induce transformation, whereas overexpression of either gene alone doesn't do this or does it weakly. It may be that overexpression of two normal cellular kinases can activate the neoplastic transformation process.[433]

Amplification of the Her-2/*neu* gene has been implicated as a factor in the progression of human cancer, particularly breast cancer, and in transgenic mice expression of *neu* driven by a mouse mammary tumor virus (MMTV) promoter induces mammary tumors.[436,437] Whereas increased tyrosine kinase activity is the presumed mechanism for the transformation of cells by the *neu* oncogene, in human breast epithelial cells transformed by transfection with *neu* exhibit an increase in specific protein tyrosine phosphatases.[435] This result is likely due to a cellular compensatory response to an increased protein tyrosine phosphate "load." Indeed, there is evidence to indicate that increased tyrosine phosphatase activity (particularly for two called LAR and PTIB) can counteract the transforming potential of tyrosine kinase oncogenes and may act as tumor suppressor genes. In support of this idea are the findings that the colorectal tumor suppressor protein DCC has some structural homology to LAR[438] and that the LAR gene maps to a region on chromosome 1p32-33 that is thought to contain a breast cancer tumor suppressor gene.[439]

For breast cancer patients, a correlation of relapse and poor survival with amplification of the HER-2/*neu* oncogene has been observed.[440] Administration of the anti-HER 2/*neu* antibody Herceptin has been an effective treatment strategy for those patients who have overexpression of the marker. Overexpression of HER-2/*neu*, determined by immunohistochemical staining of tissues or in situ hybridyation techniques, has also been shown to be associated with poor survival in advanced ovarian cancer[441] and gastric cancer.[442] Overexpression of HER-2/*neu* mRNA is also a marker for intrinsic drug resistance in non–small cell lung carcinoma cell lines.[443]

Other Growth Factor or Growth Factor Receptor Oncogenes

A number of other oncogenes that have growth factor or growth factor receptor actions have been discovered, and it is likely that many more will be found. Indeed, whenever there is a mutation, translocation, amplification, or other means of overexpression, it is possible, perhaps even highly likely, that control of cell proliferation will be deregulated, leading to a hyperplastic proliferation of cells with the concomitant increased chance of malignancy due to the genetic instability that can follow rapid cell division. A summary of characteristics of some growth factor or growth factor receptor–like oncogenes is listed below.

FMS

The v-*fms* oncogene is contained in feline sarcoma viruses. It and its c-*fms* proto-oncogene counterpart represent different forms of the hematopoietic colony-stimulating factor M-CSF. c-*fms* codes for the normal receptor and the v-*fms* protein product has scattered point mutations and deletions and substitutions in the C-terminus that activate its transforming, tyrosine kinase activity.[444]

KIT

The c-*kit* proto-oncogene is the normal cellular homologue of v-*kit* found in the H-Z4 feline sarcoma virus. It codes for a transmembrane tyrosine kinase and is the receptor for stem cell factor (SCF). c-*kit* plays a key role in hematopoiesis.[445] Mutant kit protein isoforms are expressed in gastrointestinal stromal tumors (GIST).

TRK

This oncogene was first discovered in a human colon carcinoma biopsy. It is also a transmembrane tyrosine kinase. It becomes activated by chromosomal rearrangement resulting in replacement of its extracellular domain by unrelated sequences or by other recombination events.[446] The product of the proto-oncogene c-trk is now known to be a crucial component of the nerve growth factor receptor. High levels of c-trk expression and normal gene copy numbers or N-myc are associated with a favorable prognosis for patients with neuroblastoma.[447]

MET

The met proto-oncogene encodes p190[met], a membrane-associated tyrosine kinase that is the receptor for hepatocyte growth factor (HGF), also known as scatter factor. HGF, when bound to its receptor, stimulates cell motility, extracellular matrix invasion, and in some cells, a cell-proliferative response.[448] In response to HGF binding to p190[met], autophosphorylation and activation of the phosphoinositol pathway and of Src-like kinases ensue. Although the function of met in human cancer is unclear, it may be postulated that inappropriate activation or overexpression of met is related to generation of an invasive and metastatic phenotype.

POKEMON

Pokemon is a recently characterized proto-oncogene that belongs to a family of genes coding for transcriptional repressors. Pokemon-encoded proteins have essential roles in development, differentiation, and oncogenesis (reviewed in Reference 449). Pokemon stands for POK erythroid myeloid ontogenic factor. It was originally identified as a protein that binds specifically to a HIV type 1 promoter element. Pokemon inactivation impairs cellular differentiation in multiple tissues and is embryonic lethal in mice. It has been shown to induce oncogenic transformation in mouse embryo fibroblasts and to act like a proto-oncogene in cooperation with other "classic" oncogenes. Pokemon acts by repressing transcription of the tumor suppressor gene arf and is aberrantly overexpressed in human T-cell and B-cell lymphomas.[450]

CELLULAR ONC GENE EXPRESSION DURING NORMAL EMBRYONIC DEVELOPMENT

As previously noted, the ubiquitousness of c-onc genes in vertebrate organisms and their conservation through eons of evolution suggest an important role in growth and development of the normal organism. These genes were almost certainly conserved, not because they can produce uncontrolled proliferation but because they play some key role in development. A variety of studies support this thesis.

The expression of eight cellular onc genes were examined during embryonic and fetal development of the mouse using four avian (v-myc, v-erb, v-myb, and v-src) probes, two murine (v-mos and v-H-ras) probes, one feline (v-fes), and one primate (v-sis) viral cDNA probe to detect homologous sequences in cellular mRNA from various stages of development.[451] Five homologous c-onc genes detected by these probes were expressed during embryonic development: c-sis expression peaked on about day 8 of prenatal development and continued to be expressed at lower levels throughout gestation; c-myc, c-erb-A, and c-src expression peaked in the latter half of fetal development; and c-H-ras was expressed throughout embryonic development of the mouse. Although sequences homologous to v-myb, v-mos, and v-fes are contained in the mouse genome, transcription of these genes was not detected during development. In some cases, expression of certain c-onc genes has been linked to the development of specific tissues. For example, c-fos expression has been detected in placenta, c-abl in development of male germ cells and lymphoid tissues, and c-H-ras in a variety of developing tissues including bone, brain, gastrointestinal tract, kidney, lung, skin, spleen, testis, and thymus;[452] these results suggest a protean role for this gene.

There is a spatial and temporal pattern to the expression of c-myc and c-sis in developing human placenta. Expression of both genes peaks in first-trimester placental tissue and declines thereafter, in parallel with the release of platelet-derived growth factor and expression of PDGF receptors on cytotrophoblast cells.[453] Both genes are expressed most abundantly in the cytotrophoblast. This suggests that human placenta has autocrine regulation, with the ability to both produce PDGF and respond to it by increasing expression of c-myc, the expression of which correlates with cell proliferation in a number of tissues.

Expression of other c-onc genes has also been linked to differentiation of specific cell types. c-fos and c-fms gene transcription is turned on in differentiating human monocytes.[454,455] In contrast, c-myc gene expression is decreased when cultured human promyelocytic leukemia cells (HL-60) are induced to differentiate in culture, and when the stimulus for differentiation is removed, c-myc mRNA is elevated again,[456] which suggests that c-myc expression correlates with the proliferative phase and c-fms and c-fos expression with a later differentiating phase of monocyte development. In support of this idea is the finding that in regenerating liver, induced to undergo rapid cell proliferation in response to partial hepatectomy, there is a rapid onset of c-myc transcription that increases 10- to 15-fold above the normal resting level within 1 to 3 hours after partial hepatectomy and rapidly declines after 4 hours posthepatectomy.[457] In the same experimental protocol, c-H-ras transcription increased 12 hours after partial hepatectomy, peaked at 36 hours, and returned to control levels by 72 hours.[458]

Continued expression of onc genes, however, can block terminal differentiation of normal cells. When the v-src gene under viral promoter control is introduced into cultures of mouse bone marrow cells, a dramatic increase occurs in the self-renewing stem cell (CFU-S) compartment, along with a decrease in the appearance of mature granulocytes.[459] Similarly, normal mouse skin keratinocytes, when infected with Kirsten or Harvey sarcoma virus, do not progress through a complete maturation program when the v-ras gene is expressed.[460] When these cells are induced to differentiate by addition of calcium ions, they progress only to an early reversible stage of differentiation; if subsequently treated with the tumor-promoting phorbol ester TPA, such cultures revert back to a less mature cell type.

DNA TUMOR VIRUSES

The oncogenic DNA viruses consist of three main groups: papovaviruses, adenoviruses, and herpesviruses. Examples of papovaviruses are the papilloma viruses of rabbits (Shope) and other species including humans (human wart virus), SV40 virus of monkeys, and polyoma virus of the mouse. Adenoviruses have been isolated from various animal species, and a number of them have been shown to be tumorigenic in newborn animals. Oncogenic viruses of the herpesvirus class include Epstein-Barr virus, suspected of causing Burkitt's lymphoma and nasopharyngeal carcinoma in humans, the virus that causes Lucké frog renal carcinoma, and a leukemogenic virus in chickens (Marek's disease).

SV40 and Polyoma

The papovaviruses are small icosahedrons containing 3 to 5×10^6 Daltons of DNA, enough to code for three to six proteins. Of this group, SV40 and polyoma virus have been studied the most. The SV40 virus was discovered in 1960 in rhesus monkey kidney cell cultures used to produce the early polio vaccine.[461] The virus was inadvertently inoculated into thousands of people before its presence became known. Later it was shown that SV40 could produce tumors after injection into weanling hamsters[462] and that it could also transform human cells in culture.[463] However, no human disease, including cancer, has been shown to be caused by SV40 virus, even in this inadvertent human experiment. Because SV40 and the other papovaviruses contain DNA, the flow of genetic information goes directly from DNA to RNA to protein without requiring reverse transcriptase, as do the RNA viruses.

A number of things are known about the molecular biology of SV40 gene expression.[464] Similar molecular events occur during infection and cellular transformation with other papovaviruses. The virus enters the cell by the action of its coat proteins. Viral DNA then enters the nucleus of the cell, and it is transcribed in two "waves" to produce "early" and "late" mRNAs. Transcription of the early SV40 genes is required for synthesis of viral proteins involved in the replication of SV40 DNA. The early region genes also contain the information needed for cell transformation and code for the intranuclear T antigen. Late mRNA is transcribed after viral DNA replication and codes for the viral structural proteins. The SV40 viral DNA is covalently integrated into transformed host cell DNA, and the integrated sequences can be portions as well as full copies of the SV40 genome. The DNA cleavage takes place before integration occurs at various nucleotide sequences in both the viral and host cell DNA, depending on the type of transformed host cell. Mature SV40 can be rescued by a variety of methods from many transformed cells that do not produce virus under usual culture conditions. Transformed cells produce early viral mRNA and T antigen, but do not replicate viral DNA and do not produce late viral mRNA or late structural viral proteins. Thus, T antigen is required for initiation of viral DNA synthesis, for the accompanying induction of host cell DNA synthesis, and for both the establishment and maintenance of the transformed state. Because T antigen appears to be the transforming protein of the transforming papovaviruses, a considerable amount of research has gone into characterizing this protein in both SV40 and polyoma virus.

Originally, the T antigen of SV40 was thought to be one protein of about 100,000 molecular weight. This would account for most of the coding capacity of the early gene region. Later it was shown that in vitro cell-free translation of early viral mRNA isolated from infected cells produced a protein of 17,000 molecular weight that was immunoprecipitable with antiserum to T antigen.[465] These forms are called *large* and *small* T antigen. The gene coding for both forms of T antigen is called the A gene. The two mRNAs that code for large and small T have the same 5′ and 3′ ends, and thus appear to have arisen from differential splicing of the A gene transcription product.[466] For transformation to occur after SV40 infection, expression of the large T antigen appears to be a crucial event.

The SV40 large T antigen in virus-infected cells regulates not only SV40 gene transcription but also the transcription of cellular genes such as thymidine kinase, ribosomal RNA genes, and a whole subset of other cellular genes, the transcripts of which are elevated in SV40-transformed mouse cells.[467] Activation of cellular genes may be a general feature of oncogenesis induced by DNA viruses, in contrast to RNA oncogenic viruses, which carry their own activated transforming genes. However, under certain circumstances, SV40 viral sequences can

transform cells. For example, when various segments of SV40 early-region DNA were linked to a retroviral vector from Moloney murine leukemia virus (MoLV), containing only the LTR region and other regulatory sequences required for MoLV viral propagation, vectors carrying SV40 large T antigen as the only SV40 sequence were able to induce morphologic transformation of primary or established mouse and rat lines with high efficiency.[468] The authors of this study argue that expression of large T antigen by itself is capable of transforming cells, and this conclusion is supported by experiments in transgenic mice. Brinster et al.[469] have microinjected fertilized mouse eggs with plasmids containing SV40 early-region genes and a fusion gene coding for metallothionein, known to be expressed in transgenic animals. SV40 T antigen mRNA was detected at high levels only in tissues showing histopathologic changes, including thymus, kidney, and brain, but the highest levels were seen in brain tumors (of the choroid plexus) that developed in these animals. Later experiments showed that large T antigen expression is sufficient to induce the choroids plexus tumors and does not require the metallothionein fusion gene.[470] In fact, when the SV40 enhancer region is present to direct T antigen expression, tumors specifically occur in the choroids plexus, but when the SV40 enhancer region is deleted and substituted by a metallothionein–human growth-hormone fusion gene, an entirely different pattern of pathology ensues: transgenic mice bearing this hybrid gene develop peripheral neuropathies, hepatocellular carcinomas, and pancreatic islet cell adenomas.[471]

The DNA from SV40 T antigen fusion gene–induced tumors, compared with the DNA of unaffected tissues from the same animals, shows structural rearrangements, changes in DNA methylation patterns, and, frequently, SV40 gene amplification.[469] These results indicate that the enhancer or promoter sequence attached to the T antigen gene has a key role in directing the tissue specificity of T antigen expression and its tumorigenic potential, probably by directing how the T antigen gene is inserted into DNA—that is, by allowing it to be placed in an active, transcribable conformation. Similar results have been obtained by Hanahan,[472] who showed that transfer of recombinant genes made up of regulatory sequences of the insulin gene, fused with sequences of SV40 large T antigen, into fertilized mouse eggs produced tissue-specific expression of large T in β cells of the transgenic mice pancreases, inducing β-cell tumors in these animals. These data strongly suggest that tissue-specific expression of viral-transforming gene sequences, directed by enhancer or promoter elements that have a tissue-specific expression, can produce very specific target-cell oncogenesis.

The transforming T antigen of polyoma virus bears some similarity to that of SV40, but its crucial elements are coded for by a separate gene, the *hr-t* gene, which maps in a position analogous to that deleted in certain early gene deletion mutants of SV40.[466,473] A third type of T antigen has been isolated from polyoma-transformed cells.[474] This has been called *middle T antigen* and has a molecular weight of 55,000. It contains peptides not found in either small or large T antigen and is altered by *hr-t* gene deletions. Cells infected with mutant polyoma viruses that synthesize large, but not middle, T antigen have a normal phenotype and do not induce tumors in vivo, indicating that middle T antigen is more important for the transformation of polyoma-infected cells.

Transformation of normal rat cells with genetic recombinant plasmids derived from polyoma virus that allow selective expression of large T, middle T, or small T antigens indicates that middle T antigen alone is sufficient to transform established lines (already immortalized), but not primary rat embryo fibroblasts, and that large T antigen lacks intrinsic oncogenic activity, but can decrease serum dependence of growth for both normal and transformed cells.[475] Polyoma large T antigen appears to increase the efficiency of the integration of polyoma virus DNA sequences into host-cell DNA and to increase the efficiency of transformation by polyoma, but fully transformed colonies can be obtained in the absence of active large T antigen.[476] Furthermore, a recombinant DNA clone consisting of a replication-defective murine leukemia virus vector, the polyoma early region promoter, and the middle T gene can transform NIH/3T3 cells,[477] indicating that polyoma middle T antigen can act like a dominant transforming gene similar to RNA retrovirus v-*onc* genes in already immortalized cells. The transforming ability of middle T

antigen appears to relate to its phosphorylation state, since introduction of a mutation that inserts a phenylalanine for a tyrosine at residue 315 in middle T decreases phosphorylation as well as transforming activity.[478]

Papovaviruses similar to SV40 and polyoma virus have been isolated from human patients.[464] These viruses (JC and BK) can also induce tumors in newborn hamsters and transform animal cells in culture. Their transforming ability also seems to depend on expression of T antigen.

The transforming large T antigen of SV40 and middle T antigen interact with a number of cellular proteins. These interactions are involved in the transforming activity of these viral antigens (reviewed in Reference 479). For example, the binding of polyoma middle T antigen to the c-Src protein increases its kinase activity about 20-fold, and middle T antigen mutants lacking the ability to bind c-Src are transformation deficient. Middle T antigen also interacts with phosphoinositol kinase, another important signal transduction system component (see Chapter 4), and this association correlates with the ability of middle T antigen to mediate transformation. SV40 large T antigen binds to under- or nonphosphorylated Rb protein, indicating a way that SV40 large T antigen could prevent this tumor suppressor protein from blocking entry of cells into the cell division cycle.[480] SV40 large T antigen also binds to the tumor suppressor protein p53, an action that appears to mediate cell transformation events. Large T antigen of SV40 interacts with transcription factors such as AP-2, an action that appears to be involved in the turning-on of a gene for nucleic acid synthesis.

Papilloma Viruses E6 and E7

Papilloma viruses also belong to the papovavirus family, but they are somewhat larger than SV40 or polyoma viruses and have a somewhat larger genome (5×10^6 Daltons).[481] They induce benign epithelial tumors in various animal species, including humans, and are sometimes known as "wart viruses." The skin and mucosal tumors induced in animals usually regress, but at least three papilloma viruses have oncogenic potential: the Shope papilloma virus, the bovine fibropapilloma virus, and the bovine alimentary tract papilloma virus. In humans, papilloma viruses are associated with skin warts, anal and genital warts (condylomata acuminata), and oral and laryngeal papillomas. Certain subtypes of human papilloma viruses are strongly associated with cervical carcinoma.

There are 67 distinct human papilloma viruses (HPVs). Of these, a subgroup of about 20 are associated with anogenital tract lesions. Some of them cause condyloma acuminata but are considered low risk (e.g., HPV-6 and -11) because they rarely cause malignancy. Others (e.g., HPV-16, -18, -31, and -33) are considered of high risk because they are associated with high-grade squamous intraepithelial lesions and invasive carcinomas of the uterine cervix. Of the high-risk HPVs, types 16 and 18 have been most intensively studied. The HPVs express transforming oncoproteins called E6 and E7, and those of the most tumorigenic types (HPV-16 and -18) have potent cell-transforming actions. In primary cervical carcinomas and cervical cancer cell lines, the viral genomes of high-risk HPV types are frequently found integrated into the host cells' genome, allowing active transcription of the E6 and E7 mRNA (reviewed in Reference 482). However, although expression of E6 and E7 from high-risk HPVs can immortalize primary epithelial cells in culture, a fully transformed phenotype is only observed after numerous cell passages. Moreover, only a relatively low percentage of women infected with high-risk HPVs develop invasive cervical cancer, although a high percentage of cervical cancers are positive for HPV.[483]

These data indicate that other factors in addition to HPV infection are important in the causation of cervical cancer. While there are some additional associated epidemiological risk factors,[483] at the level of the cell, what appears to happen is a series of progressive events involving genetic instability of cells transformed by high-risk HPVs. One way this could happen is by association of the oncoproteins E6 and E7 with the tumor suppressor genes p53 and Rb. E6 proteins translated from high-risk HPV-16 and -18 E6 genes bind to p53 and cause its degradation by a ubiquitin-mediated process.[484] Since normal p53 is involved in protecting cells from genetic damage from a variety of DNA-damaging agents such as irradiation or chemicals by causing cell cycle arrest and allowing time for DNA repair

(see Chapter 4), it seems logical that destruction or inactivation of p53 could account for the neoplastic progression seen with chronic HPV infection. It has, in fact, been shown that the HPV-16 E6 gene transfected into human cervical epithelial cells disrupts p53-mediated cellular response to DNA damage induced by actinomycin D.[482]

While both E6 and E7 have cell-transforming properties, expression of both is required for efficient immortalization of cells.[484] Thus, a one-two punch appears to be needed. The second punch is provided by E7's ability to bind to and disrupt the action of another tumor suppressor protein, Rb, which is involved in cell cycle regulation. In this regard E7 shares a property with SV40 T antigen (see above) and adenovirus oncoprotein E1A (see below). There are regions of amino acid sequence similarity between these three proteins that are involved in binding to Rb.

It has also been shown that expression of HPV-16 E6 and E7 oncogenes in transgenic mice causes a high incidence of preneoplastic skin lesions and subsequent development of skin carcinomas.[485] Moreover, infection of nonmetastatic mouse tumor cell lines with a retrovirus bearing inserted E6 and E7 genes from HPV-16, but not HPV-6, converted these cells into metastatic ones.[486] This is consistent with the finding that HPV-16 DNA is frequently found in sites of cervical carcinoma metastasis.

The recent development of a vaccine against papilloma viruses may turn out to be a major step for the prevention of this cancer.

Adenoviruses E1A and E1B

The oncogenicity of adenoviruses was first observed by Trentin et al.[487] in 1962, who showed that adenovirus type 12 could produce tumors on inoculation into newborn hamsters. Of the 31 adenovirus serotypes isolated from humans, 3 (type 12, 18, and 31) are highly oncogenic in newborn rodents, 5 (types 3, 7, 14, 16, and 21) are less oncogenic, producing fewer tumors after a longer latent period, and other types (e.g., 1, 2, 5, and 6) do not induce tumors by direct inoculation into animals but can transform cultured rodent cells that produce tumors upon injection into animals.[488] As in the case of the papovaviruses, at least part of the adenovirus genome becomes incorporated into the host genome during transformation, and expression of a virus-induced nuclear T antigen is required for transformation. Another similarity is that the transcription of integrated viral DNA preferentially involves "early" sequences, and correlates with production of the mRNA molecules detected in infected cells before the onset of viral DNA synthesis. Thus, the process of virus-induced transformation, involving transcription of early DNA into early mRNA, which in turn is translated into a T antigen involved in the initiation and maintenance of the transformed state, is common to the oncogenic papovaviruses and adenoviruses. The adenovirus E1A early-region gene can induce immortalization of cells in culture and act in concert with a transforming c-*ras* gene or the polyoma middle T gene to transform cultured primary diploid cells.[489] The adenovirus E1 gene region encodes the E1A and E1B proteins responsible for the oncogenic properties of these viruses, although the E4 region of adenovirus 9 is involved in the production of mammary fibroadenomas, as shown in mice infected with a recombinant virus containing the E4 gene region of that virus.[490] Expression of the E1A region alone can immortalize primary cultures of rodent cells, but co-expression of E1B is required for complete transformation. Activated H-*ras* or polyoma middle T antigen can substitute for E1B to complement E1A in transformation assays, and polyoma large T antigen, members of the *myc* family, or mutated p53 can replace E1A to complement E1B in similar assays (reviewed in Reference 491).

Cellular targets for the E1A and E1B proteins have been identified. E1A binds to and inactivates Rb, and E1B complexes with and disrupts the action of p53. Whyte et al.[491] have shown that the regions of the E1A gene product that bind Rb are precisely the ones required for E1A-mediated cell transformation, strongly suggesting that inactivation of Rb by E1A accounts in a crucial way for the cell transforming activity of E1A. However, E1A is a multifunctional protein: it acts as a transcriptional activator for a number of genes, stimulates DNA synthesis, and induces the production of an epithelial cell growth factor. Hence, its biochemical effects on cells are multifactorial, and a number of these actions could be involved in the loss of growth control seen in E1A-expressing cells.

Hepatitis B Virus

Human HBV infects live cells and causes acute and chronic hepatitis. Chronic HBV infection is a high risk factor for developing hepatocellular carcinoma. The small DNA genome of this virus encodes four genes. The product of a gene called HBVx codes for a protein, pX, that is a transcriptional activator of viral and cellular genes, including N-*myc* and NF-κB.[492,493] The pX protein itself doesn't appear to be able to bind DNA directly but acts via complex formation with the transcription factors CREB and ATF-2.[493] This action as a component of a transcriptional activation event may account in part for the transforming ability of HBV.

Herpes Viruses

The other class of DNA viruses with oncogenic potential are the herpesviruses.[494] These viruses are larger than the papovaviruses and adenoviruses and have a genome that contains information for at least 50 proteins. Hence, discerning which of these gene products is the transforming protein(s) has been difficult. Herpesviruses infect humans and nearly all animal species investigated so far. Humans are subject to infection with five viruses of this class: herpes simplex virus 1 (HSV-1), herpes simplex virus 2 (HSV-2), herpes zoster virus (HZV), cytomegalovirus (CMV), and Epstein-Barr virus (EBV). Human herpesviruses have been under suspicion for some time as causative agents for certain cancers: EBV has been implicated as the responsible agent in Burkitt's lymphoma and nasopharyngeal carcinoma, and HSV-1 and HSV-2 have been suspected as contributing to the cause of cancer of the uterine cervix and possibly of other urogenital and oropharyngeal tumors.

Epstein-Barr virus can immortalize B-lymphoid cells in culture, and in so doing expresses a variety of EBV-determined nuclear antigens (EBNA 1–6). The EBNA-2 protein is involved in the immortalization of B lymphocytes and is localized in the cell nucleus where it functions as a transcription factor to enhance the expression of several viral and host genes.

Expression of the EBNA-2 protein blocks the antiproliferative effect of α-interferon on B cells (reviewed in Reference 495). Interferons may be acting as tumor suppressor genes for B cells, an action which is overcome by EBNA-2.

TUMOR SUPPRESSOR GENES

Historical Perspectives

In the section above, the role of activated oncogenes in causing malignant transformation of cells was discussed. The excitement surrounding this research dominated the scene in cancer cell biology for a number of years. It was simple, relatively clear, and a satisfying way to explain cancer. It also unified a number of theories about how chemicals, irradiation, and viruses could cause cancer. They all converged into one theme: damage to DNA causes point mutations, chromosomal rearrangements, translocations, or amplifications, all of which can lead to the activation of cellular proto-oncogenes that could take over and dominate a cell's behavior turning it into a cell programmed to survive and proliferate.

Thus, the idea was that cancer genes, once activated, were dominant genes and caused a dominant genetic change in cells. There were only one or two flies in the ointment. Back in 1969, Henry Harris and colleagues[496] showed that when malignant cells were fused with nonmalignant cells, most of the hybrid cells were nontumorigenic. If cancer was due to a dominant genetic event, this result didn't make sense at all, hence this observation was virtually ignored for almost 20 years. Another wrinkle in the prevailing theory of cancer causation was Alfred Knudson's report in 1971 of a hereditary form of the eye tumor retinoblastoma, in which some gene carriers acquired bilateral eye tumors, some had unilateral disease, and a small minority had no tumors (reviewed in Reference 497). Moreover, only three to four tumor loci per affected patient were observed. Since there are more than one million cells in a retina, it is a rare cell indeed that actually becomes cancerous even though all the cells carry the defective gene. This observation strongly suggested a second genetic event, the inherited mutation by itself not being sufficient. This led to the "two-hit hypothesis" of Knudson: in hereditary retinoblastoma, one defective gene is inherited as a germline mutation and a second mutation, occurring after conception, is necessary to induce a tumor, whereas in the nonheritary form of the disease, both mutations occur as somatic, post-conception events (see Chapter 2).

There were also other unsettling findings that didn't fit the dominant oncogene theory of cancer. For example, in solid human tumors, in contrast to leukemias and lymphomas, chromosomal deletions were commonly observed. Even when investigators began to be able to detect oncogene mutations and amplifications by sensitive molecular genetic techniques, they could only be found in 15% to 30% of human cancers.[498,499] Thus, a number of investigators began to think more seriously that loss of some inhibitory or regulatory gene function was involved in causing cancer.

A big advance in this theory was made when introduction by microcell transfer of a single human chromosome 11 from a normal human fibroblast into HeLa cells or Wilms' tumor cells resulted in suppression of the ability of these cells to induce progressive tumors in nude mice (reviewed in Reference 500). Subsequent studies have shown deletions in specific regions of chromosomes in a number of human cancers, suggesting that the loss of "tumor suppressor" genetic information is a common event in human malignant disese (Table 5–11). The presence of genes to inhibit uncontrolled cell proliferation helps to explain why human beings only have about a 25% chance of developing a full-blown cancer, even though we experience 10^{16} cell mitoses in a lifetime.

The first tumor suppressor gene cloned was the *rb* gene, the defective gene in retinoblastoma. Cavenee et al.[501] used restriction fragment length polymorphisms (RFLPs) to map the defective gene to chromosome 13q14 and showed that a loss of heterozygosity at this locus in the tumor was due to loss of the normal allele from the unaffected parent.[502] This indicated a germline mutation, uncovered by the loss of heterozygosity, and helped substantiate the Knudson hypothesis. The *rb* gene was subsequently cloned by Friend et al.[503] It is now known that a variety of other human cancers have inactivated *rb* alleles, including sarcomas, small-cell lung, bladder, and

Table 5–11. Evidence of Loss of Genetic Information in Human Cancers°

Tumor Type	Chromosome Region(s) Involved
Wilm's tumor, sporadic	11p13, 11p15
Wilm's tumor, familial	Unknown
Retinoblastoma	13q14
Osteogenic sarcoma[†]	13q14, 17p
Soft tissue sarcoma[†]	13q14
Neuroblastoma	1p, 14q, 17
Glioblastoma multiforme (Astrocytoma)	10, 17p
Bladder carcinoma	9q, 11p, 17p
Breast carcinoma	1q, 11p, 13q, 17p
Colorectal carcinoma	5q, 17p, 18q
Renal cell carcinoma	3p
Multiple endocrine neoplasia type 1	11q
Multiple endocrine neoplasia type 2	1p, 10, 22
Tumors associated with bilateral acoustic neurofibromatosis	22q
Uveal melanoma	2
Melanoma	1, 6
Myeloid leukemia	5q
Small cell lung cancer	3p, 13q, 17p
Non–small cell lung cancer	3p, 11p, 13q, 17p

°Data derived from cytogenetic and RFLP analyses.

[†]Second malignancies in familial retinoblastoma patients.

(From Stanbridge[500])

a few breast carcinomas (reviewed in Reference 504).

A number of other tumor suppressor genes or candidate tumor suppressor genes have been cloned and characterized,[505] and more continue to be discovered as more is learned about cancer cell genetics and the map of the human genome. A single mutation can be sufficient to activate an oncogene (e.g., *ras*); a second is not crucial because there wouldn't necessarily be any particular selective pressure to sustain it. Mutations in *onc* genes are gain-of-function events and lead to increased cell proliferation and decreased cell differentiation. Oncogenes are mutated in a wide variety of human cancers (e.g., *ras, myc*). In contrast, tumor suppressor gene inactivations are loss-of-function events, usually requiring a mutational event in one allele followed by loss or inactivation of the other allele. Some of these mutations may be inherited through the germline. One point of similarity is that somatic mutational events can occur in both oncogenes and tumor suppressor genes, and the number of mutational events may accumulate over a lifetime.

A point should be made about the terms *dominant* and *recessive*. In the classical Mendelian sense, these terms refer to an inheritance pattern resulting from the interplay between one paternal and one maternal allele in a diploid offspring. In cancer cells, this principle often doesn't hold. As noted above, chromosomal duplications, loss, and rearrangements often occur, leading to aneuploidy. Thus, a cancer cell may often be something other than diploid. It is clear from experimental studies that the balance between oncogene expression and tumor suppressor gene expression is a gene dosage effect.[506] For example, hybrid cell formation between a normal fibroblast and a malignant cell will usually produce a nontumorigenic hybrid if one malignant chromosome set is present, but not if there are two malignant sets. Furthermore, hybrids containing two copies of a chromosome bearing a tumor suppressor gene show more stable suppression of the malignant phenotype than cells having only one copy. The finding of "dominant" oncogenes is really a cell culture phenomenon, resulting from the neoplastic transformation of cells like mouse 3T3 cells after transfection with an activated oncogene. This sort of transformation event is seldom seen if normal diploid cells are used. Moreover, when malignant cells expressing a known oncogene are fused with normal diploid fibroblasts, malignancy is usually suppressed even though the oncogene continues to be expressed.[506] (For this reason the term *tumor suppressor gene* is preferred to *antioncogene*).[500] Thus, the terms *dominant* and *recessive* do not retain the classical Mendelian meaning in cancer.

A word about the mechanisms of action of the tumor suppressor genes is warranted here, even though this will be discussed in more detail below under each gene. Some of the suppressor gene products are localized in the cell nucleus and act as transcription factors. Some occur at the cell membrane and act in signal transduction, cell adhesion, or production of a normal extracellular matrix. Others appear to act as conduits for cell membrane–cytoskeleton interactions. Some are involved in DNA repair.

Thus, tumor suppressor genes, functionally, "come in many flavors."[507] In addition, they may act differently in different cell types, depending on the gene dosage of various positive and negative regulators. Although the Knudson two-hit model appears to apply to a number of cancer types, it is not always necessary that both alleles of a tumor suppressor gene be knocked out to generate a malignant phenotype. The state of "haploinsufficiency" may be enough to abrogate a tumor suppressor function. For example, in both people and mice, it has been observed that a heterozygous mutation that inactivates only one allele of a tumor suppressor gene produces an increased incidence of tumors, some of which develop without loss or mutation of the second allele (reviewed in Reference 507). Haploinsufficiency of a tumor suppressor gene can also increase risk of cancer in individuals who may already carry a heritable heterozygous mutation in a separate suppressor gene and thus comply with the Knudson model. A second activating mutation of an oncogene could also do the trick. Cancers that arise due to haploinsufficiency usually have a later age of onset than those that have lost function of both tumor suppressor alleles. As noted earlier, some tumor suppressor genes are "gatekeepers," such as those involved in regulating cell cycle control, signal transduction, or cell adhesion, and some are "caretakers" involved in DNA repair or chromosomal segregation during mitosis. In some situations, individuals carrying a heterozygous defect in one tumor suppressor allele may be at risk for a different type of tumor than that for which individuals having a loss of both alleles are at risk. For example, ataxic telangiectasia (AT) patients who have homozygous truncating or null mutations of the *atm* gene are at risk for developing lymphoid malignancies, whereas AT patients who are heterozygous carriers of mutations that interfere with the function of the remaining wild-type allele have an increased risk of breast cancer (reviewed in Reference 507).

It is worth noting, as Henry Harris does,[508] that Nature didn't design oncogenes to cause cancer and tumor suppressor genes to repress cancer. It is much more likely that "oncogenes" are the genes functional during rapid cell proliferation phases of development when tissues are growing, expanding, and beginning to differentiate into adult organs, and "tumor suppressor genes" are really differentiation genes that put the brakes on cell division to allow differentiation to occur (without carrying along any gene defects that avidly dividing cells might be prone to carry forward). This means that cancer is primarily a disease of faulty differentiation and not of unbridled cell proliferation.

Properties of Individual Tumor Suppressor Genes

rb

CHARACTERIZATION OF THE RB PROTEIN

The *rb*-1 gene, about 200 kilobases in length, is located on chromosome 13q14 and has 27 exons coding for a protein of 105–110 kDa, depending on the species in which it is produced (reviewed in Reference 509). It is a nuclear protein and acts to regulate the cell cycle. Mutations in the *rb* gene have been detected in retinoblastomas, osteosarcomas, bladder, small-cell lung, prostate, breast, and cervical carcinomas, and some types of leukemia. In contrast to hereditary retinoblastomas, mutations of *rb* in these other cancers appear to be somatic rather than germline because children with the inherited mutant allele may later develop osteosarcomas but only rarely get the other tumors mentioned, even though all the cells in their bodies must bear the mutation. Hence, different cell types respond differently to a germline *rb* mutation. Malignant transformation in the tissues must require additional mutations, probably because cells have redundant means to provide cell cycle regulation. The frequency of *rb* mutations detected in various tumor types also varies. *rb* mutations or deletions are seen in most if not all retinoblastomas, 80% of small-cell lung carcinomas, 20%–30% of non–small cell lung cancers, and, to a much lesser extent, in other tumor types.[510]

A number of types of *rb* gene mutations have been detected in various tumor types, including frameshift and chain termination mutations, deletions of entire exons, and point mutations. Many of these mutations affect domains between amino acids 393 to 572 and 646 to 772, which are involved in binding of viral proteins, such as SV40 large T antigen, adenovirus E1A, or human papilloma virus E7, and cell cycle regulatory proteins. Some *rb* gene mutations also decrease the ability of the Rb protein to be phosphorylated.[511]

There are three members of the Rb protein family, Rb itself and two Rb-related proteins, p107 and p130. The p130 gene maps to chromosome 16q12.2-13, a region often altered in human cancers. The p107 gene locus is on chromosome 20q11.2, an area not frequently found to be involved in cancer (reviewed in Reference 512).

INTERACTIONS OF RB PROTEINS

Rb, p130, and p107 interact with many proteins, but their central role in cell cycle regulation involves their inhibitory binding in their unphosphorylated state to the E2F family of transcription factors (Fig. 5–18), phosphorylation of Rb by cdk2 releases Rb from E2F. Part of this action involves Rb-complex recruitment of HDACs and other chromatin factors to E2F-responsive promoters.[510] Other Rb interactions include binding the transforming proteins of three oncogenic DNA viruses, SV40, adenovirus (type 5) EIA, and human papilloma viruses type 16 and 18 E7 protein, which bind avidly to the p105 Rb protein and to the Rb-related protein p107 (reviewed in References 509 and 513). Interestingly, SV40 large T antigen, adenovirus E1B, and HPV E6 bind p53, another cell cycle regulatory protein. These findings suggest that oncogenic DNA viruses have captured this mechanism to work their will on the replicative machinery of the cell to make sure that the enzymes for nucleotide synthesis, DNA polymerases, and other processes are there to foster their own replication.

SV40 T antigen, adenovirus E1A, and HPV E7 contain homologous regions of amino acids that are involved in p105 and p107 binding. If these regions are altered or mutated, binding of

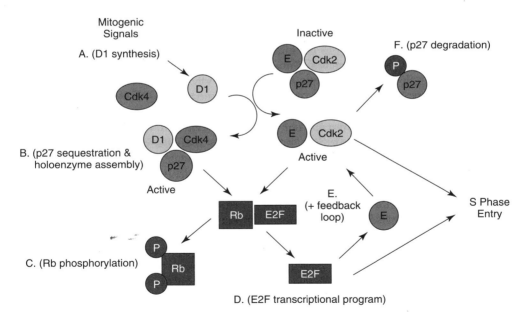

Figure 5–18. A G_1 phase regulatory cascade. Cyclin D1 synthesis (step *A*) and assembly (step *B*) in response to mitogenic signals sequesters Cip/Kip proteins (p27^{Kip1} is shown) and relieves cyclin E–Cdk2 from their constraint. Both G_1 cyclin-dependent kinases then collaborate to sequentially phosphorylate RB family proteins (only RB is shown) (step *C*). This frees E2Fs from inhibition and leads to the activation of genes required for S-phase entry (step *D*). Among the known E2F target genes is cyclin E, whose transcriptional up-regulation provides feedback to drive cells into S phase (step *E*). One substrate of cyclin E–Cdk2 is p27Kip1, whose phosphorylation triggers its ubiquitination and degradation as cells enter S phase (step *F*). (From Sherr and McCormick,[510] reprinted with permission from Elsevier.)

these DNA virus oncoproteins to these Rb-type proteins is inhibited and their transforming ability is diminished. This effect strongly suggests that ability to bind these cell cycle regulatory proteins is *de rigueur* for their ability to induce a malignant phenotype.

ROLE OF RB IN REVERSING THE MALIGNANT PHENOTYPE

Introduction of a wild-type (WT), nonmutated *rb* gene by retroviral- or transfection-mediated gene transfer into a variety of human cancer cells that have an inactivated *rb* gene results in reversion to a more normal phenotype, including reversal of morphological transformation, growth rate, growth in soft agar, and tumorigenicity in nude mice.[514] Such reversal has been noted for retinoblastoma, osteosarcoma, bladder, and prostate carcinoma cells. These data demonstrate that normal function of Rb is crucial for maintenance of cell growth control.

REQUIREMENT OF A FUNCTIONAL *RB*-1 GENE IN DEVELOPMENT

Surprisingly, gene knockout of the *rb*-1 gene by homologous recombination in mouse embryonic stem (ES) cells, followed by microinjection of the ES cells into blastocysts and implantation into foster mothers, has shown that the embryos survive until about 14 to 15 days and then die from massive cell death in the developing central nervous system and lack of hematopoiesis, particularly of erythroid cells.[514–517]

This result is surprising, because if all cell types require *rb* gene expression for regulation of the cell cycle, then how can embryos survive for 14 to 15 days, a time during which a number of cell lineages have already developed? Furthermore, why is the major defect only noted in two tissues? It is also puzzling that heterozygous mice, developed by only knocking out one *rb* allele, survived for up to 11 months; however, some of these animals developed pituitary adenocarcinomas, but none developed retinoblastomas.[516] In these pituitary carcinomas, the remaining WT allele was lost, so in this case the two-hit hypothesis for tumor development held up. It should also be noted that transfer of a normal human *rb* mini-transgene into the mutant mice corrected the developmental defects.[515] One can only conclude from these data that not

all cell lineages rely exclusively on *rb* for control of cell proliferation and differentiation and that there are species differences in the target cells for neoplastic transformation after abrogation of *rb* function.

There is, however, a report that expression of SV40 large T antigen, driven by a luteinizing hormone β gene promoter, in transgenic mice produced heritable ocular tumors similar to human retinoblastoma, and an association between SV40 T antigen and Rb p105 was shown in the tumor tissues.[518] These latter results suggest that if Rb protein function is disrupted in a specific way, similar tissue tropism for a carcinogenic effect of *rb* knockout can occur across different species.

CELL CYCLE REGULATION BY RB

As noted above, the Rb family of proteins plays a key role in cell cycle regulation, and their activity in this role is determined by their phosphorylation state. Rb proteins are kept in a hypophosphorylated state by the actions of the INK4 cell cycle regulatory proteins that inhibit cyclin D–dependent kinases through most of the "resting" G_1 phase of the cell cycle. There are at least 10 serine/threonine phosphorylation sites on Rb, and it is a substrate for cyclin-dependent kinases. Hyperphosphorylated Rb binds less tightly to its "nuclear anchor" binding site that keeps it in the nucleus, thus its binding to E2F is decreased, and it is released so it can activate genes involved in progression through the cell cycle. Hypophosphorylated Rb dampens this activation signal by "sequestering" E2F so it can't bind to its promoter–enhancer DNA sites. As noted above, the growth inhibitory function of Rb can also be down-regulated by binding to viral oncoproteins such as SV40 T antigen, E1A, or E7, which bind to hyposphorylated Rb, or by mutations that alter the ability of pRb to bind to its nuclear anchor protein(s). Any of these three events then— hyperphosphorylation, binding to oncoproteins, or mutations—could have the same end result, namely, the inability of Rb to inhibit cell cycle progression. Since normal cells don't usually carry SV40 T antigen, E1A, or E7 oncoproteins, there is presumaably a normal cellular pRb-binding protein, whose binding is displaced by the viral proteins. At least two such genes that encode Rb-binding proteins (RBP-1 and RBP-2) have been cloned.[519] The binding of the viral proteins, in contrast to the normal Rb-binding proteins, may then displace Rb from its nuclear anchor, causing it to be lost from the nucleus. Normal cell cycle progression, as opposed to viral or oncogene induced cell cycle progression, is mediated through normal mitogenic signals that turn on cyclins and cyclin-dependent kinases that phosphorylate Rb. Microinjection of Rb into cycling osteosarcoma cells in culture caused cell cycle arrest, and co-injection of Rb with c-Myc but not H-Ras, c-Jun, or c-Fos inhibited the ability of Rb to arrest the cell cycle.[519]

INTERACTIONS OF RB PROTEIN WITH TRANSCRIPTION FACTORS AND DNA REGULATORY ELEMENTS

In addition to negatively regulating cell proliferation by inactivating E2F, Rb has some more direct actions as a transcriptional regulator. Rb has ability to bind DNA itself and it has been shown to repress c-*fos* expression and AP-1 transcriptional activity in cycling 3T3 cells. Thus, Rb appears to be able to bind to its own *cis*-acting control element, termed RCE.[520] The Rb protein also induces TGF-β1 gene expression in epithelial cells, in which TGF-β is a growth-inhibiting factor, and represses TGF-β1 expression in fibroblasts, in which TGF-β1 can act as a growth promoter.[521] Furthermore, Rb has been found to activate expression of TGF-β2 in epithelial cells via an action at ATF-2 transcriptional regulatory elements.[522]

p53

CHARACTERIZATION OF P53 AND ITS MUTATIONS

Originally, p53 was thought to be an oncogenic protein. This 53 kDa protein was first detected as a complex with SV40 T antigen in SV40-transformed cells.[523] A similar complex was found between p53 and E1B protein in adenovirus-infected cells. The p53 protein was subsequently found in a variety of transformed mouse cell lines, cultured human tumor cells, and in virally, chemically, or radiation-induced murine tumors. Even more indicting was the fact that transfection of the p53 gene was found to immortalize and transform cells and to cooperate with *ras* in

inducing transformation.[524] It was also noted that the cellular half-life of p53 was increased in SV40-transformed cells, an effect that at the time was thought, for the wrong reasons, to foster the transforming action of p53. This latter observation will resurface again later in the story. Only gradually, over about 5 years time from the observations of the transforming ability of p53, did it become clear that the transforming effects of p53 were due to a mutant protein and that the nonmutated, WT p53 negatively controlled cellular proliferation and suppressed cell transformation and tumorigenesis.[525]

The p53 gene is located on chromosome 17p13 in the human genome. The gene contains 10 coding exons and is expressed in all cells of the body, although at low levels in most tissues. The human protein is 393 amino acids long and contains at least nine potentially phosphorylated serine residues, one of which, serine 316, is phosphorylated by a cyclin-dependent kinase (reviewed in Reference 509). There is a nuclear translocation domain near the Cdk phosphorylation site, suggesting a cell cycle–dependent signal for nuclear translocation of p53. Nuclear localization is important for p53 to function as a negative regulator of cell proliferation and as a tumor suppressor gene. In some human cancers, a defect in p53 function relates to its sequestration in the cell cytoplasm and inability to be transported to the cell nucleus.

There is high sequence homology for p53 among animal species; for example, there is about 56% amino acid homology from frogs (*Xenopus*) to humans, with 90% to 100% homology in some regions of the protein. Interestingly, these regions are most often found to contain mutations in human cancer, strongly implicating these regions of the protein as important to its regulatory functions.

Mutations of p53 are the most common genetic alterations observed in human cancers (50%–60% have some type of p53 alteration),[509] and there are several hot spots for these mutations (Fig. 5–19). Most of the mutations are missense point mutations in carcinomas, whereas in sarcomas, deletions, insertions, and rearrangements are more common and point mutations are rare. Some sarcomas contain an amplification of an oncogene called *mdm*2, whose protein product inactivates p53 (see below). Different

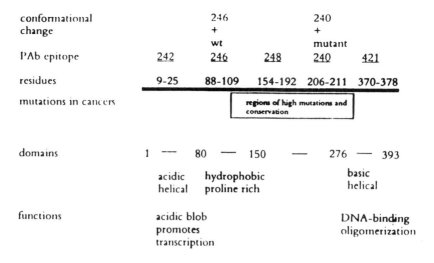

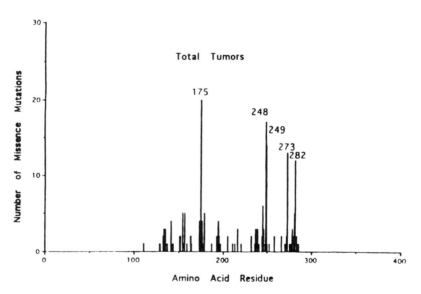

Figure 5–19. A. Schematic representation of the domains of structure of the p53 protein. B. The positions of p53 missense mutations in the p53 gene from 191 human cancers. The codon numbers or amino acid residue numbers are indicated on a linear representation of the protein. The height of the line at each codon indicates the number of independent times a mutation occurred at the codon. (From Levine,[509] with permission.)

mutational hot spots of the *p53* gene are seen in different tissues. For example, 53% of liver cancers in high-endemic exposure areas for hepatitis B infection and aflatoxin B1 have mutations in codon 249.

Germline mutations of p53 are also observed in some families with a high incidence of cancer.

The Li-Fraumeni syndrome is one such case. Many members of these families have missense and nonsense mutations in one p53 allele and tend to get osteosarcomas, adrenal corticol carcinomas, breast carcinomas, or brain cancers, often at an early age.[526,527] Curiously, colon carcinoma is not prevalent in these families, even though p53 mutations are often seen in colon cancer. This observation suggests that germline mutations tend to make certain tissues more susceptible to later somatic mutation than other tissues or some tissues have additional mechanisms for regulating cell proliferation that must be knocked out before p53 mutations become important for the cell's economy.

Mutations in the p53 protein can have at least three phenotypic effects: (1) loss of function, in which a missense mutation abrogates p53's ability to block cell division or reverse a transformed phenotype; (2) gain of function, as demonstrated by the introduction of a mutant *p53* gene into cells lacking WT p53, which induces a tumorigenic phenotype; and (3) *trans*-dominant mutation, seen when a mutant p53 allele is introduced into cells bearing a WT p53 allele, resulting in an overriding of the normal inhibitory function of p53. This latter effect is sometimes called a *dominant-negative effect*. As noted above, the cellular half-life of p53 in transformed cells is often longer than that of WT p53 in normal cells, because there are conformational differences in the mutant protein that render it less susceptible to degradation.[528] This longer half-life of the mutant form may play a role in the dominant-negative effect.

MUTAGENESIS OF P53

As discussed above, the types of mutations of p53 vary with cell type, as do the hot spots for mutations in different tumor types. Lung tumors contain both base transition and transversion mutations, but colon tumors contain primarily base transitions, often C→T. CpG dinucleotides are frequent sites of mutation, which this raises the question of whether tissue-specific methylation patterns of C in CpG sites could play a role in the types of p53 mutations observed, since methylated C residues in CpG doublets are known to have a higher mutation rate than nonmethylated C. The type of carcinogen to which different tissues may be exposed is also an important factor. The instance of liver cancer, as an example, has already been mentioned. More than half of hepatocellular carcinomas (HCC) from high aflatoxin B1 (AFB1) exposure areas have G→T transversions in the third position of codon 249(AGG), which results

in replacement of arginine by serine.[529] This can also be shown by exposure of human hepatocytes exposed to AFB1 in culture.[530] HCC tissue taken from patients in low AFB1 exposure areas, by contrast, do not usually display G→T transversions of the *p53* gene, and the observed mutations are found in other regions of the gene.[531] Thus, different carcinogens can apparently mutate p53 in different segments of the gene, but the end result is the same, i.e., inactivation of p53 function.

G→T transversions of p53 occur at high frequency in tobacco-related human cancers, including small cell and non–small cell lung cancers, esophageal carcinomas, and squamous cell carcinomas of the head and neck (reviewed in Reference 532). Benzo[a]pyrene, a component of cigarette smoke, produced a high incidence (70%) of G→T transversions in BP-induced murine skin carcinomas, whereas 7,12-dimethylbenz(a)anthracene-induced skin tumors had a similar p53 mutation frequency but a low rate of G→T transversions. These data support the concept that different carcinogens attack the p53 gene differently. Furthermore, ultraviolet B radiation–induced mouse skin carcinomas contained a prevalence of C→T transitions.[533]

An additional important fact should be noted here. Most p53 mutations occur in the non-transcribed strand of DNA. Since the non-transcribed strand is more slowly repaired, there is a potential for these errors to be passed on to daughter cells.[534]

ABILITY OF P53 TO REVERSE CELLULAR TRANSFORMATION AND TUMORIGENESIS

In several diverse cell systems and tumor cell types, introduction of the *p53* gene into cells growing in culture usually blocks cell proliferation and hangs the cells up at the G_1/S transition point in the cell cycle. Moreover, suppression of the neoplastic phenotype in culture and of tumorigenicity in nude mice is usually observed. Such effects have been observed in human colorectal, lung, and prostate carcinoma cells, glioblastomas, osteosarcomas, and acute lymphoblastic leukemia cells (reviewed in Reference 535). Introduction of a mutated *p53* gene, by contrast, does not block cell proliferation or tumorigenicity and may, in fact, enhance them.

ROLE OF P53 IN CELL CYCLE PROGRESSION AND IN INDUCING APOPTOSIS

It is now clear that WT p53 not only has anti-proliferative and anti-transforming activity but also possesses the ability to induce programmed cell death (apoptosis) after exposure of cells to DNA-damaging agents such as γ-irradiation or anticancer drugs.[536] The concept that p53 is a growth regulatory protein fits with its short half-life (5 to 20 minutes in normal mouse cells and 1 to 2 hours in normal human cells), its nuclear location and transcription factor activity (see below), and its increased synthesis in DNA-damaged cells (reviewed in Reference 537). WT p53 regulates the transcription of a number of cell replication–associated genes. Growth arrest induced by WT p53 blocks cells prior to or near the restriction point in late G_1 phase and produces a decrease in the mRNA levels for genes involved in DNA replication and cell proliferation such as histone H3, proliferating cell nuclear antigen (PCNA), DNA polymerase α, and b-myb.[538] To carry out these gene regulatory events, WT p53 has to assume a certain conformational structure, apparently modulated by its phosphorylation state, and oligomerize so that it can bind to DNA.[537] Mutant p53 cannot achieve the appropriate conformation and can block WT p53 function by forming oligomers with it.

Not all types of apoptosis, however, are mediated by p53. For example, whereas induction of apoptosis in thymocytes by γ-irradiation or the DNA-damaging drug etoposide is via a p53-dependent pathway, that induced by glucocorticoids in thymocytes is not.[539,540] WT p53 is required for the response to DNA damage; cells having mutant or no p53 fail to respond appropriately. In fact, WT p53 enhances sensitivity to ionizing irradiation[541] and anticancer drugs such as 5-fluorouracil, etoposide, and doxorubicin,[542] whereas p53 mutations increase resistance to ionizing radiation.[543] Thus, the absence of or mutation of p53 leads to an increase in cellular resistance to these agents, implying that cancer cells in patients can acquire resistance to chemotherapeutic agents or irradiation through mutations or loss of p53.

These effects may seem somewhat paradoxical, but they are understandable if one thinks of p53 as a protector or "molecular policeman" monitoring the integrity of the genome.[544] When DNA is damaged, p53 accumulates and stops DNA replication and cell division until DNA has time to repair itself. If this is not possible or the DNA repair mechanisms fail, p53 triggers a cell suicide response. Thus, in the case of massive damage to DNA, for which DNA repair is not possible, the cell dies. If p53 is mutated or lost, the cell goes on its merry way, replicating its damaged DNA, passing on mutations to daughter cells, and giving cells a survival advantage in the face of DNA damage. Cells that do this are genetically less stable and accumulate mutations and gene rearrangements, leading to the generation of an ever-increasing malignant state. This sort of event could partly explain the increased rate of mutation ("mutator phenotype") seen in tumor progression.

The above findings indicate that WT p53 acts as a "checkpoint" control protein that stops the cell cycle before S phase when DNA damage is present. Thus, p53 is analogous to the *RAD9* gene of yeast that inhibits cell cycle progression following DNA damage.[545] Loss of RAD9 or p53 causes cells to undergo a greater frequency of mutations and gene amplifications.[545–547] For example, when fibroblasts from patients with Li-Frameni syndrome are passaged in vitro, they may lose the remaining WT p53 allele, and when they do, they have a greatly increased ability to amplify drug resistance genes in response to a drug called PALA.[546,547] Introduction of a WT *p53* gene back into these cells via a retroviral vector restored cell cycle control and reduced the frequency of gene amplification to background levels.[547] Other factors, however, may also allow gene amplification to occur in tumor cells, since tumor cells with functional p53 can still amplify genes.[546]

A possible therapeutic result may be gained by taking advantage of p53's ability to induce apoptosis in tumor cells. For example, when spheroids of human lung cancer cells grown in culture are treated with a retroviral vector containing a WT *p53* gene, apoptosis was induced in the cells.[548]

MECHANISMS OF P53'S ACTIONS

Even though a lot is known about the biological actions of p53, e.g., the ability to induce G_1

arrest, to induce apoptosis following DNA damage, to inhibit tumor cell growth, and to preserve genetic stability, the way in which it does all this isn't totally clear. As noted avove, the p53 pathway is disarmed in a majority of human solid tumors at some stage in the progression pathway. Inactivation of the p53 pathway can occur by a variety of mechanisms (Table 5–12). In about 50% of tumors, p53 is inactivated by mutations and in the rest it is inactivated indirectly through binding to viral proteins or to other proteins upregulated in cancer cells. It is estimated that there are over 10,000 different tumor-associated mutations in the *p53* gene, including those observed in lower organisms and humans.[549]

The p53 network can be activated by at least three mechanisms (reviewed in Reference 549). The first is DNA strand breaks triggered by ionizing radiation or other DNA-damaging agents. This mechanism is dependent on activation of the ATM (ataxia telangiectasia-mutated) protein, Chk2, or other kinases. Interestingly, mice that are deficient in p53 function and in the ability to repair DNA double-strand breaks because of a failure in nonhomologous end-joining (NHEJ) repair develop highly aggressive pro-B-cell lymphomas.[550] The second mechanism is overexpression or aberrant expression of growth factor signals such as those turned on by oncogene proteins Ras or Myc. This occurs via activation of p14^{Arf}, which in turn inhibits MDM2's ability to stimulate degradation of p53. Finally,

cellular stress is induced by chemotherapeutic drugs, ultraviolet light, or protein kinase inhibitors. This pathway appears to involve kinases such as ATR (ataxia telangiectasia related) protein and casein kinase II.

The level of p53 proteins in cells is regulated by MDM2, which causes ubiquitination of 53, thus targeting p53 for degradation. It is the rate of degradation rather than the rate of synthesis that determines the intracellular level of p53. Full functionality of p53 as a transcription factor requires phosphorylation of serines and/or acetylation of lysines near the C-terminus. Such modifications alter the conformation of p53 and enhance its binding to DNA. It should be noted that similar modification of the folded state of the p53 protein by antibodies or small molecules could be a way to enhance its function. Similarly, small molecules that inhibit the binding of MDM2 to p53 have been designed, and their use could be a way to stabilize p53 levels in cancer cells.[551] MDM2 is overexpressed in a number of human cancers. Phosphorylation of the N-terminus of p53 diminishes its binding to MDM2 and hence its degradation. Both MDM2 and p53 contain nuclear transport signals that regulate their entry and export from the cell's nucleus. The import mechanism apparently is altered in some tumors where the nuclear and cytoplasmic levels of p53 are lower than normal. In unstressed cells, p53 is maintained at low levels by the action of MDM2.

Table 5–12. Ways in Which p53 May Malfunction in Human Cancers

Mechanism of Inactivating p53	Typical Tumors	Effect of Inactivation
Amino acid–changing mutation in the DNA-binding domain	Colon, breast, lung, bladder, brain, pancreas, stomach, esophagus and many others	Prevents p53 from binding to specific DNA sequences and activating the adjacent genes
Deletion of the carboxyl-terminal domain	Occasional tumors at many different sites	Prevents the formation of tetramers of p53
Multiplication of the *MDM2* gene in the genome	Sarcomas, brain	Extra MDM2 stimulates the degradation of p53
Viral infection	Cervix, liver, lymphomas	Products of viral oncogenes bind to and inactivate p53 in the cell, in some cases stimulating p53 degradation
Deletion of the *p14*^{ARF} gene	Breast, brain, lung and others, especially when p53 itself is not mutated	Failure to inhibit MDM2 and keep p53 degradation under control
Mislocalization of p53 to the cytoplasm, outside the nucleus	Breast, neuroblastoma	Lack of p53 function (p53 functions only in the nucleus)

From Vogelstein et al.,[549] reprinted with permission from Macmillan Publishers Ltd.

p53 acts as a transcription factor for several genes, including a number of genes involved in cell-cycle control, apoptosis, genetic stability, and angiogenesis (reviewed in Reference 549). One function of p53 is to keep the cell cycle in check. The p53 protein regulates the cell division cycle by stimulating expression of p21$^{WAF1/ CIP1}$ and of protein 14-3-3σ, which inhibit cyclin-dependent kinases (see section on cell cycle regulation in Chapter 4). This action inhibits both G_1 to S and G_2 to mitosis transitions. p53 activates a number of genes involved in inducing apoptosis, including Bax, NOXA, p53A1P1, and PUMA (p53 up-regulated modulator of apoptosis).[549,552]

The role of p53 in maintaining genetic stability appears to involve induction of genes that stimulate nucleotide excision repair, chromosomal recombination, chromosome segregation, and induction of the gene for ribonucleotide reductase. p53 also stimulates the expression of genes that inhibit angiogenesis.

One might ask how p53 regulates so many genes and why it is such a key alteration in cancer cells. The answer is that the p53 gene is at the center of so many cell regulatory networks. It is like the main control circuit breaker on an electrical panel. Thus, mutation or inactivation of p53 function by various means disrupts so many interconnecting pathways that once that central control point is breeched, numerous downstream regulators become dysfunctional, setting the stage for tumor progression.

Some other interesting clinical observations about p53 have been made. For example, specific p53 gene mutations in the urinary bladder epithelium have been observed in survivors of the Chernobyl nuclear power plant accident in the Ukraine in 1986. Urinary bladder epithelium biopsied from 45 males with a diagnosis of benign prostatic hypertrophy living in the contaminated area showed a high incidence of urothelial dysplasia, carcinoma in situ, irradiation cystitis, and one case of transitional cell carcinoma.[553] Because the incidence of urinary bladder cancer in the Ukraine increased between 1986 and 1996, alterations of p53 may have been a prodrome for such malignancies.

Another interesting observation with potentially high clinical impact is related to the well-known fact that full-term pregnancy early in reproductive life is a protection against breast cancer. Pregnancy also provides protection in rodents from carcinogen-induced breast cancer. These effects have been postulated to be due to an increased level of differentiation of breast tissue during pregnancy such that breast stem cells became less susceptible to the proliferative- and hyperplasia-inducing effects of reproductive hormones and of other potentially carcinogenic agents. The molecular mechanism of this effect has not been elucidated; however, Sivaraman et al.[554] have shown that there is an increased and sustained level of p53 in the nucleus of mammary tissue in rats exposed to pregnancy-simulating hormonal manipulation and then challenged with the mammary cancer–inducing agent methylnitrosourea. Whether this is the mechanism for the breast cancer protective effect of pregnancy in women isn't clear, but it is an enticing and logical hypothesis.

Wilms' Tumor Suppressor Gene wt-1

Wilms' tumor is a renal cancer called *nephroblastoma* that occurs in children and in some cases has a genetic predisposition. In about 10% of cases, the tumors are bilateral. It is a rare tumor, occurring in about 1 in 10,000 children, and is associated with alterations at distinct loci on chromosome 11. Certain distinct clinical syndromes have been associated with chromosome 11 abnormalities (reviewed in References 509 and 555). A deletion in the short arm of chromosome 11 at band 13p is associated with the WAGR syndrome (Wilms' tumor, aniridia, genitourinary malformation, and mental retardation). Genetic mapping of this region in tumor and normal tissue from these patients led to the identification of a gene called *wt*-1, which has mutations and loss of heterozygosity in Wilms' tumor tissue.[556] Thus, *wt*-1 is a gene locus that has the properties of a tumor suppressor gene that plays an important role in urogenital development.

Other syndromes have also been found in association with chromosome 11 defects, including the Backwith-Wiedemann syndrome, in which the 11p15 locus is involved. These patients also have a high incidence of Wilms' tumors, implicating a second tumor suppressor gene locus on chromosome 11, called *wt*-2. Both the *wt*-1 and *wt*-2 loci have tumor suppressor activity when introduced into Wilms' tumor cell lines and tested for tumorigenicity in nude mice.[557]

Abnormalities of chromosome locus 11p have been observed in several different human cancer types. Loss of heterozygosity for DNA markers at 11p have been seen for rhabdomyosarcoma, hepatoblastoma, hepatocellular, bladder, breast, non–small cell lung, ovarian, and testicular carcinomas as well as for Wilms' tumor (reviewed in Reference 558). Deletions of the short arm of chromosome 11 are associated with hepatitis B virus integration, and human fibroblasts with 11p deletions are sensitive to transformation by HBV and BK viruses. Introduction of human chromosome 11 into BK virus–transformed mouse cells suppresses their tumorigenicity.[558]

The 50 kb *wt*-1 gene at 11p13 has 10 exons and codes for a 3 kb mRNA, but a number of mRNA splice variants have been detected. The WT-1 gene product is a 46–49 kDa, proline/glutamine-rich protein containing four zinc-finger DNA-binding motifs (reviewed in Reference 555). These domains have homology to the early growth response (EGR) family of transcription factors, but WT-1 binding to at least one of the EGR response elements (EGR-1) represses transcription rather than stimulating it.[559] In the mouse, WT-1 is first expressed at day 8 of gestation in intermediate mesoderm and subsequently in differentiating mesothelium, spinal cord, brain, and the urogenital ridge. Expression peaks at day 17 of gestation (which is 21 days long in the mouse) and is low in adult tissues, indicating a key role for WT-1 in development. This is borne out in gene knockout experiments in transgenic mice. Homozygous loss of the *wt*-1 gene results in embryonic death about day 11, with abnormal development of the kidneys, gonads, heart, lungs, and mesothelium.[555]

In addition to LOH at the 11p13 locus, point mutations and small deletions of one allele of *wt*-1 have been observed as germline defects in some children with genetic predisposition to develop Wilms' tumor. The tumors from these children have loss of the remaining WT allele by chromosomal nondisjunction or recombination events (reviewed in Reference 560), fulfilling Knudson's two-hit hypothesis. One observed point mutation in the *trans*-activation domain of *wt*-1, found in a Wilms' tumor patient, converts the encoded protein from a transcriptional re-

pressor to an activator of the EGR-1 promoter.[560]

In addition to EGR-1 sites, WT-1 also can act as a transcriptional repressor of other growth-related genes, including IGF-II and PDGF A-chain. However, under some circumstances WT-1 can activate these genes.[561] WT-1 can activate or suppress transcription from similar response elements depending on how it is bound. It does this by acting through separate functional domains. Suppression of growth-related genes occurs by binding to two independent binding sites 5' or 3' relative to the transcription start site. WT-1 functions as a transcriptional activator when it only binds at the 5' or 3' site, but not both.[562] Amino acids 84 to 179 are required for transcriptional suppression, whereas the domain containing amino acids 180 to 294 mediates transcriptional activation. A second WT-1 DNA-binding site has been found that is also involved in transcriptional suppression of growth-related genes including PDGF-A, K-*ras*, EGF receptor, insulin receptor, c-*myc*, and tumor growth factor β3.[561]

The way in which the opposing roles of WT-1 are modulated in cells isn't clear. It may be that certain mutations or deletions disrupt the ability of WT-1 to bind to the DNA sites required for transcriptional repression, leaving only the activation signal or no signal at all left in the protein. Another possibility is that interaction with other cellular proteins determines the availability of WT-1 binding sites in a cell context–dependent manner. For example, WT-1 has been shown to form complexes with p53, and this interaction modulates the ability of these two tumor suppressor proteins to regulate their responsive genes.[563] In the absence of p53, WT-1 is a *trans*-activator rather than a repressor of EGR-1 genes. Furthermore, WT-1 binding to p53 enhances p53's ability to *trans*-activate the muscle creatine kinase promoter. It is also possible that dominant-negative mutations may occur in WT-1 that prevent the normal function of a remaining normal *wt*-1 allele. Such a mutation, involving a deletion of the third zinc finger, has been observed.[563] It has also been shown that WT-1 represses transcription of the human telomerase reverse transcriptase gene (*htert*),[564] which may also contribute to its tumor suppressor activity by preventing the overexpression of *htert* seen in a number of cancers.

Adenomatous Polyposis Coli (apc) Gene

Familial adenomatous polyposis (FAP) is an autosomal-dominant disease that occurs in 1 out of 10,000 individuals in the United States, Europe, and Japan and accounts for about 10% of colerectal cancers.[509] These patients develop thousands of colonic polyps during the second to third decade of life, and a small percentage of them become cancerous. However, these people are highly likely to develop colorectal cancer during their lifetime unless treated (usually by colectomy). The gene involved in this disorder was found on chromosome 5q21 and cloned.[565,566] This gene, called *apc*, was found to contain point mutations in the germline of patients with FAP. Frameshift, nonsense, and missense mutations, clustered in the first third of the structural gene, have also been found in these individuals. LOH of this genetic locus has been observed in 35% to 45% of colorectal cancers in patients who don't have FAP, indicating that the *apc* gene has important tumor suppressor function in colorectal tissue. The *apc* gene has the information for a very large protein of 2843 amino acids, but no clear function for the *apc* gene product has been assigned. The protein does, however, bind to α- and β-catenins that are associated with and important for the function of the adhesion molecule cadherin.[567,568] This finding suggests a role for *apc* in cytoskeletal–extracellular matrix (ECM) interactions that control cell growth and differentiation. It should be noted that another tumor suppressor protein, NF-2, also plays a role of cell cytoskeleton–ECM interactions (see below).

Deleted in Colorectal Cancer (dcc) Gene

Loss of heterozygosity and allelic loss at chromosome 18q are common in colorectal cancer, occurring in more than 70% of carcinomas and about 50% of large adenomas. Since this defect is much less frequent in small, early-stage adenomas, it is thought to contribute to tumor progression more than initiation and to be altered by somatic mutational events. A gene deleted at 18q21 in colorectal cancer has been cloned and called *dcc*.[569] *dcc* mRNA is reduced or absent in more than 85% of colorectal cancer cell lines studied, and it is found at low levels in several tissue types, including normal colonic mucosa. In contrast to the *apc* gene, which is only found mutated in colorectal but not other human cancers, the *dcc* gene shows LOH or loss of expression in colorectal, gastric, esophageal, pancreatic, and prostatic carcinomas (reviewed in Reference 570). The DCC protein has significant amino acid sequence homology with the neural cell adhesion molecule N-CAM, thus it may have a role in cell–extracellular matrix interactions, the loss of which might be involved in tumor invasion and metastasis.

Hereditary Non-polyposis Colorectal Cancer (hnpcc) Gene

The HNPCC syndrome, also known as the Lynch syndrome, occurs in about 1 of every 200 people and increases the risk of developing colon, ovarian, uterine, and kidney cancers, often before 50 years of age.[571] Studies of affected families indicated a linkage to a chromosome 2p locus. This predisposition to cancer is inherited in an autosomal-dominant manner. It was thought that the *hnpcc* gene would turn out to be a tumor suppressor gene, but unlike many of them, both alleles of the affected chromosome 2p locus were retained in HPCC tumors, whereas other tumor suppressor genes are usually lost or inactivated during tumorigenesis (reviewed in Reference 572). Studies of HNPCC colorectal cancers and a subset of sporadic colorectal cancers with a similar pathologic pattern revealed alterations in microsatellite DNA involving abnormal dinucleotide or trinucleotide repeats (insertions or deletions). These and other data indicated that HNPCC and a subset of sporadic colorectal tumors were related to a heritable defect producing replication errors of microsatellite sequences, so-called RER^+ cells.

The fact that the RER^+ phenotype was reminiscent of some mismatch repair defects in bacteria and yeast was a serendipitous clue that led to characterization of the gene, once it was cloned.[572–574] There are at least three ways in which mismatched nucleotides arise in DNA:

(1) deamination of methyl C to T, creating a G-T mispair, (2) misincorporation of a nucleotide during DNA replication, e.g., at an apurinic site; and (3) genetic recombination producing heteroduplexes with mismatched bases (reviewed in Reference 573). All organisms from *E. coli* to humans have enzyme systems to repair such defects.

The affected locus in *hnpcc* maps to chromosome 2p22-21 and contains the human homolog (*hmsh2*) of a bacterial gene *MutS* that is responsible for mismatch recognition in methyl-directed mismatch repair and of a yeast gene *msh 2*, mutants of which in yeast cause instability of dinucleotide repeat sequences. The mutation rate of $(CA)_n$ repeats was directly measured in RER^+ human tumor cells and shown to be at least 100-fold that of RER^- cells.[524] This increased mutation rate was due to a defect in strand-specific mismatch repair. The finding that $(CA)_n$ repeats are unstable in RER^+ cancer cells and defective in the gene known to stabilize repetitive sequences in *E. coli* and *S. cerevisiae* supports the idea of a mutator phenotype, since these defects would be expected to accumulate with time and cause genetic instability. The *hnpcc* gene is the first example of a DNA repair defect being associated with a tumor suppressor function. A second such defect to a gene called *hmlh1* has been reported, and more are likely to be discovered, since DNA repair is so important for maintaining the integrity of the genome.

Neurofibromatosis Genes nf-1 and nf-2

Neurofibromatosis (Von Recklinghausen's diseasè) varies from a mild form with café-au-lait spots on the skin to a severe form with large, disfiguring neurofibromas resulting from the tremendous overproliferation of Schwann cells. The syndrome affects 1 out of every 3500 people, and in 50% of cases there is an inherited defective gene or genes. One of these, *nf-1*, maps to chromosome 17 q11,[575] and like some other tumor suppressor genes such as *rb*, *wt-1*, and *p53*, can act in a dominant-negative fashion, reflecting inheritance of one defective allele and subsequent loss or inactivation of the remaining normal allele. The *nf-1* gene encodes a protein with significant sequence homology to GTPase-activating proteins (GAP) that modulate the function of the Ras oncoprotein.[576] Indeed, the NF-1 protein has been shown to bind to human Ras p21 and stimulate GTPase activity.[577] These data suggest that loss of NF-1 GAP activity would keep Ras in its active Ras-GTP state and prolong the signal for cell proliferation. Somewhat curiously, although NF-1 is expressed in all tissues of the body, mutations have only been found in neurofibromas and not other cancers. The fact that both GAP and NF-1 regulate the function of Ras suggests that there is redundancy in regulation of Ras and that tissues susceptible to carcinogenic transformation by loss of NF-1 activity have little regulatory control of Ras by GAP, leaving NF-1 as the key regulator.[578]

A second neurofibromatosis susceptibility gene, *nf-2*, has been cloned and it appears to connect the cell membrane to the internal cytoskeleton. Its loss of function may cause cytoskeletal disorganization that leads to abnormal cell proliferation.[579,580] *nf-2* maps to chromosome 22q12 and mutations found in tumors (usually vestibular schwannomas or meningiomas) often result in truncated protein products.

Von Hippel-Lindau Syndrome and Renal Cell Carcinoma Gene

The Von Hippel-Landau (VHL) syndrome is dominantly inherited and predisposes carriers to develop one or more of three types of cancer: brain hemangioblastomas, pheochromocytomas, or renal cell carcinomas.[578] The *vhl* gene maps to chromosome 3p25, and its protein product is a cell surface molecule that, like NF-2 and DCC, appears to be involved in cell surface–ECM interactions and/or signal transduction mechanisms.[581] LOH and translocation of other chromosome 3p markers in renal cell carcinomas strongly suggest that there are other tumor suppressor genes to be found on this chromosome.

As noted in the section on hypoxia inducible factors (HIF) in Chapter 4, the VHL gene product targets HIF for ubiquitination and degradation. Restoration of VHL function in $vhl^{-/-}$ knockout renal carcinoma cells suppresses the ability of such cells to form tumors in nude mice, and tumor suppression by the VHL protein can be overridden by a variant HIF protein not subject to ubiquitination by VHL.[582] Thus, the tumor suppressor effects of VHL appear to be mediated by regulation of HIF levels.

BRCA1 and BRCA2

Mutations in the tumor suppressor proteins BRCA1 and BRCA2 greatly increase the susceptibility of individuals to develop breast or ovarian cancer. The overall lifetime risk for a woman to develop breast cancer is about 10%, and the lifetime risk for ovarian cancer is 1.8%.[583] However, the lifetime risk of breast cancer for women carrying BRCA1 and BRCA2 mutations is 82%. For ovarian cancer, the risk is 54% for BRCA1 mutations and 23% for BRCA2 mutations.[583] These data were obtained in a study of Ashkenazi Jewish women, who as a group have a high incidence of BRCA1 and BRCA2 mutations. An earlier meta-analysis estimated that breast cancer risk by age 70 was 65% for BRCA1 mutation carriers and 45% for BRCA2 carriers (reviewed in Reference 584). The difference between these two data sets may be a result of the meta-analysis depending on a statistical model to predict the genetic status of relatives, whereas the data from the study of Askenazi women was based on genetically confirmed carriers. Of interest was the observation in the King et al. study[583] that mutation carriers born before 1940 had a 24% risk of developing breast cancer, but the incidence in carriers born after 1940 was 67%, suggesting that lifestyle differences between the two populations is a major factor. Such risk factors likely to be increased in women born after 1940 include earlier age at menarche and later stage at menopause, obesity, low physical exercise, delayed childbearing, and other hormonal exposure effects.

Both BRCA1 and BRCA2 appear to be involved in DNA repair pathway networks, although their mechanisms for this are not clear. It has been found that BRCA1 contains a peptide domain called BRCT that appears to be a common motif in other proteins involved in DNA repair (reviewed in Reference 585). BRCT domains bind phosphopeptides in protein-binding partners and typically occur as 80–100 amino acid sequences present as tandem repeats in BRCA1. These sequences recognize substrates phosphorylated by the DNA repair kinases ATM and ATR in response to γ-irradiation.[586] A mutation in the BRCT domain of BRCA1, which

prevents binding to phosphopeptides, may explain why this mutation predisposes women to breast and ovarian cancer. Yu et al.[587] futher showed that the BRCA1 BRCT domain binds a phosphorylated, BRCA-associated, DNA repair helicase. This interaction is cell cycle regulated and required for DNA damage–induced checkpoint control of the G_2 to M cell cycle phase transition. These authors[587] suggest that BRCT domain–containing proteins are a family involved in DNA repair and cell cycle checkpoint control.

Identification of Tumor Suppressor Genes

One way to identify new tumor suppressor genes is to ask via cell hybridization and chromosome transfer experiments which human chromosomes can suppress the malignant phenotype. As noted at the beginning of this section, the original finding that led to the tumor suppressor gene hypothesis was that the tumorigenic phenotype could be suppressed when malignant cells are fused with normal cells. Although this was a big advance in our knowledge, it was difficult to determine specifically which chromosomes harbored the tumor suppressor gene or genes. To get around this difficulty a technique to transfer single chromosomes into cells was used.[588] The technique involves isolating chromosomes from colcemid-treated cells, which prevents mitotic spindle formation, allowing each chromosome to condense as an individual unit within its own nuclear membrane. The cells are then enucleated by cytochalasin B treatment and centrifugation, producing microcells that can then be fused to recipient cells. To identify which microcell contains which chromosome(s), a gene marker, sometimes artificially introduced, such as a bacterial drug resistance gene like *neo*, is required. Using this technique, the presence of putative tumor suppressor genes can be located on a given chromosome. In most instances, transfer of a single copy of a normal chromosome is sufficient to induce growth inhibition in cell culture and/or suppression of tumor growth in nude mice. Specificity is demonstrated by the observation that random chromosomes not carrying a tumor suppressor gene do not suppress cell proliferation or tumor growth. Given the fact that a chromosome has a tumor suppressor function, the next task is to find the gene and characterize its mechanism.

Another method of finding tumor suppressor genes is to pharmacologically "unmask" them by turning back on epigentically silenced genes in cancer cells. Yamashito et al.[589] used such a method to stimulate re-expression of tumor suppressor genes in human esophageal squamous cell carcinoma (ESCC) cell lines. They did this by treating cells with the DNA methyltransferase inhibitor 5-aza-2-deoxycytidine and trichostatin A, a histone deacetylase inhibitor, and then using cRNA microarrays to analyze the epigenetically silenced genes that were turned back on. Some of these turned-on genes were inactivated in tumors and displayed tumor suppressor–like activity in gene transfection experiments. Of the 58 genes identified by this approach, 44 (76%) contained dense CpG islands in their promoters, and of these, a number had their promoter sequences hypermethylated in both primary ESCC tumors and ESCC cell lines.[589]

MECHANISMS OF GENE SILENCING

In earlier sections of this chapter, I discussed how chromatin packaging, DNA methylation, and histone and nonhistone protein modification can regulate gene expression. Here I will discuss other mechanisms, some natural cellular mechanisms and some not, for silencing gene expression.

Antisense

Antisense oligonucleotides can be synthesized that will base pair by conventional Watson-Crick base-pairing with gene transcripts. The antisense molecules are usually about 15–20 nucleotides in length and are usually made with a non-natural phosphodiester linkage, such as a phosphorothioate, which provides additional stability. They are designed to inhibit the function of mRNA by one of several mechanisms: degradation of the antisense-mRNA complex by RNase H, inhibition of mRNA splicing, or disruption of ribosome assembly (reviewed in Reference 590). Because antisense oligos can be designed to bind to mRNA in a sequence-specific manner, they can be used to specifically block synthesis of a protein involved in a disease state such as cancer.

For example, antisense oligos have been used to block the JNK2 kinase but not JNK1 or other kinases.[590] This sort of specificity has been a boon in the study of signal transduction pathways and in target validation for inhibition of such pathways. Antisense oligos have also been employed clinically; one is on the market for treatment of cytomegalovirus-induced retinitis, often associated with AIDS. There are, however, a number of limitations to therapeutic use of antisense molecules. These include (1) the pharmacologic effect may be slow in onset or ineffectual if the protein whose RNA is targeted has a long half-life; (2) proteins whose functionality is primarily regulated by post-translational mechanisms may not be good targets for antisense approaches; (3) most antisense oligos used clinically to date have immunostimulating side effects; (4) design of therapeutically effective antisense oligos is largely empirical because the conformation of mRNA targets and the most effective binding sequences are not usually known ahead of time; (5) in vivo delivery may be a problem (parenteral formulations seem to work for a number of indications but oral absorption is problematic).

The pharmacology of antisense oligonucleotides has been well characterized.[591] Phosphorothioate oligonucleotides have been the most widely studied. They bind to serum albumin and α_2-macroglobulin, which gives the oligos a long plasma half-life (40–60 hours). Since the serum protein binding is of relatively low affinity and is saturable, intact antisense oligos may be recovered in the urine. Absorption after parenteral delivery from a number of routes of administration (subcutaneous, intradermal, topical, and inhalation) is generally good, and systemic bioavailability may approach 90%. Phosphorothioates are widely distributed and highest accumulations are in liver, kidney, bone marrow, skeletal muscle, and skin. Very little crosses the blood–brain barrier. The bulk of an absorbed dose is eliminated by nuclease degradation.

Antisense oligos have been employed in a number of clinical trials for a number of indications, including retinitis (the only approved indication; see above), psoriasis, Crohn's disease, ulcerative colitis, and cancer.[591] Toxic side effects include complement activation, inflammatory conditions, inhibition of clotting, and flu-like symptoms. Potential genotoxic effects must be considered because of the potential for integration of the oligos into the genome (insertional mutagenesis) and for degradation of antisense oligos into toxic or carcinogenic metabolites.

Antisense oligonucleotides are also produced endogenously in cells ranging from prokaryotes to plants to humans. Thus, this appears to be an ancient mechanism of gene regulation. Endogenous antisense RNA transcripts that result from transcription of paired sequences on both strands of DNA have been reported in many organisms. Surprisingly, such antisense transcripts often code for proteins involved in disease biological functions (reviewed in Reference 592). A large number of noncoding antisense transcripts have also been identified. Their role appears to be mainly regulatory, and they play a role in genomic imprinting. They can also affect control of gene expression through a variety of mechanisms including transcription and mRNA processing, splicing, stability, transport, and translation. It is estimated that greater than 8% of the genes in the human genome produce sense–antisense transcripts.[592] This is most likely an underestimate because the methods used in the study by Yelin et al.[592] would not have detected transcripts without poly A tails, which many antisense transcripts do not have, nor would they have found sense–antisense transcripts that were not in public databases, that spanned introns, or that were trans-encoded rather than cis-encoded (i.e., transcribed from both DNA strands in the same gene locus).[593]

The antisense transcripts anneal with the sense mRNA, and these double-stranded (ds) RNAs are targets for dsRNA-specific nucleases and dsRNA adenosine deaminase. It is now known that these dsRNAs can also produce small interfering RNAs (siRNAs) through the action of a ribonuclease called "Dicer" to produce the phenomenon of RNA interference (RNAi; see below). In sum, it is evident that the intracellular production of antisense transcripts and their small dsRNA nuclease products are major contributors to gene regulation in prokaryotic and eukaryotic, including human, cells.

Ribozymes

Ribozymes are RNA molecules that have catalytic enzyme activity. By definition they include ribo-

somes, which catalyze sequence specific peptide bond formation, self-slicing group I and II introns involved in tRNA processing, and "hammerhead" and "hairpin" RNA-cleaving ribozymes. The latter were originally identified in plant virus satellite RNA.[594] The focus in this discussion will be on the hammerhead and hairpin ribozymes (Fig. 5–20).[595]

The specificity of the hammerhead and hairpin ribozymes for cleavage of an mRNA target is base pairing between the ribozyme and the target. For example, the hammerhead ribozyme cleaves after UX dinucleotides, where U = uridine and X = any ribonucleotide except guanosine (the best cleavage rate is where X = cytosine). In addition to the complementary base-pairing sequence on the ribozyme that base pairs to a target sequence such as GUC, CUC, or UUC, hammerhead ribozymes have a 22-nucleotide catalytic domain (Fig. 5–20A) and a base-pairing sequence flanking the susceptible 3′, 5′-phosphodiester bond.[595] The cleavage site is 3′ to the recognition sequence and the reaction forms a terminus containing a 2′, 3′ cyclic phosphodiester and a 5′ hydroxyl terminus on the 3′ fragment. The hairpin ribozyme (Fig. 5–20B) has four helices and five loop regions formed between a 50-base catalytic sequence and a 14-base target RNA sequence. The target recognition motif is a BNGUC sequence where B = G, C, or U and N = any nucleotide. Cleavage occurs 5′ to the guanosine nucleotide.

Design of ribozymes for therapeutic indications is still largely empirical, because a target sequence may have a high mutation rate (e.g., HIV) or the secondary and tertiary structures of RNA and its binding to proteins that may obscure the target site in vivo are not evident a priori. There are alogrithims that can simulate in vivo secondary structure that may help predict the availability of a target sequence, but this still needs to be conformed experimentally.

For all oligonucleotide and gene therapy approaches, delivery of the therapeutic nucleic acid is key to a successful therapeutic outcome. As has often been said, the three most important issues here are "delivery, delivery, delivery." In addition, other key issues are extracellular (plasma, interstitial fluid) and intracellular stability of the oligonucleotide, target accessibility, colocalization of target and ribozyme in cells, and optimal

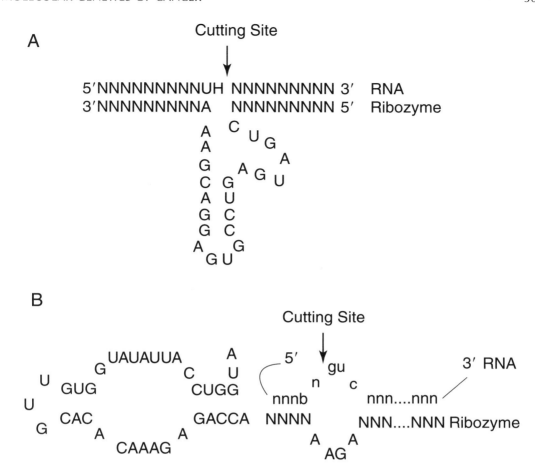

Figure 5–20. Structures of hammerhead (A) and hairpin ribozymes (B). (From Sun et al.,[595] with permission.)

catalytic activity and specificity of the ribozyme. Because RNA and its oligonucleotides have a very short half-life in body fluids, they must be protected by approaches such as derivatizing the 3' end of the ribozyme and 2' position of pyrimidines or packaging the ribozyme into cationic lipids (liposomes). If a long-term suppression of gene expression is desirable, for HIV or cancer applications, delivery of the ribozyme as a ribozyme-coding gene may be desirable.

A gene-encoded ribozyme targeting the *tat* gene of HIV has been tested in clinical trials. The approach was to transfect CD34$^+$ stem cells ex vivo with a gene encoding a *tat*-targeted ribozyme using a murine leukemia virus vector.[595a] Phase I trials showed that patients who had their autologous CD34$^+$ cells transfected in this way had long-lived ribozyme-expressing cells in their peripheral blood. The concept behind this treatment is to protect CD4$^+$ T lymphocytes, downstream progeny of the CD34$^+$ stem cells, from HIV-mediated destruction. Phase I data have demonstrated the safety of this approach.

Ribozymes have also been developed to target expression of oncogenes in cancer cells. Cancer-related targets include *bcr-abl*, the *flt*-1 gene encoding the Flt-1 receptor for VEGF (reviewed in Reference 594), the *erbB*-4 gene in estrogen receptor–positive human breast cancer cells,[596] and hepatocyte growth factor and c-*met* genes in human glioblastoma cells.[597] Most of the studies testing ribozymes as inhibitors have only been performed in cell culture systems.

Another intriguing approach is the use of a *trans*-splicing ribozyme to repair mutant p53 transcripts.[598] In this case, the ribozyme cleaves the mutant transcript, releases the downstream RNA sequence containing the mutated bases, and replaces the sequence with a 3' exon that encodes the wild-type sequence. This was successfully achieved in cultured human osteosarcoma cells.

DNAZYMES

DNAzymes are single-stranded oligodeoxy-nucleotides with enzymatic activity similar to ribozymes in that DNAzymes can also base pair with specific target mRNA sequences and cleave them. An advantage of DNAzymes over ribozymes is that the former are easier to synthesize and are more stable in body fluids such as serum (reviewed in Reference 599). DNAzymes also have greater substrate target flexibility than that of hammerhead ribozymes and can cleave effectively between almost any RNA sequence that has an unpaired purine–pyrimidine dinucleotide. This means that they can cleave sequences like the AUG translation start codon of mRNA. Since the translation start site has less secondary structure than other parts of an mRNA molecule, it is easier to predict target substrate sites for DNAzymes than for ribozymes.

The most widely studied DNAzyme is called the 10-23 DNAzyme, because it was derived from the 23rd clone of the 10th cycle of an in vitro selection.[595] A number of cancer-related gene transcripts have been successfully targeted in cell culture systems. These include c-*myc*, *bcr-abl*, the human papilloma virus 16 E6 and E7 genes (reviewed in Reference 595), and the VEGF receptor 2.[599]

RNAI

Over the past two decades various approaches have been employed to target gene expression, either to inhibit it or provide a new gene expression profile in cells (reviewed in Reference 600). Gene-targeted therapy had its first wave of enthusiasm when it was shown that a base sequence could be synthesized that would bind by Watson-Crick base-pairing in cells to block specific mRNA translation, so-called antisense therapy. This therapry has had its ups and downs as a clinical approach as noted above.

A second wave of enthusiasm occurred with the application of catalytic RNAs (ribozymes) for gene-targeted therapy. This approach had an advantage over antisense approaches in that it was easier to deliver functional ribozyme genes to cells via plasmids or viral vectors that could be controlled by promoter-based expression. In addition, ribozymes and their "partners," DNAzymes, are catalytic and, like an enzyme, can be reused over and over to chew up a specific mRNA species in cells.

The next wave of interest in gene-targeted therapy came with the discovery of RNA interference (RNAi). The relative advantages and disadvantages of these three approaches to gene-targeted therapy are shown in Table 5–13. As was noted above for antisense and ribozymes, one of the disadvantages of all three approaches is the potential for off-target effects that result from blockade of expression of genes required for normal cell metabolism.

The phenomenon of RNAi was first discovered in *C. elegans* in 1998 as a response to injected or fed (in growth medium) double-stranded RNA that triggered gene silencing.[601] This turned out to be a very potent effect: it only took a few dsRNA molecules and the gene silencing was observed in first-generation progeny (reviewed in Reference 602). This phenomenon has subsequently been observed in a wide variety of organisms, including mammals. It turns out that RNAi is an evolutionarily conserved mechanism elicited as a defense mechanism to control expression of foreign genes such as those introduced by viral infection. The mechanism for production of the dsRNA that causes the RNAi response isn't totally clear, but it may result from production of dsRNA by transcription of a viral RNA sequence by RNA-directed RNA polymerase that recognizes aberrant RNA transcripts such as those expressed by an invading virus. The evidence for this idea is that these polymerases have been shown to be essential for the RNAi response.[602] The dsRNA causes sequence-specific mRNA degradation. The mediators of this process are small, interfering RNA duplexes called siRNAs, which are produced from longer dsRNAs by cleavage with a specific nuclease, Dicer (see below). The siRNAs are about 21 nucleotides in length and are base-paired RNA duplexes with 3' end overhangs. As may be predicted, the dsRNA

Table 5–13. Relative Strengths and Weaknesses of Antisense Technologies

Approach	Advantages	Disadavantages
Antisense ODNs	Can be modified to improve selectivity and efficacy Can be targeted to introns Easy to make	Can induce interferon (if long and has CpG) Can bind proteins (aptamer activity) Only exogenous delivery possible (synthetic) Off-target effects
Ribozymes	Can discriminate single-base polymorphisms Can be used to correct defects Sequences can be appended to change target specificity Simple catalytic domain Can target introns/subcellular compartments	Requires GUC triplet—limits choice of target Binds proteins (aptamer activity)
DNAzymes	Inexpensive to make Good catalytic properties Can be modified for systemic delivery	Only exogenous activity Off-target effects?
RNAi	Effective at low concentrations Bypasses interferon pathway Can be delivered by multiple pathways Tissue-specific expression possible Nontoxic? Lasts longer?	Cannot target nuclear RNAs or introns No option for improving if target refractory Some reports of off-target effects

ODN, Oligodeoxynuceotides; RNAi, RNA interference.

(From Scherer and Rossi,[600] reprinted by permission from Macmillan Publishers Ltd.)

structure has to be unwound to base pair with mRNAs, which are single-stranded structures. Part of the excitement about this phenomenon is that dsRNA can be introduced into cells directly or by a gene transfection that leads to potent, heritable inhibition of expression of a specific gene.

The way in which this process occurs in a cell is illustrated in Figure 5–21 (see color insert). Double-stranded RNAs can be presented to or introduced into cells by replicating viruses, transfection with exogenous genes, or synthetically produced dsRNAs. These are recognized and processed into siRNAs by the ribonuclease Dicer. The double stranded siRNAs are bound into an RNA-induced silencing complex (RISC). The RISC complex is activated by ATP to process and unwind the siRNA, which in turn can base pair with the target mRNA and degrade it and/or prevent it from being translated. In addition, the processed siRNA can induce DNA methylation and chromatin remodeling to block active transcription sites. Moreover, in some cell types such as plants, siRNAs can be amplified by an RNA-directed RNA polymerase. Thus, the RNAi mechanism has a number of advantages: it has multiple mechanism for inhibiting gene transcription and translation; it is a potent, natural (endogenous) mechanism for blocking gene expression; it can be induced directly or indirectly by introducing dsRNAs or genes that produce them; and it is heritable by daughter cells. Because of these advantages, siRNA is a powerful tool to examine the function of a cellular genome by gene silencing. For example, this has led to a complete definition of genome function in such simple organisms as *C. elegans*, in which the functional analysis of all its approximately 19,000 genes was carried out (reviewed in Reference 603).

This use has also led to high interest in using siRNA as a method to block gene expression in clinical settings, such as the treatment of cancer, viral diseases, and age-related macular degeneration. There are, however, some limitations to gene silencing by transfected siRNAs. These include (1) the transient nature of expression of siRNA in transduced cells due to dilution as cells divide; (2) the decreased ability to block expression of proteins with long half-lives; (3) the difficulty of delivering siRNA directly into cells in vivo because of its instability from RNAses in body fluids; and (4) the expense of siRNAs, which have to be chemically or enzymatically synthesized.

There are a number of phenomena related to the RNAi mechanism. These include "transitive" RNAi, microRNA (miRNA), small temporal RNA (stRNA), and short hairpin RNA (shRNA). Each of these are described below.

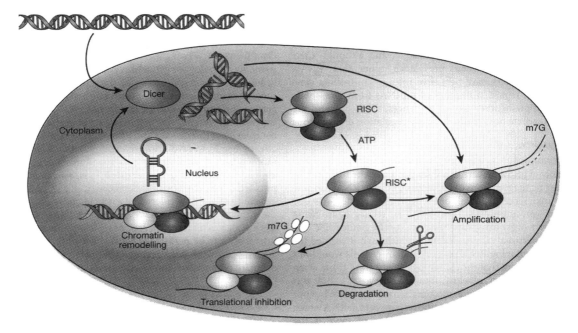

Figure 5–21. A model for the mechanism of RNAi. Silencing triggers in the form of double-stranded RNA may be presented in the cell as synthetic RNAs, replicating viruses, or may be transcribed from nuclear genes. These are recognized and processed into small interfering RNAs by Dicer. The duplex siRNAs are passed to RISC (RNA-induced silencing complex), and the complex becomes activated by unwinding of the duplex. Activated RISC complexes can regulate gene expression at many levels. Almost certainly, such complexes act by promoting RNA degradation and translational inhibition. However, similar complexes probably also target chromatin remodeling. Amplification of the silencing signal in plants may be accomplished by the siRNAs priming RNA-directed RNA polymerase (RdRP)-dependent synthesis of new dsRNA. This could be accomplished by RISC-mediated delivery of an RdRP or by incorporation of the siRNA into a distinct, RdRP-containing complex. (From Hannon,[602] reprinted with permission from Macmillan Publishers Ltd.)

Transitive RNAi

In lower organisms such as plants and *C. elegans*, the RNAi process, when triggered even by a small amount of dsRNA, can silence genes throughout the whole organism. This phenomenon is called *transitive RNAi* and refers to the movement of a silencing signal along a particular gene and from cell to cell (reviewed in Reference 602). When this response is triggered, siRNAs complementary to regions of a transcript upstream from the sequence targeted directly by the siRNA are produced and thus may silence other genes that have complementary sequences. For spreading of the RNAi response through an organism to occur, there must be some means for cell-to-cell transmission of the dsRNA or siRNA signal. In plants, this appears to occur via cytoplasmic bridges called *plasmodesmata*, which allow movement of RNA and proteins from cell to cell. In addition, the silencing signal must be able to be passed over a longer distance through the plant vasculature. In *C. elegans*, a transmembrane protein, Sid 1, may act as a channel for cellular uptake of a silencing signal. Although Sid 1 is not present in *Drosophila*, a Sid 1 homolog is found in mammalian cells, suggesting the possibility that some RNAi signals could be transmitted from cell to cell.[602]

The phenomenon of transitive RNAi does not occur in mammalian cells, most likely because dsRNAs, once they reach a certain cellular level, induce the production of interferon. This leads to a shutdown of translation, induction of RNAse L, and apoptosis.[604] This group of effects is one of the downsides of dsRNA therapy because vigorous induction of interferon could produce unwanted systemic effects on the immune system.

Micro-RNA

Micro-RNAs (miRNAs) are also small RNAs that have gene silencing activity. Unlike siRNAs, which are derived from dsRNAs produced from aberrant gene expression such as genes from viruses that have infected cells, miRNAs are transcribed from noncoding genes in the genome (what used to be called "junk-DNA"). Some estimates are that introns and other noncoding RNAs make up 98% of the transcriptional output of the human genome. There is speculation that this large amount of miRNA provides the functional regulator that makes humans so different from mice, with which we share about 95% of the same genes. Like siRNA, miRNA is processed by Dicer into about 22 nucleotide long sequences (Fig. 5–22; see color insert).[605]

Also like siRNA, miRNAs can bind specific mRNAs and degrade them or block their translation into proteins. These miRNAs could also be involved in diseases such as cancer by aberrantly regulating gene expression. The choice between cleaving an mRNA and blocking its translation appears to be governed by the degree of base-pairing match between the miRNA and its target mRNA. Degradation of mRNA is the choice when there is the best match.[605] Since miRNA can inhibit translation of even imperfectly matched mRNA targets, it is likely that a single miRNA can target the expression of multiple genes.

The expression of miRNAs correlates to a cell's developmental lineage and stage of differentiation and also reflects the differentiation state of tumors.[606] In general, a down-regulation of miRNAs in cancers compared to normal tissues has been observed. However, of the 200+ miRNAs described in humans, some clusters are overexpressed in some cancers. For example, a cluster of miRNAs derived from the *mir*-17-92 miRNA gene locus is overexpressed in human B-cell lymphomas.[607] Similarly, in breast can-

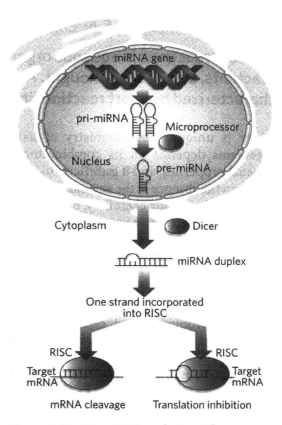

Figure 5–22. Micro RNA production. The precursor of an miRNA (pri-miRNA) is transcribed in the nucleus. It forms a stem-loop structure that is processed to form another precursor (pre-miRNA) before being exported to the cytoplasm. Further processing by the Dicer protein creates the mature miRNA, one strand of which is incorporated into the RNA-induced silencing complex (RISC). Base pairing between the miRNA and its target directs RISC to either destroy the mRNA or impede its translation into protein. The initial stem-loop configuration of the primary transcript provides structural clues that have been used to guide searches of genomic sequence for candidate miRNA genes. (From Meltzer,[605] reprinted with permission from Macmillan Publishers Ltd.)

cer, two miRNAs, *mir*-21 and *mir*-155, are up-regulated.[608] Three miRNAs are down-regulated in breast cancer: *mir*-10b, *mir*-125b, and *mir*-145. Thus, it appears that miRNAs can act either as oncogenes or tumor suppressor genes. There is evidence of interaction between miRNA expression and the *ras* and *myc* oncogenes.[605,609] c-Myc has been shown to activate expression of a cluster of six miRNAs on human chromosome 13, and two of these miRNAs, *mir*-17-5p and *mir*-20a, negatively regulate E2F1, which is one of the pro-mitogenic genes turned on by c-Myc.[609] This is an apparent negative feedback mechanism for c-Myc's proliferation stimulating effects. The sum of all these data makes it clear that there is still a lot to learn about the role of miRNAs in the regulation of cellular function.

Small Temporal RNA

Small temporal RNA (stRNA) was identified as transcripts from mutant developmental genes *lin*-4 and *lin*-7 in *C. elegans*. These gene loci encode a 70-nucleotide RNA that is processed by Dicer into 21 nucleotide forms called stRNAs because of their transient expression during specific developmental phases. They do not degrade mRNAs but block their translation.

Short Hairpin RNA

These shRNA forms are modeled on miRNAs and are endogenously produced in plants, *C. elegans*, *Drosophila*, and trypanosomes in the form of large (about 500 base pair) hairpin structures that are also processed by Dicer into siRNA molecules. ShRNAs can either be synthesized chemically or introduced into cells by plasmids or viral vectors. The advantage of shRNAs is that they are more stable in vivo[602] and thus may have an advantage for therapeutic uses.

GENE THERAPY

Gene therapy has a checkered history, to say the least. The original wave of enthusiasm has been dampened by some disappointing clinical data and some tragic results. The first wave of enthusiasm came with the treatment of a severe immune deficiency syndrome in children associated with an inherited genetic defect in adenosine deaminase (ADA). These clinical trials began in 1990 and initial trial results after 4 years were published in 1995 (reviewed in Reference 610).

ADA deficiency is due to absent ADA enzyme activity by deletion or inactivation of the gene or to a mutation that leads to production of a faulty protein. Clinically, this leads to decreased production of lymphocyte precursors in the thymus and low levels of circulating T lymphocytes because of decreased production and poor in vivo survival. The therapeutic approach, then, was to try to replace the defective gene by introducing the ADA gene into T lymphocytes. The alternative, partially successful approach had been (and still is) to treat patients with the enzyme itself, formulated as a polyethylene glycol-coupled protein (PEG-ADA). The gene therapy approach was attractive because, if successful, it would provide a much longer therapeutic effect.

After demonstration of the effectiveness of ADA gene–transduced T lymphocytes injected intraperitoneally into immunodeficient mice, a clinical trial at the National Institutes of Health (NIH) was designed and initiated (reviewed in Reference 610). The idea was to transduce ex vivo peripheral blood T cells of patients who were already on PEG-ADA therapy, using a Moloney murine leukemia virus–based retroviral vector containing a copy of the normal human ADA gene. The cells were first stimulated by exposure to anti-CD3 antibody and IL-2. In the initial trial, two children received monthly or bimonthly infusion of transduced T cells. One child had long-term persistence of ADA-transduced T lymphocytes (30% of peripheral T cells), but the other patient only had 1% of transduced cells present after several months. Although both patients had improved immune function, they had remained on PEG-ADA therapy, so the trial was only a qualified success. Nevertheless, the study did demonstrate that it was possible to transduce human peripheral blood T lymphocytes, and get them, or more likely their progeny, to persist in vivo for several years. Subsequent clinical trials of gene therapy for ADA deficiency have used transduction of hematopoietic stem cells derived from bone marrow or umbilical cord blood.[610] These trials have also shown some success, but again the patients were continued on PEG-ADA therapy.

Subsequent to these ground-breaking studies, a large number of clinical trials have been carried out. Over 300 clinical protocols involving gene therapy have been approved worldwide and more than 3500 patients have been treated with experimental gene therapies, mostly in phase I/II trials (reviewed in Reference 611). Most of these trials have been in cancer patients, but other indications have been for inherited immunodeficiency disorders, cystic fibrosis, infectious diseases (e.g., AIDS), hematopoietic disorders such as hemophilia, peripheral vascular disease, and rheumatoid arthritis. Most of these studies have involved ex vivo transduction of autologous cells and then reinjection into the donor patient. The majority of these trials have used retroviral vectors to deliver the gene, but a wide variety of viral and nonviral delivery mechanisms have been employed. Viral delivery systems include adenovirus, adeno-associated virus, poxvirus, and herpes virus vectors; nonviral delivery systems have involved naked DNA plasmids, lipid conjugates, "gene gun" delivery, electroporation (reviewed in References 611 and 612), ultrasound-enhanced transduction,[613] DNA delivery from polymer matrices,[614] cell-targeted viral vectors,[615] and gene-encoding ribozymes.[595]

The first real success for gene therapy came in the treatment of children with an inherited X-linked, severe combined immunodeficiency (SCID) syndrome.[616] This syndrome is due to a mutation in the gene encoding the common γ (γc) chain that is an essential component of five cytokine receptors, all of which are necessary for the development of T lymphocytes and natural killer (NK) cells. In patients lacking the γc chain, there is a complete absence of mature T and NK cells. Untreated, this condition is usually fatal during the first year of life because of severe recurrent infections.

Some alleviation of the deficiency can be achieved by transplantation of HLA-matched hematopoietic stem cells. However, most transplanted patients continue to have deficient B-lymphocyte function, requiring lifelong immune globulin replacement therapy, and many have persistent T-cell deficiencies as well. This devastating unmet medical need provided the rationale for a gene therapy approach. Hacein-Bey-Abina et al.[616] transduced ex vivo autologous CD34+ bone marrow cells from SCID children with a retroviral vector containing the γc gene, reinjected the transduced cells, and followed immune system function for up to 2.5 years posttransplant. The remarkable results were that transduced T and NK cells appeared in the blood of four of the five original patients; T-cell responses were mainly normal; and serum immunoglobulin levels were sufficient to correct or prevent infectious, allowing the patients to live an essentially normal life without the need for immunoglobulin therapy. All in all, 9 of 10 children treated with this protocol were cured.

Unfortunately, there has been a cloud over this result. Two of the successfully treated children developed a T-cell leukemia due to an insertion of the retroviral vector into the promoter of the proto-oncogene LM02.[617] This effect, while it might have been predicted as a potential risk due to insertional mutagenesis, was considered to be remote, and had not been observed in previous clinical trials with retroviral vectors. It has now been shown that transcription start sites in the human genome are favored targets for retroviral gene integration,[618] so this could occur more frequently than previously predicted, thus having important ramifications for gene therapy with such vectors.

The question is, why hasn't this phenomenon been observed in the hundreds of patients previously treated with retroviral vectors? There may be several reasons why it was first observed in the SCID trial. First of all, it only occurred in the two youngest children, at a time when their hematopoietic systems were presumably still developing. Secondly, the LM02 gene codes for a transcription factor required for hematopoiesis, and disruption of that gene during key phases of hematopoietic stem cell proliferation and differentiation could lead to an unbalanced proliferative effect. Third, these children already have an inborn defect in hematopoietic cell differentiation and function, and overexpression of the LM02 gene could lead to a more pronounced aberrant gene regulation defect than in a normal adult hematopoietic system.

Gene Therapy for Cancer

Gene therapy for cancer accounts for the majority of gene therapy clinical trials. Targets for this include replacement of tumor suppressor genes, "suicide genes" to activate prodrugs, antiangiogenic gene therapy, cytokine-based gene transfer, and delivery of drug resistance genes to hematopoietic stem cells to protect them from the bone marrow toxicity of chemotherapeutic agents.[619]

The tumor suppressor gene p53 is the most commonly mutated gene in human cancers. Hence, it is a good target for gene replacement therapy. Re-expression of p53 in human colon cancer cell lines bearing a mutated gene inhibits tumor cell proliferation. In a murine model of p53-mutated colon cancer, injection of an adeno-viral vector encoding the WT *p53* gene into tumors resulted in tumor regression and enhanced survival (reviewed in Reference 620). Clinical trials with *p53* gene replacement have been initiated for a number of cancers including colon and head and neck cancers. In general, the procedures were well tolerated. The main side effects were fever and transient liver enzyme abnormalities. One key question for this and other gene therapy approaches to cancer is how many cells in a tumor need to be transfected to get a therapeutic effect? There is some evidence for "bystander" effects from p53 transfection, probably due to an antiangogenesis effect.[620] Other potential tumor suppressor gene targets for which there is at least preclinical demonstration of efficacy are PTEN, E-cadherin, C-CAM, BRCA-1, and pHyde (reviewed in Reference 621).

Suicide gene therapy approaches include transfection of tumor cells with herpes simplex virus-thymidine kinase (HSV-TK) to activate the prodrug ganciclovir and cytosine deaminase to convert the nontoxic compound 5-fluor-ocytosine into cytotoxic 5-fluorouracil (reviewed in Reference 619). Suicide gene clinical trials have been carried out in prostate cancer, mesothelioma, and glioblastoma. Cytokine gene therapy approaches have included gene transfer via tumor-homing lymphocytes bearing genes encoding interleukins (IL-1B, IL-2, IL-4, IL-12), GM-CSF, and interferon-γ (IFN-γ). Clinical trials include IL-12 delivered by a vaccinia virus vector for mesothelioma and GM-CSF and IFN-γ delivered in retroviral vectors for melanoma. Some clinical responses have been observed in these trials with small numbers of patients.[619]

Gene therapy for drug resistance is aimed at protecting normal hematopoietic stem cells from the suppressive effects of cytotoxic drugs. Such approaches include transfer of the multidrug resistance gene MDR-1, dihydrofolate reductase gene variants that have reduced affinity for methotrexate, and forms of O^6-alkylguanine-DNA alkyltransferase that remove cytotoxic alkyl lesions from guanine alkylated on the O^6 position by drugs like BCNU.[619]

PERSONALIZED MEDICINE AND SYSTEMS BIOLOGY

Much has been made of the potential break-through in medicine that completion of the human genome sequence will provide. So far, only a glimmer of the advantage for human health has been seen. The potential impact is huge. Once fully realized, this knowledge will enbale prediction early in life of who is likely to get a certain disease and allow institution of chemoprevention or lifestyle changes to delay or circumvent the worst sequelae of such diseases. It will be used to develop pharmacogenomic profiles predicting who will and who won't respond to a certain drug and who will and who won't be likely to suffer severe side effects for a drug. Knowledge of the human genome sequence will provide a complete profile of the genetic alterations involved in the pathophysiology of various diseases such as cardiovascular disease, diabetes, autoimmune diseases, and cancer in individual patients.[622] Moreover, it will allow for a complete genetic and biochemical profile of the diseased cells themselves. The latter project is already being done to profile individual cancers in people. Most likely, the biggest impact of characterization of the diseased cell will be realized first in oncology. In sum, the ability to do all these things is leading to a new age of enlightenment called "personalized medicine."

Personalized medicine has as its principle that for every given human disease, the molecular changes that occur in patients' tissues, the rate and nature of disease progression, and the way each person responds to drugs is unique. This is not taken to mean that each of the four billion people who populate the earth will require separate diagnostic and separate therapies, but that individuals can be stratified into subgroups based on their genetic profiles and that of their disease. It is not easy for big pharmaceutical companies to come to grips with this prospect, because the age of the blockbuster drug is evolving into the age of orphan or sub-orphan drugs.

To realize all these advantages, current technology will have to evolve to provide rapid, cost-effective, and readily available procedures. Some of this technology is already at hand or close to being realized. It requires further refinement and scale-up to make it practical.

Here are some of the challenges:

1. There are three billion bases in the human genome. When each of the 10^{14} cells in the average human divides, every base has to be copied perfectly or a potential disease-causing mutation could occur. Of course, in most people, this is a rare phenomenon and each of us has robust DNA editing and repair systems to keep this from happening, but all it takes is for one of these base changes in a key gene to sneak through. We also have mechanisms that recognize mutated cells and kill them off by detecting significant DNA damage, leading to apoptosis. In addition, our bodies have mechanisms that recognize altered cell surface molecules and kill off aberrant cells by immune defense systems.

2. More than two million single nucleotide polymorphisms have been detected in the human genome database. Many of these are involved in determining an individual's susceptibility to disease, response to environmental toxins and drugs, and other parameters of general health and longevity.

3. Many of the technologies for rapid, high-throughput, cost-effective analyses of genomic, proteomic, and metabolomic profiles are in their infancy.

4. The science of systems biology is revealing that the interactions among DNA, RNA, proteins, carbohydrates, lipids, and, indeed, all the components of cells and tissues is extremely more complex than had been realized previously.

So just what is "systems biology"? Fundamentally, it is a conceptual framework to study, to think about, and to quantify the types of biological information contained in cells, tissues, organisms, and populations of individuals. It is these interacting networks that regulate and modulate life. Systems biology begins at the level of a cell's components, and this is where much of the focus is currently. This includes the study of genomics, proteomics, and other cellular elements and the interactions among them. For example, systems biology attempts to do the following:

1. Analyze "biological systems by measuring steady-state and dynamic relationships of a system in response to genetic or environmental perturbations across their developmental or physiological time dimensions;

2. Define protein biomodules (e.g., groups of proteins that execute a particular phenotypic function such as galactose and glucose metabolism or protein synthesis) and the protein networks of life (e.g., the skeletal framework of cells and their signal transduction pathways);

3. Delineate gene regulatory networks that govern the expression patterns of proteins across developmental or physiological time spans; and

4. Delineate the cells' effective integration of the protein and gene regulatory networks" (from Reference 623).

As noted above, the goal of systems biology is to integrate biological information across several hierarchical levels, including DNA, RNA, protein, protein–protein interactions, gene regulatory networks, cellular communication systems, tissue and organ interactions (e.g., hormonal signaling), and ecological systems. A clue as to how complex this will be can be seen from the incredibly complex genetic and protein–protein interaction networks in lower organisms. For example, global mapping of a yeast genetic interaction network containing 1000 genes revealed over 4000 interactions.[624] A single large network of 1548 proteins in yeast showed 2538 interactions.[625] Seventy-two percent of 1393 characterized proteins with at least one partner of known function predicted 364 previously uncharacterized functions. In C. elegans, more than 4000 protein–protein interactions were identified in a subset and the current version of the "worm interactions" contains over 5500 interactions.[626]

A picture is always worth a thousand words; the protein–protein interaction map of Drosophila (Fig. 5–23; see color insert) gives one some idea of the biological complexity that systems biology is trying to define. In Drosophila, a total of 10,623 predicted gene transcripts that were isolated and screened against DNA libraries produced a map of 7048 proteins and 20,405 interactions.[627] Statistical modeling of the networks

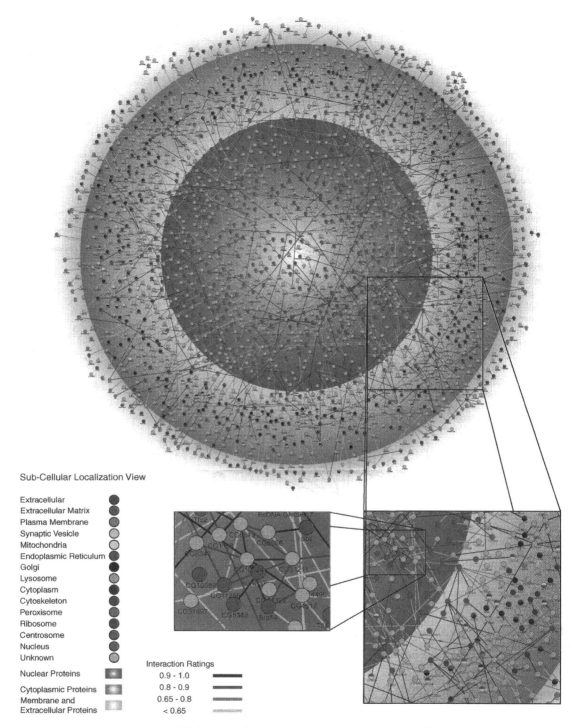

Figure 5–23. Global views of the protein interaction map: subcellular localization view. This view shows the fly interacton map with each protein colored by its gene ontology cellular component annotation. This map has been filtered by only showing proteins with less than or equal to 20 interactions and with at least one gene ontology annotation (not necessarily a cellular component annotation). Proteins for all interactions with a confidence score of 0.5 or higher are shown. This results in a map of 2346 proteins and 2268 interactions. (From Giot et al.,[627] with permission from the American Association for the Advancement of Science.)

showed two levels of organization: short-range organization, most likely corresponding to more localized pathways, and a global organization, presumably corresponding to broader, more complex connecting pathways. Analysis of these interactions detected known pathways, extended pathways, and previously unknown pathway components.

These data provide some insights into how complex a problem it will be to define the systems biology of human beings. Nevertheless, recent technological advances allow some approaches to this issue. Interactions among biologists, chemists, physicists, engineers, computer scientists, and mathematicians will be required to figure all this out. The technologies of gene expression arrays, proteomics, molecular imaging, electrical engineering, nanotechnology, and microfluidics will all be involved in developing the "lab-on-a-chip" or the "nanolab" of the future.[623]

One of the neat new technologies involves the use of nanowire sensors, of nanometer (10^{-9} meters) or less diameter, coated with a probe molecule to sense a particular signature of gene or protein expression. These nanowires can also have built-in mechanisms that produce an electrochemical signal that can detect, with great sensitivity, molecular interactions. A visionary's view of this is that this scale of instrumentation can lead to hand-held, microfluidics-based systems to detect single-cell genomic or proteomic expression, enabling a physician to analyze a patient's blood sample of a few microliters (obtained by a finger prick) to assess up to 10,000 functions. The impact of this sort of technology on the future of medicine is mind boggling indeed.

References

1. T. Boveri: *Zur Frage der Erstehung maligner Tumoren.* Jena: Fisher, 1914.
2. P. C. Nowell and D. A. Hungerford: Chromosome studies on normal and leukemic human leukocytes. *J Natl Cancer Inst* 25:85, 1960.
3. J. D. Rowley: A new consistent chromosomal abnormality in chronic myelogenous leukemia identified by quinacrine flourescence and Giemsa staining. *Nature* 243:290, 1973.
4. A. A. Sandberg and D. K. Hossfeld: Chromosomes in the pathogenesis of human cancer and leukemia. In J. F. Holland and E. Frei III, eds.: *Cancer Medicine.* Philadelphia: Lea & Febiger 1973, pp. 151–177.
5. T. Caspersson, L. Zech, and C. Johanssen: Differential binding of alkylating fluorochromes in human chromosomes. *Exp Cell Res* 60:315, 1970.
6. J. P. Chaudhuri, W. Vogel, I. Voiculescu, and U. Wolf: A simplified method of demonstrating Giemsa band pattern in human chromosomes. *Humangenetik* 14:83, 1971.
7. J. D. Rowley: Mapping of human chromosomal regions related to neoplasia: Evidence from chromosomes 1 and 17. *Proc Natl Acad Sci USA* 74:5729, 1977.
8. B. Weir, X. Zhao, and M. Meyerson: Somatic alterations in the human cancer genome. *Cancer Cell* 6:433, 2004.
9. J. A. Downs and S. P. Jackson: Protective packaging for DNA. *Nature* 424:732, 2003.
10. R. J. DeLange and E. L. Smith: Histone function and evolution as viewed by sequence studies. *Ciba Found Symp* 28:59, 1975.
11. S. C. R. Elgin and H. Weintraub: Chromosomal proteins and chromatin structure. *Annu Rev Biochem* 44:725, 1975.
12. T. A. Langan, S. C. Rall, and R. D. Cole: Variation in primary structure at a phosphorylation site in lysine-rich histones. *J Biol Chem* 246:1942, 1971.
13. L. Hong, G. P. Schroth, H. R. Matthews, P. Yau, and E. M. Bradbury: Studies of the DNA binding properties of histone H4 amino terminus. *J Biol Chem* 268:305, 1993.
14. B. D. Strahl and C. D. Allis: The language of covalent histone modifications. *Nature* 403:41, 2000.
15. S. L. Berger: The histone modification circus. *Science* 292:64, 2001.
16. D. Krylov, S. Leuba, K. van Holde, and J. Zlatanova: Histones H1 and H5 interact preferentially with crossovers of double-helical DNA. *Proc Natl Acad Sci USA* 90:5052, 1993.
17. G. H. Goodwin, C. Sanders, and E. W. Johns: A new group of chromatin-associated proteins with a high content of acidic and basic amino acids. *Eur J Biochem* 38:14, 1973.
18. J. M. Walker, G. H. Goodwin, and E. W. Johns: The similarity between the primary structures of two non-histone chromosomal proteins. *Eur J Biochem* 62:461, 1976.
19. J. M. Pash, P. J. Alfonso, and M. Bustin: Aberrant expression of high mobility group chromosomal protein 14 affects cellular differentiation. *J Biol Chem* 268:13632, 1993.
20. P. Byvoet, G. R. Shepherd, J. M. Hardin, and B. J. Noland: The distribution and turnover of labeled methyl groups in histone fractions of cultured mammalian cells. *Arch Biochem Biophys* 148:558, 1972.
21. B. G. T. Pogo, A. O. Pogo, V. G. Allfrey, and A. E. Mirsky: Changing patterns of histone acetylation and RNA synthesis in regeneration of the liver. *Proc Natl Acad Sci USA* 59:1337, 1968.
22. Y. Zhang: No exception to reversibility. *Nature* 431:637, 2004.

23. D. Y. Lee, J. J. Hayes, D. Pruss, and A. P. Wolffe: A positive role of histone acetylation in transcription factor access to nucleosomal DNA. *Cell* 72:73, 1993.

24. J. Sommerville, J. Baird, and B. M. Turner: Histone H4 acetylation and transcription in amphibian chromatin. *J Cell Biol* 120:277, 1993.

25. J. I. Nakayama, J. C. Rice, B. D. Strahl, et al.: Role of histone H3 lysine 9 mehtylation in epigenetic control of heterochromatin assembly. *Science* 292:110, 2001.

26. H. Wang, Z. Q. Huang, L. Xia, et al.: Methylation of histone H4 at arginine 3 facilitating transcriptional activation by nuclear hormone receptor. *Science* 293:853, 2001.

27. J. C. Rice and C. D. Allis: Gene regulation: Code of silence. *Nature* 414:258, 2001.

28. M. Grunstein: Histone acetylation in chromatin structure and transcription. *Nature* 389:349, 1997.

29. A. Verdel and S. Khochbin: Identification of a new family of higher eukaryotic histone deacetylases. *J Biol Chem* 274:2440, 1999.

30. Y. Wei, L. Yu, J. Bowen, et al.: Phosphorylation of histone H3 is required for proper chromosome condensation and segregation. *Cell* 97:99, 1999.

31. P. Sassone-Corsi, C. A. Mizzen, P. Cheung, et al.: Requirement of Rsk-2 for epidermal growth factor–activated phosphorylation of histone H3. *Science* 285:886, 1999.

32. R. A. DePinho: The cancer–chromatin connection. *Nature* 391:533, 1998.

33. R. J. Lin, L. Nagy, S. Inoue, et al.: Role of the histone deacetylase complex in acute promyelocytic leukemia. *Nature* 391:811, 1998.

34. S. C. R. Elgin and H. Weintraub: Chromosomal proteins and chromatin structure. *Annu Rev Biochem* 44:725, 1975.

35. R. Balhorn, R. Chalkley, and D. Granner: Lysine-rich histone phosphorylation. A positive correlation with cell replication. *Biochemistry* 11:1094, 1972.

36. T. A. Langan: Phosphorylation of liver histone following the administration of glucagon and insulin. *Proc Natl Acad Sci USA* 64:1274, 1969.

37. L. J. Kleinsmith: Phosphorylation of non-histone proteins in the regulation of chromosome structure and function. *J Cell Physiol* 85:459, 1975.

38. V. G. Allfrey, A. Inoue, J. Karn, E. M. Johnson, and G. Vidali: Phosphorylation of DNA-binding nuclear acidic proteins and gene activation in the HeLa cell cycle. *Cold Spring Harbor Symp Quant Biol* 38:785, 1973.

39. L. E. Rikans and R. W. Ruddon: Partial purification and properties of a chromatin-associated phosphoprotein kinase from rat liver nuclei. *Biochem Biophys Acta* 422:73, 1976.

40. G. M. Walton and G. N. Gill: Identity of the in vivo phosphorylation site in high mobility group 14 protein in HeLa cells with the site phosphorylated by casein II kinase in vitro. *J Biol Chem* 258:4440, 1983.

41. S.-I. Tanuma and G. S. Johnson: ADP-ribosylation of nonhistone high mobility group proteins in intact cells. *J Biol Chem* 258:4067, 1983.

42. D. J. Tremethick and H. R. Drew: High mobility group proteins 14 and 17 can space nucleosomes in vitro. *J Biol Chem* 268:11389, 1993.

43. R. D. Kornberg and Y. Lorch: Twenty-five years of the nucleosome, fundamental particle of the eukaryote chromosome. *Cell* 98:285, 1999.

44. A. L. Olins and D. E. Olins: Spheroid chromatin units (*v* bodies). *Science* 183:330, 1974.

45. G. Felsenfeld: Chromatin. *Nature* 271:115, 1978.

46. S. Weisbrod, M. Groudine, and H. Weintraub: Interaction of HMG 14 and 17 with actively transcribed genes. *Cell* 19:289, 1980.

47. J. T. Finch and A. Klug: Solenoidal model for superstructure in chromatin. *Proc Natl Acad Sci USA* 73:1897, 1976.

48. K. W. Adolph, S. M. Chevy, J. R. Paulson, and U. K. Laemmli: Isolation of a protein scaffold from mitotic HeLa cell chromosomes. *Proc Natl Acad Sci USA* 74:4937, 1977.

49. L. Manuelidis: A view of interphase chromosomes. *Science* 250:1533, 1990.

50. P. J. Horn and C. L. Peterson: Chromatin higher order folding: Wrapping up transcription. *Science* 297:1824, 2002.

51. K. E. van Holde, D. E. Lohr, and C. Robert: What happens to nucleosomes during transcription? *J Biol Chem* 267:2837, 1992.

52. C. C. Adams and J. L. Workman: Nucleosome displacement in transcription. *Cell* 72:305, 1993.

53. T. E. O'Neill, J. G. Smith, and E. M. Bradbury: Histone octamer dissociation is not required for transcript elongation through arrays of nucleosome cores by phage T7 RNA polymerase in vitro. *Proc Natl Acad Sci USA* 90:6203, 1993.

54. G.-C. Yuan, Y.-J. Liu, M. F. Dion, M. D. Slack, L. F. Wu, et al.: Genome-scale identification of nucleosome positions in *S. cerevisiae. Science* 309:626, 2005.

54a. P. B. Becker: A finger on the mark. *Nature* 442:31, 2006.

54b. J. Wysocka, T. Swiget, H. Xiao, T. A. Milne, S. Y. Kwon et al.: A PHD finger of NURF couples histone H3 lysine 4 trimethylation with chromatin remodeling. *Nature* 442: 86, 2006.

54c. X. Shi, T. Hong, K. L. Walter, M. Ewalt, E. Michishita et al.: ING2 PHD domain links histone H3 lysine 4 methylation to active gene repression. *Nature* 442: 96, 2006.

55. M. L. DePamphilis: Origins of DNA replication in metazoan chromosomes. *J Biol Chem* 268:1, 1993.

56. A. P. Wolffe and J. C. Hansen: Nuclear visions: Functional flexibility from structural instability. *Cell* 104:631, 2001.

57. K. L. Wilson, M. S. Zastrow, and K. K. Lee: Lamins and disease: Insights into nuclear infrastructure. *Cell* 104:647, 2001.

58. R. A. Hegele: The envelope, please: Nuclear lamins and disease. *Nat Med* 6:136, 2000.

59. K. Tsutsui, K. Tsutsui, S. Okada, S. Watarai, S. Seki, T. Yasuda, and T. Shohmori: Identification and characterization of a nuclear scaffold protein that binds the matrix attachment region DNA. *J Biol Chem* 268:12886, 1993.

60. N. Stuurman, A. M. L. Meijne, A. J. van der Pol, L. deJong, R. van Driel, and J. van Renswoude: The nuclear matrix from cells of different origin. *J Biol Chem* 265:5460, 1990.

61. A. W. Partin, R. H. Getzenberg, M. J. CarMichael, D. Vindivich, J. Yoo, J. I. Epstein, and D. S. Coffey: Nuclear matrix protein patterns in human benign prostatic hyperplasia and prostate cancer. *Cancer Res* 53:744, 1993.

62. P. S. Khanuja, J. E. Lehr, H. D. Soule, S. K. Gehani, A. C. Noto, S. Choudhury, R. Chen, and K. J. Pienta: Nuclear matrix proteins in normal and breast cancer cells. *Cancer Res* 53:3394, 1993.

63. S. K. Keesee, M. D. Meneghini, R. P. Szaro, and Y.-J. Wu: Nuclear matrix proteins in human colon cancer. *Proc Natl Acad Sci USA* 91:1913, 1994.

64. T. E. Miller, L. A. Beausang, L. F. Winchell, and G. P. Lidgard: Detection of nuclear matrix proteins in serum from cancer patients. *Cancer Res* 52:422, 1992.

65. D. Zink, A. H. Fischer, and J. A. Nickerson: Nuclear structure in cancer cells. *Nat Rev Cancer* 4:677, 2004.

66. H. Weintraub and M. Groudine: Chromosomal subunits in active genes have an altered conformation. *Science* 193:848, 1976.

67. S. Weisbrod: Active chromatin. *Nature* 297:289, 1982.

68. G. E. Crawford, I. E. Holt, J. C. Mullikin, D. Tai, E. D. Green, et al.: Identifying gene regulatory elements by genome-wide recovery of DNase hypersensitive sites. *Proc Natl Acad Sci USA* 101:992, 2004.

69. C. W. M. Roberts and S. H. Orkin: The SWI/SNF complex—Chromatin and cancer. *Nat Rev Cancer* 4:133, 2004.

69a. T. I. Lee, R. G. Jenner, L. A. Boyer, M. G. Grunther, S. S. Levine et al.: Control of developmental regulators by Polycomb in human embryonic stem cells. *Cell* 125: 301, 2006.

69b. B. E. Bernstein, T. S. Mikkelsen, X. Xie, M. Kamal, D. J. Huebert et al.: A bivalent chromatin structure marks key developmental genes in embryonic stem cells. *Cell* 125: 315, 2006.

70. D. Nathans: Restriction endonucleases, simian virus 40, and the new genetics. *Science* 206:903, 1979.

71. W. Gilbert: Why genes in pieces? *Nature* 271:501, 1978.

72. R. Breathnach, C. Benoist, K. O'Hare, F. Gannon, and P. Chambon: Ovalbumin gene: Evidence for a leader sequence in mRNA and DNA sequences at the exon-intron boundaries. *Proc Natl Acad Sci USA* 75:4853, 1978.

73. H. Schwartz and J. E. Darnell: The association of protein with the polyadenylic acid of HeLa cell messenger RNA: Evidence for a "transport" role of a 75,000 molecular weight polypeptide. *J Mol Biol* 104:833, 1975.

74. E. J. Sontheimer and J. A. Steitz: The U5 and U6 small nuclear RNAs as active site components of the spliceosome. *Science* 262:1989, 1993.

75. P. A. Sharp: Splicing of messenger RNA precursors. *Science* 235:766, 1987.

76. P. S. Perlman and R. A. Butow: Mobile introns and intron-encoded proteins. *Science* 246:1106, 1989.

77. A. M. Lambowitz and M. Belfort: Introns as mobile genetic elements. *Annu Rev Biochem* 62:587, 1993.

78. J. W. Tamkun, J. E. Schwarzbauer, and R. O. Hynes: A single rat fibronectin gene generates three different mRNAs by alternative splicing of a complex exon. *Proc Natl Acad Sci USA* 81:5140, 1984.

79. M. Periasamy, E. E. Strehler, L. I. Garfinkel, R. M. Gubits, N. Riuz-Opazo, and B. Nadal-Ginard: Fast skeletal muscle myosin light chains 1 and 3 are produced from a single gene by a combined process of differential RNA transcription and splicing. *J Biol Chem* 259:13595, 1984.

80. R. A. Young, O. Hagenbüchle, and U. Schibler: A single mouse α-amylase gene specifies two different tissue specific mRNAs. *Cell* 23:451, 1981.

81. Y. Fukumaki, P. K. Ghosh, E. J. Benz, Jr., V. B. Reddy, P. Lebowitz, B. G. Forget, and S. M. Weissman: Abnormally spliced messenger RNA in erythroid cells from patients with β⁺-thalassemia and monkey cells expressing a cloned β⁺-thalassemic gene. *Cell* 28:585, 1982.

82. B. Dujon: Mutants in a mosaic gene reveal functions for introns. *Nature* 282:777, 1979.

83. J. M. Johnson, J. Castle, P. Garrett-Engele, Z. Kan, P. M. Loerch, et al.: Genome-wide survey of human alternative pre-mRNA splicing with exon junction microarrays. *Science* 302:2141, 2003.

84. Z. Wang, H. S. Lo, H. Yang, S. Gere, Y. Hu, et al.: Computational analysis and experimental validation of tumor-associated alternative RNA splicing in human cancer. *Cancer Res* 63:655, 2003.

85. I. Wickelgren: Spinning junk into gold. *Science* 300:1646, 2003.

86. G. Lev-Maor, R. Sorek, N. Shomron, and G. Ast: The birth of an alternatively spliced exon: 3′ splice-site selection in *Alu* exons. *Science* 300: 1288, 2003.

87. P. Borst and D. R. Greaves: Programmed gene rearrangements altering gene expression. *Science* 235:658, 1987.

88. P. D. Sadowski: Site-specific genetic recombination: Hops, flips, and flops. *FASEB J* 7:760, 1993.

89. R. D. Camerini-Otero and P. Hsieh: Parallel DNA triplexes, homologous recombination, and other homology-dependent DNA interactions. *Cell* 73:217, 1993.

90. T. H. Morgan: An attempt to analyze the constitution of the chromosomes on the basis of

sex-limited inheritance in *Drosophila. J Exp Zool* 11:365, 1911.

91. B. McClintock: The significance of responses of the genome to challenge. *Science* 226:792, 1984.

92. C. W. Schmid and W. R. Jelinek: The Alu family of dispersed repetitive sequences. *Science* 216:1065, 1982.

93. Y. Meng-Chao, J. Choi, S. Yokoyama, C. F. Austerberry, and C.-H. Yao: DNA elimination in Tetrahymena: A developmental process involving extensive breakage and rejoining of DNA at defined sites. *Cell* 36:433, 1984.

94. M.-C. Yao and M. A. Gorovsky: Comparison of the sequence of macro- and micronuclear DNA of Tetrahymena puriformis. *Chromosoma* 48:1, 1974.

95. A. M. Campbell: Episomes. *Adv Genet* 11:101, 1962.

96. C. Coleclough: Chance, necessity, and antibody gene dynamics. *Nature* 303:23, 1983.

97. E. L. Mather, K. J. Nelson, J. Haimovich, and R. P. Perry: Mode of regulation of immunoglobulin μ- and δ-chain expression varies during B-lymphocyte maturation. *Cell* 36:329, 1984.

98. G. E. Taccioli, G. Rathbun, E. Oltz, T. Stamato, P. A. Jeggo, and F. W. Alt: Impairment of V(D)J recombination in double-strand break repair mutants: *Science* 260:207, 1993.

99. B. A. Dombroski, A. F. Scott, and H. H. Kazazian, Jr.: Two additional potential retrotransposons isolated from a human L1 subfamily that contains an active retrotransposable element. *Proc Natl Acad Sci USA* 90:6513, 1993.

100. P. R. Gross: Biochemistry of differentiation. *Annu Rev Biochem* 37:631, 1968.

101. D. V. de Cicco and A. C. Spradling: Localization of a *cis*-acting element responsible for the developmentally regulated amplification of *Drosophila* chorion genes. *Cell* 38:45, 1984.

102. F. W. Alt, R. E. Kellems, J. R. Bertino, and R. T. Schimke: Selective multiplication of dihydrofolate reductase genes in methotrexate-resistant variants of cultured murine cells. *J Biol Chem* 253:1357, 1978.

103. R. T. Schimke: Gene amplification in cultured animal cells. *J Biol Chem* 262:5989, 1988.

104. T. D. Tlsty, P. C. Brown, and R. T. Schimke: UV radiation facilitates methotrexate resistance and amplification of the dihydrofolate reductase gene in cultured 3T6 mouse cells. *Mol Cell Biol* 4:1050, 1984.

105. S. Lavi and S. Etkin: Carcinogen-mediated induction of SV40 DNA synthesis in SV 40 transformed Chinese hamster embryo cells. *Carcinogenesis* 2:417, 1981.

106. A. Varshavsky: Phorbol ester dramatically increases incidence of methotrexate-resistant colony-forming mouse cells: Possible mechanisms and relevance to tumor promotion. *Cell* 25:561, 1981.

107. G. Levan and A. Levan: Transitions of double minutes into homogeneously staining regrions and C-bandless chromosomes in the SEWA tumor. In R. T. Schimke, ed.: *Gene Amplification.* New York: Cold Spring Harbor Laboratory, 1982, p. 91.

108. J. M. Roberts, L. B. Buck, and R. Axel: A structure for amplified DNA. *Cell* 33:53, 1983.

109. Y. N. Osheim and O. L. Miller, Jr.: Novel amplification and transcription activity of chorion genes in *Drosophila melanogaster* follicle cells. *Cell* 33:543, 1983.f

110. J. Sebat, B. Lakshmi, J. Troge, J. Alexander, J. Young, et al.: Large-scale copy number polymorphism in the human genome. *Science* 305:525, 2004.

110a. H. Pearson: What is a gene? *Nature* 441:399, 2006.

111. W. S. Dynan: Modularity in promoters and enhancers. *Cell* 58:1, 1989.

112. S. McKnight and R. Tijan: Transcriptional selectivity of viral genes in mammalian cells. *Cell* 46:795, 1986.

113. C. Benoist and P. Chambon: In vivo sequence requirements of the SV40 early promoter region. *Nature* 290:304, 1981.

114. P. Gruss, R. Dhar, and G. Khoury: Simian virus 40 tandem repeated sequences as an element of the early promoter. *Proc Natl Acad Sci USA* 78:943, 1981.

115. H. R. Schöler and P. Gruss: Specific interaction between enhancer-containing molecules and cellular components. *Cell* 36:403, 1984.

116. G. Khoury and P. Gruss: Enhancer elements. *Cell* 33:313, 1983.

117. D. R. Herendee, G. A. Kassavetis, and E. P. Geiduschek: A transcriptional enhancer whose function imposes a requirement that proteins track along DNA. *Science* 256:1298, 1992.

118. K. E. Cullen, M. P. Kladde, and M. A. Seyfred: Interaction between transcription regulatory regions of prolactin chromatin. *Science* 261:203, 1993.

119. M. Ptashne: How eukaryotic transcriptional activators work. *Nature* 335:683, 1988.

120. A. D. Frankel and P. S. Kim: Modular structure of transcription factors: Implications for gene regulation. *Cell* 65:717, 1991.

121. K. Struhl: Helix-turn-helix, zinc-finger, and leucine-zipper motifs for eukaryotic transcriptional regulatory proteins. *Trends Biochem Sci* 14:137, 1989.

122. R. Tjian and T. Maniatis: Transcriptional activation: A complex puzzle with few easy pieces. *Cell* 77:5, 1994.

123. R. Schleif: DNA binding by proteins. *Science* 241:1182, 1988.

124. N. S. Foulkes and P. Sassone-Corsi: More is better: Activators and repressors from the same gene. *Cell* 68:411, 1992.

125. M. Levine and J. L. Manley: Transcriptional repression of eukaryotic promoters. *Cell* 59:405, 1989.

126. D. J. Kessler, M. P. Duyao, D. B. Spicer, and G. E. Sonenshein: NF-κB-like factors mediate interleukin 1 induction of c-myc gene transcription in fibroblasts. *J Exp Med* 176:787, 1992.

127. A. Schrivastava, S. Saleque, G. V. Kalpana, S. Artandi, S. P. Goff, and K. Calame: Inhibition of transcriptional regulator Yin-Yang-1 by association with c-myc. *Science* 262:1889, 1993.

128. R. G. Roeder: The eukaryotic transcriptional machinery: complexities and mechanisms unforeseen. *Nat Med* 9:1239, 2003.

129. L. Weis and D. Reinberg: Transcription by RNA polymerase II: Initiator-directed formation of transcription-competent complexes. *FASEB J* 6:3300, 1992.

130. P. A. Sharp: TATA-binding protein is a classless factor. *Cell* 68:819, 1992.

131. G. Gill and R. Tjian: A highly conserved domain of TFIID displays species specificity in vivo. *Cell* 65:333, 1991.

132. A. Barberis, C. W. Müller, S. C. Harrison, and M. Ptashne: Delineation of two functional regions of transcription factor TFIIB. *Proc Natl Acad Sci USA* 90:5628, 1993.

133. A. K. P. Taggart, T. S. Fisher, and B. F. Pugh: The TATA-binding protein and associated factors are components of pol III transcription factor TFIIIB. *Cell* 71:1015, 1992.

134. K. Hisatake, R. G. Roeder, and M. Horikoshi: Functional dissection of TFIIB domains required for TFIIB-TFIID-promoter complex formation and basal transcription activity. *Nature* 363:744, 1993.

135. J. W. Lillie and M. R. Green: Activator's target in sight: *Nature* 341:279, 1989.

136. S. G. E. Roberts, I. Ha, E. Maldonado, D. Reinberg, and M. R. Green: Interaction between an acidic activator and transcription factor TFIIB is required for transcriptional activation. *Nature* 363:741, 1993.

137. C. J. Fry and P. J. Farnham: Context-dependent transcriptional regulation. *J Biol Chem* 274:29583, 1999.

138. N. C. Jones, P. W. J. Rigby, and E. B. Ziff: *Trans*-acting protein factors and the regulation of eukaryotic transcription: Lessons from studies on DNA tumor viruses. *Genes Dev* 2:267, 1988.

139. T. Hai and T. Curran: Cross-family dimerization of transcription factors Fos/Jun and ATF/CREB alters DNA binding specificity. *Proc Natl Acad Sci USA* 88:3720, 1991.

140. C. Abate, S. J. Baker, S. P. Lees-Miller, C. W. Anderson, D. R. Marshak, and T. Curran: Dimerization and DNA binding alter phosphorylation of Fos and Jun. *Proc Natl Acad Sci USA* 90:6766, 1993.

141. M. I. Diamond, J. N. Miner, S. K. Yoshinaga, and K. R. Yamamoto: Transcription factor interactions: Selectors of positive or negative regulation from a single DNA element. *Science* 249:1266, 1990.

142. R. Schüle, P. Rangarajan, S. Kliewer, L. J. Ransone, J. Bolado, N. Yang, I. M. Verma, and R. M. Evans: Functional antagonism between oncoprotein c-Jun and the glucocorticoid receptor. *Cell* 62:1217, 1990.

143. C. M. Pfarr, F. Mechta, G. Spyrou, D. Lallemand, S. Carillo, and M. Yaniv: Mouse JunD negatively regulates fibroblast growth and antagonizes transformation by ras. *Cell* 76:747, 1994.

144. Z. Dong, M. J. Birrer, R. G. Watts, L. M. Matrisian, and N. H. Colburn: Blocking of tumor promoter–induced AP-1 activity inhibits induced transformation in JB6 mouse epidermal cells. *Proc Natl Acad Sci USA* 91:609, 1994.

145. M. R. Briggs, J. T. Kadonaga, S. P. Bell, and R. Tjian: Purification and biochemical characterization of the promoter-specific transcription factor, SP1. *Science* 234:47, 1986.

146. A. J. Courey, D. A. Holtzman, S. P. Jackson, and R. Tjian: Synergistic activation by the glutamine-rich domains of human transcription factor Sp1. *Cell* 59:827, 1989.

147. A. J. Udvadia, K. T. Rogers, P. D. R. Higgins, Y. Murata, K. H. Martin, P. A. Humphrey, and J. M. Horowitz: Sp-1 binds promoter elements regulated by the RB protein and Sp-1-mediated transcription is stimulated by RB coexpression. *Proc Natl Acad Sci USA* 90:3265, 1993.

148. M. H. Rosner, M. A. Vigano, P. W. J. Rigby, H. Arnheither, and L. M. Staudt: Oct-3 and the beginning of mammalian development. *Science* 253:144, 1991.

149. C. Abate-Shen: Homeobox genes and cancer: New OCTaves for an old tune. *Cancer Cell* 4:329, 2003.

150. W. Wahli and E. Martinez: Superfamily of steroid nuclear receptors: Positive and negative regulators of gene expression. *FASEB J* 5:2243, 1991.

151. K. Umesono, K. K. Murakami, C. C. Thompson, and R. M. Evans: Direct repeats as selective response elements for the thyroid hormone, retinoic acid, and vitamin D_3 receptors. *Cell* 65:1255, 1991.

152. H. Weintraub, R. Davis, S. Tapscott, M. Thayer, M. Krause, R. Benezra, T. K. Blackwell, D. Turner, R. Rupp, S. Hollenberg, Y. Zhuang, and A. Lassar: The myoD gene family: Nodal point during specification of the muscle cell lineage. *Science* 251:761, 1991.

153. D. G. Edmondson and E. N. Olson: Helix-loop-helix proteins as regulators of muscle-specific transcription. *J Biol Chem* 268:755, 1993.

154. A. B. Lassar, R. L. Davis, W. E. Wright, T. Kadesch, C. Murre, A. Voronova, D. Baltimore, and H. Weintraub: Functional activity of myogenic HLH proteins requires hetero-oligomerization with E12/E47-like proteins in vivo. *Cell* 66:305, 1991.

155. R. Benezra, R. L. Davis, D. Lockshon, D. L. Turner, and H. Weintraub: The protein Id:

A negative regulator of helix-loop-helix DNA binding proteins. *Cell* 61:49, 1990.

156. E. Bengal, L. Ransone, R. Scharfmann, V. J. Dwarki, S. J. Tapscott, H. Weintraub, and I. M. Verma: Functional antagonism between c-Jun and MyoD proteins: A direct physical association. *Cell* 68:507, 1992.

157. D. B. Mendel and G. R. Crabtree: HNF-1, a member of a novel class of dimerizing homeodomain proteins. *J Biol Chem* 266:677, 1991.

158. M. S. Kapiloff, Y. Farkash, M. Wegner, and M. G. Rosenfeld: Variable effects of phosphorylation of Pit-1 dictated by the DNA response elements. *Science* 253:786, 1991.

159. M. A. Van Dijk, P. M. Voorhoeve, and C. Murre: Pbx1 is converted into a transcriptional activator upon aquiring the N-terminal region of E2A in pre-B-cell acute lymphoblastoid leukemia. *Proc Natl Acad Sci USA* 90:6061, 1993.

160. M. J. Lenardo and D. Baltimore: NF-κB: A pleiotropic mediator of inducible and tissue-specific gene control. *Cell* 58:227, 1989.

161. M. N. Treacy, X. He, and M. G. Rosenfeld: I-POU: a POU-domain protein that inhibits neuron-specific gene activation. *Nature* 350:577, 1991.

162. I. Kola, S. Brookes, A. R. Green, R. Garber, M. Tymms, T. S. Papas, and A. Seth: The Ets1 transcription factor is widely expressed during murine embryo development and is associated with mesodermal cells involved in morphogenetic processes such as organ formation. *Proc Natl Acad Sci USA* 90:7588, 1993.

163. J. W. Pendleton, B. K. Nagai, M. T. Murtha, and F. H. Ruddle: Expansion of the Hox gene family and the evolution of chordates. *Proc Natl Acad Sci USA* 90:6300, 1993.

164. M. T. Murtha, J. F. Leckman, and F. H. Ruddle: Detection of homeobox genes in development and evolution. *Proc Natl Acad Sci USA* 88:10711, 1991.

165. W. McGinnis and R. Krumlauf: Homeobox genes and axial patterning. *Cell* 68:283, 1992.

166. W. J. Gehring: The homeobox in perspective. *Trends Biochem Sci* 200:277, 1992.

167. F. S. Jones, B. D. Holst, O. Minowa, E. M. De Robertis, and G. M. Edelman: Binding and transcriptional activation of the promoter for the neural cell adhesion molecule by HoxC6 (Hox-3.3). *Proc Natl Acad Sci USA* 90:6557, 1993.

168. D. A. Grueneberg, S. Natesan, C. Alexandre, and M. Z. Gilman: Human and *Drosophila* homeodomain proteins that enhance the DNA-binding activity of serum response factor. *Science* 257:1089, 1992.

169. K. Takeshita, J. A. Bollekens, N. Hijiya, M. Ratajczak, F. H. Ruddle, and A. M. Gewirtz: A homeobox gene of the antennapedia class is required for human adult erythropoiesis. *Proc Natl Acad Sci USA* 90:3535, 1993.

170. M. Hatano, C. W. M. Roberts, M. Minden, W. M. Crist, and S. J. Korsmeyer: Deregulation of a homeobox gene, HOX11, by the t(10;14) in T cell leukemia. *Science* 253:79, 1991.

171. T. N. Dear, I. Sanchez-Garcia, and T. H. Rabbitts: The HOX11 gene encodes a DNA-binding nuclear transcription factor belonging to a distinct family of homeobox genes. *Proc Natl Acad Sci USA* 90:4431, 1993.

172. J. M. Raser and E. K. O'Shea: Noise in gene expression: origins, consequences, and control. *Science* 309:2010, 2005.

173. K. D. Robertson: DNA methylation and human disease. *Nat Rev Genet* 6:597, 2005.

174. P. A. Jones and S. B. Baylin: The fundamental role of epigenetic events in cancer. *Nat Rev Genet* 3:415, 2002.

175. A. Razin and A. D. Riggs: DNA methylation and gene function. *Science* 210:604, 1980.

176. J. D. McGhee and G. D. Ginder: Specific DNA methylation sites in the vicinity of the chicken β-globin genes. *Nature* 280:419, 1979.

177. J. L. Mandel and P. Chambon: DNA methylatin: Organ specific variations in methylation pattern within and around ovalbumin and other chicken genes. *Nucl Acid Res* 7:2081, 1979.

178. P. W. Laird: The power and the promise of DNA methylation markers. *Nat Rev Cancer* 3:253, 2003.

179. A. Razin, C. Webb, M. Szyf, J. Yisraeli, A. Rosenthal, T. Navek-Many, N. Sciaky-Gallili, and H. Cedar: Variations in DNA methylation during mouse cell differentiation in vivo and in vitro. *Proc Natl Acad Sci USA* 81:2275, 1984.

180. A. Levine, G. L. Cantoni, and A. Razin: Methylation in the preinitiation domain suppresses gene transcription by an indirect mechanism. *Proc Natl Acad Sci USA* 89:10119, 1992.

181. A. Bird: The essentials of DNA methylation. *Cell* 70:5, 1992.

182. A. P. Bird and A. P. Wolffe: Methylation-induced repression—Belts, braces, and chromatin. *Cell* 99:451, 1999.

183. M. Okuwaki and A. Verreault: Maintenance DNA methylation of nucleosome core particles. *J Biol Chem* 279:2904, 2004.

184. M. Okano, D. W. Bell, D. A. Haber, and E. Li: DNA methyltransferases Dnmt3a and Dnmt3b are essential for de novo methylation and mammalian development. *Cell* 99:247, 1999.

185. N. Beaulieu, S. Morin, I. C. Chute, M.-F. Robert, H. Nguyen, et al.: An essential role for DNA methylation DNMT3B in cancer cell survival. *J Biol Chem* 277:28176, 2002.

186. S. K. Bhattacharya, S. Ramchandani, N. Cervoni, and M. Szyf: A mammalian protein with specific demthylase activity for mCpG DNA. *Nature* 397:579, 1999.

187. I. Keshet, J. Lieman-Hurwitz, and H. Cedar: DNA methylation affects the formation of active chromatin. *Cell* 44:535, 1986.

188. M. Higurashi and R. D. Cole: The combination of DNA methylation and H1 histone binding

inhibits the action of a restriction nuclease on plasmid DNA. *J Biol Chem* 266:8619, 1991.

189. S. M. Taylor and P. A. Jones: Multiple new phenotypes induced in 10T1/2 and 3T3 cells treated with 5-azacytidine. *Cell* 17:771, 1979.

190. E. Li, T. H. Bestor, and R. Jaenisch: Targeted mutation of the DNA methyltransferase gene results in embryonic lethality. *Cell* 69:915, 1992.

191. J. G. Herman and S. B. Baylin: Gene silencing in cancer in association with promoter hypermethylation. *N Engl J Med* 349:2042, 2003.

192. N. Sato, A. Maitra, N. Fukushima, N. Tjarda van Heek, H. Matsubayashi, et al.: Frequent hypomethylation of multiple genes overexpressed in pancreatic ductal adenocarcinoma. *Cancer Res* 63:4158, 2003.

193. A. P. Feinberg, C. W. Gehrke, K. C. Kuo, and M. Ehrlich: Reduced genomic 5-methylcytosine content in human colonic neoplasia. *Cancer Res* 48:1159, 1988.

194. F. Gaudet, J. G. Hodgson, A. Eden, L. Jackson-Grusby, J. Dausman, et al.: Induction of tumors in mice by genomic hypomethylation. *Science* 300:489, 2003.

195. S. A. Belinsky, D. M. Klinge, C. A. Stidley, J.-P. Issa, J. G. Herman, et al.: Inhibition of DNA methylation and histone deacetylation prevents murine lung cancer. *Cancer Res* 63:7089, 2003.

196. H. M. Müller, A. Widschwender, H. Fiegl, L. Ivarsson, G. Goebel, et al.: DNA methylation in serum of breast cancer patients: an independent prognostic marker. *Cancer Res* 63:7641, 2003.

197. M. L. Frazier, L. Xi, J. Zong, N. Viscofsky, A. Rashid, et al.: Association of the CpG Island methylator phenotype with family history of cancer in patients with colorectal cancer. *Cancer Res* 63:4805, 2003.

198. M. Esteller, M. Sanchez-Cespedes, R. Rosell, D. Sidransky, S. B. Baylin, et al.: Detection of aberrant promoter hypermethylation of tumor suppressor genes in serum DNA from non–small cell lung cancer patients. *Cancer Res* 59:67, 1999.

199. S. Zöchbauer-Müller, K. M. Fong, A. K. Virmani, J. Geradts, A. F. Gazdar, et al.: Aberrant promoter methylation of multiple genes in non–small cell lung cancers. *Cancer Res* 61:249, 2001.

200. I. D. C. Markl, J. Cheng, G. Liang, D. Shibata, P. W. Laird, et al.: Global and gene-specific epigenetic patterns in human bladder cancer genomes are relatively stable in vivo and in vitro over time. *Cancer Res* 61:5875, 2001.

201. N. Sato, N. Fukushima, A. Maitra, H. Matsubayashi, C. J. Yeo, et al.: Discovery of novel targets for aberrant methylation in pancreatic carcinoma using high-throughput microarrays. *Cancer Res* 63, 3735, 2003.

202. C. Battagli, R. G. Uzzo, E. Dulaimi, I. Ibanex de Caceres, R. Krassenstein, et al.: Promoter hypermethylation of tumor suppressor genes in urine from kidney cancer patients. *Cancer Res* 63:8695, 2003.

203. G. H. Kang, Y.-H. Shim, H.-Y. Jung, W. H. Kim, J. Y. Ro, et al.: CpG island methylation in premalignant stages of gastric carcinoma. *Cancer Res* 61:2847, 2001.

204. S. Baylin and T. H. Bestor: Altered methylation patterns in cancer cell genomes: cause or consequence? *Cancer Cell* 1:299, 2002.

205. D. P. Barlow: Methylation and imprinting: From host defense to gene regulation? *Science* 260:309, 1993.

206. S. M. Tilghman: The sins of the fathers and mothers: genomic imprinting in mammalian development. *Cell* 96:185, 1999.

207. Z. Xin, M. Tachibana, M. Guggiari, E. Heard, Y. Shinkai, et al.: Role of histone methyltransferase G9a in CpG methylation of the Prader-Willi syndrome imprinting center. *J Biol Chem* 278:14996, 2003.

208. R. I. Gregory, L. P. O'Neill, T. E. Randall, C. Fournier, S. Khosla, et al.: Inhibition of histone deacetylases alters alleli chromatin conformation at the imprinted *U2af1-rs1* locus in mouse embryonic stem cells. *J Biol Chem* 277: 11728, 2002.

209. E. U. Selker: DNA methylation and chromatin structure: A view from below. *Trends Biochem Sci* 15:103, 1990.

210. H. Cui, I. L. Horon, R. Ohlsson, et al.: Loss of imprinting in normal tissue of colorectal cancer patients with microsatellite instability. *Nat Med* 4:1276, 1998.

211. J. Barletta, S. Rainer, and A. Feinberg: Reversal of loss of imprinting in tumor cells by 5-aza-2-deoxycytidine. *Cancer Res* 57:48, 1997.

212. G. Christofori, P. Naik, and D. Hanahan: Deregulation of both imprinted and expressed alleles of the insulin-like growth factor 2 gene during beta-cell tumorigenesis. *Nat Genet* 10:196, 1995.

213. J. D. Ravenel, K. W. Broman, E. J. Perlman, E. L. Niemitz, T. M. Jayawardena, et al.: Loss of imprinting of insulin-like growth factor-II (IGF2) gene in distinguishing specific biologic subtypes of Wilms tumor. *J Natl Cancer Inst* 93:1698, 2001.

214. H. Nakagawa, R. B. Chadwick, P. Peltomäki, C. Plass, Y. Nakamura, et al.: Loss of imprinting of the insulin-like growth factor II gene occurs by biallelic methylation in a core region of *H19*-associated CTCF-binding sites in colorectal cancer. *Proc Natl Acad Sci USA* 98:591, 2001.

215. A. P. Feinberg: Cancer epigenetics takes center stage. *Proc Natl Acad Sci USA* 98:392, 2001.

216. H. Cui, E. L. Niemitz, J. D. Ravenel, P. Onyango, S. A. Brandenburg, et al.: Loss of imprinting of insulin-like growth factor-II in Wilms' tumor commonly involves altered methylation but not mutations of *CTCF* or its binding site. *Cancer Res* 61:4947, 2001.

217. H. Cui, P. Onyango, S. Brandenburg, Y. Wu, C.-L. Hsieh, et al.: Loss of imprinting in

colorectal cancer linked to hypomethylation of *H19* and *IGF2*. *Cancer Res* 62:6442, 2002.

217a. A. Kaneda and A. P. Feinberg: Loss of imprinting of *IGF-2*: A common epigenetic modifier of intestinal tumor risk. *Cancer Res* 65: 11236, 2005.

218. A. P. Feinberg and B. Tycko: The history of cancer epigenetics. *Nat Rev Cancer* 4:143, 2004.

219. K. Yamashita, T. Dai, Y. Dai, F. Yamamoto, and M. Perucho: Genetics supersedes epigenetics in colon cancer phenotype. *Cancer Cell* 4:121, 2003.

220. E. Solomon, R. Voss, V. Hall, et al.: Chromosome 5 allele loss in human colorectal carcinomas. *Nature* 328:616, 1987.

221. E. R. Fearon and B. Vogelstein: A genetic model for colorectal tumorigenesis. *Cell* 61:759, 1990.

222. L. Hayflick:: The limited in vitro lifetime of human diploid cell strains. *Exp Cell Res* 37:614, 1965.

223. E. H. Blackburn: Switching and signaling at the telomere. *Cell* 106:661, 2001.

224. J. W. Shay and S. Bacchetti: A survey of telomerase in human cancer. *Eur J Cancer* 33:787, 1997.

225. R. Hodes: Molecular targeting of cancer: Telomeres as targets. *Proc Natl Acad Sci USA* 98: 7649, 2001.

226. A. G. Bodnar, M. Ouellette, M. Frolkis, et al.: Extension of life-span by introduction of telomerase into normal human cells. *Science* 279: 349, 1998.

227. J. M. Gualberto, L. Lamattina, G. Bonnard, J.-H. Weil, and J.-M. Grienenberger: RNA editing in wheat mitochondria results in the conservation of protein sequences. *Nature* 341:660, 1989.

228. P. S. Covello and M. W. Gray: RNA editing in plant mitochondria. *Nature* 341:662, 1989.

229. L. F. Landweber and W. Gilbert: RNA editing as a source of genetic variation. *Nature* 363:179, 1993.

230. R. D. Klausner and J. B. Harford: *Cis-trans* models for post-transcriptional gene regulation. *Science* 246:870, 1989.

231. M. Kozak: Structural features in eukaryotic mRNAs that modulate the initiation of translation. *J Biol Chem* 266:19867, 1991.

232. C. F. Calkhoven, C. Müller, and A. Leutz: Translational control of gene expression and disease. *Trends Mol Med* 8:577, 2002.

233. M. Kozak: Bifunctional messenger RNAs in eukaryotes. *Cell* 47:481, 1986.

234. A. Lazaris-Karatzas, K. S. Montine, and N. Sonenberg: Malignant transformation by a eukaryotic initiation factor subunit that binds to mRNA 5' cap. *Nature* 345:544, 1990.

235. J. L. Hargrove and F. H. Schmidt: The role of mRNA and protein stability in gene expression. *FASEB J* 3:2360, 1989.

236. L. E. Maquat and G. G. Carmichael: Quality control of mRNA function. *Cell* 104:173, 2001.

237. E. Solomon, J. Borrow, and A. D. Goddard: Chromosome aberrations and cancer. *Science* 254: 1153, 1991.

238. J. D. Rowley: Molecular cytogenetics: Rosetta stone for understanding cancer—Twenty-ninth G.H.A. Clowes memorial award lecture. *Cancer Res* 50:3816, 1990.

239. E. M. Rego and P. P. Pandolfi: Reciprocal products of chromosomal translocations in human cancer pathogenesis: key players or innocent bystanders? *Trends Mol Med* 8:396, 2002.

240. R. Dalla-Favera, M. Bregni, J. Erikson, D. Patterson, R. C. Gallo, and C. M. Croce: Human c-*myc* onc gene is located on the region of chromosome 8 that is translocated in Burkitt lymphoma cells. *Proc Natl Acad Sci USA* 79:7824, 1982.

241. R. Taub, I. Kirsch, C. Morton, G. Lenoir, D. Swam, S. Aaronson, and P. Leder. Translocation of the c-*myc* gene into the immunoglobulin heavy chain locus in human Burkitt lymphoma and murine plasmacytoma cells. *Proc Natl Acad Sci USA* 79:7837, 1982.

242. A. Hagemeijer, D. Bootsma, N. K. Spurr, N. Heisterkamp, J. Groffen, and J. R. Stevenson: A cellular oncogene is translocated to the Philadelphia chromosome in chronic myelocytic leukemia. *Nature* 300:765, 1982.

243. N. Heisterkamp, J. R. Stephenson, J. Groffen, P. F. Hansen, A. deKlein, C. R. Bartram, and G. Grosveld: Localization of the c-able oncogene adjacent to a translocation breakpoint in chronic myelocytic leukemia. *Nature* 306:239, 1983.

244. L. C. Chan, K. K. Karhi, S. I. Rayter, N. Heisterkamp, S. Eridani, R. Powles, S. D. Lawler, J. Graffen, J. G. Foulkes, M. F. Greaves, and L. M. Wiedemann: A novel *abl* protein expressed in Philadelphia chromosome positive acute lymphoblastic leukemia. *Nature* 325:635, 1987.

245. T. G. Lugo, A.-M. Pendergast, A. J. Muller, and O. N. Witte: Tyrosine kinase activity and transformation potency of *bcr-abl* oncogene products. *Science* 247:1079, 1990.

246. Y. Tsujimoto, J. Yunis, L. Onorato-Showe, J. Erickson, P. C. Nowell, and C. M. Croce: Molecular cloning of the chromosomal breakpoint of B-cell lymphomas and leukemias with the t(11;14) chromosome translocation: *Science* 224:1403, 1984.

247. Y. Haupt, W. S. Alexander, G. Barri, S. P. Klinken, and J. M. Adams: Novel zinc finger gene implicated as *myc* collaborator by retrovirally accelerated lymphomagenesis in Eμ-*myc* transgenic mice. *Cell* 65:753, 1991.

248. Y. Tsujimoto, J. Gorham, J. Cossman, E. Jaffe, and C. M. Croce: The t(14;18) chromosome translocations involved in B-cell neoplasms result from mistakes in VDJ joining. *Science* 229:1390, 1985.

249. E. C. Collins and T. H. Rabbitts: The promiscuous *MLL* gene links chromosomal translocations to cellular differentiation and tumor tropism. *Trends Mol Med* 8:436, 2002.

250. G. Russo, M. Isobe, R. Gatti, J. Finan, O. Batuman, K. Huebner, P. C. Nowell, and

C. M. Croce: Molecular analysis of a t(14;14) translocation in leukemic T-cells of an ataxia telangiectasia patient. *Proc Natl Acad Sci USA* 86:602, 1989.

251. J. Limpens, D. de Jong, J. H. J. M. van Krieken, C. G. A. Price, B. D. Young, G.-J. B. van Ommen, and P. M. Kluin: *Bcl-2*/J_H rearrangements in benign lymphoid tissues with follicular hyperplasia. *Oncogene* 6:2272, 1991.

252. S. A. Tomlins, D. R. Rhodes, S. Perner, S. M. Dhanasekaran, R. Mehra, et al.: Recurrent fusion of *TMPRSS2* and ETS transcription factor genes in prostate cancer. *Science* 310:644, 2005.

253. J. J. Yunis: The chromosomal basis of human neoplasia. *Science* 221:227, 1983.

254. E. R. Fearon, A. P. Feinberg, S. H. Hamilton, and B. Vogelstein: Loss of genes on the short arm of chromosome 11 in bladder cancer. *Nature* 318:377, 1985.

255. J. J. Wasmuth, C. Park, and R. E. Ferrell: Report of the committee on the genetic constitution of chromosome 5. *Cytogenet Cell Genet* 51:137, 1989.

256. R. Sager, I. K. Gadi, L. Stephens, and C. T. Grabowy: Gene amplification: An example of an accelerated evolution in tumorigenic cells. *Proc Natl Acad Sci USA* 82:7015, 1985.

257. O. S. Frankfurt, J. L. Chin, L. S. Englander, W. R. Greco, J. E. Pontes, and Y. M. Rustum: Relationship between DNA ploidy, glandular differentiation, and tumor spread in human prostate cancer. *Cancer Res* 45:1418, 1985.

258. F. A. Klein, M. W. Herr, W. F. Whitmore, Jr., P. C. Sogani, and M. R. Melamed: Detection and follow-up of carcinoma of urinary bladder by flow cytometry. *Cancer* 50:389, 1982.

259. B. Tribukait, H. Gustafson, and P. L. Esposti: The significance of ploidy and proliferation in the clinical and biological evolution of bladder tumors: A study of 100 untreated cases. *Br J Urol* 54:130, 1982.

260. H. S. Smith, L. A. Liotta, M. C. Hancock, S. R. Wolman, and A. J. Hackett: Invasiveness and ploidy of human mammary carcinomas in short-term culture. *Proc Natl Acad Sci USA* 82:1805, 1985.

261. N. Wake, J. Isaacs, and A. A. Sandberg: Chromosomal changes associated with progression of the Dunning R-3327 rat prostatic adenocarcinoma system. *Cancer Res* 42:4131, 1982.

262. L. S. Cram, M. F. Bartholdi, F. A. Ray, G. I. Travis, and P. M. Kraemer: Spontaneous neoplastic evolution of Chinese hamster cells in culture: Multistep progression of karyotype. *Cancer Res* 43: 4828, 1983.

263. R. P. Stock and H. Bialy: The sigmoidal curve of cancer. *Nat Biotechnol* 21:13, 2003.

264. G. Pihan and S. J. Doxsey: Mutations and aneuploidy: Co-conspirators in cancer? *Cancer Cell* 4:89, 2003.

265. E. R. Fearon and B. Vogelstein: A genetic model for colorectal tumorigenesis. *Cell* 61:759, 1990.

266. H. Bialy: Aneuploidy and cancer—The vintage wine revisited. *Nat Biotechnol* 19:22, 2001.

267. I. Henry, C. Bonaiti-Pellié, V. Chehensse, C. Beldjord, C. Schwartz, G. Utermann, and C. Junien: Uniparental paternal disomy in a genetic cancer-predisposing syndrome. *Nature* 351:665, 1991.

268. A. C. Ferguson-Smith, B. M. Cattanach, S. C. Barton, C. V. Beechey, and M. A. Surani: Embryological and molecular investigations of parental imprinting on mouse chromosome 7: *Nature* 351:667, 1991.

269. B. Alberts, D. Bray, J. Lewis, M. Raff, K. Roberts, and J. D. Watson: *Molecular Biology of the Cell*. New York: Garland Publishing, 1983, pp. 214–215.

270. R. I. Richards and G. R. Sutherland: Dynamic mutations: A new class of mutations causing human disease. *Cell* 70:709, 1992.

271. E. J. Kremer, M. Pritchard, M. Lynch, S. Yu, K. Holman, E. Baker, S. T. Warren, D. Schlessinger, G. R. Sutherland, and R. I. Richards: Mapping of DNA instability at the fragile X to a trinucleotide repeat sequence p(CCG)n. *Science* 252:1711, 1991.

272. Y.-H. Fu, D. P. A. Kuhl, A. Pizzuti, M. Pieretti, J. S. Sutcliffe, S. Richards, A. J. M. H. Verkerk, J. J. A. Holder, R. G. Fenwick, Jr., S. T. Warren, B. A. Oostra, D. L. Nelson, and C. T. Caskey: Variation of the CGG repeat at the fragile X site results in genetic instability: Resolution of the Sherman paradox. *Cell* 67:1047, 1991.

273. The Huntington's Disease Collaborative Research Group: A novel gene containing a trinucleotide repeat that is expanded and unstable on Huntington's disease chromosomes. *Cell* 72:971, 1993.

274. M. Pieretti, F. Zhan, Y.-H. Fu, S. T. Warren, B. A. Oostra, C. T. Caskey, and D. L. Nelson: Absence of expression of the FMR-1 gene in fragile X syndrome. *Cell* 66:817, 1991.

275. C. J. Yee, N. Roodi, C. S. Verrier, and F. F. Parl: Microsatellite instability and loss of heterozygosity in breast cancer. *Cancer Res* 54:1641, 1994.

276. A. Merlo, M. Mabry, E. Gabrielson, R. Vollmer, S. B. Baylin, and D. Sidransky: Frequent microsatellite instability in primary small cell lung cancer. *Cancer Res* 54:2098, 1994.

277. V. Shridhar, J. Siegfried, J. Hunt, M. del Mar Alonso, and D. I. Smith: Genetic instability of microsatellite sequences in many non–small cell lung carcinomas. *Cancer Res* 54:2084, 1994.

278. M. Gonzalez-Zulueta, J. M. Ruppert, K. Tokino, Y. C. Tsai, C. H. Spruck III, N. Miyao, P. W. Nichols, G. G. Hermann, T. Horn, K. Steven, I. C. Summerhayes, D. Sidransky, and P. A. Jones: Microsatellite instability in bladder cancer. *Cancer Res* 53:5620, 1994.

279. N. M. Mironov, M. A.-M. Aguelon, G. I. Potapova, Y. Omori, O. V. Gorbunov, A. A. Klimenkov, and H. Yamasaki: Alterations of (CA)$_n$ DNA repeats and tumor suppressor genes in human gastric cancer. *Cancer Res* 54:41, 1994.

280. H. T. Lynch, T. C. Smyrk, P. Watson, S. J. Lanspa, J. F. Lynch, P. M. Lynch, R. J. Cavalieri, and C. R. Boland: Genetics, natural history, tumor spectrum and pathology of hereditary nonpolyposis colorectal cancer: An updated review. *Gastroenterology* 104:1535, 1993.

281. R. Fishel, M. K. Lescoe, M. R. S. Rao, N. G. Copeland, N. A. Jenkins, J. Garber, M. Kane, and R. Kolodner: The human mutator gene homolog MSH2 and its association with hereditary nonpolyposis colon cancer. *Cell* 85:1027, 1993.

282. R. Parson, G.-M. Li, M. J. Longley, W.-H. Fang, N. Papadopoulos, J. Jen, A. de la Chapelle, K. W. Kinzler, B. Vogelstein, and P. Modrich: Hypermutability and mismatch repair deficiency in RER$^+$ tumor cells. *Cell* 75:1227, 1993.

283. W. H. Brown: A case of pluriglandular syndrome—Diabetes of bearded women. *Lancet* 2:1022, 1928.

284. G. W. Liddle, J. R. Givens, W. E. Nicholson, and D. P. Island: The extopic ACTH syndrome. *Cancer Res* 25:1057, 1965.

285. L. H. Rees and J. G. Ratcliffe: Ectopic hormone production by non-endocrine tumours. *Clin Endocrinol* 3:263, 1974.

286. R. S. Yalow, C. E. Eastridge, G. Higgins, Jr., and J. Wolf: Plasma and tumor ACTH in carcinoma of the lung. *Cancer* 44:1789, 1979.

287. A. B. Gutman, T. L. Tyson, and E. B. Gutman: Serum calcium, inorganic phosphorus and phosphatase activity. *Arch Intern Med* 57:379, 1936.

288. T. B. Connor, W. C. Thomas, Jr., and J. E. Howard: The etiology of hypercalcemia associated with lung carcinoma. *J Clin Invest* 35:697, 1956.

289. A. H. Tashjian, Jr., L. Levine, and P. L. Munson: Immunochemical identification of parathyroid hormone in non-parathyroid neoplasms associated with hypercalcemia. *J Exp Med* 119:467, 1964.

290. L. M. Sherwood, J. L. H. O'Riordan, G. D. Aurbach, and J. T. Potts, Jr.: Production of parathyroid hormone by non-parathyroid tumors. *J Clin Endocrinol Metab* 27:140, 1967.

291. W. A. Winkler and O. F. Crankshaw: Chloride depletion in conditions other than Addison's disease. *J Clin Invest* 17:1, 1938.

292. W. B. Schwartz, W. Bennett, S. Curelop, and F. C. Barrter: Syndrome of renal sodium loss and hyponatremia probably resulting from inappropriate secretion of antidiuretic hormone. *Am J Med* 23:529, 1957.

293. R. L. Reeves, H. Tesluk, and C. E. Harrision: Precocious puberty associated with hepatoma. *J Clin Endocrinol Metab* 17:1651, 1959.

294. J. L. Vaitukaitis, G. T. Ross, G. D. Braunstein, and P. L. Rayford: Gonadotropins and their subunits: basic and clinical studies. *Recent Prog Horm Res* 32:289, 1976.

295. M. R. Blackman, B. D. Weintraub, S. W. Rosen, I. A. Kourides, K. Steinwascher, and M. H. Gail: Human placental and pituitary glycoprotein hormones and their subunits are tumor markers: A quantitative assessment. *J Natl Cancer Inst* 65:81, 1980.

296. R. W. Ruddon, C. Anderson, K. S. Meade, P. H. Aldenderfer, and P. D. Neuwald: Content of gonadotropins in cultured human malignant cells and effects of sodium butyrate treatment on gonadotropin secretion by HeLa cells. *Cancer Res* 39:3885, 1979.

297. Y. Yoshimotom, A. R. Wolfson, and W. D. Odell: Human chorionic gonadotropin-like substance in nonendocrine tissues of normal subjects. *Science* 197:575, 1977.

298. T. Ono and R. C. Cutler: Age-dependent relaxation of gene repression: Increase of endogenous murine leukemia virus-related and globin-related RNA in brain and liver of mice. *Proc Natl Acad Sci USA* 75:4431, 1978.

299. A. G. E. Pearse: Common cytochemical and ultrastructural characteristics of cells producing polypeptide hormones (the APUD series) and their relevance to thyroid and ultimobranchial C cells and calcitonin. *Proc R Soc Lond B* 170:71, 1968.

300. L. Zech, U. Hagland, K. Nilsson, and G. Klein: Characteristic chromosomal abnormalities in biopsies and lymphoid cell lines from patients with Burkitt and non-Burkitt lymphomas. *Int J Cancer* 17:47, 1976.

301. F. Mitelman, P. G. Nilsson, L. Brandt, G. Alimena, K. Gastaldi, et al.: Chromosome pattern, occupation, and clinical features in patient with acute nonlymphocytic leukemia. *Cancer Genet Cytogenet* 4:197, 1981.

302. R. Knapp, G. Dewald, and R. Pierre: Cytogenetic studies in 174 consecutive patients with preleukemic and myelodysplastic syndrome. *Mayo Clin Proc* 60:507, 1985.

303. FIWCL Fourth International Workshop on Chromosomes in Leukemia, 1982: Clinical significance of chromosomal abnormalities in acute nonlymphoblastic leukemia. *Cancer Genet Cytogenet* 11:332, 1984.

304. E. P. Cronkite: Chemical leukemogenesis: Benzene as a model. *Semin Hematol* 24:2, 1987.

305. H. Goldberg, E. Lusk, J. Moore, P. C. Nowell, and E. C. Besa: Survey of exposure to genotoxic agents in primary myelodysplastic syndrome: Correlation with chromosome patterns and data on patients without hematological disease. *Cancer Res* 50:6876, 1990.

306. L. F. Parada, H. Land, R. A. Weinberg, D. Wolf, and V. Rotter: Cooperation between gene encoding p53 tumor antigen and *ras* in cellular transformation. *Nature* 312:649, 1984.

307. I. Marcillac, F. Troalen, J.-M. Bidart, P. Ghillani, V. Ribrag, et al.: Free human chorionic gonadotropin β subunit in gonadal and nongonadal neoplasms. *Cancer Res* 52:3901, 1992.

308. H. M. Temin: The DNA provirus hypothesis: The establishment and implications of RNA-

directed DNA synthesis. *Science* 192:1075, 1976.

309. D. Baltimore: Viruses, polymerases, and cancer. *Science* 192:632, 1976.

310. H. M. Temin and S. Mizutani: RNA-dependent DNA polymerase in virions of Rous sarcoma virus. *Nature* 226:1211, 1970.

311. D. Baltimore: Viral RNA-dependent DNA polymerase. *Nature* 226:1209, 1970.

312. P. E. Neiman: Rous sarcoma virus nucleotide sequences in cellular DNA: Measurement by RNA-DNA hybridization. *Science* 178:750, 1972.

313. M. Hill and J. Hillova: Virus recovery in chicken cells tested with Rous sarcoma cell DNA. *Nat New Biol* 237:35, 1972.

314. P. H. Duesberg and P. K. Vogt: RNA species obtained from clonal lines of avian sarcoma and from avian leukosis virus. *Virology* 54:207, 1973.

315. D. Stehelin, R. V. Guntaka, H. E. Varmus, and J. M. Bishop: Purification of DNA complementary to nucleotide sequences required for neoplastic transformation of fibroblasts by avian sarcoma viruses. *J Mol Biol* 101:349, 1976.

316. D. H. Spector, K. Smith, T. Padgett, P. McCombe, D. Roulland-Dussoix, C. Moscovici, H. E. Varmus, and J. M. Bishop: Uninfected avian cells contain RNA related to the transforming gene of avian sarcoma viruses. *Cell* 13:371, 1978.

317. R. A. Krzyzek, A. F. Lau, A. J. Faras, and D. H. Spector: Post-transcriptional control of avian oncornavirus transforming gene sequences in mammalian cells. *Nature* 269:175, 1977.

318. J. S. Brugge and R. L. Erikson: Identification of transformation-specific antigen induced by an avian sarcoma virus. *Nature* 269:346, 1977.

319. A. F. Purchio, E. Erikson, J. S. Brugge, and R. L. Erikson: Identification of a polypeptide encoded by the avian sarcoma virus src gene. *Proc Natl Acad Sci USA* 75:1567, 1978.

320. M. S. Collett and R. L. Erikson: Protein kinase activity associated with the avian sarcoma virus src gene product. *Proc Natl Acad Sci USA* 75:2021, 1978.

321. A. D. Levinson, H. Oppermann, L. Levintow, H. E. Varmus, and J. M. Bishop: Evidence that the transforming gene of avian sarcoma virus encodes a protein kinase associated with a phosphoprotein. *Cell* 15:561, 1978.

322. R. L. Erikson, M. S. Collett, E. Erikson, and A. F. Purchio: Evidence that the avian sarcoma virus transforming gene product is a cyclic AMP-independent protein kinase. *Proc Natl Acad Sci USA* 76:6260, 1979.

323. T. Hunter and B. M. Sefton: Transforming gene product of Rous sarcoma virus phosphorylates tyrosine. *Proc Natl Acad Sci USA* 77:1311, 1980.

324. P. Madaule and R. Axel: A novel ras-related gene family. *Cell* 41:31, 1985.

325. R. W. Ellis, D. De Feo, T. Y. Shih, M. A. Gonda, H. A. Young, N. Tsuchida, D. R. Lowry, and E. M. Scolnick: The p21 src genes of Harvey and Kirsten sarcoma viruses originate from divergent members of a family of normal vertebrate genes. *Nature* 292:506, 1981.

326. D. De Feo-Jones, E. M. Scolnick, R. Koller, and R. Dhar: ras-Related gene sequences identified and isolated from *Saccharomyces cerevisiae*. *Nature* 306:707, 1983.

327. T. Kataoka, S. Powers, S. Cameron, O. Fasano, M. Goldfarb, J. Broach, and M. Wigler: Functional homology of mammalian and yeast RAS genes. *Cell* 40:19, 1985.

328. T. Toda, I. Uno, T. Ishikawa, S. Powers, T. Kataoka, D. Broek, S. Cameron, J. Broach, K. Matsumoto, and M. Wigler: In yeast, RAS proteins are controlling elements of adenylate cyclase. *Cell* 40:27, 1985.

329. D. De Feo-Jones, K. Tatchell, L. C. Robinson, I. S. Sigal, W. C. Vass, D. R. Lowy, and E. M. Scolnick: Mammalian and yeast ras gene products: Biological function in their heterologous systems. *Science* 228:179, 1985.

330. G. L. Temeles, J. B. Gibbs, J. S. D'Alonzo, I. S. Sigal, and E. M. Scolnick: yeast and mammalian ras proteins have conserved biochemical properties. *Nature* 313:700, 1985.

331. R. A. Weinberg: A molecular basis of cancer. *Sci Am* 249:126, 1983.

332. G. M. Cooper: Cellular transforming genes. *Science* 218:801, 1982.

333. O. T. Avery, C. M. Mac Leod, and M. McCarty: Studies on the chemical nature of the substance inducing transformation of pneumococcal types. Induction of transformation by a desoxyribonucleic acid fraction isolated from Pneumococcus type III. *J Exp Med* 79:137, 1944.

334. C. Shih, B. Shilo, M. P. Goldfarb, A. Dannenberg, and R. A. Weinberg: Passage of phenotypes of chemically transformed cells via transfection of DNA and chromatin. *Proc Natl Acad Sci USA* 76:5714, 1979.

335. C. Shih, L. C. Padhy, M. J. Murray, and R. A. Weinberg: Transforming genes of carcinomas and neuroblastomas introduced into mouse fibroblasts. *Nature* 290:261, 1981.

336. C. S. Cooper, M. Park, D. G. Blain, M. A. Tainsky, K. Huebner, C. M. Croce, and G. F. Vande Woude: Molecular cloning of a new transforming gene from a chemically transformed human cell line. *Nature* 311:29, 1984.

337. C. Shih and R. A. Weinberg: Isolation of a transforming sequence from a human bladder carcinoma cell line. *Cell* 29:161, 1982.

338. G. Goubin, D. S. Goldman, J. Luce, P. E. Neiman, and G. M. Cooper: Molecular cloning and nucleotide sequence of a transforming gene detected by transfection of chicken B-cell lymphoma DNA. *Nature* 302:114, 1983.

339. M. Goldfarb, K. Shimizu, M. Prucho, and M. Wigler: Isolation and preliminary characterization of a hman transforming gene from T24 bladder carcinoma cells. *Nature* 296:404, 1982.

340. B. Z. Shilo and R. A. Weinberg: Unique transforming gene in carcinogen-transformed mouse cells. *Nature* 289:607, 1981.

341. L. F. Parada, C. J. Taabin, C. Shih, and R. A. Weinberg: Human EJ bladder carcinoma oncogene is homologue of Harvey sarcoma virus *ras* gene. *Nature* 297:474, 1982.

342. C. J. Der, T. G. Krontiris, and G. M. Cooper: Transforming genes of human bladder and lung carcinoma cell lines are homologous to the *ras* genes of Harvey and Kirsten sarcoma viruses. *Proc Natl Acad Sci USA* 79:3637, 1982.

343. E. Santos, S. R. Tronick, S. A. Aaronson, S. Pulciani, and M. Barbacid: T24 human bladder carcinoma oncogene is an activated form of the normal human homologue of BALB- and Harvey-MSV transforming genes. *Nature* 298:343, 1982.

344. S. Pulciani, E. Santos, A. V. Lauver, L. K. Long, S. A. Aaronson, and M. Barbacid: Oncogenes in solid human tumours. *Nature* 300:539, 1982.

345. K. Shimizu, M. Goldfarb, Y. Suard, M. Perucho, Y. Li, T. Kamata, J. Feramisco, E. Stavnezer, J. Fogh, and M. H. Wigler: Three human transforming genes are related to the viral *ras* conogenes. *Proc Natl Acad Sci USA* 80:2112, 1983.

346. E. H. Chang, M. E. Furth, E. M. Scolnick, and D. R. Lowry: Tumorigenic transformation of mammalian cells induced by a normal human gene homologous to the oncogene of Harvey murine sarcoma virus. *Nature* 297:479, 1982.

347. R. G. Greig, T. P. Loestler, D. L. Trainer, S. P. Corwin, L. Miles, T. Kline, R. Sweet, S. Yokoyama, and G. Poste: Tumorigenic and metastatic properties of "normal" and ras-transfected NIH/3T3 cells. *Proc Natl Acad Sci USA* 82:3698, 1985.

348. J. M. Bishop: Molecular themes in oncogenesis. *Cell* 64:235, 1991.

349. T. Hunter: Cooperation between oncogenes. *Cell* 64:249, 1991.

350. A. A. Adjei: Blocking oncogenic Ras signaling for cancer therapy. *J Natl Cancer Inst* 93:1062, 2001.

351. M. Malumbres and M. Barbacid: *RAS* oncogenes: The first 30 years. *Nat Rev Cancer* 3:459, 2003.

352. T. G. Bivona, I. Pérez de Castro, I. M. Ahearn, T. M. Grana, V. K. Chiu, et al.: Phospholipase Cγ activates Ras on the Golgi apparatus by means of RasGRP1. *Nature* 424:694, 2003.

353. K. Brown, A. Buchmann, and A. Balmain: Carcinogen-induced mutations in the mouse c-Ha-*ras* gene provide evidence of multiple pathways for tumor progression. *Proc Natl Acad Sci USA* 87:538, 1990.

354. M. A. Nelson, B. W. Futscher, T. Kinsella, J. Wymer, and G. T. Bowden: Detection of mutant Ha-ras genes in chemically initiated mouse skin epidermis before the development of benign tumors. *Proc Natl Acad Sci USA* 89:6398, 1992.

355. R. W. Ruddon: *Cancer Biology*, 3rd ed. Chapter 7, New York: Oxford University Press, 1995.

356. T. C. Thompson, J. Southgate, G. Ketchener, and H. Land: Multistage carcinogenesis induced by *ras* and *myc* oncogenes in a reconstituted organ. *Cell* 56:917, 1989.

357. D. A. Greenhalgh, D. J. Welty, A. Player, and S. H. Yuspa: Two oncogenes, v-*fos* and v-*ras*, cooperate to convert normal keratinocytes to squamous cell carcinoma. *Proc Natl Acad Sci USA* 87:643, 1990.

358. J. B. Weitzman and M. Yaniv: Rebuilding the road to cancer. *Nature* 400:401, 1999.

359. J. L. Bos: *ras* oncogenes in human cancer: A review. *Cancer Res* 49:4682, 1989.

360. M. A. Knowles and M. Williamson: Mutation of H-*ras* is infrequent in bladder cancer: Confirmation by single-strand conformation polymorphism analysis, designed restriction fragment length polymorphisms, and direct sequencing. *Cancer Res* 53:133, 1993.

361. T. Enomoto, M. Inoue, A. O. Perantoni, N. Terakawa, O. Tanizawa, and J. M. Rice: K-ras activation in neoplasms of the human female reproductive tract. *Cancer Res* 50:6139, 1990.

362. D. Sidransky, T. Tokino, S. R. Hamilton, K. W. Kinzler, B. Levin, P. Frost, and B. Vogelstein: Identification of *ras* oncogene mutations in the stool of patients with curable colorectal tumors. *Science* 256:102, 1992.

363. M. Tada, M. Omata, S. Kawai, H. Saisho, M. Ohto, R. K. Saiki, and J. J. Sninsky: Detection of ras gene mutations in pancreatic juice and peripheral blood of patients with pancreatic adenocarcinoma. *Cancer Res* 53: 2472, 1993.

364. J. L. Maldonado, J. Fridlyand, H. Patel, A. N. Jain, K. Busam, et al.: Determinants of BRAF mutations in primary melanomas. *J Natl Cancer Inst* 95:1878, 2003.

365. 70.P. H. Duesberg, K. Bister, and P. K. Vogt: The RNA of avian acute leukemia virus MC29. *Proc Natl Acad Sci USA* 74:4320, 1977.

366. G. J. Kato and C. V. Dang: Function of the c-Myc oncoprotein. *FASEB J* 6:3065, 1992.

367. G. Klein: The role of gene dosage and genetic transpositions in carcinogenesis. *Nature* 294:313, 1981.

368. R. Taub, I. Kirsch, C. Morton, G. Lenoir, D. Swan, S. Tronick, S. Aaronson, and P. Leder: Translocation of the c-*myc* gene into the immunoglobulin heavy chain locus in human Burkitt lymphoma and murine plasmacytoma cells. *Proc Natl Acad Sci USA* 79:7837, 1982.

369. R. Dalla-Favera, M. Bregni, J. Erikson, D. Patterson, R. C. Gallo, and C. M. Croce: Human c-*myc* onc gene is located on the region of chromosome 8 that is translocated in Burkitt lymphoma cells. *Proc Natl Acad Sci USA* 79:7824, 1982.

370. J. M. Adams, S. Gerondakis, E. Webb, L. M. Corcoran, and S. Cory: Cellular *myc* oncogene is altered by chromosome translocation to an immunoglobulin locus in murine plasmo-

cytomas and is rearranged similarly in human Burkitt lymphomas. *Proc Natl Acad Sci USA* 80:1982, 1983.

371. R. Dalla-Favera, F. Wong-Staal, and R. C. Gallo: *onc* gene amplification in promyelocytic leukaemia cell line HL-60 and in primary leukaemic cells of the same patient. *Nature* 299:61, 1982.

372. K. Alitalo, M. Schwab, C. C. Lin, H. E. Varmus, and J. M. Bishop: Homogeneously staining chromosomal regions contain amplified copies of an abundantly expressed cellular oncogene (c-*myc*) in malignant neuroendocrine cells from a colon carcinoma. *Proc Natl Acad Sci USA* 80:1707, 1983.

373. C. D. Little, M. M. Nau, D. N. Carney, A. F. Gazdar, and J. D. Minna: Amplification and expression of the c-*myc* oncogene in human lung cancer cell lines. *Nature* 306:194, 1983.

374. B. K. Suchy, M. Sarafoff, R. Kerler, and H. M. Rabes: Amplification, rearrangements, and enhanced expression of c-*myc* in chemically induced rat liver tumors in vivo and in vitro. *Cancer Res* 49:6781, 1989.

375. J. Brennan, T. O'Connor, R. W. Makuch, A. M. Simmons, E. Russell, R. I. Linnoila, R. M. Phelps, A. F. Gazdar, D. C. Ihde, and B. E. Johnson: myc family DNA amplification in 107 tumors and tumor cell lines from patients with small cell lung cancer treated with different combination chemotherapy regimens. *Cancer Res* 51:1708, 1991.

376. M. F. Melhem, A. I. Meisler, G. G. Finley, W. H. Bryce, M. O. Jones, I. I. Tribby, J. M. Pipas, and R. A. Koski: Distribution of cells expressing myc proteins in human colorectal epithelium, polyps, and malignant tumors. *Cancer Res* 52:5853, 1992.

377. E. M. J. J. Berns, J. G. M. Klijn, W. L. J. van Putten, I. L. van Staveren, H. Portegen, and J. A. Foekens: c-myc amplification is a better prognostic factor than HER2/neu amplification in primary breast cancer. *Cancer Res* 52:1107, 1992.

378. D. D. Von Hoff, J. R. McGill, B. J. Forseth, K. K. Davison, T. P. Bradley, D. R. Van Devanter, and G. M. Wahl: Elimination of extrachromosomally amplified *MYC* genes from human tumor cells reduces their tumorigenicity. *Proc Natl Acad Sci USA* 89:8165, 1992.

379. E. M. Blackwood and R. N. Eisenman: Max: a helix-loop-helix protein that forms a sequence-specific DNA binding complex with Myc. *Science* 251:1211, 1991.

380. G. C. Prendergast, D. Lawe, and E. B. Ziff: Association of Myn, the murine homolog of Max, with c-Myc stimulates methylation-sensitive DNA binding and Ras cotransformation. *Cell* 65:395, 1991.

381. L. Kretzner, E. M. Blackwood, and R. N. Eisenman: Myc and Max proteins possess distinct transcriptional activities. *Nature* 359:426, 1992.

382. W. Gu, K. Cechova, V. Tassi, and R. Dalla-Favera: Opposite regulation of gene transcription and cell proliferation by c-Myc and Max. *Proc Natl Acad Sci USA* 90:2935, 1993.

383. B. Amati, M. W. Brooks, N. Levy, T. D. Littlewood, G. I. Evan, and H. Land: Oncogenic activity of the c-Myc protein requires dimerization with Max. *Cell* 72:233, 1993.

384. M. T. Hemann, A. Bric, J. Teruya-Feldstein, A. Herbst, J. A. Nilsson, et al.: Evasion of the p53 tumor surveillance network by tumour-derived *MYC* mutants. *Nature* 436:807, 2005.

385. H. Ogawa, K.-I. Ishiguro, S. Ganbatz, D. M. Livingston, and Y. Nakatani: A complex with chromatin modifiers that occupies E2F- and Myc-responsive genes in G_0 cells. *Science* 296:1132, 2002.

386. T. Hunter: A tail of two src's: Mutatis Mutandis. *Cell* 49:1, 1987.

387. S. A. Courtneidge: Activation of the pp60c-src kinase by middle T antigen bending or by dephosphorylation. *EMBO J* 4:1471, 1985.

388. J. A. Cooper and B. Howell: The when and how of src regulation. *Cell* 73:1051, 1993.

389. M. Susa, M. Missbach, and J. Green: Src inhibitors: drugs for the treatment of osteoporosis, cancer, or both? *Trends Pharmacol Sci* 21:489, 2000.

390. C. Bjelfman, F. Hedborg, I. Johansson, M. Nordenskjöld, and S. Pahlman: Expression of the neuronal form of pp60[c-scr] in neuroblastoma in relation to clininal stage and prognosis. *Cancer Res* 50:6908, 1990.

391. C. A. Cartwright, A. I. Meisler, and W. Eckhart: Activation of the pp60[c-src] protein kinase is an early event in colonic carcinogenesis. *Proc Natl Acad Sci USA* 87:558, 1990.

392. T. Pawson and G. D. Gish: SH2 and SH3 domains: From structure to function. *Cell* 71:359, 1992.

393. A. J. Muller, A.-M. Pendergast, K. Parmar, M. H. Havlik, N. Rosenberg, and O. N. Witte: En bloc substitution of the SRC homology region 2 domain activates the transforming potential of the c-Abl protein tyrosine kinase. *Proc Natl Acad Sci USA* 90:3457, 1993.

394. J. X. Zou, Y. Liu, E. B. Pasquale, and E. Ruoslahti: Activated *Src* oncogene phosphorylates R-Ras and suppresses integrin activity. *J Biol Chem* 277:1824, 2002.

395. Y. Maki, T. Bos, C. Davis, M. Starbuck, and P. Vogt: Avian sarcoma virus 17 carries the *jun* oncogene. *Proc Natl Acad Sci USA* 84:2848, 1987.

396. J. Schütte, J. D. Minna, and M. J. Birrer: Deregulated expression of human c-jun transforms primary rat embryo cells in cooperation with an activated c-Ha-*ras* gene and transforms *rat*-1a cells as a single gene. *Proc Natl Acad Sci USA* 86:2257, 1989.

397. V. R. Baichwar and R. Tijan: Control of c-jun activity by interaction of a cell-specific inhibitor

with regulatory domain δ: Differences between v- and c-jun. *Cell* 63:815, 1990.

398. J. Schütte, J. Viallet, M. Nau, S. Segal, J. Fedorko, and J. Minna: jun-B inhibits and c-fos stimulates the transforming and *trans*-activating activities of c-jun. *Cell* 59:987, 1989.

399. M. P. Finkel, C. A. Reilly, Jr., and B. O. Biskis: Viral etiology of bone cancer. *Front Radiat Theor Oncol* 10:28, 1975.

400. T. Jenuwein and R. Müller: Structure–function analysis of fos protein: A single amino acid change activates the immortalizing potential of v-fos. *Cell* 48:647, 1987.

401. K. Macleod, D. Leprince, and D. Stehelin: The *ets* gene family. *Trends Biochem Sci* 251, 1992.

402. A. Seth and A. Papas: The c-*ets*-1 proto-oncogene has oncogenic activity and is positively autoregulated. *Oncogene* 5:1761, 1990.

403. S. P. Goff, E. Gilboa, O. N. Witte, and D. Baltimore: Structure of the Abelson murine leukemia virus genome and the homologous cellular gene: Studies with cloned viral DNA. *Cell* 22:777, 1980.

404. O. N. Witte: Role of the BCR-ABL oncogene in human leukemia: Fifteenth Richard and Hilda Rosenthal Foundation award lecture. *Cancer Res* 53:485, 1993.

405. G. Q. Daley, R. A. Van Etten, and D. Baltimore: Induction of chronic myelogenous leukemia in mice by the P210$^{bcr/abl}$ gene of the Philadelphia chromosome. *Science* 247:824, 1990.

406. J. Willem Voncken, S. Griffiths, M. F. Greaves, P. K. Pattengale, N. Heisterkamp, and J. Groffen: Restricted oncogenicity of BCR/ABL p10 in transgenic mice. *Cancer Res* 52:4534, 1992.

407. C. L. Sawyers, J. McLaughlin, A. Goga, M. Havlik, and O. Witte: The nuclear tyrosine kinase c-Abl negatively regulates cell growth. *Cell* 77: 121, 1994.

408. A. M. Pendergast, L. A. Quilliam, L. D. Cripe, C. H. Bassing, Z. Dai, N. Li, A. Batzer, K. M. Rabun, C. J. Der, J. Schlessinger, and M. L. Gishizky: BCR-ABL-induced oncogenesis is mediated by direct interaction with the SH2 domain of the GRB-2 adaptor protein. *Cell* 75:175, 1993.

409. C. L. Sawyers: Disabling Abl—Perspectives on Abl kinase regulation and cancer therapeutics. *Cancer Cell* 3:13, 2002.

410. B. Lüscher, E. Christenson, D. W. Litchfield, E. G. Krebs, and R. N. Eisenman: Myb DNA binding inhibited by phosphorylation at a site deleted during oncogenic activation. *Nature* 344:517, 1990.

411. D. Venturelli, M. T. Mariano, C. Szczylik, M. Valtieri, B. Lange, W. Crist, M. Link, and B. Calabretta: Down-regulated c-myb expression inhibits DNA synthesis of T-leukemia cells in most patients. *Cancer Res* 50:7371, 1990.

412. S. Haldar, C. Beatty, Y. Tsujimoto, and C. M. Croce: The bcl-2 gene encodes a novel G protein. *Nature* 342:195, 1989.

413. M. Katsumata, R. M. Siegel, D. C. Louie, T. Miyashita, Y. Tsujimoto, P. C. Nowell, M. I. Green, and J. C. Reed: Differential effects of Bcl-2 on T and B cells in transgenic mice. *Proc Natl Acad Sci USA* 89:11376, 1992.

414. Z. N. Oltvai, C. L. Milliman, and S. J. Korsmeyer: Bcl-2 heterodimerizes in vivo with a conserved homolog, Bax, that accelerates programed cell death. *Cell* 74:609, 1993.

415. S. J. Weintraub, S. R. Manson, and B. E. Deverman: Resistance to antineoplastic therapy: the oncogenic tyrosine kinase-Bcl-x$_L$ axis. *Cancer Cell* 5:3, 2004.

416. R. Sen and D. Baltimore: Multiple nuclear factors interact with the immunoglobulin enhancer sequences. *Cell* 46:705, 1986.

417. S. Gosh, A. M. Gifford, L. R. Viviere, P. Tempst, G. P. Nolan, and D. Baltimore: Cloning of the p50 DNA binding subunit of NF-κB: Homology to rel and dorsal. *Cell* 62:1019, 1990.

418. M. Kieran, V. Blank, F. Logeat, J. Vanderkerckhove, F. Lottspeich, O. Le Ball, M. B. Urban, P. Kourisky, P. A. Baeuerie, and A. Israel: The DNA binding subunit of NF-κB is identical to factor KBF1 and homologous to the rel oncogene product. *Cell* 62:1007, 1990.

419. T. Lernbecker, U. Müller, and T. Wirth: Distinct NF-κB/rel transcription factors are responsible for tissue-specific and inducible gene activation. *Nature* 365:767, 1993.

420. T. Graf and H. Beug: Role of the v-erbA and v-erbB oncogenes of avian erythroblastosis virus in erythroid cell transformation. *Cell* 34:7, 1983.

421. K. Damm, C. C. Thompson, and R. M. Evans: Protein encoded by v-erbA functions as a thyroid-hormone receptor antagonist. *Nature* 339:593, 1989.

422. C. Desbois, D. Aubert, C. Legrand, B. Pain, and J. Samarut: A novel mechanism of action for v-ErbA: Abrogation of the inactivation of transcription factor AP-1 by retinoic acid and thyroid hormone receptors. *Cell* 67:731, 1991.

423. M. Sharif and M. L. Privalsky: v-erbA oncogene function in neoplasia correlates with its ability to repress retinoic acid receptor action. *Cell* 66:885, 1991.

424. R. F. Doolittle, M. W. Hunkapiller, L. E. Hood, S. G. Devare, K. C. Robbins, S. A. Aaronson, and H. N. Antoniades: Simian sarcoma virus *onc* gene, v-sis, is derived from the gene (or genes) encoding a platelet-derived growth factor. *Science* 221:275, 1983.

425. M. D. Waterfield, G. T. Scrace, N. Whittle, P. Stoobant, A. Johnson, A. Wasteson, B. Westermark, C.-H. Heldin, J. S. Huang, and T. F. Deuel: Platelet-derived growth factor is structurally related to the putative transforming protein p28^{sis} of simian sarcoma virus. *Nature* 304:35, 1983.

426. H. N. Antoniades and M. W. Hunkapiller: Human platelet-derived growth factor (PDGF):

Amino terminal amino acid sequence. *Science* 220:963, 1983.

427. S. G. Devare, E. P. Reddy, J. D. Law, K. C. Robbins, and S. A. Aaronson: Nucleotide sequence of the simian sarcoma virus genome: Demonstration that is acquired cellular sequence encode the transforming gene product p28$^{v\text{-}sis}$. *Proc Natl Acad Sci USA* 80:731, 1983.

428. T. F. Deuel, J. S. Huang, S. S. Huang, P. Stroobant, and M. D. Waterfield: Expression of a platelet-derived growth factor–like protein in simian virus transformed cells. *Science* 221:1348, 1983.

429. J. S. Huang, S. S. Huang, and T. F. Keuel: Transforming protein of simian sarcoma virus stimulates autocrine growth of SSV-transformed cells through PDGF cell-surface receptors. *Cell* 39:79, 1984.

430. T. Hunter: Oncogenes and growth control. *Trends Biochem Sci* July 1985, p. 275.

431. V. B. Lokeshwar, S. S. Huang, and J. S. Huang: Intracellular turnover, novel secretion, and mitogenically active intracellular forms of v-sis gene product in simian sarcoma virus–transformed cells. *J Biol Chemistry* 265:1665, 1990.

432. J. Downward, Y. Yarden, E. Mayes, G. Scrace, N. Totty, P. Stockwell, A. Ullrich, J. Schlessinger, and M. D. Waterfield: Close similarity of epidermal growth factor receptor of v-erb-B oncogene protein sequences. *Nature* 307:521, 1984.

433. Y. Kokai, J. N. Myers, T. Wada, V. I. Brown, C. M. LeVea, J. G. Davis, K. Dobashi, and M. I. Greene: Synergistic interaction of p185c-neu and the EGF receptor leads to transformation of rodent fibroblasts. *Cell* 58:287, 1989.

434. C. I. Bargmann and R. A. Weinberg: Increased tyrosine kinase activity associated with the protein encoded by the activated neu oncogene. *Proc Natl Acad Sci USA* 85:5394, 1988.

435. Y.-F. Zhai, H. Beittenmiller, B. Wang, M. N. Gould, C. Oakley, W. J. Esselmann, and C. W. Welsch: Increased expression of specific protein tyrosine phosphatases in human breast epithelial cells neoplastically transformed by the neu oncogene. *Cancer Res* 53:2272, 1993.

436. L. Bouchard, L. Lamarre, P. J. Tremblay, and P. Jolicoeur: Stochastic appearance of mammary tumors in transgenic mice carrying the MMTV/c-neu oncogene. *Cell* 57:931, 1989.

437. C. T. Guy, M. A. Webster, M. Schaller, T. J. Parsons, R. D. Cardiff, and W. J. Muller: Expression of the neu proto-oncogene in the mammary epithelium of transgenic mice induces metastatic disease. *Proc Natl Acad Sci USA* 89:10578, 1992.

438. P. Devilee, P. Van Vliet, N. Kuipers-Dijlesho-ony, P. L. Pearson, and C. J. Cornelisse: Somatic genetic changes on chromosome 18 in breast carcinomas: it the *DCC* gene involved? *Oncogene* 6:311, 1991.

439. M. Streuli, N. X. Krueger, P. D. Ariniello, M. Tang, J. M. Munro, W. A. Blattler, D. A.

Adler, C. M. Disteche, and H. Saito: Expression of the receptor-linked protein tyrosine phosphatase LAR: Proteolytic cleavage and shedding of the CAM-like extracellular regions. *EMBO J* 11:897, 1992.

440. D. J. Slamon, G. M. Clark, S. G. Wong, W. J. Levin, A. Ullrich, and W. L. McGuire: Human breast cancer: Correlation of relapse and survival with amplification of the HER-2/neu oncogene. *Science* 235:177, 1987.

441. A. Berchuck, A. Kamel, R. Whitaker, B. Kerns, G. Olt, R. Kinney, J. T. Soper, R. Doge, D. L. Clarke-Pearson, P. Marks, S. McKenzie, S. Yin, and R. C. Bast, Jr.: Overexpression of HER-2/neu is associated with poor survival in advanced epithelial ovarian cancer. *Cancer Res* 50:4087, 1990.

442. Y. Yonemura, I. Ninomiya, A. Yamaguchi, S. Fushida, H. Kimura, S. Ohoyama, I. Miyazaki, Y. Endou, M. Tanaka, and T. Sasaki: Evaluation of immunoreactivity of erbB-2 protein as a marker of poor short-term prognosis in gastric cancer. *Cancer Res* 51:1034, 1991.

443. C.-M. Tsai, K.-T. Chang, R.-P. Perng, T. Mitsudomi, M.-H. Chen, C. Kadoyama, and A. F. Gazdar: Correlation of intrinsic chemoresistance of non–small-cell lung cancer cell lines with HER-2/neu gene expression but not with ras gene mutations. *J Natl Cancer Inst* 85:897, 1993.

444. M. F. Roussel, J. R. Downing, C. W. Rettenmier, and C. J. Sherr: A point mutation in the extracellular domain of the human CSF-1 receptor (c-fms proto-oncogene product) activates its transforming potential. *Cell* 55:979, 1988.

445. M. Z. Ratajczak, S. M. Luger, K. DeRiel, J. Abrahm, B. Calabretta, and A. M. Gewirtz: Role of the KIT proto-oncogene in normal and malignant human hematopoiesis. *Proc Natl Acad Sci USA* 89:1710, 1992.

446. R. Oskam, F. Coulier, M. Ernst, D. Martin-Sanca, and M. Barbacid: Frequent generation of oncogenes by in vitro recombination of TRK proto-oncogene sequences. *Proc Natl Acad Sci USA* 85:2964, 1988.

447. A. Nakagawara, M. Arima-Nakagawara, N. J. Scavarda, C. G. Azar, A. B. Cantor, and G. M. Brodeur: Association between high levels of expression of the TRK gene and favorable outcome in human neuroblastomas. *N Engl J Med* 328:847, 1993.

448. S. Giordano, Z. Zhen, E. Medico, G. Gaudino, F. Galimi, and P. M. Comoglio: Transfer of motogenic and invasive response to scatter factor/hepatocyte growth factor by transfection of human MET proto-oncogene. *Proc Natl Acad Sci USA* 90:649, 1993.

449. T. Maeda, R. M. Hobbs, and P. P. Pandolfi: The transcription factor *Pokemon*: A new key player in cancer pathogenesis. *Cancer Res* 65:8575, 2005.

450. T. Maeda, R. M. Hobbs, T. Merghoub, I. Guernah, A. Zelent, et al.: Role of the

proto-oncogene *Pokemon* in cellular transformation and *ARF* repression. *Nature* 433:278, 2005.

451. D. J. Slamon and M. J. Cline: Expression of cellular oncogenes during embryonic and fetal development of the mouse. *Proc Natl Acad Sci USA* 81:7141, 1984.

452. R. Müller, D. J. Slamon, J. M. Tremblay, M. J. Cline, and I. M. Verma: Differential expression of cellular oncogenes during pre- and postnatal development of the mouse. *Nature* 299:640, 1982.

453. A. S. Goustin, C. Betsholtz, S. Pfeifer-Ohlsson, H. Persson, J. Rydnert, M. Bywater, G. Holmgren, C.-H. Heldin, B. Westermark, and R. Ohlsson: Coexpression of the *sis* and *myc* proto-oncogenes in developing human placenta suggests autocrine control of trophoblast growth. *Cell* 41:301, 1985.

454. R. L. Michell, L. Zokas, R. D. Schreiber, and I. M. Verma: Rapid induction of the expression of proto-oncogene *fos* during human monocytic differentiation. *Cell* 40:209, 1985.

455. E. Sariban, T. Mitchell, and D. Kufe: Expression of the c-*fms* proto-oncogene during human monocytic differentiation. *Nature* 316:64, 1985.

456. P. H. Reitsma, P. G. Rothberg, S. M. Astrin, J. Trial, Z. Bar-Shavit, A. Hall, S. L. Teitelbaum, and A. J. Kahn: Regulation of *myc* gene expression in HL-60 leukaemia cells by a vitamin D. metabolite. *Nature* 306:492, 1983.

457. R. Makino, K. Hayashi, and T. Sugimura: c-*myc* transcript is induced in rat liver at a very early stage of regeneration or by cycloheximide treatment. *Nature* 310:697, 1984.

458. M. Goyette, C. J. Petropoulos, P. R. Shank, and N. Fausto: Expression of a cellular oncogene during liver regeneration. *Science* 219:510, 1983.

459. D. Boettiger, S. Anderson, and T. M. Dexter: Effect of src infection on long-term marrow cultures: Increased self-renewal of hemopoietic progenitor cells without leukemia. *Cell* 36:763, 1984.

460. S. H. Yuspa, A. E. Kilkenny, J. Stanley, and U. Lichti: Keratinocytes blocked in phorbol ester–responsive early stage of terminal differentiation by sarcoma viruses. *Nature* 314:459, 1985.

461. B. H. Sweet and M. R. Hilleman: The vacuolating virus, SV40. *Proc Exp Biol Med* 105:420, 1960.

462. B. E. Eddy, G. S. Borman, G. E. Grubbs, and R. D. Young: Identification of the oncogenic substance in rhesus monkey cell cultures as simian virus 40. *Virology* 17:65, 1962.

463. H. M. Shein and J. F. Enders: Transformation of human renal cell cultures. I. Morphology and growth characteristics. *Proc Natl Acad Sci USA* 48:1164, 1962.

464. G. Khoury, C.-J. Lai, D. Solomon, M. Israel, and P. Howley: The human papovaviruses and their potential role in human diseases. In H. H. Hiatt, J. D. Watson, and J. A. Winsten, eds.: *Origins of Human Cancer*. Cold Spring Harbor, NY: Cold Spring Harbor Laboratory, 1977, pp. 971–988.

465. C. Prives, E. Gilboa, M. Revel, and E. Winocour: Cell-free translation of simian virus 40 early messenger RNA coding for viral T-antigen. *Proc Natl Acad Sci USA* 74:457, 1977.

466. P. Rigby: The transforming genes of SV40 and polyoma. *Nature* 282:781, 1979.

467. M. R. D. Scott, K.-H. Westphal, and P. W. J. Rigby: Activation of mouse genes in transformed cells. *Cell* 34:557, 1983.

468. M. Kriegler, C. F. Perez, C. Hardy, and M. Botchan: Transformation mediated by the SV40 T antigens: Separation of the overlapping SV40 early genes with a retroviral vector. *Cell* 38:483, 1984.

469. R. L. Brinster, H. Y. Chen, A. Messing, T. van Dyke, A. J. Levine, and R. D. Palmiter: Transgenic mice harboring SV40 T-antigen genes develop characteristic brain tumors. *Cell* 37:367, 1984.

470. R. D. Palmiter, H. Y. Chen, A. Messing, and R. L. Brinster: SV40 enhancer and large-T antigen are instrumental in development of choroid plexus tumors in transgenic mice. *Nature* 316:457, 1985.

471. A. Messing, H. Y. Chen, R. D. Palmiter, and R. L. Brnster: Peripheral neuropathies, hepatocellular carcinomas, and islet cell adenomas in transgenic mice. *Nature* 316:461, 1985.

472. D. Hanahan: Heritable formation of pancreatic β-cell tumours in transgenic mice expressing recombinant insulin/simian virus 40 oncogenes. *Nature* 315:115, 1985.

473. J. Feuteun, L. Sompayrac, M. Fluck, and T. Benjamin: Localization of gene functions in polyoma virus DNA. *Proc Natl Acad Sci USA* 73:4169, 1976.

474. Y. Ito, J. R. Brocklehurst, and R. Dulbecco: Virus-specific proteins in the plasma membrane of cells lytically infected or transformed by polyoma virus. *Proc Natl Acad Sci USA* 74:4666, 1977.

475. M. Rassoulzadegan, A. Cowie, A. Carr, N. Glaichenhaus, R. Kamen, and F. Cuzin: The roles of individual polyoma virus early proteins in oncogenic transformation. *Nature* 300:713, 1982.

476. G. Della Valle, R. G. Fenton, and C. Basilico: Polyoma large T antigen regulates the integration of viral DNA sequences into the genome of transformed cells. *Cell* 23:347, 1981.

477. D. J. Donoghue, C. Anderson, T. Hunter, and P. L. Kaplan: Transmission of the polyoma viurs middle T gene as the oncogene of a murine retrovirus. *Nature* 308:748, 1984.

478. G. Carmichael, B. S. Schaffhausen, G. Mandel, T. J. Liang, and T. L. Benjamin: Transformation by polyoma virus is drastically reduced by substitution of phenylalanine for tyrosine at residue 315 of middle-sized tumor antigen. *Proc Natl Acad Sci USA* 81:679, 1984.

479. A. A. Schreier and J. Gruber: Viral T-antigen interactions with cellular proto-oncogene and

anti-oncogene products. *J Natl Cancer Inst* 82:354, 1990.

480. J. W. Ludlow, J. A. DeCaprio, C.-M. Huang, W.-H. Lee, E. Paucha, and D. M. Livingston: SV40 large T antigen binds preferentially to an underphosphorylated member of the retinoblastoma susceptibility gene product family. *Cell* 56:57, 1989.

481. G. Orth, F. Breitburd, M. Favre, and O. Croissant: Papillomaviruses: Possible role in human cancer. In H. H. Hiatt, J. D. Watson, and J. A. Winsten, eds.: *Origins of Human Cancer.* Cold Spring Harbor, NY: Cold Spring Harbor Laboratory, 1977, pp. 1043–1068.

482. T. D. Kessis, R. J. Slebos, W. G. Nelson, M. B. Kastan, B. S. Plunkett, S. M. Han, A. T. Lorincz, L. Hedrick, and K. R. Cho: Human papillomavirus 16 E6 expression disrupts the p53-mediated cellular response to DNA damage. *Proc Natl Acad Sci USA* 90:3988, 1993.

483. M. H. Schiffman, H. M. Bauer, R. N. Hoover, A. G. Glass, D. M. Cadell, B. B. Rush, D. R. Scott, M. E. Sherman, R. J. Kurman, S. Wacholder, C. K. Stanton, and M. M. Manos: Epidemiologic evidence showing that human papillomavirus infection causes most cervical intraepithelial neoplasia. *J Natl Cancer Inst* 85:958, 1993.

484. M. Scheffner, B. A. Werness, J. M. Hulbregtse, A. J. Levine, and P. M. Howley: The E6 oncoprotein encoded by human papillomavirus types 16 and 18 promotes the degradation of p53. *Cell* 63:1129, 1990.

485. P. F. Lambert, H. Pan, H. C. Pitot, A. Liem, M. Jackson, and A. E. Griep: Epidermal cancer associated with expression of human papillomavirus type 16 E6 and E7 oncogenes in the skin of transgenic mice. *Proc Natl Acad Sci USA* 90:5583, 1993.

486. L. Chen, S. Ashe, M. C. Singhal, D. A. Galloway, I. Hellström, and K. E. Hellström: Metastatic conversion of cells by expression of human papillomavirus type 16 E6 and E7 genes. *Proc Natl Acad Sci USA* 90:6523, 1993.

487. J. J. Trentin, Y. Yabe, and G. Taylor: The quest for human cancer viruses. *Science* 137:835, 1962.

488. J. K. McDougall, L. B. Chen, and P. H. Gallimore: Transformation in vitro by adenovirus type 2—A model system for studying mechanisms of oncogenicity. In H. H. Hiatt, J. D. Watson, and J. A. Winsten, eds.: *Origins of Human Cancer.* Cold Spring Harbor, NY: Cold Spring Harbor Laboratory, 1977, pp. 1013–1025.

489. H. E. Ruley: Adenovirus early region 1A enables viral and cellular transforming genes to transform primary cells in culture. *Nature* 304:602, 1983.

490. R. Javier, K. Raska, Jr., and T. Shenk: Requirement for the adenovirus type 9 E4 region in production of mammary tumors. *Science* 257:1267, 1992.

491. P. Whyte, N. M. Williamson, and E. Harlow: Cellular targets for transformation by the adenovirus E1A proteins. *Cell* 56:67, 1989.

492. G. Fourel, C. Trepo, L. Bougueleret, B. Henglein, A. Ponzetto, P. Tiollais, and M-A. Buendia: Frequent activation of N-*myc* genes by hepadnavirus insertion in woodchuck liver tumors. *Nature* 347:294, 1990.

493. H. F. Maguire, J. P. Hoeffler, and A. Siddiqui: HBV X protein alters the DNA binding specificity of CREB and ATF-2 by protein–protein interactions. *Science* 252:842, 1991.

494. B. Roizman, N. Frenkel, E. D. Kieff, and P. G. Spear: The structure and expression of human hepersvirus DNAs in productive infection and in transformed cells. In H. H. Hiatt, J. D. Watson, and J. A. Winsten, eds.: *Origins of Human Cancer.* Cold Spring Harbor, NY: Cold Spring Harbor Laboratory, 1977, pp. 1069–1111.

495. P. Lengyel: Tumor-suppressor genes: News about the interferon connection. *Proc Natl Acad Sci USA* 90:5893, 1993.

496. H. Harris, O. J. Miller, G. Klein, P. Worst, and T. Tachibana: Suppression of malignancy by cell fusion. *Nature* 223:363, 1969.

497. A. G. Knudson, Jr.: Hereditary cancer, oncogenes, and antioncogenes. *Cancer Res* 45:1437, 1985.

498. R. E. Hollingsworth and W.-H. Lee: Tumor suppressor genes: New prospects for cancer research. *J Natl Cancer Inst* 83:91, 1991.

499. J. M. Bishop: The molecular genetics of cancer. *Science* 235:305, 1987.

500. E. J. Stanbridge: Human tumor suppressor genes. *Annu Rev Genet* 24:615, 1990.

501. W. K. Cavenee, T. P. Dryja, R. A. Phillips, W. F. Benedict, R. Godbout, et al.: Expression of recessive alleles by chromosomal mechanisms in retinoblastoma. *Nature* 305:779, 1983.

502. W. K. Cavenee, M. F. Hansen, M. Nordenskjold, E. Kock, I. Maumenee, et al.: Genetic origin of mutations predisposing to retinoblastoma. *Science* 228:501, 1985.

503. S. H. Friend, R. Bernards, S. Rogelj, R. A. Weinberg, J. M. Rapaport, et al.: A human DNA segment with properties of the gene that predisposes to retinoblastoma and osteosarcoma. *Nature* 323:643, 1986.

504. C. J. Marshall: Tumor suppressor genes. *Cell* 64:313, 1991.

505. J. Marx: Learning how to suppress cancer. *Science* 261:1385, 1993.

506. H. Harris: The analysis of malignancy by cell fusion: The position in 1988. *Cancer Res* 48:3302, 1988.

507. R. Fodde and R. Smits: A matter of dosage. *Science* 298:761, 2002.

508. H. Harris: Putting on the brakes. *Nature* 427:201, 2004.

509. A. J. Levine: The tumor suppressor genes. *Annu Rev Biochem* 62:623, 1993.

510. C. J. Sherr and F. McCormick: The RB and p53 pathways in cancer. *Cancer Cell* 2:103, 2002.

511. D. J. Templeton, S. H. Park, L. Lanier, and R. A. Weinberg: Nonfunctional mutants of the retinoblastoma protein are characterized by defects in phosphorylation, viral oncoprotein association, and nuclear tethering. *Proc Natl Acad Sci USA* 88:3033, 1991.

512. M. G. Paggi and A. Giordano: Who is the boss in the retinoblastoma family? The point of view of *Rb2/p130*, the little brother. *Cancer Res* 61:4651, 2001.

513. N. Dyson, K. Buchkovich, P. Whyte, and E. Harlow: The cellular 107K protein that binds to adenovirus E1A also associates with the large T antigens of SV40 and JC virus. *Cell* 58:249, 1989.

514. R. Takahashi, T. Hashimoto, H.-J. Xu, S.-X. Hu, T. Matsui, et al.: The retinoblastoma gene functions as a growth and tumor suppressor in human bladder carcinoma cells. *Proc Natl Acad Sci USA* 88:5257, 1991.

515. E. Y.-H. P. Lee, C.-Y. Chang, H. Nanpin, Y.-C. J. Wang, et al.: Mice deficient for Rb are nonviable and show defects in neurogenesis and haematopoiesis. *Nature* 359:288, 1992.

516. T. Jacks, A. Fazeli, E. M. Schmitt, R. T. Bronson, M. A. Goodell, and R. A. Weinberg: Effects of an Rb mutation in the mouse. *Nature* 359:295, 1992.

517. A. R. Clarke, E. R. Maandag, M. van Roon, N. M. T. van der Lugt, M. van der Valk, et al.: Requirement for a functional Rb-1 gene in murine development. *Nature* 359:328, 1992.

518. J. J. Windle, D. M. Albert, J. M. O'Brien, D. M. Marcus, C. M. Disteche, et al.: Retinoblastoma in transgenic mice. *Nature* 343:665, 1990.

519. D. Defeo-Jones, P. S. Huang, R. E. Jones, K. M. Haskell, G. A. Cuocolo, et al.: Cloning of cDNAs for cellular proteins that bind to the retinoblastoma gene product. *Nature* 352:251, 1991.

520. P. D. Robbins, J. M. Horowitz, and R. C. Mulligan: Negative regulation of human c-*fos* expression by the retinoblastoma gene product. *Nature* 346:668, 1990.

521. S.-J. Kim, H.-D. Lee, P. D. Robbins, K. Busam, M. B. Sporn, and A. B. Roberts: Regulation of transforming growth factor $\beta 1$ gene expression by the product of the retinoblastoma-susceptibility gene. *Proc Natl Acad Sci USA* 88:3052, 1991.

522. S.-J. Kim, S. Wagner, F. Liu, M. A. O'Reilly, P. D. Robbins, and M. R. Green: Retinoblastoma gene product activates expression of the human TGF-$\beta 2$ gene through transcription factor ATF-2. *Nature* 358:331, 1992.

523. D. P. Lane and L. V. Crawford: T antigen is bound to a host protein in SV40 transformed cells. *Nature* 278:261, 1979.

524. L. F. Parada, H. Land, R. A. Weinberg, D. Wolf, and V. Rotter: Cooperation between gene encoding p53 tumour antigen and *ras* in cellular transformation. *Nature* 312:649, 1984.

525. G. P. Zambetti, D. Olson, M. Labow, and A. J. Levine: A mutant p53 protein is required for maintenance of the transformed phenotype in cells transformed with p53 plus *ras* cDNAs. *Proc Natl Acad Sci USA* 89:3952, 1992.

526. D. Malkin, F. P. Li, L. C. Strong, J. F. Fraumeni, Jr., C. E. Nelson, et al.: Germ-line p53 mutations in a familial syndrome of breast cancer, sarcomas, and other neoplasms. *Science* 250:1233, 1990.

527. T. Frebourg, J. Kassel, K. T. Lam, M. A. Gryka, N. Barbier, et al.: Germ-line mutations of the p53 tumor suppressor gene in patients with high risk for cancer inactivate the p53 protein. *Proc Natl Acad Sci USA* 89:6413, 1992.

528. C. A. Finlay, P. W. Hinds, T.-H. Tan, D. Eliyahu, M. Oren, and A. J. Levine: Acitvating mutations for transformation by p53 produce a gene product that forms an hsc70-p53 complex with an altered half-life. *Mol Cell Biol* 8:531, 1988.

529. I. Hsu, R. Metcalf, T. Sun, J. Welsh, N. Wang, and C. Harris: Mutational hotspot in the p53 gene in human hepatocellular carcinomas. *Nature* 350:427, 1991.

530. F. Aguilar, S. P. Hussain, and P. Cerutti: Aflatoxin B$_1$ induces the transversion of G→T codon 249 of the p53 tumor suppressor gene in human hepatocytes. *Proc Natl Acad Sci USA* 90:8586, 1993.

531. S. Kress, U.-R. Jahn, A. Buchmann, P. Bannasch, and M. Schwarz: p53 mutations in human hepatocellular carcinomas from Germany. *Cancer Res* 52:3220, 1992.

532. B. Ruggeri, M. DiRado, S. Y. Zhang, B. Bauer, T. Goodrow, and A. J. P. Klein-Szanto: Benzo[a]pyrene-induced murine skin tumors exhibit frequent and characteristic G to T mutations in the p53 gene. *Proc Natl Acad Sci USA* 90:1013, 1993.

533. S. Kress, C. Sutter, P. T. Strickland, H. Mukhtar, J. Schweizer, and M. Schwarz: Carcinogen-specific mutational pattern in the p53 gene in ultraviolet B radiation-induced squamous cell carcinomas of mouse skin. *Cancer Res* 52:6400, 1992.

534. S. Kanjilal, W. E. Pierceall, K. K. Cummings, M. L. Kripke, and H. N. Ananthaswamy: High frequency of p53 mutations in ultraviolet radiation-induced murine skin tumors: Evidence for strand bias and tumor heterogeneity. *Cancer Res* 53:2961, 1993.

535. R. W. Ruddon: *Cancer Biology*, 3rd ed. Chapter 8. New York: Oxford University Press, 1995.

536. M. B. Kastan, O. Onyekwere, D. Sidransky, B. Vogelstein, and R. W. Craig: Participation of p53 protein in the cellular response to DNA damage. *Cancer Res* 51:6304, 1991.

537. S. J. Ullrich, C. W. Anderson, W. E. Mercer, and E. Appella: The p53 tumor suppressor protein, a modulator of cell proliferation. *J Biol Chem* 267:15259, 1992.

538. D. Lin, M. T. Shields, S. J. Ullrich, E. Appella, and W. E. Mercer: Growth arrest induced by wild-type p53 protein blocks cells prior to or near the restriction point in late G_1 phase. *Proc Natl Acad Sci USA* 89:9210, 1992.

539. S. W. Lowe, E. M. Schmitt, S. W. Smith, B. A. Osborne, and T. Jacks: p53 is required for radiation-induced apoptosis in mouse thymocytes. *Nature* 362:847, 1993.

540. A. R. Clarke, C. A. Purdie, D. J. Harrison, R. G. Morris, C. C. Bird, et al.: Thymocyte apoptosis induced by p53-dependent and independent pathways. *Nature* 362:849, 1993.

541. P. M. O'Connor, J. Jackman, D. Jondle, K. Bhatia, I. Magrath, et al.: Role of the p53 tumor suppressor gene in cell cycle arrest and radiosensitivity of Burkitt's lymphoma cell lines. *Cancer Res* 53:4776, 1993.

542. S. W. Lowe, H. E. Ruley, T. Jacks, and D. E. Housman: p53-dependent apoptosis modulates the cytotoxicity of anticancer agents. *Cell* 74:957, 1993.

543. J. M. Lee and A. Bernstein: p53 mutations increase resistance to ionizing radiation. *Proc Natl Acad Sci USA* 90:5742, 1993.

544. D. P. Lane: p53, guardian of the genome. *Nature* 358:15, 1992.

545. T. A. Weinert and L. Hartwell: Characterization of RAD9 of *Saccharomyces cerevisiae* and evidence that its function acts post-translationally in cell cycle arrest after DNA damage. *Mol Cell Biol* 10:6554, 1990.

546. L. R. Livingstone, A. White, J. Sprouse, E. Livanos, T. Jacks, and T. D. Tlsty: Altered cell cycle arrest and gene amplification potential accompany loss of wild-type p53. *Cell* 70:923, 1992.

547. Y. Yin, M. A. Tainsky, F. Z. Bischoff, L. C. Strong, and G. M. Wahl: Wild-type p53 restores cell cycle control and inhibits gene amplification in cells with mutant p53 alleles. *Cell* 70:937, 1992.

548. T. Fujiwara, E. A. Grimm, T. Mukhopadhyay, D. W. Cai, L. B. Owen-Schaub, and J. A. Roth: A retroviral wild-type p53 expression vector penetrates human lung cancer spheroids and inhibits growth by inducing apoptosis. *Cancer Res* 53:4129, 1993.

549. B. Vogelstein, D. Lane, and A. J. Levine: Surfing the p53 network. *Nature* 408:307, 2000.

550. R. S. Maser and R. A. De Pinho: Take care of your chromosomes lest cancer takes care of you. *Cancer Cell* 3:4, 2003.

551. D. P. Lane and P. M. Fischer: Turning the key on p53. *Nature* 427:789, 2004.

552. J. Yu and L. Zhang: No PUMA, no death: implications for p53-dependent apoptosis. *Cancer Cell* 4:248, 2003.

553. S. Yamamoto, A. Romanenko, M. Wei, C. Masuda, W. Zaparin, et al.: Specific *p53* gene mutations in urinary bladder epithelium after the Chernobyl accident. *Cancer Res* 59:3606, 1999.

554. L. Sivaraman, O. M. Conneely, D. Medina, and B. W. O'Malley: p53 is a potential mediator of pregnancy and hormone-induced resistance to mammary carcinogenesis. *Proc Natl Acad Sci USA* 98:12379, 2001.

555. J. A. Kreidberg, H. Sariola, J. M. Loring, M. Maeda, J. Pelletier, et al.: WT-1 is required for early kidney development. *Cell* 74:679, 1993.

556. K. M. Call, T. Glaser, C. Y. Ito, A. J. Buckler, J. Pelletier, et al.: Isolation and characterization of a zinc finger polypeptide gene at the human chromosome 11 Wilms' tumor locus. *Cell* 60:509, 1990.

557. S. F. Dowdy, C. L. Fasching, D. Araujo, K.-M. Lai, E. Livanos, et al.: Suppression of tumorigenicity in Wilms tumor by the p15.5-p14 region of chromosome 11. *Science* 254:293, 1991.

558. M. Negrini, A. Castagnoli, J. V. Pavan, S. Sabbioni, D. Araujo, et al.: Suppression of tumorigenicity and anchorage-independent growth of BK virus-transformed mouse cells by human chromosome 11. *Cancer Res* 52:1297, 1992.

559. S. L. Madden, D. M. Cook, J. F. Morris, A. Gashler, V. P. Sukhatme, et al.: Transcriptional repression mediated by the WT1 Wilms tumor gene product. *Science* 253:1550, 1991.

560. S. Park, G. Tomlinson, P. Nisen, and D. A. Haber: Altered *trans*-activational properties of a mutated WT1 gene product in a WAGR-associated Wilms' tumor. *Cancer Res* 53:4757, 1993.

561. Z.-Y. Wang, W.-Q. Qiu, K. T. Enger, and T. F. Deuel: A second transcriptionally active DNA-binding site for the Wilms' tumor gene product, WT1. *Proc Natl Acad Sci USA* 90:8896, 1993.

562. Z.-Y. Wang, Q.-Q. Wiu, and T. F. Deuel: The Wilms' tumor gene product WT1 activates or suppresses transcription through separate functional domains. *J Biol Chem* 268:9172, 1993.

563. S. Maheswaran, S. Park, A. Bernard, J. F. Morris, F. J. Rauscher, III, et al.: Physical and functional interaction between WT1 and p53 proteins. *Proc Natl Acad Sci USA* 90:5100, 1993.

564. S. Oh, Y. Song, J. Yim, and T. K. Kim: The Wilms' tumor 1 tumor suppressor gene represses transcription of the human telomerase reverse transcriptase gene. *J Biol Chem* 274:37473, 1999.

565. K. W. Kinzler, M. C. Nilbert, B. Vogelstein, T. M. Bryan, D. B. Levy, et al.: Identification of a gene located at chromosome 5q21 that is mutated in colorectal cancers. *Science* 251:1366, 1991.

566. G. Joslyn, M. Carlson, A. Thivers, H. Albertsen, L. Gelbert, et al.: Identification of deletion mutations and three new genes at the familial polyposis locus. *Cell* 66:601, 1991.

567. B. Rubinfeld, B. Souza, I. Albert, O. Müller, S. H. Chamberlain, et al.: Association of the

APC gene product with β-catenin. *Science* 262:1731, 1993.

568. L.-K. Su, B. Vogelstein, and K. W. Kinzler: Association of the APC tumor suppressor protein with catenins. *Science* 262:1734, 1993.

569. E. R. Fearon, K. R. Cho, J. M. Nigro, S. E. Kern, J. W. Simons, et al.: Identification of a chromosome 18q gene that is altered in colorectal carcinoma. *Science* 247:49, 1990.

570. X. Gao, K. V. Honn, D. Grignon, W. Sakr, and Y. Q. Chen: Frequent loss of expression and loss of heterozygosity of the putative tumor suppressor gene *DCC* in prostatic carcinomas. *Cancer Res* 53:2723, 1993.

571. H. T. Lynch, T. C. Smyrk, P. Watson, S. J. Lanspa, J. F. Lynch, et al.: Genetics, natural history, tumor spectrum, and pathology of hereditary nonpolyosis colorectal cancer: An updated review. *Gastroenterology* 104:1535, 1993.

572. F. S. Leach, N. C. Nicolaides, N. Papadopoulos, B. Liu, J. Jen, et al.: Mutations of a *mut*S homolog in hereditary nonpolyposis colorectal cancer. *Cell* 75:1215, 1993.

573. R. Fishel, M. K. Lescoe, M. R. S. Rao, N. G. Copeland, N. A. Jenkins, et al.: The human mutator gene homolog MSH2 and its association with hereditary nonpolyposis colon cancer. *Cell* 75:1027, 1993.

574. R. Parsons, G.-M. Li, M. J. Longley, W.-H. Fang, N. Papadopoulos, et al.: Hypermutability and mismatch repair deficiency in RER+ tumor cells. *Cell* 75:1227, 1993.

575. M. R. Wallace, D. A. Marchuk, L. B. Anderson, R. Letcher, H. M. Odeh, et al.: Type 1 neurofibromatosis gene: Identification of a large transcript disrupted in three NF1 patients. *Science* 249:181, 1990.

576. G. Xu, P. O'Connell, D. Viskochil, R. Cawthon, M. Robertson, et al.: The neurofibromatosis type 1 gene encodes a protein related to GAP. *Cell* 62:599, 1990.

577. G. A. Martin, D. Viskochil, G. Bollag, P. C. McCabe, W. J. Crosier, et al.: The GAP-related domain of the neurofibromatosis type 1 gene product interacts with ras p21. *Cell* 63:843, 1990.

578. A. G. Knudson: Antioncogenes and human cancer. *Proc Natl Acad Sci USA* 90:10914, 1993.

579. J. A. Trofatter, M. M. MacCollin, J. L. Rutter, J. R. Murrell, M. P. Duyao, et al.: A novel moesinezrin-, radixin-like gene is a candidate for the neurofibromatosis 2 tumor suppressor. *Cell* 72:791, 1993.

580. G. A. Rouleau, P. Merel, M. Luchman, M. Sanson, J. Zucman, et al.: Alteration in a new gene encoding a putative membrane-organizing protein causes neuro-fibromatosis type 2. *Nature* 363:515, 1993.

581. F. Latif, K. Tory, J. Gnarra, J. Yao, F.-M. Duh, et al.: Identification of the von Hippel-Landau disease tumor suppressor gene. *Science* 260:1317, 1993.

582. K. Kondo, J. Klco, E. Nakamura, M. Lechpammer, and W. G. Kaelin, Jr.: Inhibition of HIF is necessary for tumor suppression by the von Hippel-Lindau protein. *Cancer Cell* 1:237, 2002.

583. M.-C. King, J. H. Marks, and J. B. Mandell: Breast and ovarian cancer risks due to inherited mutations in *BRCA1* and *BRCA2*. *Science* 302:643, 2003.

584. E. Levy-Lahad and S. E. Plon: A risky business—Assessing breast cancer risk. *Science* 302:574, 2003.

585. K. W. Caldecott: The BRCT Domain: signaling with friends? *Science* 302:579, 2003.

586. I. A. Manke, D. M. Lowery, A. Nguyen, and M. B. Yaffe: BRCT repeats as phosphopeptide-binding modules involved in protein targeting. *Science* 302:636, 2003.

587. X. Yu, C. C. S. Chini, M. He, G. Mer, and J. Chen: The BRCT domain is a phosphorprotein binding domain. *Science* 302:639, 2003.

588. M. J. Anderson and E. J. Stanbridge: Tumor suppressor genes studied by cell hybridization and chromosome transfer. *FASAB J* 7:826, 1993.

589. K. Yamashita, S. Upadhyay, M. Osada, M. O. Hoque, Y. Xiao, et al.: Pharmacologic unmasking of epigenetically silenced tumor suppressor genes in esophageal squamous cell carcinoma. *Cancer Cell* 2:485, 2002.

590. E. Koller, W. A. Gaarde, and B. P. Monia: Elucidating cell signaling mechanisms using antisense technology. *Trends Pharmacol Sci* 21:142, 2000.

591. S. T. Crooke: Progress in antisense technology. *Annu Rev Med* 55:61, 2004.

592. R. Yelin, D. Dahary, R. Sorek, E. Y. Levanon, O. Goldstein, et al.: Widespread occurrence of antisense transcription in the human genome. *Nat Biotech* 21:379, 2003.

593. Ø. Røsok and M. Sioud: Systematic identification of sense–antisense transcripts in mammalian cells. *Nat Biotech* 22:104, 2004.

594. A. S. Lewin and W. W. Hauswirth: Ribozyme gene therapy: applications for molecular medicine. *Trends Mol Med* 7:221, 2001.

595. L. Q. Sun, M. J. Cairns, E. G. Saravolac, A. Baker, and W. L. Gerlach: Catalytic nucleic acids: from lab to applications. *Pharmacol Rev* 52:325, 2000.

595a. R. G. Amado, R. T. Mitsyasic, J. D. Rosenblatt, F. K. Ngok, A. Bakker et al.: Anti-human immunodeficiency virus hematopoietic progenitor cell-delovered ribozyme in a Phase I Study: Myeloid and lymphoid reconstitution in human immunodeficiency virus type-1-infected patients. *Human Gene Ther* 15:251, 2004.

596. C. K. Tang, X.-Z. Wu Concepcion, M. Milan, X. Gong, E. Montgomery, et al.: Ribozyme-mediated down-regulation of ErbB-4 in estrogen receptor–positive breast cancer cells inhibits

proliferation both in vitro and in vivo. *Cancer Res* 59:5315, 1999.

597. R. Abounader, S. Ranganathan, B. Lal, K. Fielding, A. Book, et al.: Reversion of human glioblastoma malignancy by U1 small nuclear RNA/ribozyme targeting of scatter factor/hepatocyte growth factor and c-met expression. *J Natl Cancer Inst* 91:1548, 1999.

598. T. Watanabe and B. A. Sullenger: Induction of wild-type p53 activity in human cancer cells by ribozymes that repair mutant *p53* transcripts. *Proc Natl Acad Sci USA* 97:8490, 2000.

599. L. Zhang, W. J. Gasper, S. A. Stass, O. B. Ioffe, M. A. Davis, et al.: Angiogenic inhibition mediated by a DNAzyme that targets vascular endothelial growth factor receptor 2. *Cancer Res* 62:5463, 2002.

600. L. J. Scherer and J. J. Rossi: Approaches for the sequence-specific knockdown of mRNA. *Nat Biotech* 21:1457, 2003.

601. A. Fire, S. Xu, M. K. Montgomery, S. A. Kostas, S. E. Driver, and C. C. Mello: Potent and specific genetic interference by double-stranded RNA in *Caenorhabditis elegans*. *Nature* 391:806, 1998.

602. G. J. Hannon: RNA interference. *Nature* 418: 244, 2002.

603. D. M. Dykxhoorn, C. D. Novina, and P. A. Sharp: Killing the messenger: Short RNAs that silence gene expression. *Nat Rev Mol Cell Biol* 4:457, 2003.

604. A Dillin: The specifics of small interfering RNA specificity. *Proc Natl Acad Sci USA* 100:6289, 2003.

605. P. S. Meltzer: Small RNAs with big impacts. *Nature* 435:745, 2005.

606. J. Lu, G. Getz, E. A. Miska, E. Alvarez-Saavedra, J. Lamb, et al.: MicroRNA expression profiles classify human cancers. *Nature* 435:834, 2005.

607. L. He, J. M. Thomson, M. T. Hemann, E. Hernando-Monge, D. Mu, et al.: A microRNA polycistron as a potential human oncogene. *Nature* 435:828, 2005.

608. M. V. Iorio, M. Ferracin, C.-G. Liu, A. Veronese, R. Spizzo, et al.: MicroRNA gene expression deregulation in human breast cancer. *Cancer Res* 65:7065, 2005.

609. K. A. O'Donnell, E. A. Wentzel, K. I. Zeller, C. V. Dang, and J. T. Mendell: c-Myc-regulated microRNAs modulate E2F1 expression. *Nature* 435:839, 2005.

610. R. Parkman, K. Weinberg, G. Crooks, J. Nolta, N. Kapoor, et al.: Gene therapy for adenosine deaminase deficiency. *Annu Rev Med* 51:33, 2000.

611. A. Mountain: Gene therapy: The first decade. *Trends Biotechnol* 18:119, 2000.

612. D. Ferber: Gene therapy: Safer and virus-free? *Science* 294:1638, 2001.

613. Y. Manome, M. Nakamura, T. Ohno, and H. Furuhata: Ultrasound facilitates transduction of naked plasmid DNA into colon carcinoma cells in vitro and in vivo. *Hum Gen Ther* 11:1521, 2000.

614. L. D. Shea, E. Smiley, J. Bonadio, and D. J. Mooney: DNA delivery from polymer matrices for tissue engineering. *Nat Biotech* 17:551, 1999.

615. K.-W. Peng: Strategies for targeting therapeutic gene delivery. *Mol Med Today* 5:448, 1999.

616. S. Hacein-Bey-Abina, F. Le Deist, F. Carlier, C. Bouneaud, C. Hue, et al.: Sustained correction of X-linked severe combined immunodeficiency by ex vivo gene therapy. *N Engl J Med* 346:1185, 2002.

617. S. Hacein-Bey-Abina, C. Von Kalle, M. Schmidt, M. P. McCormack, N. Wulffraat, et al.: *LMO2*-associated clonal T cell proliferation in two patients after gene therapy for SCID-X1. *Science* 302:415, 2003.

618. X. Wu, Y. Li, B. Crise, and S. M. Burgess: Transcription start regions in the human genome are favored targets for MLV integration. *Science* 300:1749, 2003.

619. P. D. Wadhwa, S. P. Zielske, J. C. Roth, C. B. Ballas, J. E. Bowman, et al.: Cancer gene therapy: scientific basis. *Annu Rev Med* 53:437, 2002.

620. G. A. Chung-Faye, D. J. Kerr, L. S. Young, and P. F. Searle: Gene therapy strategies for colon cancer. *Mol Med Today* 6:82, 2000.

621. M. S. Steiner, X. Zhang, Y. Wang, and Y. Lu: Growth inhibition of prostate cancer by an adenovirus expressing a novel tumor suppressor gene, pHyde. *Cancer Res* 60:4419, 2000.

622. J. S. Ross and G. S. Ginsburg: The integration of molecular diagnostics with therapeutics. *Am J Clin Pathol* 119:26, 2003.

623. J. R. Heath, M. E. Phelps, and L. Hood: Nanosystems biology. *Mol Imag Biol* 5:312, 2003.

624. A. Hin Yan Tong, G. Lesage, G. D. Bader, H. Ding, H. Xu, et al.: Global mapping of the yeast genetic interaction network. *Science* 303: 808, 2004.

625. B. Schwikowski, P. Uetz, and S. Fields: A network of protein–protein interactions in yeast. *Nat Biotech* 18:1257, 2000.

626. S. Li, C. M. Armstrong, N. Bertin, H. Ge, S. Milstein, et al.: A map of the interactome network of the metazoan *C. elegans*. *Science* 303:540, 2004.

627. L. Giot, J. S. Bader, C. Brouwer, A. Chaudhuri, B. Kuang, et al.: A protein interaction map of *Drosophila melanogaster*. *Science* 302:1727, 2003.

6

Tumor Immunology

HISTORICAL PERSPECTIVES

All the cells in the body have antigenic determinants on their cell surface that reflect the expression of the major histocompatibility complex (MHC) genes of that organism. In the human, this gene complex, called the *HLA complex*, is located on chromosome 6. The letters HLA stand for *human leukocyte antigens*, reflecting the cells in which the expression of these genes was initially determined. These genes determine recognition of self from non-self and are involved in the rejection of transplanted tissue from a foreign host. They also play a role in other aspects of the immune response system. When a normal cell becomes transformed into a malignant cell, it undergoes biochemical changes that often result in the production of new cellular antigens. These new antigens may be recognized by the host organism as foreign. Although new antigenic determinants may be present in other parts of a cancer cell, the ones that are most important in cancer cell recognition are most likely located on the cell surface, where they are "perceived" by interactions with host cells or shed into the bloodstream where they are recognized as foreign. In addition, peptides shed from cancer cells as they undergo cell death may be generated by proteolytic degradation and be found in the blood and urine. New antigens are found in tumors induced by chemicals, viruses, or irradiation in experimental animals. For reasons that will be discussed later, the antigenicity of tumors that arise spontaneously in animals (i.e., those that arise in certain animal species

without experimental induction or interference) and of many human tumors is low—so low, in fact, that tumor cells escape detection by the host or are able to circumvent a relatively weak reaction by the host.

The history of tumor immunology has been an up-and-down affair for almost a century. There have been times of extreme optimism, almost hubris, about the importance of the immune system in moderating or even rejecting tumor cell growth. These highs have been followed by lows in which all but a few stalwarts gave up on the idea that the immune system was capable of mounting any meaningful response against cancer cells growing in the body. This early checkered history of tumor immunology has been reviewed by Scott.[1]

The crux of the problem with the early studies on tumor rejection lay in the difficulty of differentiating true tumor rejection from rejection of a foreign tissue by a genetically incompatible host. Even today this is a problem because experimental tumors are often transplanted again and again and may change over time to the extent that they may no longer be perfectly "syngeneic" with their host animal strain.[1] Thus, the very term *syngeneic* is in a sense a misnomer because if a tumor that arises in a strain of mice, for example, is passaged to other inbred mice to achieve histocompatibility, and then subsequently shown to evoke a rejection response, it is by strict definition not syngeneic with the host animal. Moreover, if an individual animal within an inbred strain becomes mutated in a histocompatibility locus, that individual would not be syngeneic with other members of the strain. Perhaps the term *autochtonous* is a better one to describe the relationship between a tumor and its own host.[1]

A few historical benchmarks can be pointed out. As early as 1910, Peyton Rous recognized that engrafted tumors could be rejected, but he raised the question of whether this was "simply one expression of a resistance to the growth of engrafted tissues in general." Leo Loeb, using one of the first inbred strains of mice, the Japanese waltzing mouse, is credited with carrying out the first successful series of tumor transplantations in 1904. He was able to obtain virtually 100% successful "takes" in the waltzing mice and no takes in unrelated white mice.

Even though it is now known that a wide range of malignant tumors in experimental animals or in humans have tumor-associated cell surface antigens, this was not generally accepted until the 1950s. Prior to the 1940s, when little was known about the existence, let alone the immense complexity, of transplantation rejection antigens (the histocompatibility antigens), it was thought that the small group of then-known transplantable tumors (e.g., Ehrlich carcinoma, Jensen sarcoma, Walker 256 carcinosarcoma, Sarcoma 180, and Sarcoma 37) could be transplanted from one animal to another animal of different genotype because they had no incompatibility for the host and thus were not rejected.[2] Even then, however, it was appreciated that the vast majority of animal tumors were rejected by allogeneic hosts (same species, different genotype). This led to the belief that tumor cells, with the exception of the few transplantable ones, carried potent antigenic determinants that caused their rejection by the host. In fact, experiments in the 1930s and 1940s showed that pre-immunization of animals with arrested tumor cells increased the host animals' resistance to tumor transplantation. This finding led to the idea that an immunologic cure for cancer was possible.

However, after highly inbred strains of mice with known genotypes became available in the 1940s, it soon became apparent that most of the tumor rejection phenomena studied previously were due to histocompatibility differences between the tissues of mice of different genotype, leading to the conclusion that tumor cells themselves had no distinguishing immunologic features. This low point was followed by another wave of enthusiasm that started with the experiments by Foley[3] in 1953. He induced sarcomas with methylcholanthrene (MC) in inbred C3H mice, grafted them into other C3H mice, and ligated the tumors to induce tumor necrosis. Subsequent challenge of these mice with the same sarcoma frequently led to rejection of the second tumor graft, whereas the tumors grew in control animals that had not received the initial graft. Mammary carcinomas arising spontaneously in the same strain of mice were not rejected when transplanted into the mice pregrafted with MC-induced sarcomas.

These experiments indicated that tumor rejection could, in fact, be due to antigenic determinants of the tumor itself rather than simply to differences in histocompatibility antigens. Prehn and Main[4] confirmed these findings by showing that skin grafts from one mouse to another of the same inbred strain were not rejected, but that MC-induced sarcomas were rejected by host animals in which the tumors were grafted and then ligated. Klein et al.[5] subsequently demonstrated that secondary tumor transplants were rejected by the same animal (autochthonous tumor host) following surgical removal of the primary tumor and antigenic stimulation by injections of tumor cells previously inactivated by irradiation. These experiments excluded the possibility that histocompatibility differences were responsible for tumor rejection in inbred strains. These findings were subsequently confirmed in several laboratories, and the principle is now well established. It is important to keep in mind that tumor-associated antigens induce only a relative degree of host resistance, depending on the tumor burden in the animal. Tumor rejection occurs if the size of the tumor challenge is within a certain threshold, but resistance is overwhelmed by a larger tumor cell burden.[2] This threshold has important therapeutic implications for the potential use of immunotherapy in cancer treatment (see below).

With further experimentation in tumor immunology, the potential role of the immune system in modulating tumor growth loomed larger and larger. In the early 1970s, the theory of "immune surveillance" became popular.[6] In brief, this theory states that tumor cells contain aberrant cell surface antigens that a host's immune system can recognize and react to as soon as the concentration of these foreign antigens reaches a certain threshold level. This process goes on all the time, and in most young, healthy animals or people it prevents the growth of aberrant cells. However, as the individual ages, the immune surveillance mechanism becomes defective and the probability that tumor cells will escape rejection increases. This theory has had several proponents as well as detractors[7,8], and it has led to the idea that nonspecific stimulation of the immune system by such bacterial agents as the *Mycobacterium bovis* strain *Bacillus Calmette-Guerin* (BCG) and *Corynebacterium parvum* (*C. parvum*) can increase the body's ability to reject tumor growth. The initial enthusiasm for this idea, however, has waned as more extensive clinical trials have been carried out.

Indeed, the only current clinical use for the nonspecific immune system stimulator BCG is for intravesical treatment of urinary bladder cancer. Nevertheless, recent experiments in experimental tumor models indicate that the so-called innate immunity response, which is the mechanism used to reject infectious agents, can also play a role in tumor rejection. For example, it has been shown that a mixture of leukocytes containing granulocytes, macrophages, and natural killer (NK) cells adoptively transferred from tumor-primed mice conferred tumor resistance in athymic nude mice deficient in T lymphocytes. This finding suggests that the innate immune system can be a means to control malignancy (reviewed in Reference 9).

Studies on cancer induction in experimental animals have shown that, in general, tumors induced with chemicals, irradiation, or physical agents (e.g., implantation of plastic films) have unique antigenic determinants. Even two different tumors induced in the same animal by the same agent are antigenically distinct. Virally induced tumors, by contrast, contain new antigens with a common (or cross-reactive) specificity for all tumors induced by the same virus, regardless of tumor cell type or animal species. Tumors initiated by viruses may, however, carry weaker, specific tumor-associated antigens. The relative antigenic strengths and cross-reactivities of animal tumors induced by various types of agents are indicated in Table 6–1.[10] Several types of cell surface antigens can arise on tumor cells, depending on the nature of the carcinogenic agent. Those antigens involved in tumor transplant rejection are called *tumor-associated transplantation antigens* (TATA). Virus-associated antigens in virally induced tumors may be intracellular (e.g., nuclear T antigen) or on the surface (e.g., viral envelope proteins), the latter type probably being those involved in tumor transplant rejection. In addition, certain embryonic or fetal antigens may reappear on various kinds of tumors, including those that arise spontaneously.

As noted above, chemically induced tumors possess TATA unique for each neoplasm.[4,5,11]

Table 6–1. Relative Antigenic Strengths of Certain Tumors

Etiologic Agents	Relative Antigenic Strength°	Cross-Reactivity of Tumors with		
		Other Primary Tumors in Same Individual Induced by Same Agent	Tumors Induced by Same Agent in Other Individuals	Tumors Induced by Other Agents in Other Individuals
DNA viruses	++	++	++	—
RNA viruses	+	+	+	—[†]
Chemical carcinogens	++	− (±)	− (±)	—
Radiation	− (±)	− (±)	− (±)	
"Spontaneous"	− (±)			

°Even when the relative antigenic strength is high (+ +), the absolute antigenic strength compared to that of tissue alloantigens may be very low.

[†]Cross-reactivity may occur between tumors induced by closely related viruses.

(Modified from Reif[10])

However, evidence for additional cross-reacting TATA from chemically induced tumors has also been obtained. For example, cross-protection has occasionally been observed in mice pre-immunized against one chemically induced sarcoma and then challenged with a different chemically induced tumor.[12,13] These studies are frequently complicated by the fact that chemically induced mouse sarcomas often express antigens of the endogenous murine leukemia virus (MuLV) on the cell surface, and some reports have indicated that immunity induced to MuLV can protect against the development of chemically induced sarcomas in mice.[14] Moreover, the MuLV envelope glycoprotein gp70 has been detected on the surface of many MC-induced murine sarcoma cells, and those chemically produced sarcomas expressing gp70 on the cell surface induce serum antibodies against viral-envelope antigens (VEA) gp70 and another VEA, p15E, whereas tumors lacking gp70 on the cell surface do not.[15] Thus, the presence of MuLV VEA on the tumor cells' surface appears to account for a major portion of the common cross-reactivity seen between chemically induced mouse sarcomas. In some instances, however, the cross-reactivity that occurs cannot be accounted for by expression of MuLV antigens.[13]

Cell surface antigens are found on the surface of cells in tumors induced by RNA or DNA oncogenic viruses. Similar antigens are found on the surface of cells lytically infected with, but not transformed by, these viruses. Which if any of these are related to the TATA of tumors induced by these viruses? In virally induced tumors, it is conceivable that virally coded proteins, including some structural proteins, could augment the immunogenicity of tumor cells that also carry on their surface additional distinct TATA resulting from the transformation process. This could account for the presence of tumor type–specific rejection as well as cross-reactivity between tumors induced by different oncogenic agents.

The phenomenon of some shared and some distinct tumor-associated antigens is also observed in human cancer. Tumor-associated antigens have been reported for a wide variety of human neoplasms (reviewed in References 1 and 16). Several methods have been used to define human tumor immunity. Antisera to human cancer cells or extracts of human cancer cells have been prepared by inoculating other species. In a number of instances, these antisera seemed to be tumor specific. However, when these antisera were more thoroughly tested, they were directed against normal cell components present in low concentration in many normal cell types or in normal tissues not initially tested. A second approach has been to screen sera from cancer patients for antibodies to selected cell lines derived from other patients' tumors of the same histologic type. The problem with this approach is similar to that encountered in animal studies before inbred strains were available, namely, the contributions of the histocompatibility system to the observed immune reactions. A third method is aimed at estimating the cell-mediated immune response (see below) more directly by determinig the ability of tumor extracts to affect the function (e.g., migration in capillary tubes) of leukocytes from patients with the same or different tumors. With this approach, the so-called leukocyte migration inhibition (LMI) assay, it has been found that some human cancers possess common antigens that appear to elicit a cell-mediated immune response that cannot be explained by simple allogeneic histocompatibility reactions.[17] A fourth approach is to study reactions of cancer patients' sera with surface antigens of their own as well as other patients' cultured cancer cells. With this method, three classes of surface antigens have been defined in a study of human melanoma, renal cancer, and astrocytoma.[18–20]

In the case of melanoma, the three classes are as follows: class I includes unique melanoma antigens expressed only on the individual patient's own (autologous) melanoma cells and not on any other cell tested. Class II is shared melanoma antigens expressed not only on autologous melanoma cells but also on other patients' (allogeneic) melanoma cell lines. These two classes of antigens are not detected on autologous, allogeneic, or xenogeneic (other species) normal cells or on nonmelanoma tumor cells. Class III comprises antigens that are not restricted to melanoma cells and that appear to be present on autologous normal cells, as well as on allogeneic and xenogeneic normal and tumor cells. Thus, these data indicate that a variety of antigenic determinants are present on malignant melanoma cells, some of which are unique to a given tumor, some of which are tumor-type specific, and some of which are present, at least to some extent, on a variety of normal cells and other types of tumor cells. It is of interest that individual patients produce antibodies to one or more of these classes of antigens. Thus, they may develop immunity to their own tumor and also to other patients' tumors. Similar data have been obtained for human renal and brain tumors.[18–20] These findings indicate the complexity of the human immune response to cancer and have significant implications for regimen designs in immunotherapy.

Mouse monoclonal antibodies (mAbs) to human melanoma cells have been developed, and with these reagents a greater refinement of the original classification reported by Shiku et al.[18] has been generated. Using the mAbs, four categories of melanoma-associated antigens can be distinguished: (1) antigens found on all cells of melanocyte origin, whether normal or malignant melanocytes; (2) antigens found on adult but not fetal or newborn melanocytes and on a subset of melanomas; (3) antigens found on fetal and newborn but not on adult melanocytes and on a subset of melanomas; and (4) antigens found on a subset of melanomas but not on fetal, newborn, or adult normal melanocytes. Thus, a number of the originally detected antigenic determinants appear to be differentiation stage–specific antigens. More recent methods, in which phage-display is used to detect peptides to which patients have developed autoantibodies, are providing another approach to detecting a human immune response to cancer (discussed below).

In summary, there is now overwhelming experimental evidence that the immune system plays a key role in modulating and controlling tumor growth. Clinical and epidemiological data in humans support this role.

For example, immunosuppressed or anergic patients frequently have more aggressive tumors and poorer prognosis. In addition, patients who are pharmacologically immune suppressed during treatment regimens for organ transplant are more prone to develop cancer. With newer immune suppressive therapy used in transplant regimens, this problem is much less often observed.

MECHANISMS OF THE IMMUNE RESPONSE TO CANCER

To understand the immune response to cancer cells, the reader will need to review the characteristics and functions of the cellular components involved in this response.

The immune system has two functional arms: innate and adaptive immunity. Innate immunity involves phagocytic cells such as neutrophils and macrophages, the complement system, natural killer (NK) cells, cytokines, and acute-phase proteins. These components recognize foreign antigenic determinants on invading micro-organisms and parasites and can mount a rapid response, without a requirement for previous priming by specific non-self antigens. Adaptive immunity involves antigen presenting cells, T and B lymphocytes, cytokines, and the MHC system. The adaptive system is stimulated by processed antigens presented on the surface of cells, is slower to respond (may take days or weeks) than the innate system, and results in the production of memory immune cells that can produce a more vigorous and rapid response upon re-exposure to the same antigenic determinants. The adaptive response is the one most involved in tumor cell recognition and rejection, although the innate system may also play an important role. Elements and functions of the innate system have been reviewed elsewhere[21,22] and will not be further described here. The components of the adaptive immune system and their role in tumor immunity are described below.

Antigen Presenting Cells

In order for cells involved in the immune response to react to foreign organisms or cells, the non-self antigens in these organisms or cells must be "presented" in a way that the immune cells can "see" them. This requires that the foreign antigens be processed into smaller bits of information and be presented to immune cells as part of a complex with cell surface MHC molecules. There are two types of MHC molecules: class I, expressed on all cells, and class II, expressed on macrophages, dendritic cells, B cells, and occasionally on other cell types. All three of these latter cell types can present antigens to T lymphocytes. However, there are two

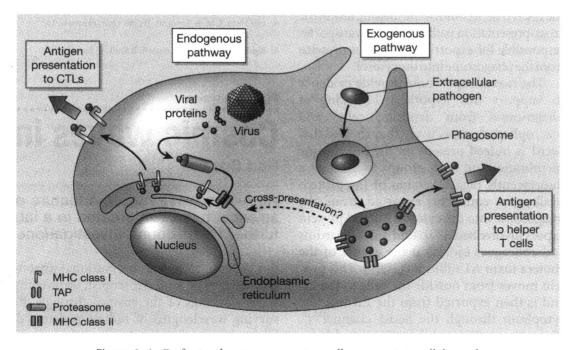

Figure 6–1. Professional antigen-presenting cells process intracellular and extracellular pathogens differently. In the endogenous pathway, proteins from intracellular pathogens, such as viruses, are degraded by the proteasome and the resulting peptides are shuttled into the endoplasmic reticulum (ER) by TAP proteins. These peptides are loaded onto MHC class I molecules and the complex is delivered to the cell surface, where it stimulates cytotoxic T lymphocytes (CTLs) that kill the infected cells. In contrast, extracellular pathogens are engulfed by phagosomes (exogenous pathway). Inside the phagosome, the pathogen-derived peptides are loaded directly onto MHC class II molecules, which activate helper T cells that stimulate the production of antibodies. But some peptides from extracellular antigens can also be "presented" on MHC class I molecules. The way in which this cross-presentation occurs has now been explained. It seems that by fusing with the ER, the phagosome gains the machinery necessary to load peptides onto MHC class I molecules. (From Roy,[23] reprinted with permission from Macmillan Publishers Ltd.)

ways that antigen loading onto MHC occurs. Intracellular antigens such as viral or cytosolic tumor peptides are complexed with MHC class I via intracellular processing pathways and presented to CD8[+] (cell surface marker) T lymphocytes (cytotoxic T cells); however exogenous antigens such as those from pathogens are processed via a different pathway and presented with MHC class II molecules to CD4[+] T cells (helper cells)[21,22] (Fig. 6–1; see color insert).[23]

Antigen-derived peptides are generated in antigen presenting cells (APCs) by one of two routes. Antigenic peptides present in the cytosol, for example, from viral-infected cells or from engulfed tumor-associated antigens, are derived by proteolytic degradation and transported into the endoplasmic reticulum where they bind to nascent MHC class I molecules.[24] In this way, T lymphocytes bearing the CD8 cell surface marker (CD8[+] cells) become stimulated to become cytotoxic T cells (CTLs). Antigenic peptides generated during endocytosis of engulfed antigenic molecules can bind to MHC class II molecules targeted to this cellular compartment on their way to be cycled back to the cell surface. Antigens presented in conjunction with MHC class II stimulate CD4[+] T cells, which become activated T-helper cells.

Macrophages scavenge dead and dying cells, engulf and kill many types of bacteria, present antigens to T cells, and are themselves effector cells in cell-mediated immune reactions and when activated can kill tumor cells. Phagocytic dendritic cells are present in many tissues where they can pick up and process antigens and then migrate to lymphoid tissue. Those that differentiate in lymphoid tissues acquire the ability to present antigens to T cells, and they are highly effective in activating CD4$^+$ cells.[25] They do not appear to have a cytotoxic effector function themselves. B cells have also been implicated as "presenters" of soluble protein antigens to CD4$^+$ cells. The interaction of T and B cells is discussed in more detail below.

Although intact antigenic proteins need to be processed to generate antigenic peptides and co-presented with MHC class I or II molecules, soluble peptides can also bind directly to empty class I or class II molecules present on the cell surface of APC cells. Such empty MHC molecules are potentially important targets for synthetic immune system–stimulatory peptides that could be anti-tumor vaccines (see below).

How Antigens Are Processed

Proteins are continually being turned over inside cells. Old, "worn-out" proteins and incomplete or misfolded proteins are degraded by the proteasomes present inside cells. Proteins are tagged for degradation by a process called *ubiquitination* that involves complexing with chains of ubiquitin molecules. The proteasome interiorizes proteins and chops them into peptides of about 15 amino acids in length. These peptides may be further clipped by amino peptidases in the cytosol and then transported to the endoplasmic reticulum by a transporter called TAP (transporter associated with antigen processing). In the endoplasmic reticulum, the peptides are further processed into peptides 8 to 10 amino acids in length that bind to peptide binding sites on MHC class I molecules (Fig. 6–2; see color insert).[26]

Surprisingly, it is the peptides of this short sequence that are recognized by receptors in CD8$^+$ T cells. More recently it has been shown that antigenic peptides can be derived from fusion of two distinct shorter sequences from the same protein.[22] This excision and splicing event appears to be catalyzed by the proteasome

during protein degradation. These fusion proteins are recognized by CD8$^+$ T cells. This finding provides support for the interesting concept that antigenic peptides do not necessarily result simply from contiguous amino acid sequences of a protein but may also from fusion of noncontiguous sequences that represent different epitopes on a protein. Similar results could be seen for proteins generated by mRNA splice variants, aberrant transcription, translation of alternative or cryptic open reading frames, translation of fusion proteins from gene translocation events, or post-translational modifications that could mask proteolytic cleavage sites. However, the finding of fusion peptides indicates that peptide splicing is another mechanism that increases the diversity of antigenic peptide presentation to T cells. This also raises the possibility that CD8$^+$ cells activated by such fusion peptides could be derived from different epitopes of normal proteins, creating a new epitope that could induce an autoimmune reaction.

Another point that should be noted here is that although carbohydrates do not bind MHC molecules and were thus thought not to be involved in activation of CD8$^+$-mediated tumor cytotoxic effects, it has now been found that both CD4$^+$ and CD8$^+$ T cells can recognize glycopeptides bearing mono- and di-saccharides in an MHC-restricted process.[28] In addition, such glycopeptide-activated T cells recognize the glycan structure with high fidelity. Since abnormal glycosylation patterns of proteins is a consistent finding on tumor cells (see Chapter 4), this may explain how tumors bearing these abnormal glycosylated glycoproteins can be rejected.

T Lymphocytes and T Cell Activation

The lymphocytic stem cell produced in the bone marrow has two pathways of differentiation. One pathway requires the thymus gland and leads to the generation of cells called *thymus-dependent*, or *T lymphocytes*, that are involved in cell-mediated or delayed immune reactions. Precursor T cells (prothymocytes) migrate to the thymus gland, where they are processed into functionally competent cells and are then released into the circulation, from which they populate the peripheral lymphoid tissue. The second pathway produces *bursal-equivalent*, or *B lymphocytes*.

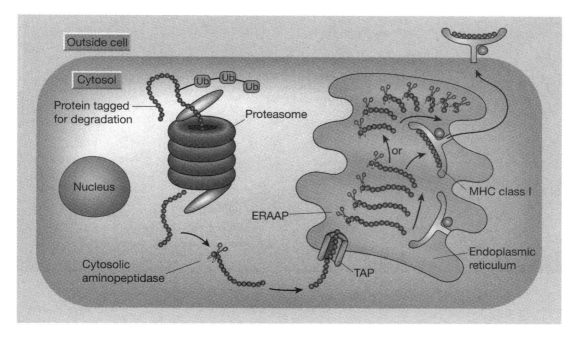

Figure 6–2. Informing the immune system. Sooner or later, every cellular protein reaches the end of its useful life and is degraded. This serves a beneficial purpose: nine–amino acid peptides representing every cellular protein are taken to the cell surface, in complexes with MHC class I molecules, and presented to the immune system. The figure shows how this happens. First, a protein is marked for degradation with chains of the ubiquitin molecule (Ub), and fed into the cell's shredder—the proteasome—to be chopped into peptides of up to 15 amino acids. Aminopeptidase enzymes in the cytosol may further shorten these peptides, some of which then enter the endoplasmic reticulum via the TAP protein. There the peptides are attacked by the enzyme ERAAP and shortened one amino acid at a time until they are completely degraded, unless an intermediate nine–amino acid peptide happens to fit into a waiting, empty MHC I molecule. (From Rammensee,[26] reprinted with permission from Macmillan Publishers Ltd.)

The term *bursal equivalent* derives from the fact that this class of cells was first clearly delineated in chickens that have a distinct bursa in which these cells are produced. In higher animals and humans, the equivalent B lymphocyte–producing tissues appear to be the lymph tissue of the gastrointestinal tract and certain areas of the spleen. Both T and B lymphocytes are present in lymph nodes and spleen, although their relative concentrations vary within these organs. Both types of cells circulate in the blood, but about 70% of circulating lymphocytes are T cells.

A number of subpopulations of human T cells have been defined on a functional basis and on the basis of their cell surface marker antigens. One of these cell populations carries a distinct surface marker called CD4$^+$. These CD4$^+$ cells constitute about 55% to 65% of peripheral T cells and are the T-helper/inducer cells of the immune response system. These cells can respond directly to antigen (although interaction with APCs is usually involved in the response, as indicated above) or to such lectins as ConA or phytohemagglutinin by undergoing a burst of cell proliferation. These cells provide T-helper function to other T cells, to B lymphocytes, and to macrophages. The effects are mediated at least in part by the release of various cytokines. The T-helper/inducer cells do not themselves have a cytolytic effector function, but they play a role in stimulating the generation of cytolytic CD8$^+$ T cells. It is the CD4$^+$ helper/inducer cells that are the target for the AIDS virus, and their loss, as observed in this disease, is devastating to the immune system.

Activated T lymphocytes also release a chemotactic factor for macrophages and polymorphonuclear leukocytes (PMNs). This factor promotes the sequestration of these phagocytic cells in the area of a cell-mediated immune reaction, and PMNs appear to aid in this reaction by the release of proteases and other lytic enzymes that promote destruction of target cells and aid in clearing the cellular debris left by target cell killing.

Although there has been some controversy about the nature of such cells, a T-cell suppressor population has been observed. These cells, called *negative regulatory* or *suppressor T cells*, are $CD4^+$, $CD25^+$ T cells that inhibit anti-tumor immune responses. These cells suppress the proliferative response of T cells to alloantigens and the production of immunoglobulin (Ig) by B lymphocytes, possibly by the release of a suppressor factor. Normally these cells function to modulate the immune response system and prevent overresponse to an antigenic stimulus. However, excessive T-suppressor cell activity can produce generalized immunosuppression and decrease the immune response to a number of foreign antigens, including those present on tumor cells. A loss of T suppressor cells has been observed in cases of excessive immune response such as occurs in certain autoimmune diseases, including systemic lupus erythematosus, hemolytic anemia, and inflammatory bowel disease. Thus, the balance of T-helper and T-suppressor activities regulates the immune response and determines the outcome of antigenic stimulation of the host.

The functional responses of lymphocytes are triggered by cell–cell interactions that produce positive or negative signals, and the balance of these signals is what drives the response. When a T cell interacts with an APC that only presents one signal, i.e., a co-stimulatory molecule, but no foreign antigens, there is no response. If, however, a T cell interacts with an APC that expresses co-stimulatory molecules and processed foreign antigens, the T cell is activated to proliferate and differentiate into an effector cell.

Co-stimulating molecules include B7-1 (CD80), B7-2 (CD86), and CD40. These are expressed on professional APCs, and their expression peaks after APCs are activated by foreign antigens such as microbial products or tumor antigens. Activated APCs produce cytokines that can further enhance T-cell responses. The APC co-stimulatory molecules (also called co-receptors) bind T-cell co-receptors CD28, CTLA-4, and CD40 ligand as part of the activation process. A key point here is that inflammatory mediators up-regulate expression of co-stimulatory molecules in APCs. Thus an inflammatory environment enhances the T-cell activation process.

T cell receptors (TCRs) on the surface of T cells are associated with the CD3 complex of molecules that aggregate with antigen bound to TCR. Aggregation of the antigen-bound receptor complex leads to activation of protein tyrosine kinase and transduction of signals to the nucleus that turn on genes for cell proliferation and differentiation (Fig. 6–3). This binding and aggregation complex is called the *immunological synapse* (see below). If co-stimulatory molecules and the antigen–TCR complex are not activated at the same time, the apoptotic pathway is activated and the T cells die.

As noted above, T cells also express negative signals that shut off responses. For example, activated T cells express a receptor called CTLA-4 (CD152) that also recognizes the same B7 co-activator molecules on APC cells that CD28 recognizes. However, the B7-CTLA-4 interaction shuts off responses by inhibiting TCR and CD28-mediated signal transduction.

The Immunological Synapse

The specialized junction between a T lymphocyte and an APC (a dendritic cell or a B lymphocyte) is called the immunological synapse[29] (Fig. 6–3).[21] It consists of a central cluster of T-cell receptors surrounded by a ring of adhesion molecules ICAM-1 and CD2 that bind to co-stimulatory molecules LFA-1 and LFA-3 on APCs. The formation of the synapse allows for a T-cell response to be mounted in response to a low level of foreign antigen and for a sustained T-cell activation response. The mature synapse structure observed when $CD4^+$ T cells are activated by antigen peptide–MHC complexes has a bulls-eye-like structure with central

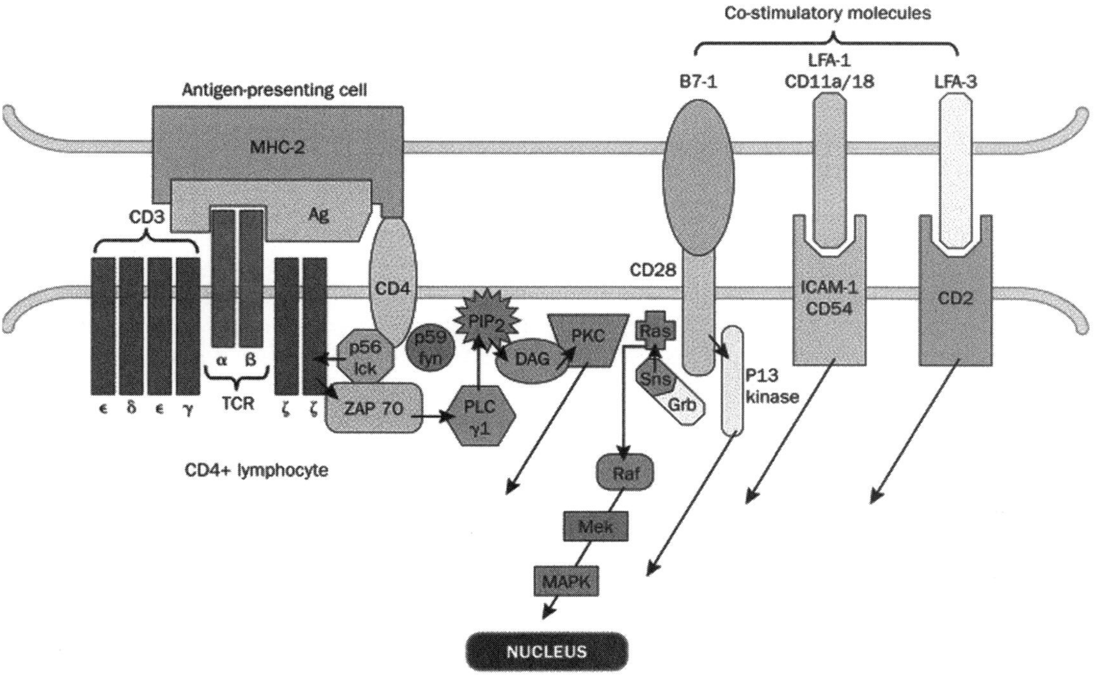

Figure 6–3. Activation of T cells. DAG, diacyl glycerol; EPK, extracellular receptor-associated kinase; Lck, lymphocyte cytoplasmic kinase; MAPK, mitogenic-associated proliferation kinase; Mek, mitogenic extracellular kinase; PKC, protein kinase C; PLC, phospholipase C; Raf, ras-associated factor; Ras, Rous adenosarcoma protein; Sos, son of sevenless; ZAP, zeta-associated protein. Cross-linking of the T-cell receptor causes aggregation with the CD3 complex containing ε, δ, γ chains together with three dimers and activation of phosphorylation and differentiation. If the co-stimulatory molecules are not activated at the same time as a different sequence of signals is activated, cell death and apoptosis results. (From Parkin and Cohen,[21] reprinted with permission from Elsevier.)

accumulation of T-cell receptors and other signaling molecules. This clustering is called the *central supramolecular activation cluster* (cSMAC). The cSMAC complex is surrounded by a ring of adhesion molecules to form a peripheral SMAC (pSMAC). Interestingly, a similar overall immunological synapse is formed at the interface of cytotoxic T cells and target tumor cells. In this case, the cSMAC contains secretory granules holding perforin and granzymes that destroy the target cells (see below).[30]

B Lymphocytes and B Cell Activation

B cells can produce antibodies to tumor cells and enhance T cell–mediated tumor cell killing, so-called antibody-dependent cell cytotoxicity (see below). Antigen recognized by the surface IgM on B cells is internalized, processed, and presented with the MHC class II molecules on the surface of B cells. Thus, B cells can also act as APCs for T cells. T cells in turn produce B-cell growth factors that induce B-cell division and maturation.

The B cells also originate in the bone marrow, but mature differently from T cells. They are the precursors of antibody-forming plasma cells. When B cells are stimulated by antigen either directly (a few antigens do not require interaction with T-helper cells to elicit antibody formation) or indirectly by T-cell interactions, B cells specifically activated by the antigenic stimulus proliferate and differentiate into immunoglobulin-producing plasma cells.

Activated T-helper cells interact with antigen-stimulated B cells in a manner analogous to that of T cell–APC interactions. Antigen, processed by B cells and presented on the cell surface together with MHC class II antigens, interacts with receptors on T cells. Contact with CD4$^+$ helper cells stimulates B cells to mature, multiply, and differentiate into antibody-secreting plasma cells as well as into a clone of memory B cells. Lymphokines secreted by CD4$^+$ cells aid in this maturation process.

The interaction of CD4$^+$ helper cells and B cells is complex. To achieve B-cell activation, T-helper cells produce a set of soluble cytokines that act at various stages in the growth and differentiation of B cells. The interaction of a T-helper cell component called CD40 with a CD40 receptor on B cells activates B cells and makes them competent to respond to soluble cytokines produced by T cells. The signal generated through the CD40 ligand–receptor interaction also plays a role in antibody switching from the more primitive IgM class to other "more sophisticated" immunoglobulins such as IgG and IgE. Cytokines produced by T-helper 2 (TH2) cells (e.g., IL-4, IL-5, IL-6, IL-10) are released locally and act locally in a tight network of T and B cells.

Individual T helper–derived cytokines have distinct actions in stimulating B cells (reviewed in Reference 29). IL-4 acts as a co-stimulant of B-cell proliferation together with anti-IgM antibodies or with the T-helper cell surface CD40 ligand. IL-4 also moderates Ig class switching from IgM to IgG4 and IgE. IL-5 acts on B cells to induce a high rate of secretion of Ig molecules and acts along with IL-4 in Ig gene switching events. IL-6 and IL-10 both strongly promote Ig secretion from differentiated B cells. IL-13 can also cause Ig class switching to IgE production.

Initially, the activated B cell looks like a primitive blast cell and produces mainly IgM-type immunoglobulin. As it matures to a plasma cell, it produces mainly IgG-type immunoglobulin. Antibody directed against the tumor antigen is released from the expanded clone of plasma cells that are specifically producing it. These antibodies can induce tumor cell killing by means of antibody-mediated, complement-dependent cell lysis. This mechanism of cell killing, however, appears to play a minor role in the immune reaction against cancer. A more active cytotoxic reaction in which antibodies participate is the so-called antibody-dependent, cell-mediated cytotoxicity (ADCC) reaction. Antibody released by plasma cells adheres to antigens on the tumor cell surface and this attracts cells that have receptors for the Fc portion of IgG. Cells with such Fc receptors on their surface include macrophages, T lymphocytes, and natural killer cells. This mechanism for cell killing is in addition to the direct killing effect of cytolytic T cells. As will be discussed later, soluble antitumor antibodies or antigen–antibody complexes, when present in high concentrations, may actually block cell-mediated cytotoxicity by binding recognition sites on cytotoxic cells.

Natural Killer Cells

In animals and humans, another population of cells is cytotoxic for tumor cells. These cells appear to belong to the lymphocyte class, but they lack surface markers that clearly place them in a specific category. They are nonphagocytic and nonadherent, they appear to possess Fc receptors after activation, and they have a low density of certain T-cell markers. Thus, these cells are thought to be derived from clones of immature pre-T lymphocytes. Their ability to kill tumor cells does not depend on prior immunization of the host and does not appear to depend on the generation of antitumor antibodies. Hence, these cells have been called natural killer (NK) cells.[31] They will kill tumor cells from both syngeneic and heterologous animal species, and in this respect are rather indiscriminate killer cells. In some animal systems, populations of NK cells have been shown to recognize certain tumor antigens and to have cytotoxic activity against normal thymus cells, macrophages, and bone marrow cells.[32] It has been postulated that NK cells can recognize several different types of specificities on cells that may have some cross-reactive antigenic determinants. This could explain the broad cytotoxic specificity that NK cells possess. NK cells are stimulated to become active killer cells by cytokines such as interferon-γ (IFN-γ), which is released by T cells as well as by activated NK cells. The latter mechanism would provide a positive feedback system even in animals lacking a functional thymus gland (e.g., nude mice). The stimulation of NK cells may partly explain the anti-tumor activity of interferon when injected into animals or cancer patients.

It has also been postulated that NK cells play a role in immune surveillance, particularly for virus-induced tumors, and some NK cell reactivity against the major envelope glycoprotein of endogenous C-type viruses has been detected in mice.[33] Mice with a genetic deficiency in NK cell production have a decreased resistance to the growth of transplanted tumors, and a disease in humans (Chediak-Higashi syndrome), characterized by a 500-fold impairment of NK cell function, is associated with a high incidence of lymphoproliferative disorders, some of which are malignant.[34] The fact that nude athymic mice lacking normal T cell–mediated immune defense mechanisms but have NK cells can still reject some heterologous tumor transplants suggests that NK cell activity does have an important immune surveillance effect against tumor cells. This effect is apparently easily overwhelmed, however, since heterologous tumor transplants generally grow much more easily in nude mice than in mice with a normal thymus gland.

While NK cells don't carry the typical T-cell markers, they may be generated from a common progenitor cell. For example, NK cells have been generated from CD34[+] hematopoietic progenitor cells cultured with IL-2. As noted above, NK cells are activated by interferons and they are also activated by IL-12 (reviewed in Reference 35). When activated, NK cells undergo a morphological change characterized by the acquisition of intracellular granules that contain lytic enzymes. The broad cell-killing specificity of NK cells appears to be mediated by the absence of normal MHC class I molecules on the surface of cells, which makes them susceptible to NK cell–mediated cytotoxicity. NK cells also produce a number of cytokines, including IFN-γ, G-CSF, CM-CSF, IL-1, and TGF-β.[35] NK cell depletion promotes metastasis in experimental animal systems, thus NK cells may play an important role in tumor cell surveillance.[35]

A diverse family of receptors that play an important role in NK-cell response to both viral infections and cancer has been found (reviewed in Reference 36). These are called *killer cell immunoglobulin-like receptors* (KIR). The KIR are members of the Ig superfamily of receptors and are encoded on chromosome 19q13.4 as part of the leukocyte receptor cluster. The diversity of these receptors provides a response mechanism for a wide variety of disease-related substrates, including HIV, hepatitis C virus, and cytomegalovirus infections, malarial infections, and cancer.

There is another species of NK cells that are a distinct lineage from T cells and that express both αβ T-cell receptors and NK-cell receptors. These cells are called *NKT cells* and they recognize a lysosomal glycosphingolipid called *isoglobotrihexosylceramide* (iGb3).[37] The production of iGb3 by tumor cells may be part of the activation of the NKT response to cancer.

Cell-Mediated Cytotoxicity

Tumor cells can be attacked in the body by a variety of mechanisms, including (1) the activation of macrophages by IFN-γ to produce tumor necrosis factor (TNF-2) and oxygen intermediates such as nitric oxide (NO), which may induce target cell killing; (2) activation of NK cells by IL-2 and other effectors to become active tumor cell killers; (3) the production of antibodies to tumor-associated antigens that results in "coating" the tumor cells and targeting them for activated macrophage or cytotoxic T lymphocytes (CTLs) (this is the induction of antibody-dependent cellular cytoxicity [ADCC] reactions); (4) anti-tumor antibody-dependent, complement-mediated tumor cell killing (this appears to be a minor mechanism); and (5) activation of CD8[+] T lymphocytes to become CTLs by an MHC-mediated cell-killing mechanism. The most important of these reactions for killing tumor cells in vivo are the cell-mediated ones involving CTLs, macrophages, and NK cells.

The mechanisms of cell-mediated cytotoxicity involve direct cell–cell contact between a killer cell and a target tumor cell (Fig. 6–4).[38] There is a cell–cell contact release of lytic enzymes from CTL cells that attach the tumor cells (reviewed in Reference 39). This process is initiated by T-cell receptor interaction with antigenic peptide plus MHC on the target cell. This triggers a Ca^{2+}-dependent pathway leading to polariza-

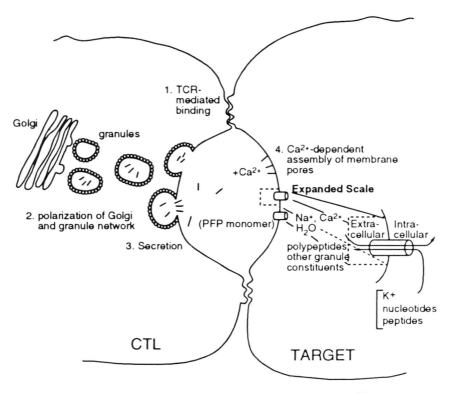

Figure 6–4. CTL-mediated cell killing. (From Young and Cohn,[38] with permission.)

tion and exocytosis of granules containing proteases ("granzymes") and the assembly of channels called *perforins* from pore-forming proteins (PFPs) in the plasma membrane of the target cell but not the activated T cell. The perforin channels allow the uptake of a variety of proteases, particularly serine proteases called *granzymes* A, B, and C. These serine proteases may be involved in activation of endonucleases that trigger the DNA fragmentation characteristic of apoptosis. Both perforin assembly and granzyme A appear to be necessary for cell-mediated cell killing because transfection of one or the other gene into inactive killer cells induced only minimal cell lysis, whereas transfection of both genes produced an active killer cell.[39] Furthermore, mice who underwent gene knockout of the perforin gene do not generate potent NK cells or CD8$^+$ CTL against virally infected cells.[39]

Another mechanism of cell-mediated cytotoxicity, not dependent on perforin or granzymes, involves activation of the cell surface receptor called *Fas* or APO-1, discussed in the apoptosis section of Chapter 4. Persistent stimulation induces T cells to express Fas, and most of the Ca^{2+}-independent component of T cell–mediated cytoxicity seems to work via the Fas pathway.

Danger Theory

Being puzzled about why newly lactating breast tissue expresses new protein products, why resident bacteria in the gastrointestinal tract aren't attacked by the body's defense mechanisms, and why tumor cells become tolerant to the host immune system, Polly Matzinger, developed the concept and subsequently provided the data to argue that the immune system doesn't distinguish self from non-self a priori, but that the immune system responds to cells that become damaged by stress or lytic cell death.[40] This theory became known as the "danger theory" of immune response. "Danger" signals are thought to activate dendritic cells to present new antigens to T lymphocytes. In the danger theory, these new antigens derive from the breakdown products of dying cells undergoing cell lysis. This may explain why resident bacteria are not attacked, since they don't cause cell death, for example.

The implications of this theory are large. In this view, transplant rejection would be initiated by the cell damage caused by surgical trauma; viral immunity would be caused by virally mediated cell death; and immune reaction to tumors would be activated by inducing local damage at the tumor site. If this theory is correct, then the tumor antigens released by tumor cell death would be key targets for development of immunologic treatment modalities. Another strategy could be to target both patients' tumor antigens and the mechanisms that generate danger signals.[40] Cellular danger signals might also be involved in development of autoimmune diseases by inflammation that causes release of tissue debris.[40]

There are a number of arguments against the danger theory of immune response. Among them is the point that cell turnover of normal cells occurs all the time in the body. Even though there is often a high rate of cell turnover in cancer tissue, the danger theory holds that it is necrotic cell death and not apoptotic cell death that is the trigger, so tumor cell death by apoptosis, which is the common mechanism of cell death after chemotherapy, may not be enough to trigger tumor immunity.

The exact cellular triggers that would trigger the danger response are not clear. It is hypothesized that released heat shock proteins are the trigger. Another idea is that intracellular molecular debris released during cell lysis is the fomenting event.[41] An opposing theory to this is the "stranger hypothesis," originally articulated by Charles Janeway, that holds that the recognition of foreign antigen by "pattern recognition receptors" (now known to be members of the Toll receptor family) stimulates an immune response (reviewed in Reference 41).

Data in favor of the danger hypothesis were reported by Shi et al.,[42] who found that uric acid, a breakdown product of nucleic acid metabolism, can stimulate dendritic cell maturation and when injected together with foreign antigen in vivo stimulates a T-cell response. The implication of

this finding is that the "stranger" and "danger" responses act in concert.

ROLE OF GENE REARRANGEMENT IN THE IMMUNE RESPONSE

The tremendous diversity of immune responses for which T and B lymphocytes are capable is determined to a large extent by these cells' rather unique ability to undergo gene rearrangement in a way that doesn't produce a malignant cell type, although sometimes this process does get "carried away," leading to a haphazard DNA recombination that can produce lymphocytic malignancies.[43]

This ability to respond to a wide variety of antigens depends on a carefully regulated rearrangement of genes encoding the antigen receptor proteins, the T cell receptors, and the immunoglobulins during lymphocytic differentiation. The genes coding for these molecules are generated by a specialized form of DNA recombination known as the *V(D)J recombinase machinery* (discussed in Chapter 5). V genes encode the variable regions of immunoglobulins and receptor proteins, D genes encode diversity regions, and J sequences are the joining gene segments. VDJ segments are flanked by DNA sequences know as recombination signal sequences (RSS) that are targeted by a protein complex containing the recombinase activating gene–encoded proteins RAG-1 and RAG-2 (reviewed in Reference 44). The binding of the recombinase complex to RSS flanking regions induces double-strand breaks in DNA. The broken ends are then rejoined by recombination events with the participation of double-strand break repair factors. Indeed, the rejoining steps resemble in many respects the steps in DNA repair after DNA double-strand breaks induced by DNA-damaging agents.

The human genome contains many cryptic recombination signal sequences that could potentially be exposed in chromatin and produce aberrant recombination events. This is carefully regulated in lymphocyte-lineage cells by cell-specific enhancer and promoter-dependent changes in chromatin structure that allow access of RAG proteins to RSS regions. This sort of chromatin remodeling appears to occur as a consequence of hyperacetylation of core histones, including histone H3.[45] This is a cell-specific phenonenom. For example, the TCR δ gene locus but not the Ig κ locus, is cleaved in isolated pro-T cells, whereas the reverse is true for isolated pro-B cell nuclei and neither locus is cleaved in fibroblast nuclei.[46] These data strongly suggest that chromatin remodeling in response to cell-specific signaling systems is what determines the unique ability of pro-lymphocytes to undergo selective gene rearrangements. Thus, TCR genes rearrange only in T cells and Ig genes rearrange completely only in B cells.

Rarely, aberrant uncovering of RSS segments in cells produces recombination events that place a proto-oncogene near an Ig gene, for example, placing the bcl-2 anti-apoptotic gene next to the antigen receptor locus.[43]

HEAT SHOCK PROTEINS AS REGULATORS OF THE IMMUNE RESPONSE

Heat shock proteins (hsps) were first identified in 1963 when it was noted that extreme temperature change caused a puffing pattern in Drosophila larvae salivary gland chromosomes and an unusual gene expression profile. The gene products encoded by heat shock–induced genes were first identified in 1974 by Tisieres et al. (reviewed Reference 47). Heat shock proteins are highly abundant proteins in all prokaryotic and eukaryotic cells and make up 5%–10% of total cellular proteins. The intracellular concentration can be induced two- to threefold by cellular insults such as heat, oxidative stress, nutritional deficits (e.g., low glucose), UV irradiation, toxic chemicals, viral infection, and ischemia-reperfusion injury (reviewed in Reference 47).

The heat shock protein family consists of small Hsps, Hsp-40, -60, -70, -90, and -110 families (Table 6–2). They are located in various intracellular compartments but can also be released into the extracellular environment under certain conditions such as inflammation and in this way can induce cytokine production and adhesion molecule expression in a range of cell types. Hsp 60 and Hsp 70 are found in the blood of healthy people. Circulating Hsp 60 levels are elevated in patients with early atherosclerosis and Hsp

70 levels are high in patients with peripheral vascular and renal disease.[47] Intracellulary, they function as chaperones for protein folding and assembly and act to prevent aggregation of unfolded or misfolded proteins, such as denatured proteins that occur in response to heat shock. They also play a key role in intracellular translocation events for many proteins, including steroid receptors and a variety of other factors that are translocated from cytoplasm to nucleus.

Inside cells, Hsps bind many peptides processed internally and can, in fact, provide a repertoire of intracellularly derived peptides. Immunity to tumor-derived peptides complexed with Hsp 70, Hsp 90, and gp 96 (another stress-induced protein) has been observed in mouse models.[48] In addition, APCs internalize gp 96-antigen peptide complexes bound to the α-macroglobulin receptor (CD 91). Such chaperoned proteins and peptides are then "pulsed" into the MHC class I–restricted processing pathway that leads to presentation of antigenic peptides (including those from tumors) to CD8+ T cells.[47]

INFLAMMATION AND CANCER

A link between chronic inflammation and cancer has been suspected for a long time on the basis of epidemiological data such as the observation that chronic inflammation often increases cancer risk in inflamed tissues and long-term use of non-steroidal anti-inflammatory drugs reduces the risk of several cancers (reviewed in Reference 49). In addition, tumors are loaded with a variety of cell types involved in inflammation such as macrophages and lymphocytes as well as cytokines that these cells produce. Deletion of certain inflammatory mediators in mouse models reduces cancer susceptibility in these animals. Key mediators for the link between inflammation and cancer are NF-κB and TNF-α (Fig. 6–5; see color insert). Supporting data for this concept come from the work of Pikarsky et al.,[50] who found that in a mdr-2 gene knockout mouse model, which spontaneously develops chronic hepatitis followed by hepatocellular carcinoma, NF-κB-induced TNF-α expression is the apparent mediator of the malignant transformation events. Treatment with an anti-TNF-α antibody, or suppression of NF-κB by induction of the

Table 6–2. Mammalian Heat Shock Protein Families and Their Intracellular Location and Function

Major Family and Membranes	Intracellular Localization	Intracellular Function
SMALL HSPS		
B-crystallin	Cytoplasm	Cytoskeletal stablization
Hsp27	Cytoplasm, nucleus	Actin dynamics
Heme oxygenase, Hsp32	Cytoplasm	Heme catabolism, antioxidant properties
HSP40		
Hsp40	Cytoplasm, nucleus	Regulates the activity of Hsp70; binds non-native proteins
Hsp47	ER	Processing of pro-collagen; processing and/or secretion of collagen
HSP60 (OR CHAPERONINS)		
Hsp60	Mitochondria	Binds to partly folded polypeptides and assist correct folding. Assembly of multimeric complexes
TCP-1	Cytoplasm	
HSP70		
Inducible: Hsp70, Hsp70hom	Cytoplasm, nucleus	Binds to extended polypeptides
Cognate or constitutive; Hsc70	Cytoplasm, peroxisome	Prevents aggregation of unfolded peptides
Grp78/BIP	ER	Dissociates some oligomers ATP Binding
MtHsp70/Grp75	Mitochondria	ATPase activity
		Hsp70 down-regulates HSF1 activity
HSP90		
Hsp90	Cytoplasm	Binds to other proteins
Grp94/gp96/Hsp100	ER	Regulates protein activity
		Prevents aggregation of refolded peptide
		Correct assembly and folding of newly synthesized protein
		Hsp90 assists in maintenance of HSF1 monomeric state in nonstressful conditions
HSP110		
Hsp110 (human)	Nucleolus, cytoplasm	Thermal tolerance
Apg-1 (mouse)	Cytoplasm	Protein refolding
Hsp105	Cytoplasm	

Apg-1, protein kinase essential for autophagy; BIP, immunoglobulin heavy-chain binding protein; ER, endoplasmic reticulum; GRP, glucose regulated protein; Hsp70hom, testis-specific Hsp70; mt, mitochondrial; TCP-1, tailless complex peptide.
(Adapted from Pockley A. G.,[47] with permission from Elsevier and also Heat shock proteins in health and disease; therapeutic targets or therapeutic agents? http://www.ermm.cbcu.cam.ac.uk/01003556h.htm (accessed Jan. 22, 2003) by permission of Cambridge University Press)

IκB suppressor of NF-κB, blocked progression to carcinoma. Because suppression of NF-κB function in young mice did not affect carcinoma development, the implication is that the promotion–progression phases of malignant transformation are the ones enhanced by inflammation, not the initiation phases.

In some human and mouse cancers, the malignant cells themselves, in addition to the inflammatory cell types, can produce the offending cytokines. On the basis of TNF-α effects on cancer-related inflammation, clinical trials of TNF-α antagonists in patients with advanced cancers are under way.[49]

IMMUNOTHERAPY

Rationale for Immunotherapy

In an immunosuppressed patient, the best way to improve immune status is to remove or destroy the bulk of the tumor. This removal by itself frequently leads to improved immune responsiveness. It is when a tumor is small that therapy aimed at stimulating the patient's immune response to a tumor is most likely to succeed. Various methods to stimulate a patient's immune system nonspecifically have been tried with minimal clinical success (e.g., with the bacterial

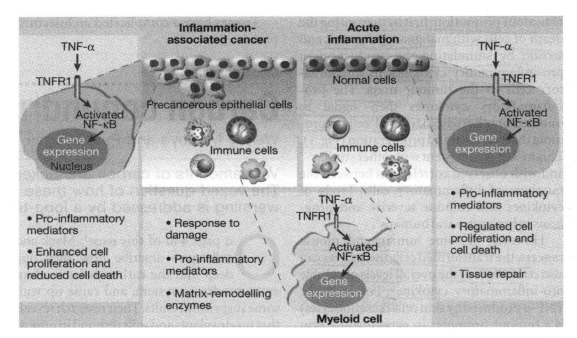

Figure 6–5. Opposing effects of the NF-κB protein in normal tissues and in cancer. Tumor-necrosis factor-α (TNF-α) acts through its receptor, TNFR1, to activate the gene-transcription factor NF-κB. During acute inflammation in "normal" epithelial cells (*right*), NF-κB activation leads to increased expression of genes that encode pro-inflammatory mediators called cytokines, and activates genes that regulate the balance between cell proliferation and cell death. In inflammatory immune cells (myeloid cells; *bottom*), NF-κB activation can also regulate cell death, but more importantly regulates short-term expression of pro-inflammatory mediators to repair the tissue damage. Ablation of NF-κB in inflammatory cells impairs the expression of pro-inflammatory cytokines. Precancerous epithelial cells (*left*) use NF-κB to enhance their survival, and their propensity to become malignant cells, by augmenting their expression of pro-inflammatory and cell-survival genes while inhibiting the death-promoting machinery. If NF-κB activation is disabled in these epithelial cells, while pro-inflammatory gene programs are maintained, cell death is favored and tumor progression is reduced. (From Balkwill and Coussens,[49] reprinted with permission from Macmillan Publishers Ltd.)

antigens BCG and *C. parvum*). Another approach is the use of monoclonal antibodies or monoclonal antibody–antitumor agent complexes directed against specific tumor antigenic determinants. Therapy aimed at overcoming a patient's immune unresponsiveness to a tumor could also be useful and could be accomplished by tumor antigen modification to render tumor cells less like normal cells and thereby less recognized as self. This therapy could conceivably be done in vitro through (1) chemical modification of tumor cell membranes by attachment of strongly antigenic chemical haptens; (2) infection of tumor cells with nonpathogenic viruses that produce altered cell surface antigens; or (3) hybridization of tumor cells to more antigenic cells.[51] Reinjection of such in vitro–altered cells (obtained originally by surgery or biopsy) back into the same patient could evoke a heightened immune response that would include antigenic recognition of the original tumor.

Over the past several years, much has changed in our understanding of tumor immunology These advances include better techniques for characterizing tumor-derived antigenic peptides, a better understanding of the role of specific cytokines in the immune response to cancer, improved knowledge of the cell–cell interactions involved in killing cancer cells, a better understanding of the mechanisms of tumor antigen processing in APCs, and development of therapeutically effective anti-tumor mAbs and radiolabeled mAbs directed to tumor cells (reviewed in References 52–54). These advances are all discussed below.

Identification and Characterization of Tumor-Derived Antigenic Peptides

As discussed above under T-lymphocyte activation, both CD8[+] cytotoxic T cells and CD4[+] T-helper cells recognize antigens presented as small peptides bound in a surface "groove" on human leukocyte antigen (HLA; the human analogue of the major histocompatibility complex, MHC). In the case of CD8[+] cells, 8 to 10 amino acid–length peptides, derived from intracellular processing of proteins, are presented to T cells via the MHC class I molecules, whereas peptide antigens recognized by CD4[+] cells are presented via MHC class II molecules. Presentation of peptides by APC cells triggers a cascade of events resulting in T cell expansion and cytokine release.

A number of techniques have been developed to identify tumor antigens. These include the following:

1. Transfection of complementary DNA (cDNA) libraries from tumor cells into target cells expressing MHC molecules and use of tumor-reactive T cells from cancer patients to identify cells expressing tumor antigens.
2. Elution of peptides from the surface of tumor cells and "pulsing" of these peptides onto APCs, followed by testing for reactivity with specific anti-tumor lymphocytes. Peptides on APCs that show reactivity can then be eluted, identified by mass spectrometry, and sequenced to identify the parent protein.
3. Exposure of T cells to cancer cells in vitro to activate T cells. If the activated T cells specifically react to the cancer cells that produce the protein, such proteins are considered tumor antigens.

4. Serologic identification by recombinant expression cloning (SEREX) to identify circulating IgG antibodies to tumor antigens. This is done by incubating diluted serum from cancer patients with bacterial cells transfected with cDNA libraries derived from cancer cells and thus expressing a profile of tumor proteins on the bacterial cell surface.
5. Use of bacteriophage libraries (phage-display) derived from cancer tissue.[55] In this case, a cDNA library is constructed from a pool of mRNA isolated from cancer tissue obtained from patients. After digestion, the cDNA library is inserted into T7 phage vectors, which are then packaged into T7 phages that express a panel of cancer-derived, cDNA-encoded tumor proteins that are then screened for reaction to sera from cancer patients. Such sera contain autoantibodies to cancer-derived peptides recognized by patients' immune systems. The phage libraries are pre-cleared by a subtraction step to remove peptide-producing phages that express peptides found in normal sera. Using this technique, Chinnaiyan and colleagues have identified autoantibody signatures specific for prostate cancer that have a much better specificity than PSA.[55]

It has been more difficult to identify MHC class II–restricted antigens than class I types because the common cDNA transfection techniques do not lead to appropriately processed peptides for MHC class II packaging. Nevertheless, a number of class II–restricted antigens have been identified. Examples of both class I– and class II–restricted antigens are shown in Table 6–3. Many of these are melanoma antigens because of the relatively high immunogenicity of melanoma cells and the resultant easier ways to identify them.

Cytokines

The term *cytokines*, used in its broadest sense, defines a large group of secreted polypeptides, released by living cells that act non-enzymatically in picomolar to nanomolar concentrations to regulate cellular functions.[56] These

Table 6–3. Examples of Human Cancer Antigens

Antigen	Antigen
I. CLASS I–RESTRICTED ANTIGENS RECOGNIZED BY CD8$^+$ LYMPHOCYTES	
Melanoma–melanocyte differentiation antigens	Cancer–testes antigens
MART-1 (Melan-A)	MAGE-1
gp100 (pmel-17)	MAGE-2
Tyrosinase	MAGE-3
Tyrosinase-related protein-1	MAGE-12
Tyrosinase-related protein-2	BAGE
Melanocyte-stimulating hormone receptor	GAGE
	NY-ESO-1
Mutated antigens	Nonmutated shared antigens overexpressed on cancers
β-catenin	α-Fetoprotein
MUM-1	Telomerase catalytic protein
CDK-4	G-250
Caspase-8	MUC-1
KIA 0205	Carcinoembryonic antigen
HLA-A2-R1701	P53
	Her-2/neu
II. CLASS II–RESTRICTED ANTIGENS RECOGNIZED BY CD4$^+$ LYMPHOCYTES	
Epitopes from non-mutated proteins	Epitopes from mutated proteins
gp100	Triosephosphate isomerase
MAGE-1	CDC-27
MAGE-3	LDLR-FUT
Tyrosinase	
NY-ESO-1	

Human cancer antigens restricted by HLA-A class I and recognized by CD8$^+$ lymphocytes fall into four general categories. (1) Melanoma–melanocyte differentiation antigens are normal, nonmutated proteins expressed exclusively on melanomas and on normal pigment-producing cells such as melanocytes. Lymphocytes reactive against these differentiation antigens can be found infiltrating into tumors. (2) Cancer–testes antigens can be widely expressed on a variety of epithelial tumors as well as on testis and placental tissue. (3) Mutated antigens represent normal proteins that contain mutations or translocation giving rise to unique epitopes. (4) Nonmutated shared antigens overexpressed on cancers. There is some evidence that overexpressed proteins, such as carcinoembryonic antigen, p53, and Her-2/neu, are tumor antigens, although evidence is controversial. As for antigens recognized by CD8$^+$ cells, epitopes recognized by CD4$^+$ cells are derived from both nonmutated and mutated proteins.

(From Rosenberg,[52] reprinted by permission from Macmillan Publishers Ltd.)

cellular functions include regulation of immune cell activity (interferons and interleukins), hematopoiesis (colony-stimulating factors, or CSFs), and regulation of proliferation and differentiation of a wide variety of cell types (peptide growth factors such as EGF, FGF, TGF-α, and TGF-β, etc.). Here we will discuss some of the cytokines that affect the immune system. The other peptide growth factors are discussed in Chapter 4. Cytokines produced by leukocytes and that have effects primarily on other leukocytes are termed *interleukins.* Some of the cytokines involved in immune system function are listed in Table 6–4.

Interferons

Interferon was discovered in 1957 by two virologists, A. Isaacs and J. Lindenmann, who were looking for a substance that blocks viral infection of cells.[57] Their research was prompted by the clinical observation that patients seldom come down with two virally induced diseases at the same time. They showed that the medium removed from influenza virus–infected chicken cells grown in culture, when added to other cultures of chicken cells, prevented infection of the second cultures by a different virus. They named this interfering substance *interferon.* Since that time, numerous studies aimed at isolating and characterizing interferons have been carried out.

Human interferons are classified into two types: type 1 (IFN-α, IFN-β) and type 2 (IFN-γ or immune interferon). IFN-α is produced by leukocytes and lymphoblastoid cells stimulated by viruses or by microbial cell components. The

Table 6–4. Cytokines

Cytokine	Source	Mode of Action
Interleukin 1	Macrophages	Immune activation; induces inflammatory response
Interleukin 2	Mainly T cells	Actives T (and natural killer) cells and supports their growth; formerly called T-cell growth factor
Interleukin 3	T cells	Mainly promotes growth of hemopoietic cells
Interleukin 4	T-helper cells	Lymphocyte growth factor; involved in IgE responses
Interleukin 5	T-helper cells	Promotes growth of B cells and eosinophils
Interleukin 6	Fibroblasts	Promotes B-cell growth and antibody production, induces acute-phase response
Interleukin 7	Stromal cells	Lymphocyte growth factor; important in development of immature cells
Interleukin 8	Mainly macrophages	Chemoattractant
Interleukin 10	CD4 cells, activated monocytes	Inhibits production of interferon, IL-1, IL-6, tumor necrosis factor, and stops antigen presentation
Interleukin 12	Monocytes and macrophages	Augments T-helper 1 responses and induces interferon
Interleukin 13	Activated T cells	Stimulates B cells
Granulocyte colony-stimulating factor	Mainly monocytes	Promotes growth of myeloid cells
Monocyte stimulating factor	Mainly monocytes	Promotes growth of macrophages
Granulocyte-macrophage colony-stimulating factor	Mainly T cells	Promotes growth of monomyelocytic cells
Interferon α	Leucocytes	Immune activation and modulation
Interferon β	Fibroblasts	Immune activation and modulation
Interferon γ	T cells and natural killer cells	Immune activation and modulation
Tumor necrosis factor α	Macrophages	Stimulates generalized immune activation and tumor necrosis; also known as cachectin
Tumor necrosis factor β	T cells	Stimulates immune activation and generalized vascular effects; also known as lymphotoxin
Transforming growth factor β	Platelets	Immunoinhibitory but stimulates connective tissue growth and collagen formation

From Parkin and Cohen,[21] reprinted with permission from Elsevier.

α interferons are members of a multigene family; they are acid stable $M_r = 16-25$ kDa polypeptides. IFN-β is a 20 kDa glycoprotein produced by fibroblasts in culture after exposure to various microorganisms, microbial components, or high-molecular-weight polyanionic oligonucleotides (e.g., poly IC). Two IFN-β genes have been identified. IFN-γ, so-called immune interferon, is produced by T lymphocytes in response to antigenic or mitogenic stimulation. There are at least two species of IFN-γ, a 20 kDa and a 25 kDa glycoprotein, and they are acid-labile. The antiviral action of the interferons appears to involve interference with viral nucleic acid and protein synthesis via an increased rate of viral RNA degradation and decreased rate of peptide chain initiation.

The anti-tumor effects of interferons appear to result from a stimulation of NK cells and macrophages, and from a direct cytotoxic effect on tumor cells. The evidence that multiple actions are involved comes from studies in tumor-bearing animals before and after depletion of NK cells. For example, in a study of the growth of tumors produced by Moloney sarcoma virus–transformed cells in mice, it was found that a mixture of IFN-α and IFN-β markedly stimulated NK-cell activity at the site of the tumor and inhibited tumor growth; however, when NK cells were depleted by in vivo treatment with an antibody to NK cells, tumor growth was still inhibited, indicating a direct effect of the interferons.[58] When the tumor load was very high, however, NK cell depletion did reduce the anti-tumor effect of IFN, a result suggesting that IFN has dual actions: direct inhibition of tumor cell multiplication and stimulation of NK cell activity, the latter of which plays a more apparent role in IFN action when the tumor burden is high.

IFN-γ, isolated initially from activated T lymphocytes and later produced by recombinant

DNA techniques, is a potent activator of tumoricidal macrophages.[59] It also has marked macrophage migration inhibitory factor activity (MIF). Thus, IFN-γ may be responsible for the immunomodulating activities previously ascribed to the lymphokines macrophage-activating factor (MAF) and MIF. One of the mechanisms by which IFN-γ increases the cellular killing effect of macrophages mediated by IgG antibodies involves induction of F_c receptors on the macrophage surface.[60] IFN-γ also enhances antigen presentation by increasing MHC class II expression of APCs.

The anti-tumor immune-modulating effect of IFN in vivo and the relatively low host toxicity of IFN led to a number of clinical trials. The greatest therapeutic usefulness of IFN has been in the treatment of a relatively rare form of leukemia called "hairy cell" leukemia, because of the spiked appearance of the cell surface. Most of the earlier clinical trials used IFN-α purified from human leukocytes and later obtained by recombinant DNA techniques. IFN-α is used in the treatment of chronic hepatitis in combination with antiviral drugs. IFN-γ has also been used in the treatment of immune deficiency diseases and some infections.

An intriguing approach to the activation of tumoricidal macrophages involves the use of liposomes to deliver macrophage-activating factors directly to these phagocytic cells.[61] Liposomes containing phosphatidylcholine and phosphatidylserine are selectively taken up by phagocytic cells, including reticuloendothelial cells in the liver, spleen, lymph nodes, and bone marrow, as well as by circulating monocytes. Incorporation of agents that activate macrophages, such as MAF, IFN-γ, and the low-molecular-weight ($M_r = 459$) synthetic compound muramyl dipeptide (N-acetylmuramyl-L-alanyl-D-isoglutamine; MDP), into liposomes provides a delivery system for these factors to macrophages. Once the carrier liposomes are engulfed, the macrophages become tumoricidal against target cells in vivo. The macrophages thus activated recognize and lyse neoplastic cells in vitro through a mechanism that requires cell-to-cell contact but apparently is independent of MHC antigens.[61] Intravenous administration of liposomes containing MAF plus MDP to nude mice who were previously injected with B16 mela-

noma cells and who had spontaneous metastases in the lungs and lymph nodes by the time of liposome treatment produced 9 of 18 250-day survivors compared to 2 of 18 250-day survivors in the liposome-only control group.[61] These data indicate that activation of macrophages in vivo is possible and support the possibility that metastatic tumor sites can also be recognized and destroyed by such activated macrophages.

Interleukins

The interleukins belong to a large family of polypeptide growth and differentiation factors called *lymphokines*. These factors, produced by lymphocytes or macrophages, stimulate the proliferation, differentiation, and function of T lymphocytes, B lymphocytes, and certain other cells involved in the immune response. Initially discovered as soluble factors present in the growth medium of cultured lymphocytes, several such activities were subsequently identified (reviewed in Reference 62). These activities were usually defined by their role in simulating an in vitro immune reaction, i.e., promoting the activation and/or proliferation of immune system cells. Accordingly, the following kinds of activities were identified: T-cell mitogenesis factor (TMF), or T-cell stimulating factor (TSF), which foster T-cell proliferation in response to added plant lectins; killer/helper factors (KHFs), which stimulate in vitro generation of antigen-specific cytolytic T cells; B-cell helper factors, which could replace T cells in fostering differentiation of B cells into antibody-producing cells; and T cell growth factor (TCGF), which is produced by mitogen-stimulated lymphocytes and promotes proliferation of antigen-activated T cells. Through the use of cloned lymphocyte populations, more extensive purification of lymphokine-containing media, and specific monoclonal antibodies, it was possible to catalogue these factors more definitively. It became clear that several of the previously described activities could be attributed to distinct polypeptides. The renaming of these factors was adopted at the Second International Lymphokine Workshop, held in Ermattingen, Switzerland in 1979. The first named were interleukin-1 (IL-1) and interleukin-2 (IL-2). The term *interleukin* was chosen because it indicates the basic property of these secreted mediators, i.e., to serve as intercellular signals between leukocytes. Many additional interleukins have now been identified. The list is now up to IL-32.

IL-1, -2, -3, and -4 have been introduced into clinical trial for a variety of malignant diseases. IL-1α and -1β have been employed to reverse bone marrow suppression due to chemotherapy or radiotherapy and to augment immunotherapy. IL-2 has been used in adoptive immunotherapy to stimulate clonal expansion of lymphokine-activated killer (LAK) cells and tumor-infiltrative lymphocytes (TIL) (see below). IL-3 has been used to stimulate bone marrow recovery in bone marrow or peripheral stem cell transplantation. IL-4 has been introduced as an immune system stimulator in various cancer treatment regimens.

Tumor Necrosis Factor

In the late 1800s it was observed that a few cancer patients had regression of their tumors after a full-blown systemic bacterial infection. This led a handful of investigators, including William Coley in the United States, to attempt to treat cancers by infecting patients with certain bacteria (reviewed in Reference 63). The infectious process proved difficult to control, although a few positive responses were noted. There was sufficient encouragement, however, from this approach to prompt Coley in 1893 to try a mixture of killed bacteria (*Streptococcus pyogenes* and *Serratia merescens*). When he injected his "Coley's toxins" directly into tumors he attained, in a few cases, some remarkable remissions of the tumor. This approach was adopted by a number of investigators and continued on an experimental basis for a number of years. In fact, Coley's toxins were considered to be the only systemic therapy for cancer until the 1930s. With the advent of radiotherapy, chemotherapy, and improved cancer surgery, this treatment was abandoned and forgotten until the early 1980s.

It has been known for a long time that filtrates of certain bacteria such as the mycobacterium BCG and the corynebacterium *C. parvum* can induce a hemorrhagic necrosis of tumors when injected into tumor-bearing mice. Later work showed that an extract of gram-negative bacteria, known as endotoxin or bacterial pyrogen, could induce similar effects. Endotoxin was subse-

quently identified as a lipopolysaccharide (LPS). Because LPS is quite toxic, it never found its way into clinical use, but BCG and *C. parvum* have undergone extensive clinical trials and were not found to be very effective. The effectiveness of BCG and LPS against certain murine tumors in vivo, however, stimulated continued interest in finding the mechanism of this effect. The evidence pointed to the fact that these agents acted by modulating the host's immune response, probably by stimulating macrophages.

Attempts to define the anti-tumor mechanism of BCG and LPS led Old and colleagues to test the serum of mice injected with BCG or LPS or both for anti-tumor activity. The serum of BCG- and LPS-treated mice produced hemorrhagic necrosis of mouse sarcomas in vivo. This effect required both BCG and LPS. Moreover, serum from BCG- and LPS-treated mice was also cytotoxic for a line of transformed mouse fibroblasts (L cells) in culture, but BCG and LPS added directly to the cultures were not, indicating that these agents were eliciting the release of some cytotoxic factor into the serum. It was also found that the in vivo necrotizing factor and the factor cytotoxic to cultured cells were one and the same. The factor was named *tumor necrosis factor* (TNF).[64]

TNF has also been shown to be identical to a cachexia-inducing factor called *cachectin*.[65] This factor, also produced by macrophages, suppresses lipogenic enzymes such as glycerol-3-phosphate dehydrogenase. It has been postulated that the cachexia associated with cancer may involve endogenous release of TNF. In addition to macrophages and certain lymphoid cell lines, there is evidence that TNF can also be produced by NK cells and endothelial cells. A variety of agents can modulate cellular release of TNF. TNF expression by producer cells is increased by IFN-γ and indomethacin and decreased by glucocorticoids and PGE$_2$. TNF-α is now known to be a mediator in a number of inflammatory and autoimmune conditions (see above).

The genes for TNF and lymphotoxin are related but not identical. There is about 30% sequence homology at the amino acid level. TNF-α has the cachectin activity, and TNF-β has lymphotoxin activity. TNF-α is produced primarily by monocyte/macrophage cells, although it is also produced by T and B lymphocytes, LAK cells, NK cells, neutrophils, astrocytes, endothelial cells, smooth muscle cells, and a variety of tumor cell lines, whereas TNF-β is produced almost exlusively by lymphoid cells (reviewed in Reference 66). TNF-α and TNF-β are considered members of the "inflammatory cytokines" and are released at sites of inflammation. They also act as immunostimulants and mediators of host resistance to infections agents and malignant cells. Overproduction of TNF-α during infection can lead to the septic shock syndrome.

The TNF-α and TNF-β genes are single-copy genes and closely linked within the MHC locus on chromosome 6 in humans. TNF-α is a trimer made up of 17,000 Da polypeptide chains. Native TNF-β is also a trimer; it consists of 25,000 Da subunits, each with one N-linked carbohydrate chain. TNF receptors of 55 kDa and 75 kDa have been identified and most cell types examined have both types. Each receptor type binds TNF-α and TNF-β with similar high affinity, and they appear to transduce signals via G protein–mediated activation of protein kinases. However, the two TNF receptors may mediate different sets of biological activity, and they have a somewhat different cellular distribution: the 55 kDa receptor is expressed on a wide range of cells and appears to be the receptor involved in direct tumor cell killing, whereas the 75 kDa receptor is expressed primarily on lymphoid and myeloid cells and may be more involved in the immune system effects of TNF.[67] The TNFs have a large number of biological activities and can induce a large number of genes in multiple target cells. These genes include transcription factors, cytokines, growth factors, adhesion molecules, inflammatory mediators, and acute-phase proteins.[66]

It is now known that the TNF family includes a large number of T-cell cytokines that includes TNF-α, TNF-β (LT-α); T–B cell recognition factors CD30, CD40, and CD27; and the apoptosis-inducing ligand Fas.[68] In fact, a family of 19 TNF ligands and 29 receptors have been identified.[67a] Many of these ligands, not including LT-α, are membrane bound and appear to produce their effects by contacting receptors on adjacent cells as they stay in the membrane bound form. Some TNF-α (and perhaps other of these ligands) is released by proteases and circulates as a soluble form. There is also a superfamily of TNF receptors, one of which is the low-affinity nerve growth factor receptor (NGFR). This suggests some evolutionary relatedness that may or may not have anything to do with a commonality of functions, but NGFRs are expressed at high levels on follicular dendritic cells of lymphoid germinal centers and TNF receptors are expressed on glial cells.

The TNF receptor family has sequence homology confined to the extracellular region. The canonical motif of all of them is the cysteine-rich repeats, each containing about six cysteines and 40 amino acids, but with considerable variation in length and number of repeats.[68] The cytoplasmic domains are all rather small and variable in sequence homology, which suggests that they may trigger different signal transduction mechanisms. None have obvious catalytic activity sequences (e.g., of a tyrosine kinase–like activity, etc.). Interaction with ligand, however, induced receptor oligomerization, as seen for other cytokines and growth factors.

The tumor cell–killing effect of TNF may be both direct and indirect. There is evidence that cells which bind TNF can be killed by induction of a cytolytic cascade of events including generation of oxygen radicals, DNA fragmentation, and apoptosis.[69] The reason why TNF tends to kill tumor cells preferentially over normal cells isn't clear, but it may be related to a higher number of TNF receptors on tumor cells, more active internalization of TNF into tumor cells, the ability of tumor cell lysosomes to more effectively process TNF into a cytotoxic form, or the lack of some protective factor in tumor cells.

Somewhat paradoxically, TNF-α release at sites of inflammation around tumors plays a tumor progression role (see above) and as a result may actually favor tumor invasion and metastasis at later stages of tumor development.

Adoptive Immunotherapy

Adoptive immunotherapy is "a treatment approach in which cells with antitumor reactivity are administered to a tumor-bearing host and mediate either directly or indirectly the regression of established tumor."[70] Several types of approaches have been tried. One idea is to take tumor cells from a patient (removed by biopsy or surgery), inactivate the cells by X-irradiation, and then use these cells plus bacterial-derived adjuvants such as BCG or *C. parvum* to attempt to induce an immune response in the patient to his or her own tumor cells.[71] Another idea is to take the patient's own lymphocytes and activate them in vitro with appropriate activating factors and then inject them back into the patient. The latter approach was taken by Rosenberg and colleagues,[70] who obtained, by leukapheresis, large quantities of patients' leukocytes, from which lymphocytes were separated. The lymphocytes were then incubated with recombinant IL-2, which stimulates a population of lymphocytes that when activated can lyse fresh, noncultured, NK cell–resistant tumor cells but not normal cells.[72] These cells were termed *lymphokine-activated killer* (LAK) cells. In earlier animal studies, Rosenberg et al. demonstrated that the adoptive transfer of LAK cells, incubated in culture with IL-2 followed by additional treatment in vivo with IL-2, induced the regression of pulmonary and hepatic metastases from a wide variety of murine tumors, including melanomas, sarcomas, a colon adenocarcinoma, and a urinary bladder carcinoma. Clinical results, however, proved to be disappointing.

Subsequently, a technique was developed to isolate T lymphocytes directly from a patient's tumor, and these tumor-infiltrative lymphocytes could also be clonally expanded in culture by adding IL-2. TIL cells bear the CD8[+] surface marker. They can kill the hosts' tumor cells in culture in an MHC class I–restricted manner and are 50–100 times more potent than LAK cells in reducing lung metastases in mouse model systems.[73] In order for TIL cells to be effective in vivo, IL-2 must also be injected, but it requires lower doses of IL-2 than for LAK cell therapy. In addition, adjuvant treatment with cyclophosphamide and fludarabine are used to immunodeplete T-suppressor cells. Such adjuvant therapy probably also reduces the tumor burden (with subsequent improved access of TIL to tumor tissue). Clinical results from TIL cell-plus-adjuvant chemotherapy in patients with metastatic melanoma has shown positive responses: of 13 patients treated, 6 had objective responses, 4 had a mixed response, and 3 had no response.[74] Subsequent studies by Rosenberg et al. showed that 18 of 35 patients with refractory metastatic melanoma had at least 50% reduction in tumor mass, including 4 complete responders.[75] The success of this study most likely resulted from the infusion of large members of activated anti-tumor lymphocytes into patients whose regulatory T cells had been depleted by pretreatment with cyclophosphamide and fludarabine.

Another approach to adoptive immunotherapy is to use tumor-target specific cytotoxic T lymphocytes. Crowley et al.[76] have generated melanoma-specific CTLs by co-culturing human peripheral blood lymphocytes with irradiated, allogeneic melanoma cells that express a restricting HLA-A region antigen, at a lymphocyte: tumor cell rate of 20:1, in the presence of IL-2. CTLs generated in this way were able to kill human melanoma cells bearing the restricting HLA-A phenotype in vitro and in a human–xenograft nude mouse model. As few as 2.5×10^6 T cells were effective in vivo and were able to produce, in combination with IL-2, 96% and 88% disease-free animals with hepatic metastases when the CTLs were injected 3 and 7, respectively, days after generation of metastases. IL-2 by itself was ineffective.

There are several potential clinical advantages of CTL adoptive immunotherapy.[76] First, CTLs are MHC restricted and can be targeted to specific HLA-A-expressing cells and can kill autologous tumor cells and allogeneic tumor cells as long as they express the HLA-A region antigen. Thus, HLA-A region–matched allogenic tumors can substitute for autologous tumor in the stimulating tumor to generate specific CTLs. Another advantage is that CTLs can be generated from peripheral lymphocytes or lymph node cells from surgical specimens, allowing generation of CTLs from patients with stage I or II disease, whereas TIL cells may only be available from larger metastatic sites. Finally, the amount of tumor needed for re-stimulation of CTLs in culture is small and if autologous tumor is unavailable for follow-up stimulation, allogenic HLA-A-matched allogeneic tumors could be used.

Still other approaches harvest peripheral blood dendritic cell (DC) precursors, expand them in culture, and pulse them with tumor antigen or transfect them with viral vectors containing tumor antigen–encoding genes. Circulating immature DC precursors make up less than 0.5% of peripheral blood mononuclear cells but they can be isolated from T cell– and monocyte-depleted peripheral blood cells after 1 to 2 days of in vitro culture in the absence of cytokines.[72] During this time DC precursors mature and change their buoyant density such that they can be isolated from other leukocytes by density-gradient centrifugation. These DC populations can then be expanded in culture in the presence of GM-CSF and IL-4. These mature, expanded cultures of DCs can take up and process protein tumor antigens and present them to T cells isolated from cancer patients and co-cultured in vitro. The DCs can also be pulsed directly with processed tumor antigen peptides or transfected with viral vectors encoding tumor antigen genes.[78] A number of clinical trials are under way using tumor antigen–activated DCs injected into patients with a variety of cancers, including lymphomas, melanoma, prostate, breast, and renal cell cancers, and multiple myeloma.[78,79]

Vaccines

Although the use of immune cell–based immunotherapy is sometimes called *vaccination*, the term is used more precisely to mean a form of specific active immunotherapy of cancer that entails immunizing patients directly with antigens that are expressed on cancer cells.[54] Vaccines include whole cancer cells (treated to make them sterile, nonproliferative cells), cancer cells transfected with cytokine or co-stimulatory encoding genes, cancer cell lysates, heat shock protein–complexed tumor antigen peptides, or cancer cells fused to APCs. Vaccination with inactivated viruses or viral antigens against viruses known to be causative or co-factors in causing human cancer could also considered as anticancer vaccination. This includes vaccination against hepatitis B virus to prevent hepatocellular carcinoma and vaccination against the E6 and E7 epitopes of human papilloma virus to prevent cervical carcinoma.

Monoclonal Antibodies

The development of monoclonal antibodies (mAbs) by Köhler and Milstein in 1975[80] opened up a new era in therapy for a number of diseases, including cancer. The development of mAbs over the years has evolved from mouse-derived immunoglobulins, which were limited in usefulness because of human anti-mouse antibody (HAMA) production in patients, to fully humanized mAbs that are less immunogenic and more efficacious. A number of mAbs have shown clinical effectiveness, including rituximab (Rituxan) against non-Hodgkin's lymphoma and trastuzumab (Herceptin) for treatment of metastatic breast cancer. Rituxan is targeted to the CD20 cell surface marker in B-cell lymphoma cells, and Herceptin is targeted against the HER 2/neu tyrosine kinase receptor overexpressed on about a 20% of metastatic breast cancers.

One of the limitations of mAbs in the treatment of cancer is that they are poorly cytocidal. Thus, their efficacy is often significantly enhanced by combination with standard chemotherapeutic drugs. Another way to address this issue is to complex a tumor-targeted mAb to a toxin, chemotherapeutic drug, or a radionuclide. These complexes can be cytocidal and not just cytostatic. mAbs complexed with yttrium-90 or iodine-131 and targeted to cell surface antigens of non-Hodgkin's lymphoma cells have proven to be clinically effective.[81]

HOW TUMOR CELLS AVOID THE IMMUNE RESPONSE

As noted in Chapter 5, the genetic instability of cancer cells means that they are continually evolving in a number of ways. One way they evolve to gain a selective advantage is to develop mechanisms to avoid the host's immune defense mechanisms. Tumors "learn" to get around these immune mechanisms in a number of ways.[53,75,82]

1. There may be insufficient numbers of CD4[+] and/or CD8[+] cells in a host. Both cell types are important to mounting a full-blown anti-tumor immune response.
2. Immune tolerance to tumors may occur through loss of co-stimulatory molecules on the tumor.
3. Down-regulation of T-cell receptor signal transduction mechanisms can occur.
4. Apoptosis of T cells rather than clonal expansion can occur because of inappropriate or lack of appropriate "danger signals."

5. Loss of MHC class 1 expression on tumor cells.
6. Production by tumors of TGF-β, which is inhibitory to the immune response.
7. Induction of T-suppressor cell proliferation.
8. Induction of T-cell suppressor mechanisms such as enhanced expression of the cytotoxic T-lymphocyte antigen CTLA-4, a negative co-stimulatory molecule.

All of these mechanisms have been observed in experimental settings, and ways to circumvent them need to be developed. Since many tumors are apparently not immunogenic enough to stimulate an effective immune response against them, increasing their antigenicity could be a way to induce a more effective host immune response. For example, one way might be to remove tumor cells from a patient, modify them with chemicals or viruses to make them more immunogenic, and then inject them back into the patient after sterilizing them by X-irradiation or cytotoxic drugs.

Tumor cell immunogenicity could also be increased by modulating expression of MHC gene products on tumor cells. Since class I MHC antigens are necessary for the presentation and recognition of tumor cell neoantigens by cytolytic T lymphocytes, their masking or absence on tumor cell surfaces may be key to their ability to escape an immune response of the host. One way to alter this is to introduce, by DNA-mediated gene transfer, for example, the genes for the missing MHC class I molecules. This has been done in adenovirus-12-transformed cells that lack expression of an H-2 class I gene product and that produce lethal tumors in syngeneic mice.[83] The induced expression of a single type of class I gene was sufficient to block the in vivo tumorigenicity of these cells. Thus, an approach to cancer therapy may be to increase or modulate expression of class I genes in tumor cells. IFN-γ can increase expression of class I antigens in certain cells.[83] Other such modulators may also be found.

Another method to boost a patient's response to a tumor may be to transfect cytokine genes into surgically removed tumor cells, followed by reimplantation or to deliver such genes by tumor-targeted gene therapy in vivo so that their rec-ognition by immune cells is increased. Several studies in murine tumor models have demonstrated increased anti-tumor responses in vivo to tumor cells transduced with cytokine genes, including IFN-γ, IL-2, IL-4, IL-6, IL-7, TNF-α, and G-CSF (reviewed in Reference 84). Clinical trials have been carried out that use TNF-α gene-transduced tumor infiltrative lymphocytes or melanoma cells transduced with TNF-α or IL-2 genes.[85]

Methods to circumvent the production of T-cell suppression are also being developed.[53] For example, antibodies against CTLA-4 have been shown to enhance anti-tumoral immunity to a GM-CSF-transduced vaccine. This method led to the regression of established transplanted syngeneic tumors.

Antibody directed to the CD25 cell surface marker on $CD4^+$ $CD25^+$ suppressor T cells has been shown to deplete such cells and lead to an enhanced immune response (the induction of CTL and NK cell cytotoxicity) and rejection of syngeneic tumors (reviewed in Reference 53). Blockade of production of the immune inhibitory factor TGF-β by inhibiting production of its inducer IL-13 by $CD4^+$ NK T cells also enhances anti-tumor responses and potentiates the efficacy of vaccines.[53]

References

1. O. C. A. Scott: Tumor transplantation and tumor immunity: personal view. *Cancer Res* 51:757, 1991.
2. G. Klein: Experimental studies in tumor immunology. *Fed Proc* 28:1739, 1969.
3. E. J. Foley: Antigenic properties of methyl cholanthrene-induced tumors in mice of the strain of origin. *Cancer Res* 13:835, 1953.
4. R. T. Prehn and J. M. Main: Immunity to methyl cholanthrene-induced sarcomas. *J Natl Cancer Inst* 18:769, 1957.
5. G. Klein, H. O. Sjogren, E. Klein, and K. E. Hellstrom: Demonstration of resistance against methylcholanthrene-induced sarcomas in the primary autochthonous host. *Cancer Res* 20:1561, 1960.
6. F. M. Burnet: Immunological surveillance in neoplasia. *Transplant Rev* 7:3, 1971.
7. R. T. Prehn: The immune reaction as a stimulator of tumor growth. *Science* 176:170, 1972.
8. G. P. Dunn, A. T. Bruce, H. Ikeda, L. J. Old, and R. D. Schreiber: Cancer immunoediting: From immunosurveillance to tumor escape. *Nat Immunol* 3:991, 2002.

9. M. J. Smyth and M. H. Kershaw: Discovery of an innate cancer resistance gene? *Mol Interventions* 3:186, 2003.

10. A. R. Reif: Evidence for organ specificity of defenses agains tumors. In H. Waters, ed.: *The Handbook of Cancer Immunology*. New York: Garland STPM Press, 1978, pp. 173–240.

11. L. J. Old and E. A. Boyse: Immunology of experimental tumors. *Annu Rev Med* 15:167, 1964.

12. M. S. Leffell and J. H. Coggin, Jr.: Common transplantation antigens on methylcholanthrene-induced murine sarcomas detected by three assays of tumor rejection. *Cancer Res* 37:4112, 1977.

13. K. E. Hellstrom, I. Hellstrom, and J. P. Brown: Unique and common-tumor specific transplantation antigens of chemically induced mouse sarcomas. *Int J Cancer* 21:317, 1978.

14. C. E. Whitmore and R. J. Huebner: Inhibition of chemical carcinogenesis by mixed vaccines. *Science* 177:60, 1972.

15. J. P. Brown, J M. Klitzman, I. Hellstrom, C. Nowinski, and K. E. Hellstrom: Antibody response of mice to chemically induced tumors. *Proc Natl Acad Sci USA* 75:955, 1978.

16. H. F. Oettgen and K. E. Hellstrom: Tumor immunology. In J. F. Holland and E. Frei III, eds.: *Cancer Medicine*. Philadelphia: Lea & Febiger, 1973, pp. 951–990.

17. J. L. McCoy, L. F. Jerome, J. H. Dean, E. Perlin, R. K. Oldham, et al.: Inhibition of leukocyte migration of tumor-associated antigens in soluble extracts of human malignant melanoma. *J Natl Cancer Inst* 55:19, 1975.

18. H. Shiku, T. Takahashi, T. E. Carey, L. A. Resnick, H. F. Oettgen, and L. J. Old: Cell surface antigens of human cancer. In R. W. Ruddon, ed.: *Biological Markers of Neoplasia: Basic and Applied Aspects*. New York: Elsevier North Holland, 1978, pp. 73–88.

19. C. L. Finstad, C. Cordon-Cardo, N. H. Bander, W. F. Whitmore, M. R. Melamed, and L. J. Old: Specificity analysis of mouse monoclonal antibodies defining cell surface antigens of human renal cancer. *Proc Natl Acad Sci USA* 82:2955, 1985.

20. F. X. Real, A. N. Houghton, A. P. Albino, C. Cordon-Cardo, M. R. Melamed, et al.: Surface antigens of melanomas and melanocytes defined by mouse monoclonal antibodies: Specificity analysis and comparison of antigen expression in cultured cells and tissues. *Cancer Res* 45:4401, 1985.

21. J. Parkin and B. Cohen: An overview of the immune system. *Lancet* 357:1777, 2001.

22. A. K. Abbas and C. A. Janeway, Jr.: Immunology: improving on nature in the twenty-first century. *Cell* 100:129, 2000.

23. C. R. Roy: Professional secrets. *Nature* 425:351, 2003.

24. A. Lanzavecchia: Identifying strategies for immune intervention. *Science* 260:937, 1993.

25. C. A. Janeway, Jr., and K. Bottomly: Signals and signs for lymphocyte responses. *Cell* 76:275, 1994.

26. H.-G. Rammensee: Survival of the fitters. *Nature* 419:443, 2002.

27. N. Vigneron, V. Stroobant, J. Chapiro, A. Ooms, G. Degiovanni, et al.: An antigenic peptide produced by peptide splicing in the proteasome. *Science* 304:587, 2004.

28. O. Werdelin, M. Meldal, and T. Jensen: Processing of glycans on glycoprotein and glycopeptide antigens in antigen-presenting cells. *Proc Natl Acad Sci USA* 99:9611, 2002.

29. W. E. Paul and R. A. Seder: Lymphocyte responses and cytokines. *Cell* 76:241, 1994.

30. D. J. Irvine: Function-specific variations in the immunological synapses formed by cytotoxic T cells. *Proc Natl Acad Sci USA* 100:13739, 2003.

31. R. B. Herberman and H. T. Holden: Natural cell-mediated immunity. *Adv Cancer Res* 27:305, 1978.

32. M. E. Nunn and R. B. Herberman: Natural cytotoxicity of mouse, rat, and human lymphocytes against heterologous target cells. *J Natl Cancer Inst* 62:765, 1979.

33. J. C. Lee and J. N. Ihle: Characterization of the blastogenic and cytotoxic responses of normal mice to ecotropic C-type gp 71. *J Immunol* 118:928, 1977.

34. K. Karre, G. O. Klein, R. Kiessling, G. Klein, and J. C. Roder: Low natural in vivo resistance to syngeneic leukaemias in natural killer–deficient mice. *Nature* 284:624, 1980.

35. W. J. Murphy, C. W. Reynolds, P. Tiberghien, and D. L. Longo: Natural killer cells and bone marrow transplantation. *J Natl Cancer Inst* 85:1475, 1993.

36. A. P. Williams, A. R. Bateman, and S. I. Khakoo: Hanging in the balance: KIR and their role in disease. *Mol Interventions* 5:226, 2005.

37. D. Zhou, J. Mattner, C. Cantu III, N. Schrantz, N. Yin, et al.: Lysosomal glycosphingolipid recognition by NKT cells. *Science* 306:786, 2004.

38. J. Ding-E Young and Z. A. Cohn: Cell-mediated killing: A common mechanism? *Cell* 46:641, 1986.

39. P. C. Doherty: Cell-mediated cytotoxicity. Cell 75:607, 1993.

40. S. Gallucci, M. Lolkema, and P. Matzinger: Natural adjuvants: endogenous activators of dendritic cells. *Nat Med* 5:1249, 1999.

41. W. R. Heath and F. R. Carbone: Dangerous liaisons. *Nature* 425:460, 2003.

42. Y. Shi, J. E. Evans, and K. L. Rock: Molecular identification of a danger signal that alerts the immune system to dying cells. *Nature* 425:516, 2003.

43. C. H. Bassing and F. W. Alt: Case of mistaken identity. *Nature* 428:29, 2004.

44. D. B. Roth and S. Y. Roth: Unequal access: regulating V(D)J recombination through chromatin remodeling. *Cell* 103:699, 2000.

45. M. Taylor McMurray and M. S. Krangel: A role for histone acetylation in the developmental

regulation of V(D)J recombination. *Science* 287: 495, 2000.

46. M. S. Schlissel: A tail of histone acetylation and DNA recombination. *Science* 287:438, 2000.

47. A. G. Pockley: Heat shock proteins as regulators of the immune response. *Lancet* 362:469, 2003.

48. H. Udono and P. K. Srivastava: Comparison of tumor-specific immunogenicities of stress-induced proteins gp96, hsp90, and hsp70. *J Immunol* 152: 5398, 1994.

49. F. Balkwill and L. M. Coussens: An inflammatory link. *Nature* 431:405, 2004.

50. E. Pikarsky, R. M. Porat, I. Stein, R. Abramovitch, S. Amit, et al.: NF-κB functions as a tumor promoter in inflammation-associated cancer. *Nature* 431:461, 2004.

51. G. Klein: Immune and non-immune control of neoplastic development: Contrasting effects of host and tumor evolution. *Cancer* 45:2488, 1980.

52. S. A. Rosenberg: Progress in human tumour immunology and immunotherapy. *Nature* 411:380, 2001.

53. T. A. Waldmann: Immunotherapy: past, present and future. *Nat Med* 9:269, 2003.

54. E. Gilboa: The promise of cancer vaccines. *Nat Rev Cancer* 4:401, 2004.

55. X. Wang, J. Yu, A. Sreekumar, S. Varambally, R. Shen, et al.: Autoantibody signatures in prostate cancer. *N Engl J Med* 353:1224, 2005.

56. C. Nathan and M. Sporn: Cytokines in context. *J Cell Biol* 113:981, 1991.

57. A Isaacs and J. Lindemann: Virus interference. I. The interferon. *Proc R Soc Ser B* 147:258, 1957.

58. K. L. Fresa and D. M. Murasko: Role of natural killer cells in the mechanism of the antitumor effect of interferon on Moloney sarcoma virus–transformed cells. *Cancer Res* 46:81, 1986.

59. L. Varesio, E. Blasi, G. B. Thurman, J. E. Talmadge, R. H. Wiltrout, and R. B. Herberman: Potent activation of mouse macrophage by recombinant interferon-γ. *Cancer Res* 44:4465, 1984.

60. Y. Akiyama, M. D. Lubeck, Z. Steplewski, and H. Koprowski: Induction of mouse IgG2a- and IgG3-dependent cellular cytotoxicity in human monocytic cells (U937) by immune interferon. *Cancer Res* 44:5127, 1984.

61. I. J. Fidler: Macrophages and metastasis—A biological approach to cancer therapy. *Cancer Res* 45:4714, 1985.

62. S. B. Mizel: The interleukins. *FASEB J* 3:2379, 1989.

63. L. J. Old: Tumor necrosis factor (TNF). *Science* 230:630, 1985.

64. E. A. Carswell, L. J. Old, R. L. Kassel, S. Green, N. Fiore, et al.: An endotoxin-induced serum factor that causes necrosis of tumors. *Proc Natl Acad Sci USA* 72:3666 1975.

65. B. Beutler and A. Cerami: Cachectin and tumor necrosis factor as two sides of the same biological coin. *Nature* 320:584, 1986.

66. J. Vilcek and T. H. Lee: Tumor necrosis factor: New insights into the molecular mechanism of its multiple actions. *J Biol Chem* 266:7313, 1991.

67. F. Balkwill: Tumor necrosis factor: Improving on the formula. *Nature* 361:206, 1993.

67a. S. R. Dillon, J. A. Gross, S. M. Ansell, and A. J. Novak: APRIL to remember: Novel TNF ligands as therapeutic targets. *Nat Rev Drug Disc* 5:235, 2000.

68. C. A. Smith, T. Farrah, and R. G. Goodwin: The TNF receptor superfamily of cellular and viral proteins: Activation, costimulation, and death. *Cell* 76:959, 1994.

69. J. W. Larrick and S. C. Wright: Cytotoxic mechanism of tumor necrosis factor-α. *FASEB J* 4:3215, 1990.

70. S. A. Rosenberg, M. T. Lotze, L. M. Muul, S. Leitman, A. E. Chang, et al.: Observations on the systematic administrations of autologous lymphokine-activated killer cells and recombinant interleukin-2 to patients with metastatic cancer. *N Engl J Med* 313:1485, 1985.

71. M. G. Hanna, J. S. Brandhorst, and L. C. Peters: Active specific immunotherapy of residual micrometastases: An evaluation of sources, doses, and ratios of BCG with tumor cells. *Cancer Immunol Immunother* 7:165, 1979.

72. J. J. Mule, S. Shu, S. L. Schwarz, and S. A. Rosenberg: Adoptive immunotherapy of established pulmonary metastases with LAK cells and recombinant interleukin-2. *Science* 225:1487, 1984.

73. S. A. Rosenberg: Immunotherapy and gene therapy of cancer. *Cancer Res* (Suppl) 51:5074s, 1991.

74. M. E. Dudley, J. R. Wunderlich, P. F. Robbins, J. C. Yang, P. Hwu, et al.: Cancer regression and autoimmunity in patients after clonal repopulation with antitumor lymphocytes. *Science* 298:850, 2002.

75. S. A. Rosenberg and M. E. Dudley: Cancer regression in patients with metastatic melanoma after the transfer of autologous antitumor lymphocytes. *Proc Natl Acad Sci USA* 101 (Suppl 2): 14639, 2004.

76. N. J. Crowley, C. E. Vervaert, and H. F. Seigler: Human xenograft-nude mouse of adoptive immunotherapy with human melanoma-specific cytotoxic T cells. *Cancer Res* 52:394, 1992.

77. J. M. Timmerman and R. Levy: Dendritic cell vaccines for cancer immunotherapy. *Annu Rev Med* 50:507, 1999.

78. C. J. Kirk and J. J. Mulé: Gene-modified dendritic cells for use in tumor vaccines. *Hum Gene Ther* 11:797, 2000.

79. J. M. Reichert and C. Paquette: Therapeutic cancer vaccines on trial. *Nat Biotech* 20:659, 2002.

80. G. Kohler and C. Milstein: Continuous cultures of fused cells secreting antibody of predefined specificity. *Nature* 256:495, 1975.

81. T. A. Davis, M. S. Kaminski, J. P. Leonard, F. J. Heu, M. Wilkinson, et al.: The radioisotope contributes significantly to the activity of radioimmunotherapy. *Clin Cancer Res* 10:7792, 2004.

82. W. Zou: Immunosuppressive networks in the tumour environment and their therapeutic relevance. *Nat Rev Cancer* 5:263, 2005.

83. K. Tanaka, K. J. Isselbacher, G. Khoury, and G. Jay: Reversal of oncogenesis by the expression of a major histocompatibility complex class I gene. *Science* 228:26, 1985.

84. M. Ogasawara and S. A. Rosenberg: Enhanced expression of HLA molecules and stimulation of autologous human tumor infiltrating lymphocytes following transduction of melanoma cells with γ-interferon genes. *Cancer Res* 53:3561, 1993.

85. S. A. Rosenberg: The immunotherapy and gene therapy of cancer. *J Clin Oncol* 10:180, 1992.

7

Cancer Diagnosis

MEDICAL AND SCIENTIFIC DRIVERS FOR EXPANDED CANCER DIAGNOSTIC TECHNIQUES

As more is learned about the genotypic, phenotypic, and metabolomic differences between cancer tissue and non-cancer tissue and about the differences between indolent and aggressive tumors, there will be an increased medical need and patient demand for such discriminatory tests. In addition, individual patient variation in response to therapeutic agents will determine which drugs and other treatments they are given. Increasing knowledge of the hereditary background of individuals and the gene alterations that determine susceptibility to getting cancer will also drive this field of research and clinical care. Moreover, identification of biomarkers that lead to early detection of malignant changes in tissues will facilitate new strategies for earlier treatment and perhaps even for prevention of early malignant changes before they progress to full-blown invasive cancer. New drug targets will emerge from the increased learning about biomarkers that reflect the molecular biology of cancer.

Some of the scientific advances that will drive this field are the following:

1. Genomic, proteomic, and epigenetic profiling. Cancer research is on the threshold of new, fundamental knowledge that will generate molecular tools to create individual profiles to direct oncology treatments and prevention.

2. Predictive biomarkers. Earlier detection, faster development, and larger oncology markets will ensue from a growing number of biomarkers, with many biomarkers finding use as surrogate end points for cancer.

3. Nanotechnology and biological micro-electro-mechanical systems (BioMEMS). Small medical devices will be developed for diagnosis and treatment over the next decade, finding applications in research as well as clinical care.

Some definitions may be helpful for topics discussed in this chapter.

Biomarker: A characteristic that is objectively measured and evaluated as an indicator of normal biological processes, pathogenic processes, or pharmacologic responses to a therapeutic intervention.

DNA methylation: A biochemical process in which methyl groups are added to certain nucleotides in genomic DNA. This affects gene expression without changing the underlying genetic sequence.

Epigenetic: Mechanisms that influence gene expression without altering the sequence of nucleotides in the DNA.

Genomic: Relating to the total set of genes carried by an individual or cell.

Genotype: The genetic constitution of an organism or cell.

Metabolomics: The study of the global metabolic profiles in cells, tissues, and organisms. These profiles are determined by the patterns of metabolites produced in cells that are typically generated by high-throughput nuclear magnetic resonance (NMR) and mass spectrometry (MS).

Nanotechnology: A branch of engineering that deals with the design and manufacture of extremely small electronic circuits and mechanical devices built at the molecular level of matter (equal to or less than 10^{-9} meters).

Pharmacogenetics: The study of how people respond differently to medicines because of their genetic inheritance. Specific focus is on the genetic differences in candidate genes that are likely to be important in drug responses. These frequently relate to drug-metabolizing genes.

Pharmacogenomics: The study of the entire complement of pharmacologically relevant genes and their variations on a molecular level and how these variations affect drug response. Specific focus is on a whole-genome approach that looks for genetic variations that could act as an individual "signature" indicating both disease susceptibility and optimal drug treatment.

Phenotype: The total characteristics displayed by an organism under a particular set of environmental factors as a result of gene expression.

Proteomics: The study of the structure and function of the proteins encoded by a genome.

Surrogate end point: A biomarker intended to substitute for a clinical end point. A surrogate end point is expected to predict clinical benefit (or harm or lack of benefit) on the basis of biochemical, epidemiologic, therapeutic, pathophysiologic, or other scientific evidence.

Better screening and diagnostic strategies will produce a "game-changer" in the way oncology is practiced in the future. For example, currently patients discover they have cancer by already having symptoms, or they are lucky enough to have the disease detected through physical exams or screenings, such as mammography or PSA tests. They may be referred to a surgeon and then wait for the results of a tissue sample biopsy. They are then told what treatment options are considered appropriate for the affected organs and their stage of cancer.

Surgery, radiation, and chemotherapy are the main options, and each carries risks and sometimes disfiguring side effects. Success in the treatment is measured by 5-year survival rates, and cancer patients must live with the worry that the disease will come back.

If cancer recurs, metastases are detected, or drug resistance develops, patients face another round of treatments. In the worst case, the treatment cycle becomes a death spiral in which patients face a difficult choice between defeat of the disease and quality of remaining life. This patient care model is shown in Figure 7–1.

In the future, patients will benefit from early detection and greater customization of treatment, thanks to new breakthroughs that lead to prolonged survival and enhanced quality of life. Cancer biomarkers and imaging devices will be used to evaluate individual tumor profiles, and patient-specific pharmacogenomics will identify the likelihood of patient response.

Large databases will help oncologists prescribe combinations of molecular-targeted cancer drugs with cytotoxic drugs and monoclonal antibodies through delivery systems that target tumor cells. The oncologist will use an electronic medical record that both tracks each patient's response to therapy and links to large data sets that guide sequencing of therapy and provide a prognosis to the doctor and patient. Physicians will treat patients as individuals. Oncologists will use the genetic characteristics of tumors along with genetic profiles of patients to predict responses to therapies and tailor each treatment. Every level of the diagnosis and treatment will be informed and transformed by the following new capabilities:

Patient ⟶ Primary Care Physician ⟶

Surgeon ⟶ Radiation Oncologist ⟶

Medical Oncologist ⟶

Empirical Selection of Drug Regimen ⟶

Drug Resistance ⟶ New Regimen... ⟶

Progression ⟶ Terminal Care

Figure 7–1. How cancer is treated now.

- Screening through biomarkers and patient genetic profiling. Individual genetic screening will determine cancer risks, and direct biomarker testing will detect pre-cancers including intraepithelial neoplasia (IEN).
- Cancer evaluation through genomic and proteomic profiling. If tumors are found, tumor profiling will allow cancers to be categorized into subtypes by genetic mutations, abnormal proteins, and other characteristics, giving greater specificity to cancer diagnosis.
- Rational selection of combination therapies based on unique tumor characteristics. Therapy will be a combination of drugs and other treatments targeted to the specific tumor. The combinations will include current and new drugs and surgical modalities, as well as complementary and alternative approaches.
- Confirmaton that drugs are compatible with the patient's pharmacogenomic profile. Drug choices will be compared with the patient's pharmacogenomic profile to determine dose and safety before expensive therapies are administered. Ethnicity is now becoming accepted as a prognosticator of clinical outcomes because of ethnic-related differences in drug metabolism.

- Continual monitoring by means of devices and panels of biomarkers. Cancer phenotypes will be monitored for potential changes. Miniaturized devices, scanners, and arrays of diagnostics such as DNA and proteomic microarrays will be integrated for surveillance of high-risk patients.
- Deployment of new therapy combinations for cancer recurrence. Treatment for recurrent cancers before clinical detection will be a combination of molecular-targeted therapies, improved drug delivery systems, and minimally invasive surgical modalities.

With the implementation of these new strategies for personalized medicine, cancer treatment in the future will follow the pathway shown in Figure 7–2, as opposed to the old paradigm show in Figure 7-1.

The pathway for future cancer care will depend on implementation of the technologies shown in Figure 7–3. This model of personalized cancer treatment depends on the assumption that science will continue to push forward the frontier of personalized treatment. Cancer will likely be the first disease for which this new model of personalized treatment gains acceptance. Molecular research and accepted cancer biomarkers will differentiate tumor types at the

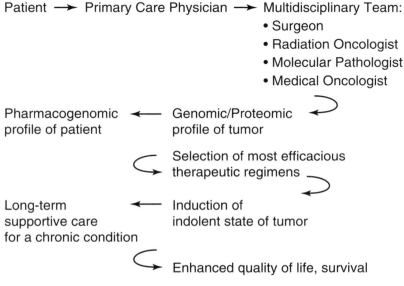

Figure 7–2. How cancer will be treated in the future.

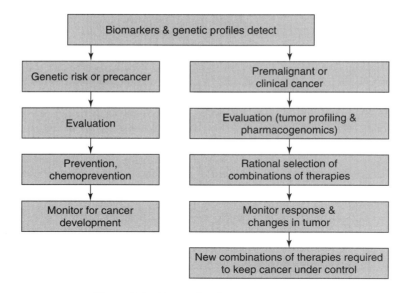

Figure 7–3. Personalized cancer treatment.

cellular level and create demand for treatments that take individual patient variability into account. This greater specificity will lead to new classification schemes that provide improved diagnosis, new targets for drug researchers, and demand for personalized treatments.

Because the molecular and clinical characteristics of cancer evolve over time and vary from patient to patient, the unique characteristics of the cancer will have to be reassessed at specific intervals. As tumors advance, the possible variations are even greater, and frequently the same cancer will consist of several phenotypes. Thus oncology will need monitoring devices that can track the progression of tumors and guide therapeutic strategy. Imaging technology, nanotechnology, and microarrays based on genomics and proteomics as noted above will combine to make the new model of personalized cancer treatment feasible for at least some cancers.

This new model of cancer treatment using personalized therapies has the potential to transform cancer from an often lethal disease into a chronic condition. In the initial stages of a cancer case, genetic risk assessments and testing for cancer biomarkers will give patients and physicians a much better ability to anticipate and identify cancers in the preclinical stage. Oncology specialists and molecular pathologists

will come to play an earlier role in diagnosis and in guiding the course of treatment. After cancers have been treated, monitoring response and recurrence by molecular imaging techniques will make the ongoing management of cancer cases increasingly important.

Key to this vision for the future of cancer diagnosis and treatment is the discovery and clinical validation of new tumor markers. This is not any easy task. Even though a number of genetic and phenotypic changes can be identified in cancer cells grown in cell culture, validation of these clinically requires long and arduous clinical trials. Some examples will serve to demonstrate this point. In 1975, Gold and Freedman[1] described the overexpression of a glycoprotein called *carcinoembryonic antigen* (CEA), so called because it was thought to be only expressed normally during fetal development and inappropriately re-expressed in colorectal cancer cells. At first it was thought to be specific for colon cancer. We now know that CEA and CEA-like antigens are expressed normally in several tissues, thus its expression is not specific for colon tissue, nor is it specific for cancer. Nevertheless, a high circulating level of CEA is still a useful clinical indicator of cancer progression, e.g., for liver metastases from colon cancer, and for recurrent breast cancer. The history of CEA has been repeated for a number of tumor markers once thought to be sensitive and specific markers of malignant disease, including the much bally-hooed and overinterpreted marker for prostate cancer, prostate-specific antigen (PSA).

The above discussion leads to two key parameters for tumor marker development and clinical interpretations: sensitivity and specificity. *Sensitivity* is a measure of a diagnostic marker's ability to detect true cases of cancer (or other disease) as defined by clinical criteria, and the level of sensitivity is the minimal amount of a marker that can be detected in a clinical sample. In other words, a sensitivity of 98% means that in 98 cases out of 100 a cancer can be detected. *Specificity*, on the other hand, is frequently the downfall of what looks like a highly sensitive cancer diagnostic in preclinical and early clinical studies. Specificity is a measure of a marker's ability to discriminate accurately between patients with true disease from those free of disease by clinical criteria. In other words, specificity defines the probability that a diagnostic test will be negative in a person free of disease (i.e., a low false-positive rate). High sensitivity tests may not have high specificity, and vice-versa. More sensitive tests, even though they may detect more cancers, are not necessarily better than lower sensitivity tests if their specificity is low. This is because false-positive tests often lead to unneeded and expensive additional tests (and sometimes unneeded treatments), not to mention a high degree of patient anxiety.

The calculation of these two parameters can be described as follows:

$$\text{Sensitivity} = \frac{\text{No. of patients with positive tests}}{\text{No. of patients with actual disease}}$$

$$\text{Specificity} = \frac{\text{No. of patients with actual disease (true positives)}}{\text{No. of total positive tests}}$$

Specificity can also be viewed as the reciprocal of the false-positive rate.

Clinical development and validation of biomarkers can be viewed as occurring in five phases: (1) preclinical exploratory; (2) demonstration that the assay detects clinical disease; (3) retrospective longitudinal studies to determine at what stage of progression cancer can be detected;

(4) prospective studies to define sensitivity and specificity; and (5) broad population studies to determine if a test can be used for screening for early cancer and what impact the test has on predicting patient survival (reviewed in Reference 3).

Some biomarkers for cancer diagnosis can also be used as surrogate markers for treatment response and in the development of mechanism-driven anticancer drugs.[4–6]

CATEGORIES OF TUMOR MARKERS

A number of types of tumor markers have been identified and employed as clinical cancer markers.[2] These include nucleic acid–based markers such as mutations, loss of genetic heterozygosity, microsatellite instability, and gene expression microarrays, as well as protein markers and protein pattern recognition profiles, circulating tumor cells, and circulating endothelial cells. Some of these markers can be detected in the circulation or in body fluids, and some require tumor tissue. Examples of some tumor markers are shown in Table 7–1.

Nucleic Acid-Based Markers

A number of genetic modifications have been detected in cancer cells (see Chapter 5), and some of these have been useful in cancer diagnosis and staging. The discovery of free DNA in plasma and urine has provided a way to assess the presence of cancer in patients. Tumors release substantial amounts of genomic DNA as cancer cells, which often have a high cell turnover rate, undergo necrosis or apoptosis. Tumor-derived DNA can be detected in plasma, urine, or stool samples.[2] Alterations in DNA can be assessed by loss of heterozygosity, mutations, microsatellite DNA alterations, and DNA methylation patterns.[7] When genetic or epigenetic alterations are detected in circulating DNA samples, they can be specific to the primary tumor of origin, but plasma DNA is a mixture of neoplastic and non-neoplastic DNA. Elevated circulating DNA levels are also seen in patients with severe infections or autoimmune diseases, for example. In addition, plasma DNA is often degraded to

Table 7–1. Selected Molecular Markers of Cancer

Cancer Type	Clinical Sample	DNA Marker°	RNA	Protein Marker[†]
Head and neck	Saliva, serum	*TP53*, microsatellite alterations, presence of HPV and EBV DNA	Cytokeratins	SCC, CD44, CYFRA, telomerase
Lung	Sputum/BAL, serum	*RAS* and *TP53* mutations, microsatellite alterations	Cytokeratins, MAGE genes, CEA	CEA, CA125, telomerase, CYFRA
Breast	Serum	Microsatellite alterations	Cytokeratins, hMAM, MAGE genes, CEA	CA15-3 (MS-1) CEA, CA125
Colon	Stool, serum	*RAS*, *APC*, and *TP53* mutations	Cytokeratins, CEA	CEA, CA19-9, CA15-3, telomerase
Pancreas	Stool, serum	*RAS* and *TP53* mutations	Cytokeratins, CEA	CA19-9
Bladder	Urine/wash, serum	*TP53* mutations, microsatellite alterations	Cytokeratins, survivin, uroplakin	CEA, CA125, CA19-9, telomerase, survivin, CD44
Prostate	Urine, serum		PSA, MAGE genes, kallikrein	PSA, free PSA, telomerase, kallikrein

°Promoter hypermethylation in DNA is listed separately in Table 7–2.

[†]Most protein markers in use are not specific enough for routine screening and are used predominantly to monitor response or disease progression. Virtually all genetic markers are still in early stages of development. Prostate-specific antigen (PSA) is widely used to screen men for prostate cancer. Cancer antigens include CA15-3, CA125, CA19-9, and carcinoembryonic antigen (CEA). Telomerase is a ribonucleoprotein and usually enzymatic activity is measured; some studies have used direct measurement of the RNA (hTR) component. Most protein markers are measured in serum but other bodily fluids such as urine, saliva, and nipple aspirates have been tested for the presence of aberrant proteins. *APC*, adenomatous polyposis coli; BAL, bronchoalveolar lavage; CYFRA, cytokeratin 19 fragment; EBV, Epstein-Barr virus; hMAM, mammaglobin; HPV, human papillomavirus; microsatellite alterations, loss of heterozygosity (LOH) and/or instability; SCC, squamous- cell carcinoma antigen.

(From Sidransky[2])

(Reprinted by permission from Macmillan Publishers Ltd.)

a variable extent, which can be a problem for assessing allelic imbalance and microsatellite markers.[7] Allelic imbalance (AI), or the loss or gain of chromosomal regions, can be detected in a variety of cancers. AI has been detected, for example, in patients with ovarian cancer and may be a useful marker for diagnosis and cancer progression, in combination with CA-125.[7]

Cancer-Associated Mutations

Detection of cancer-associated mutations in body fluids were first observed when fragments of the *p53* gene were detected in the urine of bladder cancer patients and *ras* gene mutations were found in stool samples from patients with colorectal cancer.[2] Other DNA mutations have subsequently been detected in cancer patients' blood, urine, or other body fluids. For example, the *myc*-N oncogene has been detected in the plasma of patients with neuroblastoma and most likely reflects the amplification of that gene in neuroblastoma tissue from which these extra gene copies are shed into the circulation.[8] K-*ras* gene mutations in codon 12 have been detected

in plasma DNA samples from patients with colorectal cancer.[9]

Loss of Heterozygosity and Microsatellite Instability

Loss of heterozygosity (LOH) occurs when one copy of an autosomal gene is lost. If this involves a tumor suppressor gene, particularly one for which a mutation has occurred in the other allele, tumorgenesis frequently ensues. LOH is a hallmark of many cancers and can be detected by polymerase chain reaction (PCR)-based methods in preneoplastic lesions and solid tumors. Microsatellite DNA markers can also be used as a method to detect LOH. Since the copy number of microsatellite DNAs is often altered in cancer cells, this can be a reflection of microsatellite DNA instability, which is a common feature of malignant cells (see Chapter 5). Microsatellite alterations have also been detected in the urine of patients with bladder cancer and have been more sensitive than using urine cytology to detect cancer.[2] Microsatellite DNA analysis in urine has also been reported to be a valuable tool for early detection of kidney cancer.[10] In addition, altered microsatellite DNA patterns have been found in the saliva of head and neck cancer patients, and in the plasma of patients with small cell lung cancers (reviewed in Reference 2).

DNA Methylation Patterns

As discussed in Chapter 5, DNA methylation patterns are developmentally regulated, often cell type specific, and often change during tumorigenesis. Since DNA methylation is often a negative regulatory controller of gene expression, alteration of the methylation of tumor suppressor gene regulatory sequences can lead to loss of expression of such genes. Examples of genes that are hypermethylated in human cancer are shown in Table 7–2. It is the hypermethylation of cytosines in GpC islands of promoter sequences that is responsible for shutting genes off. This phenomenon has been observed in a wide variety of cancers. PCR-based assays have been developed to accurately detect DNA methylation patterns in cancer tissue and in body fluids. Such DNA methylation pattern analysis can be used also to distinguish malignant from benign tumors. Interestingly, methylation of a gene that codes for an enzyme involved in detoxification mechanisms has been observed: increased methylation of the gene encoding the enzyme glutathione S-transferase placental enzyme 1 (GSTP1) has been found in prostate cancer cells and appears to be more specific than PSA in distinguishing prostate cancer from benign tumors.[11] Whether this enzyme plays a key role in metabolizing an agent that facilitates prostate tissue carcinogenesis isn't clear, but it may be a marker for this.

Mitochondrial DNA Mutations

Defects in oxidative metabolism in cancer cells were detected by Warburg many years ago (see Chapter 4). It is now clear that such defects are due to mitochondrial dysfunction, often related to mutations in mitochondrial DNA. Since mitochondrial DNA is repaired less efficiently than nuclear DNA, mitochondrial genes are thought to be more susceptible to carcinogenic agents. Tumor-associated mutations in mitochondrial DNA have been found in colorectal, lung, head and neck, bladder, and breast cancers.[12] Because mitochondrial DNA is much smaller than the nuclear genome, it is easier to detect mutations in mitochondrial DNA samples. Such alterations can also be found in body fluid samples; for example, Fliss et al.[12] reported that in patients with lung cancer, analysis of bronchoalveolar lavage samples detected 200-fold more mitochondrial DNA mutations than $p53$ gene mutations, the latter being one of the more common nuclear DNA mutations found in lung cancer.

Viral DNA

Viral DNA can be detected in viral-associated human tumors, for example, human papilloma virus is in cervical carcinoma samples. Thus, the detection of HPV DNA is useful to detect

Table 7–2. Examples of Genes That Are Hypermethylated in Cancer

Tumor Type	Primary Tumor°	Body Fluid
Colon	CDKN2A, MGMT, MLH1, DAPK, TIMP-3, APC	Serum (MLH1, CDKN2A)
Breast	CDKN2A, BRCA1, GSTP1, CDH1, TIMP-3, RASSF1A	Nipple aspirate (CCND2, RARβ)
Lung	CDKN2A, MGMT, DAPK, TIMP-3, APC, RASSF1A	Serum and sputum/BAL (CDKN2A, MGMT)
Head and neck	CDKN2A, MGMT, DAPK, RASSFIA	Serum and saliva (CDKN2A, MGMT, DAPK)
Bladder	APC, RASSF1A, CDH1, CDH3, FHIT, RARβ	Urine (RASSFIA, RARβ)
Pancreas	CDKN2A, MGMT, APC	None
Prostate	CDKN2A, GSTP1, ER, CH1, CD44, EDNRB	Serum and urine (GSTP1, CD44)

°Genes found to be methylated in more than 10% of primary tumors or tested in DNA isolated from a body fluid. *APC*, adenomatous polyposis coli; BAL, bronchoalveolar lavage; *BRCA1*, breast and ovarian cancer-1; *CCND2*, gene that encodes for cyclin D2; *CD44*, cluster designation 44; *CDH1*, E-cadherin-1; *CDH3*, E-cadherin-3; *CDKN2A*, cyclin-dependent kinase inhibitor-2A; *DAPK*, death-associated protein kinase; *EDNRB*, endothelin receptor B; *ER*, estrogen receptor; *FHIT*, fragile histidine triad; *GSTP1*, glutathione S-transferase protein 1; *MLH1*, Mut Lhomologue 1; *MGMT*, methylguanine-DNA methyltransferase; *RARβ*, retinoic acid receptor-β; *RASSFIA*, human RAS association domain family 1A; *TIMP-3*, tissue inhibitor of metalloproteinases-3.

(From Sidransky,[2] reprinted with permission from Macmillan Publishers Ltd.)

women at risk for developing this form of cancer.[13] Another example is the detection of Epstein-Barr virus (EBV) DNA in the plasma and serum of patients with nasopharyngeal carcinoma, and the level of EBV DNA could be used to monitor response to therapy and tumor recurrence.[14]

GENE EXPRESSION MICROARRAYS

The technology of assaying gene expression with microarrays goes back at least to 1987, when Augenlicht et al. described a method to assay transcripts of 4000 cloned cDNA sequences in a two-dimensional array on strips of nitrocellulose. This technique was used to analyze radioactively labeled cDNA derived from the mRNA of human colon carcinoma and non-tumor colon cells (reviewed in Reference 15). The investigators used this technique to compare tumor with non-tumor expression profiles. A computer-scanning and imaging system was developed to analyze the results based on the intensity of the radioactive signal derived from the binding of the radioactively labeled cDNA (from the reverse-transcribed mRNA) to the DNA gene sequences arrayed on the nitrocellulose strips. These analyses have now been modified and improved to the point where over 40,000 genes can be microarrayed by inkjet printer techniques that spot samples on glass slides the size of a standard microscope slide.[16] The spot on the slide can be cDNA or a synthetic oligonucleotide representing a portion of a gene sufficient to hybridize with a labeled cDNA sample derived from mRNA of a tissue. The tissue cDNA sequence can be amplified by the PCR to obtain sufficient samples. Fluorescent detection has now replaced radioactive detection as a means to ascertain gene expression levels. For example, mRNA from a tumor sample and reference RNA from a pooled sample representing non-tumor tissue are reverse transcribed and labeled with different fluorescent dyes. Hybridization of the fluorescent labeled cDNAs is then allowed to take place overnight, the unhybridized cDNA is washed away, and the microarrays are scanned at two wavelengths to detect the level of fluorescence over each hybridized spot on the microarray

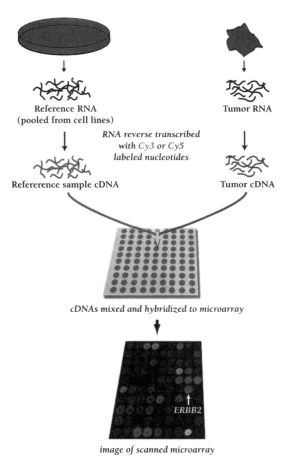

Figure 7–4. Schematic of microarray technique. RNA from a tumor sample and reference RNA (made commercially from pooled cell cultures to represent the majority of known genes) are reverse transcribed and labeled with different fluorescent dyes. The mixture is hybridized overnight to a microarray. The hybridized microarray is then scanned at two wavelengths and the intensities of red and green fluorescence are measured at each spot on the microarray. The red-to-green ratio reveals the abundance of RNA expressed by the tumor sample relative to the reference sample for every one of the 42,000 cDNA clones on the array. This technique provides a comparative measure of the global gene expression of the tumor sample. (From Jeffrey et al.,[16] reprinted with permission from ASPET.)

slide. This procedure allows comparison of gene expression in a tumor with a "normal" gene expression array (Fig. 7–4; see color insert). In this example, a red spot represents overexpression by tumor, green means lower expression in the tumor, and yellow means approximately equivalent expression.

Since the data obtained by such microarrays produces a huge amount of information, there needs to be a way to analyze and categorize such data. One method to do this is *hierarchical clustering*.[16] In this method a mathematical algorithim is used to cluster genes according to whether they are overexpressed or under-expressed in a tumor sample. Genes that have a similar expression pattern are clustered along one axis and experimental samples with similar expression patterns can be clustered along a second axis (performed by the computer algorithim). In this way, subtypes of cancers for a given diagnosis can be clustered together (Fig. 7–5; see color insert).

Hierarchial clustering done without any a prior knowledge of which genes are expected to be up- or down-regulated according to previous data is called *unsupervised clustering*. *Supervised clustering* takes into consideration previous observations and clusters genes by expected expression levels. For example, a gene set known to be overexpressed in a subtype of non-Hodgkin's lymphoma can be used to cluster data sets from uncategorized lymphomas. In this way, data from previous studies of gene expression related to pathological subtypes, clinical prognosis, and drug responsiveness can be used to categorize a new set of tumor samples. This sort of analysis can also be used to identify tumors of unknown primary origin or type.[17] DNA microarrays can also be used to screen for mutations known to be associated with human cancers, for example, the *BRCA1* gene mutations associated with breast cancer. In this case, what is probed for in the sample DNA are the known human polymorphisms or mutations in the gene, and what is arrayed on the gene chips are four areas, or fields, with oligonucleotides representing about a 20-base sequence around the suspected base mutation. In effect, this procedure poses the question of whether there is an adenine, thymine, cytosine, or guanine at that site. The DNA extracted from an individual's cells is reversed transcribed and tagged with a fluorescent probe, overlayed on the chip, and allowed to hybridize as described above. The chip tests each base position in the targeted gene to see which base is there, and which is an index of base-pairing affinity. In other words, where the individual's sample and the arrayed probe base-pair per-fectly (i.e., A-T, G-C), there will be tight binding. Thus, when the individual's sample binds tightly to the field with the mutated base and with less affinity to the site with a "normal" base sequence, the person will have the mutated base for that portion of the gene.

Laser-Capture Microdissection

Another powerful technique to enchance the selectivity of gene expression in selected cell types is laser-capture microdissection. This technique can be coupled with gene expression microarray analysis to provide a cell type–specific gene expression profile. For example, such techniques have been used to differentiate between the gene expression patterns in large and small dorsal root ganglion cells in the central nervous system[18] and to characterize gene expression patterns in lung adenocarcinomas and patterns that could delineate lung tissue patterns in smokers from those of nonsmokers.[19]

Of course, an ultimate goal of the Human Genome Project and all its attendant spin-off techniques and sophisticated analyses is to understand the *function* of all the genes contained in the human genome and their role in human development, to determine susceptibility to disease and disease progression, and to aid in the design of "individualized medicine" (see Pharmacogenomics, below). This new field of endeavor is often called "functional genomics."

As with many new, sophisticated techniques, gene expression profiling is not without its pitfalls. These include a lack of precise definition of what "normal" gene expression is, the cellular heterogeneity of cells present in tissue samples used to extract mRNA, chip-to-chip variation in preparation of arrays, lab-to-lab variation in mRNA extraction and labeling, and the vast amounts of data generated.[20,21] Nevertheless, although the technical issues are complex and the data analysis daunting, the promise for advancing biomedical research and the future practice of medicine is enormous.

Comparative Genome Hybridization

Two other techniques should be mentioned here: comparative genome hybridization and tissue arrays. In the technique of comparative

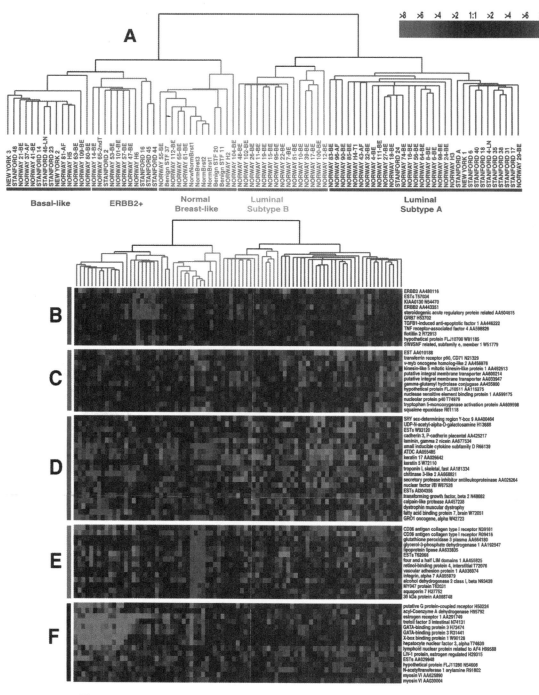

Figure 7–5. Gene expression patterns of 85 breast samples. Seventy-eight carcinomas, three benign tumors, and four normal breast tissues cluster into five subtypes: luminal A (estrogen-receptor [ER] positive, favorable survival); luminal B (ER positive, poor survival); normal breast–like; ERBB2 amplicon; basal epithelial–like cluster. Tumor clusters are represented by branched dendrograms that indicate degree of similarity between samples. Genes are clustered by rows according to similarity of expression. (From Sorlie et al., *Proc Natl Acad Sci* 98:10869–10874. Copyright 2001, National Academy of Sciences USA, in Jeffrey et al.,[16] with permission.)

genome hybridization, arrays representing entire sections of chromosomes can be arrayed on glass surfaces. These can then be probed with fluorescently tagged cDNA or protein complexes to look for protein binding sites on genes, and to determine genetic abnormalities such as gene deletions and amplifications in cancer cells.[22] This technique can be used as a more sensitive alternative to the older technique of metaphase chromosome spreads.

Tissue Arrays

Tissue arrays are analogous to DNA microarrays in the sense that up to 1000 small tissue samples, such as those obtained with a 6-mm punch biopsy, can be arrayed on a slide and probed with fluorescent cDNAs, proteins, or antibodies to look for gene and protein expression patterns at the cellular level. For example, mRNA levels can be probed by fluorescent in situ hybridization (FISH); gene abnormalities by comparative genomic hybridization (CGH); and protein expression by immunohistochemistry (IHC). The fact that this technique can be used with fixed, stored tissues allows the analysis of tissues in tissue banks. Tissue arrays have been used to analyze for Her-2/neu expression in human breast cancer samples and for progression and hormone responsiveness of human prostate cancer.[23]

Gene Expression Microarrays in Individual Cancer Types

The powerful technique of DNA microarrays has enabled investigators to look inside cancer cells and ask which genes are turned on or off in cancer cells and how cancer cells differ from the normal cells in their tissue of origin. This type of information has enabled subtyping of cancers of a given cell type, staging of cancers, estimations of prognosis, propensity of cells to metastasize, and response to chemotherapy. In addition, DNA microarrays are providing information on potential new targets for therapeutic attack and biomarkers for diagnosis and screening. Through the techniques of PCR-based cDNA subtraction and cDNA microarrays in the analysis of a human breast cancer, lung squamous cell cancer, lung adenocarcinoma, and renal cell cancer and a panel of 16 normal tissues, Amatschek et al.[24]

identified 130 genes (based on expressed sequence tags) that were up-regulated in many of the cancers. Although many genes were up-regulated in multiple tumor types, e.g., cell cycle regulatory genes such as cyclin D1, genes involved in bone matrix mineralization, and a gene (EGLN3) involved in regulation of hypoxia-inducible factor (HIF), hierarchical clustering clearly distinguished the different tumor types. Moreover, the expression of 42 genes that were up-regulated in breast cancers correlated with overall survival. Gene expression profiles either have been done or are being done for many if not most human cancers. Some examples of these data are described below.

A valuable extension of gene microarray technology is the finding that quantitative gene expression profiling can be carried out with formalin-fixed, paraffin-embedded tissues, enabling investigators to go back and review old tissue slides that were collected for histopathologic evaluation. Such review of "legacy data" allows evaluation of retrospective data accumulated over time and for which clinical data are often still available. (A caveat to this procedure is that use of such data must be compliant with federal regulations specified by the Health Insurance Portability and Accountability Act [HIPAA] and other state and federal regulations regarding appropriate consent and privacy issues.)

The method for use of formalin-fixed samples is called cDNA-*mediated annealing, selection, extension and ligation* (DASL). Using the DASL assay system, highly reproducible and cancer-related gene expression profiles can be obtained with 50 nanograms of RNA isolated from fixed and embedded tissues stored for up to 10 years.[25] The accuracy of this method was shown by comparing the gene expression data from fresh frozen tissue and formalin-fixed tissue.

Lymphoma

One of the first reports on the use of gene expression profiling to distinguish subtypes of malignancies within a given tumor type was for diffuse large B-cell lymphoma (DLBCL).[26] These investigators identified two distinct forms of DLBCL that had gene expression profiles relating to different stages of B lymphocyte differentiation. Patients with one of the profiles, indicative of germinal center B-like cells, had a significantly better overall survival than a second group of patients who had a gene expression pattern indicative of an activated B-like cell type. Subsequent studies[27] identified a third subtype of DLBCL, called type 3 diffuse large–B cell lymphoma. Of the two common oncogenic events found in unclassified DLBCL, bcl-2 translocation and c-rel amplification were found only in germinal center B cell–like groups. A profile of expression of 17 genes could be used to predict overall survival after chemotherapy.[27] This gene-based analysis was found to be an independent predictor, compared to the clinical international prognostic index, of 5-year survival. In the future, as these analyses became more detailed and are correlated with expression of specific target proteins, it should be possible to dictate which chemotherapeutic agents should be used in each subset of patients.

Leukemia

One of the distinctive features of leukemia is the high percentage of cases that have chromosomal translocations. Some are fairly typical, for example, the t(9;22) translocation in chronic myeloid leukemia (CML) and the t(8;21) translocation in acute myeloid leukemia (AML). In many leukemias, the fused genes resulting from a translocation produce an oncogenic protein, such as bcr/abl in CML. Although chromosomal translocations have been used to identify leukemia patients with distinct clinical outcomes and drug responses, these translocations do not account for all the clinical behavior of leukemias. Additional gene expression abnormalities not related to a translocation can occur, and these may accumulate over time and after chemotherapy. For example, flt 3 receptor tyrosine kinase gene mutations occur without a translocation event and are a bad prognostic sign for patients with AML. In adult AML, gene expression profiling can be used as a predictor that can identify different chromosomal translocations and a trisomic 8 condition (reviewed in Reference 28). Thus, DNA microarrays can be used to diagnose chromosomal abnormalities. In pediatric patients, gene expression profiles have been used to identify leukemia subtypes ALL,

MLL, and AML.[29] MLL was identified as a distinct entity based on its gene expression profile that is related to altered patterns of HOX gene expression. HOX 11 gene expression abnormalities are seen in T-cell ALLs, a rarer form than the usual B-cell ALLs.

In CLL, there are two overall subtypes based on mutations in immunoglobulin genes and clinical outcomes. Patients with somatic mutations in immunoglobulin genes have slowly progressing disease, whereas those without immunoglobulin gene mutations have more aggressive disease. Expression profiles that examined about 160 genes were able to distinguish between these two subtypes (reviewed in Reference 28).

Breast Cancer

Microarray technology, with its ability to interrogate up to 40,000 genes in a single sample, has been used for the molecular classification of human breast cancers. These data can be used to correlate estrogen receptor status with gene expression, clinical outcomes, likelihood of response to drugs, and metastatic potential of breast cancer cells.[30] Two of the earlier reports in the use of microarray techniques for breast cancer classification are those of Perou et al.[31] and Sgroi et al.[32] The first authors used algorithms to identify and compare clusters of genes expressed in cultured human breast cancer cell lines and resected breast cancer samples from patients. Some clusters of genes were expressed in both the tumor cell lines and resected tumor specimens. These included a "proliferation cluster" of genes, interferon-regulated genes, and cell cycle genes. As might be expected, the primary tumor samples from patients also contained lymphocytes and stromal cells. These could be identified by gene expression patterns related to these cell types, e.g., immunoglobin genes in the case of B lymphocytes and collagen type Iα genes in the case of stromal cells. Gene expression patterns in two tumor samples from the same patient were more similar to each other than to any other patients' tumors,[33] a finding suggesting that each tumor type has distinct patterns of lymphocyte infiltration and stromal cell content that may signify distinct characteristics of the total milieu of individual tumors, not just of the cancer cells themselves.

Sgroi et al.[32] took a different tack. They used laser capture microdissection to generate DNA microarrays for monitoring gene expression levels in purified normal, invasive, and metastatic breast cell populations from a single patient. These tumor cell–selected gene expression patterns were confirmed by quantitative PCR and immunohistochemical detection of the expressed proteins. Some of the genes whose expression was fourfold or more overexpressed in invasive cancer cells compared to normal breast cells were apolipoprotein D, tissue factor precursor, heat shock protein 1, annexin 1, the SWI/SNF transcriptional activator complex, and the adrenergic β receptor kinase 1.

A gene expression profile that is a predictor of breast cancer patient survival has also been reported.[34] In this study, distinct gene expression signatures were used to classify patients on the basis of lymph node involvement, metastasis, and overall survival. One of the surprising results from this study of 295 patients (180 with a poor prognosis signature and 115 with a good one) was that lymph node–positive patients, for whom a poorer prognosis is usually assumed, could be subclassified into long-term survivors and short-term survivors, indicating that lymph node status is separable from distant metastasis as an indicator of survival. Moreover, these data suggest that "the ability to metastasize to distant sites is an early and inherent genetic property of breast cancer"[34] rather than the widely held concept that metastatic potential is acquired over a sequence of genetic events that occur over multiple stages of tumor progression (see Chapter 4).

As noted above, estrogen receptor (ER) status in breast cancer is associated with a distinct gene expression signature. Employing an expression profile of 100 genes, ER^+ or ER^- tumors could be identified, indicating that they are two very distinct types of tumors.[35] Genes more highly expressed in ER^- tumors include P-cadherin, the transcription factor C/EBP β, lipocalin 2, and ladinin. Genes more highly expressed in ER^+ tumors are GATA3, TFF3, cyclin D1, and carbonic anhydrase XII. Some metastases from ER^+ primary tumors had an ER^- gene profile, a finding suggesting that there is a small percentage of aggressive ER^- cells in the primary tumor cell population or that the ER^- pheno-

type evolves over progression in a metastatic milieu.

Wang et al.[36] carried out a study similar to that of Van de Vijver et al.[34] and also found that gene expression profiles could be used to subclassify lymph node–negative breast cancer patients into high-risk and low-risk groups. The gene sets used in the two analyses, however, were quite different. There was only a three-gene overlap between the two signatures: cyclin E2, origin recognition complex, and TNF superfamily protein. This result suggests that different sets of genes can lead to breast cancer progression—the "all roads lead to Rome" idea.

Wang et al. also identified a 76-gene expression signature consisting of 60 genes for ER^+ patients and 16 genes for ER^- patients. This signature showed 93% sensitivity and 48% specificity in a subsequent analysis of samples from 171 lymph node–negative patients. The 76-gene signature also was a prognostic factor for development of metastasis in subgroups of 84 premenopausal and 87 postmenopausal patients, and in patients with small tumors (10–20 millimeters in diameter).

One of the concerns about these two studies is the virtually complete lack of overlap in the genes studied in the two data sets and the relatively few patients included in the two studies.[37] Nevertheless, if validated in larger studies, the data point to a very powerful approach to a critical clinical issue. About 60%–70% of women with lymph node–negative breast cancer at diagnosis can have long-term survival or cure when treated by local or regional excision (usually followed by irradiation), whereas 85%–90% lymph node–negative patients are routinely recommended for adjuvant systemic chemotherapy and are subject to all its attendant sequelae. Thus, a significant number of women are being treated unnecessarily. For example, based on the data of Wang et al.,[36] their 76-gene signature would have led to a recommendation for adjuvant chemotherapy for only 52% of patients compared to 90% by National Institutes of Health (NIH) guidelines.

Gene expression profiles have also been determined that correlate with hereditary breast cancers: different genes are expressed in breast cancers with BRCA1 mutations compared to BRCA2 mutations.[38] Gene expression profiles that predict aggressive, metastatic behavior of breast cancer cells have also been identified.[39] Genes coding for motility machinery proteins such as those that regulate β-actin polymerization and chemotaxis are up-regulated in invasive cells, and genes that inhibit localization of β-actin, such as ZBP1, are down-regulated.[40] In some cases, sufficient cells[41] or DNA[42] can be isolated from peripheral blood to detect the presence of breast cancer. This information may evolve into a method for the early detection of disseminated breast cancer cells.

Ovarian Cancer

Using oligonucleotide microarrays complementary to about 6,000 human genes, Welsh et al.[43] found an expression profile that correlated with ovarian cancer and distinguished cancer from normal tissue. A number of the observed up-regulated genes are also overexpressed in other epithelial cancers. These commonly overexpressed genes include CD9; CD24; cytokeratins 7, 8, 18, and 19; and Muc 1. One gene that is overexpressed in a significant fraction of ovarian cancers but only expressed at low levels in other cancers is HE4, a gene encoding a secreted extracellular protease inhibitor, which may turn out to be a diagnostic marker for ovarian cancer. A number of genes are coordinately overexpressed in ovarian cancer.[44] For example, STAT1 expression was correlated with the expression of Ep-CAM/GA733-2, Kop, Timp-3, FR1, SLP1, ApoE, and ceruloplasmin. These data suggest that these genes are targets of a common signaling pathway. Expression of other ovarian cancer–associated genes (clusterin, IGFBP-α, MGP, and S100A2) did not correlate with any other expression pattern, which suggests that the expression of these genes is related to different molecular pathways. These in turn may be related to different tumor subtypes and may require differently targeted therapies.

Prostate Cancer

Prostate cancer is the third-most common cancer in men worldwide and amounts for 6% of cancer deaths in men worldwide. In Western countries, the incidence rate is higher than in other parts of the world. In the United States,

it is the most frequently diagnosed cancer in men and the second leading cause of death in males, accounting for 10% of all cancer deaths in men.[45] The incidence and mortality rates are higher in African Americans than in Caucasians. Even though prostate cancer is a high-incidence malignancy, it is frequently, indeed most often, an indolent, slow-growing tumor. Autopsy data indicate that about 60% of men over age 60 have evidence of malignant cells in their prostate but did not die of the disease, and this rate continues to climb at about 10% per decade. Even among those men diagnosed with prostate cancer during their lifetime, only 20% will eventually die of the disease.[46]

There was a huge spike in the incidence of prostate cancer from 1989 to 1995 in the United States.[45] This was clearly due to the great increase in the use of prostate-specific antigen (PSA) as a screening tool for prostate cancers, and yet the mortality rate remained essentially unchanged over that time frame. Clearly, many of the men diagnosed with prostate cancer did not die of the disease, nor is there any evidence for some cataclysmic exposure or lifestyle change that could account for this spike; it is a PSA-related phenomenom. PSA is known to be elevated in individuals with benign prostatic hypertrophy (BPH) and prostatitis and thus is not cancer specific.

A number of other diagnostic markers of prostate cancer have been identified over the years. These include prostate acid phosphatase, prostate-specific membrane antigen (PSMA), prostate inhibin peptide, PCA-1, PR92, prostate-associated glycoprotein complex, protein-mucin antigen, 12-lipoxygenase, and p53 (reviewed in Reference 47).

Gene expression microarrays and other genetic techniques such as sequencing of expressed sequence tags (ESTs) and serial analysis of gene expression (SAGE) have provided powerful additional methods to diagnose and stage prostate cancer.[48] A number of differentially expressed genes have been found in prostate cancer tissue compared to normal prostate tissue, through cDNA library substraction and microarray. Three such genes, P503S, P504S, and P510S, were found to be overexpressed in cancerous but not normal prostate tissue.[47] Luo et al.[46] performed gene expression profiling of BPH and prostate cancer using DNA microarrays consisting of

6500 human genes. They identified 210 genes with statistically significant differences in expression between BPH and prostate cancer. One of these genes, *hepsin*, which codes for a transmembrane serine protease, had not been previously reported as one that was overexpressed in prostate cancer.

Gene expression analyses have also been used to predict prostate cancer recurrence.[49] The genes assessed included a number of metabolically related genes such as those involved in polyamine metabolism (e.g., ornithine decarboxylase [ODC], ODC antizyme, and spermidine/spermine *N*-acetyltransferase), adenosylmethionine decarboxylase, histone H3, growth arrest–specific gene 1, and glyceraldehyde 3-phosphate dehydrogenase. Analysis of expression of these genes, together with the Gleason score, lymph node involvement, and prostate volume and PSA at the time of diagnosis, resulted in correct prediction of recurrence in about 96% of patients.

It is also important to be able to subcategorize patients' tumors into those that are likely to be aggressive and those that are likely to be indolent. This distinction would prevent the large number of men who are unlikely to have progressive disease from having needless and expensive additional workups and surgery, which is often associated with high morbidity. A start has been made on this issue, with Lapointe et al.[50] using gene expression profiling to identify clinically relevant subtypes of prostate cancer. For example, expression of mucin 1 (MUC1) at both the mRNA and protein level marked a subgroup of tumors as aggressive, based on correlation with clinical–pathological features and risk of recurrence, whereas expression of the *AZGP1* gene (encoding zinc-α_2-glycoprotein) characterized a subgroup of patients with decreased risk of recurrence.

Colorectal Cancer

Progression of colorectal cancer from adenomatous polyps to invasive malignancy has been one of most well-studied models of human cancer and has led to the so-called Vogelgram model (based on the work of Vogelstein and colleagues) for cancer. This model proposes that cancer starts with a small number of genetic alterations at the state of benign adenomas and, via various stages of genetic "drift" during which additional oncogenes are turned on and various tumor suppressor genes are turned off, a full-blown invasive cancer arises. While the applicability of this model generally to human cancer is in some dispute (see Chapter 2), it has been a useful paradigm for thinking about tumor initiation, promotion, and progression. For many human solid cancers (e.g., breast, prostate, colon, lung), the time from initiation to progression takes many years. During this time, additional genetic changes (or clonal expansion of cells present in small numbers that are present in early tumors, i.e., tumor stem cells) must take place if the tumor is to become invasive. It would be very beneficial to know what genetic alterations are key to this process and how to detect them. A number of studies have been aimed at doing this.

Notterman et al.[51] used oligonucleotide arrays containing sequences complementary to 3200 full-length human cDNAs and 3400 ESTs. They examined colon adenomas, adenocarcinomas, and paired normal colon tissue obtained from each patient. Nineteen transcripts demonstrated 4- to 10-fold higher mRNA expression in carcinomas compared to normal tissue, and 47 transcripts had at least a 4-fold lower expression in tumors than in normal tissues. Some of these differences were also observed between premalignant adenomas and normal tissues (supporting the idea that some of these genetic alterations presage malignant cancer). Some of these differentially expressed genes have also been observed in other human cancers. The genes overexpressed in colon adenocarcinomas compared with paired normal tissue included melanoma growth stimulatory activity (*MGSA*), human metalloproteinase (*HMP*), and some cell cycle checkpoint genes.

In another study comparing adenocarcinomas with normal tissue, the most frequently altered genes belonged to various functional categories.[52] These were genes related to metabolism (22% of those with altered expression) and transcription and translation (11%), and nuclear genes coding for mitochondrial proteins.

A number of overexpressed genes in colorectal tumors code for secreted or cell surface proteins, which suggests diagnostic and therapeutic possibilities.[53] Gene expression profiling has also been used to determine responsiveness to 5-fluorouracil chemotherapy.[54]

Approximately 160,000 individuals are diagnosed with colorectal cancer in the United States each year and about 60,000 patients die of the disease.[55] Surgical resection of patients with early (Dukes' A) disease is highly effective. Surgery is also an effective treatment for patients with Dukes' B disease; however, 25%–30% of these patients develop recurrence and die from their disease. Thus, Dukes' B patients can often benefit from adjuvant chemotherapy with drugs such as fluoropyrimidines, irinotecan, and oxaliplatin, but the patients who would benefit from such therapy have been difficult to predict. Wang et al.[56] have identified a 23-gene signature that predicts recurrence in Dukes' B patients, which was validated in 36 independent patients after the training set of genes was established. Thirteen of 18 patients who relapsed and 15 of 18 patients who remained disease-free after surgery were correctly predicted. Thus the gene expression signature predicted a 13-fold increase in risk of relapse and identified a subset of patients who would be candidates for adjuvant chemotherapy.

Lung Cancer

There are four main histological categories of lung cancer, based on microscopic tumor cell morphology. These are squamous cell carcinomas (30%), small cell lung carcinomas (SCLC; 18%), adenocarcinomas (30%), and large cell carcinomas (10%).[57] Because these tumors are treated differently and have a different response to drugs and a different prognosis, it is important to get the diagnosis right. There is a relative consensus among pathologists for the diagnostic characteristics of SCLC. These tumors often produce neuroendocrine factors and are characterized by good initial response to chemotherapy. A typical course for these patients is (1) initial response to drugs followed by several months of complete remission; (2) recurrence associated with the tumor's development of drug resistance; and (3) death caused by systemic dissemination.

Non–small cell lung carcinomas (NSCLC) include adenocarcinomas, squamous cell carcinomas, and large cell (undifferentiated) carci-

nomas. There is much less consensus among pathologists on the subcategorization of adenocarcinomas. In one study, pathologists agreed on the subclassification only 41% of the time.[58] This lack of consensus is important to note, because for the bronchiolalveolar carcinoma subtype, the prognosis is more favorable than the other subtypes of adenocarcinomas. Another reason to have good markers for lung tumors is that many cancers metastasize to the lung, and it is important to determine the tumor site of primary origin for a lung lesion because the treatment and prognosis is often different from that of primary lung adenocarcinomas. The development of microarray methods has made strides in the ability to make these diagnostic judgments.

A microarray study correlating gene expression profiles with clinical outcome in a cohort of patients with lung adenocarcinomas has identified specific genes that predict survival among patients with stage I disease.[59] Even though most patients with NSCLC present with advanced disease and therapeutic outcomes and long-term survival are generally dismal, there is a significant subset of patients (about 25%–30%) with stage I NSCLC who will benefit from surgery alone, without other interventions. Among these patients, 35%–50% will relapse within 5 years, but the rest have a better prognosis. Again, it is important to be able to tell the difference between these levels of risk. Beer et al.[59] have identified, from six groups of stage I adenocarcinoma patients, high-risk and low-risk patients who differ in survival rates. The high-risk groups could benefit from additional therapy such as adjuvant chemotherapy. The differentiating genes fall into "all the usual suspects" categories: apoptosis, cell adhesion and structure, cell cycle and growth regulators, signal transduction, growth factor receptors, proteases, kinases and phosphodiesterases, and transcription- and translation-related genes.[59]

Renal Cancer

Renal cell carcinoma (RCC) is the most common kidney cancer and represents 2% of all cancer deaths globally.[60] By clinical and histopathologic characteristics, it is a heterogenous disease, but overall survival is dismal for all patients with

RCC. At diagnosis, about 30% of all RCC patients have metastatic disease and have an average life expectancy of 12 months. Even among those patients without clinical evidence of metastatic disease at diagnosis, most of whom are treated by nephrectomy, the relapse rate is 30%. Some patients, however, have slowly progressive disease and may live for years. Gene expression profiling has now been used to predict prognosis and survival.[60,61] A significant distinction in gene expression profiles between patients with a 100% 5-year survival rate and those with an average survival of 25.4 months (and 0% 5-year survival) was observed.[60] Similar to the studies in other cancers, a relatively small number of genes could be used to make predictions that correlated with clinical outcome; in this case, it was 40 genes, a number of which have already been found to correlate with invasion and metastasis. Vasseli et al.[61] found two patterns of gene expression: one that correlated with longer survival and one with poor survival. The vascular cell adhesion molecule–1 (*VCAM*-1) gene was the gene most predictive for survival.

Hepatic Cancer

Hepatocellular carcinoma (HCC) is an aggressive cancer with high prevalence in Asia and Africa, although incidence rates are rising in the United States and United Kingdom. HCC has been associated with hepatitis B virus infection and with exposure to parasites and certain fungal toxins (e.g., aflatoxin; see Chapter 3). Most HCC patients are diagnosed with advanced disease and have little chance of survival. Surgical resection is still the primary therapeutic option, but recurrence rates are high and intrahepatic metastases are frequent. Ye et al.[62] have analyzed gene expression profiles in HCC samples from patients with or without intrahepatic metastases. A gene signature was found that distinguished patients with metastatic disease from those without it and that also correlated with survival. Interestingly, the gene signature signifying metastases was found both in the primary HCC tumor as well as its metastases, indicating that metastatic genes were already turned on in primary tumors that would ultimately metastasize. This finding supports the concept noted above, that the pathway for metastasis may be turned on

almost from the get-go in more aggressive tumors and does not require a stepwise progression with pattern X → pattern Y → pattern Z over time. One gene, osteopontin, was identified as the "lead gene" for the metastatic profile. Intriguingly, an osteopontin-specific antibody blocked pulmonary metastases of HCC cells in nude mice, thus osteopontin may act as both a diagnostic marker and a therapeutic target.[62]

Other Cancers and Cancer-Related Phenotypes

As the reader has by now surmised, genetically based tests, using PCR, EST analyses, and gene expression arrays, are becoming a standard research approach to classifying human cancers and predicting prognosis, tumor metastatic potential, survival, and response to therapy. Cervical,[63] esophageal,[64] and central nervous systems tumors[65] are among those that have been added to the list. One can predict the day in the not too distant future when such genetic analyses will be routine for the diagnostic and therapeutic workup for cancer patients.

One question that might be asked is whether there is a common "metastatic program" of genes that is turned on in all invasive, metastatic human cancers. The answer to this question is not clear; however, the metastatic process involves the expression of certain common functions— i.e., the ability to migrate (motility), to invade tissue matrices (proteases), to survive in the bloodstream and lymphatic system, and to invade and survive in a new tissue site. Attempts have been made to use gene expression patterns to define the metastatic phenotype in human tumors.[66] A number of the genes whose over- and underexpression correlates with metastases fit the above enumerated functions. However, a number of genes identified as being associated with the metastatic phenotype have no known function,[66] which means that there is a lot of work yet to be done.

St. Croix et al.[67] have used EST, SAGE analyses, and in situ hybridization to gain an understanding of how tumor endothelium differs from normal endothelium. Endothelial cells are those that line the blood vessels (arteries, veins, capillaries, etc.) of tissues. If it could be shown that tumor vasculature differs from normal vasculature in its cell surface markers and cellular phenotypes, this difference could be used to target specifically the tumor vasculature with anti-angiogenic therapies. These authors found 170 gene transcripts that were predominantly expressed in endothelium, including 46 of which are specifically elevated in tumor-associated endothelium. Several of these genes code for extracellular matrix proteins, but the function of many these genes is unknown. Most of the tumor endothelial cell markers were also expressed in multiple other tumor types. Of some concern is the observation that the tumor-associated expression profiles were similar to that observed during normal wound healing and corpus luteum formation. These data remind us that tumor tissue often recapitulates a less well-differentiated phenotype seen normally in developing or remodeling tissues.

Chi et al.[68] employed DNA microarrays to determine the gene expression profile of endothelial cells from different tissues and different types of blood vessels. They found both tissue- and vessel-type differences in expression patterns. Expression of some endothelial cell genes was related to left–right symmetry in developing organs, and some genes were preferentially expressed in venous over arterial endothelial cells, a finding suggesting "coordination between vascular differentiation and body plan development."[68]

Several microarray studies have been designed to correlate tumor drug sensitivity or development of drug resistance with gene expression profiles. These include expression profiles to determine (1) doxorubicin resistance in cultured breast cancer cells;[69] (2) sensitivity of human esophageal cancer tissue to adjuvant chemotherapy[70]; (3) response of follicular lymphomas to the monoclonal antibody rituximab;[71] (4) therapeutic response to docetaxel in patients with breast cancer;[72] and (5) multidrug resistance in human tumor cell lines by expression profiling of ATP-binding cassette transporter genes.[73]

PROTEOMICS

The field of proteomics, in the sense of global cellular detection and functional interactions of proteins, is a relatively new field of research. It has been made possible by recent developments of sensitive molecular biological and analytical techniques. One only has to review the literature from 2000 onward to realize the newness of the field and its burgeoning scope.

Protein identification is, of course, not a new field and has been a mainstay of biochemistry since early in the twentieth century. Introduction of the Edman degradation method in 1950 and the subsequent development of the protein sequenator in 1967 enabled protein biochemists to systematically sequence proteins that could be isolated in pure form (reviewed in Reference 74). Introduction of the two-dimensional electrophoresis technique over 30 years ago led to the ability to resolve total protein extracts from cells into about 5000 individual protein spots. Two-dimensional electrophoresis has been a mainstay of the field ever since. Even this method, revolutionary as it was at the time, only detects high-abundance proteins and their modified counterparts in cells. It was not until the development of sensitive mass spectrometric techniques that the field of proteomics was really born. This has now advanced to the state where the true identification of sensitive and specific biomarkers for cancer in individual patients is becoming possible.

The term *proteome* was coined in 1994 to define the concept of the complete set of proteins expressed, and modified following expression, by the entire genome of an organism (reviewed in Reference 75). This concept is sometimes taken to mean the proteins expressed during the entire lifetime of cell types in an organism or the complement of proteins and their modification produced by a cell at a point in time. As we shall discover below, the latter definition is more meaningful for defining cellular function in real life and in various disease states. In addition, alternative mRNA splicing, rapid post-translational modification such as phosphorylation in response to changes in cellular environment, and protein turnover are constantly changing the protein profile. While coupling of gene expression patterns with protein expression profiles is a useful thing to do, they do not always have a one-to-one relationship.[74] It is, after all, the proteins of the cell that are the workhorses of cellular function, and in that sense they are a more accurate reflection of the metabolic state and health of a cell. (Besides that, protein chemists are more modest than gene expression gurus; the former speak of "consensus sequences" and the latter speak of "canonical sequences.") Moreover, the functional state of a protein depends on its folding into its three-dimensional, native form.

Thus, the term *proteomics* has come to mean different things to different people. In its broadest sense, it includes the following: (1) the complete set of proteins expressed by the entire genome; this has several subsets such as the nuclear proteome, the mitochondrial proteome, the plasma proteome, and the tissue (e.g., liver) proteome; (2) the protein–protein interactions in a cell (the *interactome*); (3) the functional proteins and their modified forms in a cell at a given time; and (4) the three-dimensional native structures of proteins that determine their function. The determination of any one of these parameters for the entire functional array of cellular proteins is more complicated than the sequencing of the human genome. It's been said that "genes were easy" compared to defining the human proteome.[75] An organization called the Human Proteome Organization (HUPO) has been formed to tackle some of these issues. It is analogous to the Human Genome Organization (HUGO). Not only is it more challenging to carry out the above determinations for proteins, there are several hundred thousand proteins, if one takes into account alternate splicing, post-translational modifications, and alterations in folded state, but only about 30,000 human genes, for which the sequence is a linear array of only four bases: A, T, G, and C.

Proteomics Methods

Two-Dimensional Electrophoresis

Originally described in 1975 by O'Farrell,[76] two-dimensional gel electrophoresis has been a mainstay for protein separation and the "founding" technology for proteomics. Proteins are separated in a first dimension based on migration to their isoelectric points in a pH gradient and then in a second dimension based on migration through the gel as determined by how large they are (molecular mass). If proteins are labeled with a radioisotope such as ^{14}C or ^{3}H as they are biosynthesized in a cell, they can be detected by radioautography. Non-labeled proteins can be detected by silver staining or other staining techniques. Although this method provided a quantum leap for protein separations at the time it was developed, only relatively high-abundance proteins can be detected, and many signal transduction signals and gene regulatory proteins escape detection by this method.

Isotope-Coded Affinity Tags (ICAT)

This method, originally described by Gygi et al.,[77] is used to label proteins with two different isotopic heavy and light forms, called *isotope-coded tags*, that couple to free thiols of cysteines in proteins. For example, the cysteines of one cell type may be labeled with an isotopically stable light isotope (e.g., ^{13}C) and the cysteines of another cell type labeled with a heavy (e.g., ^{2}H) form. The extracts of the two cell types can be mixed together, digested with trypsin, and separated by mass spectrometry. The mass spectrometer separates the heavy and light peptide forms and can determine their abundance. Thus, the relative amounts of peptides produced in one cell type (e.g., normal cells) can be compared with another cell type (e.g., cancer cells). A disadvantage of the ICAT technique is that it only detects peptides from those proteins that contain cysteines. Other isotope tagging chemistries can be used to tag proteins in cells to identify individual protein subtypes. These include tagging phosphate ester groups, N-linked carbohydrates, and active sites for serine and cysteine hydrolases.[78]

Mass Spectrometry–Based Proteomics

Mass spectrometry (MS) is the next frontier for proteomic methods. Several modifications of increasing sensitivity have been developed in recent years. An illustration of the general approach is shown in Figure 7–6 (see color insert). One of the first MS techniques developed for this is matrix-assisted laser desorption and ionization (MALDI). This involves precipitation of protein samples with an excess of a matrix material such as α-cyrano-4-hydroxycinnamic acid or dihydroxy benzoic acid, followed by desorption and ionization from the matrix by laser pulses, generation of peptide fragments, usually by trypsin diges-

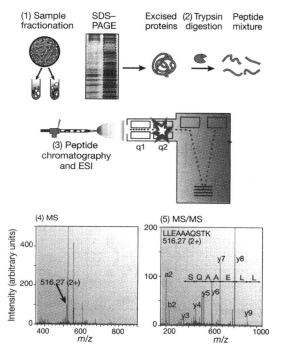

Figure 7–6. Generic mass spectrometry (MS)–based proteomics experiment. The typical proteomics experiment consists of five stages. In stage 1, the proteins to be analyzed are isolated from cell lysate or tissues by biochemical fractionation or affinity selection. This often includes a final step of one-dimensional gel electrophoresis, and defines the "sub-proteome" to be analyzed. MS of whole proteins is less sensitive than peptide MS and the mass of the intact protein by itself is insufficient for identification. Therefore, proteins are degraded enzymatically to peptides in stage 2, usually by trypsin, leading to peptides with C-terminally protonated amino acids, providing an advantage in subsequent peptide sequencing. In stage 3, the peptides are separated by one or more steps of high-pressure liquid chromatography in very fine capillaries and eluted into an electrospray ion (ESI) source where they are nebulized into small, highly charged droplets. After evaporation, multiply protonated peptides enter the mass spectrometer and, in stage 4, a mass spectrum of the peptides eluting at this time point is taken (MS_1 spectrum, or "normal mass spectrum"). The computer generates a prioritized list of these peptides for fragmentation and a series of tandem mass spectrometric or "MS/MS" experiments ensues (stage 5). These consist

tion, and separation by time-of-flight or other analyzer (see below). The MS method measures the mass-to-charge ratio (m/Z) of the peptides. Proteins are identified by comparison of their peptide patterns with protein databases. A major limitation of this is the requirement that the protein sequence of the unknown sample be present in a protein database. However, the presence of over 1.2 million human expressed sequence tags in genome databases allows for an additional method to deduce which peptide sequences relate to which gene products. For example, a 10–amino acid peptide corresponds to 30 base pairs of a gene sequence, which is often sufficient to identify a gene product. In addition, advances in liquid chromatography–mass spectrometry (see below) allow identification of peptides in the low-femtomolar range from high-perfomance liquid chromatography (HPLC) separation of complex mixtures. Peptide ions of interest can be retained on an ion trap on the basis of their m/Z and further fragmented to obtain amino acid sequence information by tandem MS/MS anaysis.[74] Thus, peptides from the first MS separation can be collected and fragmented further before being subjected to a second MS analysis. In this way, for example, one can ask whether a peptide of 1200 Daltons gives a fragmentation pattern that further identifies its parent protein in the database.

Another approach uses electrospray ionization (ESI), which ionizes the protein analytes out of a solution, and thus can be coupled to liquid-based chromatographic and electrophoretic separation techniques. MALDI-MS is usually used to analyze less complex peptide mixtures, whereas ESI-MS is usually better for complex samples. The various types of MS instruments used in proteomics research are reviewed by Aebersold

of isolation of a given peptide ion, fragmentation by energetic collision with gas, and recording of the tandem or MS/MS spectrum. The MS and MS/MS spectra are typically acquired for about 1 second each and stored for matching against protein sequence databases. The outcome of the experiment is the identity of the peptides and, therefore, the proteins making up the purified protein population. (From Aebersold and Mann,[78] reprinted with permission from Macmillan Publishers Ltd.)

and Mann.[78] The four basic types of mass analysers used in proteomics are time-of-flight (TOF), ion traps, quadropole, and Fourier transform ion cyclotron (FT-MS). These analyzers can also be put in tandem, such as a MALDI-TOF-TOF.

Protein Chips

Although for a number of technical reasons protein microarrays are more difficult to prepare and interpret than DNA microarrays, protein microarrays can have utility for identifying protein–protein, protein–nucleic acid, protein–antibody, and protein–drug interactions. The latter are being used in the pharmaceutical industry for high-throughput screening of chemical libraries. Typically, a specific protein is spotted by cross-linking to a glass surface in a grid-like format, and samples are spread over the microarray chip to detect interacting moieties.[79] Proteins could be immobilized on a microarray by reacting lysine side-chain amino groups with aldehyde-modified glass slides. These "protein chips" can be used to determine protein–protein binding interactions and kinase-mediated phosphorylation of immobilized proteins. Zhu et al.[80] made a protein chip of about 6000 yeast proteins and overlaid this with glutathione S-transferase/polyhistidine–tagged fusion proteins. This technique enabled purification of the proteins and their immobilization on the chip. Using this procedure, they were able to identify new classes of calmodulin- and phospholipid-binding proteins.

Detection of binding of proteins, peptides, or other moieties to a protein array usually depends on incorporation of a fluorescent or radioactive tag on the sample "poured" over the array. Theoretically, a protein array can contain any number of "bait" molecules to probe for things that bind it. Such bait molecules can be an antibody, a receptor, a recombinant protein or peptide, a cell or phage lysate, or a nucleic acid. The array can be queried with a probe such as a fluorescently tagged antibody, cell lysate, or serum sample to look for binding. There are a number of caveats, however, to the interpretation of such binding data.[79,81] These include (1) the broad range of protein concentrations that need

to be detected—the dynamic range of critical regulatory proteins may be in the femtomolar (10^{-15}) range; (2) protein concentrations in a given cell, tissue, or plasma extract can vary by a factor of 10^{10} among high-abundance and low-abundance proteins, making the signal-to-noise problem very difficult; (3) no PCR-like direct amplification method exists for proteins, making amplification of low-abundance proteins or peptides much more difficult than for nucleic acids; (4) accurate determination of binding affinities will usually require that the protein signal on the microarray be in the correctly folded, native state, making preparation of such arrays a complex and expensive proposition; (5) post-translational modifications such as phosphorylation and glycosylation can significantly alter the activities and binding properties of proteins, requiring the preparation of multiple forms of each so modified protein; and (6) attachment of membrane-bound proteins such as receptors and signal transduction elements is difficult because these proteins exist in a lipid bilayer environment that is necessary to maintain their native structure, and the lipid interface requires special procedures to get the complex to stick to array plates.

In spite of these difficulties, protein chip microarrays are being developed that can provide high-throughput screening techniques for protein–protein interactions, drug–receptor binding, and cancer biomarker detection. One powerful extension of such technology is the elution of bound proteins or peptides from the protein array and separation by MALDI-TOF mass spectrometry.[78]

Surface-Enhanced Laser Desorprtion/ Ionization (SELDI)

SELDI is a technique developed by Ciphergen Biosystems, Inc., that uses a surface-binding matrix and a variety of binding parameters to fractionate proteins of various chemical subtypes, based, for example, on hydrophobicity, ionic charge, phosphorylation state, and antibody binding. The SELDI protein chip uses mass spectrometry as the detection device. It has the advantage of being rapid, sensitive, and readily adaptable to a high-throughput diagnostic format. It has the disadvantage of not being quantitative

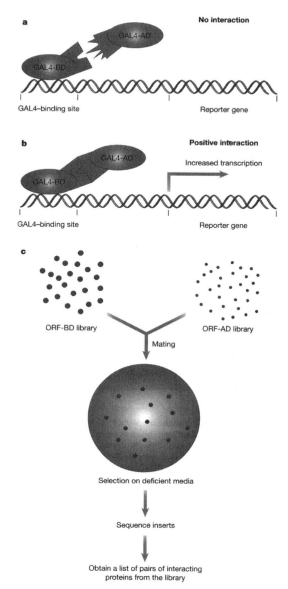

and is only useful for separating low-molecular-weight proteins (25 kDa or less).

Yeast Two-Hybrid System

Another approach to analyzing protein–protein interactions is the yeast two-hybrid system (reviewed in Reference 82). This technique uses a "bait and prey" system in which the gene for a bait protein is expressed and synthesized as a fusion protein with one component of a gene transcription factor (for example Gal 4) in one yeast strain, and the gene for a prey protein is expressed as a fusion protein between the proteins to be tested for interaction with the bait and the second component of the transcription factor in a second yeast strain. These two different yeast strains are then mated and those clones in which the bait and prey proteins interact are identified by turning on a reporter gene that codes for production of a color or a survival factor. These cells can than be separated and libraries formed from the clones (Fig. 7–7; see color insert).

A modification of the yeast two-hybrid system has been used to generate a human protein–protein interaction network (*interactome*).[83] In this sytem, the pairwise interactions among the products of about 8100 human cloned open reading frames were tested and about 2800 interactions were detected. This represents a step in the direction of determining a comprehensive human interactome map and identifying those protein connections that are disease related.

Phage Display

In this method, bacteriophage are designed to express protein or peptide products of cDNA libraries in such a way that the proteins or peptides are fused to a capsid or a coat protein on the phage-containing bacteria. This displays the

Figure 7–7. The yeast two-hybrid system. *a*. Different ORFs are expressed as fusion proteins to either the GAL4 DNA-binding domain (GAL4-BD) or its activation domain (GAL4-AD). If the proteins encoded by the ORFs do not interact with each other, the fusion proteins are not brought into close proximity and there is no activation of transcription of the reporter gene containing the upstream GAL4-binding sites. *b*. If the ORFs encode proteins that interact with each other, the fusion proteins are assembled at the GAL4-binding site of the reporter gene, which leads to activation of transcription. *c*. Library-based yeast two-hybrid screening method. In this strategy, two different yeast strains containing two different cDNA libraries are prepared. In one case, the ORFs are expressed as GAL4-BD fusions and in the other case they are expressed as GAL4-AD fusions. The two yeast strains are then mated and diploids selected on deficient media. Thus, only the yeast cells expressing interacting proteins survive. The inserts from both the plasmids are then sequenced to obtain a pair of interacting genes. (From Pandey and Mann,[82] reprinted with permission from Macmillan Publishers Ltd.)

protein or peptide on the surface of the bacterial cell. In this way, very large libraries of protein–peptide products can be screened for peptide epitopes that bind certain antibodies, protein- and peptide-binding ligands, enzyme substrates, or single-chain antibody fragments.[82]

Organelle Proteomics

One way to focus on specific subsets of proteins in a cell is to isolate the organelle in which they are contained. This enriches the source of such proteins, cuts down on background noise, and allows a more focused approach to detecting protein modifications and mutations in an organelle's protein repertoire. Protein profiles from a number of cellular organelles have been obtained by mass-spectrometric analyses (reviewed in Reference 78). The limiting factor in such studies is the ability to obtain purified, homogeneous samples of organelles. Mitochondria and the nucleolus are examples of two organelles for which the proteome has been fairly well established.[78,84]

Plasma Proteome

The Human Proteome Organization (HUPO) has established an international consortium to determine the human plasma proteome. This undertaking will be a significant challenge because the dynamic range of plasma proteins is extremely large and several physiological and genetic variables will affect what is detected. For example, the plasma proteome will vary depending on circadian rhythms, hormonal levels, diet, metabolic state, gender, ethnic background, and disease state. In addition, the plasma proteome is likely to vary over very short time frames, perhaps seconds, which will require kinetic analyses during various times of the day and during various metabolic and hormonal states (e.g., time of the menstrual cycle). Nevertheless, if baseline parameters can be established for "normal" vs. disease states such as cancer, this information could be a very useful diagnostic, prognostic, and therapy response tool. The challenges remain large. Plasma is estimated to contain hundreds of thousands of polypeptides, spanning a concentration range of up to 10 orders of magnitude—only about 500 plasma proteins have been clearly identified and very few have been quantified (reviewed in Reference 78). In addition, lab-to-lab variation in detection and quantitation makes it currently difficult to systematize the data. There is a great need for standardization and correlation of techniques to establish the true scope of the human plasma proteome.

Tissue Proteomics: Imaging Mass Spectrometry

Imaging mass spectometry joins the techniques of immunohistochemistry and fluorescence microscopy with mass spectrometry to analyze protein expression in mammalian tissues. In a typical experiment, frozen sections of a tissue are mounted on a stainless steel target plate that is then coated with a matrix solution (e.g., sinapinic acid) and dried. An MS imaging program is used to position a laser over consecutive laser spots. Material from each spot is desorbed and ionized by the laser and then delivered to MALDI-TOF MS for analysis. Each spot produces a mass spectrum obtained from molecules present within the laser-desorbed spot. Up to 30,000 spots can be sampled, and the intensity of molecular ions desorbed at each spot, in a molecular weight range of 500 Da to 80 kDa, can be determined. Over 200 protein and peptide peaks can be seen in the mass spectrum from each spot. Through use of a color-coding algorithm, the MS peaks can be integrated and displayed on a color scale.

An example of this methodology, used on brain tissue, is illustrated in Figure 7–8 (see color insert). Similar technology could be used to image proteins and peptides in specific regions of tumors and used to identify those proteins and peptides that are overexpressed compared to normal tissue. This technique could also be used to determine tumor heterogeneity and response to therapy, given the observation in the brain analysis that some protein signals were found to be highly specific for a given brain region, even though many protein signals were common to all areas of the brain.[85]

Using direct tissue matrix–assisted laser desorption ionization mass spectrometry, Caprioli and colleagues[86] have analyzed glioma tumor tissue from 108 patients and identified two patient populations—a short-term and a long-term

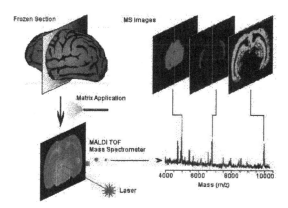

Figure 7–8. Methodology developed for spatial analysis of tissue by MALDI mass spectrometry. Frozen sections are mounted on a metal plate, coated with an UV-absorbing matrix, and placed in the mass spectrometer. A pulsed UV laser desorbs and ionizes analytes from the tissue and their m/z values are determined using a time-of-flight analyzer. From a raster over the tissue and measurement of the peak intensities over thousands of spots, mass-spectrometric images are generated at specific molecular weight values. (From Stoeckli et al.,[85] reprinted with permission from Macmillian Publishers Ltd.)

survival group—based on the tissue protein profiles. These profiles served as an independent indicator of patient survival.

Pattern Recognition

In the February 16, 2002 issue of *The Lancet*, a very provocative paper entitled "Use of Proteomic Patterns in Serum to Identify Ovarian Cancer" appeared.[87] A lot of excitement was generated by this article, because a relatively simple blood test was suggested for detecting a cancer that is hard to diagnose and is often well advanced by the time of first diagnosis. One of the intriguing aspects was that only a pattern of protein peaks obtained by SELDI-TOF mass spectrometry was needed to discriminate between normal individuals and cancer patients. A training set of mass spectra derived from analysis of serum from 50 normal and 50 ovarian cancer patients was analyzed by an iterative searching algorithm that reportedly discriminated cancer from non-cancer on the basis of a proteomic pattern, a process called *pattern recognition*. This discriminatory pattern was claimed to correctly identify 50 of 50 ovarian cancer patients and 63

of 66 cases without cancer, yielding a sensitivity of 100% and a specificity of 95%. Using similar SELDI-TOF MS techniques and iterative learning algorithms, protein pattern recognition profiles were also reported to discriminate prostate cancer from non-cancer with a high degree of sensitivity and specificity.[88,89]

Subsequent to these reports a number of other groups have reanalyzed the data and found a number of discrepancies among data sets from different experiments.[90] Moreover, there was concern that a number of the MS peaks used to discriminate cancer from non-cancer had m/Z values < 500, which is below the 2000 m/Z range considered an accurate value for a true MS peak. In a more recent study, the authors have used more sensitive instrumentation with better reproducibility, and using this approach, have reported 100% sensitivity in detecting early stage I ovarian cancers.[91]

The Unfolded Protein Response

For proteins to function properly they need to fold inside the cell into a "native" conformation that is the active state. Most folding of proteins occurs in the endoplasmic reticulum (ER), and the folding mechanisms are carried out by a series of steps involving protein-folding guide molecules called *chaperones*. Some of these belong to the heat shock family of proteins (HSPs) described earlier. This essential guardian function for correct protein folding is often subverted during oncogenesis,[92,93] leading to a logjam of the protein degradation machinery (the proteasome). This piling up of misfolded or partially degraded proteins causes cells to undergo apoptosis. Cancer cells appear to be more sensitive to this effect than normal cells, making proteasome inhibitors such as bortezomib (Velcade) useful therapeutic agents for cancers such as multiple myeloma (reviewed in Reference 94).

When eukaryotic cells are exposed to a variety of adverse physiological conditions such as changes in oxygen tension, pH, lack of nutrients, or exposure to drugs, the protein-folding machinery in the ER can be disrupted. This is called the *unfolded protein response* (UPR).[95] This response can lead to cell growth arrest and, if prolonged, to apoptosis. Induction of the UPR activates three ER stress sensors: IRE1, PKR-like ER kinase (PERK), and ATF6. As unfolded proteins accumulate in the ER, the endonuclease activity of IRE1 is turned on, activating the transcription of the *hac*1 gene. *Hac*1 is a transcription factor for elements in gene promoters that turn on UPR downstream responses. This leads to a complex array of responses that ultimately produce up-regulation of ER chaperones, the cell's attempt to resolve the pile-up of unfolded proteins by driving them back onto a normal folding pathway. If this response isn't sufficient, such as when ER stress conditions are not resolved, apoptotic pathways are initiated (see Chapter 4).

A number of studies have reported that the UPR is activated in various tumors (reviewed in Reference 95). This activation can have either apoptotic or anti-apoptotic effects on tumors, depending on the stage of tumor development, although the conditions favoring one response over the other are not well understood. Another conundrum is the effect of tumor-activated UPR on anticancer drug sensitivity. For example, activation of the UPR in cancer cells renders these cells resistant to topoisomerase II inhibitors but more sensitive to DNA cross-linking agents such as cisplatin. There is also evidence that activation of the UPR in hepatoma cells up-regulates expression of the multidrug resistance P-glycoprotein (reviewed in Reference 95).

Proteomics in Cancer Diagnosis

Earlier in this chapter, the use of nucleic acid diagnostics in cancer was discussed. It is important to note, however, that mRNA expression data alone are insufficient to predict a stage of disease or the functional status and degree of aggressiveness of a cancer cell. For example, gene expression information does not address the issue of activation state, post-translational modification, or localization of the corresponding gene-encoded proteins. In addition, as noted above, there is often a disparity between mRNA transcript and protein expression levels. Thus, in the best situation, gene expression data and proteomic analysis of cancer tissue need to be put together to provide a complete picture of the malignant state.

An example of the disparity between mRNA and protein expression was observed by Nishizuka and Charbonneau,[96] who analyzed gene and protein expression levels in 60 human cancer cell lines. For structural proteins, the mRNA and protein levels were highly correlated, but for nonstructural proteins they were not. This finding is perhaps not surprising, because the nonstructural proteins would consist of more rapidly turning over proteins such as metabolic, gene-regulatory, and signal transduction proteins whose dynamic range would be more controlled by regulation at the translational and post-translational level.

Parallel gene expression and proteomic analyses of stage I lung adenocarcinomas showed that 11 out of 27 mRNAs associated with patient survival were also represented in the profile of proteins that were survival associated and that numerous components of the glycolysis pathway were associated at either the mRNA or protein level with poor prognosis.[97] One of these, phosphoglycerate kinase I, was also detected in the serum of lung cancer patients, and increased levels correlated with poor outcome. The up-regulation of glycolytic enzymes correlates with up-regulation of hypoxia-inducible factor-1α (HIF-1α) in cancer tissue (reviewed in Reference 98).

An additional point to consider is that the "information flow" in cancer cells is mediated by protein–protein interactions,[99] and it will be important to determine these interconnecting networks of protein information to characterize the complete "wiring" diagram in cancer cells and the way in which that changes in malignant transformation and cancer progression. Moreover, the cancer cell signaling pathways do not function in isolation but as part of a complex system of cell–microenvironment interactions. For example, cellular–extracellular matrix interactions affect intracellular gene regulation and signal transduction events through ECM–cell skeleton interactions, co-regulation of growth factor expression, and regulation or angiogenesis, to name some of these interactions. An understanding of such intracellular and extracellular network interactions could lead to a new way to approach therapeutics, in that one could think of targeting an entire set of these interactions rather than a single molecular target.

Proteomic-based identification of biomarkers is already being employed for a number of human cancers. Some examples in which different proteomics techniques are used are presented here.

Lung Cancer

A proteomic approach was used to identify proteins that induce an antibody response in lung cancer. Sera from 64 patients with various types of lung cancer, 99 patients with other types of cancers, and 71 non-cancer controls were analyzed for antibody-based reactivity against lung adenocarcinoma proteins.[100] A reactive protein called protein gene product 9.5 (ubiquitin carboxyl-terminal esterase L1) was identified in 9 of 64 lung cancer patients but in only 1 of the other cancer patients and 1 of the controls, results suggesting that PGP 9.5 is a fairly specific biomarker for lung cancer.

In another study, MALDI-TOF MS was used to analyze protein profiles from 1-millimeter regions of single frozen sections of surgically resected lung tumors.[101] Fifteen distinct MS peaks distinguished proteomic patterns of cancer patients with poor prognosis from those with good prognosis.

Howard et al.[102] employed MALDI-TOF MS to analyze serum proteins of patients with or without lung cancer. They identified a peak containing serum amyloid A as a distinguishing feature of cancer patient sera: serum amyloid A levels were over eightfold higher in the sera of cancer patients than the serum levels of individuals without cancer.

Ovarian Cancer

In a five-center case–control study, serum proteomic analyses were carried out using the SELDI technique. Serum proteomic profiles were obtained on 153 patients with invasive epithelial ovarian cancer, 42 patients with other types of ovarian cancer, 166 with benign pelvic masses, and 142 healthy women.[103] Protein identification followed by serum immunoassay was carried out to validate identified biomarkers. Three were identified: apolipoprotein A1 (down-regulated in

cancer), a truncated form of transthyretin (also down-regulated), and a cleavage fragment of inter-α-trypsin inhibitor heavy chain H4 (up-regulated in cancer). Multivariate analysis of a combination of the three markers showed better sensitivity and specificity than CA-125 to detect early stage invasive epithelial ovarian cancers.

Breast Cancer

Wulfkuhle et al.[104] used two-dimensional gel separation techniques to analyze proteome patterns in normal breast tissue and ductal carcinoma in situ (DCIS) tissue, from either whole tissue sections or laser capture microdissected epithelial cells. Protein spots on the two-dimensional gels were excised and subjected to mass spectrometry sequencing. Fifty-seven proteins were differentially expressed between normal ductal epithelium and DCIS. Of these, 14 were confirmed by immunohistochemical analysis. Many of the proteins so identified had not previously been associated with breast cancers, including some involved in intracellular trafficking, cytoskeletal architecture, chaperone function, and genome instability.

Tissue arrays and immunohistochemistry have been combined to determine protein expression profiles that identify subclasses of breast cancer and predict patient prognosis.[105] A set of 21 proteins was found to closely correlate with 5-year metastasis-free survival (MFS). Among 552 patients, the 5-year MFS was 90% for patients classified in the good-prognosis subclass and 61% for those in the poor-prognosis subclass. This discriminator did not change when lymph node status, estrogen receptor status, or type of therapy were factored in.

Prostate Cancer

A quantitative proteomic technique that incorporates isotope-coded affinity tag reagents and tandem MS was used to identify secreted and cell surface proteins from human prostate cancer cells grown in culture.[106] Proteomic analyses were carried out in cells grown in the presence or absence of androgen. Analysis of the conditioned medium in which these cells grew indicated 600 protein peaks, of which 524 could be identified. The secretion of a number of these varied according to whether androgen was present in the growth medium. Some of the androgen-mediated secreted proteins appeared to be protein degradation products, thus androgen stimulation of the secreted polypeptides may have been more related to activation of proteases than to modification of transcriptional or translational control mechanisms. These secreted polypeptides may turn out to be biomarkers for prostate cancer.

Pancreatic Cancer

Two-dimensional gel eletrophoresis and MALDI-TOF MS were used to identify differentially expressed proteins in pancreatic cancer tissue, adjacent normal tissue from the same patients, pancreatitis tissue, and normal pancreas.[107] Forty differentially expressed proteins were identified, including antioxidant enzymes, chaperones, calcium-binding proteins, proteases, signal transduction proteins, and extracellular matrix proteins. Nine of these were specifically overexpressed in pancreatic cancers compared to pancreatitis or normal tissues: annexin A4, cyclophilin A, cathepsin D, galectin-1, 14-3-3, α-enolase, peroxiredoxin-I, TM2, and S100A8. Overexpression of these proteins was confirmed by Western blot of protein extracts and/or by immunohistochemical analysis of tissue sections.

CIRCULATING EPITHELIAL CELLS

There is evidence that primary cancers begin to shed malignant cells into the circulation before evidence of metastasis, suggesting that this is an early event in cancer progression. If this event can be validated, it could provide a way to detect cancers before they are large enough to be detected by standard clinical techniques such as computed tomography (CT), magnetic resonance imaging (MRI), or other imaging techniques. Such detection would be a powerful addition to the diagnostic armamentarium because even the best imaging techniques are hard pressed to detect tumors much smaller than 1 centimeter in diameter, at which time they can contain one billion cancer cells. Moreover, using sensitive genomic and proteomic techniques to analyze circulating tumor cells could provide identification

of the tissue of origin of the tumor as well as its aggressiveness and sensitivity to drugs and radiotherapy. A rate-limiting step for this sort of analysis is the number of cells it would take to carry out genomic and proteomic analyses. Nevertheless, evolving techniques such as microfluidics are making analysis of a few or even one cell more feasible.

A number of research groups have been using various methods to detect circulating cancer cells, including immunocytology, flow cytometry, and PCR (reviewed in Reference 108). Cytokeratins have been used to detect epithelial cells among the vast array and numbers of hematopoietic cells in the peripheral blood. Because cytokeratins are expressed on epithelial cells but not on blood cells, this provides a good marker to detect them in the large pool of blood cells. Cytokeratin 19 mRNA, detected by reverse-transcriptase PCR of cells found in peripheral blood, has been used to detect circulating cancer cells.[108] The detection limit of this method was 1 cancer cell in 10^7 peripheral mononuclear cells. However, low levels of CK19 mRNA are expressed by peripheral blood mononuclear cells, reducing the specificity of this method. Nevertheless, Peck et al.[108] were able to detect 40% of lung adenocarcinomas, 41% of squamous lung cancers, and 27% of small cell lung cancers. Only one non-cancer sample, from a patient with pneumonia, had circulating epithelial cells in the peripheral blood (out of a sample size of 33 pneumonia patients). In addition, these authors were able to show by serial sampling of peripheral blood for circulating epithelial cells that they could track tumor burden and response to therapy.

Jonathan Uhr and colleagues[109] have developed a sensitive assay combining immunomagnetic enrichment with flow cytometry and immunocytochemistry to detect and quantify circulating carcinoma cells in the blood. In this method, an iron-tagged (ferrofluid) monoclonal antibody to epithelial cell adhesion molecule (EPCAM) is used. Blood is incubated with the EPCAM-coated ferrofluid and then placed in a magnetic field to separate EPCAM antibody-bound cells. The magnetized cells are eluted and resuspended with a membrane permeabilization solution and further purified by interaction with anti-cytokeratin and anti-CD45 monoclonal antibodies, reacted with a fluorescent dye that stains nucleic acids, and then separated and analyzed by a fluorescence-activated cell sorter This assay can detect one epithelial cell in 1 milliliter of blood.

In the study by Racila et al.,[109] 10–20 ml of peripheral blood from 30 patients with breast cancer, 3 with prostate cancer, and 13 control individuals were subjected to the above procedure. The number of circulating epithelial cells, as defined by being positive for nucleic acids and cytokeratins and negative for CD45 (which would be present on hematopoietic stem cells), was determined. Higher numbers of circulating epithelial cells were found in the cancer patients' blood than in the controls, and by morphology, these were cancer cells. In the case of breast cancer, staining for mucin-1 (Muc-1) was also seen in a number of cases. Eight patients with breast cancer were followed for up to 10 months, and there was a correlation between changes in level of circulating tumor cells in the blood, response to therapy, and clinical status. In a subsequent study of prostate cancer patients, according to immunomagnetic selection of circulating epithelial cells,[110] the number of circulating epithelial cells correlated with disease progression. Interestingly, there was no diurnal variation observed in the number of circulating epithelial cells, which is important to know for future studies.

It has been shown by some investigators that biopsy of the prostate gland or prostatectomy releases a "shower" of epithelial cells into the peripheral blood. Since these cells may be neoplastic-like, this is further evidence for something that has been suspected for a long time—i.e., surgical removal of cancerous tissue can release cells into the peripheral circulation. While this indicates a caution for using the circulating epithelial cell assay after biopsy or surgery, it should be noted that these cells appear to have minimal viability in that they have been very difficult to grow in cell culture.

In two landmark studies, one published in the *New England Journal of Medicine*[111] and the other in *Journal of Clinical Oncology*,[112] it was shown that the number of circulating epithelial cells (now clearly defined as having cancer cell characteristics and called CTCs, for circulating tumor cells) obtained before treatment is an independent predictor of progression-free survival and overall survival in patients with metastatic breast cancer.

Global gene expression profiling of circulating tumor cells from patients with prostate cancer has also been useful to define cancer from normal cells. CTCs from 74 patients with metastatic prostate cancer were examined by gene expression profiling and compared to cells from normal donors.[113] CTC-specific expression profiles were identified that discriminated cancer from normal cells and that differentiated among separate metastatic cancers.

CIRCULATING ENDOTHELIAL CELLS AND CIRCULATING ENDOTHELIAL PROGENITOR CELLS

The process of angiogenesis was described in Chapter 4. Here the role of circulating endothelial cells (CECs) and circulating endothelial precursor cells (CEPs) and their role in cancer diagnosis and evaluation of anti-angiogenic therapy are discussed.

Originally it was thought that tumor neoangiogenesis occurred via two mechanisms: cooption and additional in-growth of capillaries from remodeled, pre-existing capillaries in the tumor environment and/or mobilization and expansion of endothelial cells from adjacent tissues. These cells could then proliferate and form new capillaries in the tumor. In either case, most neovascular tumor blood vessels differ from normal vasculature. In contrast to normal capillary beds, tumor neovasculature is dilated, leaky, and consists of a disorganized array of pericytes and smooth muscle cells. Endothelial cells from tumor neovasculature can break away and circulate in the peripheral blood. As measured by FACS analysis and quantification of peripheral blood cells by staining with endothelial cell markers such as CD34, FLK, CD105, VE cadherin, and CD31, these CECs were found to be significantly increased in the blood of patients with untreated lymphoma and breast cancer.[114] In lymphoma patients achieving complete remission after chemotherapy and in breast cancer patients after surgery, CECs were reduced to normal background levels observed in healthy controls.

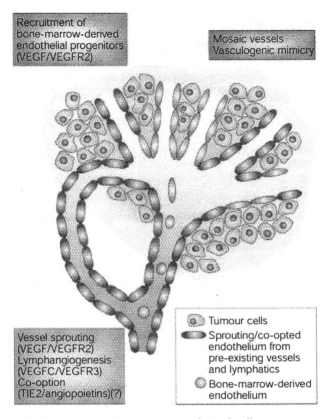

Figure 7–9. Contribution of bone marrow–derived cells to tumor angiogenesis. Angiogenic factors released by tumor cells recruit endothelial cells. Vascular and lymphatic endothelial cells incorporate into the tumor vasculature by co-option, migration, or proliferation of pre-existing neighboring vessels (sprouting). Some tumor cells are disguised as functional endothelial cells, giving rise to mosaic vessels. In addition to these processes, bone marrow–derived circulating endothelial progenitor cells (CEPs) can incorporate into tumor vasculature. The relative contribution of each of these processes to tumor angiogenesis is probably dictated by the cytokine repertoire and matrix components of each tumor cell type. Vascular endothelial growth factor (VEGF) signals through VEGF receptor (VEGFR)-2 to contribute to vessel sprouting. VEGFC signals through VEGFR3 to contribute to lymphangiogenesis, and TIE2 and angiopoietins might regulate co-option. VEGFR1 signaling supports tumor angiogenesis, probably through the release of angiogenic chemokine and cytokines by the hematopoietic cells. (From Raffii et al.,[115] reprinted with permission from Macmillan Publishers Ltd.)

There is now substantial evidence that vasculogenesis in tumors is not solely dependent on proliferation and in-growth of mature endothelial cells to generate new tumor blood vessels, but is also dependent on the mobilization and arrest in tumor sites of circulating endothelial progenitor cells (reviewed in Reference 115 and 116) (Fig. 7–9). These circulating cells (CEPs) appear to come mostly from the bone marrow.

This was initially shown in murine tumor models in which tumors failed to grow if bone marrow only contained precursor cells that lacked transcription factors modulating bone marrow endothelial cell differentiation. In this case, tumor angiogenesis did not occur. In contrast, when mice were transplanted with wild-type bone marrow cells, tumor angiogenesis and progression occurred. It has also been reported that CEPs are increased in the peripheral blood of cancer patients (reviewed in Reference 116).

In order for bone marrow–derived CEPs to contribute to tumor vasculogenesis, they have to be mobilized into the peripheral blood, migrate to the tumor site, and proliferate in the tumor environment (reviewed in Reference 115). Mobilization from the bone marrow requires activation of a number of signals, including activation of metalloproteases and release of cytokines from bone marrow stromal cells.

Mobilization of CEPs from the bone marrow occurs as a result of vascular trauma and in response to granulocyte and monocyte growth factors (GM-CSF and G-CSF as well as VEGF-A). Chemotherapeutic agents that depress the bone marrow also induce hematopoietic progenitor cell proliferation and mobilization. In addition, tumor cells release factors that favor progenitor cell mobilization. For example, tumor cell production of matrix metalloproteinase-9 (MMP-9) promotes release of ECM-bound or cell surface–bound cytokines in addition to VEGF, in turn leading to CEP mobilization. Vascular endothelial growth factor (VEGF)-mediated activation of MMP-9 promotes release of the stem cell growth-stimulatory cytokine-soluble Kit ligand, which also promotes hematopoietic cell motility.

Mobilized CEPs have a number of cell surface markers that distinguish them, including VEGF-receptor 2, CD34, PECAM, CXR4 (which they share with more mature endothelial cells) and CD133, which is only expressed in CEPs.

Although it isn't totally clear how tumor cells recruit CEPs to the tumor bed, there is evidence that CEP proliferation factors are released locally by tumors themselves, by macrophages present in the tumor, and by the tumor-supporting stroma. Tumors release a placental growth factor called PlGF that promotes angiogenesis.

Tumor endothelium–associated myelomonocytic cells release VEGF, brain-derived neurotrophic factor (BDNF), and platelet-derived growth factor (PDGF), which promote proliferation and differentiation of endothelial progenitor cells. Tumor-associated macrophages produce TNF-α and thrombospondin-1, which regulate angiogenesis. Macrophage-released factors also induce lymphangiogenesis. In vitro experiments demonstrate that addition of VEGF, fibroblast

growth factor-2 (FGF-2), and insulin-like growth factor (IGF) stimulate differentiation of CEPs into mature endothelial cells. However, the role that these factors play in vivo in this process is unclear. The tumor extracellular matrix proteins fibronectin and collagen promote differentiation of CEPs into functional tumor endothelium. It is also likely that direct cellular contact of CEPs with stromal cells is required for their proliferation and differentiation. Thus, the role of CEPs in tumor angiogenesis is a complex process, and not all the questions concerning how this occurs and its importance have been answered. Such questions include the following:

1. Which of the mechanisms for tumor vasculogenesis are most important for maintaining long-term blood supply to tumors and what role do they play in tumor progression?
2. Which tumor types rely primarily on local recruitment of mature endothelial cells, which depend on CEPs, and which depend on both (or other) mechanisms? For example, CEPs appear to play an essential role in the development of tumor vasculature of lymphomas and colon cancers but not in the initial vascularization of an experimental Lewis lung tumor model.[115]
3. What cytokines, tissue-specific stromal factors, and tumor-released factors are crucial for the mobilization, capture, and proliferation and differentiation of CEPs?
4. What new chemotherapeutic targets can be developed, based on a better understanding and validation of the factors involved in CEP-mediated vasculogenesis?

MOLECULAR IMAGING

Molecular imaging uses recent advances in visualization technologies to probe biological events inside cells and in tissues in vivo in a noninvasive manner. This is a powerful way to ask questions about protein–protein interactions, gene expression (including oncogene expression), ligand–receptor binding, drug responses, drug resistance, angiogenesis, and an almost limitless list of biological events that can be translated into fluorescent, radiographic, or ultrasound-detectable signals. These imaging techniques also allow for real-time analysis of events in cells or tissues. In general, four different noninvasive in vivo imaging technologies have been developed: (1) magnetic resonance imaging (MRI); (2) nuclear imaging by γ-camera, positron emission tomography (PET), or single-photon emission computed tomography (SPECT); (3) optical imaging by bioluminescence or fluorescence imaging; and (4) ultrasound imaging. Noninvasive imaging techniques can be direct or indirect, or use surrogate markers (reviewed in Reference 117).

Direct molecular imaging involves molecular probe–target interactions producing an image of probe localization and intensity. Examples of this include ligand–receptor occupancy, inhibitor–enzyme binding, antigen–antibody interaction on cell surfaces, probes activated by proteases, and antisense or aptamer oligonucleotide probes that specifically hybridize to a target mRNA or protein.

Indirect imaging strategies use pretargeting components that function as molecular genetic sensors—for example, the use of a PET reporter gene and a PET reporter probe (see below).

Surrogate imaging uses downstream effects of some cellular molecular process such as one that affects a signal transduction pathway with multiple steps and networks. This technique depends on identifying a biological marker that closely corresponds to an event related to oncogene activation, drug response, or tumor progression. An example is the use of ^{18}F-fluorodeoxyglucose (FDG) as a marker for response of gastrointestinal stromal tissues (GIST) to Gleevec. These tumors have a high glucose utilization and tumors in patients treated with drug have a rapid and sustained decrease in ^{18}FDG uptake as determined by PET scan.[117]

Protein–Protein Interactions

Post-transcriptional events such as translational regulation, protein–protein interactions, protein processing, and protein degradation can be imaged by fusing a reporter gene activated in response to a given protein–protein interaction or post-translational modification, similar to the yeast two-hybrid technique described above. One way to image such events is by energy transfer techniques such as Forster resonance energy transfer (FRET) and bioluminescence resonance energy transfer (BRET). These latter two techniques are powerful ones for detecting protein–protein interactions in intact cells, but have not been adapted for use in living animals or patients.[118]

Protein Degradation

Proteasomal activity can be measured by an ubiquitin-luciferase bioluminescence imaging reporter that fuses the amino terminus of firefly luciferase to four copies of a mutant ubiquitin (reviewed in Reference 118). This tetraubiquitin fusion product degrades proteins in cultured cells, but the mutant molecule has a glycine-to-valine substitution at the carboxy terminus that limits cleavage to those proteins specifically degraded by ubiquitin hydrolases. It was shown that the bioluminescent imaging of protease function in tumor xenografts was blocked by the proteasome inhibitor bortezomib (Velcade).

Imaging Gene Expression In Vivo

In vivo imaging of gene expression can be detected for either genes transduced into cells or tissues or for endogenous genes. Imaging of transgenes can be useful to monitor the expression of specific genes in response to cellular perturbation by drugs or chemical agents, or to measure the effect of a gene therapy approach to altering cellular function. Imaging of endogenous gene expression can be used to study the expression of genes during development; aging; responses to drugs, hormones, or other environmental stimuli; and alteration of gene expression during neoplastic transformation or development of other disease states.[119]

To image transgene expression, genes are transferred into cells via a viral vector, liposome, or as naked DNA by electroporation. For this, a PET reporter gene (PRG) and PET probe (PRP) can be used. The PRG is linked to a "therapy" gene that can be activated by a constitutive promoter such as the cytomegalovirus promoter. When the therapeutic gene is activated, the PRG gene is also expressed. The mRNA from the PRG is translated into a protein product that is the target

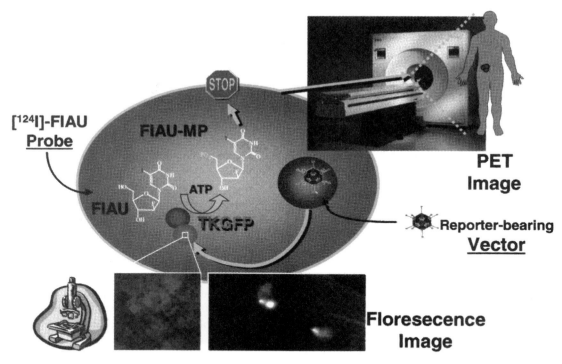

Figure 7–10. TKGFP dual-modality reporter gene imaging. This schema illustrates the steps involved for dual-modality imaging of HSV1-tk/GFP (green fluorescent proteins) reporter gene expression. The HSV1-*tk/GFP* fusion gene is transfected into target cells by a vector. Inside the transfected cell, the gene is transcribed to mRNA and then translated to a fusion protein, HSV1-TK/GFP, that retains both HSV1-TK enzymatic activity and GFP fluorescence. After administration of a radiolabeled probe (e.g., (124I)-FIAU or (18F)-FHBG) and its transport into the cell, the probe is phosphorylated by the HSV1-TK component of the reporter gene product. The phosphorylated radiolabeled probe does not readily cross the cell membrane and is "trapped" within the cell. Thus, the magnitude of probe accumulation in the cell (level of radioactivity) reflects the level of HSV1-TK enzyme activity and level of HSV1-tk gene expression. The GFP component of the reporter gene product retains fluorescence and can be used for in vivo and in situ imaging as well as FACS analysis of transduced tumor cells. (From Tjuavjev and Blasberg,[117] reprinted with permission from Elsevier.)

of the PET probe. An example of this is the use of the herpes simplex virus thymidine kinase (HSV-TK) as the PRG and the ^{18}F-labeled prodrug 8-fluoroganciclovir (FGCV) as the PRP.[120]

When FGCV is injected into animals in which the transgene reporter is expressed, it is phosphorylated and retained only in the HSV-TK-expressing cells, and the rest is washed out of the system with time. The FGCV-labeled tissues can then be imaged by PET scanning. Such an approach could also be used for imaging of gene expression in patients undergoing gene therapy or who have received ex vivo gene–transduced

cells such as T cells for immunologic therapy (Fig. 7–10; see color insert).

Bioluminescent Detection

Bioluminescence (BLI) can also be used to image gene expression or other biological events in cells and tissues. This method has an advantage of low background signals, ease of use, and low cost compared with PET imaging, but it is limited by light scattering and signal absorption when tissues deep in the body are to be imaged.[121] Photoproteins called *luciferases* are often used in

BLI. These enzymes oxidize substrates (e.g., D-luciferin) to an electronically excited state that generates an emission signal in the 400–620 nm wavelength range. An example of this is detection of tumor cells transduced with a CMV promoter-driven luciferase gene and then exposed to D-luciferin (e.g., injected by tail vein into a tumor xenograft-bearing mouse). This technique has been used by Rehemtulla et al.[122] to measure the kinetics of glioma tumor growth and response to the drug BCNU.

Another light-emitting detection mechanism is the use of fluorescent proteins of different colors to color-code cancer cells of a specific genotype or phenotype. This can also be used to discriminate cancer cells from normal cells and highly metastatic cells from low metastatic cell types.[123] For example, green fluorescent protein (GFP) can be used to label cancer cells and red fluorescent protein (RFP) to label nonmalignant cells. Genes for this protein can be delivered via viral vectors or a transgenic mouse expressing GFP in all its cells can be transplanted with tumor cells expressing RFP. Thus, the progression and metastatic sites of the tumor cells can be tracked in real time in a host animal. In vivo imaging can also be done at a single-cell level using this technique. Reversible skin flaps are employed to visualize deeper organs. Fluorescent imaging of this type can also be used to evaluate in vivo drug responses.

The application of these techniques to humans is a work in progress, but one can surmise that this could be done through a targeted delivery of fluorescent protein genes to tumors via ligand or antibody-targeting liposomes and then using fluorescent endoscopy to image tumor cells in the body.

Magnetic Resonance Spectroscopy

High-resolution magic angle spinning proton magnetic resonance spectroscopy is a method that can be used for intact tissue analysis. This technology allows identification of individual metabolites in tissues while preserving histopathologic characteristics—in effect a "metabolomics" profile for cancer.

Cheng et al.[124] have used this method to obtain metabolic profiles from prostate cancers with different Gleason scores and were able to delineate subsets of less aggressive tumors and predict tumor perineural invasion within a subset that were more aggressive. These results predict that magnetic resonance spectroscopy metabolite profiles can be used to assess more indolent from more aggressive tumors, a key issue for prostate cancer diagnostics.

Ultrasound Imaging

Imaging by ultrasound is another method to detect biologic events in tissues in vivo. An example here is the injection of ultrasound contrast microbubbles targeted to tumor microvasculature by linkage to the arginine-arginine-leucine (RRL) tripeptide.[125] Microbubbles linked to RRL were shown to specifically adhere to tumor angiogenic endothelium compared to normal vasculature. Such selective adhesion could be detected ultrasonically. Targeted microbubbles may therefore provide a noninvasive imaging techniques for imaging of tumor vasculature and for determining response to anti-angiogenesis drugs.

NANOTECHNOLOGY

In one sense, *nanotechnology* is nothing more than a new name for things that exist on a small scale, the size of groups of atoms, or, in other words, chemistry. The term, however, has taken on the meaning for things manufactured by human ingenuity that are on the scale of molecules that nature invented. The term derives from the Greek word *nanos* for dwarf. In its technical usage, *nano-* is a prefix for something that is a one billionth part (10^{-9}) of a specified unit, e.g., nanometer, nanosecond, etc. In the fields of chemistry and physics, the term is usually used to define particles of 1–100 nanometers in diameter. This is equivalent to about the size of 200 gold atoms assembled together. The Nobel Laureate Physicist Richard Feynman may be the one who originally articulated the nanotechnology vision, in a lecture at the California Institute of Technology in 1959, entitled "There Is Plenty of Room at the Bottom." In this lecture, Feynman envisioned that materials as small as molecules used by biological systems could be manufactured and that information storage could be so miniaturized that all

25,000 pages of the 1959 edition of the *Encyclopedia Britannica* could be stored in an area the size of a pinhead.

Indeed, one of the first practical uses of nanotechnology has been in the design and manufacture of transistors for microprocessors in computers. In the future, nanoscale transistors only a few atoms wide will be used to store up to 10,000 times more information than is currently possible on microprocessors.

The field of nanotechnology now encompasses several fields, including physics, chemistry, engineering (e.g., nanomaterials, nanoelectronics, microfluidics), computer science, biology, and medicine. In the field of medicine, all of these disciplines will need to work together to harness the new breakthroughs in diagnostic techniques and therapeutics that will drive the future of medical practice.

Many of the newer developments in nanobiology go back to the development by Leo Esaki of artificial inorganic nanostructures that act as semiconductor quantum devices in which the tunneling of electrons can be controlled and converted to measurable signals (reviewed in Reference 126). This work has led to the development of colloidal "quantum dots" of the size of a typical protein. These are being employed in a wide range of applications such as biological labeling of cells, tracking of cell movement, in vivo imaging, DNA detection, and production of multiplexed beads for optical coding of biomolecules (reviewed in Reference 126). Using such "nano-devices," molecular events can be sensed in cells through optical, electrical, or magnetic detection. Nanostructures such as semiconductor nanowires and carbon nanotubes can be used as sensitive and selective electrical detectors of biological events (see below).

Carbon nanotubes and nanowires are being used to detect specific DNA sequences and proteins.[127] Up to 1000 nanowire detectors (about 8 nm in diameter) can be condensed into an area about the size of a single cell. Potentially, each nanowire could contain a different antibody or oligonucleotide to detect a protein or mRNA sequence (Fig. 7–11a; see color insert).[128] On one chip, it would be possible to carry out 1000 single-cell experiments (the "lab-on-the-chip"). As an example, single-strand DNA can be bound to a nanowire and the binding of complementary mRNA will cause a signal to be generated, indicating a specific hybridization has occurred. Thus, this technique can be used to detect gene expression patterns in a few cells and potentially in a single cell when coupled to microfluidic techniques that can separate and detect single cells in a fluid sample.[127] Another method for this is the so-called nanocantilever array (Fig. 7–11b; see color insert). Here, the biomarker proteins bind to antibodies attached to cantilevers, and this binding causes the cantilevers to deflect, producing a signal that can be detected by laser beam or electronically.

One can envision using similar nanodevices for molecular detection in vivo. For example, implantable sensors could be designed to emit a signal that could be detected outside the body. A big challenge for this endeavor is the potential for nonspecific binding of tissue and serumproteins on the sensing surfaces; however, methods to circumvent such "biofouling" can be developed. One nifty futuristic concept is the coupling of nanosensing devices with a drug delivery system that could be implanted on the same minichip into a tumor vascular bed. For example, a sensor that detects the presence of a tumor biomarker or tumor physiological parameter such as low pH or low oxygen tension could produce a signal that would trigger the connected drug delivery platforms. One can also envision a multifunctional nanoparticle, which would have on its surface multiple tumor-targeting moieties such as antibodies to cell surface markers like EGF receptors and other up-regulated growth factor receptors.[128]

Various "nanovectors" can be designed to deliver drugs to targeted tissues. These include liposomes with surface-complexed antibodies to tumor antigens, "nanoshells" composed of gold over a silica core, and various polymer-coated particles such as dendritic polymers (see below). Several types of nanoparticles can also be used for in vivo molecular imaging by, for example, enhancing MRI contrast or ultrasound imaging. Not only can multiple targeting modalities be put on the surface of nanovectors, they could also be loaded with multiple drugs, each with a different time-release "minipump" such as an ultrasound or electronically signaled release mechanism. An example of this is the design of nanoparticle-RNA aptamer bioconjugates that target prostate-

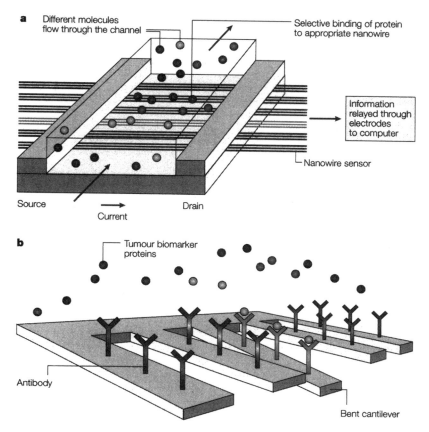

Figure 7–11. Nanowires and nanocantilevers. *a*. Nanowires deployed within a microfluidic system. Different colors indicate that different molecules (colored circles) absorb or affinity-bind to different nanowire sensors. The binding causes a change in conductance of the wires, which can be electronically and quantitatively detected in real time. The working principle is that of a (biological gated) transistor and is illustrated in the insert. The charges of the binding protein disrupt electrical conduction in the underlying nanowire. The nano size of the wire is required to attain high signal-to-noise ratios. *b*. Nanocantilever array. The biomarker proteins are affinity-bound to the cantilevers and cause them to deflect. The deflections can be directly observed with lasers. Alternatively, the shift in resonant frequencies caused by the binding can be electronically detected. As for nanowire sensors, the breakthrough potential in nanocantilever technology is the ability to sense a large number of different proteins at the same time, in real time. (From Ferrari,[128] reprinted with permission from Macmillan Publishers Ltd.)

specific membrane antigen on the surface of prostate cancer cells and that are composed of a controlled-release polymer that releases an RNA aptamer targeted to block a tumor growth factor or other tumor-related process.[129]

Some of the caveats to all these ventures include the issue of potential antigenicity of nanovectors and the difficult regulatory path that such multicomponent, multidrug devices might face at the U.S. Food and Drug Administration.

Baker and colleagues have designed dendritic polymers that function as multifunctional delivery devices.[130] The dendritic polymer-nanoparticle targets intracellular folate and selectively delivers methotrexate intracellularly. It also emits an optical imaging signal through attachment of flourescein to the nanovector. In cell culture systems, the nanovector-delivered methotrexate killed 100-fold more cancer cells than free methotrexate added to the culture medium.

This same research group has also used the folate-targeted dendrimer coupled to methotrexate and fluorescein 6-carboxytetramethylrhodamine (the latter as an imaging agent) to inject intravenously into immunodeficient mice bearing a human KB tumor.[131] Confocal microscopy confirmed internalization of the dendrimer–drug conjugates into the tumor cells. The dendrimer–drug complex had increased anti-tumor activity and lower toxicity compared with free, unconjugated methotrexate.

Gray Goo

The explosion in nanotechnology methods and applications and the burgeoning interest in industry and federal agencies in funding this research have led to a backlash of cautionary issues, including the use of nanotechnology in medical practice. This reaction spans the realms of science fiction and speculative environmental horror stories to more realistic concerns about in vivo toxicity. Some of the fictional accounts are about self-replicating nanorobots taking over the world (the "gray goo" effect). There is concern, however, about the effects on tissues of nanoparticles delivered for diagnostic or therapeutic purposes. In addition, the release of nanoparticles used in manufacturing could conceivably result in a type of chemical pollution. At this point, not much is known about the health and environmental impacts of nanomaterials. Some cell toxicity in bacterial cells and in rodents has been reported.[132] For example, a suspension of carbon nanotubes flushed into the lungs of rats caused tissue damage, respiratory problems, and even death in some animals. One can argue whether this is a legitimate model for potential human exposure, since this is unlikely to be a method of delivery for therapeutic or diagnostic purposes to humans; nevertheless, issues of in vivo toxicity need to be clearly documented. In the case of cancer diagnosis and treatment, the risk–benefit ratio would be calculated quite differently than for environmental exposure in a healthy population.

PHARMACOGENOMICS AND PHARMACOGENETICS

Pharmacogenomics can be defined as the study of the role of genetic variation in drug and xenobiotic response of individuals. Such inherited individual differences can, in addition, be modified by age, gender, disease, hormonal status, and drug interactions. Historically, the field has developed from observations in the 1950s that related to individual differences in metabolism of drugs (reviewed in Reference 133). The classic example is the prolonged muscle paralysis after succinylcholine injection that results from an inherited deficiency in plasma cholinesterase (reviewed in Reference 134). Other examples include hemolysis after antimalarial drug therapy related to the low inherited level of glucose-6-phosphate dehydrogenase activity, and peripheral neuropathy induced by the anti-tuberculosis drug isoniazed in individuals with low acetylation activity ("slow acetylators"). The molecular genetic basis for such inherited traits began to be unraveled in the 1980s when the human gene encoding the drug-metabolizing enzyme debrisoquin hydroxylase was cloned (reviewed in Reference 135). This enzyme is now known as the cytochrome P-450 isozyme CYP2D6. Originally the inherited traits linked to drug metabolism defects were all monogenic. It is now known, however, that the overall pharmacologic effects of drugs and xenobiotics are not monogenic, but rather are determined by the interaction of several gene products. In addition to drug-metabolizing enzymes, these genetic polymorphisms affect drug transporters, drug receptors, drug absorption and excretion, and drug–drug interactions.[136]

The identification of the genes involved as determinants of drug and xenobiotic effects has led to the field of pharmacogenomics, which is really the stepchild of pharmacogenetics. The two terms are often used synonymously, however, *pharmacogenomics* is usually used for the study of genome-wide approaches to identify the entire spectrum of genes that define differences in drug response and that can be used to define the population category in which an individual fits. *Pharmacogenetics* often refers more to the drug-metabolism differences among individuals.

Recent reviews of the topic of individual variation in the therapeutic effects of drugs indicate that the efficacy rate of drugs varies from 25% to 80%. Thus, up to 75% of patients may not have an effective response to a drug, and the response rate is lowest, on average, for anticancer drugs (25%). Of the 1232 drugs on the market as of 2001, 193 (16%) are associated with severe adverse events requiring a "black box" warning on the product label.[137] It should be noted that, while many adverse events have a dose–response relationship and with careful dosing and patient selection are avoidable, about 50% are idiosyncratic and often result from drug-allergic reactions that are difficult to predict a priori. Nevertheless, a meta-analysis of adverse-event reporting found that 1.8 million people were hospitalized for drug-adverse events in the United States in 1994, with over 100,000 deaths. Although such data may be an exaggeration of actual events, even if they are off by an order of magnitude, this represents a serious problem, one that use of pharmacogenomic and pharmacogenetic analyses could have greatly diminished. As one might predict from these sort of numbers, the FDA is keenly interested in having pharmacogenomic and pharmacogenetic data accompany IND submissions and has developed guidelines for this.[138]

Importance of Pharmacogenomics in Cancer

The selection of drugs to treat cancer patients is still largely empirical and based on population response data for various types of cancers, not on individual variations that affect drug efficacy and toxicity. This situation is beginning to change, but is still a clinical science in its early stages. Genetic polymorphisms have been recognized as a major cause of drug toxicity for pyrimidine analogs (e.g., 5-fluorouracil), purines (6-mercaptopurine), folate antagonists (methotrexate), anthracyclines (doxorubicin), and camptothecins (reviewed in Reference 139). Genes involved in drug metabolism and detoxification of anticancer drugs include dihydropyrimidine dehydrogenase, thiopurine methyltransferase (TPMT; see below), and UDP-glucuronosyl-transferase. In addition, genetic polymorphisms or muta-

tions in the targets of anticancer drugs can also have pronounced effects on their responses in individual patients. Recent examples are the development of resistance to Gleevec in patients who have alterations in the *bcr/abl* tyrosine kinase that is the target for the drug, and the low response rate to EGF receptor antagonists in individuals who have the "wild-type" sequence in the receptor (explaining the approximate 10% response rate in lung cancer patients). There are also ethnic differences that affect response rates to anticancer drugs, just as there are for acetylator phenotypes. For example, it has been observed that there are large differences in response and toxicity to docetaxel between Asians and Caucasians, with Asians having higher response but also increased toxicity. Polymorphisms have also been observed in the multidrug resistance gene *MDR1*, which may explain differences in drug responses.

One of the most well-studied of the enzymes that affect anticancer drug response is TPMT. 6-Mercaptopurine (6-MP) is one of the thiopurines that is inactivated by TPMT. This enzyme has genetic polymorphisms that have been observed in all populations studied. About 90% of individuals have high activity, 10% are heterozygotes with intermediate activity, and 0.3% have low or no detectable activity because they inherit two nonfunctional alleles.[139] TPMT-deficient patients are at high risk for severe, sometimes fatal, myelosuppression and thus require a dose reduction to avoid toxicity. There is one wild-type allele and eight variant alleles.[141] The genetic basis for TPMT polymorphisms involves four mutant alleles: TPMT-2, TPMT-3A, TPMT-3C, and TPMT-4. TPMT genotyping is commercially available and many clinics do it routinely before treating patients (frequently children with ALL) with 6-MP.

Interestingly, some genes involved in modulating drug and xenobiotic metabolism are also involved in cancer susceptibility. This may be related to the fact that some of these enzymes also modulate environmental chemical or hormonal metabolism. Examples of this include (1) increased breast cancer risk associated with genetic polymorphisms of estrogen-metabolizing cytochrome P-450s CYP17 and CYP1A1 and catechol-*O*-methyltransferase (COMT);[142] (2)

increased breast cancer risk associated with CYP19, glutathione S-transferase-1 (GSTP1), GST-M1, and p53 polymorphisms or mutations;[143] (3) association of the homozygous variant of CYP3A4 in African Americans with a higher risk for having advanced prostate cancer;[144] and (4) association of a genetic polymorphism in the 5, 10-methylene-tetrahydrofolate-reductase (MTHFR) gene and colorectal adenomas, which may reflect a gene–environment interaction.[145]

HAPLOTYPE MAPPING

The International HapMap Project is aimed at defining the single nucleotide polymorphisms (SNPs) in the human genome. So far, more than one million SNPs have been found by examining 269 DNA samples from four population groups.[146] These HapMap data can guide the analysis of genetic association studies, reveal recombination events, and identify genetic loci that appear to have undergone evolutionary selection.

The HapMap is based on haplotypes, which are stretches of DNA inherited as whole blocks of sequences. This haplotype basis makes identification of SNPs much faster and easier than brute-force sequencing of whole genomes. It also allows investigators to more rapidly hone in on disease genes, particularly if they are relatively rare hereditary events that correlate with an identifiable familial inheritance. HapMap data will most likely be useful for identifying monogenic-related diseases, rather than multigenic diseases or diseases that have multiple-associated mutations, such as cancer mutations like p53 or BRCA1. Nevertheless, the haplotype data will allow an investigator to rapidly scan these databases to look for specific disease-related gene mutations.

References

1. P. Gold and S. O. Freedman: Tests for carcinoembryonic antigen. Role in diagnosis and management of cancer. JAMA 234:190, 1975.
2. D. Sidransky: Emerging molecular markers of cancer. Nat Rev Cancer 2:210, 2002.
3. M. Sullivan Pepe, R. Etzioni, Z. Feng, J. D. Potter, M. L. Thompson, et al.: Phases of biomarker development for early detection of cancer. J Natl Cancer Inst 93:1054, 2001.
4. G. J. Kelloff, R. C. Bast, Jr., D. S. Coffey, A. V. D'Amico, R. S. Kerbel, et al.: Biomarkers, surrogate end points, and the acceleration of drug development for cancer prevention and treatment: An update. Clin Cancer Res 10:3881, 2004.
5. J. W. Park, R. S. Kerbel, G. J. Kelloff, J. C. Barrett, B. Z. Chabner, et al.: Rationale for biomarkers and surrogate end points in mechanism-driven oncology drug development. Clin Cancer Res 10:3885, 2004.
6. F. Vande Woude, G. J. Kelloff, R. W. Ruddon, H.-M. Koo, C. C. Sigman, et al.: Reanalysis of cancer drugs: old drugs, new tricks. Clin Cancer Res 10:3897, 2004.
7. H.-W. Chang, S. M. Lee, S. M. Goodman, G. Singer, S. K. R. Cho, et al.: Assessment of plasma DNA levels, allelic imbalance, and CA 125 as diagnostic tests for cancer. J Natl Cancer Inst 94:1697, 2002.
8. V. Combaret, C. Audoynaud, I. Iacono, M.-C. Eavrot, M. Schell, et al.: Circulating MYCN DNA as a tumor-specific marker in neuroblastoma patients. Cancer Res 62:3646, 2002.
9. M. S. Kopreski, F. A. Benko, D. J. Borys, A. Khan, T. J. McGarrity, et al.: Somatic mutation screening: Identification of individuals harboring K-ras mutations with the use of plasma DNA. J Natl Cancer Inst 92:918, 2000.
10. C. F. Eisenberger, M. Schoenberg, C. Enger, S. Hortopan, S. Shah, et al.: Diagnosis of renal cancer by molecular urinalysis. J Natl Cancer Inst 91:2028, 1999.
11. C. Jeronimo, H. Usadel, R. Henrique, J. Oliveira, C. Lopes, W. G. Nelson, and D. Sidransky: Quantitation of GSTP1 methylation in nonneoplastic prostatic tissue and organ-confined prostate adenocarcinoma. J Natl Cancer Inst 93:1747, 2001.
12. M. S. Fliss, H. Usadel, O. L. Cabellero, L. Wu, M. R. Buta, et al.: Facile detection of mitochondrial DNA mutations in tumors and bodily fluids. Science 287:2017, 2000.
13. M. E. Sherman, M. H. Schiffman, A. T. Lorincz, R. Herrero, M. L. Hutchinson, et al.: Cervical specimens collected in liquid buffer are suitable for both cytologic screening and ancillary human papillomavirus testing. Cancer 81:89, 1997.
14. Y. M. Dennis Lo, L. Y. S. Chan, A. T. C. Chan, S.-F. Leung, K.-W. Lo, et al.: Quantitative and temporal correlation between circulating cell-free Epstein-Barr virus DNA and tumor recurrence in nasopharyngeal carcinoma. Cancer Res 59:5452, 1999.
15. G. Zweiger: Knowledge discovery in gene-expression-microarray data: Mining the information output of the genome. Trends Biotechnol 17:429, 1999.
16. S. S. Jeffrey, M. J. Fero, A.-L. Borresen-Dale, and D. Botstein: Expression array technology in

the diagnosis and treatment of breast cancer. *Mol Interv* 2:101, 2002.

17. J. L. Dennis, J. K. Vass, E. C. Wit, W. N. Keith, and K. A. Oien: Identification from public data of molecular markers of adenocarcinoma characteristic of the site of origin. *Cancer Res* 62:5999, 2002.

18. L. Luo, R. C. Salunga, H. Guo, A. Bittner, K. C. Joy, et al.: Gene expression profiles of laser-captured adjacent neuronal subtypes. *Nat Med* 5:117, 1999.

19. K. Miura, E. D. Bowman, R. Simon, A. C. Peng, A. I. Robles, et al.: Laser capture microdissection and microarray expression analysis of lung adenocarcinoma reveals tobacco smoking- and prognosis-related molecular profiles. *Cancer Res* 62:3244, 2002.

20. H. C. King and A. A. Sinha: Gene expression profile analysis by DNA microarrays. *JAMA* 286:2280, 2001.

21. J. Knight: When the chips are down. *Nature* 410:860, 2001.

22. D. J. Lockhart and E. A. Winzeler: Genomics, gene expression and DNA arrays. *Nature* 405:827, 2000.

23. L. Bubendorf, M. Kolmer, J. Kononen, P. Koivisto, S. Mousses, et al.: Hormone therapy failure in human prostate cancer: Analysis by complementary DNA and tissue microarrays. *J Natl Cancer Inst* 91:1758, 1999.

24. S. Amatschek, U. Koenig, H. Auer, P. Steinlein, M. Pacher, et al.: Tissue-wide expression profiling using cDNA subtraction and microarrays to identify tumor-specific genes. *Cancer Res* 64:844, 2004.

25. M. Bibikova, D. Talantov, E. Chudin, J. M. Yeakley, J. Chen, et al.: Quantitative gene expression profiling in formalin-fixed, paraffin-embedded tissues using universal bead arrays. *Am J Pathol* 165:1799, 2004.

26. A. A. Alizadeh, M. B. Eisen, R. E. Davis, C. Ma, I. S. Lossos, et al.: Distinct types of diffuse large B-cell lymphoma identified by gene expression profiling. *Nature* 403:503, 2000.

27. A. Rosenwald, G. Wright, W. C. Chan, J. M. Connors, E. Campo, et al.: The use of molecular profiling to predict survival after chemotherapy for diffuse large B-cell lymphoma. *N Engl J Med* 346:1937, 2002.

28. L. M. Staudt: Molecular diagnosis of the hematologic cancers. *N Engl J Med* 348:1777, 2003.

29. S. A. Armstrong, J. E. Staunton, L. B. Silverman, R. Pieters, M. L. den Boer, et al.: *MLL* translocations specify a distinct gene expression profile that distinguishes a unique leukemia. *Nat Genet* 30:41, 2002.

30. B. Weigelt, J. L. Peterse, and L. J. van't Veer: Breast cancer metastasis: markers and models. *Nat Rev Cancer* 5:591, 2005.

31. C. M. Perou, S. S. Jeffrey, M. van de Ruin, C. A. Rees, M. B. Eisen, et al.: Distinctive gene expression patterns in human mammary epithelial cells and breast cancers. *Proc Natl Acad Sci USA* 96:9212, 1999.

32. D. S. Sgroi, S. Teng, G. Robinson, R. LeVangie, J. R. Hudson, Jr., et al.: In vivo gene expression profile analysis of human breast cancer progression. *Cancer Res* 59:5656, 1999.

33. C. M. Perou, T. Sørlie, M. B. Eisen, M. van de Rijn, S. S. Jeffrey, et al.: Molecular portraits of human breast tumours. *Nature* 406:747, 2000.

34. M. J. can de Vijver, Y. D. He, L. J. van't Veer, H. Dai, A. A. M. Hart, et al.: A gene-expression signature as a predictor of survival in breast cancer. *N Engl J Med* 347:1999, 2002.

35. S. Gruvberger, M. Ringnér, Y. Chen, S. Panavally, L. H. Saal, et al.: Estrogen receptor status in breast cancer is associated with remarkably distinct gene expression patterns. *Cancer Res* 61:5979, 2001.

36. Y. Wang, J. G. M. Klijn, Y. Zhang, A. M. Sieuwerts, M. P. Look, et al.: Gene-expression profiles to predict distant metastasis of lymph-node-negative primary breast cancer. *Lancet* 365:671, 2005.

37. T.-K. Jensen and E. Hovig: Gene-expression profiling in breast cancer. *Lancet* 365:634, 2005.

38. I. Hedenfalk, D. Duggan, Y. Chen, M. Radmacher, M. Bittner, et al.: Gene-expression profiles in hereditary breast cancer. *N Engl J Med* 344:539, 2001.

39. D. A. Zajchowski, M. F. Bartholdi, Y. Gong, L. Webster, H.-L. Liu, et al.: Identification of gene expression profiles that predict the aggressive behavior of breast cancer cells. *Cancer Res* 61:5168, 2001.

40. W. Wang, S. Goswami, K. Lapidus, A. L. Wells, J. B. Wyckoff, et al.: Identification and testing of a gene expression signature of invasive carcinoma cells within primary mammary tumors. *Cancer Res* 64:8585, 2004.

41. K. J. Martin, E. Graner, Y. Li, L. M. Price, B. M. Kritzman, et al.: High-sensitivity array analysis of gene expression for the early detection of disseminated breast tumor cells in peripheral blood. *Proc Natl Acad Sci USA* 98:2646, 2001.

42. J. M. Silva, G. Dominguez, J. M. Garcia, R. Gonzalez, M. J. Villanueva, et al.: Presence of tumor DNA in plasma of breast cancer patients: clinicopathological correlations. *Cancer Res* 59:3251, 1999.

43. J. B. Welsh, P. P. Zarrinkar, L. M. Sapinoso, S. G. Kern, C. A. Behling, et al.: Analysis of gene expression profiles in normal and neoplastic ovarian tissue samples identifies candidate molecular markers of epithelial ovarian cancer. *Proc Natl Acad Sci USA* 98:1176, 2001.

44. C. D. Hough, K. R. Cho, A. B. Zonderman, D. R. Schwartz, and P. J. Morin: Coordinately

up-regulated genes in ovarian cancer. *Cancer Res* 61:3869, 2001.

45. A. Jemal, T. Murray, E. Ward, A. Samuels, R. C. Tiwari, et al.: Cancer statistics, 2005. *CA Cancer J Clin* 55:10, 2005.

46. J. Luo, D. J. Duggan, Y. Chen, J. Sauvageot, C. M. Ewing, et al.: Human prostate cancer and benign prostatic hyperplasia: molecular dissection by gene expression profiling. *Cancer Res* 61:4683, 2001.

47. J. Xu, J. A. Stolk, X. Zhang, S. J. Silva, R. L. Houghton, et al.: Identification of differentially expressed genes in human prostate cancer using subtraction and microarray. *Cancer Res* 60:1677, 2000.

48. A. Waghray, M. Schober, F. Feroze, F. Yao, J. Virgin, et al.: Identification of differentially expressed genes by serial analysis of gene expression in human prostate cancer. *Cancer Res* 61: 4283, 2001.

49. S. Bettuzzi, M. Scaltriti, A. Caporali, M. Brausi, D. D'Arca, et al.: Successful prediction of prostate cancer recurrence by gene profiling in combination with clinical data: A 5-year follow-up study. *Cancer Res* 63:3469, 2003.

50. J. Lapointe, C. Li, J. P. Higgins, M. van de Rijn, E. Bair, et al.: Gene expression profiling identifies clinically relevant subtypes of prostate cancer. *Proc Natl Acad Sci USA* 101:811, 2004.

51. D. A. Notterman, U. Alon, A. J. Sierk, and A. J. Levine: Transcriptional gene expression profiles of colorectal adenoma, adenocarcinoma, and normal tissue examined by oligonucleotide arrays. *Cancer Res* 61:3124, 2001.

52. K. Birkenkamp-Demtroder, L. Lotte Christensen, S. Harder Olesen, C. M. Frederiksen, P. Laiho, et al.: Gene expression in colorectal cancer. *Cancer Res* 62:4352, 2002.

53. P. Buckhaults, C. Rago, B. St. Croix, K. E. Romans, S. Saha, et al.: Secreted and cell surface genes expressed in benign and malignant colorectal tumors. *Cancer Res* 61:6996, 2001.

54. P. A. Clarke, M. L. George, S. Easdale, D. Cunningham, R. I. Swift, et al.: Molecular pharmacology of cancer therapy in human colorectal cancer by gene expression profiling. *Cancer Res* 63:6855, 2003.

55. P. G. Johnston: Of what value genomics in colorectal cancer? Opportunities and challenges: Editorial. *J Clin Oncol* 22:1, 2004.

56. Y. Wang, T. Jatkoe, Y. Zhang, M. G. Mutch, D. Talantov, et al.: Gene expression profiles and molecular markers to predict recurrence of Dukes' B colon cancer: Original report. *J Clin Oncol* 22:1, 2004.

57. M. E. Garber, O. G. Troyanskaya, K. Schluens, S. Petersen, Z. Thaesler, et al.: Diversity of gene expression in adenocarcinoma of the lung. *Proc Natl Acad Sci USA* 98:13784, 2001.

58. A. Bhattacharjee, W. G. Richards, J. Staunton, C. Li, S. Monti, et al.: Classification of human lung carcinomas by mRNA expression profiling reveals distinct adenocarcinoma subclasses. *Proc Natl Acad Sci USA* 98:13790, 2001.

59. D. G. Beer, S. L. R. Kardia, C.-C. Huang, T. J. Giordano, A. M. Levin, et al.: Gene-expression profiles predict survival of patients with lung adenocarcinoma. *Nat Med* 8:816, 2002.

60. M. Takahashi, D. R. Rhodes, K. A. Furge, H. Kanayama, S. Kagawa, et al.: Gene expression profiling of clear cell renal cell carcinoma: gene identification and prognostic classification. *Proc Natl Acad Sci USA* 98:9754, 2001.

61. J. R. Vasselli, J. H. Shih, S. R. Iyengar, J. Maranchie, J. Riss, et al.: Predicting survival in patients with metastatic kidney cancer by gene-expression profiling in the primary tumor. *Proc Natl Acad Sci USA* 100:6958, 2003.

62. Q.-H. Ye, L.-X. Qin, M. Forgues, P. He, J. W. Kim, et al.: Predicting hepatitis B virus–positive metastatic hepatocellular carcinomas using gene expression profiling and supervised machine learning. *Nat Med* 9:416, 2003.

63. Y. Chen, C. Miller, R. Mosher, X. Zhao, J. Deeds, et al.: Identification of cervical cancer markers by cDNA and tissue microarrays. *Cancer Res* 63:1927, 2003.

64. M. J. Roth, N. Hu, M. R. Emmert-Buck, Q.-H. Wang, S. M. Dawsey, et al.: Genetic progression and heterogeneity associated with the development of esophageal squamous cell carcinoma. *Cancer Res* 61:4098, 2001.

65. S. L. Pomeroy, P. Tamayo, M. Gaasenbeek, L. M. Sturia, M. Angelo, et al.: Prediction of central nervous system embryonal tumour outcome based on gene expression. *Nature* 415: 436, 2002.

66. A. Nestl, O. D. Von Stein, K. Zatloukal, W.-G. Theis, P. Herrlich, et al.: Gene expression patterns associated with the metastatic phenotype in rodent and human tumors. *Cancer Res* 61:1569, 2001.

67. B. St. Croix, C. Rago, V. Velculescu, G. Traverso, K. E. Romans, et al.: Genes expressed in human tumor endothelium. *Science* 289:1197, 2000.

68. J.-T. Chi, H. Y. Chang, G. Haraldsen, F. L. Jahnsen, O. G. Troyanskaya, et al.: Endothelial cell diversity revealed by global expression profiling. *Proc Natl Acad Sci USA* 100:10623, 2003.

69. K. Kudoh, M. Ramanna, R. Ravatn, A. G. Elkahloun, M. L. Bittner, et al.: Monitoring the expression of profiles of doxorubicin-induced and doxorubicin-resistant cancer cells by cDNA microarray. *Cancer Res* 60:4161, 2000.

70. C. Kihara, T. Tsunoda, T. Tanaka, H. Yamana, Y. Furukawa, et al.: Prediction of sensitivity of esophageal tumors to adjuvant chemotherapy by cDNA microarray analysis of gene-expression profiles. *Cancer Res* 61:6474, 2001.

71. S. P. Bohen, O. G. Troyanskaya, O. Alter, R. Warnke, D. Botstein, et al.: Variation in gene expression patterns in follicular lymphoma and the response to rituximab. *Proc Natl Acad Sci USA* 100:1926 2003.

72. J. C. Chang, E. C. Wooten, A. Tsimelzon, S. G. Hilsenbeck, M. C. Gutierrez, et al.: Gene expression profiling for the prediction of therapeutic response to docetaxel in patients with breast cancer. *Lancet* 362:362, 2003.

73. J.-P. Gillet, T. Efferth, D. Steinbach, J. Hamels, F. de Longueville, et al.: Microarray-based detection of multidrug resistance in human tumor cells by expression profiling of ATP-binding cassette transporter genes. *Cancer Res* 64:8987, 2004.

74. P. R. Srinivas, M. Verma, Y. Zhao, and S. Srivastava: Proteomics for cancer biomarker discovery. *Clin Chem* 48:1160, 2002.

75. C. Ezzell: Proteins rule. *Sci Am* 286:40, 2002.

76. P. H. O'Farrell: High-resolution two-dimensional electrophoresis of proteins. *J Biol Chem* 250:4007, 1975.

77. S. P. Gygi, B. Rist, S. A. Gerber, F. Turecek, M. H. Gelb, et al.: Quantitative analysis of complex protein mixtures using isotope-coded affinity tags. *Nat Biotechnol* 17:994, 1999.

78. R. Aebersold and M. Mann: Mass spectrometry–based proteomics. *Nature* 422:198, 2003.

79. Y.-S. Lee and M. Mrksich: Protein chips: from concept to practice. *Trends Biotechnol* 20:S14, 2002.

80. H. Zhu, M. Bilgin, R. Bangham, D. Hall, A. Casamayor, et al.: Global analysis of protein activities using proteome chips. *Science* 293:2101, 2001.

81. L. A. Liotta, V. Espina, A. I. Mehta, V. Calvert, K. Rosenblatt, et al.: Protein microarrays: meeting analytical challenges for clinical applications. *Cancer Cell* 3:317, 2003.

82. A. Pandey and M. Mann: Proteomics to study genes and genomes. *Nature* 405:837, 2000.

83. J.-F. Rual, K. Venkatesan, T. Hao, T. Hirozane-Kishikawa, A. Dricot, et al.: Towards a proteome-scale map of the human protein–protein interaction network. *Nature* 437:1173, 2005.

84. M. Verma, J. Kagan, D. Sidransky, and S. Srivastava: Proteomic analysis of cancer-cell mitochondria. *Nat Rev Cancer* 3:789, 2003.

85. M. Stoeckli, P. Chaurand, D. E. Hallahan, and R. M. Caprioli: Imaging mass spectrometry: A new technology for the analysis of protein expression in mammalian tissues. *Nat Med* 7:493, 2001.

86. S. A. Schwartz, R. J. Weil, R. C. Thompson, Y. Shyr, J. H. Moore, et al.: Proteomic-based prognosis of brain tumor patients using direct-tissue matrix-assisted laser desorption ionization mass spectrometry. *Cancer Res* 65:7674, 2005.

87. E. F. Petricoin, A. M. Ardekani, B. A. Hitt, P. J. Levine, V. A. Fusaro: Use of proteomic patterns in serum to identify ovarian cancer. *Lancet* 359:572, 2002.

88. B.-L. Adam, Y. Qu, J. W. Davis, M. D. Ward, M. A. Clements, et al.: Serum protein fingerprinting coupled with a pattern-matching algorithm distinguishes prostate cancer from benign prostate hyperplasia and healthy men. *Cancer Res* 62:3609, 2002.

89. E. F. Petricoin, D. K. Ornstein, C. P. Paweletz, A. Ardekani, P. S. Hackett, et al.: Serum proteomic patterns for detection of prostate cancer. *J Natl Cancer Inst* 94:1576, 2002.

90. E. Check: Running before we can walk? *Nature* 429:496, 2004.

91. E. F. Petricoin and L. A. Liotta: Proteomic approaches in cancer risk and response assessment. *Trends Mol Med* 10:59, 2004.

92. L. Whitesell and S. L. Lindquist: HSP90 and the chaperoning of cancer. *Nat Rev Cancer* 5:761, 2005.

93. A. J. L. Macario and E. Conway de Macario: Sick chaperones, cellular stress, and disease. *N Engl J Med* 353:1489, 2005.

94. J. Adams: The proteasome: a suitable antineoplastic target. *Nat Rev Cancer* 4:349, 2004.

95. Y. Ma and L. M. Hendershot: The role of the unfolded protein response in tumour development: friend or foe? *Nat Rev Cancer* 4:966, 2004.

96. S. Nishizuka and L. Charboneau: Proteomic profiling of the NCI-60 cancer cell lines using new high-density reverse-phase lysate microarrays. *Proc Natl Acad Sci USA* 100:14229, 2003.

97. G. Chen, T. G. Gharib, H. Wang, C.-C. Huang, R. Kuick, et al.: Protein profiles associated with survival in lung adenocarcinoma. *Proc Natl Acad Sci USA* 100:13537, 2003.

98. S. Hanash: Integrated global profiling of cancer. *Nat Rev Cancer* 4:638, 2004.

99. E. F. Petriocoin, K. C. Zoon, E. C. Kohn, C. Barrett and L. A. Liotta: Clinical proteomics: translating benchside promise into bedside reality. *Nat Rev Drug Discov* 1:683, 2002.

100. F. Brichory, D. Beer, F. Le Naour, T. Giordano, and S. Hanash: Proteomics-based identification of protein gene product 9.5 as a tumor antigen that induces a humoral immune response in lung cancer. *Cancer Res* 61:7908, 2001.

101. K. Yanagisawa, Y. Shyr, B. J. Xu, P. P. Massion, P. H. Larsen, et al.: Proteomic patterns of tumour subsets in non-small-cell lung cancer. *Lancet* 362:433, 2003.

102. B. A. Howard, M. Z. Wang, M. J. Campa, C. Corro, M. C. Fitzgerald, et al.: Identification and validation of a potential lung cancer serum biomarker detected by matrix-assisted laser desorption/ionization–time of flight spectra analysis. *Proteomics* 3:1720, 2003.

103. Z. Zhang, R. C. Bast, Y. Yu, J. Li, L. J. Sokoll, et al.: Three biomarkers identified from serum

proteomic analysis for the detection of early stage ovarian cancer. *Cancer Res* 64:5882, 2004.

104. J. D. Wulfkuhle, D. C. Sgroi, H. Krutzsch, K. McLean, K. McGarvey, et al.: Proteomics of human breast ductal carcinoma in situ. *Cancer Res* 62:6740, 2002.

105. J. Jacquemier, C. Ginestier, J. Rougemont, V.-J. Bardou, E. Charafe-Jauffret, et al.: Protein expression profiling identifies subclasses of breast cancer and predicts prognosis. *Cancer Res* 65:767, 2005.

106. D. B. Martin, D. R. Gifford, M. E. Wright, A. Keller, E. Yi, et al.: Quantitative proteomic analysis of proteins released by neoplastic prostate epithelium. *Cancer Res* 64:347, 2004.

107. J. Shen, M. D. Person, J. Zhu, J. L. Abbruzzese, and D. Li: Protein expression profiles in pancreatic adenocarcinoma compared with normal pancreatic tissue and tissue affected by pancreatitis as detected by two-dimensional gel electrophoresis and mass spectrometry. *Cancer Res* 64:9018, 2004.

108. K. Peck, Y.-P. Sher, J.-Y. Shih, S. R. Roffler, C.-W. Wu, et al.: Detection and quantitation of circulating cancer cells in the peripheral blood of lung cancer patients. *Cancer Res* 58:2761, 1998.

109. E. Racila, D. Euhus, A. J. Weiss, C. Rao, J. McConnell, et al.: Detection and characterization of carcinoma cells in the blood. *Proc Natl Acad Sci USA* 95:4589, 1998.

110. J. G. Morena, S. M. O'Hara, S. Gross, G. Doyle, H. Fritsche, et al.: Changes in circulating carcinoma cells in patients with metastatic prostate cancer correlate with disease status. *Urology* 58:386, 2001.

111. M. Cristofanilli, G. T. Budd, M. J. Ellis, A. Stopeck, J. Matera, et al.: Circulating tumor cells, disease progression, and survival in metastatic breast cancer. *N Engl J Med* 351:781, 2004.

112. M. Cristofanilli, D. F. Hayes, G. T. Budd, M. J. Ellis, A. Stopeck, et al.: Circulating tumor cells: a novel prognostic factor for newly diagnosed metastatic breast cancer. *J Clin Oncol* 23:1420, 2005.

113. D. A. Smirnov, D. R. Zweitzig, B. W. Foulk, M. C. Miller, G. V. Doyle, et al.: Global gene expression profiling of circulating tumor cells. *Cancer Res* 65:4993, 2005.

114. S. Monestiroli, P. Mancuso, A. Burlini, G. Pruneri, C. Dell'Agnola, et al.: Kinetics and viability of circulating endothelial cells as surrogate angiogenesis marker in an animal model of human lymphoma. *Cancer Res* 61:4341, 2001.

115. S. Rafii, D. Lyden, R. Benezra, K. Hattori, and B. Heissig: Vascular and haematopoietic stem cells: novel targets for anti-angiogenesis therapy? *Nat Rev Cancer* 2:826, 2002.

116. F. Bertolini, S. Paul, P. Mancuso, S. Monestiroli, A. Gobbi, et al.: Maximum tolerable dose and low-dose metronomic chemotherapy have

opposite effects on the mobilization and viability of circulating endothelial progenitor cells. *Cancer Res* 63:4342, 2003.

117. J. Gelovani Tjuvajev and R. G. Blasberg: In vivo imaging of molecular-genetic targets for cancer therapy. *Cancer Cell* 3:327, 2003.

118. S. Gross and D. Piwnica-Worms: Spying on cancer: Molecular imaging in vivo with genetically encoded receptors. *Cancer Cell* 7:5, 2005.

119. M. E. Phelps: Positron emission tomography provides molecular imaging of biological processes. *Proc Natl Acad Sci USA* 97:9226, 2000.

120. M. Doubrovin, V. Ponomarev, T. Beresten, J. Balatoni, W. Bornmann, et al.: Imaging transcriptional regulation of p53-dependent genes with positron emission tomography *in vivo*. *Proc Natl Acad Sci USA* 98:9300, 2001.

121. L. Wu, M. Johnson, and M. Sato: Transcriptionally targeted gene therapy to detect and treat cancer. *Trends Mol Med* 9:421, 2003.

122. A. Rehemtulla, L. D. Stegman, S. J. Cardozo, S. Gupta, D. E. Hall, et al.: Rapid and quantitative assessment of cancer treatment response using in vivo bioluminescence imaging. *Neoplasia* 2:491, 2000.

123. R. M. Hoffman: The multiple uses of fluorescent proteins to visualize cancer in vivo. *Nat Rev Cancer* 5:796, 2005.

124. L. L. Cheng, M. A. Burns, J. L. Taylor, W. He, E. F. Halpern, et al.: Metabolic characterization of human prostate cancer with tissue magnetic resonance spectroscopy. *Cancer Res* 65:3030, 2005.

125. G. E. R. Weller, M. K. K. Wong, R. A. Modzelewski, E. Lu, A. L. Klibanov, et al.: Ultrasonic imaging of tumor angiogenesis using contrast microbubbles targeted via the tumor-binding peptide arginine-arginine-leucine. *Cancer Res* 65:533, 2005.

126. P. Alivisatos: The use of nanocrystals in biological detection. *Nat Biotechnol* 22:47, 2004.

127. J. R. Heath, M. E. Phelps, and L. Hood: Nanosystems biology. *Mol Imag Biol* 5:312, 2003.

128. M. Ferrari: Cancer nanotechnology: opportunities and challenges. *Nat Rev Cancer* 5:161, 2005.

129. O. C. Farokhzad, S. Jon, A. Khademhosseini, T.-N. T. Tran, D. A. LaVan, et al.: Nanoparticle-aptamer bioconjugates: A new approach for targeting prostate cancer cells. *Cancer Res* 64:7668, 2004.

130. A. Quintana, E. Raczka, L. Piehler, I. Lee, A. Mhyc, et al.: Design and function of a dendrimer-based therapeutic nanodevice targeted to tumor cells through the folate receptor. *Pharmacol Res* 19:1310, 2002.

131. J. F. Kukowska-Latallo, K. A. Candido, Z. Cao, S. S. Nigavekar, I. J. Majoros, et al.: Nanoparticle targeting of anticancer drug improves therapeutic response in animal model of human epithelial cancer. *Cancer Res* 65:1, 2005.

132. R. F. Service: Nanotechnology grows up. *Science* 304:1732, 2004.

133. R. Weinshilboum: Inheritance and drug response. *N Engl J Med* 348:529, 2003.

134. W. W. Weber: The legacy of pharmacogenetics and potential applications. *Mutat Res* 479:1, 2001.

135. W. E. Evans and M. V. Relling: Pharmacogenomics: translating functional genomics into rational therapeutics. *Science* 286:487, 1999.

136. W. E. Evans and H. L. McLeod: Pharmacogenomics—Drug disposition, drug targets, and side effects. *N Engl J Med* 348:538, 2003.

137. B. B. Spear, M. Heath-Chiozzi, and J. Huff: Clinical application of pharmacogenetics. *Trends Mol Med* 7:201, 2001.

138. L. J. Lesko and J. Woodcock: Translation of pharmacogenomics and pharmacogenetics: a regulatory perspective. *Nat Rev Drug Discov* 3:763, 2004.

139. R. Danesi, F. de Braud, S. Fogli, T. Martino de Pas, A. di Paola, et al.: Pharmacogenetics of anticancer drug sensitivity in non–small cell lung cancer. *Pharmacol Rev* 55:57, 2003.

140. J. A. Johnson and W. E. Evans: Molecular diagnostics as a predictive tool: genetics of drug efficacy and toxicity. *Trends Mol Med* 8:300, 2002.

141. R. M. Weinshilboum, D. M. Otterness, and C. L. Szumlanski: Methylation pharmacogenetics: Catechol-*O*-methyltransferase, thiopurine methyltransferase, and histamine *N*-methyltransferase. *Annu Rev Pharmacol Toxicol* 39:19, 1999.

142. C.-S. Huang, H.-D. Chern, K.-J. Chang, C.-W. Cheng, S.-M. Hsu, et al.: Breast cancer risk associated with genotype polymorphism of the estrogen-metabolizing genes *CYP17*, *CYP1A1*, and *COMT*: A multigenic study on cancer susceptibility. *Cancer Res* 59:4870, 1999.

143. A. M. Dunning, C. S. Healey, P. D. P. Pharoah, M. D. Teare, B. A. J. Ponder, et al.: A systematic review of genetic polymorphisms and breast cancer risk. *Cancer Epidemiol Biomarkers Prev* 8:843, 1999.

144. P. L. Paris, P. A. Kupelian, J. M. Hall, T. L. Williams, H. Levin, et al.: Association between a *CYP3A4* genetic variant and clinical presentation in African-American prostate cancer patients. *Cancer Epidemiol Biomarkers Prev* 8:901, 1999.

145. C. M. Ulrich, E. Kampman, J. Bigler, S. M. Schwartz, C. Chen, et al.: Colorectal adenomas and the C677T *MTHFR* polymorphism: Evidence for gene–environment interaction? *Cancer Epidemiol Biomarkers Prev* 8:659, 1999.

146. The International HapMap Consortium: A haplotype map of the human genome. *Nature* 437:1299, 2005.

8

Sequelae of Cancer
and Its Treatment

PATIENT–TUMOR INTERACTIONS

A malignant tumor growing in vivo produces a number of effects on the host (Table 8–1). The end result of one or more of these effects is what ultimately proves fatal. The effects of a growing cancer in a patient may include fever, anorexia (loss of appetite), weight loss and cachexia (body wasting), infection, anemia, and various hormonal and neurologic symptoms. These may occur out of proportion to the size of the tumor. A relatively small tumor may cause many symptoms, whereas another tumor may produce few symptoms and remain occult until it is far advanced and has metastasized. Malignant tumors affect host functions by compression, invasion, and destruction of normal tissues as well as by the elaboration of substances that circulate in the bloodstream. The effects of tumor-produced factors are called collectively the *paraneoplastic syndromes*. Approximately 15% of patients hospitalized with advanced malignancy will have clinically apparent systemic effects in organ systems distant from the primary neoplasm, even though there is no evidence of metastasis to the affected organ.[1] From 50% to 75% of cancer patients eventually experience a paraneoplastic syndrome. A common form of paraneoplastic syndrome is related to ectopic hormone production by growing tumors.

Pain

Pain is frequently associated with malignant disease. Cancer-induced pain is often unre-

lenting and difficult to treat, requiring the use of addicting narcotic drugs. It is one of the most difficult problems for the patient, the physician, and the cancer patient's family. The cause of pain in a cancer patient may be destruction of tissue by the tumor, infection, stretching of internal organs, pressure, or obstruction. An example of pain that results from tissue destruction is the bone pain from invasive or metastatic cancers that cause periosteal irritation, pressure in the medullary space of bone, or fractures.

Infection can result from decreased immunity or from an obstruction that decreases drainage from a tissue. Malignant neoplasms originating in mucous membranes, such as those of the oral cavity, vagina, or rectum, may ulcerate early in the course of the disease and produce inflammation and infection at the site of ulceration. A painful infection that often occurs in cancer patients with decreased immunity is herpes zoster, which produces a "nerve pain," or causalgia, which follows the distribution of the affected nerve.

Growth of tumors in areas that have minimal room for expansion results in pain due to pressure. A typical example is a brain tumor, either primary or metastatic, enlarging in the cranial vault. Pain from a brain tumor results from the pressure on blood vessels and membraneous septa and may be *referred pain* in the sense that it will occur over the distribution of one of the cranial nerves rather than at the primary site of the tumor. Neoplasms growing in a nasal sinus or an eye produce pain by similar mechanisms.

Table 8–1. Effects of Malignant Neoplasm on Host Functions

Effect	Manifestations	Causes
Pain	Pressure, colic, headache, etc., depending on location of tumor	Tissue destruction by tumor, obstruction, pressure on organs, and infection
Cachexia	Muscle wasting, loss of body fat, and generalized weakness	Anorexia, nausea, malabsorption, involvement of digestive organs, metabolic demands of tumor and host's defense system. TNF-α and various cytokines (IL-6, IL-1B, IFN-γ) show ubiquitin-proteasome activation
Anemia	Pallor, weakness, and fatigue	Invasion of bone marrow and crowding out of normal hematopoietic cells; hemolysis, hemorrhage, and decreased erythrocyte production and survival
Leukopenia	Infections	Crowding out of normal marrow cells; chemotherapy, radiotherapy, and trapping of cells in spleen
Thrombocytopenia	Petechia, purpura, and internal hemorrhaging	Crowding out of marrow, chemotherapy, radiation therapy, immune destruction, and hypercoagulability of blood
Blood hypercoagulability	Throbophlebitis, disseminated intravascular coagulation, hemorrhage, renal failure, and shock	Release of thromboplastin-like substances by tumor
Fever	Sweating, malaise, and confusion	Infection, pryogenic substances released by tumor
Hormone release	Examples include Cushing's syndrome (ACTH), water retention (ADH), gynecomastia or precocious puberty (hCG), hypercalcemia (PTH), and hypoglycemia (insulin-like substances)	Ectopic production by tumor, tumor involving organ-producing hormone
Hypercalcemia	Lethargy, weakness, nausea, confusion, coma, decreased gastrointestinal tract motility, cardiac arrhythmias, and kidney stones	Tumor invasion or metastasis to bone, ectopic PTH-related protein release, and tumor production of osteolytic substances, including prostaglandins and osteoclast-stimulating substance
Neurologic syndromes	Muscle weakness, decreased reflexes, and muscle atrophy	Cachexia, "toxohormones"
Dermatologic involvement	Hyperpigmentation, erythema, petechiae, purpura, hirsutism, pruritus, and infection (e.g., *Herpes zoster*)	Hormone release by tumor, thrombocytopenia, invasion of skin by tumor, allergic reactions, decreased immunity

Involvement of visceral organs may produce pain when the cancerous growth causes stretching of an organ, pressure on an adjacent structure, or obstruction. However, the slow enlargement of a tumor may not cause significant pain, and for this reason many cancers of internal organs, such as the stomach, colon, pancreas, and liver, remain occult until they are far advanced. Pain from visceral organs is frequently referred pain because the stimuli are carried over sympathetic nerves and may enter the spinal cord at an area distant from the site of the tumor. For example, pain from the esophagus is sometimes referred to the shoulder, and stomach pain may be felt more in the chest than in the abdomen. Obstruction of a hollow organ, such as the stomach, small intestine, bile duct, or colon, may produce cramps and colicky pain that may become severe if complete obstruction occurs.

Some cancers produce a characteristic pattern of pain. For example, advanced breast cancer is painful if an inflammatory reaction distends the breast or if an extensive infiltration of the chest wall occurs. Swelling of involved lymph nodes in the axilla may cause severe shoulder or arm pain. Metastases to the liver, ovary, or other visceral organs may produce referred pain and metastatic involvement of bone may produce pain in the affected part. Metastases to the brain can produce headache. In lung cancer, pain is a late syndrome and indicates local invasion or distant metastases. It may be manifest as severe shoulder or arm pain or bone pain due to periosteal stretching. Extensive skeletal pain may indicate multiple myeloma or advanced prostatic carcinoma. Diffuse bone pain and joint discomfort due to leukemia may simulate arthritic disease.

Nutritional Effects

Nutritional disorders and malnutrition are frequently the most disabling effects of cancer. These effects are manifest by weight loss, hypermetabolism in body tissues, and ultimately body wasting, or cachexia. This tissue wasting affects predominantly muscle and fat, but probably involves all organs, with the exception of the heart, liver, and brain.[2] The symptoms of malnutrition are not strictly correlated with the size of the tumor or the rate of tumor growth because some patients with widespread tumors have minimal symptoms until very late in their disease. Furthermore, the amount of cachexia is usually out of proportion to the expected metabolic demands based on the size of the tumor. Ultimately, almost all cancer patients will experience cachexia. The cachexia seen in cancer patients is caused by a variety of factors including reduced appetite (anorexia), decreased digestive functions, metabolic demands of the tumor and of the host's defense systems, and factors released by tumors.

Many patients with advanced cancer have anorexia. For example, at the time of diagnosis, 80% of patients with upper gastrointestinal tract cancers and 60% of patients with lung cancer have already had substantial weight loss (reviewed in Reference 3). Cachexia has been implicated as the main cause of death in over 20% of patients. In addition, abnormalities in taste sensation that make certain foods less palatable have been reported.[4] Certain patients with liver involvement may have a postprandial hyperglycemia that signals the "glucostat" in the central nervous system to reduce appetite.[5] Liver dysfunction and wasting of tissues also produce elevated circulating levels of amino acids and fatty acids that may decrease appetite, probably by a central nervous system–mediated mechanism.[2,5]

Although anorexia is a major contributing factor to the cachexia of advanced cancer, it cannot fully explain the progressive weight loss associated with malignant disease. In certain patients, malabsorption of nutrients from the gastrointestinal (GI) tract may occur if the GI tract is obstructed or ulcerated. Direct involvement of the liver, bile duct, or pancreas can reduce the production of bile salts or digestive enzymes that are needed for digestion of foodstuffs. But even in patients who have no evidence of malabsorption or direct involvement of the digestive system, weight loss frequently occurs. Thus, there must be something related to the cancerous growth itself that contributes to weight loss and cachexia. Observations in experimental animals supports this.[6,7] After the inoculation of a tumor into an animal, an initial increase in carcass weight may occur for several days because of fluid retention. Thereafter, the weight of the animal's carcass progressively decreases as the tumor grows. Normal tissues other than the liver, brain, and heart lose weight and have a decreased nitrogen content. With continued tumor growth the animals go into a negative nitrogen balance, yet the tumor continues to gain weight. Thus, the tumor and the host's tissues appear to comprise two separate metabolic compartments.[6] A tumor continues to grow and sequester nutrients even in the face of starvation of the host organism: host tissues are, in fact, dismantled, by depletion of protein, fat, and carbohydrate, to feed the tumor. Mider et al.[8] have proposed the idea that malignant tumors are nitrogen "traps" in that the tumor continues to have a positive nitrogen balance in the face of protein loss from the host animal's tissues, and the nitrogen sequestered by the tumor does not become available to the host. In tumor-bearing animals that are force-fed, tumor growth is stimulated, but no sustained weight gain of the animal occurs.[7] However, it has been found that the relative growth of tumor and host tissues depends on the type of nutrient supplied.[9] If mammary carcinoma–bearing rats are fed carbohydrate alone, neither host nor tumor tissue growth is stimulated.

Amino acids alone, given in amounts adequate to improve host nitrogen balance, stimulate tumor growth. Adequate amino acids and carbohydrates given together induce optimal tumor growth and maintain host tissues. In contrast, when a diet that is isocaloric with the carbohydrate–amino acid diet, but consists of both fat and amino acids, is fed intravenously to the animals, host tissue is maintained and no stimulation of tumor growth is observed. These data suggest that the host's normal tissues and the mammary carcinoma have different primary mechanisms for energy metabolism. If these tumors have a high rate of anaerobic glycolysis, for example, they would be expected to consume large amounts of glucose or gluconeogenic precursors (at the expense of the host) during active growth. Thus, when animals are fed amino acids alone or amino acids plus glucose, tumor growth is stimulated. However, when calories are provided as a mixture of a nongluconeogenic substrate (i.e., fat) and amino acids, the host tissues, which can use nongluconeogenic substrates more efficiently than the tumor, regain some advantage for maintenance of their own growth. These kinds of results hold out the hope that appropriate alimentation of cancer patients could maintain their nutritional balance without stimulating tumor growth. In a study of patients with small cell lung cancer, however, only short-term gains in weight and caloric intake were achieved when these patients were placed on total parenteral nutrition for 4 weeks; no long-term differences in nutritional status were observed, nor was there any significant improvement in response rates to chemotherapy or in overall survival.[10]

Experiments in tumor-bearing animals indicate a number of mechanisms by which a tumor growing in a host can cause increased use of energy and body wasting even in the face of a normal caloric intake. A study of sarcoma-bearing rats, before the onset of cachexia, has shown that there is an increased rate of glucose turnover, secondary to both an increased rate of gluconeogenesis and an increased rate of recycling of glucose, mostly from normal tissues to tumor.[11] The tumor-bearing animals had significantly lower plasma glucose and higher blood lactic acid levels than those of non-tumor-bearing controls, but this abnormality was not due to changes in serum insulin or glucagon levels. The sarcoma-bearing rats also had a higher rate of glucose production than controls, most likely because of increased glucose synthesis from lactate in the host's liver and gluconeogenesis from amino acid precursors.

The explanation for these findings is that the tumor has a high rate of glucose utilization because of its high rate of glycolysis. This increased consumption of glucose by the tumor forces the host to increase endogeneous production of glucose to maintain glucose levels. To do this, the host increases the rate of both glucose recycling (from lactate) and gluconeogenesis. These are energy-inefficient processes, and along with the high metabolic demands of the tumor itself, place metabolic strains on the host that, even if normal caloric intake is maintained, can lead to weight loss and eventually to cachexia.

Other animal studies support these conclusions. For example, Cameron and Ord[12] found that rats bearing Morris hepatomas relied on gluconeogenesis to maintain glucose levels, mobilized more liver glycogen, catabolized more of their muscle proteins, and had higher blood lactate levels than normal control animals. Lindmark et al.[13] found that sarcoma-bearing mice had an increased fat oxidation and loss of body lipids as well as a significantly higher energy expenditure in relation to their food intake compared with pair-fed controls.

Studies in cancer patients also bear out the findings with tumor-bearing animals. Noncachectic lung cancer patients, under conditions of constant caloric and nitrogen intake, had a significantly higher turnover rate of total body protein, as measured by continuous infusion of [^{14}C]lysine, than did control individuals. In addition, muscle catabolism rates, determined by 3-methylhistidine/creatinine excretion rates, were elevated in the lung cancer patients. Glucose production rates were also higher in the patient group, but serum levels of ACTH, insulin, and glucagon were normal, as were 24-hour urinary cortisol levels. Thus, abnormalities in the levels of these hormones could not explain the increased protein turnover, glucose production, and muscle catabolism noted in these patients. Increased synthesis and breakdown of whole-body protein has also been observed in children with newly diagnosed leukemia or lymphoma.[15] In another study, comparing malnourished cancer patients with malnourished patients without cancer, it was found that the cancer patients had a doubling of glucose turnover, indicating that cancer patients have an increased glucose drain, which could account for a loss of about 0.9 kg of body fat per

30-day period.[16] In cachectic patients with colo-rectal cancer, elevated rates of glucose produc-tion and recycling by means of lactate, compared with those of a group of age-matched normal subjects, have been observed.[17]

In addition to the great nutrient demands of growing tumor tissue because of high growth fraction and their inefficient use of substrates (e.g., high rate of glycolysis), the demands placed on the body's defense systems may ex-pend a significant amount of energy. Maintain-ing cells with a high turnover rate, such as cells of the bone marrow involved in the production of granulocytes, lymphocytes, and monocytes, requires a tremendous amount of energy. Even under normal circumstances, the bone marrow produces several billion cells a day, and the in-creased demands placed on that system by a chronic illness such as cancer may be very sig-nificant. Decreased protein synthesis and loss of protein in the urine may also account for ni-trogen loss in some patients. For example, pa-tients with advanced cancer often have low al-bumin levels in their blood, and this is at least partly due to decreased albumin synthesis in the liver.[2] Many cancer patients also have significant proteinuria[18] and lose protein that way. Thus, the greatly increased demand on the cancer pa-tient's energy stores produces a generalized hy-permetabolic state with an increased rate of turnover of normal tissue components.

The production and release of certain sub-stances from tumors have sometimes been in-voked as the mechanism of cachexia in cancer patients. Nakahara and Fukuokia[19] observed that the decreased hepatic catalase activity often seen in tumor-bearing animals and patients[20] could be produced by injection of water-soluble, thermo-stable, and enthanol-precipitable material that was extracted from human gastric or rectal car-cinoma. They called this "toxohormone." This substance appears to be a polypeptide, and when it is injected into normal animals it produces, in addition to decreased liver catalase activity, de-creased levels of plasma iron, liver ferritin, and diphosphopyridine nucleotides. It also causes involution of the thymus, an enlarged liver and spleen, and an increased level of liver protopor-phyrin. Evaluation of data from a number of studies supports the concept that proteolytically generated peptides derived from tumors play a central role in producing cancer-related ca-chexia.[20] In addition, substances that uncouple oxidative phosphorylation in normal liver mito-chondria have been found in the rat Novkioff hepatoma[21] and in the serum of sarcoma-bearing rats.[22] A cytotoxic "diffusible polypeptide" that inhibits the growth of a variety of cells has been found in ascites or pleural fluids from tumor-bearing animals and cancer patients.[23]

The implication of these data is that tumors secrete substances that can have a far-reaching effect on host functions, and these substances may contribute significantly to the aberrant metabolism and cachexia observed in tumor-bearing animals. It remains to be proven, how-ever, that such substances cause these phe-nomena in the cancer patient.

Although the molecular pathways that lead to cachexia are still not clearly defined, current studies provide strong evidence for the role of inflammatory cytokines in this tissue wasting syn-drome. These include tumor necrosis factor α (TNF-α), interleukin-1, interleukin-6, interfer-on-α, and interferon-γ. It is the combination of such factors that appears to be involved, since these cytokines, by themselves, may not cause the profound cachexia seen in cancer patients. It is also now known that protein turnover medi-ated by the ubiquitin-proteasome system plays a key role (see below). Additional mechanisms that appear to play a central role in the cancer-related anorexia–cachexia syndrome are the presence of a chronic inflammatory state, cir-culating tumor-derived lipolytic and proteolytic factors, and oxidative stress leading to produc-tion of reactive oxygen species (reviewed in Ref-erence 3).

The progressive loss of skeletal muscle is a common event in many types of advanced can-cer. Muscle wasting is primarily due to activa-tion of the ubiquitin-proteasome system, which leads to the degradation of myosin and other muscle proteins. Kwak et al.[24] have shown that the E3 ubiquitin ligase E3α-II is up-regulated in two different animal models of cancer-induced cachexia, one of them a murine xenograft of the human C-26 colon cancer cell line. E3α-II is dif-ferentially activated by TNF-α or IL-6, which, as noted above, are major proinflammatory cyto-kines known to be involved in development of cachexia.

Cancer patients with weight loss also excrete in their urine a lipid-mobilizing factor (LMF) that stimulates lipolysis in isolated murine epididymal adipocytes.[25] This bioactivity was not detected in the urine of cancer patients without weight loss or in the urine of normal individuals. Purification of LMF showed it to be Zn-α_2-glycoprotein. These data, as well as similar data from a murine cachexia model, suggest that Zn-α_2-glycoprotein production by tumors is a key factor responsible for the lipid catabolism seen in cancer patients.

There are also data linking production of reactive oxygen species (ROS) to cancer-induced cachexia (reviewed in Reference 3). For example, in a murine model of muscle wasting and cachexia, the cachexia-inducing agent TNF-α induces ROS production, and this could be prevented by treatment with antioxidants in animal models of cachexia. Moreover, serum levels of an ROS protecting enzyme, glutathione peroxidase, were lower than normal in cancer patients and decreased further as cancer progression occurred, at a time when levels of TNF-α and IL-6 serum levels were increased.[3]

Hematologic Effects

Hematologic complications occur frequently in patients with cancer, particularly in those with disseminated cancer. Depression of the hematopoietic tissues of the bone marrow can occur as a result of direct invasion by cancer cells or the abnormal expansion of the immature bone marrow cell compartment that occurs in leukemia. Metastases to the bone marrow occur in patients with carcinomas of the breast, prostate, lung, adrenal, thyroid, and kidney; malignant melanoma; neuroblastoma; Hodgkin's disease; reticulum cell sarcoma; and other lymphomas. Bone marrow depression also occurs as a complication of chemotherapy or radiotherapy. As in the case of the nutritional abnormalities of cancer patients, indirect effects of cancer growing elsewhere than in the bone marrow can also be observed, possibly because of the remote effects of tumor-released products.

Erythropoiesis

The most common hematologic effect of cancer is anemia. Anemia can develop in cancer patients through a number of mechanisms, including malnutrition, blood loss due to neoplastic invasion of tissues, autoimmune hemolysis, and decreased red blood cell production and survival.[26] The decreased nutritional status of cancer patients can lead to a secondary anemia resulting from decreased levels of folic acid and other essential nutrients. Folic acid deficiency is common in patients who have difficulty eating, such as those with head and neck cancers, and in cachectic patients. Malabsorption of folate has been reported in patients with intestinal lymphoma, reticulum cell sarcoma, and leukemia and also in patients who have had bowel resections or extensive radiation to the abdomen. Decreased vitamin B_{12} absorption, coupled with anemia, is a common finding in patients with gastric carcinoma. Anemia secondary to ulcerating lesions is commonly seen in cancers of the gastrointestinal tract, head and neck, urinary bladder, and uterus. This blood loss may be occult and leads in time to an iron-deficiency anemia. Iron-deficiency anemia can also occur in cancer patients as a result of malabsorption or hemolysis. An immune-type hemolytic anemia resulting from production of antibodies directed against the patient's own erythrocytes is seen in certain types of cancer, particularly those involving the lymphatic and reticuloendothelial systems (e.g., chronic lymphocytic leukemia, Hodgkin's disease, and other malignant lymphomas). The mechanism of this apparent autoimmunity is not well understood. A number of cancer patients have a decreased total plasma iron-binding capacity and an increased rate of removal of iron from the plasma. In many of these patients, however, erythropoiesis is not decreased and, in fact, may be higher than normal, but the rate of erythrocyte destruction is elevated.[27] The decreased survival of erythrocytes appears to be due to their damage or destruction as they pass through the tumor mass[28] or to production of a hemolytic substance by the tumor[29] or to a combination of these factors.

The opposite of anemia, namely, a marked increase in circulating erythrocytes (*erythrocytosis*), occurs in some cancer patients, most notably those with renal tumors, cerebellar hemangioblastomas, and hepatocellular carcinomas, and less frequently in patients with uterine fibroma, pheochromocytoma, adrenal adenoma, ovarian carcinoma, or carcinoma of the lung.[26]

This erythrocytosis appears to be caused by the ectopic production of erythropoietin by the tumors.

Leukopoiesis

Leukopenia can occur in cancer patients as part of the bone marrow depression seen with marrow metastases. The mechanism is thought to be a "crowding out" of the hematopoietic cells of the marrow. A decrease in circulating white blood cells is also sometimes observed in cancer patients with tumor-involved enlarged spleens, presumably as a result of increased trapping of these cells by the spleen. The most common cause of leukopenia in cancer patients, however, is associated with the bone marrow–depressing effects of cancer chemotherapeutic agents and radiation therapy.

Increased circulating leukocytes (*leukocytosis*) is, of course, seen in leukemias, in which the nature of the disease process itself results in an overproduction of immature leukocytes, but it is also observed in patients with other cancers, usually metastatic ones. The stimulus for this leukocytosis is not understood, but may involve the release of colony-stimulating factors (CSFs) or other leukocyte-stimulating growth factors by tumors that favor leukopoiesis similar to the erythropoietic effect noted previously. Because many cancer patients have chronic infections of one sort or another, the leukocytosis seen in many patients may be secondary to the infectious process.

Platelets

Depression of the number of circulating platelets (thrombocytopenia) is the most common platelet abnormality observed in cancer patients. A decreased platelet count produces bleeding due to the accompanying blood coagulation defect. The most common cause of thrombocytopenia in cancer patients is impaired production of megakaryocytes, the platelet progenitor cell, in the bone marrow. The mechanism for this is similar to that described for leukopenia (see above)—that is, marrow infiltration by tumor and increased sequestration of platelets in the spleen. An immune type of thrombocytopenia, similar to that described for immune hemolysis and re-

sulting in increased platelet destruction, has also been described.[30] Increased platelet consumption also leads to a lowered platelet count; blood hypercoagulability, observed in some cancer patients, is responsible.

In some cases, hypercoagulability is widespread and leads to a phenomenon known as disseminated intravascular coagulation (DIC), characterized by an extensive activation of the blood coagulation system within the circulatory system and the deposition of fibrin clots in small blood vessels. It has been postulated that DIC is due to the release of thromboplastin-like substances by the tumor.[31,32] Some studies indicate that a low-grade DIC occurs in many cancer patients.[33] The intravascular coagulation that typifies this syndrome consumes clotting factors, particularly factors I (fibrinogen), II, V, VIII, and XIII, as well as platelets. Secondarily, the increased fibrin deposition activates the plasminogen system, eliciting an increased activity of the major fibrinolysin, plasmin, which in turn leads to the presence of fibrin degradation products in the blood. The most common clinical manifestation of DIC is bleeding because of the consumption of coagulation factors and platelets. The bleeding may be intermittent or continuous and tends to occur in the urinary tract, gastrointestinal tract, lungs, and skin. Organ failure (usually renal) secondary to localized obstruction of the microcirculation and shock due to generalized obstruction of the microcirculation are other clinical effects of DIC.

Thrombosis

Venous thromboembolism is a common occurrence in cancer patients and can be related to release of coagulation factors from the tumor, vascular stasis or injury, and effects of chemotherapeutic agents. Cancer patients are reported to have a sixfold higher risk of venous thrombosis compared to normal individuals, and cancer patients with this condition have a poor prognosis.[34] The reasons for the poor prognosis are not clear, but may relate to fatal pulmonary embolism or to the point that thrombosis is a marker for tumor aggressiveness and metastasis.

Tumor cells secrete procoagulant molecules, the best characterized of which are tissue factor (TF) and "cancer procoagulant."[35] TF is a single-chain, 263–amino acid transmembrane protein of 47 kDa molecular weight. It is a surface receptor and cofactor for activated coagulation protease Factor VII, which when bound to TF activates coagulation factors FXa and FIIa (thrombin). Some cancer chemotherapeutic agents as well as the chemopreventive agent tamoxifen and the supportive agent erythropoietin have been shown to be associated with increased risk of thrombosis (reviewed in Reference 35).

Fever and Infection

Fever, a common manifestation of malignant neoplastic disease, results from the systemic effects of the malignant process itself or from infection, although the latter is the most frequent cause. The incidence of fever is high in patients with advanced disease: about 70% of cancer patients experience febrile episodes during hospitalization.[36,37] The incidence of fever is related to the type of cancer. Patients with acute leukemia have elevated temperatures during about 50% of their hospitalization, and patients with Hodgkin's disease have elevated temperatures about 26% of the time. Other diseases with a high incidence of fever are lymphosarcoma and reticulum cell sarcoma. Fever is also common in patients with widely disseminated solid cancers, such as metastatic carcinomas of lung, kidney, pancreas, and gastrointestinal tract, and particularly in those patients with liver involvement.[36] The presence of fever often correlates with a poor prognosis. The mechanism of tumor-induced fever is not well established, but there is evidence for a fever-inducing (pyrogenic) substance in the urine of febrile patients with Hodgkin's disease[38] and in tissue culture fluids of lymphoid cell cultures derived from lymphoma patients.[39]

Patients with cancer have an increased incidence of infection with such pathogenic organisms as staphylococci, streptococci, and pneumococci, which frequently cause infections in patients without cancer. In addition, cancer patients become infected with "opportunistic" micro-organisms that are normally held in check by host defense mechanisms. These latter organisms include bacteria of the *Pseudomonas* genus, fungi, such as *Candida albicans*, and viruses such as herpes zoster. Infection is present at the time of death in 70% to 80% of patients with

leukemia or lymphoma and in 15% to 40% of patients with metastatic carcinoma.[36]

The type of infection depends to some extent on the nature of the impairment of the host's defense mechanisms (Table 8–2).[40] For example, patients with marked granulocytopenia most frequently develop infections of the respiratory track and gastrointestinal tract with organisms such as *Escherichia coli*, *Klebsiella pneumonia*, *Pseudomonas aeruginosa*, or *Staphylococcus aureus*. Patients whose primary deficiency is in cell-mediated immunity tend to develop infections with *Listeria*, *Salmonella*, and *Mycobacterium* species, for example. Patients with a primary deficiency in humoral immunity are at increased risk for infections by *Streptococcus pneumoniae* and *Hemophilus influenzae*.

Lung, skin, gastrointestinal tract, and urinary tract are the most common sites of infection in cancer patients. The site of infection is often determined by the location of the primary tumor or its metastases. For example, pneumonia and abscess formation in the lung occur peripherally to tumors blocking major bronchi, and urinary tract infections occur in patients whose tumors obstruct the ureter or urinary bladder. The increased susceptibility of cancer patients to infection, however, is most often related to depression of normal host defense mechanisms induced by the disease process itself or by chemotherapy or irradiation. Depression of granulocytes, the humoral immune response, and cell-mediated immunity all occur in cancer patients as a result of the disease process (e.g., bone marrow involvement) or immunosuppressive therapy. Granulocytopenia is responsible for many of the bacterial infections seen in cancer patients, especially those with hematologic malignancies or metastatic involvement of the marrow. Patients with chronic lymphocytic leukemia often have low levels of immunoglobulins, and bacterial infections occur in a high percentage of these patients. Cell-mediated immunity plays the key role in the body's defense against certain bacteria (e.g., *Mycobacterium tuberculosis*), fungi (e.g., *Cryptococcus*), and viruses (e.g., herpes), and infections with these organisms are frequently seen in patients with malignant lymphoma (e.g., Hodgkin's disease).

Before the widespread use of antibiotics, pneumococci, streptococci, and staphylococci

Table 8–2. Factors Predisposing to Infection among Patients with Cancer and Organisms Commonly Involved

---

GRANULOCYTOPENIA (E.G., ACUTE LEUKEMIA)

Usually with associated damage to body barriers (especially alimentary canal mucosa, respiratory tract ciliary function, and integument)

Common organisms

Gram-negative bacilli: *Pseudomonas aeruginosa*, *Klebsiella pneumoniae*, and *Escherichia coli*

Gram-positive cocci: *Staphylococcus aureus* and *Staphylococcus epidermis*

Yeasts: *Candida* species, *Torulopsis glabrata*

Fungi: *Asperigillus* species, *Mucor*

CELLULAR IMMUNE DEFICIENCY (E.G., LYMPHOMA)

Common organisms

Bacteria: *Listeria monocytogenes*, *Salmonella* species, *Mycobacterium* species, *Nocardia asteroids*, and *Legionella pneumophilia*

Viruses: varicella-zoster, herpes simplex, and cytomegalovirus (CMV)

Fungi: *Cryptococcus neoformans*, *Histoplasma capsulatum*, and *Coccidioides immitus*

Protozoa: *Pneumocystis carinii* and *Toxoplasma gondii*

Helminth: *Strongyloides stercoralis*

HUMORAL IMMUNE DYSFUNCTION (E.G., MULTIPLE MYELOMA)

Common organisms: *Streptococcus pneumoniae* and *Hemophilus influenzae*

OBSTRUCTION TO NATURAL PASSAGES (E.G., SOLID TUMORS)

Common sites: respiratory tract, biliary tract, and urinary tract

Common organisms: locally colonizing forms

CENTRAL NERVOUS SYSTEM DYSFUNCTION (E.G., BRAIN TUMORS)

Common sites: pneumonitis and urinary tract infection

Common organisms: locally colonizing forms

INFECTIONS ASSOCIATED WITH MEDICAL PROCEDURES

Procedures: intravascular catheters, urinary catheters, and respiratory assist devices

Common organisms: locally colonizing forms

From Pizzo and Schimpff[40]

were major causes of infection in cancer patients. More recently the biggest problem has been with such gram-negative bacilli such as *Pseudomonas aeruginosa*, *Klebsiella* species, and *E. coli*. About 25% of such infections are caused by *P. aeruginosa*, with this organism accounting for about 35% to 50% of the fatal septicemias in patients with acute leukemia.[36] The lung is the most common site of infection with *Pseudomonas* organisms. Yeast infections, particularly with *Candida* and *Aspergillus*, have become increasingly prevalent in hospitalized cancer patients, especially those with hematoligic malignancies or lymphomas. *Candida* infections are the most common; they usually occur in the mouth and gastrointestinal tract, but the infection may become disseminated. The incidence of viral infections is also high in cancer patients. For example, herpes zoster, which is rare in the general population over 50 years of age, occurs in 3% to 15% of patients with lymphoma, multiple myeloma, or chronic lymphocytic leukemia.[36] Because of the weakened condition of patients with advanced cancer and their markedly compromised defense mechanisms, their infections are extremely difficult to control and are often the ultimate cause of death. This problem is compounded by large increase in drug-resistant bacteria in recent years. In patients with hematologic malignancies and lymphomas, infection is the most common lethal event.

Despite the advances in therapy and supportive care for infections in cancer patients, emerging bacterial fungal and viral pathogens remain a large problem. Among the emerging bacterial pathogens are a number of drug-resistant species, including β-latamase-producing *Enterobacter* species, *Escherichia coli*, *Klebsiella pneumoniae*, and *Stentotrophomonas maltophilia* as well as multidrug-resistant *Pseudomonas aeruginosa* and *Acinetobacter* species.[41] Also of concern is the emergence of vancomycin-resistant *Staphylococcus aureus* and *Enterococcus* species. Treatment for some of these drug-resistant organisms is becoming increasingly problematic. In addition, the hospital environment is now a source for some of these drug-resistant organisms. The expanding spectrum of viral and fungal infections is an evolving challenge, particularly in immunosuppressed patients.

Common and emerging viral pathogens that infect cancer patients include the Herpes virus family (*Herpes simplex*, *Varicella zoster*, cytomegalovirus, human herpes virus-6, and Epstein-Barr virus), influenza, parainfluenza, respiratory syncytial virus, metapneumovirus, adenovirus species, and polyomavirus species.[41] Although there are antiviral drugs available to treat some of these viral infections, effective treatments are quite limited.

Candida and *Aspergillus* remain the most common fungal infections seen in cancer patients. Although *Candida albicans* is still the most common *Candida* pathogen infection observed, other *Candida* species infections are increasing in frequency.[42] *Aspergillus* infections are acquired through the respiratory tract, in contrast to *Candida*'s infectious route, which is via the gastrointestinal tract. *Aspergillus* infections are a major complication for patients with prolonged neutropenia. New antifungal agents are being developed, but other measures are being investigated, including neutrophil transfusion for neutropenic patients and various cytokines (G-CSF, GM-CSF, and γ-interferon).[42]

Hormonal Effects

In Chapter 4, the production of ectopic hormones was discussed. Since the biochemical form of the hormone released by the tumor may have low biologic activity (e.g., "big" ACTH secreted by lung tumors), the incidence of clinical syndromes related to ectopic hormones underestimates the actual frequency of this phenomenon. Nevertheless, a number of instances of clinical syndromes related to ectopic hormone production have been clearly defined. In addition, the clinical symptoms exhibited by patients with ectopic humoral syndromes are often more complex than would be expected from the overproduction of a single hormone. This is probably because complex host–tumor interactions may modify or mask the hormonal effects. For example, in a patient with severe infection and cachexia, abnormal physiologic events resulting from ectopic hormone production may be attributed to other causes. Also, some tumors produce multiple hormones, some of which may have different actions.[43,44] A clinical rule of thumb, however, is that "if a patient has overproduction of a hormone, look for a tumor, and if a patient has a tumor, look for evidence of hormone overproduction."[45]

Hypercalcemia

Elevated plasma calcium occurs in up to 20%–30% of patients with disseminated cancer[46] and is the most frequent paraneoplastic syndrome. This condition affects multiple organ systems, and the resulting events may be more immediately life threatening than the cancer itself. Manifestations of hypercalcemia include central nervous system effects (e.g., confusion, psychotic behavior, coma), gastrointestinal effects (e.g., decreased motility, ulceration), renal failure and salt loss by the kidney, and cardiac arrythmias and cardiovascular system collapse. Hypercaliemia in a cancer patient indicates a very poor prognosis: about 50% of patients die within 30 days.[46]

A number of mechanisms may operate in cancer patients to produce hypercalcemia. An obvious one is the direct invasion of bone by metastatic tumor cells. Carcinomas of the breast, lung, kidney, thyroid, ovary, and colon are tumors that commonly have bony metastases and produce hypercalcemia. This may lead to direct destruction of bone, with release of calcium into the bloodstream. Hypercalcemia has been observed, however, in patients in whom there is no evidence of bony metastasis, and may account for at least 15% of cases of malignant hypercalcemia.[47] In these cases, there is no obvious correlation between clinically detectable bone involvement and the degree of hypercalcemia.[48]

There are four types of hypercalcemia associated with cancer (Table 8–3):[46] (1) osteoclastic bone resorption due to bony metastasis or tumors growing in the marrow space, e.g., multiple myeloma; (2) humoral hypercalcemia of malignancy caused by systemic secretion of parathyroid hormone–related protein (PTHrP) by a cancer; PTHrP increases bone resorption and retention of calcium by the kidney; (3) secretion of vitamin D (1, 25,(OH_2) D) by lymphomas; vitamin D causes hypercalcemia as a result of bone resorption and enhanced intestinal absorption of calcium; and (4) ectopic secretion by tumors of parathyroid hormone (rare).

The cancers associated with hypercalcemia are usually large and most likely have already metastasized. An exception to this is the small neuroendocrine tumors such as islet cell tumors and pheochromocytomas.

Diagnosis of hypercalcemia in a cancer patient should be viewed as a medical emergency. A rapid increase in serum calcium can lead to marked neurological dysfunction and cardiovascular failure. Treatment includes hydration, diuretics,

Table 8–3. Types of Hypercalcemia Associated with Cancer°

Type	Frequency (%)	Bone Metastases	Causal Agent	Typical Tumors
Local osteolytic hypercalcemia	20	Common, extensive	Cytokines, chemokines, PTHrP	Breast cancer, multiple myeloma, lymphoma
Humoral hypercalcemia of malignancy	80	Minimal or absent	PTHrP	Squamous cell cancer (e.g., of head and neck, esophagus, cervix, or lung), renal cancer, ovarian cancer, endometrial cancer, HTLV-associated lymphoma, breast cancer
1,25(OH)$_2$D-secreting lymphomas	<1	Variable	1,25(OH)$_2$D	Lymphoma (all types)
Ectopic hyperparathyroidism	<1	Variable	PTH	Variable

Abbreviations: HTLV, human T-cell Lymphotrophic Virus; PTH, parathyroid hormone; PTH$_r$P, parathyroid hormone–related protein; 1,25(OH)$_2$D, 1,25 dihydroxyvitamin D.

(From Stewart,[46] reprinted with permission from the Massachusetts Medical Society.)

phosphate repletion, and intravenous bisphosphonates.[46] Second-line treatments include mithramycin and calcitonin. Bisphosphonates should be given as soon as hypercalcemia is diagnosed because it takes 2 to 4 days to obtain a therapeutic response. Mithramycin, which was the mainstay of treatment for acute hypercalcemia but has adverse effects such as bone marrow depression similar to other anticancer drugs, is now second-line therapy.

Neurologic Effects

Patients with neoplastic disease may have disorders of the nervous or muscular systems that are related to the presence of a malignant tumor, but not directly due to invasion or metastases of these systems. A term used to describe these effects is *carcinomatous neuromyopathy*. These disorders have been reported to occur in about 7% of cancer patients, most often in those with carcinomas of the lung, ovary, stomach, prostate, breast, colon, or cervix, in that order.[49] Although some of the neuromuscular effects can be explained on the basis of cachexia, many cannot. The most common abnormality is muscle wasting and weakness with decreased tendon reflexes, out of proportion to the degree of cachexia.[49] This is seen more often in patients with extensive local or metastatic disease. Actual degeneration and demyelinization of peripheral nerves is observed in some patients. A wide variety of paraneoplastic syndromes that affect the nervous system have been reported in cancer patients.[50] These include syndromes that affect the brain and cranial nerves (e.g., encephalitis and cerebellar degeneration), the spinal cord (e.g., necrotizing myelopathy), dorsal root ganglia (e.g., sensory neuronopathy), peripheral nerves (e.g., autonomic neuropathy), neuromuscular junction (e.g., myasthenia gravis), and muscle (e.g., polymyositis).

The cause of many of these effects is unknown, but they may be due to the release of toxic biologically active substances by malignant neoplasms. A syndrome known as paraneoplastic neurological degeneration (PND) is known to be caused by an autoimmune response to neuronal antigens expressed in cancer cells.[51] These PND antigens were identified by using antisera from patients with a PND to screen expression complementary DNA libraries. The PND immune response is characterized by the presence of PND antigen-specific CD8$^+$ cytotoxic cells (CTCs) in the blood of patients with PND. The activation of CD8$^+$ CTCs occurs in lymph nodes and is dependent on the presence of CD4$^+$ T-helper cells. This indicates that antigenic tumor-derived polypeptides produced by cancers are shared by central nervous system (CNS) cells (so-called onconeural antigens) and that the activated CTCs can enter the CNS to produce PND. They may do so via areas where tumors have compromised the blood–brain barrier.

Patients with PND are typically unaware that they have cancer, and their first clue may be the appearance of a neurological symptom such as imbalance, memory loss, muscle weakness, or vision loss. Most commonly, PNDs are observed in patients with breast, ovarian, or small cell lung cancers.

Dermatologic Effects

Cutaneous manifestations are frequent concomitants of certain forms of cancer.[52] Examples of this include hyperpigmentation of the skin, reddening or flushing (erythema), bleeding into the skin (purpura), abnormal hair growth (hirsutism), and itching (pruritus). Skin pigmentation can occur as a result of ectopic release of ACTH or melanocyte-stimulating hormone (MSH) or as a result of excess corticosteroid release by adrenal gland invasion. An unusual dark hyperpigmentation of the skin associated with skin hyperplasia, called acanthosis nigricans, occurs in the axillae, lower back, neck, groin, and antecubital spaces of some patients with carcinomas of the gastrointestinal tract, uterus, prostate, ovary, or kidney or with lymphomas. Erythema tends to occur more frequently in patients with lymphomas, leukemias, and carcinomas of the gastrointestinal tract, breast, lung, and cervix. Bleeding into the skin may take the form of small, pinpoint purplish red spots (petechiae) or larger, more diffuse areas of hemorrhage (purpura). These occur in patients with thrombocytopenia resulting from marrow involvement or as a result of chemotherapy or radiation therapy. Hirsutism occurs with some ovarian cancers and adrenal tumors that secrete male hormones. Pruritus is common in Hodgkin's disease and other lymphomas, and also occurs occasionally in patients with various carcinomas (e.g., of the pancreas, stomach, brain).

In some cases, the causes of the dermatologic effects of cancer are clearly defined, for example, as in the case of the known biologic effects of excess hormonal production, or the results of decreased platelet count, but in others (e.g., pruritus, acanthosis nigricans, erythema) the cause is not known.

Kaposi's sarcoma is a malignancy of endothelial cell origin that is manifested by vascular, raised or flat, pink, brown, or blue lesions that can arise anywhere on the skin or in the mucosa of the mouth or gastrointestinal tract. The lesions can resemble bruises or nevi and occasionally they appear first in enlarged lymph nodes. Although this used to be a rare tumor, usually seen in elderly men of Mediterranean descent, it occurs with increased frequency in AIDS patients.

Fatigue

Fatigue is a prevalent and distressing symptom for cancer patients. It affects 70%–100% of cancer patients, and cancer survivors report fatigue as a problem months to years after treatment ends (reviewed in Reference 53). In one survey, a third of patients reported severe and persistent fatigue 3 years after diagnosis. The National Comprehensive Cancer Network (NCCN) defines cancer-related fatigue as "a persistent, subjective sense of tiredness related to cancer or cancer treatment that interferes with usual functioning."[53] The specific mechanisms involved in cancer-related fatigue are in many cases unknown. In some patients, anemia, cachexia, tumor burden, and release of certain cytokines appear to be causative, but this is not an explanation for the symptom in a fairly large number of individuals. For example, many patients who are not anemic report high levels of fatigue. Even patients who have regained their normal weight and are not cachetic complain of severe fatigue. Although the extent of tumor burden seems like a logical explanation for those patients with residual disease, there is often no correlation. In some instances, cytokines that may contribute to fatigue by exerting effects on the endocrine system or on neurotransmitter release are elevated in patients' sera. However, there is no clear relationship here either.

Treatment options include treatment with erythropoietin (Epo) or psychostimulants. Epo treatment has shown benefit for cancer-related anemia and fatigue, but as noted above, anemia isn't always the problem. There are only limited data for the use of stimulants for reducing cancer-related fatigue. The clearest positive data are for exercise, as a way to increase energy and decrease the symptoms of fatigue. Eleven published reports show significantly lower levels of fatigue in individuals who exercised than in those who

Table 8-4. Recognized Late Effects of Cancer Treatment

Reproductive problems: infertility, premature menopause
Endocrine changes: thyroid disease, vasomotor symptoms
Osteoporosis
Sexual dysfunction
Decreased energy
Cognitive complaints
Pain syndromes secondary to surgery and radiation
Psychosocial concerns: increased sense of vulnerability
  and worry, hypervigilance
Socioeconomic: job discrimination, insurance concerns
Cardiac injury
Pulmonary fibrosis
Renal failure
Pregnancy loss
Low birth weight of offspring
Second primary cancers

survivors now make up about 3.5% of the U.S. population (reviewed in Reference 54) and now number about 10 million individuals. This includes survivors of breast cancer (22% of the total number of survivors), prostate cancer (17%), colorectal cancer (11%), gynecologic cancer (10%), hematologic malignancies (7%), genitourinary cancers (7%), and melanoma (6%). However, these successes are not without some secondary effects that can be of significant consequence to patients (Table 8–4). These late effects can include infertility, endocrine changes, cognitive complaints, pain, job discrimination, and secondary late cancers. There may also be significant sequelae of treatments, including cardiac injury, pulmonary fibrosis, and renal failure.

A fair number of anticancer drugs have cardiotoxic effects (Table 8–5).[55] These effects can be exacerbated by combination chemotherapy employing those drugs with molecular-targeted agents such as Herceptin and other ERB receptor inhibitors (e.g., EGFR antagonists).

Nausea and vomiting is one of the most common side effects of chemotherapy, but can be significantly moderated in most patients by the newer antinausea drugs. What bothers many patients the most, particularly women, is hair loss (alopecia). Many anticancer drugs cause hair loss because these drugs target cells with

didn't (reviewed in Reference 53). Most of these studies employed moderate exercise such as walking for 30 minutes four to five times a week.

SEQUELAE OF CANCER TREATMENT

For many cancers, 5-year or longer survival has increased significantly in the last two decades, despite the fact that overall mortality rates for several of the major cancers, e.g., breast, colon, lung, and prostate, haven't changed very much for the past 30 years (see Chapter 3). Cancer

Table 8–5. Cardiotoxicity of Select Chemotherapeutic Agents

Drug	Toxic Dose Range	Toxicities
Doxorubicin	400–550 mg/m^2 (emerging data implicate lower doses, particularly in setting of radiation)	Arrythmia, pericarditis–myocarditis syndrome, myocardial infarction, sudden cardiac death, cardiomyopathy, congestive heart failure
Mitoxantrone	>100–140 mg/m^2	Congestive heart failure, decreased left ventricular ejection fraction, myocardial infarction, ECG changes, arrhythmia
Cyclophosphamide	>100–120 mg/kg (over 2 days)	Hemorrhagic cardiac necrosis, reversible systolic dysfunction, ECG changes, congestive heart failure
Ifosfamide		ECG changes, congestive heart failure, arrhythmias
Cisplatin	Standard dose	Myocardial ischemia, Raynaud's phenomenon, ECG changes
Fluorouracil	Standard dose	Myocardial infarction, angina, cardiogenic shock, sudden death, dilated cardiomyopathy
Trastuzumab		Ventricular dysfunction, congestive heart failure, cardiomyopathy
Paclitaxel	Standard dose	Sudden death, bradyarrhythmia, myocardial dysfunction, myocardial infarction

From Hale et al.,[55] reprinted with permission from the American Society of Clinical Oncology.

a high division rate. Unfortunately, a number of normal tissues also have high cell division rates, including the bone marrow and hair follicles. Some approaches to preventing chemotherapy-induced alopecia have been tried over the years, such as sulfhydryl reagents that react with and inactivate alkylating agents, but none have proven very effective. A newer approach is the use of inhibitors of cyclin-dependent kinase 2 (CDK2). Davis et al.[56] applied a CDK2 inhibitor, formulated in dimethyl sulfoxide, topically to newborn rats treated with etopside or a cyclophosphamide-doxorubicin combination. This treatment prevented alopecia in 33%–50% of the animals. In another set of experiments, they transplanted human scalp hair onto immunodeficient mice. In this case, CDK2 inhibitor reversibly inhibited hair follicle division, suggesting that the inhibitor could slow cell division of human hair follicles and take them out of the cell division cycle long enough to protect them from an infusion of cytotoxic anticancer drugs. Importantly, the investigators did not detect any interference with the ability of anticancer drugs to kill cancer cells in tumor-bearing animal models, perhaps because the CDK2 inhibitor is applied topically, which would limit its systemic absorption.

References

1. T. C. Hall: The paraneopiastic syndromes. In P. Rubin, ed.: Clinical Oncology, 4th ed. Rochester: American Cancer Society, 1974, pp. 119–128.
2. C. Waterhouse: Nutritional disorders in neoplastic disease. J Chron Dis 16:637, 1963.
3. G. Mantovani, C. Madeddu, A. Macciò, G. Gramignano, M. R. Lusso, et al.: Cancer-related anorexia/cachexia syndrome and oxidative stress: An innovative approach beyond current treatment. Cancer Epidemiol Biomarkers Prev 13:1651, 2004.
4. W. DeWys: Metastases and disseminated cancer. In P Rubin, ed.: Clinical Oncology, 4th ed. Rochester: American Cancer Society, 1974, pp. 508–525.
5. W. DeWys: Working conference on anorexia and cachexia of neoplastic disease. Cancer Res 30:2816, 1970.
6. S. D. Morrison: Partition of energy expenditure between host and tumor. Cancer Res 31:98, 1971.
7. A. Theologides: Pathogenesis of cachexia in cancer: A review and a hypothesis. Cancer 29:484, 1972.
8. G. B. Mider, H. Tesluk, and J. J. Morton: Effect of walker carcinoma 256 on food intake, body weight, and nitrogen metabolism in growing rats. Int J Cancer 6:409, 1948.
9. G. P. Buzby, J. L. Muller, T. P. Stein, E. E. Miller, C. L. Hobbs, et al.: Host–tumor interaction and nutrient supply. Cancer 45:2940, 1980.
10. W. K. Evans, R. Makuck, G. H. Clamon, R. Feld, R. S. Weiner, et al.: Limited impact of total parenteral nutrition on nutritional states during treatment for small cell lung cancer. Cancer Res 45:3347, 1985.
11. M. E. Burt, S. F. Lowry, C. Gorshboth, and M. E. Brennan: Metabolic alterations in a noncachectic animal tumor system. Cancer 47:2138, 1981.
12. I. L. Cameron and V. A. Ord: Parenteral level of glucose intake on glucose homeostasis, tumor growth, gluconeogenesis, and body consumption in normal and tumor-bearing rats. Cancer Res 43:5228, 1983.
13. L. Lindmark, S. Edstrom, L. Ekman, I. Karlberg, and K. Lundholm: Energy metabolism in non-growing mice with sarcoma. Cancer Res 43:3649, 1983.
14. D. Heber, R. T. Chlebowski, D. E. Ishibashi, J. N. Herroid, and J. B. Block: Abnormalities in glucose and protein metabolism in noncachectic lung cancer patients. Cancer Res 43:4815, 1982.
15. C. L. Kien and B. M. Camitta: Increased whole body protein turnover in sick children with newly diagnosed leukemia or lymphoma. Cancer Res 43:5586, 1983.
16. E. Edén, S. Edström, K. Bennegard, T. Scherstén, and K. Lundholm: Glucose flux in relation to energy expenditure in malnourished patients with and without cancer during periods of fasting and feeding. Cancer Res 44:1718, 1984.
17. C. P. Holroyde, C. L. Skutches, G. Boden, and G. A. Reichard: Glucose metabolism in cachectic patients with colorectal cancer. Cancer Res 44:5910, 1984.
18. D. Rudman, R. K. Chawla, D. W. Nixon, W. R. Vogler, J. W. Keller, and R. C. MacDonnell: A system of cancer-related urinary glycoproteins: biochemical properties and clinical applications. Trans Assoc Am Physicians 90:286, 1977.
19. W. Nakahara: Toxhormone. In H. Busch, ed.: Methods in Cancer Research, Vol. II. New York: Academic Press, 1967, pp. 203–237.
20. H. Rubin: Cancer cachexia: Its correlations and causes. Proc Natl Acad Sci USA 100:5384, 2003.
21. T. M. Devlin and M. P. Pruss: Oxidative phosphorylation in transplanted Novikoff hepatoma of the rat. Fed Proc 17:211, 1958.
22. G. Nanni and A. Casu: In vitro uncoupling of oxidative phosphorylation in normal liver mitochondria by serum of sarcoma bearing rats. Experientia 17:402, 1961.
23. B. Sylvén and B. Holmberg: On the structure and biological effects of a newly discovered cytotoxic polypeptide in tumor fluid. Eur J Cancer 1:199, 1965.

24. K. S. Kwak, X. Zhou, V. Solomon, V. E. Baracos, J. Davis, et al.: Regulation of protein catabolism by muscle-specific and cytokine-inducible ubiquitin ligase E3α-II during cancer cachexia. *Cancer Res* 64:8193, 2004.

25. P. T. Todorov, T. M. McDevitt, D. J. Meyer, H. Ueyama, I. Ohkubo, et al.: Purification and characterization of a tumor lipid-mobilizing factor. *Cancer Res* 58:2353, 1998.

26. W. B. Kremer and J. Laszio: Hematologic effects of cancer. In J. F. Holland and E. Frei III, eds.: *Cancer Medicine*. Philadelphia: Lea & Febiger, 1973, pp. 1065–1099.

27. G. A. Hyman: Anemia in malignant neoplastic disease. *J Chron Dis* 16:645, 1963.

28. V. E. Price, R. E. Greenfield, W. R. Sterling, and R. C. MacCardle: Studies on the anemia of tumor-bearing animals. III. Localization of erythrocyte iron within the tumor. *J Natl Cancer Inst* 22:877, 1959.

29. K. Oh-Uti, S. Inoue, and S. Minato: Purification of an anemia-inducing factor from human placenta and its application to diagnosis of malignant neoplasms. *Cancer Res* 40:1686, 1980.

30. S. Ebbe, B. Wittels, and W. Damesbek: Autoimmune thrombocytopenic purpura ("TTP") type with chronic lymphocytic leukemia. *Blood* 19:23, 1962.

31. R. B. Davis, A. Theologides, and B. J. Kennedy: Comparative studies of blood coagulation and platelet aggregation in patients with cancer and nonmalignant diseases. *Ann Intern Med* 71:67, 1969.

32. S. P. Miller, J. Sanchez-Avalos, T. Stefanski, and L. Zuckerman: Coagulation disorders in cancer. I. Clinical and laboratory studies. *Cancer* 20:1452, 1967.

33. S. D. Peck and C. W. Reiquam: Disseminated intravascular coagulation in cancer patients: Supportive evidence. *Cancer* 31:1114, 1973.

34. J. A. Heit, M. D. Silverstein, D. N. Mohr, et al.: Predictors of survival after deep vein thrombosis and pulmonary embolism: A population-based cohort study. *Arch Intern Med* 159:445, 1999.

35. M. N. Levine, A. Y. Lee, and A. K. Kakkar: Thrombosis and cancer. *Proc Am Soc Clin Oncol: Education Book*, p. 748, 2005.

36. G. P. Bodey: Infections in patients with cancer. In J. F. Holland and E. Frei III, eds.: *Cancer Medicine*. Philadelphia: Lea & Febiger, 1973, pp. 1135–1165.

37. R. T. Silver: Fever and host resistance in neoplastic diseases. *J Chron Dis* 16:677, 1963.

38. J. E. Sokal and K. Shimaoka: Pyrogen in the urine of febrile patients with Hodgkin's disease. *Nature* 215:1183, 1967.

39. P. Bodey: Pyrogen production in vitro by lymphoid tissue in malignant lymphoma. *J Clin Invest* 51:11a, 1972.

40. P. A. Pizzo and S. C. Schimpff: Strategies for the prevention of infection in the myelosuppressed

41. or immunosuppressed cancer patient. *Cancer Treat Rep* 67:223, 1983.

41. T. J. Walsh, P. Lin, and K. Cortez: Emerging bacterial and viral pathogens in pediatric patients with cancer: Approaches to management. *Proc Am Soc Clin Oncol: Education Book*, p. 848, 2005.

42. N. L. Seibel: New developments in diagnosis and treatment of pediatric fungal infections. *Proc Am Soc Clin Oncol: Education Book*, p. 844, 2005.

43. L. H. Rees, G. A. Bloomfield, G. M. Rees, B. Corrin, L. M. Franks, et al.: Multiple hormones in a bronchial tumor. *J Clin Endocrinol Metab* 38:1090, 1974.

44. Y. Hirata, S. Matsukura, H. Imura, T. Yakura, S. Ihjima, et al.: Two cases of multiple hormone-producing small cell carcinoma of the lung: Coexistence of tumor ADH, ACTH, and β-MSH. *Cancer* 38:2575, 1976.

45. G. W. Liddle and J. H. Ball: Manifestations of cancer mediated by ectopic hormones. In J. F. Holland and E. Frei III, eds.: *Cancer Medicine*. Philadelphia: Lea & Febiger 1973, pp. 1046–1057.

46. A. F. Stewart: Hypercalcemia associated with cancer. *N Engl J Med* 352:373, 2005.

47. D. Chopra and E. P. Clerkin: Hypercalcemia and malignant disease. *Med Clin North Am* 59:441, 1975.

48. A. Besarab and J. F. Caro: Mechanisms of hypercalcemia in malignancy. *Cancer* 41:2276, 1978.

49. E. P. Richardson: Neurological effects of cancer. In J. F. Holland and E. Frei III, eds.: *Cancer Medicine*. Philadelphia: Lea and Febiger, 1973, pp. 1057–1067.

50. R. B. Darnell and J. B. Posner: Paraneoplastic syndromes involving the nervous system. *N Engl J Med* 349:1543, 2003.

51. M. L. Albert and R. B. Darnell: Paraneoplastic neurological degenerations: keys to tumour immunity. *Nat Rev Cancer* 4:36, 2004.

52. W. L. Dobes and R. R. Kierland: Dermatologic effects of cancer. In J. F. Holland and E. Frei III, eds.: *Cancer Medicine*. by Philadelphia: Lea and Febiger, 1973, pp. 1067–1074.

53. K. Ahlberg, T. Ekman, F. Gaston-Johansson, and V. Mock: Assessment and management of cancer-related fatigue in adults. *Lancet* 362:640, 2003.

54. P. A. Ganz: Late effects of cancer in adult survivors: What are they and what is the oncologist's role in follow-up and prevention? *Proc Am Soc Clin Oncol: Education Book*, p. 724, 2005.

55. E. R. Hale, S. E. Lipshultz, and L. S. Constine: Latent cardiac injury following the double-edged sword of chemotherapy and radiation. *Proc Am Soc Clin Oncol: Education Book*, p. 739, 2005.

56. S. T. Davis, B. G. Benson, H. N. Bramson, D. E. Chapman, S. H. Dickerson, et al.: Prevention of chemotherapy-induced alopecia in rats by CDK inhibitors. *Science* 291:134, 2001.

9

Cancer Prevention

To think about cancer prevention, one needs to consider the causes of cancer (Chapter 2) and the epidemiology of cancer (Chapter 3) as well as how tumors progress from an early, indolent, or less aggressive type into a full-blown invasive, metastatic cancer. We live in a "sea" of carcinogens in modern society. This is a fact of life. We are continually exposed to environmental agents that can potentially cause cancer in susceptible individuals. A key question then is: Can we adopt a lifestyle or take minimally toxic agents that lessen this risk? The answer to the first part of this question is clearly "yes." The dangers of cigarette smoking are clearly defined. Obesity is linked to increased risk of certain cancers. Some dietary elements are suggested to help, based on epidemiological data, although some of these data are still controversial. Early and prolonged exposure to estrogen or other hormones plays a role in breast cancer. Thus, there are some things that people can do to help decrease risk. The answer to the second part of the question relating to effective chemopreventive agents is less clear, but there are some encouraging data that suggest this is a possibility.

One of the key "causes" of cancer is aging. Unfortunately, none of us can avoid it, but the accumulation of molecular insults over a lifetime must play an important role, because the incidence of cancer is clearly age related. There are, of course, a number of childhood malignancies and some hereditary susceptibility genes that cause cancer to occur at an earlier age, but by and large, the most common cancers such as colon, lung, breast, and prostate tend to occur later in life.

Thus, one of the ways to think about prevention is to consider the mechanisms that lead to aging and how they may contribute to causing cancer.

MOLECULAR MECHANISMS OF AGING AND ITS PREVENTION

There are a number of proposed mechanisms for the aging process, which are discussed below.

Somatic Mutation

Numerous investigations have shown that DNA damage and somatic mutations occur in aging animals and during cancer progression in humans. As noted in Chapter 3, it may take 15–20 years for a tumor to progress to an invasive, metastatic phenotype in humans. During this time, a series of chromosomal derangements and DNA mutations occur. These data suggest that the capacity for DNA repair is an important determinant of age-related conditions, including cancer. Indeed, there is a positive relationship between longevity and DNA repair capability[1]—that is, people who live longer have better and longer-lasting DNA repair functions.

Telomere Loss

There is a decline in cellular-division capacity with age that can be detected in human somatic tissues and cells taken from aged animals. This decline correlates in many instances with a shortening of telomeres, which get progressively shorter with continuous cell division, leading to cellular senescence. However, oxidative stress–induced DNA damage, such as that caused by reactive oxygen species, appears to be a more important cause of telomere loss than the DNA–end replication effect.[1] Telomere shortening is significantly accelerated in cells exposed to oxidative stress.

Mitochondrial Damage

There is evidence for an age-related increase in cytochrome c oxidase–deficient cells that is associated with mitochondrial DNA mutations in human muscle, brain, and gastrointestinal tract (reviewed in Reference 1). This increase leads to an age-dependent decline in mitochondrial integrity and function. The amount of damage to mitochondrial DNA can be assessed by the presence of 8-oxo-2'-deoxyguanosine in the DNA, which may result from inefficient electron exchange of the aging mitochondrial.[2]

Formation of Oxygen-Free Radicals

Harman proposed the "free radial theory" about 50 years ago.[3] The theory is based on the idea that aging and its associated degenerative diseases are caused by the effects of free radicals generated in cells by metabolic functions on cellular components. These free radicals come from "the interaction of the respiratory enzymes involved in the direct utilization of molecular oxygen."[3] The aerobic metabolic pathway does indeed generate reactive oxygen species (ROS) that can attack DNA and other cellular components; however, cells have protective enzymes such as superoxide dismutase and catalase that can scavenge ROS. Although it isn't clear precisely how ROS generation relates to aging and cancer, there is a lot of circumstantial evidence that it does (reviewed in Reference 2).

The vast majority of ROS are produced in mitochondria during the process of oxidative phosphorylation (Fig. 9–1). This is the energy-generating machinery of the cell essential for life in higher organisms. Contrary to many popular theories, a higher rate of oxidative phosphorylation and oxygen consumption does not necessarily lead to more ROS generation. In this case, the oxidative system can actually operate more

efficiently and produce less ROS,[2] which may account for some marathon runners being so long-lived. However, lowering the ambient oxygen concentration can extend the generation time span of cells (Hayflick number), as does increasing the intracellular levels of superoxide dismutase. Interestingly, some of the metabolic genes in the TCA (Krebs) cycle can act as tumor suppressor genes, and genes that slow aging decrease the development of chronic degenerative diseases (reviewed in Reference 2).

Cell Senescence

Replicative senescence, as defined by Hayflick and colleagues, describes the number of cellular doublings that occur before cells stop dividing in culture. This is related to the shortening of telomeres with each cell division. In vivo, however, cellular senescence in renewable tissues occurs in response to diverse stress responses, DNA damage (which turns on p53, rB, and other tumor suppressor genes), overexpressed mitogenic signals, chromatin alterations, and other phenotypic changes that occur in damaged cells. It is also apparent that a number of oncogenic stimuli induce a cell senescence response,[4] including unrepaired DNA strand breaks and other types of unrepaired DNA damage as well as epigenetic changes to chromatin organization. Generation of ROS also appears to play a role in this, since the Ras mitogenic pathway, for example, stimulates ROS production, which is in turn responsible for at least part of the Ras effect on mitogenesis and induction of cell senescence (reviewed in Reference 4).

The accumulation of senescent cells may hinder tissue renewal capabilities. Senescent fibroblasts, for example, secrete high levels of matrix metalloproteinases, epithelial growth factors, and inflammatory cytokines.[4] This phenotype is associated with a cancer-like stroma that facilitates the progression of carcinomas. Thus, an understanding of how this senescent phenotype is regulated could provide another therapeutic target.

DNA Repair and Genome Stability

There is a significant amount of data supporting the conclusion that DNA damage and mutations accumulate with age in mammals (reviewed in

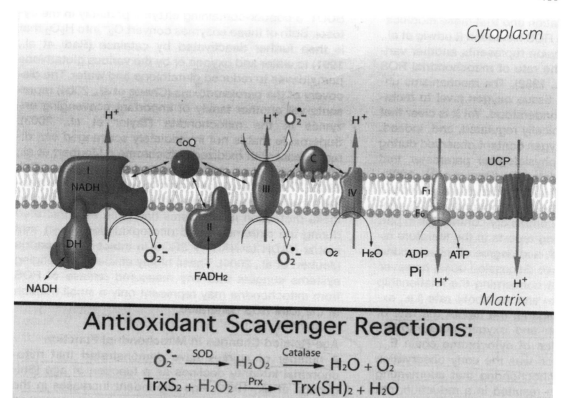

Figure 9–1. Schematic model of ROS generation in the mitochondria. The major production sites of superoxide anions at sites I and III are identified along with the major ROS scavenging pathways. Antioxidant enzymes include various isoforms of peroxiredoxin (Prx), superoxide dismutase (SOD), and glutathione peroxidase (GP). The scavenging reaction of the peroxiredoxin family requires other cellular dithiol proteins such as thioredoxin (TrxS$_2$). Similarly, the enzymatic action of GP requires reduced glutathione (GSH). Specific family members of SOD, GP, and Prx are found inside the mitochondria, while other family members localize to the cytosol or extracellular space. The different complexes of oxidative phosphorylation are color coded with regard to the magnitude of E$_{ox}$ for reducing oxygen, with red (dehydrogenases [DH] and site I) having the highest potential and pink (site IV) the lowest potential. The family of uncoupling protein (UCP), here denoted in green, reduces the overall mitochondrial membrane potential ($\Delta\Psi$). This is believed to result in a generalized decrease in E$_{ox}$ for both sites I and III and hence a reduction in ROS formation. (From Balaban et al.,[2] reprinted with permission from Elsevier.)

Reference 5). In addition, chromosome derangements such as translocations, insertions, dicentric chromosomes, and acentric fragments increase with age in mammalian cells. These DNA and chromosomal alterations vary from tissue to tissue, most likely reflecting their rate of cell division and decreased activity of DNA repair systems.

Damage to cells' genomic apparatus comes from both external and intrinsic sources. The former includes exposure to irradiation, cytotoxic chemicals, sunlight, tobacco smoke, etc. Intrinsic sources of damage include ROS, genetic susceptibility mutations, and DNA base–deamination reactions. Both intrinsic and extrinsic damage would be expected with age.

Studies of genetic mutations that cause premature aging support the concept that accumulated damage in aging cells can lead to malignant transformation. Some of the mutations associated with premature aging that increase the incidence rate of cancer and the cellular processes that these mutations affect are listed in Table 9–1.[5]

Caloric Restriction

It has been known for over 70 years that limiting food consumption in rodents increases their life span (reviewed in Reference 6). This effect is also true for yeast, round worms (*C. elegans*), fruitflies (*Drosophila*), and most likely primates. Since nature has a way of creating an evolutionary continuum, it is probably also true for humans. This observation relates to the number of calories in the diet; hence the term *caloric restriction* (CR) is used to define the phenomenon. In general, caloric-restricted diets contain 60%–70% of what animals would eat ad libitum. The phenotype that a CR diet produces includes lower body temperature, blood glucose, and insulin levels, reduced body fat, and lower total body weight.[6] Interestingly, although the size of organs in such CR-fed animals is lower, brain size is not reduced. CR animals are also more resistant to temperature and oxidative stress. Evolutionarily this makes sense, because an organism that could survive times of food scarcity would have a reproductive advantage.

Caloric-restriction diets not only slow the aging process but also prevent the onset of late-onset diseases, including cancer. An example

Table 9–1. Mutations Associated with Premature Aging

Mutant Gene	Process Affected
Atm	DNA double-strand break repair
BRACA1$^{\Delta11/\Delta11}$/*p53*$^{+/-}$	Double-strand break and other DNA repair mechanisms
DNA-dependent protein kinase catalytic subunit	Nonhomologous end-joining DNA repair
Terc/Wrn	Telomere maintenance and DNA repair
Terc/Wrn/Blm	Telomere maintenance and DNA repair

from animal studies is that CR extended the life span of tumor suppressor–deficient mice (e.g., p53$^{-/-}$), who have a high frequency of cancers and die early (reviewed in Reference 6).

The genes that play a role in regulating the CR response include the *sir2* (silent information regulator 2) family of genes. *sir2* was first found in yeast where it mediates gene-silencing events. The *sir2* ortholog in mammals is called *sirt1*, and it appears to mediate physiological events that result from a CR diet (reviewed in Reference 6). *sir2* also has an ortholog in *C. elegans* and its expression is a determinant of life span in that organism. Since yeast and *C. elegans* diverged about one billion years ago, the presence of this ortholog suggests evolutionary conservation of this process.

The *sir2* gene product (called sirtuin) is a NAD-dependent histone deacetylase and the *sirt2* product in mammalian cells deacetylates histones and nonhistone substrates. Since histone deacetylation is a mechanism to shut down expression of some genes (see Chapter 5), this suggests a way that the "metabolic thermostat" could be turned down, allowing cells to survive longer. Support for the role of the *sir2*-related genes in CR comes from experiments in which the *sir2* gene was deleted in yeast (CR did not extend life span in this case) and in which CR was shown to increase the silencing activity of *sir2*.[7] Moreover, a group of compounds known as STACs (sirtuin-activating compounds) extend the replicative life span of yeast, *C. elegans*, and *Drosphila* as well as of human cells in culture.[8] One of the STACs, resveratrol, that does this is a naturally occurring polyphenol antioxidant found in raspberries, blueberries, peanuts, grapes, grapeskins, and red wine. Anecdotal epidemiological data indicate that one or more glasses of wine a week may lower the risk of upper digestive tract cancers and cardiovascular disease. Dark chocolate also contains similar anti-oxidants, making it and wine an attractive dietary regimen for wine and chocolate lovers.[9,10]

Another mechanistic link of *sirt1* activity to cancer prevention comes from studies showing that *sirt1* is a negative regulator of PPARγ activity. PPARγ overexpression is associated with aging changes and cancer progression in colon, bladder, breast, and prostate (reviewed in Reference 6). Thus, inhibition of PPARγ activity may be another mechanism of cancer prevention by *sirt*1 activation.

DIET AND CANCER PREVENTION

A number of studies support the hypothesis that a diet high in fat and low in carbohydrates, fruits, fruit and vegetable fiber increases the risk of colorectal[11] and others types of cancer.[12] A considerable amount of evidence from both animal experiments and human epidemiological studies suggests that high-fat and high-caloric diets increase the risk of cancer.[13] Conversely, caloric restriction has been shown to reduce the incidence of chemically induced tumor development in the skin[14] and pancreas[15] of experimental animals.

There are a number of potentially cancer-preventive ingredients in foods (Fig. 9–2)[16] that can act at various stages in tumor initiation and promotion (Fig. 9–3). This information and additional data on naturally occurring food constituents and vitamins (see below) have led to the idea that diet modification can be a way to prevent cancer.[17] One such nutrition intervention trial has been carried out in a region in China where the mortality rates of gastric and esoph-ageal cancer are among the highest in the world.[18] Dietary supplementation with vitamin A (retinol) and zinc reduced the incidence of gastric cancer, and study participants who received supplements of β-carotene, vitamin E, and selenium had a reduced incidence of esophageal cancer. Overall cancer mortality was reduced over a 5.25-year period for those receiving β-carotene, vitamin E, and selenium.

Selenium has also been shown to have chemopreventive effects for prostate cancer.[18a] High consumption of milk and calcium is associated with a lower risk of colorectal cancer.[18b]

β-carotene is one of the natural dietary sources of vitamin A (retinol). It is found in a variety of plant sources (including certain green and yellow vegetables) but is not synthesized by animals or humans. The other usual dietary source of vitamin A is the alcohol and aldehyde forms and their esters found in milk, eggs, and meat.[19] β-carotene is converted to retinol during intestinal absorption, from where it and preformed vitamin A from animal sources are transported to liver and fat tissue and stored. Retinol, from β-carotene or retinol esters (the storage form of vitamin A, which is converted to retinol upon mobilization from tissues), is transported in the blood as part of a complex with retinol-binding

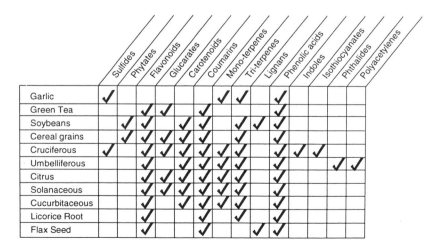

Figure 9–2. Qualitative distribution of major food plant phytochemicals. Fourteen classes of phytochemicals are known or believed to possess cancer-preventive properties. They are believed to appear in greatest abundance in the foods and ingredients included in this diagram. (From Caragay,[16] with permission.)

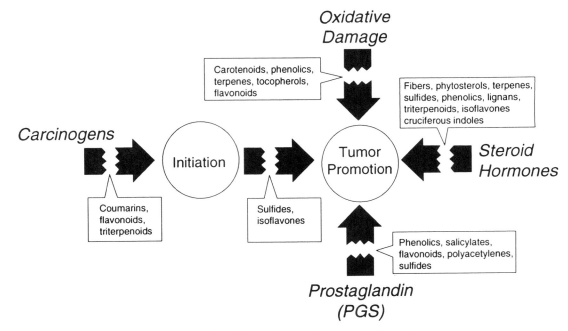

Figure 9–3. Dietary phytochemicals can affect metabolic pathways associated with breast cancer. Certain phytochemicals are known or believed to block specific pathways that lead to the development of breast cancer. (From Caragay,[16] with permission.)

protein, transthyretin, and thyroxine.[19] Vitamin A has an important role in growth, reproduction, and epithelial differentiation.

β-carotene, vitamin A, and synthetic retinoids have attracted a lot of attention as possible chemopreventive agents. β-carotene is thought to function as an electron-scavenging antioxidant (see below), whereas retinol and retinoids are thought to enhance cellular differentiation by acting via their specific nuclear receptor proteins RAR and RXR, which regulate gene expression, cell proliferation, and tissue differentiation.[20] A number of studies have been designed to test the ability of β-carotene to act as a chemopreventive agent for cancer (see below under Retinoids).

There is also experimental evidence supporting a role of vitamin D in preventing colorectal cancer (reviewed in Reference 21). Vitamin D has been reported to regulate cellular proliferation and differentiation and to inhibit angiogenesis. These properties may be the mechanisms for its preventive actions and have been attributed primarily to the active metabolite 1, 25-dehydroxyvitamin D, [1, 25(OH)$_2$D]. The conversion of natural vitamin D [25(OH)D] to its active metabolite can be carried out in the kidney and in other tissues, including the colon.

The concept that vitamin D could prevent colorectal cancer originated with the epidemiological observation that colon cancer mortality rates were lower in states with the highest mean solar radiation and has been further supported by disease incidence studies and prospective studies reporting that colorectal cancer risk was 67% lower in women in the highest quintile of vitamin D intake over time (reviewed in Reference 21).

These studies and others have suggested that exposure to sunlight, which is the main source for production of vitamin D in situ, can prevent or slow the onset of other cancers such as lymphomas. Thus, exposure to sunlight in moderate doses appears to be a good thing. There is some controversy about what the term "moderate" means, and dermatologists warn that more than a million cases of skin cancers annually in the United States are attributed to sun exposure, including 54,000 cases of melanoma.

A large number of observational epidemiologic studies have reported an inverse relationship between dietary intake of fruits, vegetables, and micronutrients and cancer incidence, including breast cancer.[22] However, there continues to be a disconnect between observational epidemiological studies and randomized clinical trials.[23] The reasons for this are probably twofold. First, the relationship between intake of fruits and vegetables and cancer is based on total intake of these dietary entities, which contains a large array of potential preventive agents, i.e., "the entire biological action package." Thus, variation in the mix of these nutrients in various studies can be a confounding issue. Second, dietary studies designed to show the protective effect of individual components in a randomized clinical trial frequently fail because they do not recognize that the preventive action is most likely due to multiple interacting agents. The development of biomarkers that connect the effect of dietary intake of various components in the carcinogenic process is critical to success in this endeavor.

CHEMOPREVENTION

The observations that certain chemicals, some of them natural dietary constituents and some not, can decrease tumorigenesis in animals and, by epidemiological implication, in humans have led to the idea that the intake of certain chemicals can either prevent cancer or slow its progression. Numerous chemicals have been tested in experimental animal studies and a number of clinical trials are ongoing.[24]

The idea that carcinogenesis could be prevented by the administration of agents that inhibit the carcinogeneric process is an attractive one. Theoretically, prevention of carcinogenesis could be accomplished by blockade of the initiation or the promotion–progression phases. Blockade of initiation events could be brought about by agents that decrease the metabolic activation of chemicals to the ultimate carcinogen, increase the detoxification of chemical carcinogens, or prevent the binding of carcinogens to their cellular targets.

Simply put, chemoprevention is the prevention of cancer with drugs. In this context, *drugs* is used in its broadest sense to include dietary supplements, vitamins, hormones, antihormones, etc., as well as "real" drugs such as aspirin, oltipraz, and other synthetic agents used for therapeutic

purposes. *Chemoprevention*, then, can be defined as "the use of intervention with pharmaceuticals, vitamins, minerals, or other chemicals to reduce cancer incidence."[25]

Cancer chemopreventive agents can be classified as antimutagens and carcinogen-blocking agents, antiproliferatives, or antioxidants (Table 9–2);[26] however, there is some overlap among agents in these categories, i.e., they may have more than one mechanism of action. The need for biomarkers to track the success or lack thereof in such clinical trials is as important here as for the dietary studies noted above. These studies take years to complete and require a large study population to obtain meaningful data. Thus, it is critical to have some biochemical or histopathologic means to track the progress of the study population. Unfortunately, there are no sure-fire

Table 9–2. Pharmacologic and Chemical Structural Classification of Promising Chemopreventive Agents

ANTIMUTAGENS AND CARCINOGEN-BLOCKING AGENTS

Phase II metabolic enzyme inducers: N-acetyl-L-cysteine, S-allyl-L-cysteine, oltipraz, phenhexyl isothiocyanate
Polyphenols: ellagic acid
Other: curcumin, DHEA, fluasterone (16-fluoro-DHEA)

ANTIPROLIFERATIVES

Retinoids and carotenoids: β-carotene, 4-HPR, 13-*cis*-retinoic acid, vitamin A
Antihormones: finasteride, tamoxifen
Anti-inflammatories: aspirin, carbenoxolone, curcumin, 18β-glycyrrhetinic acid, ibuprofen, piroxicam, sulindac
G6PDH inhibitors: DHEA, fluasterone
ODC inhibitors: N-acetyl-L-cysteine, aspirin, carbenoxolone, curcumin, DFMO, 18β-glycyrrhetinic acid, 4-HPR, ibuprofen, piroxicam, 13-*cis*-retinoic acid, sulindac, vitamin A
Protein kinase C inhibitors: carbenoxolone, 18β-glycyrrhetinic acid, 4-HPR, tamoxifen
Other: calcium

ANTIOXIDANTS

Anti-inflammatories: see under Antiproliferatives (above)
Antioxidants: N-acetyl-L-cysteine, β-carotene, curcumin, ellagic acid, fumaric acid
Phase II metabolic enzyme inducers: see under Antimutagens and Carcinogen-Blocking Agents (above)
Thiols: N-acetyl-L-cysteine, S-allyl-L-cysteine, oltipraz
Retinol
Vitamin C
Vitamin E

DFMO, 2-difluoromethylornithine; DHEA, dehydroepiandrosterone; G6PDH, glucose-6-phosphate dehydrogenase; ODC, ornithine decarboxylase; 4-HPR, all-*trans*-N-(4-hydroxyphenyl)-retinamide.
(Modified from Kelloff et al.[26])

markers to do this for most types of cancer. For cancers on external surfaces or areas that can be visualized by endoscopic techniques, changes in the degree of dysplasia can be observed (e.g., oral cavity, uterine cervix, and urinary bladder). However, for many other cancers this is not possible. Hence, noninvasive biochemical tests are needed to track these cases.

The clinical strategy for chemoprevention is threefold: (1) "to block or reverse carcinogens before the development of invasive cancer,"[27] (2) to prevent disease progression, and (3) to prevent the occurrence of second primary tumors. In some situations, chemopreventive agents are also used as adjuvants to chemotherapy or surgery. Clinical cancer chemoprevention is based on two concepts derived from experimental and epidemiological studies of cancer. The first is that carcinogenesis is a multistep process beginning with premalignant changes that progress ultimately to invasive cancer. The second concept is "field carcinogenesis," in which "carcinogen exposure diffusely damages the epithelium and predisposes the entire carcinogen-exposed field to the development of multiple independent cancers."[27] Exposure of the skin to sunlight, the colon to fatty acids, or the lung to cigarette smoke are examples of carcinogenic insults that can widely injure epithelial tissues and produce multiple primary tumors by field carcinogenesis.

Molecular Targets for Chemoprevention

In addition to the targets noted above (metabolic processes, production of ROS, etc.), the increasing knowledge of the molecular steps in cancer initiation and progression has provided a new rationale for design of molecular-targeted agents for chemoprevention, just as it has for development of anticancer therapeutic agents. The molecular hallmarks of cancer development are often the same or overlapping in both cancer and premalignancy. These include evasion of apoptosis, overexpression of growth factors, enhanced cell proliferation, and angiogenesis.[24]

As a model for study, the early phase of cancer called *intraepithelial neoplasia* (IEN) is probably the best one for clinical evaluation of chemopreventive agents. IEN is an early, noninvasive lesion that is in an intermediate state between normal epithelium and invasive cancer and has many of the some genetic and phenotypic characteristics of more advanced cancers. Thus the targets for prevention and therapy overlap. With more sensitive methods of cancer detection, more cancers are diagnosed in the IEN stage.

The overlapping prevention and therapy targets include epidermal growth factor receptor (EGFR), HER-2/neu, p53, β-catenin, peroxisome proliferators activated receptor delta (PPAR-δ), cyclooxygenase-2 (COX-2), selective estrogen receptor modulators (SERMs), and aromatase inhibitors.[24] However, the end points of prevention trials and therapeutic trials are somewhat different. In a phase I therapeutic trial, the primary outcome is the maximum tolerated dose (MTD). In a prevention trial, the primary outcome is usually inhibition of progression from IEN to invasive cancer or occurrence of a second tumor (e.g., breast cancer in the contralateral breast). In addition, a prevention trial may be aimed at preventing conversion of a predisposing condition to cancer, such as inhibition of formation of colon polyps. Prevention trials have a much longer time to end point. This means that the agents used or the doses of agents used must be tolerable over a prolonged dosage period, in some cases perhaps for life. Prevention trials are thus expensive and harder to evaluate. Big Pharma tends to shy away from such trials. Abbruzzese and Lippman[24] propose a potential solution for this: if overlapping prevention and therapy targets are known, convergent trials could be designed to include IEN patients and patients with more advanced cancers. Since a number of patients undergo biopsy and subsequent surgery, tissue samples could be available to see if progression of IEN or early cancer was slowed and if the target was being hit by the drug. White blood cells obtained from plasma could also be used to validate targets for both types of agents.

Antimutagens and Carcinogen-Blocking Agents

Isothiocyanates

Organic isothiocyanates (R-N = C = S) are widely distributed in plants and show up in the human food chain in a variety of "Mikey will eat it" type vegetables such as broccoli, brussels sprouts, cabbage, cauliflower, and kale. Broccoli and other cruciferous vegetables are not only good for building character but may also decrease the risk of getting cancer. Allyl isothiocyanates are found in mustard seeds and are responsible for the pungent flavor and odor of mustard. In addition to their characteristic flavors and odors, isothiocyanates have a variety of pharmacologic properties such as antibacterial, antifungal, and antiprotozoal actions. A keen interest has developed in them because of a number of studies showing that they have anticarcinogenic activity (reviewed in Reference 28).

Organic isothiocyanates block the production of tumors in rodents by a wide variety of carcinogens, primarily by blocking their activation to ultimate carcinogens. The carcinogenic actions of polycyclic aromatic hydrocarbons, azo dyes, ethionine, N-2-fluorenylacetamide, and nitrosamines have been shown to be decreased by isothiocyanates.[28] Active agents in this class include α-naphthyl, β-naphthyl, and phenyl-, benzyl-, phenethyl-, and other arylalkyl-isothiocyanates. Tumor development in the liver, lung, mammary gland, forestomach, and esophagus has been inhibited by feeding these compounds to animals. The anticarcinogenic action of isothiocyanates appears to be mediated by decreasing the activity of the P-450 isozymes involved in carcinogen activation and by inducing phase 2 xenobiotic metabolizing enzymes that detoxify and inactivate electrophilic intermediates generated from chemical carcinogens. These enzymes include glutathione transferases and NADPH: quinone reductase. The induction of phase 2 enzymes appears to be the most likely mechanism of anticarcinogenic action of isothiocyanates.[28] Potent inducers of phase 2 enzymes such as sulforaphane (isolated from broccoli) and similar synthetic analogs block the formation of mammary tumors in rats treated with dimethylbenzanthracene (DMBA). In addition to the metabolic actions of sulforaphane, it can also suppress proliferation of cancer cells in culture and in vivo by inhibiting cell cycle progression and inducing apoptosis. This effect is related to sulforaphane's ability to induce the production of the pro-apoptotic proteins Bax and Bak.[29] Since many of these agents are found in high levels in human diets and have thus proven to be safe to take, they are ideal candidates for the develop-

ment of chemoprotective compounds. Indeed, epidemiological studies have suggested that individuals ingesting diets rich in cruciferous vegetables have a lower incidence of cancer.

Oltipraz

Oltipraz is a synthetic dithiolthione analog, similar to some dithiolthiones found in cruciferous vegetables, that was developed in the late 1970s as an effective antischistosomal drug. During studies of its mechanism of action, Bueding et al.[30] observed increased levels of phase 2 enzymes and glutathione in tissues of rodents receiving the drug. Wattenberg and Bueding[31] first established the anticarcinogenic effects of oltipraz in inhibiting diethylnitrosamine-, benzo[a]-pyrene-, or uracil mustard-induced carcinomas in the lungs and forestomachs of mice. Subsequent studies have shown the chemoprotective properties of oltipraz against various carcinogen-induced tumors of the breast, bladder, skin, trachea, and liver in rodents (reviewed in References 32 and 33). This broad range of anticarcinogenic activity and relative low toxicity prompted the testing of the agent in phase I clinical trials. In these trials, only low-grade (I/II) toxicities were observed.[33] Phase II trials were designed to test anticarcinogenic activity against liver cancer development in high aflatoxin-ingestion regions, using the production of aflatoxin-N^7 guanine adducts as a biomarker.[34]

Other Organosulfur Compounds

The organosulfur compounds include the isothiocyanates and dithiolthiones discussed above as well as diallyl sulfides present in garlic and onions and compounds endogenously formed in the body such as N-acetylcysteine and taurine (2-aminoethanesulfonate). These compounds have been shown to induce phase 2 enzymes glutathione S-transferase, NADPH-dependent quinone reductase, and UDP-glucuronosyl transferase in liver and colon tissue and to inhibit chemically induced tumors in the colon and other tissues of rodents (reviewed in Reference 35). Diallyl sulfide has also been shown to inhibit P450 2EI activity[36] and N-acetylcysteine to inhibit formation of DNA adducts in organs of rats treated with benzo[a]pyrene or exposed to cigarette smoke.[37] These activities may contribute to the anticarcinogenic actions of organosulfur compounds. N-acetylcysteine, which also has antioxidant activity, has undergone phase I trials in patients in remission with oral, laryngeal, or lung cancer and was shown to have a low frequency of side effects.[38]

Ellagic Acid

Ellagic acid is related to the coumarin class of lactones and is found in a variety of fruits and vegetables in the human diet. It has been shown to inhibit carcinogen-induced tumors in rodent skin, mammary gland, and forestomach (reviewed in Reference 39). It appears to act by preventing formation of the activated forms of polycyclic aromatic hydrocarbons.

Dehydroepiandrosterone (DHEA)

DHEA is an adrenal steroid hormone precursor. Low serum levels have been found in women with breast cancer and in bladder cancers of both sexes (reviewed in References 39 and 40). DHEA has also shown chemopreventive activity in animal models for skin, breast, colon, and lung carcinogenesis. Its anticarcinogenic effects appear to be due to its ability to inhibit P450-mediated activation of carcinogens[39] and/or its ability to inhibit isoprenylation (and hence membrane targeting) of $p21^{ras40}$ (see Chapter 4). Other synthetic analogs of DHEA such as 16-fluoro-DHEA have been synthesized and may be better to use in human trials because they have less ability to be converted to testosterone or estrogen.[39]

Antiproliferative Agents

Retinoids and β-Carotene

Retinoids are synthetic or natural analogs of vitamin A, the fat-soluble vitamin first recognized in 1909[41] and first named in 1920.[42] It was noted early on that diets deficient in vitamin A led to keratinization of epithelium and squamous metaplasia (reviewed in Reference 19). In a classic paper published in 1925, Wolbach and Howe[43] described gastrointestinal, respiratory, and urogenital epithelial metaplasia in rats fed a

vitamin A–deficient diet. In 1926, Fujumaki[44] observed that such a diet led to gastric carcinomas in rats. A paper in 1941 by Abels et al.[45] described an association of human cancer with vitamin A–deficient diets. The first demonstration that a vitamin A analog could suppress premalignant epithelial lesions was by Lasnitzke,[46] who showed that such lesions could be reversed by addition of retinyl acetate to the diet.

Retinoids play a key role in regulating differentiation and proliferation of a number of tissues, including normal epithelium and connective tissues as well as preneoplastic and neoplastic tissues. They act through a family of receptors (see below) found ubiquitously on cells. During animal development, retinoids have been shown to affect tissue determination, regional polarization of the embryo, and limb bud development. For this reason, retinoic acid has been thought to act as a "morphogen" in embryonic pattern formation (see Chapter 4).

The term *retinoid* has been used to include a large family of natural and synthetic compounds that encompasses vitamin A (retinol) and its esters, β-all-*trans* retinoic acid (tretinoin), 13-*cis*-retinoic acid (isotretinoin or Accutane), an aromatic ethyl ester derivative of retinoic acid (etretinate), the retinamides such as N-(4-hydroxyphenyl) retinamide (4HPR or Fenretinide), and retinoidal benzoic acid derivatives such as TTNPB.[19] Over 2000 retinoid analogs have been synthesized, and the structures of some of them are shown in Figure 9–4.

The activity of retinoids in reversion or slowing the growth of preneoplastic and neoplastic cells in vitro and in vivo has led to a number of clinical trials of agents in this class, with the goal of determining whether such agents could slow tumor progression or prevent the recurrence of cancers treated by other means. Clinical trials of this group of agents have included use in head and neck, lung, cervical, prostate, bladder, breast, and skin cancers, as well as in acute promyelocytic leukemia (APL).[27,47]

One large U.S.-based multicenter, double-blind randomized trial called CARET was designed to test whether oral administration of β-carotene plus retinyl palmitate could decrease the incidence of lung cancer in heavy smokers and asbestos workers.[20] This and other such studies were based on observations that

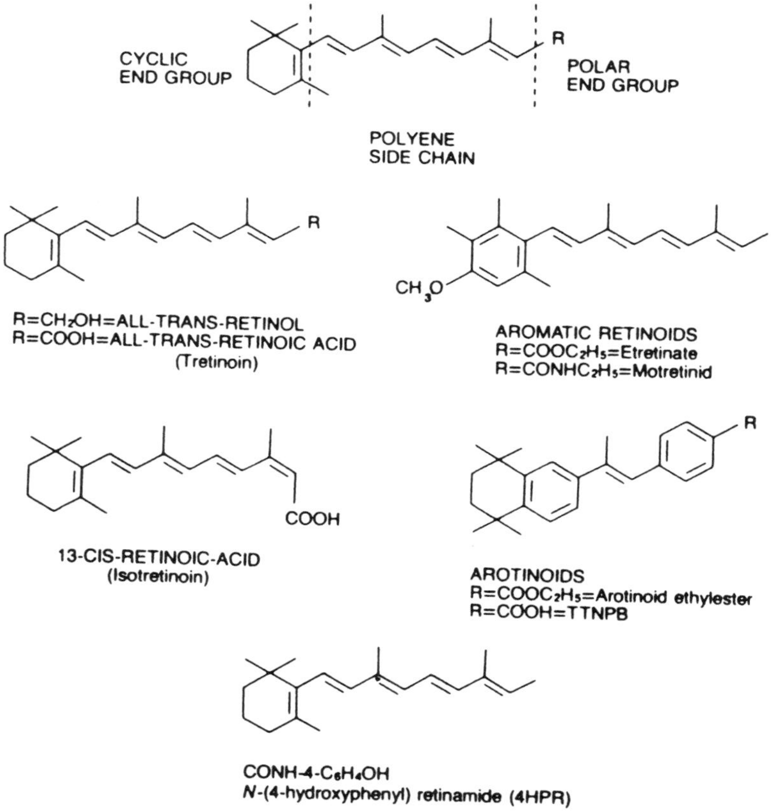

Figure 9–4. Chemical structures of retinoids. TTNPB = (E)-4-[2-(5,6,7,8-tetrahydro-5.5.8,8-tetramethyl-2-naphthalenyi)-1-propenyl] benzoic acid. (From Lippman et al.,[19] with permission.)

retinoids have reversed cigarette smoking–induced preneoplastic bronchial lesions[48] and that retinyl palmitate increased the time to relapse or development of new primary tumors in patients with stage I lung cancer.[49] Another study found a decreased risk of developing lung cancer in *nonsmokers*, who had been consuming diets rich in β-carotene and raw fruits and vegetables.[50] Other studies have shown that oral ingestion of β-carotene induces regression of oral leukoplakia, a premalignant lesion for oral cancer (reviewed in Reference 51). One study done in Finland has reported no decrease in the incidence of lung cancer in heavy smokers who were placed on β-carotene, α-tocopherol (vitamin E), or both and then followed for 5 to 8 years.[52] Moreover, this study also found that there was a small (8%) but statistically significant increase in mortality among the group taking β-carotene. The conclusions of this study are somewhat

surprising (and controversial) because they are contradictory to a number of other studies' findings. The reason for this discrepancy may be that the Finnish study population consisted of relatively heavy smokers (5 or more cigarettes a day) who had been smoking for several years (the study population was 50 to 69 years of age). Thus, one could argue that the lung tissue in these individuals had already been "initiated" and undergone several years of "promotion" at the time of entry into the study. Nevertheless, these results point out an important caveat for all chemoprevention studies: once tissues have undergone initiation and several stages of promotion–progression, it is unlikely that the tumorigenesis process can be stopped—slowed perhaps, but not stopped.

A number of studies have suggested that retinoids can slow or prevent skin cancer progression. For example, isotretinoin has been reported to reverse preneoplastic skin lesions and to prevent invasive skin cancer in high-risk patients with xeroderma pigmentosum, and in combination with interferon α-2a (IFN-α2a), to induce remissions in advanced squamous cell carcinomas of the skin (reviewed in Reference 53).

Hormonal Chemoprevention

A large body of evidence indicates that hormones have an important role in the causation of cancer in women and men. For example, female hormones have been linked to the development of cancers of the breast, uterine endometrium, and ovary. Male hormones are implicated in the cause of prostate cancer. This type of hormone-induced carcinogenesis appears to result from the ability of hormones to stimulate cell division in hormone target organs. This may in turn lead to an increased probability of accumulating genetic errors over time. It is estimated that hormone-related cancers account for at least 20% of all male and 40% of all female cancers in the United States.[54] Thus, chemoprevention of cancer by the manipulation of hormone levels or by the use of "antihormones" is a legitimate strategy. Several hormonal chemopreventive agents have been shown to be effective as chemopreventive agents and others are under intense clinical study (Table 9–3).

ORAL CONTRACEPTIVES

The findings that sequential oral contraceptives (estrogen alone, followed by progesterone alone) and that estrogen replacement therapy (at the earlier used higher doses) increased the risk of endometrial cancer led to the "unopposed estrogen" hypothesis for the cause of this form of uterine cancer (reviewed in Reference 54). In contrast, the now commonly used combination oral contraceptives (COCs), which contain lower amounts of estrogen plus progesterone, have markedly decreased the risk of endometrial and ovarian cancer. Epidemiological studies show an 11.7% per year decrease in endometrial cancer and a 7.5% per year decrease in ovarian cancer with the increased use of COCs.[54]

Breast cancer is a different story, however. In breast tissue, both estrogen and progesterone stimulate cell proliferation. Thus, the combina-

Table 9–3. Some Hormone Chemopreventive Agents in Clinical Use

Chemopreventive Agents	Cancer Site	Mechanism of Action
Oral contraceptives	Endometrium Ovary	Anti-estrogen Inhibit ovulation
GnRH agonists	Breast, Endometrium	Inhibit ovarian steroid hormone production
	Ovary	Inhibit ovulation
Progestogens (HRT)	Endometrium	Anti-estrogen
Tamoxifen	Breast	Anti-estrogen
Raloxifene	Breast	Anti-estrogen
Anastrozole	Breast	Aromatase inhibitor
Finasteride	Prostate	5α-reductase inhibitor

GnRH, Gonadotropin-releasing hormone; HRT, hormone replacement therapy.

tion of the two appears to have a greater stimulatory effect on breast cell division than estrogen alone, leading to the "estrogen augmented by progesterone" theory of breast cancer etiology. The protective effects of late menarche, early menopause, and early child-bearing fit this concept. Studies of breast cancer risk in COC users have not produced a clear answer. A 3.1% increase in breast cancer risk per year of COC use has been reported for women diagnosed under age 45, whereas no increased risk with COC use has been seen in women diagnosed over age 45.[54]

GONADOTROPIN-RELEASING HORMONE ANALOGS (GnRHAs)

GnRHAs reversibly inhibit ovulation and reduce production of ovarian steroid hormones. Clinical trials are under way to test whether such agents can act as contraceptives and also reduce the risk of breast, ovarian, and endometrial cancer.

HORMONE REPLACEMENT THERAPY

Hormone replacement therapy (HRT) is being used to prevent the symptoms of menopause and prevent osteoporosis associated with menopause. As noted above, the use of estrogen by itself is associated with an increased risk of endometrial cancer that is related to both dose and duration of therapy. This increased risk is calculated to be about 3.5-fold after 5 years of unopposed estrogen therapy. However, the addition of progestogens to the regimen and lowering the estrogen dose decreased the overall risk of endometrial cancer between 1973 and 1977 from 27.9% to 14.4%.[54]

The effect of HRT on breast cancer risk is less clear and more controversial. However, lowering the estrogen dose appears to lower risk. Lowering the daily dose from 1.25 mg to 0.625 mg of the commonly used conjugated equine estrogen, used by itself, decreased the risk of breast cancer from 3.1% per each year of estrogen-alone replacement therapy to less than 2% per year of use.[55]

TAMOXIFEN, RALOXIFENE, AND AROMATASE INHIBITORS

Tamoxifen has been a mainstay in the therapy of breast cancer for many years. It has been clearly established that tamoxifen decreases the risk of appearance of a second tumor in the opposite breast of women who have had a primary tumor in one breast.[54,56] These data plus the well-documented epidemiological evidence that the lifetime duration of exposure to estrogen (i.e., early menarche, late menopause, nulliparity) increases breast cancer risk led to the concept that an antiestrogen such as tamoxifen could be a chemopreventive agent for women at high risk to develop breast cancer. As a result, large-scale clinical trials have been carried out or are ongoing in the United States and Europe to test this hypothesis. Such trials are not without controversy. After all, is it wise to give a hormonal-type agent to high-risk but otherwise healthy woman for many years? A compelling argument to do so is the dramatically lower risk (35% decrease) for development of a cancer in the contralateral breast after a primary in one breast. Also, this treatment has a relatively mild list of side effects, including hot flashes, nausea and vomiting, menstrual irregularities, and vaginal bleeding and discharge. The greatest objection to the long-term use of tamoxifen as a chemopreventive agent is the small but real increased risk of endometrial cancer. Even though it acts as an antiestrogen in breast tissue, tamoxifen has weak estrogen agonist action on the uterine endometrium. Thus, it can act as an unopposed estrogen in this tissue. Clearly, the risk–benefit ratio has to be considered in such clinical trials.

Raloxifene is another estrogen receptor inhibitor that has shown evidence of its preventative potential for breast cancer.[57]

There is also a lot of interest in aromatase inhibitors such as anastrozole, exemestane, and letrozole for breast cancer prevention among postmenopausal women. A trial comparing aromatase inhibitors to tamoxifen has shown a greater reduction in recurrence rates and new contralateral tumors for the aromatase inhibitors.[58] In some trials, patients have initially been treated with tamoxifen for 2–3 years and then with an aromatase inhibitor.

ANTIANDROGENS

Dihydrotestosterone (DHT) is a metabolic product derived from testosterone by the action of an enzyme called 5α-reductase. DHT stimulates prostatic cells to proliferate and its blood levels appear to correlate with prostate cancer risk; for example, DHT levels are lower in Japanese men, a low-risk group, than in U.S. black males, a high-risk group.

The drug finasteride (Proscar) inhibits the enzyme testosterone 5α-reductase and thereby lowers the production of DHT and prevents hyperplasia of the prostatic stroma. The Prostate Cancer Prevention Trial, which enrolled over 18,000 men, reported a 24.8% decrease in prostate cancer in the finasteride treatment compared to the placebo group; however, in the patients who received finasteride, the prostate cancer that did develop had a higher Gleason score.[59]

Anti-Inflammatory Agents

A number of studies, but not all, suggest that chronic aspirin intake lowers the incidence and mortality of colon cancer (reviewed in Reference 60). Aspirin is one of a family of drugs known as nonsteroidal anti-inflammatory drugs (NSAIDs) that inhibit the cyclooxygenase arm of the arachidonic acid cascade, which produces prostaglandins. Other members of the NSAID class include indomethacin, ibuprofen, piroxicam, and sulindac.

In rodent models of colon carcinogenesis, NSAIDs have been shown to reduce the number of tumor-bearing animals and the number of tumors per animals.[60] Waddell and Loughry[61] first reported that sulindac reduced the size and number of rectal polyps in individuals with familia polyposis but that tumors recurred when treatment was stopped.[62] Additional studies have also shown a reversible regression of polyps in familial polypsis patients taking sulindac (reviewed in Reference 60). While the mechanism of the anti-tumor effects of the NSAIDs is not clear, it may relate to inhibition of prostaglandin-mediated stimulation of cell proliferation and/or to the effects of NSAIDs on reversing immune suppression[60] and decreasing inflammation.

Cyclooxygenase-2 Inhibitors

Cyclooxygenase-2 (COX-2) is an inducible enzyme that, in collaboration with its constitutively expressed partner, COX-1, is responsible for synthesis of prostaglandins. COX-2 is upregulated by oxidative stress, growth factors such as EGF, inflammatory cytokines like interleukin-6 (IL-6), tumor necrosis factor-α (TNF-α), and estrogen (reviewed in Reference 63). COX-2 activity is generally low in normal nonproliferative epithelial tissues, but overexpressed in a number of cancers, including colon and breast carcinomas. The selective COX-2 inhibitor celecoxib has been shown to reduce the number and size of polyps in patients with familial adenomatous poloyposis.[64]

COX-2 is also overexpressed in breast hyperplastic lesions, ductal carcinoma in situ, and invasive breast cancer. This heightened activity is associated with promotion and progression of breast cancer through its positive effects on estrogen production, cell proliferation, protease activity, and angiogenesis, and its anti-apoptotic activity.[63]

Studies in rats have shown a dose-dependent decrease in growth of carcinogen-induced estrogen receptor–positive breast cancers and HER-2-overexpressing breast tumors in transgenic mice. Celecoxib appeared to be more effective than nonselective COX-1/COX-2 inhibitors like the NSAIDs in the rat studies. These data, together with the initial clinical studies suggesting less gastrointestinal tract side effects with selective COX-2 inhibitors, led to the initiation of a number of breast cancer chemoprevention trials.[63] Unfortunately, because of an observed increased risk of cardiovascular toxicity and deaths, these and other COX-2 chemoprevention trials have been put on hold. They may remain in limbo for a long time, and it is unclear when the chemopreventive effects of such trials will be known.

Ornithine Decarboyxlase Inhibitors

Hämäläinen[65] was the first to study in detail the levels of polyamines in human neoplasms. An increased polyamine content has since been found in a wide variety of human and animal neoplasms. Polyamines play an important role in cell proliferation and differentiation, and ornithine decarboylase (ODC) is an essential enzyme in their biosynthetic pathway. ODC activity is increased in proliferating tissues and in tumors. Several ODC inhibitors can inhibit tumor formation in rodent models. One such inhibitor, D, L-α-difluoromethylornithine (DFMO), has been reported to inhibit chemical carcinogenesis in skin, tongue, liver, colon, breast, urinary bladder, kidney, and brain (reviewed in Reference 66). Inhibitory effects of DFMO on the growth of bladder and kidney cancers in humans have also been reported.[67] Such data have led to clinical trials of DFMO as a chemopreventive agent for human cancer.[68]

Antioxidants

A number of chemopreventive agents discussed above also have the ability to scavenge oxygen radicals and to act as antioxidants. These include the organic sulfur-containing compounds, retinol, β-carotene, ellagic acid, oltipraz, and certain of the anti-inflammatory agents. Since oxygen-radical damage to DNA is a well-recognized effect of a number of carcinogenic agents, including a number of chemical carcinogens and irradiation, the ability to block the production of oxygen radicals could potentially provide a major protective effect against carcinogenesis.

The damage produced by endogenously generated oxygen radicals has been suggested to be a major factor in aging and in such aging-related diseases as heart disease, mental senility, and cancer. Endogenous oxidants are produced continuously by normal cellular metabolism, and when not kept in check by cellular protective factors such as glutathione can result in extensive damage to proteins, lipids, and DNA.

Oxidative damage to DNA, on the basis of urinary excretion of DNA adducts such as 8-hydroxyguanosine, is estimated to be about 10 "hits" per cell per day in humans.[69]

A number of vitamins and minerals have antioxidant actions. These include vitamins A, C, and E, carotenoids, and selenium. Vitamins E and C and carotenoids trap free radicals and reactive oxygen molecules. Micronutrient elements such as selenium, zinc, copper, iron, and manganese are essential cofactors for antioxidant enzymes. In addition, vitamin C can protect against cancer by inhibiting nitrosation of secondary amines and N-substituted amides to form nitrosamines and nitrosamides in acidic conditions such as those found in the stomach.[70] There is epidemiological evidence that these nitroso compounds contribute to the etiology of cancers of the stomach, esophagus, and nasopharynx (reviewed in Reference 70). Hence, vitamin C may be an effective chemopreventive agent in those parts of the world where the incidence of these cancers is high, such as China.

A number of clinical and epidemiological studies have suggested a role for antioxidants in preventing human cancer. For instance, in the Iowa Women's Health Study, a high intake of vitamin E in the diet correlated with a decreased risk for colon cancer, especially in women under 65 years of age.[71] In contrast to these findings, a prospective study of breast cancer risk in over 89,000 women enrolled in the Nurses' Health Study showed no protective effect of diets rich in vitamin C or E.[72]

Protease Inhibitors

Certain protease inhibitors have been shown to suppress carcinogenesis in in vitro and in vivo assay systems. One of these, the Bowman-Birk inhibitor (BBI) derived from soybeans, has both antitrypsin and antichymotrypsin activity. BBI can block cell transformation of cells in culture exposed to chemical carcinogens and in several tissues (e.g., colon, liver, lung, esophagus, and oral cavity) in rodents exposed to various carcinogens (reviewed in Reference 73). While the mechanism of anticarcinogenic action of BBI and similar protease inhibitors is not clear, several observed effects may be involved. For in-

stance, protease inhibitors have been observed to inhibit the expression of the oncogenes c-*myc* and c-*fos* and to block carcinogen-induced gene amplification.[73]

There is circumstantial epidemiological evidence for a role of soybean protease inhibitors in preventing human cancer. For example, populations who have a high intake of soybeans or soybean-derived products (e.g., Japanese and Seventh Day Adventists) have lower risk of many cancers common in countries of the Western world.[73]

Histone Deacetylase Inhibitors

As noted in Chapter 5, the acetylation state of histones bound to chromatin regulates gene expression. Acetylation of histones "opens" chromatin and activates gene expression. A number of genes, including some tumor suppressor genes, are shut off if the histone components regulating their expression are deacetylated, hence, the rationale for the use of histone deacetylase (HDAC) inhibitors in cancer therapy. HDACs play a role in cell-cycle progression and differentiation and their dysregulation has been observed in several cancer cell types. Inhibitors of HDACs such as trichostatin A and suberoylanilide hydroxamic acid have antitumor effects in mouse models. The latter agent induces growth arrest, cell differentiation, and apoptosis in a variety of cancer cell types and inhibits tumor growth in vivo in animal models (reviewed in Reference 74). Suberoylanilide hydroxamic acid also has low toxicity, and clinical trials as a cancer chemotherapeutic agent have been initiated. Since the drug also induces growth arrest, differentiation, and apoptosis in cultured human colon cancer cells, it is a candidate for a chemopreventative agent in colorectal cancer.[74]

Statins

Statins are a class of drugs in wide use to lower lipid levels via their inhibition of 3-hydroxy-3-methylglutaryl coenzyme A (HMG-CoA) reductase. Since proliferating cancer cells require a high amount of cell membrane lipid biosynthesis, inhibition of HMG-CoA reductase, which is a key enzyme in the biosynthesis of cellular lipids, is a good target. Inhibitors of this enzyme inhibit tumor cell proliferation and induce apoptosis in cancer lines in vitro. In clinical studies of the lipid-lowering effects of statins in patients prone to cardiovascular disease, there was some evidence for a decreased incidence of colorectal cancer, although the results from various trials were inconsistent, perhaps because the studies were not sufficiently powered to detect such differences. The small number of cancers observed in these studies limited their statistical power to detect a clear association between statin use and cancer risk. Poynter et al.[75] carried out a population-based, case–control study of patients who had been diagnosed with colorectal cancer, comparing statin use and cancer incidence. Statin use was associated with a 47% relative reduction in colorectal cancer risk. There was significant risk reduction due to statin use even after adjusting for use or non-use of aspirin or other NSAIDs; the presence or absence of physical activity, hypercholesterolemia, or family history of colorectal cancer; ethnic groups; and level of vegetable consumption.

Multiagent Chemoprevention

The rationale for multiagent chemoprevention is the same as that for combination chemotherapy, namely, additive $(1+1=2)$ or synergistic $(1+1=3)$ effects may be achieved if the single agents used in the regimen (1) have different mechanisms of action, (2) do not have significantly overlapping toxicities, (3) have demonstrated activity as single agents, and (4) do not significantly interfere with one another's pharmacokinetic or pharmacodynamic profiles. Several examples of effective multiagent chemoprevention have been obtained in animal and human studies. These include combinations of (1) a peroxisome proliferator–activator receptor–γ agonist and a histone deacetylase inhibitor in inhibiting adenocarcinoma cell growth; (2) ursodeoxycholic acid and low-dose sulindac in preventing intestinal adenomas in Min mice (an animal model of familial adenomatous polyposis with a high incidence of colon carcinomas); (3) combination of the NSAID piroxicam and difluoromethylornithine in reducing carcinogen-induced colon carcinomas in

rats; (4) combination of the statin lovastatin with the COX-2 inhibitor celecoxib in inhibiting colon cancer in Min mice; and (5) combination of β-carotene, vitamin E, and selenium in reducing stomach cancer deaths in China.[76,77]

These sorts of studies have led to a number of multiagent clinical trials for chemoprevention. However, the cost and duration of such trials is daunting. Not only does the absence of definitive surrogate biomarkers make it difficult to assess efficacy, but the need to obtain single-agent efficacy and toxicity data to satisfy FDA requirements can be a show stopper. Fortunately, if the combination includes a vitamin or food supplement "generally regarded as safe" (GRAS designation), a combination trial can often proceed with data from the other non-GRAS agents alone.[77]

References

1. T. B. L. Kirkwood: Understanding the odd science of aging. *Cell* 120:437, 2005.
2. R. S. Balaban, S. Nemoto, and T. Finkel: Mitochondria, oxidants, and aging. *Cell* 120:483, 2005.
3. D. Harman: Aging: A theory based on free radical and radiation chemistry. *J Gerontol* 11:298, 1956.
4. J. Campisi: Senescent cells, tumor suppression, and organismal aging: Good citizens, bad neighbors. *Cell* 120:513, 2005.
5. D. B. Lombard, K. F. Chua, R. Mostoslavsky, S. Franco, M. Gostissa, et al.: DNA repair, genome stability, and aging. *Cell* 120:497, 2005.
6. L. Guarente and F. Picard: Calorie restriction— The *SIR2* connection. *Cell* 120:473, 2005.
7. S. J. Lin, M. Kaeberlein, A. A. Andalis, L. A. Sturtz, P. A. Defossez, et al.: Calorie restriction extends *Saccharomyces cerevisiae* life span by increasing respiration. *Nature* 418:344, 2002.
8. J. G. Wood, B. Rogina, S. Lavu, K. Howtiz, S. L. Helfand, et al.: Sirtuin activators mimic caloric restriction and delay ageing in metazoans. *Nature* 430:686, 2004.
9. J. Schultz: Resveratrol may be a powerful cancer-fighting ally. *J Natl Cancer Inst* 96:1497, 2004.
10. J. Gertner: Eat chocolate, live longer? *NY Times*, October 10, 2004.
11. R. S. Sandler, C. M. Lyles, L. A. Peipins, C. A. McAuliffe, J. T. Woosley, et al.: Diet and risk of colorectal adenomas: Macronutrients, cholesterol, and fiber. *J Natl Cancer Inst* 85:884, 1993.
12. R. Doll and R. Peto: The causes of cancer: Quantitative estimates of avoidable risks of cancer. *J Natl Cancer Inst* 66:1191, 1981.

13. R. Weindruch, D. Albanes, and D. Kritchevsky: The role of calories and caloric restriction in carcinogenesis. *Hematol Oncol Clin North Am* 5: 79, 1991.

14. D. F. Birt, H. J. Pinch, T. Barnett, A. Phan, and K. Dimitroff: Inhibition of skin tumor promotion by restriction of fat and carbohydrate calories in SENCAR mice. *Cancer Res* 53:27, 1993.

15. B. D. Roebuck, K. J. Baumgartner, and D. L. MacMillan: Caloric restriction and intervention in pancreatic carcinogenesis in the rat. *Cancer Res* 53:46, 1993.

16. A. B. Caragay: Cancer-preventive foods and ingredients. *Food Technol* 46:65, 1992.

17. C. Ip, D. J. Lisk, and J. A. Scimeca: Potential of food modification in cancer prevention. *Cancer Res* 54:1957s, 1994.

18. P. R. Taylor, B. Li, S. M. Dawsey, J. Li, C. S. Yang, et al.: Prevention of esophageal cancer: the nutrition intervention trials in Linxian, China. *Cancer Res* 54:2029s, 1994.

18a. P. R. Taylor, H. L. Parnes, and S. M. Lippman: Science peels the onion of selenium effects on prostate carcinogenesis. *J Natl Cancer Inst* 96:645, 2004.

18b. E. Cho, S. A. Smith-Warner, D. Spiegelman, W. L. Beeson, P. A. van den Brandt, et al.: Dairy foods, calcium, and colorectal cancer: A pooled analysis of 10 cohort studies. *J Natl Cancer Inst* 96:1015, 2004.

19. S. M. Lippman, J. F. Kessler, and F. L. Meyskens, Jr.: Retinoids as preventive and therapeutic anticancer agents (part I). *Cancer Treat Rep* 71:391, 1987.

20. G. S. Omenn, G. Goodman, M. Thornquist, J. Grizzle, L. Rosenstock, et al.: The β-carotene and retinol efficacy trial (CARET) for chemoprevention of lung cancer in high-risk populations: Smokers and asbestos-exposed workers. *Cancer Res* 54:2038s, 1994.

21. D. Feskanich, J. Ma, C. S. Fuchs, G. J. Kirkner, S. E. Hankinson, et al.: Plasma vitamin D metabolites and risk of colorectal cancer in women. *Cancer Epidemiol Biomarkers Prev* 12:1502, 2004.

22. M. M. Gaudet, J. A. Britton, G. C. Kabat, S. Steck-Scott, S. M. Eng, et al.: Fruits, vegetables, and micronutrients in relation to breast cancer modified by menopause and hormone receptor status. *Cancer Epidemiol Biomarkers Prev* 13: 1485, 2004.

23. F. L. Meyskens, Jr. and E. Szabo: Diet and cancer: The disconnect between epidemiology and randomized clinical trials. *Cancer Epidemiol Biomarkers Prev* 14:1366, 2005.

24. J. L. Abbruzzese and S. M. Lippman: The convergence of cancer prevention and therapy in early-phase clinical drug development. *Cancer Cell* 6:321, 2004.

25. D. E. Brenner: Multiagent chemoprevention: An overview. *Proc Am Soc Clin Oncol Education Book*, pp. 124, 2005.

26. G. J. Kelloff, C. W. Boone, V. E. Steele, J. A. Crowell, R. Lubet, et al.: Progress in cancer chemoprevention: Perspectives on agent selection and short-term clinical intervention trials. *Cancer Res* 54:2015s, 1994.

27. S. M. Lippman, S. E. Benner, and W. K. Hong: Retinoid chemoprevention studies in upper aerodigestive tract and lung carcinogenesis. *Cancer Res* 54:2025s, 1994.

28. Y. Zhang and P. Talalay: Anticarcinogenic activities of organic isothiocyanates: Chemistry and mechanisms. *Cancer Res* 54:1976s, 1994.

29. S. Choi and S. V. Singh: Bax and bak are required for apoptosis induction by sulforaphane, a cruciferous vegetable–derived cancer chemopreventive agent. *Cancer Res* 65:2035, 2005.

30. E. Bueding, P. Dolan, and J. P. Leroy: The antischistosomal activity of oltipraz. *Res Commun Chem Pathol Pharmacol* 37:293, 1982.

31. L. W. Wattenberg and E. Bueding: Inhibitory effects of 5-(2-pyrazinyl)-4-methyl-1,2-dithiol-3-thione (oltipraz) on carcinogenesis induced by benzo[a]pyrene, diethyl-nitrosamine and uracil mustard. *Carcinogenesis (Lond)* 7:1379, 1986.

32. M. G. Bolton, A. Munoz, L. P. Jacobson, J. D. Groopman, Y. Y. Maxuitenko, et al.: Transient intervention with oltipraz protects against aflatoxin-induced hepatic tumorigenesis. *Cancer Res* 54:3499, 1993.

33. A. B. Benson: Oltipraz: A laboratory and clinical review. *J Cell Biochem* 17f:278, 1993.

34. J. D. Groopman, G. N. Wogan, B. D. Roebuck, and T. W. Kensler: Molecular biomarkers for aflatoxins and their application to human cancer prevention. *Cancer Res* 54:1907s, 1994.

35. B. S. Reddy, C. V. Rao, A. Rivenson, and G. Kelloff: Chemoprevention of colon carcinogenesis by organosulfur compounds. *Cancer Res* 53: 3493, 1993.

36. C. S. Yang, T. J. Smith, and J. Hong: Cytochrome P-450 enzymes as targets for chemoprevention against chemical carcinogenesis and toxicity. *Cancer Res* 54:1982s, 1994.

37. A. Izzotti, F. D'Agostini, M. Bagnasco, L. Scatolini, A. Rovida, et al.: Chemoprevention of carcinogen–DNA adducts and chronic degenerative diseases. *Cancer Res* 54:1994s, 1994.

38. N. De Vries and S. De Flora: *N*-acetyl-*l*-cysteine. *J Cell Biochem* 17F:270, 1993.

39. C. W. Boone, G. J. Kelloff, and W. E. Malone: Identification of candidate cancer chemopreventive agents and their evaluation in animal models and human clinical trials: A review. *Cancer Res* 50:2, 1990.

40. S. Schulz, R. C. Klann, S. Schonfeld, and J. W. Nyce: Mechanisms of cell growth inhibition and

cell cycle arrest in human colonic adenocarcinoma cells by dehydroepiandrosterone: Role of isoprenoid biosynthesis. *Cancer Res* 52:1372, 1992.

41. W. Steppe: Versuche uber Futterung mit lipoidfreier Nahrung. *Biochem Z* 22:452, 1909.

42. J. C. Drummond: The nomenclature of the so-called accessory food factors (vitamins). *Biochem J* 14:660, 1920.

43. S. B. Wolbach and P. R. Howe: Tissue changes following deprivation of fat-soluble A vitamin. *J Exp Med* 42:753, 1925.

44. Y. Fujumaki: Formation of gastric carcinoma in albino rats fed on deficient diets. *J Cancer Res* 10:469, 1926.

45. J. C. Abels, A. T. Gorham, G. T. Pack, C. P. Rhoads: Metabolic studies in patients with cancer of the gastrointestinal tract. I. Plasma vitamin A levels in patients with malignant neoplastic disease, particularly of the gastrointestinal tract. *J Clin Invest* 20:749, 1941.

46. I. Lasnitzki: The influence of A hypervitaminosis on the effect of 20-methylcholanthrene on mouse prostate glands grown in vitro. *Br J Cancer* 9:434, 1955.

47. K. Slawin, D. Kadmon, S. H. Park, P. T. Scardino, M. Anzano, et al.: Dietary fenretinide, a synthetic retinoid, decreases the tumor incidence and the tumor mass of *ras+myc*-induced carcinomas in the mouse prostate reconstitution model system. *Cancer Res* 53:4461, 1993.

48. J. L. Misset, G. Mathe, G. Santelli, J. Gouveia, J. P. Homasson, et al.: Regression of bronchial epidermoid metaplasia in heavy smokers with etretinate treatment. *Cancer Detect Prev* 9:167, 1986.

49. U. Pastorino, M. Infante, G. Chiesa, M. Maioli, M. Clerici, et al.: Lung cancer chemoprevention. In U. Pastorino and W. K. Hong, eds.: *Chemoimmuno Prevention of Cancer.* New York: Thieme Medical Publishers, pp. 147–159, 1991.

50. S. T. Mayne, D. T. Janerich, P. Greenwald, S. Chorost, C. Tucci, et al.: Dietary β-carotene and lung cancer risk in U.S. nonsmokers. *J Natl Cancer Inst* 86:33, 1994.

51. H. S. Garewal: β-carotene cancer and vitamin E in oral cancer prevention. *J Cell Biochem* 17F:262, 1993.

52. The α-tocopherol, β-carotene cancer prevention study group: The effect of vitamin E and β-carotene on the incidence of lung cancer and other cancers in male smokers. *N Engl J Med* 330:1029, 1994.

53. S. M. Lippman, D. R. Parkinson, L. M. Itri, R. S. Weber, S. P. Schantz, et al.: 13-*cis* retinoic acid and interferon α-2a: Effective combination therapy for advanced squamous cell carcinoma of the skin. *J Natl Cancer Inst* 84:235, 1992.

54. B. E. Henderson, R. K. Ross, and M. C. Pike: Hormonal chemoprevention of cancer in women. *Science* 259:633, 1993.

55. M. C. Pike, L. Bernstein, and D. V. Spicer: In J. E. Niederuber, ed.: *Current Therapy in Oncology.* St. Louis: Dekker, Mosby-Yearbook, 1993, pp. 292–303.

56. S. G. Nayfield, J. E. Karp, L. G. Ford, F. A. Door, and B. S. Kramer: Potential role of tamoxifen in prevention of breast cancer. *J Natl Cancer Inst* 83:1450, 1991.

57. S. Martino, J. A. Cauley, G. Barrett-Conor, T. J. Powles, J. Mershon, et al.: Continuing outcomes relevant to Evista: Breast cancer incidence in postmenopausal osteoporotic women in a randomized trial of raloxifene. *J Natl Cancer Inst* 96:1751, 2004.

58. J. M. Cuzick: New directions in risk assessment and the prevention of breast cancer. *Proc Am Soc Clin Oncol Education Book*, p. 145, 2005.

59. A. S. Tsao, E. S. Kim, and W. K. Hong: Chemoprevention of cancer. *CA Cancer J Clin* 54:150, 2004.

60. L. J. Marnett: Aspirin and the potential role of prostaglandins in colon cancer. *Cancer Res* 52: 5575, 1992.

61. W. R. Waddell and R. W. Loughry: Sulindac for polyposis of the colon. *J Surg Oncol* 24:83, 1983.

62. W. R. Waddell, G. F. Gasner, E. J. Cerise, and R. W. Loughry: Sulindac for polyposis of the colon. *Am J Surg* 157:175, 1989.

63. C. J. Fabian and B. F. Kimler: COX-2 inhibitors for breast cancer prevention: a current understanding. *Proc Am Soc Clin Oncol Education Book*, p. 139, 2005.

64. G. Steinbach, P. M. Lynch, R. K. Phillips, et al.: The effect of celecoxib, a cyclooxygenase-2 inhibitor, in familial adenomatous polyposis. *N Engl J Med* 342:1946, 2000.

65. R. Hämäläinen: Über die quantitative Bestimmung des Spermins in Organismus und sein Verkommen in menschlichen Geweben und Körperflussigkeiten. *Acta Soc Med Fenn Duodecim Ser A* 23:97, 1947.

66. T. Kojima, T. Tanaka, T. Kawamori, A. Hara, and M. Mori: Chemopreventive effects of dietary D,L-α-difluoromethylornithine, and ornithine decarboxylase inhibitor, on initiation and postinitiation stages of diethylnitrosamine-induced rat hepatocarcinogenesis. *Cancer Res* 53:3903, 1993.

67. U. Dunzendorfer: The effect of α-difluoro methylornithine on tumor growth, acute phase reactants, β-2-microglobulins and hydroxyproline in kidney and bladder carcinomas. *Urol Invest* 36:128, 1981.

68. R. R. Love, P. P. Carbone, A. K. Verma, D. Gilmore, P. Carey, et al.: Randomized phase I chemoprevention dose-seeking study of α-difluoromethylornithine. *J Natl Cancer Inst* 85: 732, 1993.

69. C. G. Fraga, P. A. Motchnik, M. K. Shigenaga, H. J. Helbock, R. A. Jacob, et al.: Ascorbic acid

protects against endogenous oxidative DNA damage in human sperm. *Proc Natl Acad Sci USA* 88:11003, 1991.

70. S. S. Mirvish: Experimental evidence for inhibition of *N*-nitroso compound formation as a factor in the negative correlation between vitamin C consumption and the incidence of certain cancers. *Cancer Res* 54:1948s, 1994.

71. R. M. Bostick, J. D. Potter, D. R. McKenzie, T. A. Sellers, L. H. Kushi, et al.: Reduced risk of colon with high intake of vitamin E: The Iowa Women's Health Study. *Cancer Res* 53:4230, 1993.

72. D. J. Hunter, J. E. Manson, G. A. Colditz, M. J. Stampfer, S. Rosner, et al.: A prospective study of the intake of vitamins C, E, and A and the risk of breast cancer. *N Engl J Med* 329:234, 1993.

73. A. R. Kennedy: Prevention of carcinogenesis by protease inhibitors. *Cancer Res* 54:1999s, 1994.

74. L. C. Hsi, X. Xi, R. Lotan, I. Shureiqi, and S. M. Lippman: The histone deacetylase inhibitor suberoylanilide hydroxamic acid induces apoptosis via induction of 15-Lipoxygenase-1 in colorectal cancer cells. *Cancer Res* 64:8778, 2004.

75. J. N. Poynter, S. B. Gruber, P. D. R. Higgins, R. Almog, J. D. Bonner, et al.: Statins and the risk of colorectal cancer. *N Engl J Med* 352:2184, 2005.

76. C. V. Rao: Multiagent regimens for colon cancer chemoprevention. *Proc Am Soc Clin Oncol Education Book*, p. 129, 2005.

77. E. Szabo: Multiagent chemoprevention—Clinical concepts. *Proc Am Soc Clin Oncol Education Book*, p. 135, 2005.

Index